Catalogus Fossilium Austriae

Ein systematisches Verzeichnis aller auf österreichischem Gebiet festgestellten Fossilien

Schriftleitung
Werner E. Piller, Graz

Band 2

Echinoidea neogenica

mit 82 Tafeln und 94 Abbildungen im Text

Von
Andreas Kroh, Wien

Verlag der
Österreichischen Akademie der Wissenschaften
Wien 2005

Kontaktadressen:

Andreas KROH
Naturhistorisches Museum, Geologisch-Paläontologische Abteilung,
A-1010 Wien, Burgring 7

Werner E. PILLER
Universität Graz, Institut für Erdwissenschaften, Bereich Geologie und Paläontologie,
A-8010 Graz, Heinrichstraße 26

ISBN 3-7001-3491-6

Verlag der Österreichischen Akademie der Wissenschaften
Postgasse 7, A-1010 Wien
Tel. +43-1-51581 / DW 3402–3406
Fax +43-1-51581 / DW 3400
E-Mail: verlag@oeaw.ac.at
http://verlag.oeaw.ac.at
http://hw.oeaw.ac.at/3491-6

Gesamtherstellung:
Crossdesign Weitzer GmbH, A-8042 Graz

Contents

Foreword

Starting in 2001 with the first issue of the volume "Bivalvia neogenica" authored by O. Schultz the Catalogus Fossilium Austriae received not only a new layout but also a distinctly improved quality. This was accomplished by including also taxa reported from outside Austria, providing extensive distributional data for all taxa and adding a complete, high-quality photographic documentation. This way of presentation necessitated the splitting of volume 1 into three issues, published 2001, 2003 and 2005.

The present volume "Echinoidea neogenica" by A. Kroh goes one step further in presenting a full taxonomic/systematic revision of all Austrian Neogene echinoids. This includes extensive synonymy lists, material listing, detailed descriptions, discussions and distributional data. One genus and two species are newly described and a nomen novum is established. The text is entirely in English to make the volume easier accessible to an international readership. Besides a complete, high-quality photo documentation of the Austrian material (82 plates) 94 figures provide drawings of specific features and a variety of diagrams with statistical data. Additionally, some of these figures present photographs comparing Austrian material with other fossil and even recent echinoids. All localities cited are documented in maps which are not restricted to Austrian territory but include also locations in Czechia, Slovakia, Hungary, Croatia, Bosnia and Herzegovina, Serbia and Montenegro, Poland, Ukraine, Romania, and Bulgaria. A comprehensive systematic index of the mentioned taxa concludes the volume. An additional special service to the reader are the identification charts which are offered on the inner sides of the front and back covers for easier use.

For providing this huge amount of information and data and for his constructive cooperation my warmest thanks go to the author of this volume, Dr. Andreas Kroh.

Graz, May 2005

Werner Piller
Editor

Introduction

The present volume combines listings of literature records with new data, critical revisions, new descriptions and illustrations. The aim of this is to present to the reader a single volume including all data on Neogene echinoids of Austria combined with sufficient information to identify echinoids from Austria or neighbouring regions.

The CFA series necessitates the inclusion of poorly resolved references and/or material as it intends to document the occurrence of fossil taxa (of any level) as completely as possible (see Kühn in Flügel & Kropfitsch-Flügel, 1965). Therefore, fragmented and poorly preserved material, as well as records with unresolved taxonomy or records based on such material are also taken into account.

Descriptions and illustrations

All of the descriptions are based on Austrian material, except where explicitly stated otherwise. The material and its repositories, as well as short remarks (e.g. which part of the animal is preserved; if the material represents type or figured material of earlier papers) is listed under the section "Material" for each taxon. This section is followed by a section entitled "Foreign material for comparison" listing specimens which were used for direct comparison with the Austrian material. These are usually specimens from the Central Paratethys or (depending on availability) material from the type area of the species discussed. Likewise the specimens figured on the plates are of Austrian provenance, with very few exceptions. When foreign material is figured (usually in text-figures) it is explicitly labelled as such (by black numbers on white circles on the plates). Plate-drawings have been made with the aid of a *camera lucida* or by tracing plate outlines on greatly enlarged photographs and SEM photos. Where not readily visible, plate-contours could often be enhanced by wetting the specimens (with a mixture of water and glycerine), or in the case of some scutellid specimens by etching and/or grinding and polishing the oral surface until the micro-canal system was revealed. Stippled lines were used in cases in which contours were ambiguous. Cross-hatched areas indicate test surface damage. In clypeasteroids and spatangoids the interambulacra are usually shaded in grey.

Species concepts

In this time of increasing importance of genetics and DNA sequencing methods in systematics, it is necessary to comment on the species concepts employed in the present study. Although it has been shown that the echinoids can exhibit considerable morphological plasticity (e.g. Marcus, 1983; McNamara, 1995 for *Schizaster compactus*; ...) and morphology can be strongly linked to substrate and other environmental parameters (Ernst, 1973a, b; Stanton et al., 1979 for *Dendraster excentricus*) the basic premise in the present paper is that echinoid species can nevertheless be identified by a phenetic, or morphospecies approach. This is in fact a necessity in palaeontology as, of course, all other data (physiological, histological, genetic and in most cases also ontogenetic data, as well as any data on fertilisation success and reproductive isolation) are lost during fossilisation. Interestingly, a large number of systematic papers by neontologists [e.g. revisions of the extant Australian species of the family Schizasteridae by McNamara & Philip (1980b), of the Australian species of the genus *Pericosmus* by McNamara (1984), of the extant species of *Mellita* by Harold & Telford (1990), of the extant species of *Dendraster* by Mooi (1997)] relied on the same assumption.

Recent molecular studies on Indo-West Pacific members of the genus *Echinometra* (Matsuoka & Hatanaka, 1991; Palumbi et al., 1997), the genus *Eucidaris* (Lessios et al., 1999), the East Pacific and Atlantic species of *Echinometra* (McCartney et al., 2000) and the members of the genus *Diadema* from all over the world (Lessios et al., 2001b) shed new light on the suitability of phenetic (morphological) approaches in the systematics of "regular" echinoids (similar works on "irregular" echinoids such as the clypeasteroids or spatangoids are largely still lacking). Their results indicate a good agreement between prior classification on the basis of morphology and the groupings obtained by molecular analysis (i.e. most species defined on morphological grounds are monophyletic groupings in the trees obtained from genetic sequences). Individual molecular clades and sibling species, however, were often not recognised prior to the molecular analysis.

Studies linking genetic and morphological analyses are rare, notable exceptions are the papers on the *Echinocardium cordatum* complex by the working group around Bruno David and Jean-Pierre Féral (e.g. Laurin et al. 1994; but see also more recent papers by David & Laurin, 1996; Féral et al., 1998; David et al., 1999; Chenuli & Féral, 2003 and references therein). According to these studies both approaches (phenetic and molecular) have their merits, but results may differ.

As stated above, a phenetic approach was followed in the present study. Initial classification of the specimens was based on character analysis and spatial and temporal distribution. Whenever possible (depending on the availability of suitable numbers of specimens and their preservation) these classifications were subsequently tested by morphometric analyses. Classification prior to testing was favoured over "blind" analysis to avoid circular argument in subsequent testing of the classification. Potentially powerful multivariate analysis (e.g. principal component analysis, landmark methods, or canonical discriminant analysis) could not be employed due to the high number of missing values in the data sets. Instead, bivariate statistics, a method successfully applied in many studies before (e.g. Wörheide, 1995 for *Echinocardium cordatum*) was favoured.

The present study attempts to provide diagnostic characters for all species encountered in the study area (Austria) and, where possible, to differentiate them from all other species occurring in the Neogene of the Central Paratethys. The species accepted were the smallest morphologically diagnosable groups. During this process many previously established species turned out to form gradational series rather than distinct groups. Previously employed species concepts (e.g. by Vadász, 1915) led to the differentiation of closely related forms based on subtle differences in size and shape. This went so far as to recognize nearly each individual as a separate species (e.g. in the genus *Clypeaster*, where c. 80 species and subspecies were reported from the shallow sublittoral carbonate sediments of

Ma	System	Series	Standard Stages	Central Paratethys regional Stages	Historical Stages (Paratethys)		Eastern Paratethys regional Stages	Planktonic Foraminifera	Calcareous Nannoplankton
5		Pliocene	Zanclean	Dacian	Pliozän	Thracische Stufe	Kimmerian	PL1	NN 15-13; NN12
6			Messinian	Pontian			Pontian		b
7					Pont	Levantinische Stufe		M14	NN11
8							Maeotian	M13b	a
9		Late	Tortonian	Pannonian				M13a	NN10
10					Pannon	Pontische Stufe	Khersonian	M12	NN9
11							Bessarabian	M11	NN8; NN7
12			Serravallian	Sarmatian	Sarmat	Sarmatische Stufe	Volhynian	M9b	
13							Konkian	M8-9a	NN6
14	NEOGENE	MIOCENE Middle		Badenian (Early; Mid.; Late)	Torton	II. Mediterranstufe (Obermediterran)	Karaganian/ Chokrakian	M7; M6	NN5
15			Langhian				Tarkhanian		
16								M5	NN4
17				Karpatian	Helvet		Kotsakhurian	M4	
18			Burdigalian	Ottnangian					NN3
19		Early			Burdigal	I. Mediterranstufe (Untermediterran)		M2	
20				Eggenburgian			Sakaraulian		
21									NN2
22			Aquitanian		Aquitan		Karadzhalgian	M1	
23									NN1
24				Egerian					
25					Chatt			P22	NP25
26		Late	Chattian				Kalmykian		
27								b	
28								P21	
29	PALAEOGENE	OLIGOCENE		Kiscellian				a	NP24
30					Rupel			P20	
31		Early	Rupelian				Solenovian	P19	NP23
32									
33					Latdorf		Pshekian	P18	NP22
34								P17	NP21
		Eocene	Priabonian	Priabonian	Priabon		Beloglinian	P16	NP20

Table 1. Chronostratigraphic time scale for the interval Late Eocene to Early Pliocene. Modified from SCHULTZ (2001: tab. 1), RÖGL et al. (2002: tab. 4), and HARZHAUSER et al. (2003: fig. 2). Eastern Paratethys regional stages taken from GONCHAROVA et al. (2001: 517, tab.). Absolute dates of stage boundavies modified according to Gradstein et. al. (2004).

the Badenian). The present approach, in contrast, is a kind of "minimum species approach" attempting to provide groupings that can be differentiated. Features that are used in the classification of related extant forms were also used here when possible.

Systematic treatment

The synonymy lists give the exact citation of the cited paper including spelling errors, punctuation and use of capitals. Clarifying remarks are added in square brackets. Spelling errors are underlined to indicate that they were present in the original publication. Critical symbols are used in the synonymy lists throughout this work. An asterisk (*) indicates references where a valid species was established; the double cross (#) indicates references where a junior synonym was established; a dot (.) indicates citations which are explicitly placed into the synonymy of the species under discussion, these are usually references accompanied by a figure and/or extensive description sufficient for revision; the letter "v" indicates that the material mentioned in the corresponding reference has been seen by the author; "pp" means than only part of the reference material of a citation belongs to the considered species; "non" means that a reference is explicitly excluded from the synonymy of the discussed species, the reasons for this are usually given in the discussion or in brackets at the end of the line; a question-mark (?) indicates that there are reasons to doubt that the reference belongs in the synonymy of the considered species; if there is no symbol the information found in the corresponding paper was insufficient to decide on the correctness of the determination but that there was also no evidence suggesting otherwise.

In the occurrence section the Austrian localities are listed first, followed by the Paratethyan ones and finally that of the Mediterranean. Within these sub-sections the localities are arranged according to their palaeogeographic position (e.g. the basins in which they are situated). The stratigraphic range of a given species in the respective areas (Austria, Paratethys and Mediterranean) is based on the references in the respective section corrected according to new stratigraphical data when available. References in this section consist of author and date only and the reader is referred to the synonymy lists to find the full citations.

Arrangement and use of higher level taxonomy follows Smith (1984a) and Smith & Wright (1989, 1990, 1993, 1996, 1999, 2000, 2003) respectively but is emended where necessary. The classification of the Clypeasteroida follows Mooi (1989), taking also into account the results of Mooi (1990c, d) and Mooi & Chen (1996). For the Cassiduloida the work of Kier (1962) forms the baseline, but new data based on morphology (Mooi, 1990a) and cladistic analyses (Suter, 1994a, b; Smith, 2001) are incorporated where necessary.

The descriptive terminology follows Melville & Durham (1966), Durham & Wagner (1966) and Smith (1984b) (see also glossary and explanatory illustrations on the inside of the frontside- and backside-cover). Ambulacral pore morphology follows Smith (1978, 1980c) with additional types from McNamara (1985b). Nomenclature for fasciole morphology follows Néraudeau et al. (1998b). Ambulacral compounding nomenclature follows Jensen (1982).

Open nomenclature is used where necessary. The use of the open nomenclature (cf., aff.) follows Bengtson (1988). Where the material was insufficient for determination to species or even generic rank, higher level taxa are used.

Descriptions are based on Austrian specimens only, reference to other material is given in the "Remarks" and "Discussion" sections respectively. All specimens referred to have been examined by the author personally, except where explicitly stated otherwise.

The results of the taxonomic revision are summarized at the end of the introduction. Table 2 shows the Central Paratethyan taxa currently considered as valid and their ranges within the Neogene of Austria, the Central Paratethys and the total known ranges. Doubtful records that could not be substantiated or rejected during the present work are summarised in Table 3. Taxa mentioned in earlier papers but rejected from the Austrian and/or Central Paratethyan echinoid faunas are listed in Tables 4 and 5, with short explanatory statements.

Geological Setting

The Austrian Neogene basins are part of the Paratethys, an epicontinental sea ranging from the Rhône Basin and western Switzerland (Lake Geneva) in the west to the Transcaspian area (Lake Aral) in the east (Laskarev, 1924). It developed in the Oligocene by separation from the Mediterranean through the newly formed land masses of the Alps, Dinarides, Hellenides and the Anatolian Massif (see Rögl & Steininger, 1983; Rögl, 1998). Its palaeogeography and connection to the Mediterranean is strongly linked to the Alpine orogenesis (Fig. 1).

The Paratethys may be subdivided into the Western, Central and Eastern Paratethys. The Western Paratethys comprises the French, Swiss and part of the Bavarian Molasse. The Central Paratethys includes the Eastern Bavarian, Upper Austrian and Lower Austrian Molasse, the Carpathian Foredeep, the Vienna Basin, the Styrian Basin, the Danube Basin, the Great Hungarian Basin, the Zala, Sáva and Dráva Basins, the Novohrad Basin, the Transcarpathian Basin and the Transylvanian Basin (ordered from west to east and north to south) (Fig. 2). The Eastern Paratethys extends from the Black Sea area to the Transcaspian area. At its largest extent the Paratethys extended for more than 55° longitude and 15° latitude (not considering the reduction due to alpine orogenesis), an area similar to or larger than the Mediterranean today. With continued raising and northwards movement of the alpine mountain chains, the connections between Mediterranean and Paratethys, as well as between Central and Eastern Paratethys became severed. At the Sarmatian-Badenian boundary most stenohaline organisms disappeared due to short-term changes of sea-water chemistry. Although normal to hypersaline conditions prevailed during most of the Sarmatian, only a few of these groups re-appeared due to restricted connections with the world ocean (Piller & Harzhauser, in press). By the time of the Pannonian (Tortonian, Late Miocene), marine sedimentation ceased in the Paratethys. Since that time the extent of the Paratethys shrank and fresh-water conditions and endemism prevailed. For further information the reader is referred to Rögl (1998; and references therein).

Stratigraphy

The history of stratigraphy in the Central Paratethys is complex. Although the faunal development of the Paratethys is closely linked to that of the Mediterranean, geodynamic processes connected with the Alpine orogenesis strongly influenced that development. Times of isolation and good connections alternated, resulting in a distinct faunal history. The similarities and differences led to the development of a regional chronostratigraphic subdivision that differs from that of the Mediterranean, but can be correlated with it in a rather straightforward way (see Tab. 1). To understand the older literature it is necessary to give a short overview of the different subdivisions previously employed for the Lower and Middle Miocene sediments of the Central Paratethys previously. Moreover, as in the international literature, Middle Miocene occurrences, species, and outcrops of the Central Paratethys are commonly attributed to the Late Miocene (e.g. Philippe, 1998). However, marine sedimentation in the Central Paratethys, however, ceased during the late Middle Miocene (see above).

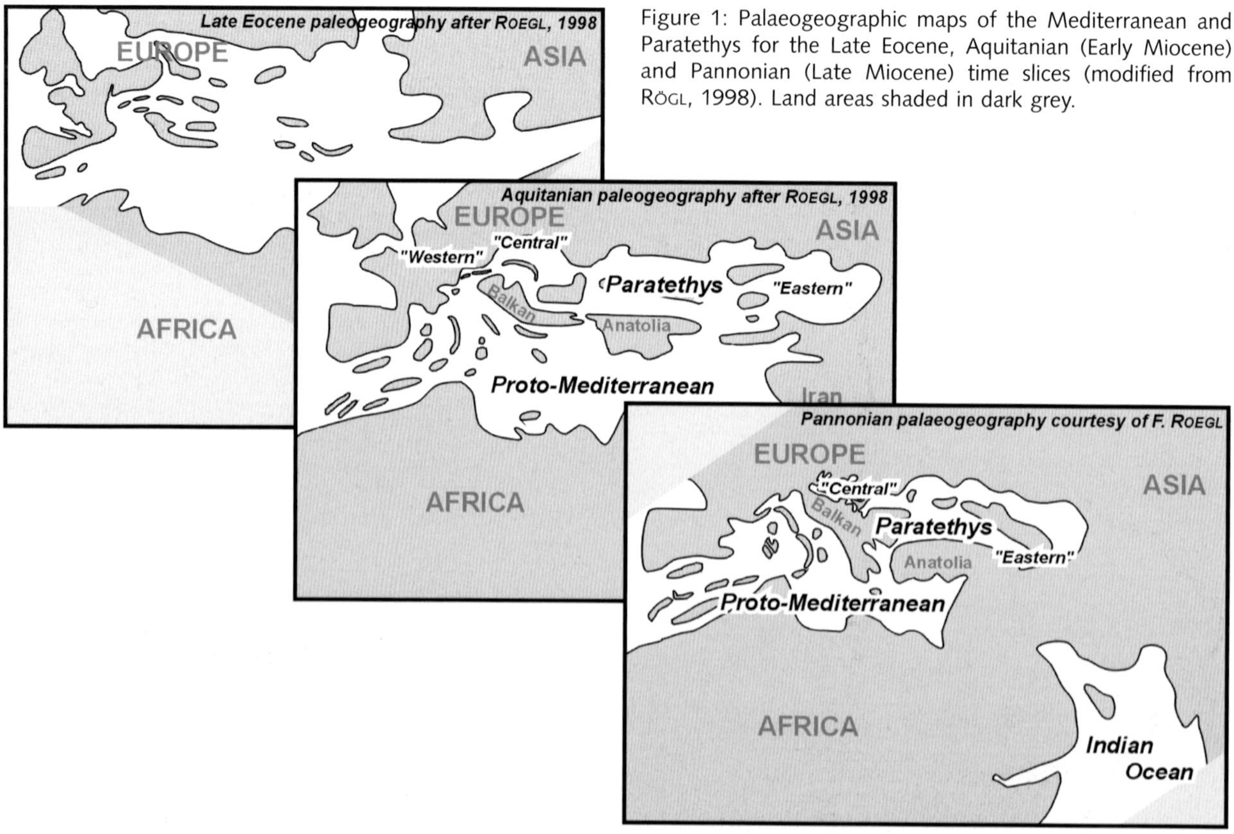

Figure 1: Palaeogeographic maps of the Mediterranean and Paratethys for the Late Eocene, Aquitanian (Early Miocene) and Pannonian (Late Miocene) time slices (modified from RÖGL, 1998). Land areas shaded in dark grey.

In the second half of the 19th and first decade of the 20th century a concept distinguishing between two stages, the "I. Mediterranstufe" or "Untermediterran" and the "II. Mediterranstufe" or "Obermediterran" was employed. The former was used mainly for Lower Miocene sediments and the latter for Middle Miocene ones. In the early 20th century this concept was replaced with a three-fold subdivision: "Burdigal – Helvet – Torton". While the "Burdigal" of this concept at least corresponds to what is known as Early Burdigalian today, the terms "Helvet" and "Torton" are misleading. "Helvet" was used for Middle Burdigalian as well as Langhian sediments. Most confusing, however, is the term "Torton" which was employed for Langhian to Lower Serravallian sediments based on an erroneous correlation with the Italian "Tortoniano" (Late Miocene) by SCHAFFER (1927b). This subdivision was in use as early as 1906 through *c.* 1975 by scientists working in the Central Paratethys, but is still present in the international literature (see PAPP et al., 1978a; and HARZHAUSER et al., 2003 for more information). The currently employed subdivision distinguishes four Early Miocene and two Middle Miocene stages (see Tab. 1). It was developed by a working group of the RCMNS around T. BALDI, I. CICHA, A. PAPP, F. RÖGL, J. SENEŠ, and F.F. STEININGER and published in the series "*Chronostratigraphie und Neostratotypen*" of the Slovak Academy of Sciences (later volumes were published by the Hungarian Academy of Sciences and Yugoslavian Academy of Sciences and Arts). This subdivision is constantly improved and substantiated by new results (e.g. BRZOBOHATÝ et al., 2003).

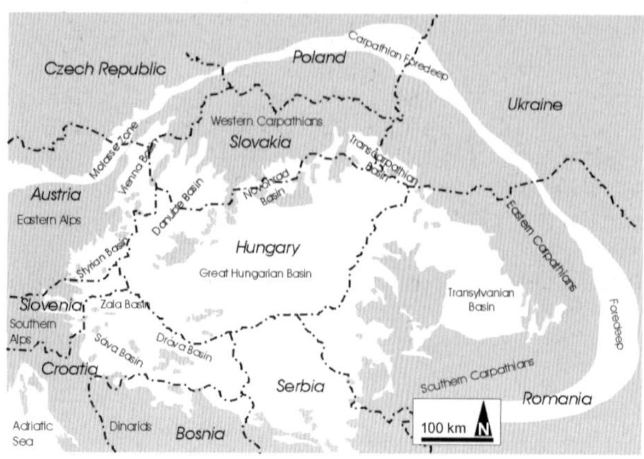

Figure 2: The Recent Carpathian-Pannonian Basin complex of the Central Paratethys showing the different sedimentary basins (modified from HARZHAUSER et al., 2003). Pre-Neogene sediments and basement shaded in grey.

Localities

At the end of the introduction the reader will find sketch maps showing the geographic position of the outcrops and historical localities mentioned in this work. The Austrian localities are shown on three maps: one for Oberösterreich (Upper Austria), one for Niederösterreich (Lower Austria), Wien (Vienna) and the northern half of Burgenland and one for Steiermark (Styria), the southern half of Burgenland and the easternmost part of Kärnten (Carinthia) (Figs. 3-5). In these maps, pre-Neogene sediments and basement are shaded in grey in these maps for easier orientation. In addition, smaller maps of other areas with Central Paratethyan sediments are provided to locate other Central Paratethyan localities (Figs. 6-9).

Note that many outcrops (mainly those in the Austrian-Czech, Austrian-Slovak, Austrian-Hungarian, and Hungarian-Romanian border regions) may be known by several different names in the geological literature. Wherever possible the currently used names are given in the "Occurrence" sections. The older names are mentioned in brackets (see also Tab. 6)

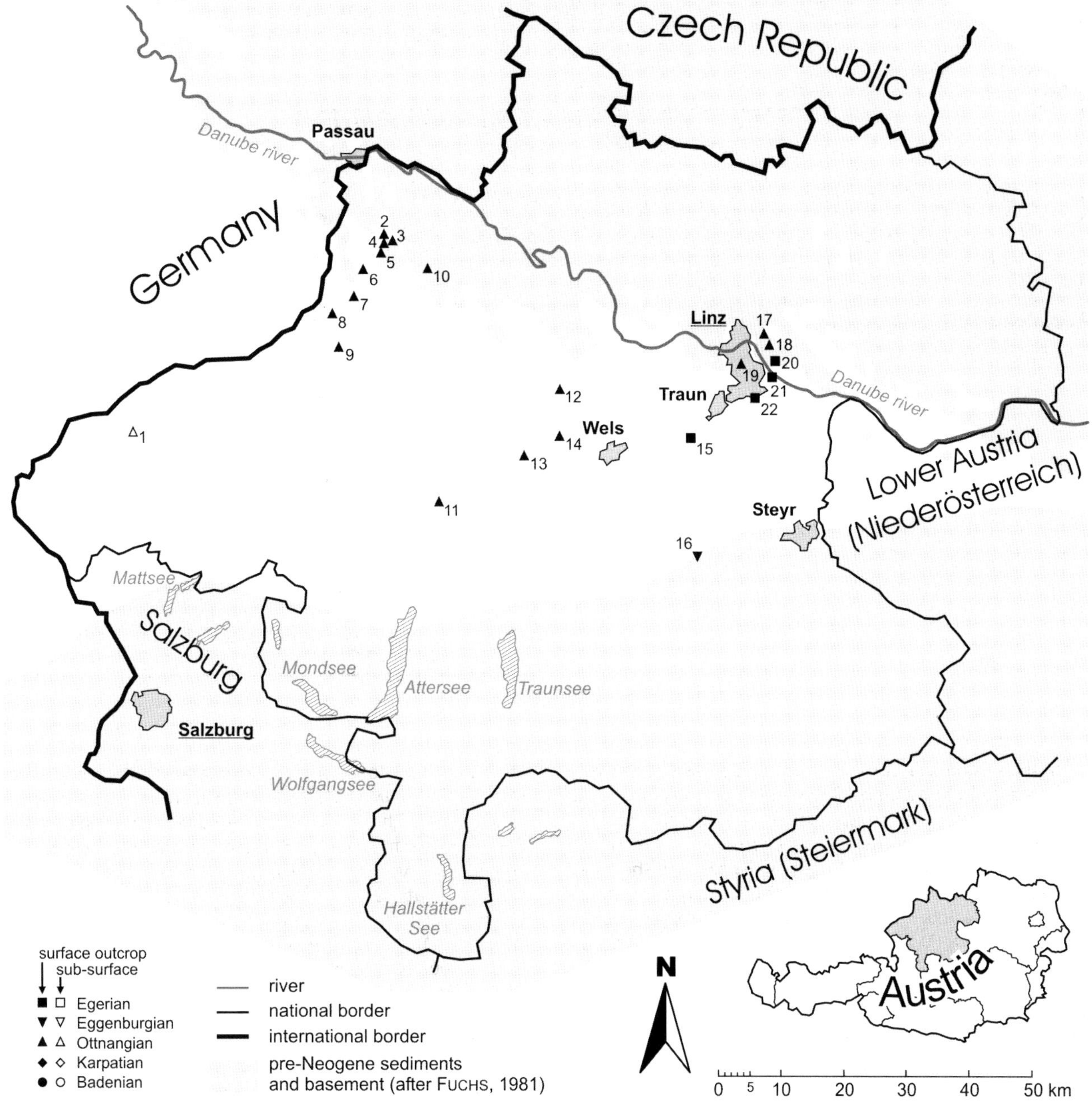

Allerding	6	Kletzenmarkt	12	Plesching (Pfennigberg)	18
Antiesenhofen	8	Linz (phosphoritic sand)	19	Plesching (phosphoritic sand)	17
borehole Eisenhub 2	1	Mitterndorf	10	Pucking (Ebelsberg Fm.)	15
Ebelsberg (Ebelsberg Fm.)	22	Offenhausen	13	Rainbach	3
Edermaning	4	Ort im Innkreis	9	Steyregg	20
Großwiesenhardt	7	Ottnang	11	Weikerlsee (Ebelsberg Fm.)	21
Haselbach	2	Pfarrkirchen (Haller Schlier)	16		
Höbmannsbach	5	Pichl bei Wels	14		

Figure 3: Echinoid-bearing localities in Upper Austria (Oberösterreich) mentioned in this paper.

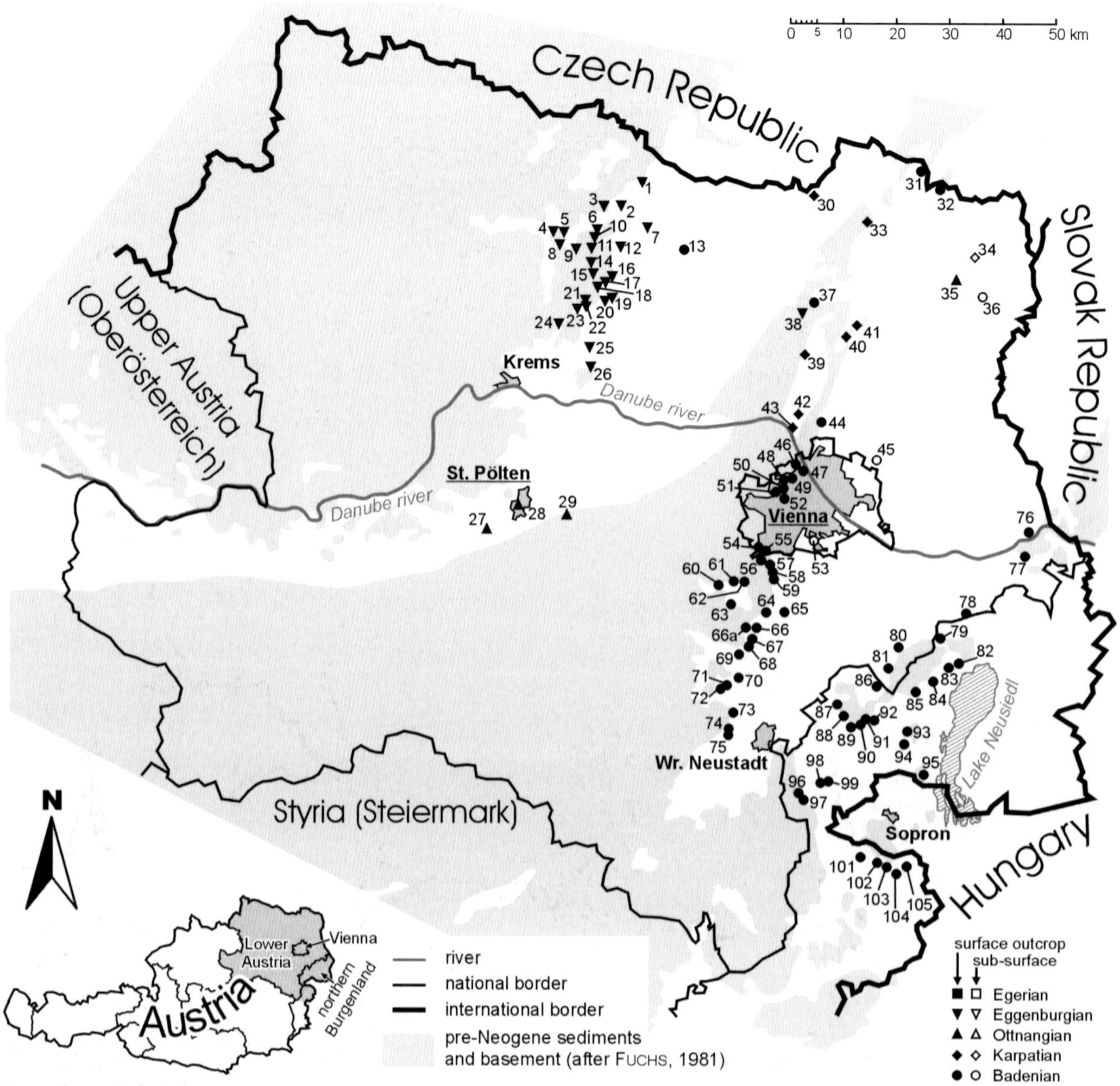

Burgenland (Bgld)

Locality	No.
Breitenbrunn	83
Deutschkreutz	105
Donnerskirchen	85
Eisenstadt	91
Forchtenau	97
Forchtenstein	96
Großhöflein	89
Haschendorf	103
Hornstein	87
Kaisersteinbruch	79
Kleinhöflein	90
Mattersburg	98
Mörbisch	95
Müllendorf	88
Neckenmarkt	102
Oslip	93
Pinkafeld	see Fig. 5
Purbach	84
Ritzing	101
Schreibersdorf	see Fig. 5
St. Georgen	92
St. Margarethen	94
Stotzing	86
Unterpetersdorf	104
Walbersdorf	99
Wiesfleck	see Fig. 5
Winden	82

Vienna (W)

Locality	No.
Dornbach	51
Grinzing	49
Kahlenberg	46
Kalksburg	54
Nußdorf	47
Oberlaa (borehole)	53
Ottakring (Wien XVI)	52
Pötzleinsdorf	50
Rodaun	55
Sievering	48

Lower Austria (NÖ)

Locality	No.
Aderklaa (borehole)	45
Bad Fischau	74
Bad Vöslau	68
Baden	66
Baden (Rauchstall-brunngraben)	66a
Bayersdorf	22
Bruck an der Leitha	78
Brunn am Steinfeld	75
Brunn am Gebirge	57
Eggenburg	11
Eggendorf	23
Eichhorn 1 (borehole)	36
Enzersdorf	33
Enzesfeld	71
Ernstbrunn	38
Fels am Wagram	26
Gaindorf	19
Gainfarn	69
Gauderndorf	10
Gösing	25
Grübern	21
Grund	13
Gumpoldskirchen	64
Hagenbrunn	44
Hainburg an der Donau	76
Hof am Leithagebirge	81
Hundsheim	77
Karnabrunn	39
Klein-Meiseldorf	5
Korneuburg 1 (borehole)	43
Kühnring	9
Laa an der Thaya	30
Leobersdorf	70
Limberg	16
Lindabrunn	72
Maissau	18
Mannersdorf	80
Maria Dreieichen (Achberg)	4
Maria-Enzersdorf	58
Mechters	29
Mödling	59
Mold	8
Möllersdorf	65
Neubau	41
Niederkreuzstetten	40
Niederleis	37
Ober-Dürnbach	17
Obernholz	24
Perchtoldsdorf	56
Platt	7
Pulkau	3
Ravelsbach	20
Reinprechtspölla	12
Roggendorf	6
Rohrendorf	2
Schrattenberg	32
Siegenfeld	63
Sittendorf	60
Sonndorf	15
Sooss	67
Sparbach	61
St. Pölten	28
Steinebrunn	31
St. Ulrich 2 (borehole)	34
Teiritzberg	42
Unternalb	1
Völlerndorf	27
Weissenbach	62
Wöllersdorf	73
Zistersdorf (Steinberg)	35
Zogelsdorf	14

Figure 4: Echinoid-bearing localities in Lower Austria (Niederösterreich) and northern Burgenland mentioned in this paper.

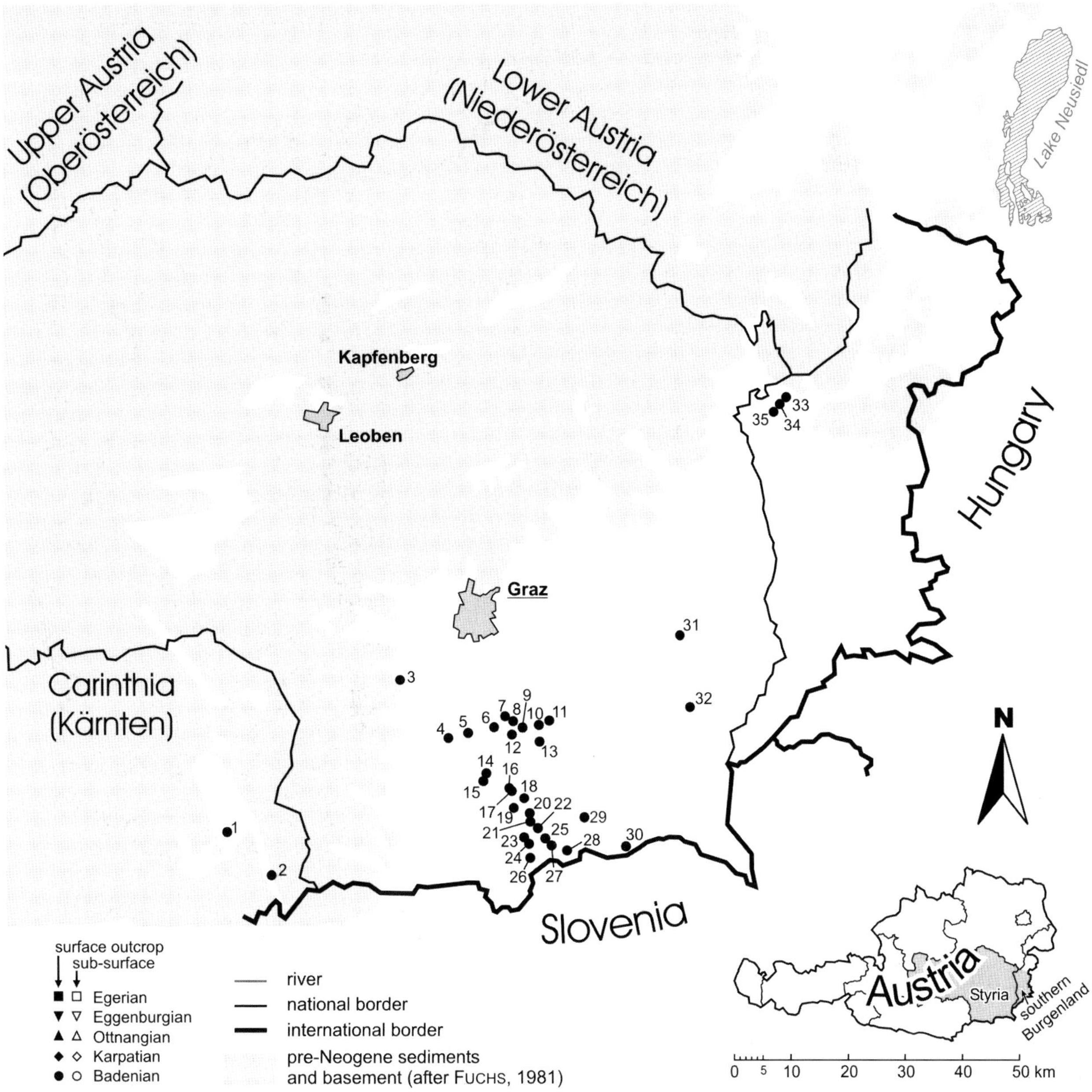

Carinthia (Ktn)

Ettendorf im Lavanttal	2
Mühldorf im Lavanttal	1

Styria (Stmk)

Aflenz an der Sulm	21
Altenberg	16
Ehrenhausen	25
Ewitsch	27
Gamlitz	24
Gamlitz (Eckberg)	26
Grubthal (Weinleiten)	23
Kainberg	17
Kollitschberg	8
Leibnitz	18
Leibnitz (Seggauberg)	19
Muggenau	14
Neurath	31
Ober-Mureck	30
Perbersdorf	29
Pesendorf (Stiefingbach)	10
Petersdorf	32
Pichla	11
Pöls an der Wieserbahn	5
Retznei (Lafarge quarry)	22
Spielfeld	28
St. Georgen	13
St. Nikolai im Sausal	15
Sukdull	9
Wagna (brickyard)	20
Weißenegg	7
Weitendorf	6
Wetzelsdorf in der Weststeiermark	4
Wildon	12
Zirknitz	3

Burgenland (Bgld)

Pinkafeld	35
Schreibersdorf	33
Wiesfleck	34

Figure 5: Echinoid-bearing localities in Styria (Steiermark), Carinthia (Kärnten) and southern Burgenland mentioned in this paper.

Březí (= Bratelsbrunn)	10	Mikulov (= Nikolsburg)	11	Služín	7
Hrušovany nad Jevišovkou (= Grusbach)	9	Mušlov	12	Strachotín	19
Kinberk (= Kienberg)	18	Nový Rybik (= Portzteich)	13	Úvaly (= Garschenthal)	14
Knínice (= Kinitz)	1	Orlová (= Orlau)	22	Valtice (= Feldsberg)	15
Kralice nad Oslavou (= Kralitz)	3	Ostrava (= Ostrau)	21	Židlochovice (= Groß-Seelowitz)	20
Květnice, near Lomnièka	4	Podivín	17	Znojmo (= Znaim)	2
Lednice (= Eisgrub)	16	Rudolticé (= Rudelsdorf)	6		
Lysice (= Lissitz)	5	Slatinky	8		

Figure 6: Czech localities mentioned in the text. Former German names used in the older literature are given in brackets.

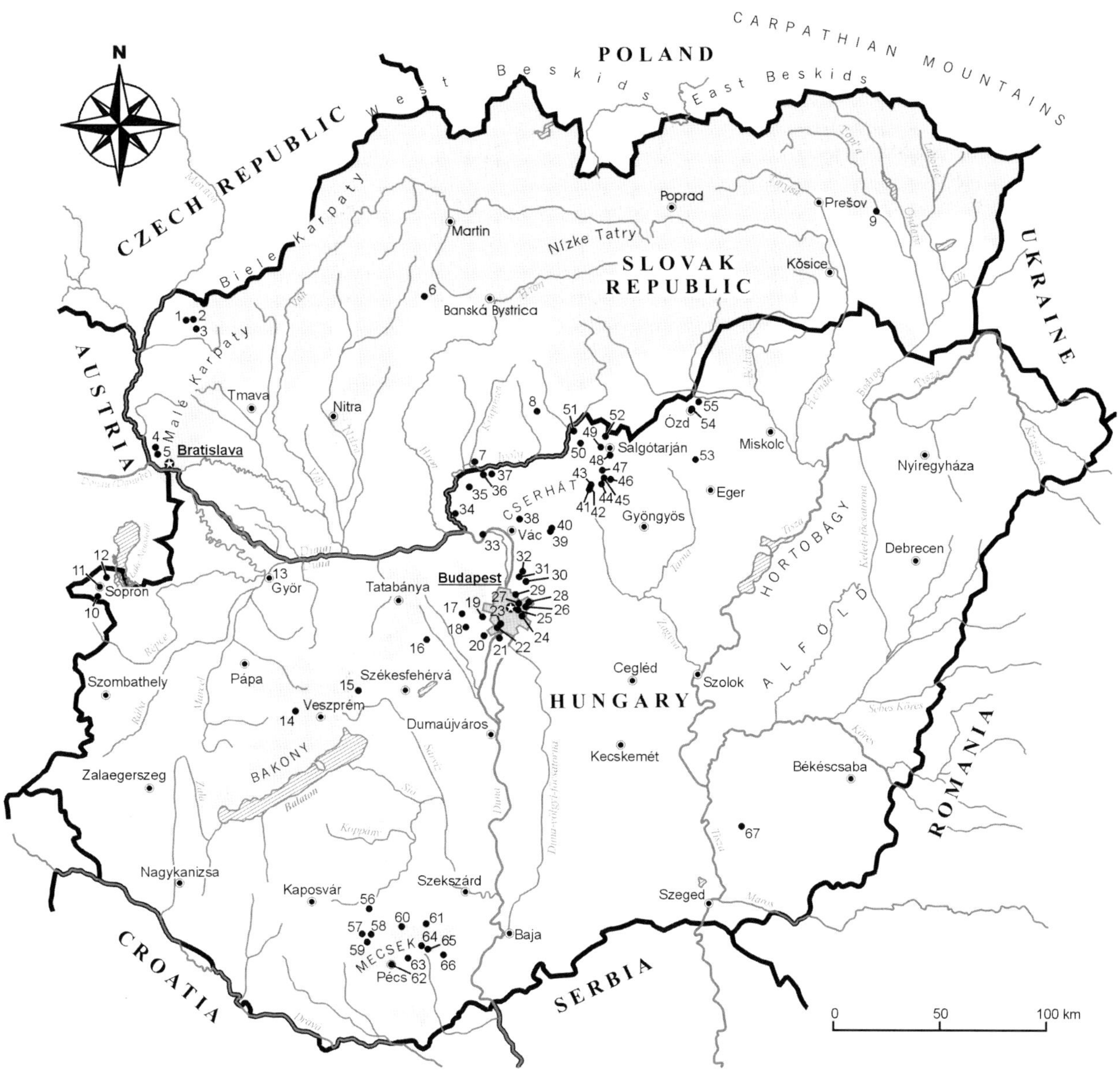

Slovak Republic	
Devínska Nová Ves (= Dévény-Újfalu, = Neudorf an der March)	4
Devínska Kobyla (= Thebener Kogel)	5
Dubovce	1
Hor. Strháre (= Felsö-Esztergály)	8
Hlinné	9
Šahy (= Ipoly-Shag)	7
Oreske	3
Radošovce	2
Vel'ká Čausa (borehole ČČ-3)	6

Hungary	
Acsa	40
Alcsútdoboz	16
Alsótold (= Alsó Told)	41
Bélapátfalva	53
Biatorbágy (= Bia)	17
Bodolyaber (= Egyházbér)	58
Budaörs	19
Budapest-Budafok	23
Budapest-Budatétény	22
Budapest-Gyakorló út	25
Budapest-Kerepesi út	25
Budapest-Rákos	29
Budapest, Xth district	24
Budapest, XIVth district	27
Cinkota	28
Csomád	32
Drégelypálank (= Drégely)	37
Érd	20
Érdliget	21
Etes (= Ettes)	49
Felsötold (= Felsö-Told)	42
Fertörákos (= Kroisbach)	12
Fót (= Fóth)	31
Garáb	43
Györ	13
Harka (= Harkau)	10
Herend-Márkó	14
Hidas (= Hidasd)	61
Hird	63
Hont	36
Karacs-Lapujtö (= Lapujtö)	52
Kemence	35
Királd	55
Kisgeresd	66
Kishajmás (= Háromház)	57
Kovácsszénnája (= Kovacéna, = Kovóicsszénaja)	59
Letkés	34
Magyaregregy	60
Márkháza	47
Mátraszölös (= Mátraszöllös)	44
Mátraverebély	46
Mátyásföld	26
Mogyoród	30
Nagymaros	33
Nagypall	65
Nógrádszakál (= Szakall)	51
Ösagárd	38
Paptelek	67
Pécs	62
Pécsvárad	64
Piliny	50
Püspökhatvan	39
Sajóvárkony	54
Sámsonháza	45
Sásd	56
Sopron (= Oedenburg)	11
Sopron-Rákos = Fertörákos	12
Sóskut	18
Szatina	57
Tétény = Budatétény	22
Tótmarokháza	47
Várpalota	15
Zagyvapálfalva (= Pálfalva)	48

Figure 7a: Slovak and Hungarian localities mentioned in the text. Former German and Hungarian names used in the older literature are given in brackets.

Slovenia

Locality	No.
"Gallenegg", near Zagorje ob Savi (= Sagor)	1
La'sko (= Tüffer)	3
Šentjernej (= St. Bartolomae)	4
Trbovlje (= Trifail)	2

Croatia

Locality	No.
Baćin Dol	21
Cepeliš Brijeg	13
Čučerje	9
Donja Velešnja, near Kostajnica	14
Donja Voča	5
Dvor, near Bosanski Novi	17
Gaj, near Trstenika	6
Giznik, near Samobor	7
Gradina	23
Grahovljani	20

Croatia (cont.)

Locality	No.
Hrastovica	13
Kaznica Potok, near Selci Đakovački	25
Kirin	12
Jelav Potok	22
Mesarski Brijeg, near Daruvar	19
Ostrožin	11
Podsused	8
Popov Dol-Poljanica	7
Samobor Kupal	7
Repušnica Potok	18
Sovsko Dola	26
Sovsko Jezero	24
Sutjeska	8
Utinje	10
Zaprešić Breg	7
Zrinj	16

Bosnia-Herzegovina

Locality	No.
Bosanska Kostajnica	15
Derventa	27
Gradačac	29
Maglajc-Poljane	30
Ugljevik	31
Vujići	28
Una river, Bosanski Novi (= Bosnisch Novi)	17

Serbia & Montenegro

Locality	No.
Bukovac	32
Despotovac	37
Golubac	38
Koceljevo	36
Leštane	35
Slankamen	33
Višnjica	34

Figure 7b: Neogene echinoderm localities in Bosnia & Herzegovina, Croatia, Serbia & Montenegro, and Slovenia. Former names used in the older literature are given in brackets.

Poland

Busko	8
Chomentów	4
Huta Lubycka	17
Huta Różaniecka	16
Karsy	3
Kików	10
Korytnica	5
Nawodzice	13
Niechobrz	14
Niskowa	1b
Pińczów	7
Rybnica	12
Skowronno	6
Świniary	11
Szczaworyz	9
Trzęsiny	15
Wieliczka	1a
Żerniki	2

Ukraine

Bilka	27
Gleboviti (= Chlebowice)	24b
Grabnik	21
Krekhiv (= Krehov, = Krechów)	20
Krzywcze Gorne (= Krzywczyce)	35
Zolochiv district (loc. Lackie małe)	28
L'viv (= Lwów)	22
Monastyrek (= Monastyrz)	19
Mykolaiv (= Mikolajów)	23
Podgortsg (= Podhorce)	29
Pod'yarkov (= Podjarków)	26
Potelych (= Potylicz)	18
Radziechów (= Radziejów)	30
Szuszkowce	32
Velyki Birky (= Borki-Wielkie)	34
Vybranovka (= Wybranówska)	24a
Zalesce (= Zalezce)	25
Zbarazh (= Zbaraz)	31
Zhukov (= Zuków, = Żukowice, = Zukowce, = Shukowze)	33

Figure 8: Polish and Ukrainian localities mentioned in the text. Former Polish names used in the older literature are given in brackets.

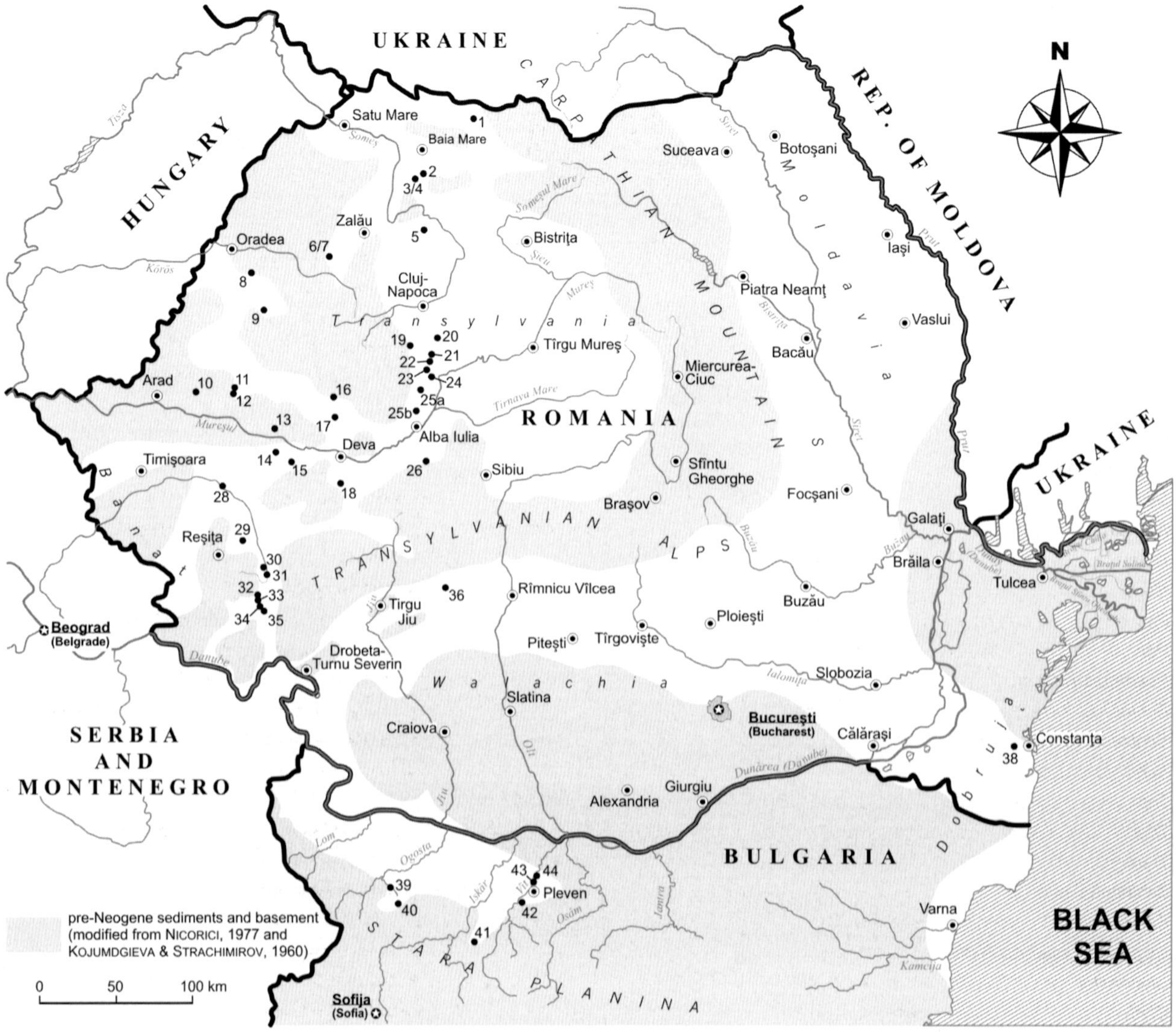

Romania

Armeniş (= Örményes)	31
Berchezoaia	3
Brăteşti	9
Buciumi	4
Buituri (= Bujtur)	18
Cacovalerii (= Cacova, = Kákova)	19
Covăşinţ (= Kovaszí, =Kovaszinc)	10
Chechiş	2
Ciceu-Hăşmaş	5
Ciolt	4
Coşteiu de Sus (=Kostéj)	14
Cricău (= Krakkó)	25b
Gârbova (= Szász-Orbó)	26
Gârbova de Sus (= Gîrbova de Sus, = Felsö-Orbó)	25a
Globu Craiovei (= Globukrajova)	34
Iablanitja (= Bélajablánc, = Jablanica)	35
Ilteu	13
Lăpugiu des Sus (=Felsölapugy, =Lapugy)	15
Livezile (= Ùrháza, = Vládháza)	24
Lugoj	28
Luncaviţa (= Lunkavica)	32
Mănăstire Polovragi (= Kloster Polowratsch)	36
Mehadica (= Mehádia, = Mehádika)	33
Minişel (= Kresztaménes)	12
Minişu des Sus (= Felménes)	11
Moldoveneşti (= Várfalva)	21
Murfatlar, Basarabi	38
Rachiş (= Oláh-Rákos, = Oláhrákos)	23
Petnek (= Petnik)	btw. 34 and 35
Pietroasa (= Csegez)	22
Ribicioara (= Ribice)	16
Rugi-Delineşti	29
Selişte (= Szelistye)	13
Sănduleşti (= Szind)	20
Sîg (= Sâg, = Şag)	6
Slatina-Timiş (= Temes-Szlatina)	30
Tărăţel (= Cerecel, = Czereczel)	17
Taşad (= Tasadfö)	8
Tusa (= Tusza)	7
Vălea Satului	3
Valeni	1

Bulgaria

Bivolare	44
Ochrid	39
Opanez	43
Palilula	41
Tarnene	42
Urovene	40

Figure 9: Romanian and Bulgarian localities mentioned in the text. Former Hungarian names used in the older literature are given in brackets.

Early Miocene: Aquitanian (Egerian), Burdigalian (Eggenburgian Early | Late, Ottnangian, Karpatian); Middle Miocene: Langhian (Badenian Early to Middle | Late), Serravall. (Sarmatian)

Species	total range (worldwide)
Prionocidaris sismondai (MAYER, 1864)	Chatt.-Langh.
Stylocidaris ? polyacantha (REUSS, 1860)	
Stylocidaris ? schwabenaui (LAUBE, 1869)	
Eucidaris zeamays (SISMONDA, 1842)	Aquit./Burd.-Serr.
Plegiocidaris ? peroni (COTTEAU, 1877)	Burd.-Serr.
Cidarinae sp. 1	
Cidaroida indet.	
Retzneiosoma jaseneki KROH, nov. sp.	
Diadematidae indet.	
Arbacina catenata (DESOR, 1846)	Burd.-Early Langh.
Arbacina cf. *macrophyma* LAMBERT, 1910	
Genocidaris sp.	
Psammechinus dubius dubius (AG., 1840)	Aquit.-Serr.
Psammechinus cf. *dubius gauthieri* COTT., 1885	
Psammechinus sp.	
Echinometra mathaei (DE BLAINVILLE, 1825)	Burd. or Tort., Plio.-rec.
Schizechinus hungaricus (LAUBE, 1869)	
Schizechinus sp.	
Tripneustes planus AGASSIZ in AG. & DES., 1846	Burd.
Tripneustes cf. *ventricosus* (LAMARCK, 1816)	Mioc.-rec.
Tripneustes sp.	
Echinacea indet.	
Echinoneus aff. *abnormalis* DE LORIOL, 1883	
Clypeaster calabrus SEGUENZA, 1880	Early(?)-Middle Mioc.
Clypeaster campanulatus (SCHLOTHEIM, 1820)	Middle Mioc.
Clypeaster folium AGASSIZ in AG. & DES., 1847	Middle Mioc.
Clypeaster latirostris MICHELIN, 1861	Aquit.-Burd.
Clypeaster scillae DES MOULINS, 1837	Burd.-Serr.
Clypeaster intermedius DES MOULINS, 1837	Aquit.-Burd., ? M. Mioc.
Clypeaster sp. (allochth.)	
Clypeasteridae indet.	
Echinocyamus transylvanicus LAUBE, 1869	
Echinocyamus pseudopusillus COTTEAU, 1895	Early-Middle Mioc.
Echinocyamus sp. A	
Parascutella gibbercula DE SERRES, 1829	Serr.-Tort.
Parascutella paulensis (AGASSIZ, 1841)	Aquit.-Burd., ? M. Mioc.
Parascutella ? sp.	
Parmulechinus hoebarthi (KÜHN, 1936)	
Amphiope bioculata (DES MOULINS, 1837)	Aquit.-Serr.
Echinolampas barcinensis LAMBERT, 1906	Burd.?, Langh.-Serr.
Echinolampas hemisphaerica (LAMARCK, 1816)	Burd.-Tort., ? Mess.
Echinolampas manzonii POMEL, 1883	Burd.-Langh.
Echinolampas aff. *manzonii* POMEL, 1883	
Echinolampas sayni LAMBERT, 1913	Burd.
Echinolampas schultzi KROH nom. nov.	
Echinolampas sp. 1	
Echinolampas sp. indet.	

Aquit. = Aquitanian, Burd. = Burdigalian, Chatt. = Chattian, Langh. = Langhian, Mess. = Messinian, Mioc. = Miocene, Oligoc. = Oligocene, Plio. = Pliocene, Serr. = Serravallian, Tort. = Tortonian

Legend: thick bar = definite occurrence within the finest temporal division used; ■ ■ ■ ■ = occurrence within the interval, but not finely resolved; thin line = whole Central Paratethys; •••••• = record dubious

Table 2. List of accepted taxa and their ranges (arranged systematically).

Early Miocene: Aquitanian (Egerian); Burdigalian (Eggenburgian Early / Late, Ottnangian, Karpatian). Middle Miocene: Langhian (Badenian Early to Middle), Serravall. (Badenian Late, Sarmatian).

Species	total range (worldwide)
Conolampas elegans (AIRAGHI, 1900)	Middle Mioc.
Conolampas subpentagonalis (GREGORY, 1891)	Middle Mioc.
Conolampas cf. *subpentagonalis* (GREGORY, 1891)	
Conolampas ? sp.	
Pliolampas vassalli (WRIGHT, 1855)	? Aquit.-Burd., Langh.
Studeria corsica ? (COTTEAU, 1877)	Late Burd.-Early Serr.
Ditremaster scillae (WRIGHT, 1855)	Aquit.-Langh.
Pericosmus latus (DESOR in AG. & DES., 1847)	Aquit.-Mess.
Schizaster (*S.*) *eurynotus* SISMONDA, 1841	Aquit.-Mess.
Schizaster (*S.*) *karreri* LAUBE, 1869	Burd.-Tort.
Schizaster (*S.*) *laubei* HOERNES, 1875	
Schizaster sp.	
Aliaster cotteauii (WRIGHT, 1855)	Chatt.-Serr.
Linthia ? *hlinnensis* (SENEŠ, 1955)	
Linthia ? *summesbergeri* nov. sp.	
Prenaster fuchsi (LAUBE, 1871)	
Pseudobrissus corsicus (COTTEAU in LOCARD, 1877)	? Oligoc., Burd.-Langh.
Brissus abeli (REIDL, 1941)	
Brissopsis ottnangensis HOERNES, 1875	? Mioc.
Brissopsis sp.	
Spatangus austriacus LAUBE, 1869	? Burd.
Spatangus cf. *austriacus* LAUBE, 1869	
Spatangus sp. 1	
Spatangus sp. 2	
Spatangus sp.	
Mariania deydieri (COTTEAU, 1897)	Burd.
Echinocardium depressum (AGASSIZ in AG. & DES., 1847)	Burd.
Echinocardium sp.	
Hemipatagus ocellatus (DEFRANCE, 1827)	Burd., ? Langh.
Spatangoida indet.	
Additional Paratethys taxa not recorded from Austria	
Prionocidaris avenionensis (DES MOULINS, 1837)	?Chatt., Burd.-Serr., ?Tort.
Clypeaster barcinensis ? LAMBERT, 1906	
Clypeaster neudorfensis LAMBERT, 1928	
Clypeaster myriophyma ? POMEL, 1887	? Middle Mioc.
Echinocyamus sp. B	
Schizaster latipetalus VADÁSZ, 1915	
Brissopsis crescenticus WRIGHT, 1855	Aquit.-Langh.
Plagiobrissus hungaricus VADÁSZ, 1915	
Meoma locardi (COTTEAU in LOCARD, 1877)	Burd.-Mess.
Spatangus hungaricus (VADÁSZ, 1915)	
Echinocardium deikei DESOR, 1858	
Echinocardium leopolitanum RADWAŃSKI & WYSOCKA, 2001	Burd.
Echinocardium peroni COTTEAU, 1877	
Hemipatagus ? *hungarica* (VADÁSZ, 1915)	
Lovenia mortenseni ČTYROKÝ, 1965	

Aquit. = Aquitanian, Burd. = Burdigalian, Chatt. = Chattian, Langh. = Langhian, Mess. = Messinian, Mioc. = Miocene, Oligoc. = Oligocene, Plio. = Pliocene, Serr. = Serravallian, Tort. = Tortonian

▬ definite occurrence within the finest temporal division used
■ ■ ■ ■ occurrence within the interval, but not finely resolved
— whole Central Paratethys
······ record dubious

Table 2. List of accepted taxa and their ranges (arranged systematically) (continued).

species	first recorded by	remarks
Austria		
"Prionechinus" felmenesensis LAMBERT & THIÉRY, 1911	LÓCZY, 1877 (as *Echinus* cf. *dux*)	probably conspecific with *Arbacina* cf. *macrophyma*
"Prionechinus" loczyi LAMBERT & THIÉRY, 1911	LÓCZY, 1877 (as *Ps.* cf. *monilis*)	only known specimen is lost; description and illustration insufficient for revision
Echinocyamus calariensis (LAMBERT, 1908)	VADÁSZ, 1915	few specimens known, which are usually poorly preserved; needs to be substantiated
"Hemiaster" kalksburgensis LAUBE, 1869	LAUBE, 1869	type material lost; description and illustration insufficient for revision
Schizaster desori WRIGHT, 1955	LAUBE, 1869	reference material lost, illustration and description insufficient; not recorded for more than 120 year from the Central Paratethys
Central Paratethys		
"Fibularia" sandalina SZÖRÉNYI, 1953	SZÖRÉNYI, 1953	type specimen (only specimen known) not seen, illustration and description are poor; needs to be re-described
Clypeaster angustus POMEL, 1887	VADÁSZ, 1915	reference material not located, description insufficient, record doubtful
Clypeaster coronalis LAMBERT, 1913	VADÁSZ, 1915	reference material not located, record doubtful
Clypeaster martini DES MOULINS, 1837	MEZNERICS, 1941	sole record; needs to be confirmed
Clypeaster megastoma mediterraneus VADÁSZ, 1915	VADÁSZ, 1915	probably a misidentified *Clypeaster scillae*
Clypeaster cf. *latirostris laganoides* AGASSIZ	MEZNERICS, 1941	sole record; needs to be confirmed
Clypeaster cf. *parvituberculatus* POMEL, 1887	VADÁSZ, 1915	based on poorly preserved material, not illustrated, description insufficient for revision
Clypeaster cf. *petalodes* POMEL, 1887	VADÁSZ, 1915	reference material not located, description insufficient, record doubtful
"Scutella" eichwaldi SZÖRÉNYI, 1953	SZÖRÉNYI, 1953	probably conspecific with *Parascutella gibbercula*
"Scutella" cf. leoganensis LAMBERT, 1903	SZÖRÉNYI, 1953	illustration and description poor, might be conspecific with *Parascutella gibbercula*
Parascutella ? guebhardi (LAMBERT, 1915)	KÓKAY, 1988b	mentioned without description or illustration, needs to be confirmed
Parascutella ? lusitanica (DE LORIOL, 1896)	KÓKAY, 1988b	mentioned without description or illustration, needs to be confirmed
Echinolampas dacica VADÁSZ	VADÁSZ, 1915	no specimens seen, but species seems to be closely related, if not synonymous with *Echinolampas hemisphaerica*
Echinolampas dacica humilis VADÁSZ	VADÁSZ, 1915	see under *Echinolampas dacica*
Echinolampas dumasi COTTEAU, 1893	VADÁSZ, 1915	record based on a single specimen, variation unknown; might be a separate species or an abnormal *E. hemisphaerica*
Echinolampas italicus LAMBERT, 1906	VADÁSZ, 1915	material not seen, difficult to differentiate from flat specimens of *E. hemisphaerica*
Echinolampas percrassus MEZNERICS, 1941	MEZNERICS, 1941	needs to be re-described; could also be a very high *E. manzonii* or a juvenile *Conolampas*
Pliolampas gauthieri (COTTEAU, 1880)	SZÖRÉNYI, 1953	unconfirmed, reported to co-occur with eight other closely related species, each represented by a single specimen (revision necessary)
"Tristomanthus" angulosus (MAZZETTI, 1885)	SZÖRÉNYI, 1953	see under *Pliolampas gauthieri*
"Tristomanthus" aremoricus (BAZIN, 1884)	SZÖRÉNYI, 1953	see under *Pliolampas gauthieri*
"Tristomanthus" fischeuri (POMEL, 1887)	SZÖRÉNYI, 1953	see under *Pliolampas gauthieri*
"Tristomanthus" meslei (GAUTHIER, 1886)	SZÖRÉNYI, 1953	see under *Pliolampas gauthieri*
"Tristomanthus" podolicus SZÖRÉNYI, 1953	SZÖRÉNYI, 1953	see under *Pliolampas gauthieri*
"Tristomanthus" subcylindricus (AGASSIZ, 1846)	SZÖRÉNYI, 1953	see under *Pliolampas gauthieri*
Hemiaster sp.	HILBER, 1882	doubtful record without description or illustration
Pericosmus airaghii LAMBERT, 1909	SENEŠ, 1955	doubtful record without description or illustration
Schizaster cf. *scillae* DES MOULINS, 1837	VADÁSZ, 1915	specimens poorly preserved, identification doubtful
Schizaster cf. *parkinsoni* (DEFRANCE)	VADÁSZ, 1915	specimens poorly preserved, identification doubtful
"Macropneustes" compressus NEMES, 1888	NEMES, 1888a	description and illustration insufficient for revision, needs to be re-described
Spatangus delphinus DEFRANCE, 1827	VADÁSZ, 1915	reference material not located; description and illustration insufficient for revision
Spatangus desmaresti MÜNSTER in GOLDFUSS, 1829	EICHWALD, 1852	species from the Oligocene of Northern Germany; reported without sufficient description or illustration; needs to be confirmed
Spatangus fabianii ? (LAMBERT, 1924)	SZÖRÉNYI, 1953	material needs to be re-examined to confirm identification
Spatangus fothiensis (STRAUSZ, 1926)	STRAUSZ, 1926	type material lost; new material needed
Spatangus hungaricus buekkensis (KUTASSY, 1928)	KUTASSY, 1928	holotype (only specimen available) poorly preserved; needs to be re-described
Spatangus peroni COTTEAU in LOCARD, 1877	VADÁSZ, 1915	record based on single poorly preserved specimen; unconfirmed
Spatangus pustulosus WRIGTH, 1864	MEZNERICS, 1941	identification doubtful, needs to be confirmed
Spatangus cf. *delphinus* DEFRANCE, 1827	MĄCZYŃSKA, 1988	description and illustration insufficient for revision
Spatangus cf. *corsicus* COTTEAU, 1877	VADÁSZ, 1915	record based on poorly preserved material; unconfirmed
Spatangus sp. 3	MAJCEN et al., 1997	mentioned as *S. austriacus*, but clearly not conspecific with any known *Spatangus* species from the Central Paratethys; provenance unknown
Echinocardium intermedium LÓCZY, 1877	LÓCZY, 1877	RADWAŃSKI & WYSOCKA (2001) questioned the generic attribution; needs to be re-described

Table 3. Doubtful records of Neogene echinoids from Austria and the Central Paratethys (arranged systematically).

species	first recorded by	reason for rejection
acuminatus DESOR, *Clypeaster*	MICHELIN, 1861	jun. syn. of *Clypeaster campanulatus*
affinis LAUBE, *Pericosmus*	LAUBE, 1869	type material lost, tentatively placed into the synonymy of *Aliaster cotteauii*
airaghii LAMBERT, *Clypeaster*	SCHOUPPÉ, 1947	unconfirmed, likely to be a misidentification
alticostatus MICHELIN, *Clypeaster*	MICHELIN, 1861	jun. syn. of *Clypeaster campanulatus*
altus LAMARCK, *Clypeaster*	KARRER, 1868	misidentified *Clypeaster campanulatus*, respectively *C. scillae* (some of the later records)
altus conicus, *Clypeaster*	QUENSTEDT, 1874	misidentified *Clypeaster campanulatus*
angulare altum KLEIN, *Scutum* [*Clypeaster*]	KNORR / WALCH, 1771	*Clypeaster campanulatus*
angulatus oblonus VADÁSZ, *Clypeaster*	TOLLMANN, 1955	jun. syn. of *Clypeaster campanulatus*
angulata MÉRIAN, *Echinolampas*	VADÁSZ, 1915	misidentified *Echinolampas manzonii*
angustistellatus LAUBE, *Echinolampas*	LAUBE, 1869	nomen dubium (type material not unambiguously identified, probably hybrid species of *Echinolampas sayni* and *E. schultzi*)
catenata (DESOR), *Genocidaris*	KROH, 2003a	misidentified; material belongs to *Genocidaris* sp.
crassicostatus AGASSIZ, *Clypeaster*	ANDRAE, 1855	nomen nudum, misidentified *Clypeaster scillae*
cucurbites MERCATI, *Echinus*	STÜTZ, 1807	pre-Linnéan taxon, mentioned without description or illustration
desori REUSS, *Diadema*	REUSS, 1860	isolated spines of diadematids, undeterminable below family level
duciei WRIGHT, *Echinus, Psammechinus, Schize.*	LAUBE, 1869	misidentified *Schizechinus hungaricus*
dux LAUBE, *Echinus, Schizechinus*	LAUBE, 1869	junior synonym of *Schizechinus hungaricus*
elliptica DESOR, *Amphiope*	LAUBE, 1869	jun. syn. of *Amphiope bioculata*
euglyphus LAUBE, *Spatangus*	LAUBE, 1869	not conspecific with type material; referred to *Spatangus* aff. *austriacus* here
extraalpinus SCHAFFER, *Psammechinus*	SCHAFFER, 1912	junior synonym of *Psammechinus dubius dubius*
floridus MERCATI, *Echinus*	STÜTZ, 1807	pre-Linnéan taxon, mentioned without description or illustration
faujasii DEFRANCE, *Scutella*	FUCHS, 1869	misidentified *Parascutella gibbercula*
gibbosus DE SERRES, *Clypeaster*	MICHELIN, 1861	misidentified *Clypeaster campanulatus*
grandiflorus BRONN, *Clypeaster*	BRONN, 1837	jun. syn. of *Clypeaster scillae*
grateloupii SISMONDA, *Schizaster*	HOERNES, 1875a, b	misidentified *Linthia ? hlinnensis*
gregoryi LAMBERT, *Clypeaster*	SCHOUPPÉ, 1947	unconfirmed, likely to be a misidentification
hemisphaericus rhodensis LAUBE, *Echinolampas*	LAUBE, 1869	Paratethyal records belong to *Echinolampas hemisphaerica*, name restricted to material from the Pliocene of Rhodes island
hirta SISMONDA, *Cidaris*	PETERS, 1857	doubtful record, Pliocene species [jun. syn. of *Histocidaris rosaria* (BRONN)]; Paratethyal material are probably misidentified *Stylocidaris ? polyacantha* spines
hungaricus VADÁSZ, *Echinolampas*	TOLLMANN, 1955	jun. syn. of *Conolampas elegans*
kalksburgensis WIESBAUR, *Scutella*	WIESBAUR, 1874	jun. syn. of *Parascutella gibbercula*
kleinii GOLDFUSS, *Echinolampas*	FUCHS, 1868	misidentified *Echinolampas schultzi*
laubei LAMBERT, *Amphiope*	LAMBERT, 1912	jun. syn. of *Amphiope bioculata*
laurillardi AGASSIZ, *Echinolampas*	LAUBE, 1869	misidentified *Echinolampas schultzi* (Eggenburgian records) or *E. hemisphaerica* (Badenian records)
laurillardi acuminata SCHAFFER, *Echinolampas*	SCHAFFER, 1912	= *Echinolampas schultzi* nom. nov. (preoccupied by *E. acuminata* ABICH, 1882, after raising to species rank)
leithanus LAUBE, *Schizaster*	LAUBE, 1869	jun. syn. of *Schizaster* (*S.*) *eurynotus*
linkii GOLDFUSS, *Clypeaster, Echinolampas*	GOLDFUSS, 1829	jun. syn. of *Echinolampas hemisphaerica*; Eggenburgian specimens referred to this species by later authors belong to *E. schultzi*
marginatus ? LESKE, *Echinanthus* [*Clypeaster*]	SEDGWICK & MURCHISON, 1831	misidentified *Clypeaster* sp.
media SCHAFFER, *Scutella*	SCHAFFER, 1962	jun. syn. of *Parascutella paulensis*
michelotti AGASSIZ, *Clypeaster*	MICHELIN, 1861	misidentified *Clypeaster campanulatus*
miocaenicus SCHAFFER, *Brissus* (*Allobrissus*)	SCHAFFER, 1961	jun. syn. of *Brissus abeli*
mirabilis NICOLET, *Psammechinus*	LAUBE, 1869	misidentified *Psammechinus* cf. *dubius gauthieri* and/or *Schizechinus hungaricus*
monilis DESMAREST, *Echinus, Psammechinus*	LAUBE, 1869	misidentified *Psammechinus* cf. *dubius gauthieri*
multiconcava SCHAFFER, *Scutella*	SCHAFFER, 1962	jun. syn. of *Parascutella gibbercula*, robust morphotype
papillata LESKE, *Dorocidaris*	MANZONI, 1880	extant species [jun. syn. of *Cidaris cidaris* (LINNÉ), not conspecific with *Stylocidaris ? schwabenaui*]; later records are misidentified *S. ? schwabenaui* or *S. ? polyacantha* spines
parkinsoni DEFRANCE, *Schizaster*	LAUBE, 1869	misidentified *Schizaster* (*S.*) karreri
partschii MICHELIN, *Clypeaster*	MICHELIN, 1861	jun. syn. of *Clypeaster campanulatus*
pernatus SCHAFFER, *Spatangus, Maretia*	SCHAFFER, 1912	jun. syn. of *Hemipatagus ocellatus*
perspicillata AGASSIZ, *Amphiope*	LAUBE, 1869	misidentified *Amphiope bioculata*
plagiosomus AGASSIZ, *Conoclypus, Echinolampas*	LAUBE, 1869	= *Conolampas subpentagonalis* (not conspecific with type material of *C. plagiosomus*); some later records are misidentified *C. elegans*
portentosus MICHELIN, *Clypeaster*	MICHELIN, 1861	jun. syn. of *Clypeaster campanulatus*
pusillus MÜLLER, *Echinocyamus*	VINASSA de REGNY, 1897	extant species, not conspecific with *Echinocyamus transylvanicus*
pyramidalis MICHELIN, *Clypeaster*	MICHELIN, 1861	jun. syn. of *Clypeaster campanulatus*
rotundus LAUBE, *Hemiaster*	LAUBE, 1869	jun. syn. of *Ditremater scillae*
rosaceus, *Echinus* [*Clypeaster*]	STÜTZ, 1807	extant species (*Clypeaster rosaceus*, Caribbean), misidentified *Clypeaster* sp.
scillae DES MOULINS, *Schizaster*	LAUBE, 1869	misidentified *Schizaster* (*S.*) *eurynotus*
serresii DES MOULINS, *Echinus, Psammechinus*	LAUBE, 1869	misidentified *Schizechinus hungaricus*
styriaca HOERNES, *Amphiope*	HOERNES, 1883	jun. syn. of *Amphiope bioculata*, large morphotype
styriaca SCHAFFER, *Scutella*	SCHAFFER, 1962	jun. syn. of *Parascutella gibbercula*
subfolium POMEL, *Clypeaster*	KAZÁR, 2002	jun. syn. of *Clypeaster folium*
subpartschi SCHAFFER, *Clypeaster*	SCHAFFER, 1912	jun. syn. of *Clypeaster intermedius*
ventiencis vindobonensis LAMBERT, *Clypeaster*	LAMBERT, 1912	jun. syn. of *Clypeaster latirostris*
ventricosus austriacus TAUBER, *Tripneustes*	TAUBER, 1951	validity of ssp. remains to be shown; *Tripneustes* cf. *ventricosus* (LAMARCK, 1816)

Table 4. List of rejected Neogene echinoids from Austria (species arranged alphabetically).

species	first recorded by	reason for rejection
vindobonensis LAUBE, *Scutella, Parascutella*	LAUBE, 1869	jun. syn. of *Parascutella gibbercula*
vindobonensis secunda SCHAFFER, *Scutella*	SCHAFFER, 1962	jun. syn. of *Parascutella gibbercula*, flat morphotype
aff. *italicus* LAMBERT, *Echinolampas*	TOLLMANN, 1955	single unconfirmed record without description or illustration
aff. *seguenzai* VADÁSZ, *Clypeaster*	TOLLMANN, 1955	misidentified *Clypeaster campanulatus*
***Amphiope* sp.**	KROH, 2002a	= *Amphiope bioculata*
***Brochopleurus* sp.**	SCHMID et al., 2001	misidentified; material belongs to *Genocidaris* sp.
***Centrostephanus* sp.**	TOLLMANN, 1955	isolated spines of diadematids, undeterminable below family level
***Echinus* sp.**	FUCHS, 1877	misidentified *Psammechinus* or more likely *Schizechinus*
***Hemiaster* sp.**	KAZÁR, 2002	misidentified *Aliaster cotteauii*

Table 4 (continued). List of rejected Neogene echinoids from Austria (species arranged alphabetically).

Acknowledgements

First of all I want to express my sincere thanks to my wife Andrea and my son Christian whose patience and moral support enabled me to complete this monograph.

Without the enormous amount of help received from numerous colleagues, librarians, and private collectors it would not have been possible to finish this study. Foremost of all others, however, I have to thank four colleagues who enabled me to start and finish this study firsthand: Werner E. PILLER (UGP), my supervisor, who arranged financial support, publication possibilities and worked hard in reviewing the first manuscript, Mathias HARZHAUSER and Heinz KOLLMANN (both NHMW), who provided access to the collections and facilities of the NHMW and continuously supported me and Ortwin SCHULTZ (NHMW), for his patience in discussing taxonomic questions.

My thanks go to: Cynthia AHEARN (USNM), Tomasz K. BAUMILLER (Univ. Michigan), Enrico BORGHI (private collector, Reggio Emilia, Italy), Wolfgang BRUNNBAUER (NHMW), Burgenländische Kreidewerke (Müllendorf, Bgld), Rostislav BRZOBOHATÝ (Univ. Brno), Burt CARTER (Georgia Southwestern College), Wolfgang DANNINGER (private collector, Andorf, OÖ), Bruno DAVID (Université de Bourgogne), Horst DOLEZAL (Mödling, Austria), Hubert DOMANSKI (Univ. Vienna), Stephen K. DONOVAN (Museum Naturalis, Leiden), Alfréd DULAI (Hungarian Natural History Museum, Budapest), Anton ENGLERT (NHMW), Christoph ERHART (UGP), Rosalinde ESSBERGER (NHMW), J. Georg FRIEBE (inatura – Erlebnis Naturschau Dornbirn), Heinz FÜRNKRANZ (private collector, Vienna), Jaume GALLEMÍ (Museo de Geologia, Barcelona), Michael GATT (private collector, Rabat, Malta), Irina A. GONCHAROVA (Russian Academy of Sciences, Moskva), Martin GROSS (Joanneum, Graz), Sheila HALSEY (BMNH), Hartmut HIDEN (Graz), Eva HÖCK (NHMW), Gudrun HÖCK (NHMW), Frank HOLMES (Heathmont, Australia), Anna ŽOLDOVÁ (SNM) John W. M. JAGT (Natuurhistorisch Museum Maastricht), Arie W. JANSSEN (Xewkija, Malta), Manfred JASENEK (private collector, Vienna), Josef KASTL (private collector, Linz), Miklós KÁZMER (Univ. Budapest), Rudolf KNÖPPEL (private collector, Berndorf-St.Veit, NÖ), Gertrude KOFLER (HM), Artúr KÖHLER (Budapest, Hungary), László KORDOS (MAFI), Johanna KOVAR-EDER (Naturkunde Mus. Stuttgart), Reinhold KUNZ (private collector, Vienna), Lafarge-Perlmoser cement company (Retznei, Styria), Christine LATAL (Univ. Graz), David N. LEWIS (BMNH), Alexander Walter LUFT (private collector, Vienna), LUKENEDER (NHMW), Oleg MANDIC (GBA), Martina MARINELLI (Univ. Vienna), Mayer Sand (Stotzing, Bgld), Kenneth J. MCNAMARA (Western Australian Museum), Ursula MENKVELD (Naturhistorisches Museum Bern), Ashley MISKELLY (North Turramurra, Australia), Rich MOOI (Californian Acad. Sci., Los Angeles), James H. NEBELSICK (Univ. Tübingen), Johann NEITZ (private collector, Lindabrunn, NÖ), Didier NÉRAUDEAU (Univ. Rennes), Christian NEUMANN (Naturkundemuseum Berlin), Therese PFISTER (Naturhistorisches Museum Bern), Benno PLÖCHINGER (GBA), Edeltraud PREIS (NHMW), Johann PREIS (NHMW), Andrzej RADWAŃSKI (Univ. Warszawa), Ula RADWAŃSKA (Univ. Warszawa), Michael RASSER (UGP), Karl RAUSCHER (Univ. Vienna), Mike REICH (Univ. Göttingen), Johannes REISCHER (GBA), Adam REITZEL (Univ. Boston), Fred RÖGL (NHMW), Reinhard ROETZEL (GBA), Jean ROMAN (Paris), Brian ROSEN (BMNH), Helmut SATTMANN (NHMW), Andrew B. SMITH (BMNH), Helga SCHMITZ (NHMW), Heinke SCHULTZ (Hemdingen, Germany), Fritz F. STEININGER (Senckenberg Museum, Frankfurt), Franz STOJASPAL (GBA), Sabine STÖHR (Swedish Museum of Natural History), Barbara STUDENCKA (Muzeum Ziemi, Warszawa), Herbert SUMMESBERGER (NHMW), Tomas SUTTNER (Univ. Vienna), Stefan SZAILER (NHMW), Harald TEMMEL (private collector, Vienna), Petra TEMPFER (Univ. Vienna), Andreas THINSCHMIDT (KM), Adam TOMAŠOVÝCH (Univ. Würzburg), Franz TOPKA (NHMW), Richard TURNER (Florida Inst. of Technology), Johannes M. TUZAR (KM), Norbert VAVRÁ (Univ. Vienna), Loïc VILLER (Univ. Marseilles), Michael WAGREICH (Univ. Vienna), Gerhard WANZENBÖCK (private collector, Gainfarn, NÖ), Anna WYSOCKA (Univ. Warszawa), Louis G. ZACHOS (Texas Memorial Museum), Kamil ZÁGORŠEK (Czech National Museum, Praha), George ZAMMIT MAEMPEL (Ghar Dalam Museum, Malta), Reinhard ZETTER (Univ. Vienna), Irene ZORN (GBA), Martin ZUSCHIN (Univ. Vienna)

Special thanks go to Alice SCHUMACHER (NHMW) who made the numerous photographs (except the SEM images) in this paper.

This study was supported by the Austrian Science Foundation (FWF), project No. P-14366-Bio to Werner E. Piller (University of Graz) and by the "Kommission für die paläontologische und stratigraphische Erforschung Österreichs" of the Austrian Academy of Sciences (OEAW).

species	first recorded by	reason for rejection
acclivis POMEL, *Clypeaster*	VADÁSZ, 1915	misidentified *Clypeaster calabrus*
acuminatus DESOR, *Clypeaster*	DESOR, 1847	jun. syn. of *Clypeaster campanulatus*
acuminatus robustus VADÁSZ, *Clypeaster*	VADÁSZ, 1915	jun. syn. of *Clypeaster campanulatus*
aequizonatus GREGORY, *Echinolampas*	MITROVIĆ-PETROVIĆ, 1969	specimen poorly preserved, probably a misidentified *Pliolampas vassalli*
agassizi SISMMONDA, *Clypeaster*	VADÁSZ, 1915	material lost, probably conspecific with *Clypeaster campanulatus*
airaghii LAMBERT, *Centrostephanus*	VADÁSZ, 1915	isolated spines of diadematids, undeterminable below family level
airaghi LAMBERT, *Clypeaster*	POLJAK, 1938	misidentified *Clypeaster scillae*
almerai LAMBERT, *Clypeaster*	VADÁSZ, 1915	misidentified *Clypeaster campanulatus*
angulatus VADÁSZ, *Clypeaster*	VADÁSZ, 1915	jun. syn. of *Clypeaster campanulatus*
angulatus oblonus VADÁSZ, *Clypeaster*	VADÁSZ, 1915	jun. syn. of *Clypeaster campanulatus*
angulosa MAZZETTI, *Milletia*	VADÁSZ, 1915	misidentified *Pliolampas vassalli*
angustipetalus VADÁSZ, *Echinolampas*	VADÁSZ, 1915	tentatively placed into the synonymy of the co-occurring *Echinolampas hemisphaerica*
atrophus podoloicus SZÖRÉNYI, *Echinolampas*	SZÖRÉNYI, 1953	junior synonym of *Echinolampas manzonii*
balillai (LOVISATO), *Clypeaster*	MIKUŽ & MITROVIĆ -PETROVIĆ, 2001	misidentified *Clypeaster folium*
barcinensis LAMBERT, *Schizaster*	VADÁSZ, 1915	misidentified *Schizaster* (*S.*) *karreri*
beaumonti SISMONDA, *Clypeaster*	POLJAK, 1938	probably a misidentified *Clypeaster folium*
calarensis COTTEAU, *Centrostephanus*	VADÁSZ, 1915	isolated spines of diadematids, undeterminable below family level
calceolus LAMBERT, *Schizaster*	VADÁSZ, 1915	poorly preserved material, probably misidentified specimens of the co-occurring *Schizaster* (*S.*) *eurynotus*
campanulatus carapezzai CHECCHIA-RISPOLI, *Cly.*	KALABIS, 1949	misidentified *Clypeaster neudorfensis* (partly)
campanulatus declinatus VADÁSZ, *Clypeaster*	VADÁSZ, 1915	type material lost, probably conspecific with *C. campanulatus*
campanulatus rotundus VADÁSZ, *Clypeaster*	VADÁSZ, 1915	type material lost, probably conspecific with *C. campanulatus*
campanulatus sphaericus VADÁSZ, *Clypeaster*	VADÁSZ, 1915	type material lost, probably conspecific with *C. campanulatus*
chateleti LAMBERT, *Schizechinus*	MĄCZYŃSKA, 1988	misidentified; referred to *Psammechinus* ? sp.
chiassi LOVISATO, *Clypeaster*	MITROVIĆ-PETROVIĆ, 1969	misidentified *Clypeaster scillae*
circularis CAPEDER, *Echinocyamus*	MĄCZYŃSKA, 1977	probably misidentified *Echinolampas transylvanicus*
consobrinus LAMBERT, *Brissopsis*	VADÁSZ, 1915	misidentified *Brissopsis ottnangensis*
convexus VADÁSZ, *Clypeaster*	VADÁSZ, 1915	type material lost, probably conspecific with *C. campanulatus*
crassus AGASSIZ, *Clypeaster*	VADÁSZ, 1915	misidentified *Clypeaster scillae* and *Clypeaster calabrus*
cruciatus AGASSIZ, *Schizobrissus*	VADÁSZ, 1915	misidentified *Meoma locardi* (but the two species are probably synonymous)
cureti LAMBERT, *Plegiocidaris*	HALMAI, 1981	jun. syn. of *P.* ? *peroni* (fide PHILIPPE, 1998)
dacicus VADÁSZ, *Clypeaster*	VADÁSZ, 1915	jun. syn. of *Clypeaster campanulatus*
danubicus VADÁSZ, *Clypeaster*	VADÁSZ, 1915	jun. syn. of *Clypeaster calabrus*
delagadoi DE LORIOL, *Clypeaster*	VADÁSZ, 1915	misidentified *Clypeaster campanulatus*
delphinus (LAMBERT), *Schizechinus*	VADÁSZ, 1915	poorly preserved, probably *Schizechinus hungaricus*
depressus VADÁSZ, *Clypeaster*	VADÁSZ, 1915	type material lost, probably conspecific with *C. campanulatus*
desmoulinsi SISMONDA, *Cidaris*	VADÁSZ, 1915	probably a misidentified *Stylocidaris* ? *schwabenaui*; also often confuse with *Eucidaris zeamays*
digitalis VADÁSZ, *Clypeaster*	VADÁSZ, 1915	jun. syn. of *Clypeaster scillae*
egregyensis SZÖRÉNYI, *Hypsoclypus*	SZÖRÉNYI, 1953	probably a junior synonym of *Conolampas subpentagonalis*
epianthus MEZNERICS, *Clypeaster*	MEZNERICS, 1941	jun. syn. of *Clypeaster calabrus*
excentricus VADÁSZ, *Clypeaster*	VADÁSZ, 1915	jun. syn. of *Clypeaster calabrus*
felmenesensis VADÁSZ, *Clypeaster*	VADÁSZ, 1915	jun. syn. of *Clypeaster campanulatus* (subadult specimen)
fontannesi (COTTEAU), *Parasalenia*	MĄCZYŃSKA, 1977	misidentification, juvenile echinaceans
gauthieri LOVISATO, *Clypeaster*	VADÁSZ, 1915	misidentified *Clypeaster campanulatus*
gombosae MIHÁLY, *Clypeaster*	MIHÁLY, 1990	jun. syn. of *Clypeaster campanulatus*
gracilis VADÁSZ, *Clypeaster*	VADÁSZ, 1915	type material lost, probably conspecific with *C. campanulatus*
grandiflorus altipetalus VADÁSZ, *Clypeaster*	VADÁSZ, 1915	jun. syn. of *Clypeaster scillae*
grandiflorus anteacutus VADÁSZ, *Clypeaster*	VADÁSZ, 1915	jun. syn. of *Clypeaster scillae*
grandiflorus basiconcava VADÁSZ, *Clypeaster*	VADÁSZ, 1915	jun. syn. of *Clypeaster scillae*
grandiflorus rhabdopetalus POMEL, *Clypeaster*	VADÁSZ, 1915	jun. syn. of *Clypeaster scillae*
gregoryi LAMBERT, *Clypeaster*	POLJAK, 1938	misidentified *Clypeaster calabrus*
guebhardi LAMBERT, *Clypeaster*	LAMBERT, 1913	misidentified *Clypeaster neudorfensis*
gustavi LOVISATO, *Clypeaster*	MITROVIĆ-PETROVIĆ, 1984	misidentified *Clypeaster campanulatus*
hemisphaericus VADÁSZ, *Clypeaster*	VADÁSZ, 1915	type material lost, probably conspecific with *C. campanulatus*
hemisphaericus depressa LAMBERT, *Echinol.*	MIHÁLY, 1969	Paratethyal records tentatively placed into the synonymy of *Echinolampas barcinensis*
hemisphaericus maxima DE LORIOL, *Echinol.*	VADÁSZ, 1915	Paratethyal records tentatively placed into the synonymy of *Echinolampas barcinensis*
herepeyi KOCH, *Clypeaster*	KOCH, 1887	jun. syn. of *Clypeaster folium*
hungarica VADÁSZ, *Scutellina*	VADÁSZ, 1915	poorly preserved juvenile scutellid; probably *Parascutella gibbercula*
hungaricus VADÁSZ, *Clypeaster*	VADÁSZ, 1915	type material lost, tentatively referred to *C. campanulatus*
hungaricus VADÁSZ, *Echinolampas, Heteroclypeus*	VADÁSZ, 1915	jun. syn. of *Conolampas elegans*
hungaricus VADÁSZ, *Schizaster*	VADÁSZ, 1915	jun. syn. of *Schizaster* (*S.*) *karreri*
hungaricus dispar VADÁSZ, *Clypeaster*	VADÁSZ, 1915	tentatively referred to *Clypeaster campanulatus*
ilottoi LAMBERT, *Schizaster*	VADÁSZ, 1915	jun. syn. of *Schizaster* (*S.*) *karreri*
inflatus VADÁSZ, *Clypeaster*	VADÁSZ, 1915	probably conspecific with *Clypeaster campanulatus*
italicus LAMBERT, *Echinolampas*	MITROVIĆ-PETROVIĆ, 1969	misidentified *Echinolampas hemisphaerica*
jacquementi LAMBERT, *Brissus*	SZÖRÉNYI, 1953	tentatively referred to *Brissus abeli*
karreri hungaricus VADÁSZ, *Schizaster*	VADÁSZ, 1906	jun. syn. of *Schizaster* (*S.*) *karreri*
kemencensis MEZNERICS, *Clypeaster*	MEZNERICS, 1941	jun. syn. of *Clypeaster calabrus*
lamberti extensus VADÁSZ, *Clypeaster*	VADÁSZ, 1915	jun. syn. of *Clypeaster campanulatus*
lecointreae LAMBERT, *Fibularia*	SZÖRÉNYI, 1950	misidentified *Echinocyamus pseudopusillus*
lecointreae LAMBERT, *Echinolampas*	MITROVIĆ-PETROVIĆ, 1969	misidentified *Echinolampas hemisphaerica*

Table 5. List of rejected Neogene echinoids from the Central Paratethys (except Austria, see Tab. 4).

species	first recorded by	reason for rejection
leognanensis LAMBERT, *Scutella*	VADÁSZ, 1915	figured specimens (Badenian, Felsö-Orbó) at the MAFI coll. are conspecific with *Parascutella gibbercula*, Karpatian specimens not located, but probably also conspecific with *P. gibbercula* (judging from the description)
linearis CAPEDER, *Echinocyamus*	MĄCZYŃSKA, 1977	misidentified *Echinocyamus* sp. B (not conspecific with CAPEDER's material)
lovisatoi COTTEAU, *Trachyaster*	VADÁSZ, 1915	jun. syn. of *Aliaster cotteauii*
lovisatoi rakosensis VADÁSZ, *Schizaster*	VADÁSZ, 1906	jun. syn. of *Schizaster* (*S.*) *karreri*
ludovici LAMBERT, *Amphiope*	MIHÁLY, 1985	jun. syn. of *Amphiope bioculata*
magnus VADÁSZ, *Clypeaster*	VADÁSZ, 1915	type material lost, probably conspecific with *C. campanulatus*
majocchii LOVISATO, *Clypeaster*	MITROVIĆ-PETROVIĆ, 1969	misidentified *Clypeaster scillae*
malladai LAMBERT, *Clypeaster*	MITROVIĆ-PETROVIĆ, 1984	misidentified *Clypeaster campanulatus*
manini LOVISATO, *Clypeaster*	MITROVIĆ-PETROVIĆ, 1969	misidentified *Clypeaster scillae*
marginatus LAMARCK, *Clypeaster*	MIKUŽ & MITROVIĆ-PETROVIĆ, 2001	misidentified *Clypeaster folium*
marioi LOVISATO, *Echinocyamus*	KOJUMDGIEVA & STRACHIMIROV, 1960	misidentified *Echinocyamus transylvanicus*
melitensis FORBES, *Cidaris*	VADÁSZ, 1915	misidentified *Eucidaris zeamays*
melitensis MICHELIN, *Clypeaster*	MICHELIN, 1861	Paratethyal specimen removed from *Clypeaster melitensis* by LAMBERT (1928) and renamed *Clypeaster neudorfensis*
michelotti DESOR, *Psammechinus*	MÁRTONFI, 1894	unclear, later attributed to *Arbacina tenera* by VADÁSZ, 1915
miocenica de LORIOL, *Echinometra*	KOJUMDGIEVA & STRACHIMIROV, 1960	jun. syn. of *Echinometra mathaei*
monilis (DESMAREST in DEFRANCE), *Arbacina*	VADÁSZ, 1915	most records are undeterminable juvenile echinaceans the others misidentifications; presence remains to be confirmed
montesiensis MAZZETTI, *Hypsoheteroclypus*	SZÖRÉNYI, 1953	misidentified *Conolampas subpentagonalis*
mostoi LOVISATO, *Clypeaster*	MITROVIĆ-PETROVIĆ, 1969	misidentified *Clypeaster scillae*
muelleri MIHÁLY, *Scutella*	MIHÁLY, 1985	juvenile of *Parascutella gibbercula*
nevianii AIRAGHI, *Echinolampas*	VADÁSZ, 1915	junior synonym of *Conolampas subpentagonalis*
olisiponensis MICHELIN, *Clypeaster*	MITROVIĆ-PETROVIĆ, 1969	misidentified *Clypeaster campanulatus*
ovatus GOLDFUSS, *Echinocyamus*	LÖRENTHEY, 1894	Oligocene taxon, identification unconfirmed
periplanus VADÁSZ, *Clypeaster*	VADÁSZ, 1915	jun. syn. of *Clypeaster campanulatus*
periplanus subtilis VADÁSZ, *Clypeaster*	VADÁSZ, 1915	type material lost, probably conspecific with *C. campanulatus*
petaliferus SEGUENZA, *Clypeaster*	VADÁSZ, 1915	misidentified *Clypeaster campanulatus*
plagiosomus corsicanus SZÖRÉNYI, *Hypsoh.*	SZÖRÉNYI, 1953	probably a junior synonym of *Conolampas subpentagonalis*
plagiosomus lamberti CHECCHIA-RISPOLI, *Hypsoh.*	SZÖRÉNYI, 1953	misidentified *Conolampas elegans*
podjarkovii SZÖRÉNYI, *Tristomanthus*	SZÖRÉNYI, 1953	tentatively placed into the synonymy of *Pliolampas vassalli*
pseudoserrata COTTEAU, *Cidaris*	TOULA, 1893	doubtful record, Eocene species, probably misidentification
pygmea KOCH, *Scutella*	KOCH, 1887	probably a juvenile *Parascutella gibbercula*
pyramidalis ABICH, *Hypsoheteroclypus*	SZÖRÉNYI, 1953	misidentified *Conolampas elegans*
rakosensis VADÁSZ, *Schizaster*	LAMBERT, 1906	jun. syn. of *Schizaster* (*S.*) *karreri*
reidii WRIGHT, *Clypeaster*	VADÁSZ, 1915	misidentified *Clypeaster campanulatus*
rhodanicus (MAYER-EYMAR), *Centrostephanus*	MĄCZYŃSKA, 1987	disarticulated diadematid material, undeterminable below family level
romani MIHÁLY, *Scutella*	MIHÁLY, 1985	juvenile of *Parascutella gibbercula*
rosaria (BRONN), *Cidaris, Histocidaris*	HALMAI, 1981	unconfirmed record without description or illustration
sardiniensis COTTEAU, *Clypeaster*	VADÁSZ, 1915	misidentified *Clypeaster campanulatus*
sardiniensis COTTEAU, *Schizaster*	VADÁSZ, 1915	probably a misidentified *Ditremaster scillae*; later records (e.g. SZÖRENYI, 1953) are misidentified *Schizaster* (*S.*) *karreri*
sardiniensis ellipticus VADÁSZ, *Clypeaster*	VADÁSZ, 1915	jun. syn. of *Clypeaster campanulatus*
scillae alienus VADÁSZ, *Clypeaster*	MAJCEN et al., 1997	misidentified *Clypeaster scillae*
scillae* cf. *alienus VADÁSZ, *Clypeaster*	MIKUŽ & MITROVIĆ-PETROVIĆ, 2001	misidentified *Clypeaster scillae*
scillae* var. *alienus VADÁSZ, *Clypeaster*	VADÁSZ, 1915	jun. syn. of *Clypeaster calabrus*
seguenzai VADÁSZ, *Clypeaster*	VADÁSZ, 1915	nom. nov. pro *Clypeaster accuminatus* SEGUENZA, 1880 non DESOR, 1847; Paratethys specimens conspecific with *C. campanulatus*
semiglobosus LAMARCK, *Echinolampas*	VADÁSZ, 1915	misidentified *Clypeaster subpentagonalis*
semseyana MIHÁLY, *Kieria*	MIHÁLY, 1985	very probably juveniles of the co-occurring *Amphiope bioculata*
stellata CAPEDER, *Fibularia*	VADÁSZ, 1915	misidentified *Echinocyamus transylvanicus* or *E. pseudopusillus*
striatula (DE SERRES), *Parascutella*	PHILIPPE, 1998	Early Miocene species of France, not conspecific with Badenian material from Hungary
studeri (SISMONDA), *Echinocyamus*	HALMAI, 1981	unconfirmed record without description or illustration
subacutus POMEL, *Clypeaster*	VADÁSZ, 1915	tentatively referred to *Clypeaster campanulatus*
subacutus depressus VADÁSZ, *Clypeaster*	VADÁSZ, 1915	jun. syn. of *Clypeaster campanulatus*
subaltus VADÁSZ, *Clypeaster*	VADÁSZ, 1915	type material lost, probably conspecific with *C. campanulatus*
subconoideus VADÁSZ, *Clypeaster*	VADÁSZ, 1915	jun. syn. of *Clypeaster campanulatus*
subconoideus VADÁSZ, *Echinolampas*	VADÁSZ, 1915	tentatively placed into the synonymy of *Echinolampas barcinensis*
subfolium POMEL, *Clypeaster*	VADÁSZ, 1915	jun. syn. of *Clypeaster folium*
sublaganoides VADÁSZ, *Clypeaster*	VADÁSZ, 1915	jun. syn. of *Clypeaster folium*
suboblongus POMEL, *Clypeaster*	VADÁSZ, 1915	misidentified *Clypeaster campanulatus*
subrotunda LAMARCK, *Scutella*	EICHWALD, 1830	probably a misidentified *Parascutella gibbercula*
subtrigona LAUBE [sic], Scutella	GHIURCA, 1996	misidentified *Parascutella gibbercula*
subularis d'ARCHIAC, *Cidaris*	TOULA, 1893	doubtful record, Eocene species, probably misidentified *Stylocidaris* ? *schwabenaui*
szoerenyiae MIHÁLY, *Scutella*	MIHÁLY, 1969	jun. syn. of *Parascutella gibbercula*, holotype has an irregular margin due to predation
tauricus DESOR, *Clypeaster*	MITROVIĆ-PETROVIĆ, 1969	misidentified *Clypeaster scillae*
tarabellianus GRATELOUP, *Clypeaster*	KORNHUBER, 1865	unconfirmed, probably misidentification
tenera DE LORIOL, *Arbacina*	VADÁSZ, 1915	jun. syn. of *Arbacina catenata*, most records are undeterminable juvenile echinaceans or misidentifications
tenuipetalus SEGUENZA, *Clypeaster*	VADÁSZ, 1915	probably misidentified *Clypeaster neudorfensis*
transylvanicus VADÁSZ, *Clypeaster*	VADÁSZ, 1915	type material lost, probably conspecific with *Clypeaster campanulatus*
transylvanicus VADÁSZ, *Echinolampas*	VADÁSZ, 1915	tentatively placed into the synonymy of *Echinolampas barcinensis*
turritus PHILIPI, *Clypeaster*	MITROVIĆ-PETROVIĆ, 1969	misidentified *Clypeaster campanulatus*
ugolinii STEFANINI, *Echinolampas*	MITROVIĆ-PETROVIĆ, 1969	specimen poorly preserved, identification at species level doubtful

Table 5 (continued). List of rejected Neogene echinoids from the Central Paratethys (except Austria, see Tab. 4).

species	first recorded by	reason for rejection
vadaszi Roman, *Echinolampas, Hypsoclypus*	Roman, 1965	jun. syn. of *Conolampas elegans*
ventiensis Lambert, *Schizaster*	Vadász, 1915	jun. syn. of *Schizaster (S.) karreri*
vicinoconoideus Szörényi, *Hypsoheteroclypus*	Szörényi, 1953	jun. syn. of *Conolampas elegans*
vindobonensis altus Mihály, *Scutella*	Mihály, 1990	jun. syn. of *Parascutella gibbercula*, high morphotype
vindobonensis planata Kókay, *Scutella*	Somos & Kókay, 1960	jun. syn. of *Parascutella gibbercula*, flat morphotype
wrighti Gregory, *Echinolampas*	Mitrović-Petrović, 1969	misidentified *Echinolampas hemisphaerica*
zanoni Checchia-Rispoli, *Clypeaster*	Mitrović-Petrović, 1969	jun. syn. of *Clypeaster scillae*
aff. *pusilla* Müller, *Fibularia*	Horusitzky, 1927	probably misidentified *Echinocyamus pseudopusillus*
cf. *balillai* (Lovisato), *Clypeaster*	Mikuž & Mitrović-Petrović, 2001	misidentified *Clypeaster folium*
cf. *bellardi* Agassiz, *Schizaster*	Vadász, 1915	misidentified *Schizaster (S.) karreri*
cf. *curtus* Pomel, *Schizaster*	Vadász, 1915	specimen indeterminable
cf. *laganoides* Agassiz, *Clypeaster*	Mikuž & Mitrović-Petrović, 2001	misidentified *Clypeaster folium*
cf. *longispinus* Philipi, *Centrostephanus*	Vadász, 1915	isolated spines of diadematids, undeterminable below family level
cf. *parmularius* Lambert, *Echinolampas*	Kochansky-Devidé, 1957	unique record without description or illustration
cf. *scutellatus* de Serres, *Clypeaster*	Mikuž & Mitrović-Petrović, 2001	misidentified *Clypeaster folium*
***Arbacina* sp.**	Vadász, 1915	undeterminable juvenile echinaceans or misidentifications
***Cyphosoma* sp.**	Toula, 1893	Late Jurassic to Eocene genus; misidentification
Parasalenia sp.	Mączyńska, 1977	misidentification, diadematid ambulacral plate and poorly preserved *Schizechinus* or *Psammechinus* corona
***Salenia* ? sp.**, genital plates	Toula, 1893	Cretaceous genus; probably a misidentification of cidaroid genital plates

Table 5 (continued). List of rejected Neogene echinoids from the Central Paratethys (except Austria, see Tab. 4).

Alsopéterfa – Unterpetersdorf
Borbolya – Walbersdorf
Császárköbánya – Kaisersteinbruch
Feketeváros – Purbach
Fertöfehéregyháza – Donnerskirchen
Fertömeggyes – Mörbisch (formerly Merwisch)
Fertöszéleskút – Breitenbrunn
Fraknó – Forchtenstein
Hasfalva – Haschendorf
Kismarton – Eisenstadt
Lajtaszék – Stotzing
Lajtaszentgyörgy – St. Georgen, near Eisenstadt
Nagyhöflány – Kleinhöflein
Nagymarton – Mattersburg
Oszlop – Oslip
Sásony – Winden
Sopronkeresztúr – Deutschkreutz
Sopronnyék – Neckenmarkt
Szárazvám – Müllendorf
Szarvkö – Hornstein
Szentmargitbánya – St. Margarethen
Ujrétfalu – Wiesfleck

Table 6. Hungarian names of Austrian localities (Burgenland) used in the literature (late 19th, early 20th century)

Abbreviations

ant.	anterior
AO	figured specimen
APP	anterior paired petals
Bgld	Burgenland, a state of Austria
BMNH	British Museum of Natural History, London
dist.	distal
CZ	Czech Republic
coll.	collection
Fm.	Formation (lithostratigraphic unit)
GBA	Geologische Bundesanstalt (Geological Survey of Austria), Wien, Austria
HM	Höbarth Museum, Horn, NÖ, Austria
HT	holotype
IPUW	Institut für Paläontologie, Universität Wien, Austria
KM	Krahuletz-Museum, Eggenburg, NÖ, Austria
LT	lectotype
LMJ	Landesmuseum Joanneum, Graz
MAFI	Magyar Állami Földtani Intézet (Geological Survey of Hungary), Budapest, Hungary
[MAFI]	in occurrence lists: material from the respective locality is housed in the MAFI collection
NHMW	Naturhistorisches Museum Wien, Austria
[NHMW]	in occurrence lists: material from the respective locality is housed in the NHMW collection
NMB	Naturhistorisches Museum Bern, Switzerland
NÖ	Niederösterreich (Lower Austria), a state of Austria
OÖ	Oberösterreich (Upper Austria), a state of Austria
pc	periproct
PLT	paralectotype(s)
PPP	posterior paired petals
prox.	proximal
ps	peristome
RCMNS	Regional Committee on Mediterranean Neogene Stratigraphy
SEM	scanning electron microscope
SK	Slovak Republic
SNM	Slovenské narodné múzeum (Slovak National Museum)
ST	syntype(s)
Stmk	Steiermark (Styria), a state of Austria
UGP	Institut für Geologie und Paläontologie, Karl-Franzens-Universität Graz, Austria
USNM	U.S. National Museum, Smithsonian Institution, Washington, USA
ÚÚG-PC	Ústředni ústav geologický (Geological Survey), Praha, Czech Republic
W	Wien (Vienna), a state of Austria
(= *locality*)	former name of a locality used in the literature
locality (*xxx*)	xxx = additional information to a locality, this may either be a more refined localisation (e.g. example hill) or a lithostratigraphic unit

Symbols used in the Plates

White numbers on black background are used for Austrian material. Black numbers in white circles are used for foreign material.

Symbols used in the synonymy lists

[ ]	in synonymy lists: remarks by the author
*	first description of a valid species
#	first description of an available species placed into the synonymy of an older species
?	attribution questionable
.	placed explicitly into the synonymy of the present taxon
<u>....</u>	exact spelling of the original publication (e.g. misspelled taxa)
"genus"	indicates that the respective name is obsolete and/or needs to be revised
aff.	affinis (see Bengtson, 1988), used for new, undescribed taxa to relate them to a known taxon
cf.	confer (see Bengtson, 1988), used for provisional identifications (i.e. in cases where critical features are not preserved)
indet.	more specific determination unclear
non	explicitly excluded from the synonymy of the present taxon
pp	pro parte – only part of the material that forms the base of the corresponding record belong into the synonymy of the present taxon
sp.	species, indicates that a specific identification is impossible
v	the specimens on which the corresponding record is based were seen by the author

Abbreviations used for the measurements

APP_ant	anterior paired petals, length of anterior poriferous zone
APP_post	anterior paired petals, length of posterior poriferous zone
APP_Cr	anterior paired petals, corresponding test radius
InfD	infundibulum depth
PeL	petalodium length
PeW	petalodium width
PPP_ant	posterior paired petals, length of anterior poriferous zone
PPP_post	posterior paired petals, length of posterior poriferous zone
PPP_Cr	posterior paired petals, corresponding test radius
PPP_W	posterior paired petals, width at the level of half the test height
PW	maximum petal width
TD	test diameter
TH	test height
TL	test length
TW	test width

(see also Figs. 24, 55)

Glossary

aboral – dorsal side of the animal, opposite of the peristome (mouth)

acetabulum – basal pit on the proximal end of a spine with which it is sitting on the mamelon (rounded topmost part of a tubercle)

adapical – towards the apical disc (→)

adapical suture – suture on the side of a plate nearest to the apical disc (→)

adoral – towards the peristome (→)

adoral suture – suture on the side of a plate nearest to the peristome (→)

adradial suture – vertical suture between ambulacral and interambulacral plates/columns

ambitus – largest circumference of the test

ambulacra (sing. ambulacrum) – five segments of (usually) biserially arranged plates bearing ambulacral pores, through which the water-vascular system extends; extending from the apical disc to the peristome and alternate with five interambulacra; designated by Roman numbers in the Lovénian system

ambulacral plates – individual plate of an ambulacral column (→)

ambulacral pore – paired (usually) perforations in the ambulacral plates (→) through which the vessels of the water-vascular system pass (→), outside of the test (→) a tube foot (→) is attached to the pore in the living animal (a terminology for the ambulacral pores was developed by Smith, 1978, 1980c)

amphisternous – type of plastron (→) in which the labrum (→) is followed by two similar sized sternal plates (→)

ampullae – A) contractile organ of the water-vascular system (→) to operate tube feet (→) (*anat.*); B) internal swelling of a coronal plate (→) below an enlarged primary tubercle (e.g. in *Lovenia*) (*skeletal*)

anal plates – A) plates on the membrane covering the periproct (→); B) paired plates of interambulacrum 5 which border the periproct (→)

anterior paired petals – petals in ambulacra II and IV

apex – highest part on the aboral (dorsal) surface of the test

apical disc – group of plates at the terminal end of the ambulacral and interambulacral columns (→); usually on apex (→) of the test (→); consists of two types of plates [genital (→) and ocular plates (→)], but may incorporate one or more complementary plates in certain groups; in "regular" echinoids five plates of each type are present forming a single or double ring around the periproct (→); in other groups individual plates may be reduced or fused

apophysis (pl. apophyses) – internal projections of the interambulacral basicoronal plates (→) for the attachment of the muscles supporting the lantern (→)

areole – marked depression around each tubercle for attachment of the muscles of the spines

Aristotle's lantern – structure of (usually) 40 ossicles forming the jaw apparatus of the animal; (usually) consists of 10 demipyramids (→), 10 epiphyses (→), 5 rotulae (→), 5 teeth (→), 10 compass elements (→); in some groups completely reduced (e.g. spatangoids and holasteroids) in others partially (e.g. clypeasteroids)

base (spine base) – specialised proximal part of a spine which it is attached to the tubercle (→) and the musculature in the living animal

basicoronal plates – first formed plates of the ambulacral (→) and interambulacral columns (→); usually bordering the peristome (→), may be reduced, occluded or incorporated in the buccal membrane in some taxa; important in the classification in some groups (e.g. clypeasteroids)

boss – part of a tubercle; cone shaped skeletal structure between areole (→) and mamelon (→)

bourrelet – inflated adoral part of the interambulacra in the cassiduloids, lie between the phyllodes (→)

camarodont – descriptive term for a lantern (→) with keeled teeth (→) and closed foramen magnum (→)

camellae – internal swelling of a coronal plates (→) below an enlarged primary tubercle (e.g. in *Lovenia*)

camellate – adj., see camellae

collar – part of the spine base; between milled ring (→) and neck (→); may be smooth or finely striated

column – vertical series of coronal plates (→), extending from apical disc (→) to peristome (→); both ambulacra (→) and interambulacra (→) usually consist of two columns each

compass – element of the lantern (→ Aristotle's lantern); slender, arched, radial rod in ambulacral position on to of the lantern; each rod consists of a simple proximal part and a bifurcating distal part

compound plates – fused ambulacral plates (2 or more) united by a single large tubercle; the number of plates in a compound plate corresponds to the number of pore pairs on the compound plate; type of compounding (→ echinoid, diadematoid,...); important for classification

confluent – descriptive term for weakly separated areoles (→) of adjacent tubercles (used mainly in cidaroids)

conjugate pores – ambulacral pores (→) where the two individual pores of each pair are connected by a groove on the external surface of the test

corona – principal skeleton of an echinoid excluding external appendages (e.g. spines), apical disc, lantern as well as all peristomal and periproctal elements; consists of all interambulacral and ambulacral plates; usually forms a more or less solid globular, cylindrical or disc-shaped structure [except in groups with imbricate (→) or weakly connected coronal plates; e.g. echinothurioids]

coronal plate – either an ambulacral or interambulacral plate; see also *corona*

cortex – dense external layer of cidaroid spines; bearing ornamentation; contains no living tissue in mature spines

crenulate – descriptive term for a tubercle (→) or acetabulum (→) with ribbed/strongly structured periphery

demiplate – term for individual ambulacral plates united in a compound (→) or pseudocompound (→) plate; may be occluded from perradial (→), adradial (→) or both sutures (→)

demipyramid – element of the lantern; one of 10 paired elements supporting the teeth; largest element of the lantern

diadematoid compound plate – compound plate with three similar sized demiplates, non of which is occluded from the adradial or perradial suture (characteristic for diadematoids) dicyclic – descriptive term for a condition in the apical disc of "regular" echinoids where all ocular plates (→) are exsert (→); i.e. only the genital plates (→) form the margin of the periproct (→)

disjunct – interrupted plate column [may occur both in ambulacra and interambulacra; occurs commonly in the plastron of some groups (→)]

echinoid compound plate – compound plate with three or more sized demiplates of markedly different size; usually only the adapical most and adoral most of the demiplate are in contact with both adradial and perradial suture, the others are in contact with the adradial suture only; the adoral most demiplate is the largest (characteristic for the order Echinoida and some Temnopleuroida)

epiphysis – element of the lantern; one of 10 paired elements sitting on top of the demipyramids (→)

episternal plates – paired plates in adoral interambulacrum 5 following the sternal plates in spatangoids

ethmolytic – descriptive term for a condition in the apical disc in spatangoids where genital plate 2 separates genital plates 3 and 4, as well as ocular plates I and V

ethmophract – descriptive term for a condition in the apical disc in spatangoids where genital plate 2 is not elongate; genital plates 3 and 4, as well as ocular plates I and V are in contact with each other

exsert – descriptive term for ocular plates in the apical disc of "regular" echinoids which are not in contact with the periproct

fasciole – specialised band of minute tubercles in holasteroids and spatangoids (for the types see Néraudeau et al. 1998b)

floscelle – star-shaped area around the peristome (→) in cassiduloids, formed by bourrelets (→) and phyllodes (→)

food grooves – sunken grooves in the ambulacra on the oral surface of most clypeasteroids; specialised system for transport of food particles from the site of collection (e.g. margin of the test) to the peristome; bottom of the groove is perforated by numerous microunipores

foramen magnum – space between the upper ends of the joined demipyramids in the lantern

frontal petal – odd petal in ambulacrum III

genital plates – primordial interradial plate in the apical disc perforate by one or more genital pores (→)

genital pores – opening in the test connected to the gonads (reproductive organs) which allows passage of the sperm or eggs

gill slits – or *buccal slits*, indentations in the peristomal margin, allowing passage of compensation or expansion sacs of peripharyngeal tissue; associated with volume control necessary for a active protrusible lantern, not with gaseous exchange (as believed earlier)

gonopores – see *genital pores*

heel – inflated protrusion below the periproct

hemicyclic – descriptive term for a condition in the apical disc of "regular" echinoids where some ocular plates (→) are insert (→) and some exsert (→); i.e. only some ocular plates are in contact with the margin of the periproct (→)

hemilytic – descriptive term for a condition in the apical disc in spatangoids where genital plate 2 separates genital plates 3 and 4, but not ocular plates I and V

holamphisternous – amphisternous (→) plastron (→) with symmetrical sternal plates (→) and symmetrical but otherwise undifferentiated episternal plates (→)

hydropores – perforations in the apical disc providing access to the water-vascular system (→)

imbricate – descriptive term for overlapping coronal plates (like roof tiles; occurs in echinothurioids)

infundibulum – sunken area around the peristome in clypeasteroids

insert – descriptive term for ocular plates in the apical disc of "regular" echinoids which form part of the periproctal margin

interambulacra (sing. interambulacrum) – five columns of (usually) biserially arranged plates extending from the apical disc to the peristome; alternating with the five ambulacral columns; designated by Arabic numbers in the Lovénian system

ambulacral plates – individual plate of an ambulacral column (→)

internal ampullae – see *camellae*

internal fasciole – type of fasciole running around the apical disc and crossing all ambulacra

interporiferous zone – space between the poriferous zones (→) in the petals (→)

interporal partition – ridge of stereom (→) separating the two pores of a pore pair (→)

interradial suture – vertical suture (→) between the two plate columns in an interambulacrum (→)

keel – raised ridge-like area of the corona (→); also used for the ridge on the internal side of the teeth

labrum – differentiated basicoronal plate of interambulacrum 5 in the holasteroids and spatangoids

lamellae – area of radially arranged stereom (→) meshwork between the cortex (→) and the medulla (→) in cidaroid spines

lantern – see Aristotle's lantern

latero-anal fasciole – type of fasciole co-occurring with a peripetalous fasciole (→) in some groups; branching off the peripetalous fasciole behind the anterior paired petals and running towards the posterior end where it crosses interambulacrum 5 below the periproct

Lovénian system (or Lovén's rule) – numbering system for the ambulacra (Roman numbers) and interambulacra (Arabic numbers) developed by Lovén (1874); often used to elucidate plate architecture and homologies in echinoids; when an "irregular" (bilaterally symmetrical) echinoid is viewed from the oral side the ambulacra are counted clockwise starting with the first ambulacrum clockwise of the periproct; interambulacra are numbered in the same fashion; as each ambulacrum comprises two plate columns Lovén named the first column encountered in the clockwise sequence "a" and the second "b", again the same may be done in the interambulacra; in the pentamerous (five-fold symmetrical) "regular" echinoids genital plate 2 is by convention the madreporite (see David et al., 1995 for the ontogenetic basis of Lovén's rule and potential trap when using it to identify homologous structures)

lunule – natural perforation of the test connecting aboral and oral surface; occurs in some clypeasteroid groups; two types: ambulacral lunules (at the perradial suture) and interambulacral lunules (at the interradial suture)

madreporite – plate (usually genital plate 2) of the apical disc perforated by hydropores which provide access to the water-vascular system (→)

mamelon – rounded, topmost part of the tubercles on which the acetabulum (→) articulates

marginal fasciole – type of fasciole (→) surrounding the test at or near the ambitus

medulla – central part of a cidaroid spine, made up of coarse stereom meshwork

mesamphisternous – amphisternous (→) plastron (→) with near-symmetrical sternal plates (→) and biserially offset episternal plates (→)

microcanal system – system of canals within the plates for the passage of the secondary branches of the water-vascular

system (→) leading to accessory tube feet; known only in clypeasteroids
microunipore – minute unipore (→)
miliary spine/tubercle – very small, non-primary spine, respectively the corresponding tubercle
milled ring – widened part of the base of a spine acting as attachment point for the musculature moving the spine
monobasal – type of apical disc with apparently fused genital plates
monocyclic – descriptive term for a condition in the apical disc of "regular" echinoids where all ocular plates (→) are insert (→); i.e. all genital (→) and all ocular (→) plates are in contact with the margin of the periproct (→)
neck – smooth, cylindrical portion of a primary spine, between collar and shaft
neural canal – small notch in the margin of the perradial pore of each pore pair for the passage of the nerve cords into the tube feet
non-conjugate (pore) – opposite of conjugate (→)
non-crenulate (tubercle) – opposite of crenulate (→)
non-perforate (tubercle) – opposite of perforate (→)
non-petaloid – descriptive term for ambulacra without adapically differentiated petal (→)
occluded – descriptive term for the a condition in which a coronal plate (usually an ambulacral plate) does not reach one or both of the two vertical sutures (→) that border it usually
ocular plates – primordial, radial plate of the apical disc, terminating the ambulacra; perforated by a single ocular pore (→)
ocular pore – perforation in the ocular plate (→), for the passage of the water-vascular system (→) into the terminal tentacle (a tube foot)
oral – ventral side of the animal which bears the peristome (→)
orthofasciole – descriptive term for a fasciole (→) band which is well differentiated from the remaining surface tuberculation and does not include primary tubercles (see Néraudeau et al. 1998b)
parapet – part of a tubercle; slope of the boss (→)
pedicellaria (pl. pedicellariae) – minute pincer-like skeletal structures on thin stalks on the surface of the animal between the spines; function ranges from sanitary to defensive; important in the classification of some extant groups, yet rarely preserved in fossils (for a terminology the reader is referred to Melville & Durham, 1966, as it is not employed here)
perforate (tubercle) – descriptive term for tubercles which show a central pit for ligaments connecting spine and tubercle (not present in all groups)
peripetalous fasciole – type of fasciole that surrounds the petalodium (→)
peri-plastronal areas – oral portions of ambulacra I and V next to the plastron (→), usually with differentiated tuberculation (in spatangoids and holasteroids)
periporal area – area of differentiated stereom for the attachment of the soft tissue of the tube feet (→)
periproct – anus
peristome – mouth
perradial suture – vertical suture (→) between the two plate columns in an ambulacrum (→)
petal – differentiated adapical part of an ambulacrum (→) with enlarged/specialised ambulacral pores facilitating tube feet (→) adapted for respiration
petaloid – descriptive term for ambulacra with adapically differentiated petal (→)
petalodium – all five (or four if ambulacrum III nonpetaloid) petals (→)
phyllode – differentiated adoral part of the ambulacra (→) with tube feet (→) adapted for handling food particles
pillar – internal support structure in clypeasteroids
plastron – differentiated adoral part of interambulacrum 5 in spatangoids and holasteroids; structure important for classification
platform – part of a tubercle; flattened part of the boss (→), just below the mamelon (→)
polyporous – descriptive term for compound plates (→) with numerous ambulacral pores (→)
pore pair – ambulacral pore partitioned by a wall of stereom (usual condition), leading to a single tube foot (→)
postbasicoronal plates – plates following the basicoronal plates (→)
posterior paired petals – petals in ambulacra I and V
preanal plates – paired plates between the episternal (→) and anal plates (→) in interambulacrum 5
primary spine/tubercle – largest spines, usually first to appear during growth; one per plate in "regular" echinoids
pseudocompound plates – group of united ambulacral plates not bound together by a single large tubercle
quinquegeminate – descriptive term for compound plates (→) with five ambulacral pores (→)
rostrate – descriptive term for a pointed posterior end of a spatangoid or holasteroid
rostrum – pointed posterior part of interambulacrum 5 in a spatangoid or holasteroid
rotula (pl. rotulae) – element of the lantern; one of 5 ossicles sitting between the paired epiphyses (→)
scrobicular tubercles – circle enlarged tubercles arranged around the areole (→) in cidaroids
secondary spine/tubercle – intermediate sized spine, occurring later during growth than primary spines
shaft – main part of the spine between base and tip
sphaeridia – minute spherically modified spines on short stalks commonly situated adorally in pits near the perradial suture; sometimes completely enclosed within the plates (in sand dollars); function as balance organ
sphaeridial pit – pit in which the sphaeridia (→) are situated
spine – moveable calcareous rods attached to special structures (→ tubercles) on the coronal plates by muscles and ligaments; mainly defensive and/or locomotory in function (depending on group), but fulfil also numerous other tasks
stereom – three-dimensional meshwork of high-Mg calcite of which all skeletal elements are composed; strongly differentiated according to the type of skeletal element, the position within that element and its function (see Smith, 1980 for details and a terminology)
sternal plates – first pair of postbasicoronal plates [following the labrum (→)] of interambulacrum 5 in spatangoids and holasteroids
sternum – see *sternal plates*
stirodont – descriptive term for a lantern (→) with keeled teeth (→) and open foramen magnum (→)
subanal fasciole – type of fasciole lying below the periproct; usually ovate or bilobed
subequal tubercles – in some "regular" echinoids the secondary spines/tubercles (→) may become as large as the primary spines/tubercles at ambital coronal plates, obscuring the differentiation in primary and secondary spines (e.g. *Schizechinus*, *Toxopneustes*, *Tripneustes*,...)
suture – narrow zone marking the contact between adjacent plates; usually represented by a thin line on surface of the test (may be made visible by wetting the test with alcohol or glycerine in extant material, respectively by wetting, weathering, etching or grinding in fossil specimens)
test – collective term for the corona and the plates of the apical disc, as well as any periproctal or peristomal ossicles

tetrabasal – type of apical disc with four separate genital plates (genital plate 5 not present)

tetrageminate – descriptive term for compound plates (→) with four ambulacral pores (→)

tooth – element of the lantern; one of 5 ossicles used for mastication; complex microstructure (useful in classification), growing continuously at the upper end

trigeminate – descriptive term for compound plates (→) with three ambulacral pores (→)

tube foot – external extension of the water-vascular system (→); usually specialised for certain tasks according to the position on the test; functions as locomotory, feeding, respiratory, sanitary, adhesive,... organs or any combination

tubercle – specialised structure on the surface of a coronal plate on which the spines (→) articulate (for a terminology see Smith, 1980b)

ultramphisternous – amphisternous (→) plastron (→) with symmetrical sternal plates (→) followed by symmetrical plate pairs of markedly dissimilar size and shape (e.g. large triangular episternal, followed by strongly differently shaped preanal and anal plates); often disjunct (→) or markedly narrowed

unipore – ambulacral pore (→) which consists only of a single perforation

water-vascular system – system of internal, radially arranged vessels, which contain water; arranged below the ambulacra and extending through the ambulacral pores into the tube feet (→); used to operate the tube feet through regulation of the water pressure inside the vessels by contractile organs (ampullae)

References

Abel, O. (1898): Studien in den Tertiärbildungen von Eggenburg. – *Beiträge zur Paläontologie und Geologie Österreich-Ungarns und des Orients* **11**: 211-226, 3 figs., Wien.

Abel, O. (1902): Zwei neue Menschenaffen aus den Leithakalkbildungen des Wiener Beckens. – *Sitzungsberichte der Kaiserlichen Akademie der Wissenschaften, mathematisch-naturwissenschaftliche Classe, Abtheilung* I **111**: 1171-1207, 1 pl., 2 figs., Wien.

Abel, O. (1904): Die Sirenen der mediterranen Tertiärbildungen Österreichs. – *Abhandlungen der kaiserlich-königlichen Geologischen Reichsanstalt* **19**/2: 1-223, 7 pls., 26 figs., Wien.

Abel, O. (1922): Lebensbilder aus der Tierwelt der Vorzeit. vii+643 pp., 1 pl., 507 figs., Jena (Gustav Fischer).

Abel, O. (1924): Landschaft und Tierleben des Wiener Beckens in der Tertiärzeit. – In: Abel, O. (ed.): *Wien – Sein Boden und seine Geschichte.* pp. 52-83, figs. 22-42, Wien (Wolfrum Verlag).

Abich, H. (1882): Geologische Forschungen in den Kaukasischen Ländern, II. Theil. Geologie des Armenischen Hochlandes. I. Westhälfte. I-X, 1-478 pp., 19 pls., 49 figs., 5 maps, Wien (Alfred Hölder Verlag).

Agassiz, A. (1863): List of the Echinoderms sent to different Institutions in Exchange for other Specimens, with Annotations. – *Bulletin of the Museum of Comparative Zoölogy at Harvard College* **1**/2: 17-28, Cambridge, MA.

Agassiz, A. (1869): Preliminary Report on the Echini and Starfishes dredged in deep water between Cuba and the Florida Reef, by L. F. de Pourtalès, Assist. U.S. Coast Survey. – *Bulletin of the Museum of Comparative Zoölogy at Harvard College* **1**/9: 253-308, 1 tab., Cambridge, MA.

Agassiz, A. (1872-1874): Revision of the Echini. – *Memoirs of the Museum of Comparative Zoology at Harvard College* **III**: pt. 1-2: i-xii, 1-378, pls. 1-49 (1872); pt. 3: 379-628 +1, pls. 50-77 (1873); pt. 4: 629-762, pls. 78-94 (1874), Cambridge, MA.

Agassiz, A. (1878): Report on the results of dredging, under the supervision of Alexander Agassiz, in the Gulf of Mexico, by the U.S. Coast Survey Steamer "Blake", Lieutenant-Commander C. D. Sigsbee, U.S.N., Commanding. Report on the Echini. – *Bulletin of the Museum of Comparative Zoölogy at Harvard College* **5**/9: 181-195, 5 pls., 2 figs., Cambridge, MA.

Agassiz, A. (1879): Preliminary report on the Echini of the Exploring Expedition of H.M.S "Challenger". – *Proceedings of the American Academy of Arts and Sciences, new series* **6**: 190-212, Cambridge, MA.

Agassiz, A. (1881): Report on the scientific results of the voyage of the H.M.S. "Challenger" during the years 1873-1876, Zoology, Vol. III, Part IX, Report on the Echinoidea. 321 pp., London (Longmans & Company).

Agassiz, A. (1883): Report on the results of dredging under the supervision of Alexander Agassiz, in the Gulf of Mexico (1877-1878), in the Caribbean Sea (1878-79), and along the Atlantic coast of the United States (1880), by the U.S. Coast Survey Steamer "Blake", Lieut.-Com. C. D. Sigsbee, U.S.N., and Commander J. R. Bartlett, U.S.N., commanding. XXIV, Part 1. Report on the Echini. – *Memoirs of the Museum of Comparative Zoölogy at Harvard College* **10**/1: i-v+1-94, 32 pls., Cambridge, MA.

Agassiz, A. (1904): The Panamic Deep Sea Echini. – *Memoirs of the Museum of Comparative Zoölogy at Harvard College* **31**: 1-243, 110 pls., Cambridge, MA.

Agassiz, A. & Clark, H.L. (1907): Preliminary report on the Echini collected 1906, from May to December, among the Aleutian Islands, in Bering Sea, and along the coasts of Kamchatka, Saghalin, Korea, and Japan, by U.S. Fish Commission steamer "Albatross", in charge of Lieut.-Commander L. M. Garrett, U.S.N., commanding. – *Bulletin of the Museum of Comparative Zoology at Harvard College* **51**/3: 107-139, Cambridge, MA.

Agassiz, L. (1835): Prodrome d'une monographie des Radiaires ou Échinodermes. – *Mémoires de la Société des Sciences Naturelles de Neuchâtel* **1**: 168-199, Neuchâtel.

Agassiz, L. (1839): Description des Échinodermes fossiles de la Suisse; première partie, Spatangoides et Clypéasteroides. – *Mémoires de la Société helvétique des Sciences naturelles* **3**: i-viii, 1-101, 14 pls., Neuchâtel.

Agassiz, L. (1840a): Description des Echinodermes fossiles de la Suisse; Seconde partie. Cidarides. – *Mémoires de la Société helvétique des Sciences naturelles* **4**: i-iv, 1-97, 9 pls., Neuchâtel.

Agassiz, L. (1840b): Catalogus systematicus Ectyporum Echinodermatum fossilium Musei Neocomiensis, secundum ordinem zoologicum dispositus; adjectis synonymis recentioribus, nec non stratis et locis in quibus reperiuntur. Sequuntur characteres diagnostici generum novorum vel minus cognitorum. 1-20 pp., Neuchâtel (Oliv. Petitpierre).

Agassiz, L. (1841a): Monographies d'Échinodermes vivans et fossiles. Échinites. Famille des Clypéasteroides. Seconde Monographie. Des Scutelles. 149 pp., 32 pls., Neuchâtel (Petitpierre).

Agassiz, L. (1841b): Observations sur les progrés récens de l'histoire naturelle des échinodermes. – In: Agassiz, L.: *Monographies d'Échinodermes vivants et fossiles.* pp. 20, Neuchâtel (Petitpierre).

Agassiz, L. & Desor, P.J.E. (1846): Catalogue raisonné des familles, des genres, et des espèces de la classe des échinodermes. – *Annales des Sciences Naturelles, Troisième Série, Zoologie* **6**: 305-374, pls. 15-16, Paris.

Agassiz, L. & Desor, P.J.E. (1847a): Catalogue raisonné des espèces, des genres, et des familles d'échinides. – *Annales des Sciences Naturelles, Troisième Série, Zoologie* **7**: 129-168, Paris.

Agassiz, L. & Desor, P.J.E. (1847b): Catalogue raisonné des espèces, des genres, et des familles d'échinides. – *Annales des Sciences Naturelles, Troisième Série, Zoologie* **8**: 5-35, 355-380, Paris.

Airaghi, C. (1898): Echinidi del Pliocene Lombardo. – *Atti della Società Italiana di Scienze Naturali e del Museo Civico di Storia Naturale in Milano* **37**: 357-377, 1 pl., Milano.

Airaghi, C. (1899): Echinidi del Bacino della Bormida. – *Bolletino della Società Geologica Italiana* **18**: 140-178, 2 pls., 1 tab., Roma.

Airaghi, C. (1900): Di alcuni Conoclipeidi. – *Bolletino della Società Geologica Italiana* **19**: 173-178, 1 pl., Roma.

AIRAGHI, C. (1901): Echinidi terziari del Piemonte e della Liguria. – *Palaeontographia Italica* **7**: 149-219, 9 pls., Pisa.

AIRAGHI, C. (1905a): Echinidi miocenici della Sardegna. – *Atti della Società Italiana di Scienze Naturali e del Museo Civico di Storia Naturale in Milano* **44**: 209-217, 2 figs., Milano.

AIRAGHI, C. (1905b): Echinodermi miocenici dei dintorni di S. Maria Tiberina (Umbria). – *Atti della Reale Accademia della Scienze di Torino* **40**: 43-54, 1 pl., Torino.

AIRAGHI, C. (1908): Revisione degli Asteroidi e degli Echinidi Lombardi. – *Rendiconti Reale Istituto Lombardo di Scienze e Lettere, Serie II* **41**: 244-259, 5 figs., Milano.

AIRAGHI, C. (1930): Echinidi del Quarternario antico dell'Isola di Rodi (Dodecaneso). – *Atti della Società Italiana di Scienze Naturali e del Museo Civico di Storia Naturale in Milano* **69**/3-4: 311-318, pl. 8, Milano.

ALCOCK, A.W. (1893): Natural History notes from H.M. Indian Marine Survey steamer "Investigator". Ser. 2, No. 9. An account of the deep-sea collection made during the season of 1892-93. – *Journal of the Asiatic Society* **62**/2-4: 169-184, pls. 8-9, Bengal. [not seen]

ALI, M.S.M. (1983): The paleogeographical distribution of *Clypeaster* (Echinoidea) during the Cenozoic Era. – *Neues Jahrbuch für Geologie und Paläontologie, Monatshefte* **1983/8**: 449-464, 4 figs., 1 tab., Stuttgart.

ALI, M.S.M. (1985): On some Pliocene echinoids from the area north of Mersa Alam, Red Sea coast, Egypt. – *Paläontologische Zeitschrift* **59**/3-4: 277-300, 12 figs., 4 tabs., Stuttgart.

ALI, M.S.M. (1998): Some Miocene Scutellina (Echinoidea, Echinodermata), from the northern Western Desert, Egypt: A preliminary study. – In: MOOI, R. & TELFORD, M. (eds.): *Echinoderms: San Francisco.- Proc. of the 9th Int. Echinoderm Conf. San Francisco/California(USA), 5-9 Aug. 1996*, pp. 541-546, 3 figs., Brookfield (A.A. Balkema).

ANDRAE, K.J. (1855): Bericht über die Ergebnisse geognostischer Forschung im Gebiete der 14., 18. und 19. Section der General-Quartiermeisterstabs-Karte von Steiermark und Illyrien während des Sommers 1854. – *Jahrbuch der kaiserlich-königlichen Geologischen Reichsanstalt* **6**: 265-304, Wien.

ANGELIS D'OSSAT, G. (1897): Contribuzione allo studio paleontologico dell'Alta valle dell'Aniene. – *Bollettino della Società Geologica Italiana* **16**: 280-318, Roma.

ANONYMOUS (1767): Auszug eines Schreibens aus Stargard in Meckl. an den hn. C. R. R. von Neustrelitz von einer unbekannten und bisher noch nicht beschriebenen Versteinerung. – *Hamburgische Nachrichten aus dem Reiche der Gelehrsamkeit* **84-85**: 701-706, Hamburg.

BACHMAYER, F. (1962): Ein neuer Aufschluß im Torton südlich von Hagenbrunn (N.-Ö.). – *Verhandlungen der Geologischen Bundesanstalt* **1962**: 377-378, Wien.

BAGDASARYAN, K.G. (1970): Molluscs from the Tarkhanian Horizon and their living conditions [in Russian]. 150 pp., 7 pls., Tbilisi (Mecniereba).

BATHER, F.A. (1908a): The Echinoid Name *Cidaris* and its Modern Application. – *The Annals and Magazine of Natural History, 8th Series* **1**: 284-288, London.

BATHER, F.A. (1908b): The Genotype of *Cidaris*. – *The Annals and Magazine of Natural History, 8th Series* **2**: 134-136, London.

BATHER, F.A. (1909): The Type of *Cidaris*. – *The Annals and Magazine of Natural History, 8th Series* **3**/13: 88, London.

BAZIN (1884): Sur les Échinides du Miocène moyen de la Bretagne. – *Bulletin de la Société Géologique de France, Série 3* **12**: 34-45, 3 pls., Paris.

BENGTSON, P. (1988): Open Nomenclature. – *Palaeontology* **31**/1: 223-227, London.

BERGGREN, W.A., KENT, D.V., SWISHER III, C.C. & AUBRY, M.-P. (1995): A revised Cenozoic geochronology and chronostratigraphy. – In: BERGGREN, W.A., KENT, D.V., AUBRY, M.-P. & HARDENBOL, J. (eds.): *Geochronology, time scales and global stratigraphic correlations: A unified temporal framework for a historical geology. – SEPM Special Publication* **54**/4 pp. 129-212, Tulsa (Society for Sedimentary Geology).

BERNHAUSER, A. (1955): Zur Kenntnis der Retzer Sande. – *Sitzungsberichte der Österreichischen Akademie der Wissenschaften Wien, mathematisch-naturwissenschaftliche Klasse, Abteilung* I **164**: 163-192, 1 pl., 12 figs., Wien.

BILLINGS, E. (1858): On the Asteriadae of the Lower Silurian rocks of Canada. – *Geological Survey of Canada, Figures and Descriptions of Canadian Organic Remains* **decade 3**: 75-85, Montreal. [not seen; cited from HOTCHKISS, 1976]

BITNER, M.A. & PISERA, A. (2000): Brachiopod fauna from the Middle Miocene deposits of Niechobrz, south-eastern Poland. – *Tertiary Research* **20**/1-4: 7-16, 2 pls., 2 figs., Leiden.

BITTNER, A. (1880): Beiträge zur Kenntnis alttertiärer Echinidenfaunen der Südalpen. – *Beiträge zur Paläontologie und Geologie Österreich-Ungarns und des Orients* **1**/1: 1-68, 8 pls., Wien.

BITTNER, A. (1882): Beiträge zur Kenntnis alttertiärer Echinidenfaunen der Südalpen. – *Beiträge zur Paläontologie und Geologie Österreich-Ungarns und des Orients* **1**/2: 73-110, 4 pls., Wien.

BITTNER, A. (1884): Die Tertiär-Ablagerungen von Trifail und Sagor. – *Jahrbuch der kaiserlich-königlichen Geologischen Reichsanstalt* **34**: 433-600, pl. 10, 7 figs., Wien.

BITTNER, A. (1891): Über *Parabrissus* und einige andere alttertiäre Echinidengattungen. – *Verhandlungen der kaiserlich-königlichen Geologischen Reichsanstalt* **25**/6: 133-144, 1 fig., Wien.

BLONDEL, T. & PHILIPPE, M. (1992): Les Echinides du Burdigalien terminal-Langhien de Tunisie. Données paléobiologiques et paléogéographiques. – *Revue de Paléobiologie* **11**/2: 433-446, 2 pls., 2 figs., Genève.

BOBIES, C.A. (1928): Das Gaadener Becken. (Ein Beitrag zur jüngsten Geschichte des Kalkalpenrandes bei Wien.) – *Mitteilungen der Geologischen Gesellschaft in Wien* **19** (1926): 41-78, 1 tab., 1 map, Wien.

BOBIES, C.A. (1930): Über bryozoenführende Sedimente des inneralpinen Wiener Beckens. – *Mitteilungen der Geologischen Gesellschaft in Wien* **21** (1928): 24-34, Wien.

BODA, J. (1964): Catalogus Originalum Fossilium Hungariae. Pars Zoologica. 229 pp., Budapest (M. Áll. Földtani Intézet).

BOGGILD, G.R. & ROSE, E.P.F. (1984): Mid-Tertiary echinoid biofacies as palaeoenvironmental indices. – *Annales Géologiques des Pays Helléniques* **32**: 57-67, 3 figs., Athènes.

BOJAR, A.-V., HIDEN, H., FENNINGER, A. & NEUBAUER, F. (2004): Middle Miocene seasonal temperature changes in the Styrian basin, Austria, as recorded by isotopic composition of pectinid and brachiopod shells. – *Palaeogeography, Palaeoclimatology, Palaeoecology* **203**: 95-105, 6 figs., 1 tab., Amsterdam.

BÖHM, A. von (1882): Über einige tertiäre Fossilien von der Insel Madura, nördlich von Java. – *Denkschriften der kaiserlichen Akademie der Wissenschaften, mathematisch-naturwissenschaftliche Classe* **45**: 359-372, 4 pls., 2 figs., Wien.

BORGHI, E. (1993): Nuove acquisizioni relative a: *Schizechinus serialus* POMEL, 1887. Echinodermi fossili emiliani – II. – *Bollettino de la Società Reggiana Scienze naturali* **13**/2: 1-11, 3 pls., 2 figs., Reggio Emilia.

Borghi, E. (1994): *Echinolampas (Macrolampas) hoffmanni* Desor, 1847. Echinodermi fossili emiliani – VI°. – *Bollettino de la Società Reggiana Scienze naturali* **14**/1: 1-11, 2 pls., 5 figs., 2 tabs., Reggio Emilia.

Borghi, E. (1997): Genus *Brissopsis* in Pliocene and Pleistocene of Emilia region (North Italy). Fossil Echinoderms – XII°. – *Bollettino de la Società Reggiana Scienze naturali* **17**/1: 1-15, 2 pls., 9 figs., Reggio Emilia.

Botto-Micca, L. (1896): Contribuzione allo studio degli Echinidi Terziarii del Piemonte (famiglia Spatangidi). – *Bolletino della Società Geologica Italiana* **15**: 341-375, Roma.

Breton, G., Néraudeau, D. & Cuenca-Boulat, C. (1992): *Gnatichnus stellarum* ichnosp. nov., trace de broutagne d'un échinide du Campanien des Charentes (France). – *Revue de Paléobiologie* **11**/1: 219-229, 1 pl., 11 figs., Paris.

Brighton, A.G. (1929): The echinoid species *Echinoneus melitensis* (Wright) and *E. cyclostomus* Leske. – *The Annals and Magazine of Natural History, 10th Series* **3**: 85-95, 13 figs., London.

Brighton, A.G. (1931): The Geology of the Farsan Islands, Gizan and Kamaran Islands, Red Sea. Part 3. – Echinoidea. – *Geological Magazine* **68**/805: 323-333, 6 figs., 1 tab., London.

Brix, F. (1980): Zur Geologie des Tertiärs und Quartärs der Umgebung von Hirtenberg (Niederösterreich). – In: *Heimatbuch der Marktgemeinde Hirtenberg*. pp. 37-79, 3 pls., 1 map., Leobersdorf (Stiepandruck).

Brix, F. (1988): Jungtertiär und Quartär. – In: Brix, F. & Plöchinger, B. (eds.): *Erläuterungen zu Blatt 76 Wiener Neustadt. Geologische Karte der Republik Österreich 1:50.000*. pp. 29-51, 2 tabs., Wien (Geologische Bundesanstalt).

Bromley, R.G. (1975): Comparative analysis of fossil and recent echinoid bioerosion. – *Palaeontology* **18**/4: 725-739, pls. 85-89, 2 figs., London.

Bromley, R.G. & Asgaard, U. (1975): Sediment structures produced by a spatangoid echinoid: A problem of preservation. – *Bulletin Geological Society of Denmark* **24**: 261-281, Copenhagen.

Brongniart, A. (1822): Mémoire sur les terrains de sédiment superieurs calcaréo-trappéens du Vicentin, et sur quelques terrains d'Italie, de France, d'Allemagne, etc., qui peuvent se rapporter à même époque. iv+86 pp., 6 pls., Paris. [not seen; cited from Fischer (1966)]

Bronn, H.G. (1835-1837a): Lethaea geognostica oder Abbildungen und Beschreibungen der für die Gebirgs-Formationen bezeichnendsten Versteinerungen. Erster Band, das Übergangs- bis Oolithen-Gebirge enthaltend. vi+544 pp., Stuttgart (E. Schweizerbart's Verlagshandlung).

Bronn, H.G. (1837b): Lethaea geognostica oder Abbildungen und Beschreibungen der für die Gebirgs-Formationen bezeichnendsten Versteinerungen. XLVII Tafeln mit Abbildungen zur Lethäa Geognostica. 16+8 pp., 47+1 pls., 2 tabs., Stuttgart (E. Schweizerbart's Verlags-Buchhandlung).

Bronn, H.G. (1838): Lethaea geognostica oder Abbildungen und Beschreibungen der für die Gebirgs-Formationen bezeichnendsten Versteinerungen. Zweiter Band, das Kreide- und Molasse-Gebirge enthaltend. 545-1346, and 4 unpaginated pages pp., Stuttgart (E. Schweizerbart's Verlagshandlung).

Bronn, H.G. & Roemer, F. (1853-1856a): H. G. Bronn's Lethaea Geognostica oder Abbildungen und Beschreibungen der für die Gebirgs-Formationen bezeichnendsten Versteinerungen. Dritte stark vermehrte Auflage. Dritter Band. viii+1130 (pt. 6) pp., Stuttgart (E. Schweizerbart's Verlagshandlung und Druckerei).

Bronn, H.G. & Roemer, F. (1850-1856b): Atlas zu H. G. Bronn's Lethaea Geognostica oder Abbildungen und Beschreibungen der für die Gebirgs-Formationen bezeichnendsten Versteinerungen. Dritte stark vermehrte Auflage. 123 pp., 63 pls., Stuttgart (E. Schweizerbart's Verlagshandlung und Druckerei).

Brzobohatý, R., Cicha, I., Kováč, M. & Rögl, F. (eds.) (2003): The Karpatian – A Lower Miocene Stage of the Central Paratethys. 360 pp., Brno (Masaryk University).

Bunzel, E. (1869): Ueber den marinen Tegel vom Porzteich bei Voitelsbrunn. – *Verhandlungen der kaiserlich-königlichen Geologischen Reichsanstalt* **3**: 81, Wien.

Cahuzac, B. & Cluzaud, A. (1998): Observations géologiques sur le carrière de Bougue (Landes) et animations scientifiques dans cette commune. – *Bulletin de la Société de Borda* **451**/4: 413-442, 3 pls., 11 figs., Dax, Landes.

Cahuzac, B. & Roman, J. (1994): Les échinoides de l'Oligocene supérieur (Chattien) des Landes (Sud-Aquitaine, France). – *Revue de Paléobiologie* **13**/2: 351-373, 3 pls., 4 figs., 3 tabs., Genève.

Cameron, R.A., Boidron-Metairon, I. & Monterrosa, O. (1985): Does the embryonic response to temperature and salinity by four species of Caribbean sea urchins parallel the reproductive synchrony. – *Proceedings of the Fifth International Coral Reef Symposium, Tahiti* **5**: 273-278. [not seen]

Capeder, G. (1906): Fibularidi del Miocene medio di S. Gavino a mare (Portotorres) Sardegna. – *Bolletino della Società Geologica Italiana* **25**: 195-534, 1 pl., Roma.

Carrasco, J.F. (2003): Presencia de la icnoespecie *Gnathichnus pentax* sobre *Dimya richei* (Bivalvia) en el Eoceno inferior de La Puebla de Roda (Huesa). – *Batalleria – Revista de Paleontologia* **11**: 37-40, 1 pl., Barcelona.

Ceranka, T. & Złotnik, M. (2003): Traces of cassid snails predation upon the echinoids from the Middle Miocene of Poland. – *Acta Palaeontologica Polonica* **48**/3: 491-496, 2 figs., 2 tabs., Warszawa.

Challis, G.R. (1979): Miocene echinoid biofacies of the Maltese Islands. – *Annales Géologiques des Pays Helléniques, tome hors série* **1**: 253-262, 1 fig., Athènes.

Challis, G.R. (1980): Palaeoecology and Taxonomy of Mid-Tertiary Maltese echinoids – unpublished Ph. D. Thesis, Department of Geology, Bedford College, University of London, 401 pp., 81 figs., 132 pls., London.

Chavanon, S. (1974): L'approche statistique dans l'interpretation de la systematique de l'evolution et de la paleoecologie des Échinides des formation du Cenozoique Bordelais. – *Memoires de l'Institut de Geologie du Bassin d'Aquitaine* **6**/1-2: 1-457, 24 pls., 51 + 269 figs., Talence.

Checchia-Rispoli, G. (1917): Gli Echinidi viventi e fossili della Sicilia. Parte quinta: Echinidi miocenici. – *Palaeontographia Italica* **23**: 55-78, 6 pls., Pisa.

Checchia-Rispoli, G. (1920): Osservazioni sulla morfologia interna dei Clipeastri e loro classificazione. – *Palaeontographia Italica* **26**: 57-74, 1 pl., 1 tab., Pisa.

Checchia-Rispoli, G. (1923): Nuove osservazioni sulla struttura interna dei Clipeastri. – *Palaeontographia Italica* **29-30**: 25-30, 1 pl., Pisa.

Checchia-Rispoli, G. (1925): Illustrazione dei Clipeastri miocenici della Calabria seguita da uno studio sulla morfologia interna e sulla classificazione dei Clipeastri. – *Memorie descrittive della carta geologica d'Italia* **9**/3: 1.75, 24 pls., 21 figs., Roma.

Checchia-Rispoli, G. (1927a): Illustrazione degli echinidi cenozoici della Cirenaica raccolti dall'Ing. C. Crema. III. – Generi *Schizaster* e *Trachyaster*. – *Bollettino del R. Ufficio Geologico d'Italia* **52**/3: 1-6, 1 pl., 3 figs., Roma.

Checchia-Rispoli, G. (1927b): Illustrazione degli echinidi cenozoici della Cirenaica raccolti dall'Ing. C. Crema. VI. – Generi *Stolonoclypus, Clypeaster, Tretodiscus*. – *Bollettino del R. Ufficio Geologico d'Italia* **52**/9: 1-18, 5 pls., 4 figs., Roma.

Chenuil, A. & Féral, J.-P. (2003): Sequences of mitochondrial DNA suggest that *Echinocardium cordatum* is a complex of several sympatric or hybridizing species: A pilot study. – In: Féral, J.-P. & David, B. (eds.): *Echinoderm Research 2001 – Proc. 6th Europ. Conf. Echinoderms Res., Banyuls/France/ 3-7 Sept. 2001.* pp. 15-21, 2 figs., 2 tabs., Lisse (Swets & Zeitlinger).

Chesher, R.H. (1968): The systematics of sympatric species in West Indian spatangoids: a revision of the genera *Brissopsis, Plethotaenia*, and *Saviniaster*. – *Studies in Tropical Oceanography* **7**: i-viii+1-168, Miami, FL.

Chesher, R.H. (1969): Contributions to the biology of *Meoma ventricosa* (Echinoidea: Spatangoida). – *Bulletin of Marine Science* **19**/1: 72-110, Coral Gables, FL.

Chesher, R.H. (1970): Evolution in the Genus *Meoma* (Echinoidea: Spatangoida) and a Description of a New Species from Panama. – *Bulletin of Marine Science* **20**/3: 731-761, 9 figs., Coral Gables, FL.

Cicha, I. (1978): 3. Faziostratotypus: Židlochovice, Karpatische Vortiefe in Mähren, Tschechoslovakei (Unterbaden, $M_{4b;}$ Obere Lagenidenzone). – In: Papp, A., Cicha, I., Seneš, J. & Steininger, F. (eds.): *M_4 Badenien (Moravien, Wielicien, Kosovien). – Chronostratigraphie und Neostratotypen, Miozän der Zentralen Paratethys 6*, pp. 168-170, fig. 42, Bratislava (VEDA – Vydavatel'stvo Slovenskej akad. vied).

Cicha, I., Ondrejíčková, A., Seneš, J., Špička, V., Tejkal, J. & Zapletalová, I. (1967): Holostratotypus und Faziostratotypen des M_3 (Karpatien). – In: Cicha, I., Seneš, J. & Tejkal, J. (eds.): *M_3 Karpatien. Die Karpatische Serie und ihr Stratotypus. – Chronostratigraphie und Neostratotypen, Miozän der Zentralen Paratethys 1*, pp. 50-99, figs. 6-13, Bratislava (VEDA – Vydavatel'stvo Slovenskej akad. vied).

Clark, A.M. & Rowe, F.W.E. (1971): Monograph of shallow-water Indo-West Pacific Echinoderms. 238 pp., 31 pls., London (Trustees of the British Museum (Natural History)).

Clark, H.L. (1908): The Type of *Cidaris*. – *The Annals and Magazine of Natural History, 8th Series* **1**: 532-533, London.

Clark, H.L. (1909): The Type of *Cidaris*. – *The Annals and Magazine of Natural History, 8th Series* **3**/13: 88, London.

Clark, H.L. (1917): Hawaiian and other Pacific Echini. The Echinoneidae, Nucleolitidae, Urechinida, Echinocorythidae, Calymnidae, Pourtalesiidae, Palaeostomatidae, Aeropside, Palaeopneustidae, Hemiasteridae, and Spatangidae. – *Memoirs of the Museum of Comparative Zoology at Harvard College* **46**/2: 81-283, 18 pls. Cambridge, MA.

Clark, H.L. (1925): A Catalogue of the Recent Sea-Urchins (Echinoidea) in the Collection of the British Museum (Natural History). 250 pp., 12 pls., London (Oxford University Press).

Claus, C.F.W. (1876): Grundzüge der Zoologie. 3rd ed. XII+1254 pp., 1 vol., Marburg & Leipzig (N. G. Elwert´sche Universitätsbuchhandlung).

Claus, C.F.W. (1880): Grundzüge der Zoologie. 4th ed. vol. 1: VII+821 pp., vol. 2: IV+522 pp., Marburg & Leipzig (N. G. Elwert´sche Universitätsbuchhandlung).

Codrea, V. (1981): Contribuţii la cunoaşterea scutelidelor Badeniene din Bazinul Beiuş. – *Nymphaea – Folia Naturae Bihariae* **8-9** (1980-1981): 139-147, 5 figs., 3 tabs., Oradea.

Comaschi Caria, I. (1955): Il Sottogenere Amphiope in Sardegna. – *Bolletino della Società Geologica Italiana* **74**: 183-194, 15 pls., Roma.

Comaschi Caria, I. (1963): La fauna miocenica della zona di *Funtanazza compressa* tra le Marine di Montevecchio ed Arbus in Sardegna. – *Rendiconti del Seminario della Facoltà di Scienze della Università di Cagliari.* **33 (1964)**/3-4: (1-74), 11 pls., Bologna.

Comaschi Caria, I. (1972): The miocene echinoids of Sardinia. 96 pp., Cagliari (Stabil. Tipogr. ed. Fossataro).

Commenda, H. (1900): Materialien zur Geognosie Oberösterreichs. Ein Beitrag für die Landeskunde in Einzeldarstellungen. – *Jahres-Bericht des Museums Francisco-Carolinum* **58**: 1-272, Linz.

Cooke, C.W. (1959): Cenozoic Echinoids of Eastern United States. – *United States Geological Survey Professional Papers* **321**: III+1-106, 43 pls., Washington, DC.

Coppi, F. (1884): Il Miocene medio nei colli modenesi; appendice alla Paleontologia Modenese. – *Bollettino del R. Comitato Geologico d'Italia* **15**/5-6: 171-201, Roma.

Cotteau, G. (1858-1880): Échinides nouveaux ou peu connus. 1re série. (Extraits de la Revue et Magasin de Zoologie). 1858: 1-2, pls. 1-2 (No. 1-6); 1859: 13-27, pls. 3-4 (No. 7-12); 1860: 29-42, pls. 5-6 (No. 13-22); 1861: 43-58, pls. 7-8 (No. 23-34); 1862: 59-76, pls. 9-10 (No. 35-42); 1863: 77-92, pls. 11-12 (No. 43-51); 1864: 93-108, pls. 13-14 (No. 52-61); 1866: 109-124, pls. 15-16 (No. 62-69); 1867: 125-139, pls. 17-18 (No. 70-77); 1869: 141-152, pls. 19-20 (No. 78-83); 1872: 153-167, pls. 21-22 (No. 84-90); 1873: 168-178, pls. 23-24 (No. 91-96); 1875: 180-191, pls. 25-26 (No. 97-103); 1876: 192-201, pls. 27-28 (No. 104-110); 1878: 202-217, pls. 29-30 (No. 111-120); 1880: 218-234, pls. 31-32 (No. 121-128) pp., Paris (J.-B. Baillière). [also published in the journal „Revue et Magasin de Zoologie", with different pagination]

Cotteau, G. (1862-1967): Terrain Crétace, Tome Septième. Échinides. – Paléontologie Française. Description des Animaux Invertébrés commencée par Alcide d'Orbigny continuée sous la direction d'un comité spécial. 1-64, pls. 1007-1018 (Feb. 1861); 65-128, pls. 1019-1030 (Jul. 1861), 129-176, pls. 1031-1043 (Nov. 1861); 177-224, pls. 1044-1052, 1080, 1088 (May 1862); 225-272, pls. 1053-1064 (Jul. 1862); 273-329, pls. 1065-1075, 1054bis (Oct. 1862), 321-368, pls. 1076-1079, 1081-1087, 1087bis (April 1863); 369-416, pls. 1089-1100 (Jul. 1863); 417-464, pls. 1101-1112 (Oct. 1863); 465-512, pls. 1113-1124 (Feb. 1864); 513-560, pls. 1125-1131, 1133-1137 (June 1864); 561-608, pls. 1132, 1138-1148 (Sept. 1864); 609-656, pls. 1149-1160 (Jan. 1865); 657-704, pls. 1161-1172 (April 1865); 705-752, pls. 1173-1184 (Oct. 1865); 753-800, pls. 1185-1196 (May 1866); 801-892, pls. 1197-1204 (Jan. 1867), Paris (V. Masson et fils).

Cotteau, G. (1863): Échinides fossiles de Pyrénées. 160 pp., 9 pls., 1 tab., Paris (F. Savy).

Cotteau, G. (1875): Description des Échinides Tertiaires des îles St-Barthélemy et Anguilla. – *Kongelige Svenska Vetenskaps-Akademiens Handlingar* **13**/6: 1-48, 8 pls., Stockholm.

Cotteau, G. (1877): Description des Echinides. – In: Locard, A. (ed.): Description des Faune des terrains Tertiaires moyen de la Corse. – *Annales de la Société d'Agriculture, Historie Naturelle et Arts utiles de Lyon* **1877**: 227-335, pls. 8-17, Paris-Genève.

Cotteau, G. (1882-1887): Échinides nouveaux ou peu connus. 2e série, articles 1-6. (Extraits du Bulletin de la Société Zoologique de France). 1882: 1-19, pls. 1-2 (No. 1-18); 1883: 21-35, pls. 3-4 (No. 19-28); 1884: 37-51, pls. 5-6 (No. 29-36); 1885: 53-66, pls. 7-8 (No. 37-45); 1886: 69-89, pls. 9-10 (No. 46-55); 1887: 91-103, pls. 11-12 (No. 56-62) pp., Paris. [also published in the journal "Bulletin de la Société Zoologique de France", with different pagination]

Cotteau, G. (1885-1889): Terrain Tertiaire, Tome I. Échinides Éocènes. Familles des Spatangidées, des Brissidées, des Échinonéidées et des Cassidulidées (pars). – Paléontologie

Française ou description des fossiles de la France continuée par une réunion de paléontologistes sous la direction d'un comité spécial, 1re Série. – Animaux Invertébrés 1-48, pls. 1-2 (Nov. 1885); 49-96, pls. 3-24 (Jan. 1886); 97-144, pls. 25-36 (Apr. 1886); 145-176, pls. 37-48 (Apr. 1886); 177-208, pls. 49-60 (Jul. 1886); 209-240, pls. 61-72 (Nov. 1886); 241-272, pls. 73-84 (Dec. 1886); 273-320, pls. 85-96 (Jan. 1887); 321-368, pls. 97-108 (March 1887); 369-416, pls. 109-120 (Jun. 1887); 417-464, pls. 121-132 (Oct. 1887); 465-512, pls. 133-144 (Dec. 1887); 513-544, pls. 145-156 (Apr. 1888); 545-576, pls. 157-168 (May 1888); 577-608, pls. 609-640, pls. 181-192 (Jan, 1889); 641-672, pls. 193-200 (Feb. 1889), Paris (G. Masson).

COTTEAU, G. (1889-1894): Terrain Tertiaire, Tome II. Échinides Éocènes. Familles des Cassidulidées (pars), Conoclypéidées, Clypéasteroidées, Scutellidées, Scutellinidées, Fibularidées et les Échinides réguliers. – Paléontologie Française ou description des fossiles de la France continuée par une réunion de paléontologistes sous la direction d'un comité spécial, 1re Série. – Animaux Invertébrés 1-48, pls. 201-212 (Nov. 1889); 49-80, pls. 213-224 (March 1890); 81-112, pls. 225-236 (May 1890); 113-160; pls. 237-248 (Nov. 1890); 161-208, pls. 249-260 (Jan. 1891); 209-256, pls. 261-272 (Apr. 1891); 157-304, pls. 273-284 (Dec. 1891); 305-352, pls. 287-296 (Jan. 1892); 353-400, pls. 297-308 (May 1892); 401-448, pls. 309-320 (Jul. 1892); 449-528, pls. 321-328 (Dec. 1892); 529-576, pls. 329-340 (March 1893); 577-624, pls. 341-352 (Aug. 1893); 625-672, pls. 353-364 (Dec. 1893), 673-720, pls. 365-376 (April 1894); 721-788, pls. 377-384 (May 1894), Paris (G. Masson).

COTTEAU, G. (1895): Description des Échinides recueillis par M. Lovisato dans le Miocène de la Sardaigne. – *Mémoires de la Société Géologique de France, Paléontologie* **13**: 5-56, 5 pls., Paris.

COTTEAU, G. (1897): Monographie des *Spatangus* du système Miocène de France. – *Bulletin de la Société de Statistique des Sciences Naturelles et des Arts Industriels du Département de l'Isére, 4me série* **3**/28: 159-196, 12 pls., Grenoble.

COTTEAU, G., PÉRON, A. & GAUTHIER, V. (1891): Échinides fossiles de l'Algérie, 10e fasc.; Étages miocène et pliocène. 273 pp., 8 pls., Paris (G. Masson).

COTTREAU, J. (1912): *Echinolampas elongatula* MILLET, 1854. – *Palaeontologia Universalis* **256**: a-b, 15 figs., Paris.

COTTREAU, J. (1913a): Les Échinides néogènes du Bassin Méditerranéen. – *Annales de l'Institut Océanographique Monaco* **6**/3: 1-192, 15 pls., 41 figs., 2 maps, Monaco.

COTTREAU, J. (1913b): *Scutella gibbercula* M. de Serres et *Clypeaster scutellatus* M. de Serres. – *Compte Rendu Sommaire des Séances de la Société Géologique de France* **1913**/3: 19, Paris.

COWPER REED, F.R. (1932): New Miocene fauna from Cyprus. – *Geological Magazine* **69**: 511-517, London.

COWPER REED, F.R. (1933): Notes on the Neogene Faunas of Cyprus. – *The Annals and Magazine of Natural History, 10th Series* **12**: 225-244, London.

CRAM, D.L. (1971): Life history studies on South African echinoids (Echinodermata) 2. *Echinolampas* (*Palaeolampas*) *crassa* (Bell) (Echinolampadidae). – *Transactions of the Royal Society of South Africa* **39**/3: 339-352, 8 figs., 1 tab., Cape Town.

ČTYROKÝ, P. (1957): Zwei Seeigel aus dem Helvet der Westslovakei. – Časopsis pro Mineralogii a Geologii **2**/3: 233-239, Praha.

ČTYROKÝ, P. (1965): *Lovenia mortenseni* n. sp., a new species of Spatangoid echinoid from the Miocene of the Vienna Basin. – *Sborník geologických vêd, paleontologie* **5**: 107-121, 6 pls., Praha.

CUTRESS, B.M. (1965): Observations on the growth in *Eucidaris tribuloides* (LAMARCK), with special reference to the origin of the oral primary spines. – *Bulletin of Marine Science* **15**/4: 797-834, 5 figs., 4 tabs., Coral Gables, FL.

CUTRESS, B.M. (1980): Cretaceous and Tertiary Cidaroida (Echinodermata: Echinoidea) of the Carribean area. – *Bulletin of American Paleontology* **77 (No. 309)**: 1-221, 14 pls., 16 tabs., 1 fig., Ithaca, NY.

CŽJŽEK, J. (1852): Geologische Verhältnisse der Umgebung von Hainburg, des Leithagebirges und der Ruster Berge. – *Jahrbuch der kaiserlich-königlichen Geologischen Reichsanstalt* **3**/4: 35-55, Wien.

d'ALESSANDRO, A. & LAVIANO, A.R.G.S.A. (1979): Il Neogene del Monte Gargano. – *Bolletino della Società Paleontologica Italiana* **18**/1: 9-118, 19 pls., 32 figs., 2 tab., Modena.

DAFNI, J. (1992): Growth rate of the sea urchin *Tripneustes gratilla elatensis*. – *Israel Journal of Zoology* **38**/1: 25-33, 2 figs., 1 tab., Jerusalem.

DAVID, B. & DE RIDDER, C. (1989): Echinodermes: Échinides irréguliers. – In: FOREST, J. (ed.): Résultats des Campagnes Musorstom, Volume 4. – *Mémoires du Muséum National d'Historire Naturelle, Série A (Zoologie)* **143**/5: 203-2227, 4 pls., 7 figs., Paris.

DAVID, B. & LAURIN, B. (1996): Morphometrics and cladistics: measuring phylogeny in the sea urchin *Echinocardium*. – *Evolution* **50**/1: 348-359, 15 figs., Lawrence, KS.

DAVID, B., LAURIN, B., CHONÉ, T. & MAGNIEZ, F. (1999): Morphological disparity in the genus *Echinocardium* (Echinoidea: Spatangoida). – In: CANDIA CARNEVALI, M.D. & BONASORO, F. (eds.): *Echinoderm Research 1998.- Proc. 5th Europ. Conf. Echinoderms/Milan/Italy/7-12 Sept.1998.* pp. 253-259, 5 figs., Rotterdam (A.A. Balkema).

DAVID, B., MOOI, R. & TELFORD, M. (1996): The ontogenetic basis of Loven's Rule clarifies homologies of the echinoid peristome. – In: EMSON, R., SMITH, A.B. & CAMPBELL, A. (eds.): *Echinoderm Research 1995. – Proc. 4th Europ. Echinoderm Colloq./London, UK/10-13 April 1995.* pp. 155-164, 7 figs., Rotterdam (A.A. Balkema).

DEFRANCE, M.J.L. in de BLAINVILLE & al. (1816-1836): *Dictionnaire des Sciences naturelles par plusieurs professeurs du Jardin du Roi et des principales Ecoles de Paris.* t. 9: 560 p. (1817), t. 18: 594 p. (1820), t37: 559 p. (1825), t. 50: 554 p. (1827), Paris. [not seen]

DEPÉRET, C. (1892): II. Description Générale, Paléontologie et Classification. – In: FONTANNES, F.: *Études stratigraphiques et paléontologiques pour servir a l'Histoire de la Période Tertiaire dans le Bassin du Rhone. (Memoire posthume réd'gé et complété par le Docteur Ch. Depéret). X. Les Terrains Tertiaires marins de la Cote de Provence. Deuxième Partie. Étage Helvétien. La Couronne – Plan-d'Aren – Istres – Miramas.* pp. 26-78, Paris (Société Géologique de France).

DESALVO, L.H., RANDALL, J.E. & CEA, A. (1988): Ecological reconnaissance of the Easter Island sublittoral marine environment. – *National Geographic Research* **4**: 451-473, Washington, DC.

DESIO, A. (1929): La Paleontologia (Molluschi – Echini – Arthropodi). Estratto dai resultati scientifica della missione alla oase di Giarabub (1926-27). – *Reale Società Geografica Italiana* **7**/3: 1-441, 47 pls., Roma.

DESMAREST, A.C. (1816): *Echinus monilis*. – In: DEFRANCE, M.J.L.: *Dictionnaire des Sciences Naturelles* **37** pp. 100, Strasbourg & Paris (F.G. Levrault).

DESOR, E. (1853): Notice sur les échinides du terrain nummulitique des Alpes, avec les diagnoses et plusieurs espèces et genres nouveaux – *Actes de la Société Helvétique des Sciences Naturelles* **38**: 270-279, Aarau.

Desor, E. (1855-1858): Synopsis des échinides fossiles lxviii+ 490 pp., 44 pls., Paris (Reinwald).

Dixon, H.L. & Donovan, S.K. (1998): Oligocene echinoids of Jamaica. – *Tertiary Research* **18**/3-4: 95-124, 10 pls., 6 figs., 3 tabs., Leiden.

Döderlein, L. (1906): Die Echinoiden der Deutschen Tiefsee-Expedition. – In: Chun, C. (ed.): *Wissenschaftliche Ergebnisse der Deutschen Tiefsee-Expedition auf dem Dampfer „Valdivia" 1898-1899* **5** (2): pp. 61-290, pls. 9-50, 46 figs., Jena (Gustav Fischer).

Dollfus, R.-P. & Roman, J. (1981): Les Échinides de la mer Rouge. – *Ministère des Universités, Comité des Travaux Historiques et Scientifiques, Mémoires de la Section des Sciences* **9**: 1-145, 41 pls., 20 figs., 11 tabs., Paris.

Donovan, S.K. (1993): Jamaican Cenozoic Echinoidea. – In: Wright, R.M. & Robinson, E. (eds.): Biostratigraphy of Jamaica. – *Memoirs of the Geological Society of America* **182**: 371-412, 19 figs., 2 tabs., New York, NY.

Donovan, S.K. (2001): Evolution of the Caribbean echinoderms during the Cenozoic: moving towards a complete picture using all of the fossils. – *Palaeogeography, Palaeoclimatology, Palaeoecology* **166**: 177-191, 3 figs., 6 tabs., Amsterdam.

Donovan, S.K. & Collins, J.S.H. (1997): Unique Preservation of an *Echinometra* Gray (Echinodermata: Echinoidea) in the Pleistocene of Jamaica. – *Caribbean Journal of Science* **33**/1-2: 123-124, 1 fig., Mayagüez.

Donovan, S.K. & Gordon, C.M. (1993): Echinoid taphonomy and the fossil record: supporting evidence from the Plio-Pleistocene of the Caribbean. – *Palaios* **8**/3: 304-306, 1 tab., Lawrence, KS.

Donovan, S.K., Lewis, D.N., Vincent, H. & Blackman, I.M. (2001): Pleistocene echinoids from Tobago, West Indies. – *Caribbean Journal of Earth Science* **35**: 1-12, 4 figs., 3 tabs., Mayagüez.

Donovan, S.K. & Veale, C. (1996): The irregular echinoids *Echinoneus* Leske and *Brissus* Gray in the Cenozoic of the Antillean region. – *Journal of Paleontology* **70**/4: 632-640, 5 figs., 2 tabs., Lawrence, KS.

Douxami, H. (1896): Étude sur les terrains tertiaires du Dauphiné, de la Savoie et de la Suisse occidentale. – *Annales de l'Université de Lyon* **27**: 1-318, 6 pls., 31 figs., Lyon. [not seen]

Dreger, J. (1894): Geologische Beschreibung der Umgebung der Städte Pettau und Friedau und des östlichen Theiles des Kollosgebirge in Südsteiermark. – *Verhandlungen der kaiserlich-königlichen Geologischen Reichsanstalt* **28**/2: 69-74, Wien.

Dreger, J. (1913): Geologische Mitteilungen aus dem Kartenblatte Wildon und Leibnitz in Steiermark. – *Verhandlungen der kaiserlich-königlichen Geologischen Reichsanstalt* **47**/2: 65-72, Wien.

Dullo, W.-C. (1983): Fossildiagenese im miozänen Leitha-Kalk der Paratethys von Österreich: Ein Beispiel für Faunenverschiebung durch Diageneseunterschiede. – *Facies* **8**: 1-112, 15 pls., 22 figs., 2 tabs., Erlangen.

Duncan, P.M. (1877): On the Echinodermata of the Australian Cainozoic (Tertiary) Deposits. – *Quarterly Journal of the Geological Society of London* **33**: 42-73, pls. 3-4, 1 fig., London.

Duncan, P.M. (1889): A revision of the genera and great groups of the Echinoidea. – *Journal of the Linnean Society, Zoology* **23**: 1-311, London.

Duncan, P.M. & Sladen, W.P. (1884): Fossil Echinoidea of western Sind and the Coast of Bilúchístán and of the Persian Gulf, from the Tertiary Formations. Fasc. III. Echinoidea from the Khirthar Series of Nummulitic Strata in Western Sind. – *Palaeontologia Indica, Ser. XIV* **I**/3 Fasc. III: 101-246, 18 pls., Calcutta.

Durham, J.W. (1953): Type Species of *Scutella*. – *Journal of Paleontology* **27**/3: 347-352, 1 pl., 1 fig., Tulsa, OK.

Durham, J.W. (1955): Classification of clypeasteroid echinoids. – *California University Publications in Geological Science* **31**/4: 73-198, pls. 3-4, 38 figs., Berkeley & Los Angeles, CA.

Durham, J.W. (1966): Clypeasteroids. – In: Moore, R.C. (ed.): *Treatise on Invertebrate Paleontology, U. Echinodermata. 3 (2)* pp. U450-U491, figs. 335-377, Boulder, CO & Lawrence, KS (GSA & Univ. Kansas Press).

Durham, J.W., Fell, H.B., Fischer, A.G., Kier, P.M., Melville, R.V., Pawson, D.L. & Wagner, C.D. (1966): Echinoids. – In: Moore, R.C. (ed.): *Treatise on Invertebrate Paleontology, U. Echinodermata. 3 (1-2)* pp. U211-U640, figs. 152-517, Boulder, CO & Lawrence, KS (GSA & Univ. Kansas Press).

Durham, J.W. & Wagner, C.D. (1966): Glossary of morphological terms applied to echinoids. – In: Moore, R.C. (ed.): *Treatise on Invertebrate Paleontology, U. Echinodermata. 3 (1)* pp. U251-U256, Boulder, CO & Lawrence, KS (GSA & Univ. Kansas Press).

Ebner, F. & Gräf, W. (1977): Die Fauna von Weitendorf. – *Landesmuseum Joanneum, Jahresberichte, Neue Folge* **6** (1976): 157-183, 10 pls., Graz.

Eichwald, E. (1830): Naturhistorische Skizze von Lithauen, Volhynien und Podolien in geognostisch-mineralogischer, botanischer und zoologischer Hinsicht. 256 pp., 3 pls., Wilna (Joseph Zawadzki).

Eichwald, E. (1852): Lethaea Rossica ou le Monde Primitif de la Russie. Troisième Atlas. Période Moderne. 1-4 pp., 14 pls., Stouttgart (E. Schweizerbart).

Eichwald, E. (1853): Lethaea Rossica ou Paléontologie de la Russie. Troisième Volume. Dernière Période. 1-533 pp., Stouttgart (E. Schweizerbart).

Ernst, G. (1970): Faziesgebundenheit und Ökomorphologie bei irregulären Echiniden der nordwestdeutschen Oberkreide. – *Paläontologische Zeitschrift* **44**/1-2: 41-62, 5 pls., 8 figs., Stuttgart.

Ernst, G. (1973a): Aktuopaläontologie und Merkmalsvariabilität bei mediterranen Echiniden und Rückschlüsse auf die Ökologie und Artumgrenzung fossiler Formen. – *Paläontologische Zeitschrift* **47**/3-4: 188-216, 2 pls., 9 figs., Stuttgart.

Ernst, G. (1973b): Evolution und ökologische Varianz bei fossilen Echiniden. – *Aufsätze und Reden der senckenbergischen naturforschenden Gesellschaft* **24**: 83-102, Frankfurt am Main.

Ertl, R. (1964): Ein Seeigel aus dem Leithakalk. – *Aufschluß* **15**: 253-255, 2 figs., Heidelberg.

Fabian, K. (1906): Das Miozänland zwischen der Mur und der Stiefing bei Graz. – *Mitteilungen des naturwissenschaftlichen Vereins für Steiermark* **1905**: 3-21, 1 map, 1 profile, 2 figs., Graz.

Fell, F.J. (1974): The Echinoids of Easter Island (Rapa Nui). – *Pacific Science* **28**/2: 147-158, 3 figs., Honolulu, HI.

Fell, H.B. (1954): Tertiary and Recent Echinoidea of New Zealand: Cidaridae. – *Paleontological Bulletin of the New Zealand Geological Survey* **23**: 1-62, 15 pls., 15 figs., Wellington.

Fell, H.B. (1964): Oligocene Echinoids from the Trelissic Basin, New Zealand. – *Transactions of the Royal Society New Zealand (Zoology)* **4**/15: 201-205, 6 figs., Wellington.

Fell, H.B. (1966a): Cidaroids. – In: Moore, R.C. (ed.): *Treatise on Invertebrate Paleontology, U. Echinodermata. 3 (1)* pp. U312-U340, figs. 235-254, Boulder, CO & Lawrence, KS (GSA & Univ. Kansas Press).

Fell, H.B. (1966b): Diadematacea. – In: Moore, R.C. (ed.):

Treatise on Invertebrate Paleontology, U. Echinodermata. 3 (1) pp. U340-U366, figs. 255-271, Boulder, CO & Lawrence, KS (GSA & Univ. Kansas Press).

Fell, H.B. (1967): Cretaceous and Tertiary surface currents of the oceans. – *Oceanography and Marine Biology: An Annual Review* **5**: 317-341, 5 figs., London.

Fell, H.B. & Pawson, D.L. (1966): Echinacea. – In: Moore, R.C. (ed.): *Treatise on Invertebrate Paleontology, U. Echinodermata. 3 (2)* **3** (2): pp. U367-U440, figs. 272-328, Boulder, CO & Lawrence, KS (GSA & Univ. Kansas Press).

Fenninger, A. & Wassermann, W. (1982): Clavagellen (Lamellibranchiata) aus den Tonmergeln von Weitendorf (Badenien). – *Mitteilungen des naturwissenschaftlichen Vereins für Steiermark* **112**: 49-58, 3 pls., Graz.

Féral, J.-P., Poulin, É. & Oubelkheir, K. (1998): Geographic and genetic differentiation of *Echinocardium cordatum* (Pennant) – The current state of a complex question: The status of *Echinocardium fenauxi* (Péquignat). – In: Mooi, R. & Telford, M. (eds.): *Echinoderms: San Francisco.- Proc. of the 9th Int. Echinoderm Conf. San Francisco/California(USA), 5-9 Aug. 1996.* pp. 647-649, 1 tab., Rotterdam (A.A. Balkema).

Fischer, A.G. (1966): Spatangoids. – In: Moore, R.C. (ed.): *Treatise on Invertebrate Paleontology, U. Echinodermata. 3 (2)* **3** (2): pp. U543-U628, figs. 427-514, 1 tab., Boulder, CO & Lawrence, KS (GSA & Univ. Kansas Press).

Flügel, H., Hauser, A. & Papp, A. (1952): Neue Beobachtungen am Basaltvorkommen von Weitendorf bei Graz. – *Sitzungsberichte der Österreichischen Akademie der Wissenschaften Wien, mathematisch-naturwissenschaftliche Klasse, Abteilung* I **161**/2-3: 173-184, 1 figs., Wien.

Flügel, H. & Heritsch, H. (1968): Das Steirische Tertiär-Becken. 2. Auflage des Geologischen Führers durch das Tertiär- und Vulkanland des Steirischen Beckens von Artur Winkler-Hermaden. – *Sammlung geologischer Führer* **47**: xi+196 pp., 8 pls., 27 figs., 1 map, 6 suppl., Berlin-Stuttgart (Gebrüder Bornträger).

Foetterle, F. (1860): ... legte Suite von Fossilien zur Ansicht vor, welche von Herrn k. k. Hofrath Ritter von Schwabenau in Oedenburg der k. k. geologischen Reichsanstalt als wertvolles Geschenk zugekommen sind.... – *Jahrbuch der kaiserlich-königlichen Geologischen Reichsanstalt, Verhandlungen* **11**: 75-76, Wien.

Fontaine, A. (1953): The shallow water echinoderms of Jamaica. Part III. The sea urchins (class Echinoidea). – *Natural History Notes of the Natural History Society of Jamaica* **6**: 3-9, 14 figs., Kingston. [not seen, cited from Rose, 1978]

Forbes, E. (1841): A history of British starfishes and other animals of the Class Echinodermata. xx+267 pp., London. [not seen]

Fourtau, R. (1900): Revision des Échinides fossiles de l'Égypte. – *Mémoires présentés a l'Institut Égyptien* **3**/8: 605-740, 4 pls., Le Caire.

Fourtau, R. (1901): Notes pour servir a l'étude des Échinides fossiles d'Égypte. – *Bulletin de la Société Géologique de France, Série 4* **1**: 623-626, Paris.

Fourtau, R. (1902): Notes sur les Échinides fossiles de l'Égypte. – *Bulletin de l'Institute Egyptien, IV. Série* **2 (1901)**: 31-117, 6 pls., Le Caire.

Fourtau, R. (1920): Catalogue des Invertébrés fossiles de l'Égypte. Terrains Tertiaires. 2de partie: Échinodermes Néogènes. – *Palaeontological Series* 4: iv+100 pp., 12 pls., Cairo (Geological Survey of Egypt).

Franzenau, Á. (1881): Adatok a rákosi (Budapest) felsö mediterrán emelet foraminifera faunájához. – *Földtani Közlöny* **11**/1-3: 31-55, 83-107, pl. 3, Budapest.

Frazer, T.K., Lindberg, W.J. & Stanton, G.R. (1991): Predation on Sand Dollars by gray Triggerfish, *Balistes capriscus*, in the northeastern Gulf of Mexico. – *Bulletin of Marine Science* **48**/1: 159-164, 2 figs., 1 tab., Coral Gables, FL.

Friebe, J.G. (1990): Lithostratigraphische Neugliederung und Sedimentologie der Ablagerungen des Badenium (Miozän) um die Mittelburgenländische Schwelle (Steirisches Becken, Österreich). – *Jahrbuch der Geologischen Bundesanstalt* **133**/2: 223-257, 29 figs., Wien.

Fuchs, T. (1868): Die Tertiärbildung der Umgebung von Eggenburg. – *Jahrbuch der kaiserlich-königlichen Geologischen Reichsanstalt* **18**: 584-598, pl. 15, 1 fig., Wien.

Fuchs, T. (1869): Der Steinbruch im marinen Conglomerate von Kalksburg und seine Fauna, mit einer Einleitung über die Darstellung von Local-Faunen. – In: Karrer, F. & Fuchs, T. (eds.): Geologische Studien in den Tertiärbildungen des Wiener Beckens. – *Jahrbuch der kaiserlich-königlichen Geologischen Reichsanstalt* **19**/2: 189-195, 1 fig., 1 tab., Wien.

Fuchs, T. (1875): Neue Brunnengrabungen in Wien und Umgebung. – In: Fuchs, T. & Karrer, F. (eds.): Geologische Studien in den Tertiärbildungen des Wiener Beckens. – *Jahrbuch der Geologischen Reichsanstalt* **25**: 19-62, Wien.

Fuchs, T. (1877): Geologische Uebersicht der jüngeren Tertiärbildungen des Wiener Beckens und des Ungarisch-Steirischen Tieflandes. – *Zeitschrift der Deutschen Geologischen Gesellschaft* **29**/4: 653-709, 1 tab., Berlin.

Fuchs, T. (1900): Beiträge zur Kenntnis der Tertiärbildungen von Eggenburg. – *Sitzungsberichte der Kaiserlichen Akademie der Wissenschaften, mathematisch-naturwissenschaftliche Classe, Abtheilung* I **109**: 859-904, 1 pl., 6 figs., Wien.

Fuchs, T. (1902): Über Anzeichen einer Erosionsepoche zwischen Leythakalk und sarmatischen Schichten. – *Sitzungsberichte der Kaiserlichen Akademie der Wissenschaften, mathematisch-naturwissenschaftliche Classe, Abtheilung* I **111**: 351-355, Wien.

Fuchs, T. (1903a): Exkursion in die Umgebung von Atzgersdorf, Baden und Vöslau. (Tagesausflug) – In: Teller, F. (ed.): *IX. Internationaler Geologen-Kongress 1903. Führer für die Exkursionen in Österreich.* pp. 1-9, Wien (Brüder Hollinek).

Fuchs, T. (1903b): Exkursion nach Eggenburg. (Tagesausflug). – In: Teller, F. (ed.): *IX. Internationaler Geologen-Kongress 1903. Führer für die Exkursionen in Österreich.* pp. 1-13, Wien (Brüder Hollinek).

Fuchs, T. & Schaffer, F. (1903): Exkursion in die Ziegeleien der Wienerberger Ziegelfabrik und Baugesellschaft bei Inzersdorf am Wiener Berg. (Nachmittagsausflug). – In: Teller, F. (ed.): *IX. Internationaler Geologen-Kongress 1903. Führer für die Exkursionen in Österreich.* pp. 1-4, Wien (Brüder Hollinek).

Fuchs, W. (1965): Geologie des Ruster Berglandes (Burgenland) – *Jahrbuch der Geologischen Bundesanstalt* **108**: 155-194, 2 pls., 3 figs., Wien.

Fuchs, W. (1981): Großtektonische Neuorientierung in den Ostalpen unter Einbeziehung plattentektonischer Gesichtspunkte. – In: Fuchs, W. & Grill, R. (eds.): Gelogische Karte von Wien und Umgebung 1:200.000., Wien (Geologische Bundesanstalt).

Gaál, S. (1905): Beiträge zur Mediterranen Fauna des Ostroski-Verpor Gebirges. – *Földtani Közlöny* **35**/6-7: 338-365, Budapest.

Gábos, L. & Ghiurca, V. (1969): Les petits échinides du faciès récifal du Tortonien de Transylvanie. – Studia Universitatis Babeş Bolyai, Series Geologica-Geographica **14**/1: 81-91, 2 pls., 2 tabs., Cluj.

Ghiold, J. & Hoffman, A. (1984): Clypeasteroid echinoids and historical biogeography. – *Neues Jahrbuch für Geologie und Paläontologie Monatshefte* **1984**/9: 529-538, 2 figs., Stuttgart.

Ghiold, J. & Hoffman, A. (1985): Vicariance and dispersal; a historical biogeographic perspective from clypeasteroid echinoids. – In: *Abstracts with Programms – The Geological Society of America; 98th annual meeting. Orlando, Florida, USA, 28-31 Oct. 1985* **17** (7): Boulder, CO, USA (Geological Society of America).

Ghiold, J. & Hoffman, A. (1986): Biogeography and biogeographic history of clypeasteroid echinoids. – *Journal of Biogeography* **13**/3: 183-206, 2 tabs., sketch maps, Oxford.

Ghiurca, V. (1996): Rumänien für Fossiliensammler. – *Fossilien* **13**/3: 188-192, 6 figs., Korb.

Giorgio, A.d. (1923): Echinidi miocenici della Sardegna. – *Atti della Società Toscana di Scienze Naturali* **35**: 116-130, 1 pl., Pisa.

Givulescu, R. & Duşa, A. (1960): Zwei für das Torton von Coştei-Nemeseşti (Bez. Hunedoara, Rumänien) neue Echiniden. – *Geologie* **9**: 935-938, 6 figs., Berlin.

Gladfelter, W.B. (1978): General ecology of the cassiduloid urchin *Cassidulus caribbearum*. – *Marine Biology* **47**: 149-160, Heidelberg.

Gmelin, J.F. (1777-1779): Vollständiges Natursystem des Mineralreichs von Linné nach der 12. lateinischen Ausgabe in einer freyen und vermehrten Uebersetzung, Nüenberg (Raspe). [not seen]

Gmelin, J.F. (1788-1793): Caroli a Linne Systema Naturae per Regna tria Naturae. Editio decimatertia, aucta, reformata, cura J. F. Gmelin. Vol. 1, Regnium Animale: pt. 1, Mammalia, pp. ix, 1-232 (1788); pt. 2, Aves, pp. 233-1032 (1789); pt. 3, Amphibia et Pisces, pp. 1033-1516 (1789); pt. 4, Insecta, pp. 1517-2224 (1790); pt. 5, Insecta, pp. 2225-3020 (1790); pt. 6, Vermes, pp. 3021-3910 (1791); pt. 7, Index, pp. 3911-4120 (1792); Vol. 2, pt. 1, Regnum Vegetabile, pp. i-xi, 1-884 (1792); pt. 2, Regnum Vegetabile, pp. 885-1661 (1792); Vol. 3, Regnum Lapideum, pp. 1-476, 3 pls. (1793), Lipsiae.

Goldfuss, A. (1826-1944): Petrefacta Germaniæ tam ea, quae in Museo Universitatis Regiae Borussicae Fridericiae Wilhelmiae Rhenanae servantur quam alia quaecunque in Museis Hoeninghusiano Muensteriano aliisque extant, Iconibus et Descriptionibus illustrata. 1(1): i-viii+1-76; pls. 1-25 (1826); 1(2): 77-164; pls. 26-50 (1829); 1(3): 165-240; pls. 51-70 (1831); 1(4): 241-252 (1833); 2(1): i-iii+1-68; pls. 71-96 (1834); 2(2): 69-140; pls. 97-121 (1835); 2(3): 141-224; pls. 122-146 (1837); 2(4): 225-312; pls. 147-165 (1840); 3(1): i-iv+1-20; pls. 166-171 (1841); 3(2): 21-28; pls. 172-195 (1844); 3(3): 29-128; pls. 196-200 (1844), Düsseldorf (Arnz & Comp.).

Goncharova, I.A., Khondkarian, S.O. & Shcherba, I.G. (2001): The Tarkhanian-Karaganian Stage in Development of the Euxinian-Caspian Basin (Eastern Patatethys). Part 1: Tarkhanian. – *Stratigraphy and Geological Correlation* **9**/5: 508-522, 3 figs., 1 tab., Moskva.

Goncharova, I.A., Khondkarian, S.O. & Shcherba, I.G. (2002): The Tarkhanian-Karaganian Stage in Development of the Euxinian-Caspian Basin (Eastern Patatethys). Part 2. – *Stratigraphy and Geological Correlation* **10**/2: 196-208, Moskva.

Gordon, I. (1991): The poor fossil record of *Echinometra* (Echinodermata: Echinoidea) in the Caribbean region. – *Journal of the Geological Society of Jamaica* **28**: 37-41, Kingston.

Gourret, P. (1892): La Faune tertiaire marine de Carry, de Sausset et de la Couronne (près de Marseille). Faciès des étages tertiaires dans la Basse-Provence. – *Bulletin de la Société Belge de Géologie, de Paléontologie et d'Hydrologie* **4** (1890): 73-143, 4 pls., Bruxelles.

Gradstein, F.M., Ogg, J.G. & Smith, A.G. (eds.) (2004): A Geological Time Scale 2004. xix+589 pp., Cambridge (Cambridge Univ. Press).

Gras, A. (1848): Description des oursins fossiles du département de l'Isère, précédée de notions élémentaires sur l'organisation & la glossologie de cette classe de Zoophytes, et suive d'une notice géologique sur les divers terrains de l'Isère. 96 pp., 6 pls., Grenoble & Paris (Ch. Vellot et comp. édit. & Victor Masson).

Gras, A. (1852): Catalogue des corps organisés fossiles qui se recontrent dans le départment de l'Isere. 54 pp., 4 pls., Grenoble (Imprimerie Maisonville).

Grateloup, J.P.S.d. (1836): Mémoire de géo-zoologie sur les Oursins fossiles, (Échinides), qui se recontrent dans terrains calcaires des environs de Dax (Département des des Landes). – *Actes de la Société Linnéenne de Bordeaux* **8**/4: 103-191, 2 pls., Bordeaux.

Gray, J.E. (1825): An attempt to divide the Echinida, or sea eggs, into natural families. – *Annals of Philosophy, new series* **10**: 423-431, London.

Gray, J.E. (1845): Description of two new invertebrated animals from Australia. – In: Eyre, E.J. (ed.): *Journals of expeditions of discovery into Central Australia and overland from Adelaide to King Georg's Sound in 1840-41. Volume 1.* pp. 435-436, pl. 6, London. [not seen]

Gray, J.E. (1851a): Description of two new genera and some new species of Scutellidae and Echinolampidae in the Collection of the British Museum. – *Proceedings of the Zoological Society London* **19**: 34-38, London.

Gray, J.E. (1851b): Descriptions of some new genera and species of Spatangidae in the British Museum. – *The Annals and Magazine of Natural History, 2nd Series* **7**: 130-134, London.

Gray, J.E. (1855): Catalogue of the Recent Echinida, or Sea Eggs, in the Collection of the British Museum. Part I.-Echinida Irregularia. 69 pp., 6 pls., London (Woodfall & Kinder).

Greenstein, B.J. (1993): Is the Fossil Record of Regular Echinoids Really So Poor? A Comparison of Living and Subfossil Assemblages. – *Palaios* **8**/6: 587-601, 18 figs., 2 tabs., 1 appendix, Lawrence, KS.

Gregory, J.W. (1891): The Maltese fossil Echinoidea and their evidence on the correlation of the Maltese rocks. – *Transactions of the Royal Society of Edinburgh* **36**: 585-639, 2 pls., 1 tab., Edinburgh.

Gregory, J.W. (1906): Fossil Echinoidea from Sinai and Egypt. – *Geological Magazine* **3**: 216-227, 246-254, 2 pls., London.

Gregory, J.W. (1911): The Fossil Echinoidea of Cyrenaica. – *Quarterly Journal of the Geological Society of London* **67**: 661-679, 3 pls., London.

Grill, R. (1935): Das Oligocänbecken von Gallneukirchen bei Linz a. D. und seine Nachbargebiete. – *Mitteilungen der Geologischen Gesellschaft in Wien* **27** (1934): 37-72, 1 map, Wien.

Grill, R. (1943): Über mikropaläontologische Gliederungsmöglichkeiten im Miozän des Wiener Beckens. – *Mitteilungen des Reichsamt für Bodenforschung, Zweigstelle Wien* **6**: 33-44, 8 pls., Wien.

Grill, R. (1953): Der Flysch, die Waschbergzone und das Jungtertiär um Ernstbrunn (Niederösterreich). – *Jahrbuch der Geologischen Bundesanstalt* **96**: 65-116, 4 pls., Wien.

Gutzwiller, A. (1883): Erster Theil. Molasse und Jüngere Ablagerungen enthalten auf Blatt IV und V des Eidgen. Atlas mit Ausnahme des Gebietes nördlich von Rhein und vom Bodensee. – In: Gutzwiller, A. & Schalch, F. (eds.): Geologische Beschreibung der Kantone St. Gallen, Thurgau und Schaffhausen. – *Beiträge zur Geologischen Karte der Schweiz* **19**/1: v-viii, 1-133, 2 pls., Bern.

Haeckel, E. (1896): Systematische Phylogenie der wirbellosen

Thiere (Invertebrata). Vol. 2. XVIII+720 pp., Berlin (Georg Reimer Verlag).

HALL, T.S. (1907): Four new echinoids from the Australian Tertiary. – *Proceedings of the Royal Society Victoria* **19**: 47-53, Victoria.

HALMAI, J. (1981): A fóti Formáció Makrofaunája. – *A Magyar Állami Földtani Intézet Évi Jelentése* **1979**: 97-113, 3 tabs., Budapest.

HÁMOR, G. (1985): A Nógrád-Cserhát Kutatási Terület Földtani Viszonyai. – *Geologica Hungarica, Series Geologica* **22**: 1-307, 86 figs., 56 tabs., Budapest.

HAROLD, A.S. & TELFORD, M. (1990): Systematics, phylogeny and biogeography of the genus *Mellita* (Echinoidea: Clypeasteroida). – *Journal of Natural History* **24**/4: 987-1026, 24 figs., 2 tabs., London.

HARZHAUSER, M. & KROH, A. (1999): Palökologie der Echinidenfaunen von Unternalb (Retz-Formation – Eggenburgium). – In: ROETZEL, R. (ed.): *Arbeitstagung Geologische Bundesanstalt 1999*. pp. 221-222, Wien (Geological Survey of Austria).

HARZHAUSER, M., MANDIC, O. & ZUSCHIN, M. (2003): Changes in Paratethyan marine molluscs at the Early/Middle Miocene transition: diversity, palaeogeography and palaeoclimate. – *Acta Geologica Polonica* **53**/4: 323-339, 11 figs., Warszawa.

HARZHAUSER, M., DAXNER-HÖCK, G., KOLLMANN, H., KOVAR-EDER, J., RÖGL, F., SCHULTZ, O. & SUMMESBERGER, H. (2004): 100 Schritte Erdgeschichte – Die Geschichte der Erde und des Lebens im Naturhistorischen Museum in Wien. 96 pp., Wien (Naturhistorisches Museum Wien).

HENDERSON, R.A. (1975): Cenozoic spatangoid echinoids from New Zealand. – *Paleontological Bulletin, New Zealand geological Survey* **46**: 1-90, 18 pls., 17 figs., 2 tabs., Wellington.

HENDLER, G., MILLER, J.E., PAWSON, D.L. & KIER, P.M. (1995): Sea Stars, Sea Urchins, and Allies. – Echinoderms of Florida and the Carribean. xi+390 pp., 191 figs., Washington, London (Smithsonian Institute Press).

HENSON, F.R.S., BROWNE, R.V. & MCGINTY, J. (1949): A synopsis of the stratigraphy and geological history of Cyprus. – *Quarterly Journal of the Geological Society of London* **55**: 1-41, 2 pls., London.

HERMITE, H. (1879): Études géologiques sur les Iles Baléares (Majorque et Minorque). – Thèses présentées a la Faculté des Sciences de Paris pour obtenir le Grade de Docteur ès-Sciences Naturelles, Série A No. 35, Vol. 423: 357 pp., 4 pls., Paris (F. Pichon & F. Savy).

HESS, H. (1975): Die fossilen Echinodermen des Schweizer Juras. – *Veröffentlichungen aus dem Naturhistorischen Museum Basel* **8**: 1-130, 48 pls., 57 figs., 2 tabs., Basel.

HESSE, E. (1900): Die Microstructur der fossilen Echinoideenstachel und deren systematische Bedeutung. – *Neues Jahrbuch für Mineralogie, Beilagenband* **13**: 185-264, 2 pls., 8 figs., Stuttgart.

HICKMAN, C.P. (1998): A Field Guide to Sea Stars and other Echinoderms of Galápagos. – Glápagos Marine Life Series 83 pp., Lexington, VA (Sugar Spring Press).

HIDEN, H. (1993): Erstfund von *Conoclypus plagiosomus* AGASSIZ aus Retznei bei Ehrenhausen. – *Der Steirische Mineralog* **4**/6: 17-18, 1 fig., Graz.

HIDEN, H.R. (1998): Paläopathologische Befunde an Echinoideen aus dem Mittelmiozän (Badenium) Ost-Österreichs. – *Mitteilungen des Referats für Geologie und Paläontologie am Landesmuseum Joanneum* **SH2**: 215-221, 1 pl., Graz.

HILBER, V. (1877): Die Miocänschichten von Gamlitz bei Ehrenhausen in Steiermark. – *Jahrbuch der kaiserlich-königlichen Geologischen Reichsanstalt* **27**: 251-270, 1 pl., Wien.

HILBER, V. (1878): Die Miocän-Ablagerungen um das Schiefergebirge zwischen den Flüssen Kainach und Sulm in Steiermark. – *Jahrbuch der kaiserlich-königlichen Geologischen Reichsanstalt.* **28**: 505-580, 1 pl., 1 map, Wien.

HILBER, V. (1882): Geologische Studien in den ostgalizischen Miocän-Gebieten. – *Jahrbuch der kaiserlich-königlichen Geologischen Reichsanstalt* **32**/2: 193-330, Wien.

HILBER, V. (1892): Fauna der Pereiraïa-Schichten von Bartolomae in Unter-Krain. – *Sitzungsberichte der Kaiserlichen Akademie der Wissenschaften, mathematisch-naturwissenschaftliche Classe, Abtheilung* I **101**: 1005-1032, 1 pl., Wien.

HILBER, V. (1894): Das Tertiärgebiet um Hartberg in Steiermark und Pinkafeld in Ungarn. – *Jahrbuch der kaiserlich-königlichen Geologischen Reichsanstalt* **44**: 389-414, Wien.

HILBER, V. (1913): Eine Diskordanz im steirischen Leithakalk. – *Mitteilungen der Geologischen Gesellschaft in Wien* **6**: 229-233, Wien.

HOCHULI, P.A. (1978): Palynologische Untersuchungen im Oligozän und Untermiozän der Zentralen und Westlichen Paratethys. – *Beiträge zur Paläontologie von Östereich* **4**: 1-132, 14 pls., 21 figs., 3 tabs., Wien.

HOERNES, R. (1875a): Die Fauna des Schlier von Ottnang. – *Jahrbuch der Geologischen Reichsanstalt* **25**/4: 333-400, pls. 10-15, Wien.

HOERNES, R. (1875b): Die Fauna des Schlier von Ottnang. – *Verhandlungen der kaiserlich-königlichen Geologischen Reichsanstalt* **9**/11: 209-212, Wien.

HOERNES, R. (1880): *Amphiope* nov. sp. vom Seckauer Berg bei Leibnitz. – *Verhandlungen der kaiserlich-königlichen Geologischen Reichsanstalt* **14**/11: 194, Wien.

HOERNES, R. (1883): Ein Beitrag zur Kenntniss der miocänen Meeres-Ablagerungen der Steiermark. – *Mittheilungen des naturwissenschaftlichen Vereines für Steiermark* **1882**: 1-50, 1 pl., Graz.

HOERNES, R. (1884): Ein Vorkommen des *Pecten denudatus* Reuss und anderer „Schlier"-Petrefacte im inneralpinen Theil des Wiener Beckens. – *Verhandlungen der kaiserlich-königlichen Geologischen Reichsanstalt* **18**/15: 305-306, Wien.

HOERNES, R. (1903): Bau und Bild der Ebenen Österreichs. – In: DIENER, C., HOERNES, R., SUESS, F.E. & UHLIG, V. (eds.): *Bau und Bild Österreichs*. pp. 915-1110, 1 pl., 27 figs., Wien-Leipzig (F. Tempsky & G. Freytag).

HOLLER, A. (1900): Die Fauna der Meeresbildungen von Wetzelsdorf bei Preding in Steiermark. – *Mitteilungen des naturwissenschaftlichen Vereins für Steiermark* **1899**: 49-71, 1 fig., Graz.

HOLMES, F.C. (1999): Australian Tertiary Apatopygidae (Echinoidea). – *Proceedings of the Royal Society Victoria* **111**/1: 51-70, 8 figs., 2 appendices, Victoria.

HOPKINS, T.S. (1988): A review of the distribution and proposed morphological groupings of extant species of the genus *Clypeaster* in the Caribbean Sea and the Gulf of Mexico. – In: BURKE, R.D., MLADENOV, P.V., LAMBERT, P. & PARSLEY, R.L. (eds.): *Echinoderm Biology. Proceed. 6th Internat. Echinoderm Conf., Victoria, 23-27 Aug. 1987* pp. 337-345, 2 figs., 1 tab., Rotterdam (A.A. Balkema).

HÖRNES, M. (1850): Bericht über die Bereisung mehrerer Fundorte von Tertiär-Petrefacten im Wiener-Becken. – *Jahrbuch der kaiserlich-königlichen Geologischen Reichsanstalt.* **1**: 662-679, Wien.

HÖRNES, M. (1853): Verzeichnis der in Ottnang vorkommenden Versteinerungen. – *Jahrbuch der kaiserlich-königlichen Geologischen Reichsanstalt.* **4**: 190, Wien.

HORUSITZKY, F. (1927): Neue Daten zur Miozän-Stratigraphie der Umgebung von Budapest. – *Földtani Közlöny* **56**: 21-30, 161-171, Budapest.

HORUSITZKY, F. (1934): Megjegyzések a Budapest környéki Burdigalien Kérdéséhez. – *Földtani Közlöny* **64**/10-12: 321-334, 1 fig., Budapest.

Hotchkiss, F.H.C. (1976): Devonian Ophiuroids from New York State: Reclassification of *Klasmura*, *Antiquaster* and *Stenaster* into the Suborder Scalarina nov., order Stenurida. – *New York State Museum Bulletin* **425**: 1-39, 10 pls., 2 figs., 2 tabs., Albany, NY.

Hughes, R.N. & Hughes, P.I. (1971): A study of the gastropod *Cassis tuberosa* (L.) preying upon sea urchins. – *Journal of experimental marine Biology and Ecology* **7**: 215-315, Amsterdam.

ICZN (1954a): Opinion 207. Designation, under the Plenary Powers, of the type species in harmony with accustomed use for the genera *Echinocyamus* van Phelsum, 1774, and *Fibularia* Lamarck, 1816 (Class Echinoidea). – *Opinions and Declarations rendered by the International Commission on Zoological Nomenclature* **3**/26: 339-352, London.

ICZN (1954b): Opinion 209. Validation of, designation of type species for, *Brissus* Gray, 1825, *Echinocardium* Gray, 1825, and *Spatangus* Gray, 1825 (Class Echinoidea) under the Plenary Powers, and designation, under those Powers, of a type species for *Schizaster* Agassiz (L.), 1836, and in so far as necessary, for *Moira* Agassiz (A.), 1872. – *Opinions and Declarations rendered by the International Commission on Zoological Nomenclature* **3**/28: 367-392, London.

ICZN (1961): Opinion 608. *Spatangus* O. F. Müller, 1776 (Echinoidea); Suppression under the Plenary Powers. – *Bulletin of Zoological Nomenclature* **18**/4: 262-264, London.

Imbesi Smedile, M. (1958): Clipeastri Aquitanian, Elveziani e Tortoniani della Calabria. – *Palaeontographia Italica* **53**: 1-47, 22 pls., 2 figs., Roma.

Jackson, R.T. (1912): Phylogeny of the Echini with a revision of Palaeozoic species. – *Memoirs of the Boston Society of Natural History* **7**: 1-491, 76 pls., 256 figs., Boston, MA.

Jagt, J.W.M. (2000): Late Cretaceous-Early Palaeogene echinoderms and the K/T boundary in the southeast Netherlands and northeast Belgium – Part 4: Echinoids. – *Scripta Geologica* **121**: 181-375, 30 pls., 23 figs., Leiden.

Janoschek, R. (1931): Die Geschichte des Nordrandes der Landseer Bucht im Jungtertiär (Mittleres Burgenland). – *Mitteilungen der Geologischen Gesellschaft in Wien* **24**: 38-132, 1 fig., 1 tab., Wien.

Janoschek, R. (1943): Das Inneralpine Wiener Becken. – In: Schaffer, F.X. (ed.): *Die Geologie der Ostmark.* pp. 427-514, Wien (Franz Deuticke).

Janoschek, R. (1951): Das Inneralpine Wiener Becken. – In: Schaffer, F.X. (ed.): *Geologie von Österreich. Zweite veränderte Auflage.* pp. 525-693, Wien (Franz Deuticke).

Jaskó, S. (1940): A Rima és Tarna Közének Oligocén Rétegei és Kövületei. – *Földtani Közlöny* **70**/10-12: 294-317, 369-373, fig. 56, tabs. 8-13, Budapest.

Jeannet, A. (1928): Contribution à l'étude des Echinides tertiaires de la Trinité et du Venezuela. – *Mémoires de la Société Paléontologique Suisse* **48**: 49 p., 6 pls., 12 figs., Neuchâtel.

Jeannet, A. & Martin, R. (1937): Ueber Neozoische Echinoidea aus dem Niederlaendisch-Indischen Archipel. – *Leidsche Geologische Mededeelingen* **8**/2: 215-308, 67 figs, Leiden.

Jensen, M. (1982): Morphology and classification of Euechinoidea Bronn, 1860 – a cladistic analysis. – *Videnskabelige Meddelelser fra Dansk Naturhistorisk Forening* **143 (1981)**: 7-99, 40 figs., Copenhagen.

Kaiser, D., Rasser, M., Nebelsick, J.H. & Piller, W.E. (2001): Late Oligocene algal limestones on a mixed carbonate-siliciclastic ramp at the southern margin of the Bohemian Massif (Upper Austria). – In: Piller, W.E. & Rasser, M. (eds.): *Paleogene of the Eastern Alps. – Schriftenreihe der Erdwissenschaftlichen Kommissionen.* **14** pp. 197-223, 2 pls., 9 figs., Wien (Österreichische Akademie der Wissenschaften).

Kalabis, V. (1935): Roentgenovy x Paprsky v Paleontologii. – Zvláštni otisk časopisu "Příroda" **28**/9: 274-278, 3 figs., Brno.

Kalabis, V. (1937a): Anomalie u největších fosilních ježovek rodu "*Clypeaster*". – Zvláštni otisk časopisu "Příroda" **30**/2: 1-4, 2 figs., Brno.

Kalabis, V. (1937b): O pobřežní facii tortonského moře u Služína nedaleko Prostějova. – Časopis Vlasteneckého Spolku Museiního v Olomouci **50**: 107-112, 1 fig., Olomouci.

Kalabis, V. (1937c): Příspěvek ke studiu miocenních "Stolonclypů". – Zvláštni otisk časopisu "Příroda" **30**/6: 1-3, 2 figs., Brno.

Kalabis, V. (1938a): Über die miozänen *Clypeaster*-Arten des Wiener Beckens. – Věstník Královské České Společnosti Nauk Třída matematicko-přírodovědecká **1938**: 1-11, Praha.

Kalabis, V. (1938b): Výsledky studia miocenních Clypeastrů z Č SR. a z pánve vídenšké s určovacim klíčem. – Věstník Královské České Společnosti Nauk Třída matematicko-přírodovědecká **1937**/7: 1-13, Praha.

Kalabis, V. (1939): O hrabavých miocenních ježovkách rodu *Schizaster* Ag. od Badenu. – Sborník Klubu Přírodovědeckého v Brně **21** (1938): 88-92, 2 figs., Brno.

Kalabis, V. (1949): Monografie Clypeasterů z Československého Miocénu. – Rospravy Státního Geologického Ústavu Republiky Československé **11**: 1-65, 8 pls., 17 figs., Praha.

Kalabis, V. (1974): Les Echinodermes du Néogène tchécoslovaque. – *Mémoires du Bureau de Recherches Géologiques et Minières* **78**/1: 313-319, Paris.

Kalabis, V. (1975): Importance stratigraphique de la répartition de quelques espèces d'échinides dans le territoire de la Paratéthys centrale au Miocène. – *Proceedings of the VIth Congress of the Regional Committee on Mediterranean Neogene Stratigraphy, Bratislava, 4-7 Sept. 1975, Vol. 1.* 175-176, Bratislava.

Kapounek, J. (1939): Geologische Verhältnisse der Umgebung von Eisenstadt. – *Jahrbuch der Geologischen Bundesanstalt* **87** (1938): 49-103, 1 map, Wien.

Karrer, F. (1867): Zur Foraminiferenfauna in Österreich. – *Sitzungsberichte der Kaiserlichen Akademie der Wissenschaften, mathematisch-naturwissenschaftliche Classe, Abtheilung* I **55**: 331-368, 3 pls., 1 tab., Wien.

Karrer, F. (1868): Ueber die Tertiärbildungen in der Bucht von Berchtoldsdorf bei Wien. – *Jahrbuch der kaiserlich-königlichen Geologischen Reichsanstalt* **18**: 567-584, pl. 15, Wien.

Karrer, F. (1877): Geologie der Kaiser Franz Josefs Hochquellen-Wasserleitung. – *Abhandlungen der kaiserlich-königlichen Geologischen Reichsanstalt* **9**: 1-420, 20 pls., 96 figs., 11 tabs., Wien.

Kazár, E. (2002): Revised phylogeny of the Physeteridae (Mammalia: Cetacea) in the light of *Placoziphius* Van Beneden, 1869 and *Aulophyseter* Kellogg, 1927. – *Bulletin de l'Institut Royal des Sciences Naturelles de Belgique, Sciences de la Terre* **72**: 151-170, 4 pls., 3 figs., Bruxelles.

Kienel, U., Rehfeld, U. & Bellas, S.M. (1995): The Miocene Blue Clay Formation of the Maltese Islands: Sequence-stratigraphy and palaeooceanographic implications based on calcareous nannofossil stratigraphy. – *Berliner geowissenschaftliche Abhandlungen* **E16** (Gundolf Ernst Festschrift): 533-557, 3 pls., 3 figs., Berlin.

Kier, P.M. (1962): Revision of the cassiduloid echinoids. –

Smithsonian Miscellaneous Collection **144**: 1-262, 44 pls., 184 figs., Washington, D.C.

KIER, P.M. (1966): Cassiduloids. – In: MOORE, R.C. (ed.): *Treatise on Invertebrate Paleontology, U. Echinodermata. 3 (2)* pp. U492-U523, figs. 378-413, Boulder, CO & Lawrence, KS (GSA & Univ. Kansas Press).

KIER, P.M. (1967): Sexual dimorphism in an Eocene echinoid. – *Journal of Paleontology* **41**/4: 988-993, 2 pls., 3 figs., Tulsa, OK.

KIER, P.M. (1969): Sexual dimorphism in fossil echinoids. – In: WESTERMANN, G.E.G. (ed.): *Sexual Dimorphism in Fossil Metazoa and Taxonomic Implications.* pp. 215-222, 1 pl., 2 figs., Stuttgart (Schweitzerbart'sche Verlagsbuchhandlung).

KIER, P.M. (1972): Tertiary and Mesozoic Echinoids of Saudi Arabia. – *Smithsonian Contributions to Paleobiology* **10**: 1-105, 67 pls., 50 figs., Washington, DC.

KIER, P.M. (1975): The Echinoids of Carrie Bow Cay, Belize. – *Smithsonian Contributions to Zoology* **206**: 1-45, 12 pls., 8 figs., Washington, DC.

KIER, P.M. (1977): The poor fossil record of the regular echinoid. – *Paleobiology* **3**: 168-174, 3 figs., 3 tabs., Lancaster, PA.

KIER, P.M. (1984): Fossil Spatangoid Echinoids of Cuba. – *Smithsonian Contributions to Paleobiology* **55**: 1-336, 90 pls., 45 figs., Washington, DC.

KIER, P.M. & GRANT, R.E. (1965): Echinoid distribution and habits, Key Largo Coral Reef Preserve, Florida. – *Smithsonian Miscellaneous Collection* **149**/6: 1-68, 16 pls., 15 figs., Washington, DC.

KIER, P.M. & LAWSON, M.H. (1978): Index of living and fossil echinoids 1924-1970. – *Smithsonian Contributions to Paleobiology* **34**: 1-182, Washington, DC.

KITTL, E. (1886): Ueber den miocenen Tegel von Walbersdorf. – *Annalen des k. k. Naturhistorischen Hofmuseums – Notizen* **1**: 19-23, 2 tabs., Wien.

KNORR, G.W. (1771): Sammlung der Merckwürdigkeiten der Natur und Alterthümern des Erdbodens oder versteinte und andere gegrabene Cörper in illuminirten Kupfertafeln. 3. Theil., 83 pls., Nürnberg (Paul Johnatan Felßecker).

KOCH, A. (1884-1887a): Die Alttertiären Echiniden Siebenbürgens. – *Mitteilungen aus dem Jahrbuch der königlichen ungarischen geologischen Anstalt* **7**: 45-132, 4 pls., Budapest.

KOCH, A. (1887b): Die Echiniden der obertertiären Ablagerungen Siebenbürgens. – *Orvos-Természettudományi Értesitö az Erdélyi Múzeum-Egylet Orvos-Természettudományi Szakosztályának II. Természettudományi Szak* **12**/2: 129-146, 255-271, pl. 5, Kolozsvárt.

KOCHANSKY-DEVIDÉ, V. (1957): O Miocenskim Ježincima i Koraljima Sjeverne Bosne. – *Geološki Vjesnik* **10** (1956): 135-136, Zagreb.

KOEHLER, R. (1914): Échinides du Musée Indien à Calcutta. I. Spatangidés. – *Echinoderma of the Indian Museum. Part 8, Echinoidea (1).* 258 pp., 20 pls., Calcutta. [not seen]

KOJUMDGIEVA, E. & STRACHIMIROV, B. (1960): Les Fossiles de Bulgarie. VII. Tortonien. 1-320 pp., 59 pls., 2 figs., 1 tab., Sofia (Académie des Sciences de Bulgarie).

KÓKAY, J. (1966): A Herend – Márkói Barnaköszénterület földtani és öslénytani vizsgálata. – *Geologica Hungarica, Series Palaeontologica* **36**: 1-149, 15 pls., Budapest.

KÓKAY, J. (1988a): Tengeri kifejlödésü Kárpati Rétegek elöfordulása Alcsútdobozon. – *Földtani Intézet Évi Jelentése* **1986**: 263-279, 6 pls., Budapest.

KÓKAY, J. (1988b): Bántapusztai Formáció, Foti Formáció – Bakony, Várpalota, Bántapuszta. – *Magyarország geológiai alapszelvényei*, 6 p. (unpaginated), 9 figs.

KÓKAY, J. (1991): Stratigraphische Revision der unter- und mittelmiozänen Bildungen des Beckens von Várpalota (Bakony-Gebirge). – In: LOBITZER, H. & CSÁZÁR, G. (eds.): *Jubiläumsfestschrift 20 Jahre Geologische Zusammenarbeit Österreich – Ungarn* **1**, pp. 101-108, 3 figs., Wien (Geological Survey of Austria).

KÓKAY, J., MIHÁLY, S. & MÜLLER, P. (1984): Bádeni korú rétegek a budapesti Örs vezér tere környéken. – *Földtani Közlöny* **114**: 285-295, 2 figs., Budapest.

KOLLMANN, K. (1965): Jungtertiär im Steirischen Becken. – *Mitteilungen der Geologischen Gesellschaft in Wien* **57**: 479-632, 6 pls., 2 figs., Wien.

KOLLMANN, K. & RÖGL, F. (1978): 2. Faziostratotypus: Sassbachtal, Steirisches Becken, Steiermark, Österreich (Boundary Stratotypus: Badenien – Sarmatien; M_{4a-d} – M_{5ab}). – In: PAPP, A., CICHA, I., SENEŠ, J. & STEININGER, F. (eds.): *M_4 Badenien (Moravien, Wielicien, Kosovien). – Chronostratigraphie und Neostratotypen, Miozän der Zentralen Paratethys 6*, pp. 158-168, figs. 37-41, Bratislava (VEDA – Vydavatel'stvo Slovenskej akad. vied).

KORNHUBER, A. (1898): Der Thebener Kobel. Ein Beitrag zu seiner Naturgeschichte. – *Verhandlungen des Vereins für Naturkunde und Heilkunde zu Presburg, Neue Folge* **10** (1897): 1-41, Pressburg.

KORNHUBER, G.A. (1865): Beiträge zur physikalischen Geographie der Presburger Gespanschaft. – *Gedenkbuch der XI. Versammlung ungarischer Ärzte und Naturforscher zu Presburg.* **1865**: i-xcv, 1 map, Pressburg.

KOTCHETOFF, B., KOTCHETOFF, Y. & VEIGA FERREIRA, O.d. (1975): Contribution à la connaissance des gisements fossilifères miocènes au Nord du Cap d'Espichel. – *Comunições Serviços Geológicos de Portugal* **59**: 59-106, 8 pls., Lisboa.

KOVAR, J. B. (1982): Eine Blätter-Flora des Egerien (Ober-Oligozän) aus marinen Sedimenten der Zentralen Paratethys im Linzer Raum (Österreich). – *Beiträge zur Paläontologie von Östereich* **9**: 1-209, 38 pls., 12 figs., 8 tabs., Wien.

KOWALEWSKI, M. & NEBELSICK, J.H. (2003): Predation on Recent and Fossil Echinoids. – In: KELLEY, P.H., KOWALEWSKI, M. & HANSEN, T.A. (eds.): *Predator – Prey Interactions in the Fossil Record.* pp. 279-302, 6 figs., 1 tab., New York, Boston, Dortrecht, London, Moscow (Kluwer Academic/Plenum Publishers).

KRAINER, B. (1987): Sedimentation und Shoshonit von Weitendorf, Badenien, Steirisches Becken. – *Mitteilungen der Österreichischen Geologischen Gesellschaft* **80**: 143-156, 4 figs., 2 tabs., Wien.

KROH, A. (2002a): Die Echiniden (Echinodermata) aus dem Karpatium des Korneuburger Beckens und der Kreuzstettner Bucht (Niederösterreich, Untermiozän). – *Beiträge zur Paläontologie* **27**: 305-315, 2 pls., 2 figs., Wien.

KROH, A. (2002b): Die Echinodermen des Badenium (Mittel-Miozän) von Gainfarn, Niederösterreich.– In: HUBMANN, B. (ed.): 9. Jahrestagung der Österreichischen Paläontologischen Gesellschaft. – *Berichte des Instituts für Geologie und Paläontologie der Karl-Franzens-Universität Graz* **5**: 11-13, Graz.

KROH, A. (2003a): The Echinodermata of the Langhian (Lower Badenian) of the Molasse Zone and the northern Vienna Basin (Austria). – *Annalen des Naturhistorischen Museums in Wien* **104A**: 155-183, 5 pls., 4 figs., Wien.

KROH, A. (2003b): Echinoderms of the Karpatian. – In: BRZOBOHATÝ, R., CICHA, I., KOVÁČ, M. & RÖGL, F. (eds.): *The Karpatian – A Lower Miocene Stage of the Central Paratethys.* pp. 247-256, 1 pl., with an appendix by Eva POSPÍCHALOVÁ, Brno (Masaryk University).

KROH, A. (2004a): *Echinoneus* & *Echinometra* – Two new records of tropical echinoids from the Miocene of Austria and their palaeoclimatic implications. – *Berichte des Institutes für Erdwissenschaften der Karl-Franzens-Universität Graz* **9**: 229-230, Graz.

KROH, A. (2004b): Closing a gap – Discovery of a rare echinothurioid (Echinodermata: Echinoidea) in the Miocene of Styria. – *Berichte des Institutes für Erdwissenschaften der Karl-Franzens-Universität Graz* **9**: 231-232, Graz.

KROH, A. (submitted): Climate Changes in the Early to Middle Miocene of the Central Paratethys and the Origin of its Echinoderm Fauna. *Palaeogeography, Palaeoclimatology, Palaeoecology*, Amsterdam.

KROH, A., (submitted): *Hemipatagus* – a misinterpreted Loveniid (Echinodermata: Echinoidea). *Journal of Systematic Palaeontology*, London.

KROH, A. & HARZHAUSER, M. (1999): An Echinoderm Fauna from the Lower Miocene of Austria: Paleoecology and Implications for Central Paratethys Paleobiogeography. – *Annalen des Naturhistorischen Museums in Wien* **101A**: 145-191, 9 pls., 4 figs., Wien.

KROH, A., HARZHAUSER, M. & PILLER, W.E. (2002): Mass accumulations of regular sea urchins in the Late Badenian (Middle Miocene) of the Leitha Platform (Eastern Austria).– In: Pangeo I Austria, 28-30/6/2002, Salzburg. – Programm & Kurzfassungen, pp. 101-102, Salzburg.

KROH, A., HARZHAUSER, M., PILLER, W.E. & RÖGL, F. (2003): The Lower Badenian (Middle Miocene) Hartl Formation (Eisenstadt – Sopron Basin, Austria). – In: PILLER, W.E. (ed.): Stratigraphia Austriaca. – *Schriftenreihe der Erdwissenschaftlichen Kommissionen* **16**: 87-109, 3 pls., 5 figs., 2 appendices, Wien.

KROH, A. & NEBELSICK, J.H. (2003): Echinoid assemblages as a tool for palaeoenvironmental reconstruction – an example from the Early Miocene of Egypt. – *Palaeogeography, Palaeoclimatology, Palaeoecology* **201**/1-2: 157-177, 8 figs., 2 tabs., Amsterdam.

KÜHN, O. (1936): Eine neue Burdigalausbildung bei Horn. – *Sitzungsberichte der Akademie der Wissenschaften Wien, mathematisch-naturwissenschaftliche Klasse, Abteilung* I **145**/1-2: 35-45, 1 pl., 3 figs., Wien.

KÜHN, O. (1952): Unsere palaeontologische Kenntnis vom österreichischen Jungtertiär. – *Verhandlungen der Geologischen Bundesanstalt Sonderheft* **C**: 114-126, 2 tabs., Wien.

KÜHN, O. (ed.) (1962): Autriche. – Lexique Stratigraphique International, Congrès Géologique International – Commission de Stratigraphie Europe, Fascicule 8, 1-646 pp., 2 maps, Paris (Centre National de la Recherche Scientifique).

KÜHN, O. (1965): Zweck und Anlage des Catalogus Fossilium Austriae. – In: FLÜGEL, H. & KROPFITSCH-FLÜGEL, M. (1965): Catalogus Fossilium Austriae. Heft VIe1: Ammonoidea palaeozoica. 31 pp., Wien (Österreichische Akademie der Wissenschaften).

KÜPPER, H. (1968): Wien. – *Verhandlungen der Geologischen Bundesanstalt, Bundesländerserie*, 1-206, 8 pls., 23 figs., 20 tabs., 20 appendices, Wien.

KÜPPER, H. & BOBIES, C.A. (1927): Das Tertiär am Ostrand des Anninger. – *Jahrbuch der Geologischen Bundesanstalt* **77**: 1-27, 1 pl., Wien.

KUTASSY, E. (1928): A Borsodmegyei Királd Barnaszénmedencéje. – *Földtani Szemle* **1**/5: 1-20, 1 pl., 1 map, Budapest.

KUTSCHER, M. (1985): Die Echinodermen des Magdeburger Grünsandes (Mittel-Oligozän). – *Abhandlungen und Berichte für Naturkunde und Vorgeschichte* **12**/6: 3-14, 3 pls., Magdeburg.

LACKHEM, H. & ROMAN, J. (1995): Les échinoïdes irréguliers (Néognathostomes et Spatangoïdes) du Messinien de Melilla (Maroc septentrional). – *Annales de Paléontologie* **81**/4: 247-278, 5 pls., 1 fig., 4 tabs., Paris.

LACOUR, D. & NÉRAUDEAU, D. (2000): Évolution de la diversité des Brissopsis (Echinoida, Spatangoida) en Méditerranée depuis la «crise messinienne»: application paléoécologique aux *B. lyrifera* intragypses de Sorbas (SE Espagne). – *Geodiversitas* **22**/4: 509-523, 10 figs., Paris.

LAMARCK, J.B.P.M. de (1801): Système des animaux sans vertèbres, ou tableau général des classes, des ordres et des genres de ces animaux [...]. VIII+432 pp., Paris (Deterville).

LAMARCK, J.B.P.M. de (1816): Histoire Naturelle des Animaux sans Vertèbres, présentant les caractères géneraux et particuliers de ces animaux, leur distribution, leur classes, leurs familles, leurs generes, et le citation des principales espèces qui s'y rapportent; précédée d'une Introduction offrant la Détermination des caractères essentiells de l'animal, sa distinction du végétal et des autres corps naturels, enfin, l'Exposition des Principes fondamentaux de la Zoologie. Tome Troisième. 586 pp., Paris (Verdière).

LAMARCK, J.B.P.M. de (1835): Histoire naturelle des animaux sans vertèbres [...]. Tome Troisième, Paris (Verdiere). [not seen]

LAMBERT, J. (1891): Note sur le genre *Echinocyamus* Van PHELSUM, 1774. – *Bulletin de la Société Géologique de France, Série 3* **19**: 749-752, Paris.

LAMBERT, J. (1897): Echinidi neogenici del Museo Parmense, par P.E. Vinassa de Regny. [annotated review] – *Revue Critique de Paléozoologie* **1**/3: 120, Paris.

LAMBERT, J. (1900): Étude sur quelques Échinides de l'Infra-Lias et du Lias. – *Bulletin de la Société des Sciences Historiques et Naturelles de l'Yonne* **52** (1899)/2: 3-57, 1 pl., 2 tabs., Auxerre.

LAMBERT, J. (1903): Echinofauna oligomiocenica della conca benacense, del Dott, C. AIRAGHI. [annotated review] – *Revue Critique de Paléozoologie* **7**/3: 173, Paris.

LAMBERT, J. (1905): Notes sur quelques Échinides éocéniques de l'Aude et de l'Hérault. – In: Catalogue descriptif des fossiles nummulitiques de l'Aude et de l'Hérault. – *Annales de l'Université de Lyon, Nouvelle Série, I. Sciences, Médecine* **17**: 129-164, pl. 6, Lyon-Paris.

LAMBERT, J. (1906a): Description des Échinides de la Province de Barcelone. fasc. 2-3 – *Mémoires de la Société Géologique de France, Paléontologie* **14**: 59-128, 5 pls., Paris.

LAMBERT, J. (1906b): Étude sur les Échinides de la Molasse de Vence. – *Annales de la Société des Lettres, Sciences et Arts des Alpes Maritimes* **20**: 1-64, 10 pls., Nice.

LAMBERT, J. (1906c): Note sur quelques Échinides de la Haute-Garonne. – *Bulletin de la Société Géologique de France, Série 4* **6**: 695-723, 3 pls., Paris.

LAMBERT, J. (1906d): Catalogue descriptif des fossiles nummulitiques de l'Aude et de l'Herault, par L. DONCIEUX. [annotated review] – *Revue Critique de Paléozoologie* **10**/1: 62-64, Paris.

LAMBERT, J. (1907a): Description des échinides fossiles des terrains miocéniques de la Sardaigne. – *Mémoires de la Société Paléontologique Suisse* **34**: 1-72, 5 pls., Genève.

LAMBERT, J. (1907b): Recherches sur le genere *Amphiope*. – *Bulletin de la Société des Sciences naturelles de Béziers* **29**: 49-62, 5 figs., 1 tab., Béziers.

LAMBERT, J. (1909): Description des échinides fossiles des terrains miocéniques de la Sardaigne. – *Mémoires de la Société Paléontologique Suisse* **35** (1908): 73-142, 6 pls., Genève.

LAMBERT, J. (1910a): Description des échinides des terrains néogènes du bassin Rhône. fasc.1 – *Mémoires de la Société Paléontologique Suisse* **37**: 1-48, 3 pls., Genève.

LAMBERT, J. (1910b): Les Échinides des Îles Snow-Hill et Seymour. – *Wissenschaftliche Ergebnisse der Schwedischen Südpolarexpedition, 1901-03* **3**/11: 1-15, 1 pl., Stockholm.

LAMBERT, J. (1911): Note sur quelques Échinides éocéniques des Corbières septentrionales. – In: DONCIEUX, L. (ed.): Catalogue descriptif des fossiles nummulitiques de l'Aude et de l'Hérault. – *Annales de l'Université de Lyon, nouvelle série, I. Sciences, Médecine* **30**: 163-199, 1 pl., Lyon.

LAMBERT, J. (1912): Description des échinides des terrains néogènes du bassin Rhône. fasc.2 – *Mémoires de la Société Paléontologique Suisse* **38**: 51-103, 4 pls., Genève.

LAMBERT, J. (1913a): Description des échinides des terrains néogènes du bassin Rhône. fasc.3 – *Mémoires de la Société Paléontologique Suisse* **39**: 105-154, 5 pls., Genève.

LAMBERT, J. (1913b): Note sur le *Scutella gibbercula* Marcel de SERRES, 1829. – *Bulletin de la Société Géologique de France, Série 4* **13**: 148-150, Paris.

LAMBERT, J. (1915a): Description des échinides des terrains néogènes du bassin Rhône. fasc.4 – *Mémoires de la Société Paléontologique Suisse* **41**: 155-240, 5 pls., Genève.

LAMBERT, J. (1915b): Revision des échinides fossiles du Bordelais. II: Échinides de l'Oligocène. – *Actes de la Société Linnéenne de Bordeaux* **69**: 13-59, Bordeaux.

LAMBERT, J. (1918): Considérations sur la classification des Échinides Atélostomes. I. Brachygnata et Procassiduloida. – *Mémoires de la Société Académique d'Agriculture des Sciences, Arts et Belles-Lettres du Département de l'Aube, Troisième Série* **55**/(Vol. 82 of the whole series): 9-54, 4 figs., 1 tab., Troyes.

LAMBERT, J. (1926): Échinides de la molasse marine du Jura neuchâtelois. – *Eclogae Geologicae Helvetiae* **19** (1925-26): 250-254, Bâle.

LAMBERT, J. (1927a): Revision des Echinides fossiles de la Catalogne. – *Memorias del Museo de Ciencias Naturales de Barcelona, Serie Geologica* **1**/2: 1-62, pls. 5-8, Barcelona.

LAMBERT, J. (1927b): Révision des Échinides fossiles du Bordelais. III. – Échinides du Miocène. – *Actes de la Société Linnéenne de Bordeaux* **79**: 71-132, 8 figs., 1 tab., Bordeaux.

LAMBERT, J. (1928): Sur deux Échinides fossiles de Cuba. – *Bulletin de la Société Géologique de France, Série 4* **28**: 19-21, 2 figs., Paris.

LAMBERT, J. (1931-1932): Étude sur les échinides fossiles du Nord de l'Arfrique. – *Mémoires de la Société Géologique de France, Nouvelle Serie* **(Mémoire No. 16)**: **7**/2: 1-108, pls. 1-4 (1931); **7**/4: 109-228, pls. 5-8 (1932), Paris.

LAMBERT, J. (1933): Échinides de Madagascar communiqués par M. H. Besairie. – *Madagascar Annales Géologiques du Service des Mines* **3**: 1-49, 4 pls., 8 figs., Tananarive.

LAMBERT, J. & JEANNET, A. (1928): Nouveau catalogue des moules d'échinides fossiles du Musée d'Histoire naturelle de Neuchâtel exécutés sous la direction de L. Agassiz et E. Desor. – *Mémoires de la Société helvétique des Sciences naturelles* **64**/mém. 2: 233 p., Zürich.

LAMBERT, J. & THIÉRY, P. (1909-1925): Essai de Nomenclature Raisonnée des Échinides. fasc. 1: i-iii, 1-80, pls. 1-2 (March 1909); fasc. 2: 81-160, pls. 3-4 (July 1910); fasc. 3: 161-240, pls. 5-6 (May 1911); fasc. 4: 241-320, pls. 7-8 (March 1914); fasc. 5: 321-384, pl. 9 (Sept. 1921); fasc. 6-7: 385-512, pls. 10-11, 14 (Dec. 1924); fasc. 7-8: 513-607, pls. 12, 13, 15 (Feb. 1925), Chaumont (L. Ferrière).

LANGER, I. (1939): Geologische Beschreibung des Bisamberges. – *Jahrbuch der Geologischen Bundesanstalt* **88** (1938): 351-393, 1 map, 4 pls., 3 figs., Wien.

LASKAREV, V. (1924): Sur les équivalents du sarmatien supérieure en Serbie. – In: *Recueil de travaux offert à M. Jovan CVIJIC par ses amis et collaborateurs à l'occasion de ses trente-cinq ans de travail scientifique.* pp. 73-85 (1-13), 1 fig., 2 tabs., Belgrade (Imprimerie d'État du Royaume des Serbes, Croates et Slovènes).

LATREILLE, P.A. (1825): Familles naturelles du régne animal. 570 pp., Paris (Baillière).

LAUBE, G.C. (1868): Ein Beitrag zur Kenntniss der Echinodermen des vicentinischen Tertiärgebietes. – *Denkschriften der kaiserlichen Akademie der Wissenschaften, mathematisch-naturwissenschaftliche Classe* **29**/2: 1-38, 7 pls., Wien.

LAUBE, G.C. (1869a): Die Echinoiden der österreichisch-ungarischen oberen Tertiärablagerungen. – *Verhandlungen der kaiserlich-königlichen Geologischen Reichsanstalt* **3**/9: 182-184, Wien.

LAUBE, G.C. (1869b): Beschreibung neuer fossiler Echinodermen aus Süd-Australien. – *Sitzungsberichte der Kaiserlichen Akademie der Wissenschaften, mathematisch-naturwissenschaftliche Classe, Abtheilung* I **59** (1869): 183-198, 1 pl., Wien.

LAUBE, G.C. (1870): Die Echinoiden der österreichisch-ungarischen oberen Tertiärablagerungen. – *Verhandlungen der kaiserlich-königlichen Geologischen Reichsanstalt* **4**/16: 312-313, Wien.

LAUBE, G.C. (1871): Die Echinoiden der oesterreichisch-ungarischen oberen Tertiaerablagerungen. – *Abhandlungen der kaiserlich-königlichen Geologischen Reichsanstalt* **5**/3: 55-74, pls. 16-19, Wien.

LAURIN, B., DAVID, B., FÉRAL, J.-P. & DERELLE, E. (1994): Polytypism in the spatangoid sea urchin *Echinocardium*: A morphological vs. molecular approach. – In: DAVID, B., GUILLE, A., FÉRAL, J.-P. & ROUX, M. (eds.): *Echinoderms through Time.- Proc. of the 8th Internat. Echinoderm Conf. Dijon/France, 6-10 Sept. 1993.* pp. 739-745, 6 figs., Rotterdam & Brookfield, IL (A.A. Balkema).

LAWRENCE, J.M. & AGATSUMA, Y. (2001): The ecology of *Tripneustes*. – In: LAWRENCE, J.M. (ed.): *Edible Sea Urchins: Biology and Ecology. Developments in Aquaculture and Fisheries Science* **32** pp. 395-413, 3 tabs., Amsterdam (Elsevier).

LAWRENCE, J.M., HERRERA, J. & COBB, J. (2004): Vertical posture of the clypeasteroid sand dollar *Encope michelini*. – *Journal of the Marine Biological Association of the United Kingdom* **84**: 407-408, 2 figs., Cambridge, MA.

LESKE, N.G. (1778): Jacobi Theodori Klein naturalis dispositio echinodermatum . . ., edita et descriptionibus novisque inventis et synonomis auctorem aucta. Addimenta ad I. T. Klein naturalem dispositionem Echinodermatum. XXII+278 pp., 54 pls., Leipzig (G. E. Beer).

LESSIOS, H.A., GARRIDO, M.J. & KESSING, B.D. (2001a): Demographic history of *Diadema antillarum*, a keystone herbivore on caribbean reefs. – *Proceedings of the Royal Society of London, Series B* **268**: 1-7, London.

LESSIOS, H.A., KESSING, B.D. & PEARSE, J.S. (2001b): Population structure and speciation in tropical seas: global phylogeography of the sea urchin *Diadema*. – *Evolution* **55**/5: 955-975, Lawrence, KS.

LESSIOS, H.A., KESSING, B.D., ROBERTSON, D.R. & PAULAY, G. (1999): Phylogeography of the pantropical sea urchin *Eucidaris* in relation to land barriers and ocean currents. – *Evolution* **53**/3: 806-817, Lawrence, KS.

LESSIOS, H.A. & PEARSE, J.S. (1996): Hybridization and introgression between Indo-Pacific species of *Diadema*. – *Marine Biology* **126**: 715-723, Berlin.

LEWIS, J.L. (1960): The fauna of rocky shores of Barbados, West Indies. – *Canadian Journal of Zoology* **38**: 391-435, Ottawa.

LINDER, G. (1990): Muscheln und Schnecken der Weltmeere: Aussehen, Vorkommen, Systematik. 256 pp., 1257 figs., München, Wien, Zürich (BLV).

LINDLEY, I.D. (2003): Echinoids of the Kairuku Formation (Lower Pliocene), Yule Island, Papua New Guinea: Regularia. – *Proceedings of the Linnean Society of New South Wales* **124**: 137-151, 9 figs., 1 tab., Kingsford, NSW.

LINNÆI, C. (1758): Systema Naturæ per Regna tria Naturæ,

secundum Classes, Ordines, Genera, Species, cum characteribus, differentiis, synonymis, locis. Edito Decima, Reformata. Tomus I. 824 pp., Holmiæ (Impensis Direct. Laurentii Salvii).

Lóczy, L. (1877): Echioniden aus den Neogen-Ablagerungen des Weissen Korösthales. – *Természetrajzi Füzetek* **1**: 61-67, 1 pl., Budapest.

Lörenthey, I. (1894): Jelentés az Erdélyi Múzeum-Egylet megbizásában 1891. Nyarán tett földtani kirándulásaimnak eredményeiröl. – *Értesitö az Erdélyi Múzeum-Egylet Orvos-Természettudományi Szakosztályából. II. Természettudományi Szak* **15** (1893)/1: 55-68, 100-102, Kolozsvárt.

de Loriol, P. (1875-1876): Echinologie Helvétique. Description des Échinides tertiaires de la Suisse. Troisième partie. Échinides de la Période Tertiaire. – *Mémoires de la Société Paléontologique Suisse* **2** (1875): 1-88, pls. 1-8; **3** (1876): 89-142, pls. 9-23, Genève.

de Loriol, P. (1880): Monographie des Échinides contenus dans les couches nummulitiques de l'Égypte. – *Mémoires de la Société de Physique et d'Histoire Naturelle de Genève* **27**/1: 1-92, 11 pls., Genève.

de Loriol, P. (1882): Description des Échinides des environs de Camerino (Toscane). – *Mémoires de la Société de Physique et d'Histoire Naturelle de Genève* **28**/1: 1-32, 3 pls., Genève.

de Loriol, P. (1883): Catalogue Raisonne des Échinodermes recueillis par M. V. Robillard a l'île Maurice. – *Mémoires de la Société de Physique et d'Histoire Naturelle de Genève* **28**/8: 1-64, 6 pls., Genève.

de Loriol, P. (1896): Description des Échinodermes Tertiaires de Portugal. Accompagnée d'un Tableau Stratigraphique par J. C. Berkeley Cotter. 50 pp., 13 pls., 1 tab., Lisbonne (Direction des Traveaux Géologiques du Portugal, Imprimerie de l'Academie Royale des Sciences Lisbonne).

de Loriol, P. (1897): Description de quelques échinodermes. Appendice à la note de M. E. Pellat sur le Burdigalien supérieur. – *Bulletin de la Société Géologique de France, Série 3* **25**/1: 115-129, 1 pl., Paris.

de Loriol, P. (1902): Notes pour servir à l'étude des échinodermes. 1re Série, 10e article. 32 pp., 4 pls., Bâle et Genève (Librairie Georg et cie.).

Lovén, S. (1874): Études sur les échinoïdées. – *Kongelige Svenska Vetenskaps-Akademiens Handlingar* **11**: 1-91, Stockholm.

Lovisato, D. (1909): *Clypeaster Pillai* Lov. – *Palaeontographia Italica* **15**: 207-302, 1 pl., Pisa.

Lovisato, D. (1910): Una parola sul *Clypeaster lovisatoi* Cottreau e specie nuove di *Clypeaster* ed *Echinolampas*. – *Palaeontographia Italica* **16**: 137-146, 3 pls., Pisa.

Lovisato, D. (1911a): Note di paleontologia miocenica della Sardegna. Specie nuove di *Clypeaster* e di *Amphiope*. – *Palaeontographia Italica* **17**: 37-48, 3 pls., Pisa.

Lovisato, D. (1911b): Nuove specie di «*Clypeaster*» del Miocene medio di Sardegna. – *Rivista Italiana di Paleontologia* **17**/3: 29-36, 2 pls., Parma.

Lovisato, D. (1912a): Altre contributo echinologica con nuove specie di «*Clypeaster*» in Sardegna. – *Bolletino della Società Geologica Italiana* **31**: 357-378, pls. 15-16, Roma.

Lovisato, D. (1912b): Da Cagliari a Thiesi – Altre Specie nuove di *Clypeaster* miocenici. – *Palaeontographia Italica* **18**: 129-140, 3 pls., Pisa.

Lovisato, D. (1913): Nuove Specie di «Clypeaster» Miocenici Sardi dal Vulcano S.Matteo di Ploaghe per Nurecci e Senis alla regione Fraos nella Planargia e all'amba del Capo della Frasca. – *Bolletino della Società Geologica Italiana* **32**: 401-438, 2 pls., Roma.

Lovisato, D. (1914a): Altre specie nuove di *Clypeaster*, *Scutella* ed *Amphiope* della Sardegna. – *Rivista Italiana di Paleontologia* **20**/3-4: 89-114, 2 pls., Parma.

Lovisato, D. (1914b): Fibularidi e Clipeastridi miocenici della Sardegna. – *Bolletino della Società Geologica Italiana* **33**: 485-564, 2 pls., Roma.

Lukeneder, A. & Harzhauser, M. (2002): Shell Accumulations of the Nautilidae *Aturia* (*Aturia*) *aturi* (Bast.) in the Lower Miocene Paratethys (Lower Austria). – In: Summesberger, H., Histon, K. & Daurer, A. (eds.): Cephalopods – Present and Past. – *Abhandlungen der Geologischen Bundesanstalt* **57**: 459-466, 1 pl., 5 figs., Wien.

Machalski, M. (1996): Fossilien sammeln in Polen: Vorkarpatensenke. – *Fossilien* **13**/1: 22-26, 5 figs., Korb.

Mączyńska, S. (1977): Echinoids from the Korytnica basin (Middle Miocene; Holy Cross Mountains, Poland). – *Acta Geologica Polonica* **27**/2: 193-200, 7 pls., Warszawa.

Mączyńska, S. (1979): Echinoids from the Miocene deposits of the Roztocze Region, South-eastern Poland. – *Prace Muzeum Ziemi* **32**: 29-32, 10 pls., Warszawa.

Mączyńska, S. (1987): A supplementary account on the echinoids from the Korytnica Basin (Middle Miocene; Holy Cross Mountains, Central Poland). – *Acta Geologica Polonica* **37**/3-4: 145-153, 8 pls., Warszawa.

Mączyńska, S. (1988): Echinoids from the Middle Miocene (Badenian) sands from southern Poland. – *Prace Muzeum Ziemi* **40**: 59-64, 6 pls., Warszawa.

Mączyńska, S. (1993): Echinoids from the Pińczów Limestones (Middle Miocene; Holy Cross Mountains, Central Poland). – *Acta Geologica Polonica* **43**/1-2: 103-114, 6 pls., Warszawa.

Mączyńska, S. (1996): Middle Miocene (Badenian) echinoids from Niechobrz near Rzeszów (southern Poland). – *Prace Muzeum Ziemi* **43**: 39-46, 2 pls., 1 fig., 1 tab., Warszawa.

Majcen, T., Mikuž, V. & Pohar, V. (1997): Okamine v Paleontološki zbirki Laškega Muzeja. – In: 13th Meeting of Slovenian Geologists. – *Geološki Zbornik* **13**: 104-118, 11 figs., Ljubljana.

Májer, I. (1915): A Börzsony hegység északi részének üledékes képzödményei. – *Földtani Közlöny* **45**: 18-40, 69-64, pl. 2, figs. 4-5, Budapest.

Manzoni, A. (1873): Il Monte Titano (territorio della Repubblica di San Marino), i suoi fossili, la sua età ed il suo modo d'origine. – *Bollettino del R. Comitato Geologico d'Italia* **4**/1-2: 1-28, 1 pl., Firenze.

Manzoni, A. (1879): Gli echinodemi fossili dello Schlier delle colline di Bologna. – *Denkschriften der kaiserlichen Akademie der Wissenschaften, mathematisch-naturwissenschaftliche Classe* **39**: 149-164, 4 pls., Wien.

Manzoni, A. (1880a): Echinodermi fossili della molassa serpentinosa e supplemento agli echinodermi dello Schlier della colline di Bologna. – *Denkschriften der kaiserlichen Akademie der Wissenschaften, mathematisch-naturwissenschaftliche Classe* **42**: 185-190, 3 pls., Wien.

Manzoni, A. (1880b): Echinodermi fossili pliocenici. – *Atti della Società Toscana di Scienze Naturali* **4**/2 (1879): 327-335, Pisa.

Manzoni, A. (1881): Spugne Silicee della Molassa Miocenica del Bolognese. – *Atti della Società Toscana di Scienze Naturali residente in Pisa, Memorie* **5**/1 (1880): 173-176, Pisa.

Marcopoulou-Diacantoni, A. (1967): La faune des échinides néogènes des Pays helléniques. – *Annales Géologiques des Pays Helléniques* **18**: 331-406, pls. 48-69, 10 figs., 1 tab., Athènes.

Marcopoulou-Diacantoni, A. (1970): [Quelques observations sur les anomalies et les irregulatités du test des échinides et notamment de ceux qui proviennent du Néogène des Pays Helléniques]. – *Annales Géologiques des Pays Helléniques* **22**: 256-262, Athènes.

Marcopoulou-Diacantoni, A. (1972): Échinides (*Clypeaster*,

Schizaster, Spatangus, Brissopsis) de l'Helvétien de l'ile de Crète centrale et Orientale. – *Annales Géologiques des Pays Helléniques* **24**: 133-160, 2 pls., 9 figs., Athènes.

MARCUS, N.H. (1983): Phenotypic variability in echinoderms. – In: JANGOUX, M. & LAWRENCE, J.M. (eds.): *Echinoderm Studies, Vol. 1* pp. 19-38, Rotterdam (A.A. Balkema).

MÁRTONFI, L. (1894): Adatok Bujtur Fossil Faunájahóz. (Jelentés az erdélyi múzeum-egylet megbizásában 1892. év nyarán végezett földtani kirándulás eredményeiröl). – *Értesitö az Erdélyi Múzeum-Egylet Orvos-Természettudományi Szakosztályából. II. Természettudományi Szak* **15**/2 (1893): 141-158, 181, Kolozsvárt.

MATSUOKA, N. & HATANAKA, T. (1991): Molecular evidence for the existence of four sibling species within the sea-urchin, *Echinometra mathaei* in Japanese waters and their evolutionary relationships. – *Zoological Science* **8**: 121-133, Tokyo.

MAYER, K. (1864): Die Tertiär-Fauna der Azoren und Madeiren. Systematisches Verzeichniss der fossilen Reste von Madeira, Porto Santo und Santa Maria, nebst Beschreibung der neuen Arten. vi+107 pp., 7 pls., Zürich (Eigenverlag).

MAZZEI, R. (1985): The Miocene Sequence of the Maltese Islands: Biostratigraphic and Chronostratigraphic References based on Nannofossils. – *Atti della Società Toscana di Scienze Naturali, Memorie* **A 92**: 165-197, 3 pls., 3 figs., 1 tab., Pisa.

MAZZETTI, G. (1882a): Echinodermi fossili di Montese. – *Annuario della Società dei Naturalisti in Modena, Serie 2* **15**: 108-126, 3 pls., Modena.

MAZZETTI, G. (1882b): Echinodermi fossili di Pantano. – *Annuario della Società dei Naturalisti in Modena, Serie 2* **15**: 127-129, Modena.

MAZZETTI, G. & PANTANELLI, D. (1883): Cenno monografico intorno allo fauna fossile di Montese. – *Atti della Società dei Naturalisti di Modena, Memorie Serie III* **2**/17: 58-96, 2 pls., Modena.

MCCARTNEY, M.A., KELLER, G. & LESSIOS, H.A. (2000): Dispersal barriers in tropical oceans and speciation in Atlantic and eastern Pacific sea urchins of the genus *Echinometra*. – *Molecular Ecology* **9**: 1391-1400, 2 figs., 2 tabs., Edinburgh.

MCCLANAHAN, T.R. & MUTHIGA, N.A. (2001): The ecology of Echinometra. – In: LAWRENCE, J.M. (ed.): *Edible Sea Urchins: Biology and Ecology. Developments in Aquaculture and Fisheries Science* **32** pp. 225-243, 2 tabs., Amsterdam (Elsevier).

MCCOY, F. (1882): Tertiary Echinodermata. – In: Prodromus of the palaeontology of Victoria; or, figures and descriptions of the Victorian organic remains, Decade VII, pp. 15-22, pls. 64-68, Melbourne (Geological Survey of Victoria).

MCKINNEY, M.L., MCNAMARA, K.J. & WIEDMAN, L.A. (1988): Echinoids from the La Meseta Formation (Eocene), Seymour Island, Antarctica. – *Geological Society of America, Memoir* **169**: 499-503, 3 figs., Boulder, CO.

MCNAMARA, K.J. (1984): Living Australian species of the echinoid *Pericosmus* (Spatangoida: Pericosmidae). – *Records of the Western Australian Museum* **11**/2: 87-100, 8 figs., Perth.

MCNAMARA, K.J. (1985a): The spatangoid echinoid *Linthia* from the Late Eocene of Southern Australia. – *Transactions Royal Society of South Australia* **109**/4: 161-165, 1 fig., Adelaide.

MCNAMARA, K.J. (1985b): Taxonomy and evolution of the Cainozoic spatangoid echinoid *Protenaster*. – *Palaeontology* **28**/2: 311-330, 3 pls., 8 figs., London.

MCNAMARA, K.J. (1995): The spatangoid echinoid *Schizaster (Schizaster) compactus* (KOEHLER, 1914) in Western Australia. – *Records of the Western Australian Museum* **17**: 315-323, 8 figs., Perth.

MCNAMARA, K.J. & BRYCE, C. (1983): Habitats of the heart urchins *Protenaster* and *Rhynobrissus*. – *Records of the Western Australian Museum* **11**/1: 73-74, Perth.

MCNAMARA, K.J. & KENDRICK, G.W. (1994): Cenozoic molluscs and echinoids of Barrow Island, Western Australia. – *Records of the Western Australian Museum Supplement* **51**: 1-50, 18 figs., Perth.

MCNAMARA, K.J. & PHILIP, G.M. (1980a): Australian Tertiary schizasteroid echinoids. – *Alcheringa* **4**: 47-65, 9 figs., Sydney.

MCNAMARA, K.J. & PHILIP, G.M. (1980b): Living Australian schizasterid schinoids. – *Proceedings of the Linnean Society of New South Wales* **104**/2 (1979): 127-146, 13 figs., 1 tab., Sydney.

MCNAMARA, K.J. & PHILIP, G.M. (1980c): Tertiary species of *Echinolampas* (Echinoidea) from southern Australia. – *Memoirs of the National Museum of Victoria* **41**: 1-14, 4 pls., 6 figs., Victoria.

MCNAMARA, K.J. & Philip, G.M. (1984): A Revision of the Spatangoid Echinoid *Pericosmus* from the Tertiary of Australia. – *Records of the Western Australian Museum* **11**/4: 319-356, 28 figs., Perth.

MCNAMARA, K.J., Philip, G.M. & Kruse, P.D. (1986): Tertiary Brissid echinoids of southern Australia. – *Alcheringa* **10**: 55-86, 17 figs., Sydney.

MCPHERSON, B.F. (1968a): Contributions to the biology of *Eucidaris tribuloides* (Lamarck). – *Bulletin of Marine Science* **18**/2: 400-443, 16 figs., 8 tabs., Coral Gables, FL.

MCPHERSON, B.F. (1968b): Feeding and oxygen uptake of the tropical sea urchin *Eucidaris tribuloides* (Lamarck). – *Biological Bulletin* **135**: 308-321, 7 figs., 3 tabs., Woods Hole, MA.

MELION, V.J. (1853): Bericht über das Vorkommen fossiler tertiärer Mollusken bei Littenschitz, Lomniczka und Rossitz. – *Jahrbuch der kaiserlich-königlichen Geologischen Reichsanstalt.* **4**: 703-710, Wien.

MELVILLE, R.V. & DURHAM, J.W. (1966): Skeletal Morphology – In: MOORE, R.C. (ed.): *Treatise on Invertebrate Paleontology, U. Echinodermata. 3 (1)* pp. U220-U257, figs. 163-196, Boulder, CO & Lawrence, KS (GSA & Univ. Kansas Press).

MENESINI, E. (1967): Gli Echinidi miocenici delle "Arenarie di Porsano" (Volterra – Prov. di Pisa). – *Palaeontographia Italica* **62**: 143-167, 6 pls., 4 figs., Pisa.

MENESINI, E. (1975): Considerazione su *Schizaster parkinsoni* Defrance del Miocene dell'Archipelago Maltese. – *Atti della Società Toscana di Scienze Naturali, Memorie, Serie A* **82**: 94-116, 2 pls., 6 figs., 8 tabs., Pisa.

MENESINI, E. (1979a): Echinidi fossili dell'Arcipelago Maltese. I. – *Atti della Società Toscana di Scienze Naturali, Memorie, Serie A* **86**: 51-64, 2 figs., Pisa.

MENESINI, E. (1979b): Maltese Fossil Echinoids. – *Annales Géologiques des Pays Helléniques, tome hors série* **2**: 799-806, Athènes.

MERCATI, M. (1717-19): Metallotheca Vaticana, opus posthuman opera et studio Johannis Mariae Lancisii illustratum. lxiv+378+18 pp., 6 pls., figs. (1717), Appendix, 54 pp., figs. (1719) pp., Romae.

MEZNERICS, I. (1936): Die Schlierbildungen des mittelsteirischen Beckens. – *Mitteilungen des naturwissenschaftlichen Vereins für Steiermark* **73**: 118-140, 1 pl., Graz.

MEZNERICS, I. (1941): Neue Stachelhäuter (Echinodermen) aus dem Miocän Ungarns. – *Annales Musei Nationalis Hungarici, Pars Mineralogica, Geologica et Palaeontologica* **34**: 83-96, 3 pls., Budapest.

MICHALÍK, J. (1977): Systematics and ecology of *Zeilleria* Bayle and other brachiopods in the Uppermost Triassic of the West Carpathians. – *Geologický Zbornik – Geologica Carpathica* **28**/2: 323-346, 17 figs., Bratislava.

MICHELIN, H. (1863): Échinides et Stéllerides recueillis par M. Maillard, ingénieur colonial de 1856 a 1860. – In: MAILLARD, L. (ed.): *Notes sur l'Ile de la Réunion (Bourbon). Deuxième édition.* Annexe A. pp. 1-7, pls. 14-16 (14-15 in color), Paris (Dentu).

MICHELOTTI, G. (1847): Description des Fossiles l'Italie septentrionale. 408 pp., 17 pls., Leiden (Arnz & Comp.). [not seen]

MIHÁLY, S. (1969): Tortonai Echinoideák a Kerepesi úti csatornázás feltárásából. – *Földtani Közlöny* **99**/3: 253-257, Budapest.

MIHÁLY, S. (1985): Felsö-Bádeni Echinoideák Budapesti új Feltárásokból. – *A Magyar Állami Földtani Intézet Évi Jelentése* **1983**: 235-262, Budapest.

MIHÁLY, S. (1990): Két új Echinoidea faj a Magyarországi Miocénböl. – *A Magyar Állami Földtani Intézet Évi Jelentése* **1987**: 237-249, 4 pls., Budapest.

MIKUŽ, V. & MITROVIĆ-PETROVIĆ, J. (2001): Miocenski morski ježki iz okolice Šentilja v Slovenskih Goricah. 1. Clypeasteridae. – *Razprave IV. Razreda Sazu Dissertationes* **42**/1: 47-77, 6 pls., 1 fig., Ljubljana.

MILLET, P.-A. (1854): Paléontologie de Maine et Loire comprenant, avec des observations et l'indication des diverses formations géologiques du department de Maine et Loire, un relevé des roches, des minéraux et des fossiles qui se rapportent à chacune d'elles. viii+187 pp., 1 tab., Angers (Imprimerie de Cosneier et Lachèse).

MILLET, P.-A. (1865): Indicateur de Maine-et-Loire, Tome II, Supplémente, Paléontographie. [not seen, cited from COTTREAU, 1912]

MILLET, P.-A. (1866): Paléontographie ou description des fossiles nouveaux du terrain tertiaire marin ou terrain miocène supérieur du départemente de Maine et Loire. [not seen, cited from COTTREAU, 1912]

MIRONOV, A.N. & SAGAIDACHNY, A.Yu. (1984): [Morphology and distribution of the recent echinoids of the Genus Echinocyamus (Echinoidea: Fibulariidae)]. – *Trudy Instituta Okeanologii* **119**: 179-203, 2 pls., 10 figs., 4 tabs., Moskva.

MISKELLY, A. (2002): Sea Urchins of Australia and the Indopacific. 180 pp., illustr., Sydney (Capricornica Publications).

MITROVIĆ-PETROVIĆ, J. (1966): Les échinides du Crétacé et du Miocène de la Serbie. [in Russian, with French summary] – *Annales Géologiques de la Péninsule Balkanique* **32**: 87-164, 7 pls., 19 figs., 3 tab., Beograd.

MITROVIĆ-PETROVIĆ, J. (1969): Les Echinides du Miocène moyen dans la Bosnie septentrionale (La vallé de la Save en Bosnie). [in Serbo-Croatian, with French summary] – *Priridislovna Istraživanja, Knjiga 36, Acta Geologica* **6**: 113-146, 30 pls., 1 fig., 2 tabs., Zagreb.

MITROVIĆ-PETROVIĆ, J. (1981): Les echinides comme les fossiles faciaux. [in Russian, with French summary] – *Annales Géologiques de la Péninsule Balkanique* **45**: 173-185, 13 figs., 1 tab., Beograd.

MITROVIĆ-PETROVIĆ, J. (1984): Importance biostratigraphique et paleoecologique du genre *Clypeaster* (Echinoidea) pour les sediments Neogenes de la Yougoslavie. – *Annales Géologiques des Pays Helléniques* **32**: 211-235, 13 pls., Athènes.

MONTENAT, C. & ROMAN, J. (1970): Échinides Neogènes d'Espagne (Provinces d'Alicante et de Murcie). – *Annales de Paléontologie (Invertébrés)* **56**/1: 89-138, 3 pls., 2 figs., 1 tab., Paris.

MOOI, R. (1989): Living and Fossil Genera of the Clypeasteroida (Echinoidea: Echinodermata): An Illustrated Key and Annotated Checklist. – *Smithsonian Contributions to Zoology* **488**: 1-51, 34 figs., Washington, DC.

MOOI, R. (1990a): Living cassiduloids (Echinodermata: Echinoidea): A key and annotated list. – *Proceedings of the Biological Society Washington* **103**/1: 63-85, 12 figs., Washington, DC.

MOOI, R. (1990b): A New "Living Fossil" Echinoid (Echinodermata) and the Ecology and Palaeobiology of Caribbean Cassiduloids. – *Bulletin of Marine Science* **46**/3: 688-700, 6 figs., Coral Gables, FL.

MOOI, R. (1990c): Paedomorphosis, Aristotle's lantern, and the origin of the sand dollars (Echinodermata: Clypeasteroida). – *Paleobiology* **16**/1: 25-48, 8 figs., Lancaster, PA.

MOOI, R. (1990d): Progenetic miniaturization in the sand dollar *Sinaechinocyamus*: implications for clypeasteroid phylogeny. – In: DE RIDDER, C., DUBOIS, P., LAHAYE, M.-C. & JANGOUX, M. (eds.): *Echinoderm Research. – Proc. 2nd European Echinoderm Conf., Brussels, 18-21 Sept. 1989.* pp. 137-143, 5 figs., Rotterdam (A.A. Balkema).

MOOI, R. (1997): Sand dollars of the genus *Dendraster* (Echinoidea: Clypeasteroida): Phylogenetic systematics, heterochrony, and distribution of extant species. – *Bulletin of Marine Science* **61**/2: 343-375, Miami, FL.

MOOI, R. & CHEN, C.-P. (1996): Weight belts, diverticula, and the phylogeny of the sand dollars. – *Bulletin of Marine Science* **58**/1: 186-195, 3 figs., 3 tabs., Miami, FL.

MOOI, R., DAVID, B., FELL, J.F. & CHONE, T. (2000): Three new species of bathyal cidaroids (Echinodermata: Echinoidea) from the Antarctic region. – *Proceedings of the Biological Society Washington* **113**/1: 224-237, 4 figs., Washington, DC.

MOORE, H.B., JUTARE, T., JONES, J.A., MCPHERSON, B.F. & ROPER, C.F.E. (1963): A contribution to the biology of *Tripneustes esculentus*. – *Bulletin of Marine Science* **13**: 267-281, 10 figs., Coral Gables, FL.

MOORE, R.C. (ed.) (1966): Treatise on Invertebrate Paleontology, U. Echinodermata 3. XXX + 695 pp., 3485 figs., Boulder, CO & Lawrence, KS (GSA & Univ. Kansas Press).

MORTENSEN, T. (1907): The Danish Ingolf-Expedition 1895-1896. Vol. 4, No. 2. Echinoidea, pt. 2. 200 pp., 19 pls., 27 figs., Copenhagen. [not seen]

MORTENSEN, T. (1909): Die Echinoiden der Deutschen Südpolar-Expedition 1901-1903. – In: Drygalski, E.v. (ed.): *Deutsche Südpolar-Expedition 1901-1903 im Auftrage des Reichsamtes des Innern, XI. Band, Zoologie III. Band, Heft I* pp. 1-114, pls. I-XIX, Berlin (Georg Reimer).

MORTENSEN, T. (1912): Echinological Notes IV. On natural hybrids of Echinoderms. – *Videnskabelige Meddelelser fra den naturhistoriske Forening i Kjøbenhavn* **63**: 58-66, 2 figs., Kjøbenhavn.

MORTENSEN, T. (1913): Die Echinoiden des Mittelmeeres. – *Mittheilungen aus der Zoologischen Station zu Neapel* **21**: 1-39, Neapel.

MORTENSEN, T. (1928): A Monograph of the Echinoidea. I. Cidaroidea. 551 pp., 88 pls., 173 figs., Copenhagen & London (C. A. Reitzel & Oxford Univ. Press).

MORTENSEN, T. (1942): New Echinoidea. Preliminary notice. – *Videnskabelige Meddelelsar Dansk Naturhistoriske Forening i Kjøbenhavn* **106**: 225-232, Kjøbenhavn.

MORTENSEN, T. (1943a): A Monograph of the Echinoidea. III, 2. Camarodonta. I. Orthopsidæ, Glyphocyphidæ, Temnopleuridæ and Toxopneustidæ. vii+553 pp., 56 pls., 321 figs., Copenhagen (C. A. Reitzel).

MORTENSEN, T. (1943b): A Monograph of the Echinoidea. III, 3. Camarodonta. II. Echinidæ, Strongylocentrotidæ, Parasaleniidæ, Echinometridæ. 446 pp., 66 pls., 215 figs., Copenhagen (C. A. Reitzel).

MORTENSEN, T. (1948a): A Monograph of the Echinoidea. IV, 1 Holectypoida, Cassiduloida. 371 pp., 14 pls., 326 figs., Copenhagen (C. A. Reitzel).

MORTENSEN, T. (1948b): A Monograph of the Echinoidea. IV, 2.

Clypeasteroida. Clypeasteridæ, Arachnoidæ, Fibulariidæ, Laganidæ and Scutellidæ. 471 pp., 72 pls., 258 figs., Copenhagen (C. A. Reitzel).

MORTENSEN, T. (1950): A Monograph of the Echinoidea. V, 1. Spatangoida I. Protoseternata, Meridosternata, Amphisternata I. Palæopneustidæ, Palæostomatidæ, Aëropsidæ, Toxasteridæ, Micrasteridæ, Hemiasteridæ. 432 pp., 25 pls., 315 figs., Copenhagen (C. A. Reitzel).

MORTENSEN, T. (1951): A Monograph of the Echinoidea. V, 2. Spatangoida II. Amphisternata II. Spatangidæ, Loveniidæ, Pericosmidæ, Schizasteridæ, Brissidæ. 593 pp., 64 pls., 286 figs., Copenhagen (C. A. Reitzel).

DES MOULINS, C. (1837): Troisième Mémoire sur les Échinides. – Synonymie générale. – *Actes de la Société Linnéenne de Bordeaux* **9**/6: 45-364, Bordeaux.

DES MOULINS, C. (1869): Spécification et noms légitimes de six Échinolampes. – *Actes de la Société Linnéenne de Bordeaux* **27**: 309-322, 3 pls., Paris-Bordeaux.

MÜLLER, A.H. (1963): Lehrbuch der Paläozoologie. Band II. Invertebraten. Teil 3. Arthropoda 2 – Stomocordata. 1-698 pp., 854 figs., Jena (Gustav Fischer Verlag).

MÜLLER, A.H. (1978): Lehrbuch der Paläozoologie. Band II. Invertebraten. Teil 3. Arthropoda 2 – Hemicordata. 2. Auflage. 1-748 pp., 852 figs., Jena (Gustav Fischer Verlag).

MÜLLER, O.F. (1776): Zoologiæ Danicæ Prodromus, seu Animalium Daniæ et Norvegiæ indigenarum Characteres, Nomina, et Synonyma Imprimis Popularium. xxxii+282 pp., Havniæ (Typis Hallageriis).

MUNIER-CHALMAS, E. (1885): Observations sur l'appareil apical de quelques Échinides Crétacés et Tertiaires. – *Comptes Rendus de l'Académie des Sciences Paris* **101**: 1074-1077, Paris. [not seen; cited from FISCHER, 1966]

MÜRLE, K. (1867): Neue Brunnenbohrung in Hainburg an der Donau. (Aus einem Schreiben an Bergrath Dr. G. Stache). – *Verhandlungen der kaiserlich-königlichen Geologischen Reichsanstalt* **1**: 332-333, Wien.

NEBELSICK, J.H. (1989): Temperate Water Carbonate Facies of the Early Miocene Paratethys (Zogelsdorf Formation, Lower Austria). – *Facies* **21**: 11-40, 10 figs., 1 tab., Erlangen.

NEBELSICK, J.H. (1991): Sand-Dollar. – In: STEININGER, F.F. & PILLER, W.E. (eds.): *Eggenburg am Meer – Eintauchen in die Erdgeschichte. Katalogreihe des Krahuletz-Museums Nr. 12* pp. 75-79, Eggenburg (Peter Schindler).

NEBELSICK, J.H. (1992): The Northern Bay of Safaga (Red Sea, Egypt): An Actuopalaeontological Approach III. Distribution of Echinoids – *Beiträge zur Paläontologie von Österreich* **17**: 5-79, 8 pls., 37 figs., 6 tabs., Wien.

NEBELSICK, J.H. (1999): Taphonomic comparison between Recent and fossil sand dollars. – *Palaeogeography, Palaeoclimatology, Palaeoecology* **149**: 349-358, 6 figs., Amsterdam.

NEBELSICK, J.H. & KROH, A. (2002): The Stormy Path from Life to Death Assemblages: The Formation and Preservation of Mass Accumulations of Fossil Sand Dollars. – *Palaios* **17**/4: 378-393, 9 figs., 2 tabs., Lawrence, KS.

NEBELSICK, J.H., ROETZEL, R., STEININGER, F.F., RUPP, C. & JENKE, Y. (1991a): F/4: Eggenburg, Brunnstube. – In: ROETZEL, R. & NAGEL, D. (eds.): *Exkursionen im Tertiär Österreichs.* pp. 86-91, figs. 23-24, Gmünd (Österreichische Paläontologische Gesellschaft).

NEBELSICK, J.H., SCHMID, B. & STACHOWITSCH, M. (1997): The encrustation of fossil and recent sea-urchin tests: ecological and taphonomical significance. – *Lethaia* **30**: 271-284, 8 figs., Oslo.

NEBELSICK, J.H., STEININGER, F.F., VÁVRA, N. & JENKE, Y. (1991b): F/11: Limberg, Steinbruch Hengl. – In: ROETZEL, R. & NAGEL, D. (eds.): *Exkursionen im Tertiär Österreichs.* pp. 117-120, figs. 33, 34, Gmünd (Österreichische Paläontologische Gesellschaft).

NEBELSICK, J.H., STEININGER, F.F., VÁVRA, N. & PILLER, W.E. (1991c): F/6: Zogelsdorf, Johannes-Steinbruch. – In: ROETZEL, R. & NAGEL, D. (eds.): *Exkursionen im Tertiär Österreichs.* pp. 97-99, figs. 23, 26, Gmünd (Österreichische Paläontologische Gesellschaft).

NEGRETTI, B. (1984): Échinides néogènes du littoral de la Nerthe (Basse Provence, France): stratigraphie, paléontologie, conclusion paléoécologiques. – *Travaux du Laboratoire de Stratigraphie et Paléoecologie, Université de Provence, Marseille, N.S.* **2**: 1-139, 10 pls., 16 figs., Marseille.

NEGRETTI, B., PHILIPPE, M., SOUDET, H.J., THOMASSIN, B.A. & OGGIANO, G. (1990): *Echinometra miocenica* LORIOL, échinide Miocène, synonymie d'*Echinometra mathaei* (BLAINVILLE), actuel: biogéographie et paléoécologie. – *Géobios* **23**/4: 445-459, 2 pls., 5 figs., 2 tabs., Lyon.

NELLI, B. (1903): Fossili miocenici del macigno di Porretta. – *Bolletino della Società Geologica Italiana* **22**: 181-252, 2 pls., Roma.

NELLI, B. (1907): Il miocene del Monte Titano nella republica di St. Marino. – *Bolletino della Società Geologica Italiana* **26**: 239-320, 3 pls., Roma.

NEMES, F.D. (1888a): Palaeontologiai tanulmányok erdély tertiaerjéröl. – *Értesitö az Erdélyi Múzeum-Egylet Orvos-Természettudományi Szakosztályából. II. Természettudományi Szak* **10 (13)**/2: 161-170, 217-220, pl. 6, Kolozsvárt.

NEMES, F.D. (1888b): Újabb adatok a bujturi mediterrán rétegek Fauná-Jának ismeretéhez. – *Értesitö az Erdélyi Múzeum-Egylet Orvos-Természettudományi Szakosztályából. II. Természettudományi Szak* **10 (13)**/1: 19-32, 118-119, Kolozsvárt.

NÉRAUDEAU, D. (1994): Hemiasterid echinoids (Echinodermata: Spatangoida) from the Cretaceous Tethys to the present-day Mediterranean. – *Palaeogeography, Palaeoclimatology, Palaeoecology* **110**: 319-344, 8 figs., 4 tabs., Amsterdam.

NÉRAUDEAU, D., BORGHI, E. & ROMAN, J. (1998a): Le genre d'échinide *Spatangus* dans les localités du Pliocène et du Pléistocénee d'Émilie (Italie du Nord). – *Annales de Paléontologie* **84**/3-4: 243-264, 5 pls., 9 figs., Paris.

NÉRAUDEAU, D., CRAME, J.A. & KOOSER, M. (2000): Upper Cretaceous Echinoids from Ross Basin, Antarctica. – *Géobios* **33**/4: 455-466, 4 figs., Lyon.

NÉRAUDEAU, D., DAVID, B. & MADON, C. (1998b): Tuberculation in spatangoid fascioles: Delineating plausible homologies. – *Lethaia* **31**: 323-334, 10 figs., 4 tabs., appendix, Oslo.

NÉRAUDEAU, D., GOUBERT, E., LACOUR, D. & ROUCHY, J.M. (2001): Changing biodiversity of Mediterranean irregular echinoids from the Messinian to Present-Day. – *Palaeogeography, Palaeoclimatology, Palaeoecology* **175**: 43-60, 9 figs., 1 tab., Amsterdam.

NÉRAUDEAU, D., ROMAN, J. & BORGHI, E. (1999): Impact of the Messinian crisis on the Mediterranean echinoid fauna. – In: CANDIA CARNEVALI, M.D. & BONASORO, F. (eds.): *Echinoderm Research 1998.- Proc. 5th Europ. Conf. Echinoderms/ Milan/Italy/7-12 Sept.1998.* pp. 355-360, Rotterdam (A.A. Balkema).

NEUMANN, C. (2000): Evidence of predation on Cretaceous sea stars from north-west Germany. – *Lethaia* **33**: 65-70, 3 figs., 1 tab., Oslo.

NEVIANI, A. (1887): Contribuzioni alla geologia del Catanzarese. – *Bollettino della Società Geologica Italiana* **6**: 168-208, Roma.

NICHOLS, D. (1959): Changes in the Chalk heart-urchin *Micraster* interpreted in relation to living forms. – *Philosophical transactions of the Royal Society of London (B)* **242**: 347-437, 1 pl., London.

NICORICI, E. (1977): Les pectinidés Badéniens de Roumanie. –

Mémoires Institut de Géologie et de Géophysique **26**: 119-159, 44 pls., 9 figs., 2 tabs., Bucarest.

ONODA, K. (1943): Notes on the development of some Japanese echinoids. Report IV. Hybridization experiments with some Japanese echinoids. – *Japanese Journal of Zoology* **10**/3: 407-426, Tokyo.

OPPENHEIM, P. (1902a): Revision der tertiären Echiniden Venetiens und des Trentino, unter Mittheilung neuer Formen. – *Zeitschrift der Deutschen Geologischen Gesellschaft* **54**/ A. Aufsätze: 159-283, 3 pls., Berlin.

OPPENHEIM, P. (1902b): Nachtrag zu meinem Aufsatz „Revision der tertiären Echiniden Venetiens und des Trentino, unter Mitteilung neuer Formen". – *Zeitschrift der Deutschen Geologischen Gesellschaft* **54**/B. Briefliche Mittheilungen: 66-71, Berlin.

d'ORBIGNY, A. (1852): Prodrome de Paléontologie stratigraphique universelle des Animaux Mollusques & Rayonnés faisant suite au cours élémentaire de Paléontologie et de Géologie stratigraphiques. Troisième Volume. 191 pp., Paris (Victor Masson).

PALUMBI, S.R., GRABOWSKY, G., DUDA, T., GEYER, L. & TACHINO, N. (1997): Speciation and population genetic structure in tropical Pacific sea urchins. – *Evolution* **51**: 1506-1517, Lawrence, KS.

PANTANELLI, D. (1882): Cenno monografico sopra la fauna miocenica di Montese [...] Giuseppe Mazzetti. [book review]. – *Atti della Società dei Naturalisti di Modena, Memorie Serie III* **1**/17: 45-48, Modena.

PAPP, A. (1952): Die Altersstellung der fossilführenden Schichten. – In: BECK-MANNAGETTA, P. (ed.): Zur Geologie und Paläontologie des Tertiärs des unteren Lavanttals. – *Jahrbuch der Geologischen Bundesanstalt* **95**: 5-28, Wien.

PAPP, A., CICHA, I. & SENEŠ, J. (1978a): Die Stellung des Badenien in der Stratigraphie der Zentralen Paratethys. – In: PAPP, A., CICHA, I., SENEŠ, J. & STEININGER, F. (eds.): *Chronostratigraphie und Neostratotypen. Miozän der Zentralen Paratethys. M_4 Badenien (Moravien, Wielicien, Kosovien).* pp. 29-30, 2 figs., 1 tab., Bratislava (VEDA – Vydavatel'stvo Slovenskej akad. vied).

PAPP, A., CICHA, I. & STEININGER, F. (1978b): 2. Charakteristische Fossilgruppen im Badenien. – In: PAPP, A., CICHA, I., SENEŠ, J. & STEININGER, F. (eds.): *M_4 Badenien (Moravien, Wielicien, Kosovien). – Chronostratigraphie und Neostratotypen, Miozän der Zentralen Paratethys,* 6 pp. 31-33, Bratislava (VEDA – Vydavatel'stvo Slovenskej akad. vied).

PAPP, A. & STEININGER, F. (1978): A. Holostratotypus des Badenien. Holostratotypus: Baden-Sooss. – In: PAPP, A., CICHA, I., SENEŠ, J. & STEININGER, F. (eds.): *Chronostratigraphie und Neostratotypen. Miozän der Zentralen Paratethys. M_4 Badenien (Moravien, Wielicien, Kosovien).* pp. 138-145, 2 figs., 1 tab., Bratislava (VEDA – Vydavatel'stvo Slovenskej akad. vied).

PARONA, C.F. (1887): Appunti per la paleontologia miocenica della Sardegna. – *Bolletino della Società Geologica Italiana* **6**: 289-358, Roma.

PAUCĂ, M. (1936): Le Bassin Neogène de Beiuş. – *Anuarul Institutului Geologic al României* **17** (1932): 133-223, 11 pls., 11 figs., 1 tab., 2 maps, Bucureşti.

PAWSON, D.L. (1978): The echinoderm fauna of Ascension Island, South Atlantic Ocean. – *Smithsonian Contributions to Marine Sciences* **2**: i-iv, 1-31, 11 figs., Washington, DC.

PEDLEY, H.M., HOUSE, M.R. & WAUGH, B. (1976): The Geology of Malta and Gozo. – *Proceedings of the Geologists' Association* **87**/3: 325-341, 4 figs., London.

PENNANT, T. (1777): British Zoology. Crustacea, Mollusca, Testacea. 4 vols., 136 pp., 93 pls., London. [not seen]

PETERS, K. (1857): Geologische Studien aus Ungarn. – *Jahrbuch der kaiserlich-königlichen Geologischen Reichsanstalt* **8**: 308-334, Wien.

PETERS, W. (1853): Über die an der Küste von Moçambique beobachteten Seeigel und insbesondere über die Gruppe der Diademen (Auszug). – *Bericht über die zur Bekanntmachung geeigneten Verhandlungen der Königlichen Preussischen Akademie der Wissenschaften zu Berlin* **1853**: 484-465, Berlin.

PHILIPPE, M. (1973): Le Burdigalien (Miocène) du plateau des Courennes (Vaucluse). Études stratigraphique et paléontologique. – *Documents des Laboratoires de Géologie de la Faculte des Sciences de Lyon* **51 (1972)**: 1-275, 50 figs., 12 tabs., Lyon.

PHILIPPE, M. (1974): Découverte de gisments à faune burdigalienne dans les "marnes bleues de Faucon" à Entrechaux (Vaucluse) et à Mollans (Drôme). – *Bulletin mensuel de la Société Linnéenne de Lyon* **43**/1: 5-8, Lyon.

PHILIPPE, M. (1984): Échinides. – In: POUYET, S. (ed.): La faune du faciès "marnes bleues", Burdigalien du Bassin de Faucon-Mollans-Malaucène (Sud-Est de la France). – *Nouvelles Archives du Muséum d'Histoire Naturelle de Lyon* **22**: 85-91, pl. 5, fig. 14, Lyon.

PHILIPPE, M. (1989): Révision des Échinides miocènes du Bassin du Rhône: résultats d'ordre systématique. – *Actes 6ème Séminaire international Echinodermes, Les Embiez, 1988; Vie Marine, Hors Serie* **10**: 24-35, Marseille.

PHILIPPE, M. (1998): Les Échinides miocènes du Bassin du Rhône: révision systématique. – *Nouvelles Archives du Muséum d'Histoire Naturelle de Lyon* **36**/1-2: 3-241, 249-441, 26 pls., 36 figs., 74 tabs., Lyon.

PHILIPPE, M., NEGRETTI, B., LARY, C., ARNAUD, M. & MONLEAU, C. (1990): Les échinides miocènes de la Nerthe (Bouches-du-Rhône), répartition et conclusions biostratigraphiques. – *Géologie méditerranéenne* **15**/4 (1988): 235-250, 5 figs., Marseille.

PHILIPPI, R.A. (1842): Ueber *Clypeaster altus* Scilla's und einige Verwandte. – *Neues Jahrbuch für Mineralogie, Geognosie, Geologie und Petrefakten-Kunde* **1842**: 52-55, pl. II, figs. 3-5, Stuttgart.

PHILIPPI, R.A. (1845): Beschreibung einiger neuer Echinodermen nebst kritischen Bemerckungen über einige weniger bekannte Arten. – *Archiv für Naturgeschichte* **11**/1: 344-359, pl. XI, Berlin.

PHILIPPI, R.A. (1851): Ueber *Clypeaster altus*, *Cl. turritus* und *Cl. Scillae*. – *Palaeontographica – Beiträge zur Naturgeschichte der Vorwelt* **1**: 324-323, pl. XXXVIII-XL, Cassel.

PICTET, F.J. (1846): Traité élémentaire de Paléontologie ou Histoire Naturelle des Animaux Fossiles considérés dans leurs rapports zoologiques et géologiques. Tome 4. xiv+ 458 pp., 20 pls., Genève (Imprimerie de Jules-Guillaume Fick).

PICTET, F.J. (1857): Traité élémentaire de Paléontologie ou Histoire Naturelle des Animaux Fossiles considérés dans leurs rapports zoologiques et géologiques. Seconde Édition.Tome 4. xiv+768 pp., Paris (J.-B. Baillière et Fils).

PIJPERS, P. (1933): Geology and Palaeontology of Bonaire (D.W.I.). – *Geographische en Geologische Mededeelingen* **8**: 1-103, 2 pls., 157 figs., Utrecht.

PILLER, W.E. (2000): Eisenstadt – Johannesgrotte. – In: SCHÖNLAUB, H.P. (ed.): *Erläuterungen zur geologischen Karte des Burgenlandes 1:200.000.* pp. 86-87, 2 figs., Wien (Geological Survey Austria).

PILLER, W.E. & HARZHAUSER, M. (1999): Badenian (Middle Miocene) Ecosystems. – In: PILLER, W.E., DAXNER-HÖCK, G., DOMNING, D.P., FORKE, H.C., HARZHAUSER, M., HUBMANN, B., KOLLMANN, H.A., KOVAR-EDER, J., KRYSTYN, L., NAGEL, D., PERVESLER, P., RABEDER, G., ROETZEL, R., SANDERS, D. & SUMMESBERGER,

H.: Palaeontological Highlights of Austria. – *Mitteilungen der Österreichischen Geologischen Gesellschaft* **92**: 220-225, 4 figs., Wien.

PILLER, W.E. & HARZHAUSER, M. (in press): The myth of the brackish Sarmatian Sea. – *Terra Nova*, 5 p., 5 figs.

PLÖCHINGER, B. & KARANITSCH, P. (2002): Faszination Erdgeschichte mit Brennpunkt Mödling am Alpenostrand. 238 pp., 458 figs., 5 tabs., Mödling (Heimat Verlag).

PODDUBIUK, R.H. & ROSE, E.P.F. (1984): Relationships between Mid-Tertiary echinoid faunas from the Central Mediterranean and eastern Carribean and their palaeobiogeographic significance. – *Annales Géologiques des Pays Helléniques* **31**: 115-127, 5 tabs., Athènes.

POLJAK, J. (1938): Prilog poznavanju miocenskih Echinoidea Hrvatske i Slavonije. – *Vesnik geološkog Instituta kraljevine Jugoslavije* **7**: 167-204, pls. 2-10, 1 tab., Beograd.

POMEL, A. (1869): Revue des échinodermes et de leur classification pour servir d'introduction a l'étude des fossiles. i-lxvii pp., Paris (Deyrolle).

POMEL, A. (1883): Classification méthodique et Genera des Échinides vivante et fossiles. – Thèses présentées a la Faculté des Sciences de Paris pour obtenir le Grade de Docteur ès Sciences Naturelles 503: 131 pp., 1 pl., 1 tab., Alger (Aldolphe Jourdan).

POMEL, A. (1885): Paléontologie ou description des animaux fossiles de l'Algérie. Zoophytes. 2e fascicule. – Échinodermes. 1re livraison. lxvii+132 pp., pls. A i-xvi, B i-xlix, C i-xii, D i-iii + 1 pls, Alger (Adolphe Jourdan).

POMEL, A. (1887): Paléontologie ou description des animaux fossiles de l'Algérie. Zoophytes. 2e fascicule. – Échinodermes. 2e livraison. 344 pp., Alger (Adolphe Jourdan).

POMEL, A. (1888): Notes d'Échinologie synonymique. – *Bulletin de la Société Géologique de France, Série 3* **16**: 441-453, Paris.

POSPÍCHALOVÁ, E. (2003): New Records of Echinoids from the Karpatian of Southern Moravia. Appendix in: KROH, A.: Echinoderms of the Karpatian. – In: BRZOBOHATÝ, R., CICHA, I., KOVÁČ, M. & RÖGL, F. (eds.): *The Karpatian – A Lower Miocene Stage of the Central Paratethys.* pp. 253, Brno (Masaryk University).

PROCHÁZKA, V.J. (1892a): [Ein Beitrag zur Kenntniss der Fauna des marinen Tegels und des diesen überlagerndn Sandsteines von Walbersdorf]. – Rozpravy České akademie Císaře Františka Josefa pro Vědy, Slovesnost a Umě ní v Praze **1892**: 729-750, Praze.

PROCHÁZKA, V.J. (1892b): Předběžná zpráva o stratigrafických a faunistických poměrech nejzazší části miocaenu západní Moravy. – Věstníka královské české společnosti nauk **1892**: 326-368, Praze.

PROCHÁZKA, V.J. (1900): Das Ostböhmische Miocaen. – *Archiv für Naturwissenschaftliche Durchforschung Böhmens* **10**/2: 3-173, 72 figs., Prag.

PUSCH, G.G. (1837): Polens Paläontologie oder Abbildung und Beschreibung der vorzüglichsten und der noch unbeschriebenen Petrefakten aus den Gebirgsformationen in Polen, Volhynien und den Karpaten nebst einigen allgemeinen Beiträgen zur Petrefaktenkunde und einem Versuch zur Vervollständigung der Geschichte des Europäischen Auer-Ochsen. xiii+218 pp., 16 pls., Stuttgart (E. Schweizerbart's Verlagshandlung).

QUENSTEDT, F.A. (1874): Atlas zu den Echiniden., pls. 62 – 89, Leipzig (Fues's Verlag (R. Reisland)).

QUENSTEDT, F.A. (1875): Petrefactenkunde Deutschlands. I. Abtheilung, 3. Band. Echinodermen. Die Echiniden. 720 pp., 28 pls., Leipzig (Fues's Verlag (R. Reisland)).

RADWAŃSKI, A. (1970): Dependence of rock-borers and burrowers on the environmental conditions within the Tortonian littoral zone of southern Poland. – In: CRIMES, T.P. & HARPER, J.A. (eds.): Trace Fossils – *Geological Journal, Special Issues* **3**: 371-390, Liverpool.

RADWAŃSKI, A. (1973): Transgresja dolnego tortonu na polidniowo-wschodnich; wschodnich stokach Gor Swietokrzyskich. – *Acta Geologica Polonica* **23**/2: 375-434, 10 pls., 13 figs., Warszawa.

RADWAŃSKI, A. & WYSOCKA, A. (2001): Mass aggregation of Middle Miocene spine-coated echinoids Echinocardium and their integrated eco-taphonomy. – *Acta Geologica Polonica* **51**/4: 295-316, 8 pls., 8 figs., Warszawa.

RADWAŃSKI, A. & WYSOCKA, A. (2004): A farewell to Świniary sequence of mass-aggregated, spine-coated echinoids *Psammechinus* and their associates (Middle Miocene; Holy Cross Mountains, Central Poland). – *Acta Geologica Polonica* **54**/3: 381-399, 10 pls., 10 figs., Warszawa.

RADWAŃSKA, U. (1999): Lower Kimmeridgian echinoids of Poland. – *Acta Geologica Polonica* **49**/4: 287-364, 49 pls., 13 figs., 1 tab., Warszawa.

RAGAINI, L. (1990): Variazioni morfologiche in *Opissaster scillae* (Wright): un esempio dal «Lower Globigerina Limestone» dell'Archipelago Maltese. – *Atti della Società Toscana di Scienze Naturali, Memorie, Serie A* **97**: 1-14, 10 figs., 4 tabs., Pisa.

RAGAINI, L. (1994): Neogene echinoids from Salento (Apulia, Southern Italy). – In: MATTEUCCI, R. & al. (eds.): Studies on Ecology and Paleoecology of Benthic Communities. – *Bollettino della Societa Paleontologica Italiana, Special Volume* **2**: 287-290, 1 fig., 2 tabs., Modena.

RAHMAN, M.A., UEHARA, T. & PEARSE, J.S. (2001): Hybrids of Two Closely Related Tropical Sea Urchins (Genus *Echinometra*): Evidence Against Postzygotic Isolating Mechanisms. – *Biological Bulletin* **200**: 97-106, 4 figs., 5 tabs., Miami, FL.

RASOUMOVSKY, G.d. (1822): Observations minéralogiques sur les environs de Vienne. 60 pp., 10 pls., Vienne (Chez Leopold Grund).

RAVN, J.P.J. (1928): De Regulære Echinider i Danmarks Kridtaflejringer. – *Det Kongelige Danske Videnskabernes Selskabs Skrifter, naturvidenskabelig og mathematisk Afdeling, 9. Række* **1**/1: 1-63, 6 pls., København.

REDLICH, K.A. (1895): Ein Beitrag zur Kenntniss des Tertiärs im Bezirke Gorju (Rumänien). – *Verhandlungen der kaiserlich-königlichen Geologischen Reichsanstalt* **29**/12: 330-334, 1 fig., Wien.

RÉGIS, M.B. & THOMASSIN, B.A. (1982): Ecologie des Échinoides réguliers dans les récifs coralliens de la région de Tuléar (S.W. de Madagascar): adaption de la microstructure de piquants. – *Annales de l'Institut Océanographique, n. s.* **58**: 117-158, Paris. [not seen]

REHFELD, U. & JANSSEN, A.W. (1995): Development of Phosphatized Hardgrounds in the Miocene Globigerina Limestone of the Maltese Archipelago, Including a Description of *Gamopleura melitensis* sp. nov. (Gastropoda, Euthecosomata). – *Facies* **33**: 91-106, 4 pls., 4 figs., Erlangen.

REIDL, G. (1937): Paläobiologische Untersuchungen im Leithakalkaufschlusz am Äuszeren Berg bei Mülendorf im Bgld. – Ph.D. thesis (unpublished), University of Vienna, Vienna.

REIDL, G. (1941): Über eine neue Spatangidenart *Plagiobrissus abeli* nov. spec. aus dem Torton von Müllendorf (ehem. Burgenland). – *Berichte der Reichsstelle für Bodenforschung* **1941**: 24-29, 2 figs., Wien.

REUSS, A.E. (1860): Die marinen Tertiärschichten Böhmens und ihre Versteinerungen. – *Sitzungsberichte der Kaiserlichen Akademie der Wissenschaften, mathematisch-naturwissenschaftliche Classe, Abtheilung* I **39**: 207-288, 8 pls., Wien.

REUSS, A.E. (1867): Die fossile Fauna der Steinsalzablagerungen von Wieliczka in Galizien. – *Sitzungsberichte der Kaiserli-*

chen Akademie der Wissenschaften, mathematisch-naturwissenschaftliche Classe, Abtheilung I **55**: 17-179, 8 pls., 1 tab., Wien.

RIEDL, R. (ed.) (1983): Fauna und Flora des Mittelmeeres. 3. Aufl. 836 pp., 298 pls., Hamburg (Paul Parey Verlag).

ROETZEL, R., STEININGER, F.F. & PILLER, W.E. (1991): E/10: Steyregg, Sandgrube Treul. – In: ROETZEL, R. & NAGEL, D. (eds.): *Exkursionen im Tertiär Österreichs.* pp. 47-49, fig. 13, Gmünd (Österreichische Paläontologische Gesellschaft).

RÖGL, F., HOCHULI, P. & MÜLLER, C. (1979): Oligocene – Early Miocene stratigraphic correlations in the Molasse Basin of Austria. – *Annales géologiques des pays Helléniques, Hors Série* **3**: 1045-1049, 1 fig., Athens.

RÖGL, F. & MÜLLER, C. (1976): Das Mittelmiozän und die Baden-Sarmat Grenze in Walbersdorf (Burgenland). – *Annalen des Naturhistorischen Museums in Wien* **80**: 221-232, 1 fig., 2 tabs., Wien.

RÖGL, F. & NAGYMAROSY, A. (2004): Biostratigraphy and correlation of the Lower Miocene Michelstetten and Ernstbrunn sections in the Waschberg Unit, Austria (Upper Egerian to Eggenburgian, Central Paratethys). – *Courier Forschungsinstitut Senckenberg* **246**: 129-151, 3 pls., 4 figs., 3 tabs., Frankfurt am Main.

RÖGL, F., SCHULTZ, O. & KOVAR, J. (1982): Tertiär. Das Ende der Tethys – In: KOLLMANN, H.A., KOVAR, J., RÖGL, F., SCHÖNLAUB, H.P., SCHULTZ, O. & SUMMESBERGER, H.: *Schätze im Boden – Österreichs Boden im Wandel der Zeit.* pp. 49-72, Horn (Ferdinand Berger & Söhne).

RÖGL, F., SPEZZAFERRI, S. & CORIC, S. (2002): Micropaleontology and biostratigraphy of the Karpatian-Badenian transition (Early-Middle Miocene boundary) in Austria (Central Paratethys). – *Courier Forschungsinstitut Senckenberg* **237**: 47-67, 2 pls., 1 fig., 4 tabs., Frankfurt am Main.

ROHATSCH, A. (1999): Gesteinsbestand des Wiener Neustädter Domes. – In: URBAN, J. (ed.): *Der Dom in neuem Glanz – Abschluß der Restaurierung des Liebfrauendomes.* pp. 21-23, 6 figs., Wiener Neustadt (Verein zur Erhaltung des Liebfrauendomes).

ROLLE, F. (1855): Über einige neue Vorkommen von Foraminiferen, Bryozoen und Ostracoden in den tertiären Ablagerungen Steiermarks. – *Jahrbuch der kaiserlich-königlichen Geologischen Reichsanstalt.* **6**: 351-354, Wien.

ROLLE, F. (1856): Die tertiären und diluvialen Ablagerungen in der Gegend zwischen Gratz, Köflach, Schwanberg und Ehrenhausen in Steiermark. – *Jahrbuch der kaiserlich-königlichen Geologischen Reichsanstalt.* **7**: 535-602, 6 figs., Wien.

ROMAN, J. (1952): Sur les structures internes des Clypéastres. – *Bulletin de la Société Géologique de France, Série 6* **2**: 403-416, 1 fig., 1 tab., Paris.

ROMAN, J. (1958): Échinides du Miocène de France avec allusion spéciale aux *Echinolampas*. – *Actes du 83ᵉ Congrès des Sociétés Savantes, Aix-en-Provence; Colloque sur le Miocène,* 247-263, Aix-en-Provence.

ROMAN, J. (1960): Échinides (*Clypeaster, Scutella, Schizaster*) de l'Helvetien du bassin de Karaman (Turquie). – *Bulletin of the Mineral Research and Exploration Institute of Turkey* **55**: 64-96, 8 pls., 1 fig., Ankara.

ROMAN, J. (1965): Morphologie et évolution des *Echinolampas* (Échinides, Cassiduloïdes). – *Mémoires du Muséum National d'Historire Naturelle Nouvelle Série, Science de la Terre, Série C* **15**: 1-341, 12 pls., 136 figs., Paris.

ROMAN, J. (1974): Les Échinides du Néogène rhodanien: répartion géographique et stratigraphique. – *Mémoires du Bureau de Recherches Géologiques et Minières* **78**/1: 329-340, 3 figs., 1 tab., Paris.

ROMAN, J. (1977): Biogéographie d'un groupe d'échinides cénozoïques (*Echinolampas* et ses sous-generes *Conolampas* et *Hypsoclypeus*). – *Géobios* **10**/3: 337-349, 3 figs., Lyon.

ROMAN, J. & SOUDET, H.J. (1990): Échinides de bassins néogenes bétiques orientaux. – *Documents et Travaux de l'IGAL* **12-13**: 53-56, 1 tab., Cergy-Pontoise.

ROMAN, J. & STROUGO, A. (1988): Échinoids de l'étage Mokattamien (Éocène moyen et superieur) de la region du Grand Caire (Égypte) – *Bulletin du Museum National d'Histoire naturelle, Section C: Sciences de la Terre: Paleontologie, Geologie, Mineralogie* **10**/2: 137-161, 3 pls., 1 fig., 1 tab., Paris.

ROSE, E.P.F. (1974a): The Miocene echinoids of Libya: a summary and review of their stratigraphical distribution. – *Mémoires du Bureau de Recherches Géologiques et Minières* **78**/1: 341-347, 3 figs., Paris.

ROSE, E.P.F. (1974b): Stratigraphical and facies distribution of irregular echinoids in Miocene limestones of Gozo, Malta, and Cyrenaica, Libya. – *Mémoires du Bureau de Recherches Géologiques et Minières* **78**/1: 349-355, 3 figs., Paris.

ROSE, E.P.F. (1975): Oligo-Miocene echinoids of the Maltese Islands. – *Proceedings of the VIᵗʰ Congress of the Regional Committee on Mediterranean Neogene Stratigraphy, Bratislava, 4-7 Sept. 1975, Vol. 1.* 75-79, 4 tabs., Bratislava.

ROSE, E.P.F. (1978): Some observations on the recent holectypoid echinoid *Echinoneus cyclostomus* and their palaeoecological significance. – *Thalassina Jugoslavica* **12**/1: 299-306, Zagreb.

ROSE, E.P.F. & PODDUBIUK, R.H. (1987): Morphological variation in the cenozoic echinoid *Clypeaster* and its ecological and stratigraphical significance. – *Annales Instituti Geologici Publici Hungarici* **70**: 463-469, 1 fig., 1 tab., Budapest.

ROSE, E.P.F., PRATT, S.K. & BENNET, S.M. (1992): Evidence for Sea-level Changes in the Globigerina Limestone Formation (Miocene) of the Maltese Islands. – *Paleontología i Evolució* **24-25**: 265-276, 6 figs., Barcelona.

ROSE, E.P.F. & WATSON, A.C. (1998): Burrowing adaptions of schizasteroid echinoids from the Globigerina Limestone (Miocene) of Malta and their evolutionary significance. – In: MOOI, R. & TELFORD, M. (eds.): *Echinoderms: San Francisco.- Proc. of the 9th Int. Echinoderm Conf. San Francisco/ California(USA), 5-9 Aug. 1996.* pp. 811-817, 5 figs. Rotterdam (A.A. Balkema).

ROTH VON TELEGD, L. (1891): Mediterrane Petrefacte von Felsö-Esztergály. – *Földtani Közlöny* **21**/4-5: 150-151, Budapest.

ROTH VON TELEGD, L. (1899): Die Randzone des siebenbürgischen Erzgebirges in der Gegend von Várfalva, Toroczkó und Hidas. – *Jahrbuch der königlichen ungarischen geologischen Anstalt* **1897**: 67-103, Budapest.

RUPP, J.H. (1973): Effects of temperature on fertilization and early cleavage of some tropical echinoderms, with emphasis on *Echinometra mathaei*. – *Marine Biology* **23**: 183-189, 5 figs., 2 tabs., Berlin – Heidelberg – New York.

RUPP, C., ROETZEL, R. & STOJASPAL, F. (1991): E/7: Ottnang-Schanze. – In: ROETZEL, R. & NAGEL, D. (eds.): *Exkursionen im Tertiär Österreichs.* pp. 38-41, fig. 10, Gmünd (Österreichische Paläontologische Gesellschaft).

RZEHAK, A. (1880): Ueber die Gliederung und Verbreitung der älteren Mediterranstufe in der Umgebung von Gr. Seelowitz in Mähren. – *Verhandlungen der kaiserlich-königlichen Geologischen Reichsanstalt* **14**/16: 300-303, Wien.

RZEHAK, A. (1902): Die Tertiärformation in der Umgebung von Nikolsburg in Mähren. I. Theil. – *Zeitschrift des mährischen Landesmuseum* **1902**: 28-61, Brünn.

Sánchez Roig, M. (1949): Los equinodermos fosile de Cuba. – *Paleontologia Cubana* **1**: 1-302, 50 pls., La Habana. [not seen; cited after Kier, 1984]

Sánchez Roig, M. (1952): Nuevos generos y especies de equinoideos fosiles cubanos. – *Torreia* **17**: 1-18, La Habana. [not seen; cited after Kier, 1984]

Sánchez Roig, M. (1953): Nuevos equinodos fosiles de Cuba. – *Anales de la Academis de Ciencias Medicas, Fisicas y Naturales de la Habana* **91**: 135-176, La Habana. [not seen; cited after Kier, 1984]

Saraiman, A. (1988): The presence of some echinidae in the Badenian from Murfatlar. – Analele ştiinţifice ale Universităţii "Al. I. Cuza" din Iaşi, Serie nouă, IIb. Geologie-Geografie **34**: 37-38, 1 pl., Iaşi.

Sauer, R., Seifert, P. & Wessely, G. (1992): Guidebook to Excursions in the Vienna Basin and the Adjacent Alpine-Carpathian Thrustbelt in Austria. – *Mitteilungen der Österreichischen Geologischen Gesellschaft* **85**: 5-239, 200 figs., Wien.

Savin, L. (1903): Catalogue raisonné des Échinides fossiles du Départment de La Savoie. – *Bulletin de la Société d'Histoire Naturelle de Savoie, Série 2* **8**: 59-249, 3 pls., Chambéry. [not seen; cited from Fischer (1966)]

Savin, L. (1907): Supplément au Catalogue des Échinides fossiles du départment de la Savoie. – *Bulletin annuel de la Société d'Histoire Naturelle de Savoie* **12** (1906): 5-58, 1 pl., Chambéry.

Schafarzik, F. (1892): A Cserhát piroxén-andezitjei. Petrografiai geologiai tanulmány. A kir. magyar természettudományi társulat megbizásából. – *A Magyar királyi Földtani Intézet évkönyve* **9** (1890-92): 175-328, pls. 7-9, 26 figs., Budapest.

Schaffer, F. (1897): Der marine Tegel von Theben-Neudorf in Ungarn. – *Jahrbuch der kaiserlich-königlichen Geologischen Reichsanstalt* **47**/3: 533-548, 7 figs., Wien.

Schaffer, F.X. (1906): Geologie von Wien. II. Teil. 242 pp., 17 pls., 25 figs., 1 map, Wien (R. Lechner – Wilh. Müller).

Schaffer, F.X. (1907): Geologischer Führer für Exkursionen im inneralpinen Becken der nächsten Umgebung von Wien. – *Sammlung geologischer Führer* **12**: i-viii, 1-127 pp., 11 figs., Berlin (Gebrüder Borntraeger).

Schaffer, F.X. (1908): Geologischer Führer für Exkursionen im Inneralpinen Wienerbecken. II. Teil. nebst einer Einführung in die Kenntnis seiner Fauna. – *Sammlung geologischer Führer* **13**: i-viii, 1-157 pp., 13 pls., Berlin (Gebrüder Borntraeger).

Schaffer, F.X. (1912a): Die Echiniden der Miocänbildung von Eggenburg. – *Abhandlungen der Geologischen Reichsanstalt* **22**/2: 187-192, pls. 58-60, 4 figs., Wien.

Schaffer, F.X. (1912b): Geologischer Anschauungsunterricht in der Umgebung von Wien. 143 pp., 43 figs., Wien (Franz Deuticke).

Schaffer, F.X. (1913a): Geologischer Führer für Exkursionen im Wiener Becken. III. Teil, nebst einer Einführung in die Kenntnis der Fauna der ersten Mediterranstufe. – *Sammlung geologischer Führer* **18**: i-x, 1-167 pp., 12 pls., 3 figs., 1 map., Berlin (Gebrüder Borntraeger).

Schaffer, F.X. (1913b): Zur Kenntnis der Miocänbildungen von Eggenburg (Niederösterreich). VI. Grunderschichten und Tegel von Gaindorf bei Ravelsbach. – *Sitzungsberichte der Kaiserlichen Akademie der Wissenschaften, mathematisch-naturwissenschaftliche Klasse, Abteilung* I **122**: 55-63, Wien.

Schaffer, F.X. (1924a): Geologischer Anschauungsunterricht in der Umgebung von Wien. Zweite, veränderte Auflage. 139 pp., 50 figs., Leipzig & Wien (Franz Deuticke).

Schaffer, F.X. (1924b): Lehrbuch der Geologie. II.Teil. Grundzüge der Historischen Geologie (Geschichte der Erde, Formationskunde). Erste bis dritte Auflage. 628 pp., 1 pl., 795 figs., Leipzig & Wien (Franz Deuticke).

Schaffer, F.X. (1927a): Geologische Geschichte und Bau der Umgebung Wiens. 112 pp., 1 pl., 157 figs., 1 map, Leipzig & Wien (Franz Deuticke).

Schaffer, F.X. (1927b): Der Bergriff der „miozänen Mediterranstufen" ist zu streichen. – *Verhandlungen der Geologischen Bundesanstalt* **1927**/2-3: 86-88, Wien.

Schaffer, F.X. (1928): Grundzüge der Geologie. 215 pp., 1 pl., 232 figs., 1 map, Leipzig & Wien (Franz Deuticke).

Schaffer, F.X. (1942): Geologischer Führer für Exkursionen in der Umgebung von Wien. Zweite veränderte Auflage. – *Sammlung geologischer Führer* **12**: i-xv, 1-170 pp., 34 figs., Berlin-Zehlendorf (Gebrüder Borntraeger).

Schaffer, F.X. (1943): Das Alpenvorland – In: Schaffer, F.X. (ed.): *Die Geologie der Ostmark.* pp. 515-553, Wien (Franz Deuticke).

Schaffer, F.X. & Grill, R. (1951): Das Alpenvorland – In: Schaffer, F.X. (ed.): *Geologie von Österreich. Zweite veränderte Auflage.* pp. 694-761, Wien (Franz Deuticke).

Schaffer, H. (1959): Der stratigraphische Wert der Scutelliden im österreichisch-ungarischen Miozän. – *Anzeiger der Österreichischen Akademie der Wissenschaften, mathemathisch-naturwissenschaftliche Klasse* **13**: 253-257, Wien.

Schaffer, H. (1961): *Brissus (Allobrissus) miocaenicus*, eine neue Echinidenart aus dem Torton von Mühlendorf (Burgenland). – *Sitzungsberichte der Österreichischen Akademie der Wissenschaften Wien, mathematisch-naturwissenschaftliche Klasse, Abteilung* I **170**/1: 149-157, 2 pls., 1 fig., Wien.

Schaffer, H. (1962): Die Scutelliden des Miozäns von Österreich und Ungarn. – *Paläontologische Zeitschrift* **36**/3/4: 135-170, 5 pls., 11 figs., Stuttgart.

Schlotheim, E.F.B.v. (1820): Die Petrefactenkunde auf ihrem jetzigen Standpunkte durch die Beschreibung seiner Sammlung versteinerter und fossiler Überreste des Thier- und Pflanzenreichs der Vorwelt erläutert. lxii + 437 pp., 15 pls., Gotha (Becker'sche Buchhandlung).

Schmid, B. (1989): Cheilostome Bryozoen aus dem Badenien (Miozän) von Nußdorf (Wien). – *Beiträge zur Paläontologie von Österreich* **15**: 1-101, 16 pls., 6 figs., Wien.

Schmid, H.-P., Harzhauser, M. & Kroh, A. (2001): Hypoxic Events on a Middle Miocene Carbonate Platform of the Central Paratethys (Austria, Badenian, 14 Ma). – *Annalen des Naturhistorischen Museums in Wien* **102A**: 1-50, 8 pls., 8 figs., Wien.

Schmid, H. (1968): Das Jungtertiär an der SE-Seite des Leithagebirges zwischen Eisenstadt und Breitenbrunn. – *Wissenschaftliche Arbeiten aus dem Burgenland* **41**: 1-74, 6 tab., 3 maps, 2 profiles, Eisenstadt.

Schopf, T.J.M. (1978): Fossilization potential of an intertidal fauna: Friday Harbor, Washington, DC. – *Paleobiology* **4**/3: 261-269, 4 tabs., 3 append., Lancaster, PA.

Schoppe, S. & Werdings, B. (1996): The boreholes of the sea urchin genus *Echinometra* (Echinodermata: Echinoidea: Echinometridae) as a microhabitat in the tropical south America. – *Marine Ecology, Pubblicazioni della Stazione Zoologia di Napoli I* **17**: 181-186, 2 figs., 2 tabs., Berlin.

Schouppé, A. (1949): Die Fauna des Steinbruchs Retznei bei Ehrenhausen. – *Mitteilungen des naturwissenschaftlichen Vereins für Steiermark* **77-78**: 142-144, Graz.

Schultz, O. (1965): Der tertiäre Grobsand zwischen Höbmannsbach und Rainbach bei Schärding am Inn, O.Ö. – In: Kühn, O.: Korallen aus dem Helvetien von Österreich. – *Sitzungsberichte der Österreichischen Akademie der Wissenschaften Wien, mathematisch-naturwissenschaftliche*

Klasse, Abteilung I **174**/7-10: 282-285, 2 pls., 1 fig., Wien.

SCHULTZ, O. (1998): Tertiärfossilien Österreichs. 159 pp., 65 pls., 7 figs., 1 tab., Freiburg (Goldschneck-Verlag).

SCHULTZ, O. (2001): Catalogus Fossilium Austriae. Bivalvia neogenica. Band 1 / Teil 1 (Nuculacea-Unionacea). 379 pp., 56 pls., 8 figs., 1 tab., Wien (Österreichische Akademie der Wissenschaften).

SCHWINNER, R. (1926): Der Säuerling von Perbersdorf (Mittelsteiermark). – *Verhandlungen der Geologischen Bundesanstalt* **1925**/11: 195-197, 1 tab., Wien.

SCILLA, A. (1670): La vana speculazione disinganuata dal senso, Lettera risponsiva circa: Corpi Mariniche Petrificata si trovano in varii luoghi terrestri. viii+168 pp., 29 pls., Napoli. [not seen]

SCILLA, A. (1752): De corporibus marinis lapidescentibus, quae defossa reperiuntur. viii+84+6 pp., 32 pls., Romae (ed. altera emendatior). [not seen]

SEDGWICK, A. & MURCHISON, R.I. (1831): A Sketch of the Structure of the Eastern Alps; with Sections through the Newer Formations on the Northern Flanks of the Chain, and through the Tertiary Deposits of Styria, &c. &c. – *Transactions of the Geological Society of London, 2. Series* **3**: 301-420, 6 pls., London.

SEGUENZA, G. (1879-1880): Le formazioni terziarie nella provincia di Reggio (Calabria). – *Memoria della Reale Accademia dei Lincei, Serie 3ª Memorie della Classe di scienze fische, matematiche e naturali* **6** (1877): 1-446, 15 pls., 2 maps, Roma.

SENEŠ, J. (1971): 15. Faziostratotypus: Vel'ká Čausa bei Prievidza – Bohrung ČČ-3, Westslovakei, Tschechoslovakei. – In: STEININGER, F.F., SENES, J. & al. (eds.): *M1 Eggenburgien. Die Eggenburger Schichtengruppe und ihr Stratotypus. – Chronostratigraphie und Neostratotypen, Miozän der Zentralen Paratethys* 2, pp. 194-202, figs. 26-27, Bratislava (VEDA – Vydavatel'stvo Slovenskej akad. vied).

SENEŠ, J. (1955): Stratigrafický a biofaciálny výskum niektorých Neogénnych sedimentov východného Slovenska na základe makrofauny. – *Geologické Práce* **40**: 1-160, 9 pls., Bratislava.

SERRES, M.d. (1829): Géognosie des Terrains tertiaires, ou Tableau des principaux animaux invertébrés des Terrains marins Tertiaires, du midi de la France. xcii+277 pp., 6 pls., Montpellier – Paris (Pomathio-Durville).

SHIGEI, M. (1986): The sea urchins of Sagami Bay collected by His Majesty the Emperor of Japan. 204 pp., Tokyo (Maruzen).

SIEBER, R. (1935): Kurze Mitteilung über die Grunder Fauna von Platt bei Zellerndorf, N. Ö. – *Verhandlungen der Geologischen Bundesanstalt* **1935**: 96-98, Wien.

SIEBER, R. (1953a): Die Fauna des Schlierbasisschuttes des Steinberggebietes von Zistersdorf (N.-Ö.). – *Verhandlungen der Geologischen Bundesanstalt* **1953**/3: 202-208, Wien.

SIEBER, R. (1953b): Die Tortonfauna von Pötzleinsdorf. – *Verhandlungen der Geologischen Bundesanstalt* **1953**/3: 184-195, 1 fig., Wien.

SIEBER, R. (1954): Verschiedene Fossilien aus dem Mittelmiozän der Umgebung von Wien. – In: GRILL, R. & KÜPPER, H. (eds.): *Erläuterungen zur geologischen Karte der Umgebung von Wien*, pl. 10, Wien (Geologische Bundesanstalt).

SIEBER, R. (1956): Die faunengeschichtliche Stellung der Makrofossilien von Ottnang bei Wolfsegg. – *Jahrbuch des Oberösterreichischen Musealvereines* **101**: 309-318, Linz.

SIEBER, R. (1958a): Die Tortonfauna von Steinabrunn bei Drasenhofen (Bez. Mistelbach, N.-Ö.). – *Verhandlungen der Geologischen Bundesanstalt* **1958**/2: 142-155, Wien.

SIEBER, R. (1958b): Paläontologisch-stratigraphische Untersuchungen im Jungtertiär des nördlichen Niederösterreichs. – *Verhandlungen der Geologischen Bundesanstalt* **1958**: 296-297, Wien.

SIEHL, A. (1957): Das Jungtertiär in der näheren Umgebung von Hornstein im Burgenland. – *Mitteilungen der Gesellschaft der Geologie- und Bergbaustudenten in Wien* **8**: 60-75, 115-116, pls. 17-18, 1 tab., Wien.

SIMONELLI, V. (1884): Il monte della Verna e i suoi fossili. – *Bolletino della Società Geologica Italiana* **2**: 235-283, Roma.

SIMPSON, G. (2001): First record of the holectypoid echinoid *Echinoneus cyclostomus* Leske from the late Pleistocene Falmouth Formation of Jamaica. – *Caribbean Journal of Earth Science* **35**: 33-36, 2 figs., Mayagüez.

SISMONDA, E. (1841): Monografia degli Echinidi Fossili del Piemonte. – *Memoria della Reale Accademia della Scienze di Torino, Serie II* **4**: 1-54, 3 pls., Torino.

SISMONDA, E. (1842): Appendice alla Monografia degli Echinidi Fossili del Piemonte. – *Memoria della Reale Accademia della Scienze di Torino, Serie II* **4**: 385-394, 1 tab., Torino.

SISMONDA, E. (1843): Memoria geo-zoologica sugli Echinidi fossili del Contado di Nizza. – *Memoria della Reale Accademia della Scienze di Torino, n. ser.* **6**/2: 341-412, 2 pls., Torino.

SISMONDA, E. (1847): Synopsis Methodica Animalium Invertebratorum Pedemontii Fossilium (exceptis speciebus ineditis). viii+62 pp., Augustae Taurinorum (Typis Regiis).

SLOAN, N.A., CLARK, A.M. & TAYLOR, J.D. (1979): The echinoderms of Aldabra and their habitats – *Bulletin of the British Museum (Natural History), Zoology Series* **37**/2: 81-128, 22 figs., London.

SMITH, A.B. (1978): A functional classification of the coronal pores of regular echinoids. – *Palaeontology* **21**/4: 759-789, 4 pls., 3 figs., appendix, London.

SMITH, A.B. (1980a): Stereom microstructure of the echinoid test. – *Special Papers in Palaeontology* **25**: 1-81, 23 pls., 20 figs., 4 tabs., London.

SMITH, A.B. (1980b): The structure and arrangement of echinoid tubercles. – *Philospohical transactions of the Royal Society of London (B)* **289**: 1-54, 22 pls., 109 figs., Appendix, London.

SMITH, A.B. (1980c): The structure, function, and evolution of tube feet and ambulacral pores in irregular echinoids. – *Palaeontology* **23**/1: 39-83, 6 pls., 15 figs., 1 tab., London.

SMITH, A.B. (1981): Implications of lantern morphology for the phylogeny of post-Palaeozoic echinoids. – *Palaeontology* **24**/4: 779-801, 10 figs., London.

SMITH, A.B. (1984a): Classification of Echinodermata. – *Palaeontology* **27**/3: 431-459, London.

SMITH, A.B. (1984b): Echinoid Palaeobiology. – *Special topics in palaeontology* **1**: x+190 pp., 81 figs., 9 tabs., London (Allen & Unwin).

SMITH, A.B. (2001): Probing the cassiduloid origins of clypeasteroid echinoids using stratigraphically restricted parsimony analysis. – *Paleobiology* **27**/2: 392-404, 4 figs., 1 appendix, Lancaster, PA.

SMITH, A.B. (2003/2004): The Echinoid Directory. – *www.nhm.ac.uk/palaeontology/echinoids.*

SMITH, A.B. & JEFFERY, C.H. (2000): Maastrichtian and Palaeocene echinoids: a key to world faunas. – *Special Papers in Palaeontology* **63**: 1-406, 152 figs., London.

SMITH, A.B. & WRIGHT, C.W. (1989): British Cretaceous echinoids. Part 1, general introduction and Cidaroida. – *Monographs of the Palaeontographical Society* **141 (578)**: 1-101, pls. 1-32, figs. 1-18, London.

SMITH, A.B. & WRIGHT, C.W. (1990): British Cretaceous echinoids. Part 2, Echinothurioida, Diadematoida and Stirodonta (1, Calycina). – *Monographs of the Palaeontographical Society* **143 (583)**: 101-198, pls. 33-72, figs. 19-65, London.

Smith, A.B. & Wright, C.W. (1993): British Cretaceous echinoids. Part 3, Stirodonta 2 (Hemicidaroida, Arbacioida and Phymosomatoida, part 1). – *Monographs of the Palaeontographical Society* **147 (593)**: 199-267, pls. 73-92, figs. 66-90, 1 tabs., London.

Smith, A.B. & Wright, C.W. (1996): British Cretaceous echinoids. Part 4, Stirodonta 3 (Phymosomatidae, Pseudodiadematidae) and Camarodonta. – *Monographs of the Palaeontographical Society* **150 (602)**: 268-341, pls. 93-114, figs. 91-125, London.

Smith, A.B. & Wright, C.W. (1999): British Cretaceous echinoids. Part 5, Holectypoida, Echinoneoida. – *Monographs of the Palaeontographical Society* **153 (612)**: 342-390, pls. 115-129, figs. 127-148, London.

Smith, A.B. & Wright, C.W. (2000): British Cretaceous echinoids. Part 6, Neognathostomata (Cassiduloids). – *Monographs of the Palaeontographical Society* **154 (615)**: 391-429, pls. 130-138, figs. 150-173, London.

Smith, A.B. & Wright, C.W. (2003): British Cretaceous echinoids. Part 7, Atelostomata, 1. Holasteroida. – *Monographs of the Palaeontographical Society* **156 (619)**: 440-568, pls. 139-182, figs. 174-238, London.

Socin, C. (1956): Fauna echinologica dell'Oligo-miocene somalo. – In: Paleontologia della Somalia. – *Palaeontographia Italica* **32**, suppl. 7: 1-22, 1 pl., Pisa.

Somos, L. & Kókay, J. (1960): Földtani megfigyélesek a Mecsekhegységi Liásban és Miocénben. – *Földtani Közlöny* **90**/3: 331-347, 2 pls., 12 figs., Budapest.

Stache, G. (1867): Die geologischen Verhältnisse der Fundstätte des *Halitherium*-Skelettes bei Hainburg an der Donau. – *Verhandlungen der kaiserlich-königlichen geologischen Reichsanstalt* **1**/7: 141-144, Wien.

Stadler, J. (1926): Geologie der Umgebung von Passau. – *Geognostische Jahreshefte* **38** (1925): 39-117, 28 figs., 1 map, München.

Stancu, I. & Andreescu, E. (1968): Fauna Tortoniană din Regiunea Rugi-Delineşti (Bazinul Caransebeşului). – Studii şi Cercetări de Geologie, Geofizică, Geografie – 2. *Geologie* **13**: 455-471, 86 figs., Bucarest.

Stanton, R.J., Dodd, J.R. & Alexander, R.R. (1979): Eccentricity in the clypeasteroid echinoid *Dendraster*: environmental significance and application in Pliocene paleoecology. – *Lethaia* **12**: 75-87, 5 figs., 2 tabs., Oslo.

Stchepinsky, V. (1939): Sivas Vilâyetinin Miosen devrine ait Faunasi. – *"METEAE" – Maden Tetkik ve Arama Enstitü sü Yayinlarindan, Seri C: Monografiler (Publications de l'Institut d'Etudes er de Recherches Minières de Turquie, Série C: Monographies)* **1**: 5-63, 10 pls., 1 map, 1 profile, Ankara.

Stefanini, G. (1907): Conoclipedi e Cassidulidi conoclipeiformi. – *Bolletino della Società Geologica Italiana* **26**: 343-376, 2 pls., Roma.

Stefanini, G. (1908a): Echini miocenici di Malta. Esistenti nel Museo di Geologia di Firenze. – *Bolletino della Società Geologica Italiana* **27**: 435-483, 1 pl., Roma.

Stefanini, G. (1908b): Echinidi del Miocene medio dell'Emilia. Parte prima. – *Palaeontographia Italica* **14**: 65-120, 4 pls., Pisa.

Stefanini, G. (1909): Echinidi del Miocene medio dell'Emilia. Parte seconda. – *Palaeontographia Italica* **15**: 1-58, 2 pls., Pisa.

Stefanini, G. (1911): Osservazioni sulla distribuzione geografica, sulla origini e sulla filogenesi degli Scutellidae. – *Bolletino della Società Geologica Italiana* **30**: 739-754, 2 figs., 1 tab., Roma.

Stefanini, G. (1919): Fossili del Neogene Veneto. Parte seconda: Brachiopoda-Echinodermata. – *Palaeontographia Italica* **25**: 127-171, 2 pls., 5 figs., Pisa.

Steininger, F. (1963): Die Molluskenfauna aus dem Burdigal (U-Miozän) von Fels am Wagram in Niederösterreich. – *Denkschriften der kaiserlichen Akademie der Wissenschaften, mathematisch-naturwissenschaftliche Klasse* **110**/5: 1-88, 13 pls., 3 figs., 2 tabs., Wien.

Steininger, F. (1969): Das Tertiär des Linzer Raumes. – In: *Geologie und Paläontologie des Linzer Raumes.* pp. 36-53, 14 pls., 1 tab., Linz (Stadtmuseum Linz & OÖ Landesmuseum).

Steininger, F. (1971a): Die Echinodermen des Eggenburgiens. – In: Steininger, F., Senes, J. & al. (eds.): *M1 Eggenburgien. Die Eggenburger Schichtengruppe und ihr Stratotypus. – Chronostratigraphie und Neostratotypen, Miozän der Zentralen Paratethys* 2, pp. 593-603, 3 pls., Bratislava (VEDA – Vydavatel'stvo Slovenskej akad. vied).

Steininger, F. (1971b): 2. Faziostratotypus: Achberg (bzw. Eichberg, s. u.) bei Maria Dreieichen E Horn in Niederösterreich (Österreich) (= Scutellensande und Lithothammnienkalke, Kühn 1936, S. 36). – In: Steininger, F., Senes, J. & al. (eds.): *M1 Eggenburgien. Die Eggenburger Schichtengruppe und ihr Stratotypus. – Chronostratigraphie und Neostratotypen, Miozän der Zentralen Paratethys* 2, pp. 115-119, fig. 13, Bratislava (VEDA – Vydavatel'stvo Slovenskej akad. vied).

Steininger, F. (1971c): 3. Faziostratotypus: Brunnstubengraben an der S-Einfahrt von Eggenburg, Niederösterreich (Österreich) (= Liegendsande Fuchs 1900b; Molassesandstein, Suess 1866 = Brunnstubensandstein, Abel 1898a; Eggenburger-Schichten, Suess 1866). – In: Steininger, F., Senes, J. & al. (eds.): *M1 Eggenburgien. Die Eggenburger Schichtengruppe und ihr Stratotypus. – Chronostratigraphie und Neostratotypen, Miozän der Zentralen Paratethys* 2, pp. 119-130, fig. 14, Bratislava (VEDA – Vydavatel'stvo Slovenskej akad. vied).

Steininger, F. (1971d): 5. Faziostratotypus: Gauderndorf bei Eggenburg (ehemalige Zottersandgruben) in Niederösterreich (Österreich) (= Liegendsande Fuchs 1900b). – In: Steininger, F., Senes, J. & al. (eds.): *M1 Eggenburgien. Die Eggenburger Schichtengruppe und ihr Stratotypus. – Chronostratigraphie und Neostratotypen, Miozän der Zentralen Paratethys* 2, pp. 134-139, fig. 16, Bratislava (VEDA – Vydavatel'stvo Slovenskej akad. vied).

Steininger, F. (1971e): 9. Faziostratotypus: Fels am Wagram, E-Krems in Niederösterreich (Österreich) (= Sande von Fels am Wagram). – In: Steininger, F., Senes, J. & al. (eds.): *M1 Eggenburgien. Die Eggenburger Schichtengruppe und ihr Stratotypus. – Chronostratigraphie und Neostratotypen, Miozän der Zentralen Paratethys* 2, pp. 157-167, fig. 19, Bratislava (VEDA – Vydavatel'stvo Slovenskej akad. vied).

Steininger, F. & Papp, A. (1978): C. Faziostratotypen des Badenien. 9. Faziostratotypus: Gross Höflein NNW, Steinbruch „FENK". – In: Papp, A., Cicha, I., Seneš, J. & Steininger, F. (eds.): *Chronostratigraphie und Neostratotypen. Miozän der Zentralen Paratethys. M_4 Badenien (Moravien, Wielicien, Kosovien).* pp. 194-199, 3 figs., Bratislava (VEDA – Vydavatel'stvo Slovenskej akad. vied).

Steininger, F.F., Rögl, F., Hochuli, P. & Müller, C. (1989): Lignite deposition and marine cycles. The Austrian Tertiary lignite deposits – A case history. – *Sitzungsberichte der Österreichischen Akademie der Wissenschaften Wien, mathematisch-naturwissenschaftliche Klasse, Abteilung* I **197**/5-10: 309-332, 3 figs., Wien.

Steininger, F.F., Seneš, J., Kleemann, K.H. & Rögl, F. (eds.) (1985): Neogene of the Mediterranean Tethys and Paratethys. Stratigraphic Correlation Tables and Sediment Distribution Maps. Vol. 1: xiv+189, Vol. 2: xxvi+536 pp., numerous figs., 10 maps, Wien (Institute of Palaeontology, University of Vienna).

Stojaspal, F. (1975): Katalog der Typen und Abbildungsoriginale der Geologischen Bundesanstalt. – 1. Teil: Wirbellose des Kaenozoikums. – *Verhandlungen der Geologischen Bundesanstalt* **1975**: A159-A193, Wien.

STRAUSZ, L. (1926): Neuere Daten zur untermediterranen Fauna von Fót. – *Földtani Közlöny* **55**: 367-369, 1 fig., Budapest.

STUR, D. (1871): Geologie der Steiermark. Erläuterungen zur geologischen Uebersichtskarte des Herzogthumes Steiermark. xxxi+654 pp., illustr. (numerous figs and tabs. without nos.), Graz (geognostisch-montanistischer Verein für Steiermark).

STUR, D. (1873): Neogen-Petrefacte aus dem in neuerer Zeit eröffneten zweiten Steinbruche bei Kalksburg. Geschenk des Steinbruchbesitzers Herrn André Kraif. – *Verhandlungen der kaiserlich-königlichen Geologischen Reichsanstalt* **7**/5: 91, Wien.

STÜTZ, A. (1807): Mineralogisches Taschenbuch enthaltend eine Oryctographie von Unterösterreich zum Gebrauche reisender Mineralogen, herausgegeben von J. G. Megerle v. Mühlfeld. [posthumous] 394 pp., Wien & Triest (Geistinger's Buchhandlung).

SUESS, E. (1866a): Untersuchungen über den Charakter der österreichischen Tertiärablagerungen. I. Über die Gliederung der tertiären Bildungen zwischen Mannhart, der Donau und dem äusseren Saume des Hochgebirges. – *Sitzungsberichte der Kaiserlichen Akademie der Wissenschaften, mathematisch-naturwissenschaftliche Classe, Abtheilung* I **54**/1: 87-152, 2 pls., Wien.

SUESS, E. (1866b): Untersuchungen über den Charakter der österreichischen Tertiärablagerungen. II. Über die Bedeutung der sogenannten „brackischen Stufe" oder der „Cerithienschichten". – *Sitzungsberichte der Kaiserlichen Akademie der Wissenschaften, mathematisch-naturwissenschaftliche Classe, Abtheilung* I **54**: 218-257, Wien.

SUESS, F.E. (1891): Beobachtungen über den Schlier in Oberösterreich und Bayern. – *Annalen des k. k. Naturhistorischen Hofmuseums* **6**: 407-429, 3 figs., Wien.

SUTER, S.J. (1994a): Cladistic analysis of cassiduloid echinoids: trying to see the phylogeny for the trees. – *Biological Journal of the Linnean Society* **53**: 31-72, 10 figs., 3 tabs., 3 appendices, London.

SUTER, S.J. (1994b): Cladistic analysis of living cassiduloids (Echinoidea), and the effects of character ordering and successive approximations weighting. – *Zoological Journal of the Linnean Society* **112**: 363-387, 5 figs., 4 tabs., 2 appendices, London.

SZÖRÉNYI, E. (1936): Cas tetralogique d'une Clypeaster miocene de Matraszöllös. – *Földtani Közlöny* **66**: 300-302, Budapest.

SZÖRÉNYI, E. (1950): Miocén-Echinidák a Mescekhegységböl. – *Földtani Közlöny* **80**/1-3: 140-148, 2 pls., Budapest.

SZÖRÉNYI, E. (1952): A *Maretia hungarica* Vadász mjocén korú faj horvatartozásának tisztázása. – *Földtani Közlöny* **82**: 302-303, 2 figs., Budapest.

SZÖRÉNYI, E. (1953): Miozäne Echiniden aus den westlichen Teilen der Ukraine. – *Geologica Hungarica, Series Palaeontologica* **23**: 1-104, 8 pls., Budapest.

TAUBER, A.F. (1951): *Tripneustes ventricosus austriacus* nov. ssp. ein tropischer Seeigel aus dem Torton des Wiener Beckens. – *Sitzungsberichte der Österreichischen Akademie der Wissenschaften Wien, mathematisch-naturwissenschaftliche Klasse, Abteilung* I **160**/3-4: 303-320, 4 figs., Wien.

TAVANI, G. (1939): Fossili del Miocene della Cirenaica (cont. e fine). – *Palaeontographia Italica* **39**: 17-76, 2 pls., Pisa.

TELFORD, M. & MOOI, R. (1986): Resource partitioning by sand dollars in carbonate and siliceous sediments: evidence from podial and particle dimensions. – *Biological Bulletin* **171**/1: 197-207, 3 figs., 1 tab., Woods Hole, MA.

TELFORD, M., MOOI, R. & HAROLD, A.S. (1987): Feeding activities of two species of *Clypeaster* (Echinoides, Clypeasteroida): further evidence of clypeasteroid resource partitioning. – *Biological Bulletin* **172**/3: 324-336, 4 figs., 3 tabs., Woods Hole, MA.

THENIUS, E. (1955): Niederösterreich im Wandel der Zeiten. – Grundzüge der Erd- und Lebensgeschichte von Niederösterreich. 114 pp., 14 pls., 22 figs., 1 tabs., 1 map, Wien (NÖ Landesmuseum).

THENIUS, E. (1962a): Niederösterreich. – *Verhandlungen der Geologischen Bundesanstalt, Bundesländerserie*: 1-125, 7 pls., 9 tabs., 1 map, 2 profiles, Wien.

THENIUS, E. (1962b): Niederösterreich im Wandel der Zeiten. – Grundzüge der Erd- und Lebensgeschichte von Niederösterreich. 2. Aufl. 114 pp., 9 pls., 23 figs., 1 tabs., Wien (G. A. Neumann).

THENIUS, E. (1970): Die Tier- und Pflanzenwelt des Wiener Raumes von einst. – In: STARMÜHLNER, F. & EHRENDORFER, F. (eds.): *Naturgeschichte Wiens. Band I (Lage, Erdgeschichte und Klima)*. pp. 191-234, Wien & München (Jugend und Volk).

THENIUS, E. (1974): Niederösterreich. 2. Auflage. – *Verhandlungen der Geologischen Bundesanstalt, Bundesländerserie*: 1-280, 48 figs., 16 tabs., Wien.

THENIUS, E. (1979): Niederösterreich zur Braunkohlenzeit. Landschaft, Klima, Tier- und Pflanzenwelt Niederösterreichs zur Tertiärzeit. – *Wissenschaftliche Schriftenreihe Niederösterreich* **43/44**: 1-64, 4 pls., 14 figs., 1 tab., St. Pölten – Wien.

THENIUS, E. (1983): Niederösterreich im Wandel der Zeiten. – Die Entwicklung der vorgeschichtlichen Tier- und Pflanzenwelt von Niederösterreich. 3. Aufl. – Katalog des Niederösterreichischen Landesmuseums. 144: 196 pp., 9 pls., 63 figs., 4 tabs., Wien (Amt der Niederösterreichischen Landesregierung).

THOMSON, C.W. (1872a): On the Echinoidea of the "Porcupine" Deep-sea Dredging-Expeditions. (Abstract). – *Proceedings of the Royal Society of London* **20**/137/19: 491-497, London. [text identical with THOMSON, 1872b]

THOMSON, C.W. (1872b): On the Echinoidea of the "Porcupine" Deep-sea Dredging-Expeditions. (Abstract). – *The Annals and Magazine of Natural History, 4th Series* **10**/58: 300-306, London. [text identical with THOMSON, 1872a]

THUM, A.B. & ALLEN, J.C. (1975): Distribution and abundance of the lamp urchin *Echinolampas crassa* (BELL) 1880 in False Bay, Cape. – *Transactions of the Royal Society of South Africa* **41**/4: 359-373, 5 figs., 7 tabs., Cape Town.

THUM, A.B. & ALLEN, J.C. (1976): Reproductive ecology of the lamp urchin *Echinolampas crassa* (BELL) 1880 from a subtidal biogenous ripple train. – *Transactions of the Royal Society of South Africa* **42**: 23-33, 4 figs., 3 tabs., Cape Town.

TOKIOKA, T. (1963): Supposed effects of the cold weather of the winter 1962-63 upon the intertidal fauna in the vicinity of Seto. – *Publications of Seto Marine Biology Laboratory* **9**: 415-424, Kyoto.

TOKIOKA, T. (1966): Recovery of the Echinometra populations in the intertidal zone in the vicinity of Seto, with a preliminary note on the mass mortality of some sea urchins in the summer season. – *Publications of Seto Marine Biology Laboratory* **14**: 7-16, Kyoto.

TOLLMANN, A. (1955): Das Neogen am Nordwestrand der Eisenstädter Bucht. – *Wissenschaftliche Arbeiten aus dem Burgenland* **10**: 1-74, 7 figs., 1 map, 2 profiles, 11 tabs., Eisenstadt.

TOLLMANN, A. (1957): Die Mikrofauna des Burdigal von Eggenburg (Niederösterreich). – *Sitzungsberichte der Akademie der Wissenschaften Wien, mathematisch-naturwissen-*

schaftliche Klasse, Abteilung I **166**: 165-213, 7 pls., 2 figs., 2 tabs., Wien.

TOLLMANN, A. (1985): Geologie von Österreich. Band II. Außeralpiner Anteil. i-xv, 1-710 pp., 286 figs., 27 tabs., Wien (Franz Deuticke).

TOMAŠOVÝCH, A. (1998): Bádenské mäkkýše z tehelne Devínska Nová Ves (Bratislava, Slovensko). – *Mineralia Slovaca* **30**: 357-386, 10 pls., 2 figs., 1 tab., Bratislava.

TOTH, G. (1942): Paläobiologische Untersuchungen über die Tortonfauna aus der Gaadener Bucht. – Palaeobiologica **7/** 5-6: 496-530, 5 figs., Wien.

TOTH, G. (1950): Zur Kenntnis des österreichischen Miozäns. – III. Nachtrag zur Tortonfauna von Sparbach. – *Annalen des Naturhistorischen Museums in Wien* **57**: 170-173, 1 fig., Wien.

TOULA, F. (1884): Ueber die Tertiärablagerungen bei St. Veit an der Triesting und das Auftreten von *Cerithium lignitarum* Eichw. – *Verhandlungen der kaiserlich-königlichen Geologischen Reichsanstalt* **1884**/12: 219-233, Wien.

TOULA, F. (1893): Die Miocänablagerungen von Kralitz in Mähren. – *Annalen des k. k. Naturhistorischen Hofmuseums* **8**: 283-293, 1 tab., Wien.

TOULA, F. (1914): Über eine kleine Mikrofauna der Ottnanger-(Schlier-) Schichten. – *Verhandlungen der kaiserlich-königlichen Geologischen Reichsanstalt* **48**/7-8: 203-217, 7 figs., Wien.

TOULA, F. (1915): Über den marinen Tegel von Neudorf an der March (Dévény – Ujfalu) in Ungarn und seine Microfauna. – *Jahrbuch der Geologischen Reichsanstalt* **64**: 635-673, 1 pl., Wien.

TOURNOUËR, R. (1869): Recensement des échinodermes de l'étage du calcaire a astéries dans le s.-o. de la France. – *Actes de la Société Linnéenne de Bordeaux* **27**: 263-308, 3 pls., Paris-Bordeaux.

TROSCHEL, F.H. (1872): Die Familie der Echinocidariden (1). – *Archiv für Naturgeschichte* **38**/1: 293-356, Berlin.

TURNER, R.L. & NORLUND, C.M. (1988): Labral morphology in heart urchins of the genus *Brissopsis* (Echinodermata: Spatangoida), with an illustrated and revised key to the western Atlantic species. – *Proceedings of the Biological Society of Washington* **101**/4: 890-897, 5 figs., Washington, DC.

VADÁSZ, E. (1906): Budapest-Rákos felsö mediterránkorú faunája. – *Földtani Közlöny* **36**: 256-283, 322-351, 1 pl., 2 figs., Budapest.

VADÁSZ, E. (1907): A Ribicei Felsö-Mediterrán Korszaki Korallpad Faunájáról. – *Földtani Közlöny* **37**/9-11: 368-373, 420-425, Budapest.

VADÁSZ, E. (1914): Regenerationserscheinungen an fossilen Echinoiden. – *Centralblatt für Mineralogie, Geologie und Paläontologie* **1914**: 283-288, 3 figs., Stuttgart.

VADÁSZ, E. (1915): Die mediterranen Echinodermen Ungarns. – *Geologica Hungarica* **1**/2: 79-253, 6 pls., 123 figs., Budapest.

VALDINUCCI, A. (1974a): «*Aliaster*» nuovo genere di echinide istituito per un piccolo gruppo di specie tratte dai generi *Trachyaster* ed *Opissaster* (Famiglia Hemiasteridae Clark). – *Bolletino del Servizio Geologico d'Italia* **94 (1973)**/2: 454-479, 1 pl., 4 figs., Roma.

VALDINUCCI, A. (1974b): Su *Trachypatagus lamberti* (Checchia Rispoli) e considerazioni sulle altre specie del genre. – *Bolletino del Servizio Geologico d'Italia* **94 (1973)**/2: 443-452, 2 pls., Roma.

VÁVRA, N. (1974): Cyclostome Bryozoen aus dem Badenien (Mittelmiozän) von Baden bei Wien (Niederösterreich). – *Neues Jahrbuch für Geologie und Paläontologie, Abhandlungen* **147**/3: 343-375, 3 figs., Stuttgart.

VEIT, E. (1943): Zur Stratigraphie des Miozäns im Wiener Becken. – *Mitteilungen des Reichsamts für Bodenforschung, Zweigstelle Wien* **6**: 2-32, 4 tabs., Wien.

VENDL, M. (1930): Die Geologie der Umgebung von Sopron. II. Teil: Die Sedimentgesteine des Neogen und des Quartär. – *Erdészeti Kisérletek* **32**: 1-175, 23 figs. 3 supplements, Sopron.

VENTURA, G. & Zanfrà, S. (2001): Stato attuale dei *Clypeaster* della Collezione Checchia Rispoli (1925) conservati presso il Servizio Geologico Nazionale. – *Bollettino del Servizio Geologico d'Italia* **116** (1997): 55-81, 10 pls., 2 figs., Roma.

VERRI, A. & Angelis d'Ossat, G. (1900): II.° contributo allo Studio del Miocene nell'Umbria. – *Bolletino della Società Geologica Italiana* **19**/1: 239-279, Roma.

VIA, L. & PADRENY, J. (1970): Dos nuevas especies de *Clypeaster* del Eoceno de Cataluña. – *Instituto de Investigaciones Geológicas de la Diputación Provincial* **24**: 89-97, 3 figs., 3 tabs., Barcelona.

de VINASSA REGNY, P. (1897): Echinidi neogenici del Museo Parmense. – *Atti della Società Toscana di Scienze Naturali, Memorie* **15**: 139-155, 2 figs., 1 tab., Pisa.

VOGL, V. (1907): Adatok a Fóti alsó-mediterran isnmeretéhez. – *Földtani Közlöny* **37**: 243-246, 303-307, 3 figs., Budapest.

WAGNER, C.D. & DURHAM, J.W. (1966): Holectypoids. – In: MOORE, R.C. (ed.): *Treatise on Invertebrate Paleontology, U. Echinodermata. 3 (2)* pp. U440-U450, figs. 329-334, Boulder, CO & Lawrence, KS (GSA & Univ. of Kansas Press).

WALCH, J.E.I. (1771): Die Naturgeschichte der Versteinerungen zur Erläuterung der Knorrischen Sammlung von Merkwürdigkeiten der Natur. 3. Theil. 6+235 pp., Nürnberg (Paul Johnatan Felßecker).

WANK, M. (1981): Fossilien aus dem Lavanttaler Tertiär. – *Carinthia II* **171/91**: 377-386, 5 pls., 2 figs., Klagenfurt.

WANK, M. & STOJASPAL, F. (1980): Neue Fossilfunde im Badenien von Mühldorf (Lavanttal). – *Carinthia II* **170/90**: 445-449, 1 fig., Klagenfurt.

WEIDERT, W.K. (2002): Burgenland. – *Fossilien* **19**/3: 141-142, 1 fig., Korb.

WEIDLINGER, J. (2003): Augen auf. – *Fossilien* **20**/4: 199, 3 figs., Weinstadt.

WESSELY, G. (1961): Geologie der Hainburger Berge. – *Jahrbuch der Geologischen Bundesanstalt* **104**: 273-349, 6 pls., 5 figs., Wien.

WIESBAUR, J. (1874): Fossile Pflanzen im marinen Tertiär Conglomerate zu Kalksburg bei Wien. – *Verhandlungen der kaiserlich-königlichen Geologischen Reichsanstalt* **8**/7: 157-165, Wien.

WIGGELSWORTH, J.C. (1964): The Tertiary Stratigraphy and Echinoid Palaeontology of Gozo, Malta. – Ph.D. thesis (unpublished), University of Durham, 133 pp., 26 pls., Durham.

WILLMANN, R. (1987): Mißverständnisse um das biologische Artkonzept. – *Paläontologische Zeitschrift* **61**/1-2: 3-15, Stuttgart.

WINKLER, A. (1926): Das kohleführende Miozänbecken in Südweststeiermark. – *Montanistische Rundschau* 1926/5: 1-8, Wien.

WINKLER, E. (1942): Tortonfauna aus einem neuen Aufschluß bei Wien-Perchtoldsdorf. – *Berichte des Reichsamts für Bodenforschung* **1942**: 103-108, Wien.

WOLF, H. (1870): Die Stadt Oedenburg und ihre Umgebung. – *Jahrbuch der kaiserlich-königlichen Geologischen Reichsanstalt* **20**: 15-64, 1 pl., 8 figs., 1 map, Wien.

WÖRHEIDE, G. (1995): Bi- und multivariate Analyse borealer und mediterraner Populationen der *Echinocardium cordatum* – Gruppe (Echinoidea; Spatangoida). – *Berliner geowissenschaftliche Abhandlungen* **E16**: 731-763, 3 pls., 18 figs., 4 tabs., Berlin.

WRIGHT, T.W. (1855): On Fossil Echinoderms from the Island of Malta; with notes on the stratigraphical distribution of the Fossil Organisms in the Maltese beds. – *The Annals and Magazine of Natural History, 2nd Series* **15**: 101-127, 175-196, 262-277, 4 pls., London.

WRIGHT, T.W. (1864): On the fossil Echinidæ of Malta. With additional notes on the Miocene beds of the Island, and the stratigraphical distribution of the species therein; by A. Leith Adams. – *Quarterly Journal of the Geological Society of London* **20**: 471-491, 2 pls., London.

ZAMMIT-MAEMPEL, G. (1979): The Indo-Pacific affinities of some Maltese Tertiary fossils. – *Central Mediterranean Naturalist* **1**/1: 1-12, Valetta.

ZITTEL, K.A.v. (1865): Die Bivalven der Gosaugebilde in den nordöstlichen Alpen. Ein Beitrag zur Charakteristik der Kreideformation in Österreich. – *Denkschriften der kaiserlichen Akademie der Wissenschaften, mathematisch-naturwissenschaftliche Classe* **24**/2: 105-178, pls. 1-10, Wien.

ZOLLIKOFER, T.v. (1859): Die geologischen Verhältnisse des Drannthales in Unter-Steiermark. – *Jahrbuch der kaiserlich-königlichen Geologischen Reichsanstalt.* **10**: 200-219, 2 pls., 10 figs., Wien.

ZORN, H. (1972): Sind morphometrische Merkmale von Organismen durch Verhältniszahlen quantitativ zu erfassen? – *Neues Jahrbuch für Geologie und Paläontologie, Abhandlungen* **140**/3: 354-377, 17 figs., Stuttgart.

ZORN, I. (2000): Das Paläogen und Neogen. – In: Schönlaub, H.P. (ed.): *Erläuterungen zur geologischen Karte des Burgenlandes 1:200.000.* pp. 15-30, 1 pl., 10 figs., 1 tab., Wien (Geologische Bundesanstalt).

ŽÍTT, J. (1985): Revision of *Meoma* (Echinoidea, Spatangoids) from Loma Caoba (Cuba, middle Eocene) with some remarks to the evolution of the group. – Časopis pro mineralogii a geologii **30**/3: 265-284, 4 pls., 8 figs., Praha.

ZITTEL, K. A. von (1879): Echinodermata. Handbuch der Paläontologie: Paläozoologie. Vol. 1. Pt. 1. 308-560 pp., figs. 212-404, München & Leipzig (R. Oldenbourg).

Class Echinoidea LESKE, 1778

Subclass Cidaroidea CLAUS, 1880
Order Cidaroida CLAUS, 1880

The use of cidaroid spines and isolated coronal plates in systematic studies is problematic. While being well characterised as a group and readily recognisable even in disarticulated ossicles generic and specific determination of such remains is very difficult. Yet the conservatism of coronal material is partly compensated for by the diversity displayed by primary spine morphology and ornamentation. In fact, MOOI et al. (2000) claim that many cidaroid species cannot be identified without information on primary spines.

Cidaroids are known to exhibit a wide range of spine morphologies on a single individual. Additionally, spine morphology is correlated (to an unknown extent) with the habitat. Nevertheless, these spines often represent the only remains of taxa (often very common) in the fossil record and hence need to be considered. Data on test morphology is scare or entirely missing. Following CUTRESS (1980) disarticulated spine and coronal material is grouped in parataxa as working-hypothesis, although the author is aware that these groupings do not represent species in a biological sense and could be partly artificial. The attribution to the different "species" is based mostly on ornamentation and microstructure. Despite the [theoretical] problems associated with this procedure the results are reproducible.

The influence of the sedimentary environment respectively habitat on the spine ornamentation as well as the extent of individual variation is unclear and needs to be investigated in extant species, where additional independent evidence/ features can be used for determination (e.g. test morphology, pedicellariae, molecular data,...). Personal observations on extant *Stylocidaris affinis* from the Mediterranean, *Eucidaris metularia*, *Prionocidaris baculosa* and *Phyllacanthus imperialis* from the Red Sea showed that spine ornamentation is very homogenous at least within a single site/region. This suggests that the method outlined above might produce reasonable biodiversity data if applied cautiously and variation due to spine position is taken into account.

Family Cidaridae GRAY, 1825
Subfamily Rhabdocidarinae LAMBERT, 1900
Genus *Prionocidaris* AGASSIZ, 1863

Type-species: *Cidarites pistillaris* LAMARCK, 1816; by original designation (AGASSIZ, 1863: 18).
Diagnosis: Corona arched or low; more or less flattened apex; primary tubercles noncrenulate adorally, noncrenulate to weakly subcrenulate aborally; areoles shallow, usually well separated, except for the most adoral which may be confluent; pores distinctly conjugate to subconjugate; primary spines usually long, tapering and ornamented with coarse thorns in longitudinal series; less commonly spines are smooth or bear large thorns arranged in whorls; oral primaries with long collar. (modified from FELL, 1966a).
Distribution: Eocene to Recent – circumtropical
Remarks: Fossil species of *Prionocidaris* (e.g. *P. avenionesis* DES MOULINS, 1837) were often attributed to the genus *Cyathocidaris* (e.g. LAMBERT, 1910b) a genus of predominantly Antarctic Late Cretaceous occurrence. First photographs of *Cyathocidaris* spines and tests are given by NÉRAUDEAU et al. (2000), whereas LAMBERT (1910a) gave only drawings. The photographs and description clearly show the different nature of the morphology in the Antarctic *Cyathocidaris* and the Miocene species attributed to this genus. Whereas the trumpet-shaped crown of e.g. *P. avenionensis* is formed by fused thorns, which are common also along the shaft (as in extant *Prionocidaris*), the crown of true *Cyathocidaris* species forms by an extension of the cortex and lateral flattening of the distal part of the spine. Thorns or large granules similar to extant and fossil prionocidarids are absent in the Antarctic cyathocidarids. Additionally, the coronas of the Antarctic species have also few similarities with the Mediterranean species.

Prionocidaris sismondai (MAYER, 1864)

(Pl. 1, Figs. 1-5; Pl. 2, Figs. 10-11)

*	1864	*Rhabdocidaris Sismondai*. MAYER. – MAYER: 12; pl. 1, fig. 6 [needs to be re-examined]
	1891	*Cidaris avenionensis*, DESMOULINS, 1837. – GREGORY: 587-589, pl. 1, fig. 1a-c
pp	1913a	*Cyathocidaris avenionensis* DESM. (*Cidarites*). – COTTREAU: 44
v	1913a	*Cyathocidaris avenionensis* DESM. (*Cidarites*). – COTTREAU: 23, 79-80; fig. 9 [poorly preserved misidentified *P. sismondai*]
non	1913a	*Cyathocidaris avenionensis* DESM. (*Cidarites*). – COTTREAU: pl. 1, figs. 1-8
pp	1913a	*L.*[*eiocidaris*] *Sismondai* MAYER (*Rhabdocidaris*). – COTTREAU: 23, 45
	1980	*Prionocidaris sismondai* (MAYER) – CHALLIS: 69-72; pl. 21, figs. b-c; pl. 22, figs. a-c
v.	2003	*Prionocidaris* sp. – KROH et al.: 92

Type-material:
The type material could not be located. According to the information in MAYER (1864) it is probably in the palaeontological collection in Zurich or in the collection of BRONN housed (at that time) in the mineralogical collection of Heidelberg. The types of this species need to be re-examined and newly illustrated. The Austrian material is compared with the only available good description given by CHALLIS (1980), which is, however, based on Maltese material.

Material:
Early Badenian (Langhian) – Eisenstadt (Hartl Fm., Hartl hill), Bgld, Austria
NHMW: 1 spine fragment (NHMW 1981/56/5)
Early Badenian (Langhian) – Stotzing (sandpit Mayer), Bgld, Austria
NHMW: 1 spine fragment (NHMW 2004z0050/0002)
Early Badenian (Langhian) – Weissenegg Fm., Retznei [Lafarge (formerly Perlmoser) quarry], Styria, Austria
NHMW: 1 spine fragment (NHMW 2004z0098/0004)
Badenian (Langhian-Early Serravallian) – Forchtenau, Bgld, Austria
NHMW: 4 spine fragments (NHMW 1853.III.256)
Badenian (Langhian-Early Serravallian) – old sandpit between

Großhöflein and Kleinhöflein, Bgld, Austria
NHMW: 14 spine fragments (NHMW 2003z0081/0051, 2004z0004/0001)
Badenian (Langhian-Early Serravallian) – Rauchstallbrunngraben, near Baden, NÖ, Austria
NHMW: 2 large spine fragments (NHMW 1981/56/8, 1997z0178/0904)

Foreign material for comparison:
Badenian (Langhian- Early Serravallian) – Mátyásföld, Budapest, Hungary
NHMW: several spine fragments (NHMW 1997z0178/2093)

Description:
Spines: Primary spines, thick, straight and very robust. The shaft is ornamented with large pointed thorns which are arranged in irregular longitudinal rows. On one side of the spines (~25 % of the circumference in cross-section) the thorns are distinctly smaller, or in some cases missing and replaced by moderately large granulae arranged in distinct vertical rows. The spines are round to slightly oval in cross-section. Preserved distal ends bear a trumpet shaped whorl of thorns (Pl. 1, Fig. 5). The collar is about the same length as the distance acetabulum – milled ring. The neck is short but well developed and lacks any ornamentation. The milled ring is prominent. The acetabulum has a noncrenulate margin.

Discussion:
The specimens considered here are placed in the genus *Prionocidaris* based on the thorny nature of the spines with a trumpet shaped whorl of thorns on their distal end and a noncrenulate acetabulum. Extant *Prionocidaris* show very similar morphologies (compare Mortensen, 1928; Fell, 1966a; Dollfuss & Roman, 1981). Specific determination is more difficult with the limited material at hand. Basically three *Prionocidaris* species are recorded in the Early and Middle Miocene of the Mediterranean: *P. avenionensis*, *P. scillae* and *P. sismondai*. The investigated material fits best with *P. sismondai*, which was re-described by Challis (1980) on base of material from the Chattian to Langhian of the Maltese Islands. Some spine fragments come also close to *P. avenionensis* (compare Philippe, 1998 [under the genus *Cyathocidaris*] and Kroh, in press), but generally the investigated material shows more coarsely ornamented spines than present in that species. Unfortunately, no coronal material could be recovered. *P. sismondai* has a narrow admedian zone in contrast to *P. avenionensis* and *P. scillae* which have broader admedian zones often ornamented by distinct grooves. Literature records of *P. avenionensis* (most under the [sub-] genus *Cyathocidaris*) from the Central Paratethys are in part based on material very similar to that documented here (summarised in Kroh, 2003b). Other specimens (e.g. material of Vadász, 1915 from the Ottnangian, Karpatian and Early Badenian of Hungary in the collection of the Hungarian Geological Survey), however, are closely similar to *P. avenionensis* and it seems possible that both species are present in the Central Paratethys. Further material (especially coronal plates) is needed to explore the possibility that part of the material attributed to *P. sismondai* here might belong to *P. avenionensis* and to check if the latter species is present in the Austrian Miocene too.

Cidaris oligocenus Gregory, 1891 from the Lower Coralline Limestone of Malta is a junior synonym of *Prionocidaris sismondai* according to Challis (1980). Cottreau (1913a) and Philippe (1998), on the other hand, place it into the synonymy of *P. avenionensis* [referred to the genus *Cyathocidaris* in that papers].

Occurrence:

Austria: Badenian (Langhian to Early Serravallian)
Vienna Basin: Rauchstallbrunngraben, near Baden, NÖ ([NHMW])
Eisenstadt-Sopron Basin: Forchtenau, Bgld ([NHMW]); Großhöflein, Bgld ([NHMW]); Hartl Fm., Hartl, Eisenstadt, Bgld (Kroh et al., 2003; [NHMW])
Styrian Basin: Retznei (Weissenegg Fm., Lafarge quarry), Styria ([NHMW])

Paratethys (non-Austrian occurrences): Badenian (Langhian to Early Serravallian)
Great Hungarian Basin (Pannonian Basin): Mátyásföld, Budapest ([NHMW])

Mediterranean: ? Chattian to Langhian
Central Mediterranean: *Globigerina* Limestone, Malta (Gregory, 1891; Cottreau, 1913a; Challis, 1980)

Atlantic Ocean: Miocene
Madeira Archipelago: Porto da Calheta, Porto Santo (Mayer, 1864)

Subfamily Cidarinae Gray, 1825
Genus ? *Stylocidaris* Mortensen, 1909

Type-species: *Cidaris affinis* Philippi, 1845; by original designation (Mortensen, 1909: 52, 54).
Diagnosis: Corona usually flattened; areoles deep or shallow, usually well separated except the most adoral two or three, which may be confluent; primary tubercles noncrenulate adorally, sometimes weakly subcrenulate aborally, especially in juveniles; primary spines long, tapering; secondary spines flattened; tridentate pedicellariae slender; large globiferous pedicellariae without end-tooth (slightly modified from Fell, 1966a).
Distribution: Miocene to Recent – circumtropical
Remarks: The extant genus *Cidaris* differs only by the presence of an end-tooth in the globiferous pedicellariae, a feature rarely preserved in fossil specimens. Nevertheless, most fossil and extant cidaroids of uncertain generic position were originally referred to that genus (which was the first cidaroid genus to be established) [Moore (in Fell, 1966a: U331-U332) proposed the use of inverted commas to distinguish between *Cidaris* sensu stricto and sensu lato].

Although it cannot be excluded that the present material belongs to the genus *Cidaris* s.s., this seems to be unlikely. Instead the material is tentatively assigned to the more diverse and more widely distributed genus *Stylocidaris* on base of the high similarities in spine and coronal morphology.

Stylocidaris ? polyacantha (Reuss, 1860)

(Pl. 1, Figs. 11-19; Pl. 2, Figs. 1-2; Pl. 3, Figs. 1-16)

v.*	1860	*Cidaris polyacantha* n. sp. – Reuss: 223; pl. 3, figs. 4, 4a
	1880	*Cidaris polyacantha* Rss. – Rzehak: 302
	1900	*Cidaris polyacanthus*, Rss. – Procházka: 74
	1900	*Cydarus polyacanthus*, Reuss. – Procházka: 146
	1942	Cidariten (ähnlich dem *Cidaris polyacanthus* Reuss) – Toth: 525
	1950	*Cidaris polyacantha* Reuss – Toth: 171
v.	1998	"*Cidaris*" sp. – Schultz: 116; pl. 52b, figs. 5a-b

Type-material:
Syntypes: 9 spine fragments (NHMW 1859.X.135)
Locus typicus: Rudoltické (= Rudelsdorf), Czech Republic
Stratum typicum: "Rudoltitzer Tegel"
Age: Badenian (Langhian-Early Serravallian), Middle Miocene

Material:
Early Badenian (Langhian) – Badener Tegel, Baden, NÖ, Austria
NHMW: 2 spine fragments (NHMW 1846.37.952), 2 spine fragments (NHMW 1862.I.241), 5 spine fragments

(NHMW 1866.I.131), 19 spine fragments (NHMW 1997z0178/0899), 1 spine fragment (NHMW 2004z0001/0010), 4 spine fragments (NHMW 2004z0001/0011), 1 spine fragment and 1 interambulacral plate (NHMW 2004z0001/0012a+b), 6 spine fragments (NHMW 2004z0001/0063a-f)

GBA: 18 spine fragments and 1 interambulacral plate (no inventory numbers)

Early Badenian (Langhian) – Badener Tegel, Vöslau, NÖ, Austria

NHMW: 65 spine fragments and 1 interambulacral plate (NHMW 2004z0001/0002), 1 spine fragment + 1 interambulacral plate (NHMW 2004z0001/0012), 1 spine fragment (NHMW 2004z0001/0014)

Early Badenian (Langhian) – Badener Tegel, Sooss, near Baden, NÖ, Austria

GBA: 17 spine fragments (no inventory numbers)

Early Badenian (Langhian) – Niederleis, NÖ, Austria

NHMW: 1 spine fragment (NHMW 2004z0001/0016c)

Early Badenian (Langhian) – Stotzing (sandpit Mayer), Bgld, Austria

NHMW: 7 spine fragments (NHMW 2004z0050/0018)

Foreign material for comparison:

Early Badenian (Langhian) – Lăpugiu des Sus (= Lapugy), Romania

NHMW: 4 spine fragments (NHMW 1859.XLV.617B.d)

Badenian (Langhian-Early Serravallian) – Hrušovany nad Jevišovkou (= Grusbach), Czech Republic

NHMW: 22 spine fragments (NHMW 1997z0178/0740), 124 spine fragments (NHMW 1981/56/3)

Badenian (Langhian-Early Serravallian) – Knínice (= Kinitz), Czech Republic

NHMW: 4 primary spine fragments and 1 secondary spine (NHMW 2004z0001/0017a-e)

Badenian (Langhian-Early Serravallian) – Lysice (= Lissitz), Moravia, Czech Republic

NHMW: 4 spine fragments (NHMW 1865.XV.Anhang.b)

Badenian (Langhian-Early Serravallian) – Rudoltićé (= Rudelsdorf), Czech Republic

NHMW: 9 spine fragments (NHMW 1859.X.135 [**type material** of *C. polyacantha* Reuss, 1860])

Description:

Interambulacral plates: Each interambulacral plate bears one large perforate primary tubercle. Aborally the tubercles are weakly crenulate in the adapical part of the platform. Ambitally and adorally the primary tubercles are noncrenulate. The areoles are not confluent, but separated only by narrow ridges with smaller scrobicular tubercles ambitally. The scrobicular tubercles are eye- or lens-shaped and are about twice as large as the remaining secondary tubercles. The adradial zone is narrow bearing a single row of secondary tubercles additionally to the scrobicular tubercles. The admedian zone is wider (about twice as wide as the adradial zone) and bears 3-4 irregular rows of secondary tubercles additionally to the scrobicular tubercles. Both adradial and admedian zones show horizontally and/or radially arranged, narrow grooves. This feature is, however, not very prominent and there are only few grooves on each plate [not as prominent as in e.g. *Prionocidaris canaliculata* (Duncan & Sladen, 1884) or *Phyllacanthus titan* Fell, 1954].

Primary spines: The spines are up to 70 mm long, straight and slightly tapering distally. Their diameter ranges from 1.5 to 5.2 mm in the proximal part, just above the neck. They are ornamented by 10 to 18 well separated, longitudinal ridges bearing short, pointed thorns (Pl. 3, Figs. 3-16). These thorns are flexed distally, producing a saw-like appearance. The number of ridges decreases towards the tip of the spines. The space between the ridges, the ridges themselves and the base of the thorns are covered by small granulae. In some cases an outer layer of cortical hair is preserved in patches, looking like a fine mesh spanning the gaps between the granulae (Pl. 3, Fig. 9). In some spines remnants of what may have been the original coloration is preserved by a pinkish stain. The thinner spines often show less closely spaced, larger and more sharply pointed thorns along the ridges in the distal part of the spine (e.g. Pl. 3, Fig. 4). Rarely thin spines occur with a flattened, flaring distal part terminating in a small "crown-like" whorl of small thorns (Pl. 3, Fig. 15). The collar and base are approximately equally long and separated by a finely striated milled ring. There is a well developed neck, which is free of ornamentation, showing a smooth, glossy surface. It is a little bit longer than collar or base. Both spines with partly crenulate acetabulum (aboral spines; Pl. 3, Figs. 10, 11) and such with noncrenulate acetabulum (ambital and oral spines; Pl. 3, Figs. 9, 12, 14) are present. Few spines with intact tips were observed, some seem to terminate in a small "crown-like" whorl of small thorns.

In cross-section the spines show a microstructure similar to *Stylocidaris*-spines, with non-specialised lamellae and a moderately thick cortex (Pl. 2, Figs. 1-2).

Differential diagnosis:

Spines of *Stylocidaris ? schwabenaui* Laube, 1869 are overall very similar to *S. ? polyacantha* but differ by their ornamentation (and microstructure). The longitudinal ridges in *S. ? schwabenaui* are usually more numerous and more closely spaced. Moreover, the ridges are usually strongly beaded, being made up of large, closely spaced nodules (or rarely short thorns). While the tips of the thorns are sharp and free of granulae in *S. ? polyacantha*, they are blunt and bear granulae except on the summit in *S. ? schwabenaui*. As a whole the spines of *S. ? schwabenaui* are more robust and lack the delicate thorny habit of *S. ? polyacantha* spines. Additionally, the "naked" neck of *S. ? schwabenaui* spines is longer but less constricted. No flattened, distally flaring spines have been observed in this species and the tubercles are always noncrenulate.

Spines of the Ottnangian cidaroid species [referred to as Cidarinae sp. 1 here] differ by their deeper grooves between the longitudinal ridges, which are again usually more numerous than in *S. ? polyacantha*. Additionally, these ridges do not really bear thorns or are made up of nodules, but are simple ridges, often undulating in their width. Moreover, some spines have a triangular to T-shaped cross-section in their distal part. Co-occurring with this spines serrated oral primary spines have been observed, which were never found in association with the other two species.

Spines of the Eggenburgian species *Plegiocidaris ? peroni* (Cotteau, 1877) differ by their more strongly crenulate acetabula, longer striated collar and short neck. The ornamentation of the spines is more "robust" and like in *S. ? schwabenaui*, the thorns are completely covered with granulae.

Discussion:

In contrast to *Stylocidaris ? schwabenaui* Laube, 1869, which occurs most abundantly in psammitic and maerl-type sediments *S. ? polyacantha* is found mainly in pelitic settings. Spine-microstructure, however, differs so strongly that the hypothesis of an ecomorphological gradient between the two morphs is rejected.

Occurrence:

Austria: Early Badenian (Langhian) (Late Badenian occurrences not ascertained)

Vienna Basin: Badener Tegel, Bad Vöslau, NÖ (Schultz, 1998 as "*Cidaris*" sp. [NHMW]); Badener Tegel, Baden, NÖ ([NHMW]); Badener Tegel, Sooss, near Baden, NÖ ([NHMW])

Gaaden Bay: roadcuts between Sittendorf-Sparbach-Weissenbach, NÖ (Toth, 1942, 1950)

Paratethys (non-Austrian occurrences): Early Badenian (Langhian) (Late Badenian occurrences not ascertained)

Molasse Zone: Hrušovany nad Jevišovkou (= Grusbach), Czech Republic ([NHMW]); Lysice (= Lissitz), Moravia, Czech Republic ([NHMW]); Rudolticé (= Rudelsdorf), Czech Republic (Reuss, 1860; Procházka, 1900; [NHMW]); Židlochovice (= Gross Seelowitz), Czech Republic (Rzehak, 1880)

Transylvanian Basin: Lăpugiu des Sus (= Lapugy), Romania ([NHMW])

Mediterranean: unknown

Stylocidaris ? schwabenaui (Laube, 1869)

(Pl. 1, Figs. 20-30; Pl. 2, Figs. 3-4; Pl. 4, Figs. 1-11)

*	1869a	*Cidaris Schwabenaui* Laube. – Laube: 182
	1870	*Cidaris Schwabenaui* Laube. – Laube: 314
	1871	*Cidaris Schwabenaui* Laube. – Laube: 58; pl. 16, fig1.
pp	1880b	*Dorocidaris papillata*, Leske. – Manzoni: 329 [Austrian records]
?	1895	*Cidaris* cf. *Schwabenaui* Laube – Redlich: 333
non	1897	*Dorocidaris papillata* Leske. – Vinassa de Regny: 144-145
	1898	*Cidaris Schwabenaui* Lbe. – Kornhuber: 31 [spines]
v	1915	*Cidaris (Dorocidaris) papillata* Leske. – Vadász: 105-106; pl. 8 (2), fig. 16
	1939	*Cidaris Schwabenaui* Laube. – Kapounek: 74

Type-material:
Holotype: figured in Laube (1871: pl. 16, fig. 1); collection of Hofrat von Schwabenau, current whereabouts unknown
Locus typicus: St. Margarethen, Bgld, Austria
Stratum typicum: Leitha limestone of St. Margarethen
Age: Late Badenian (Early Serravallian), Middle Miocene
Remarks: Hofrat von Schwabenau was living in Linz, OÖ in the year 1864 and was probably employed at the Museum of Linz. In any case, he had a private collection and had also access to the collection at the Museum of Linz (compare Zittel, 1865: 106). Some of the specimens of his collection are now located in the systematic collection of the Naturhistorisches Museum Wien (at that time k. & k. Hofmuseum), but the type specimen is not among them.

Material:
Badenian (Langhian-Early Serravallian) – Großhöflein, Bgld, Austria
NHMW: 1 interambulacral plate (NHMW 2004z0004/0002, determination tentative)
Badenian (Langhian-Early Serravallian) – Kahlenberg, Vienna, Austria
NHMW: 5 spine fragments (NHMW 1981/56/10a)
Badenian (Langhian-Early Serravallian) – Perchtoldsdorf (Brunnergasse, Highway-ramp), NÖ, Austria
NHMW: 12 spine fragments (NHMW 1981/56/9a)
Badenian (Langhian-Early Serravallian) – Steinebrunn (= Steinabrunn), NÖ, Austria
NHMW: 4 spine fragments (NHMW 1859.XLV.634B.b)
Badenian (Langhian-Early Serravallian) – Wöllersdorf, NÖ, Austria
NHMW: 6 spine fragments (NHMW 1981/56/6)
Late Badenian (Early Serravallian) – Walbersdorf, Bgld, Austria
NHMW: 1 spine (NHMW 2004z0001/0001)
Late Badenian (Early Serravallian) –Winden am Neusiedlersee (Nirgl quarry), Bgld, Austria
NHMW: 5 spine fragments (NHMW 1859.L.833)
Late Badenian (Early Serravallian) – Winden [old quarry, N of the Ludlloch (cave), N of the village], Bgld, Austria
NHMW: test fragment consisting of 3 interambulacral plates with attached ambulacral plates (NHMW 1997z0178/1683), 6 test fragments (NHMW 2003z0082/0009, ../0036), 107 spine fragments (NHMW 2003z0082/0008), several hundred primary spine fragments (NHMW 2003z0082/0031-32, ../0035, ../0037-39, 2004z0004/0002)

Foreign material for comparison:
Late Badenian (Early Serravallian) – Sandberg, Devínska Nová Ves (= Neudorf an der March), Slovak Republic
NHMW: 46 spine fragments and 1 interambulacral plate with attached ambulacral plates (NHMW 1997z0178/1939a-b)
Late Badenian (Early Serravallian) – Devínska Nová Ves (= Neudorf an der March; = Dévény-Újfalu), Slovak Republic
NHMW: 78 spine fragments (NHMW 1857.XLV.48), 165 spine fragments (NHMW 1868.VIII.27a-h)
MAFI: 16 spine fragments (MAFI Ech 295 [material referred to as *Cidaris (Dorocidaris) papillata* Leske by Vadász, 1915: 105-106])

Interambulacral plates: Each interambulacral plate bears one large perforate noncrenulate primary tubercle. The areoles are not confluent, but separated by a narrow ridge bearing only scrobicular tubercles. The scrobicular tubercles are eye- or lens-shaped and are about twice as large as the remaining secondary tubercles. The adradial zone is narrow bearing up to two rows of secondary tubercles additionally to the scrobicular tubercles. The admedian zone is bearing 3-4 irregular rows of secondary tubercles additionally to the scrobicular tubercles. There is no naked area along the interradial suture and the suture is not depressed.
Ambulacral plates: The ambulacra are sinuous. Each plate bears a single large, oblique partitioned isopore (P1 to P2 type). Medially a single vertical row of moderately large marginal tubercles and a vertical row of inner tubercles is present in each half ambulacrum. The inner tubercles are situated in the perradial/adoral corner of each ambulacral plate and are much smaller than the marginal tubercles. Both marginal tubercles and pore pairs take up the full height of the ambulacral plates. There is no naked area along the perradial suture. 12 to 13 ambulacral plates per interambulacral plate ambitally.
Primary spines: The largest spine fragments are more than 50 mm long, a maximum spine length of up to 70 mm or more seems likely. The spines are straight and taper slightly distally. Spine diameter ranges from *c.* 1.0 to 4.0 mm, most fragments, however, have a diameter between 2.5 and 3.1 mm. The shaft is ornamented by up 12 to 20 beaded longitudinal ridges. These ridges may be relatively straight (Pl. 4, Fig. 8) to strongly beaded (Pl. 4., Fig. 10) and sometimes may bear short, blunt thorns (Pl. 4, Fig. 6). Contrary to *S. ? polyacantha* these thorns are never pointed and are wholly covered by small granulae (not only the bases as in *S. ? polyacantha*). The ridges themselves and the narrow grooves in-between are covered with small granulae as well. In some cases an outer layer of cortical hair is preserved in patches, looking like a fine mesh spanning the gaps between those granulae (Pl. 4, Fig. 10). The collar is striated and moderately long. The neck is up to twice as long as the collar, free of ornamentation and not conspicuously constricted. The acetabulum has a noncrenulate margin. Few distal spine fragments preserved, showing a blunt, worn tip.

In cross-section the spines show a microstructure similar to extant *Stylocidaris*-spines, with non-specialised lamellae and a moderately thick cortex (Pl. 2, Fig. 3-4).

Differential diagnosis:
For the differences to *S. ? polyacantha* see above under that species.

Spines of the Ottnangian cidaroid species [referred to as Cidarinae sp. 1 here] are generally similar to those of *S. ? schwa-*

benaui. Spine ornamentation differs, however, in its microstructure: whereas there are pronounced longitudinal ridges with undulating width in the Ottnangian spines, the ridges of *S. ? schwabenaui* are more like longitudinal rows of thorns than true ridges. Moreover the strongly serrated oral primary spines that are typical of the Ottnangian species were never observed in *S. ? schwabenaui*

Spines of the Eggenburgian species *Plegiocidaris ? peroni* (COTTEAU, 1877) differ by their strongly crenulate acetabula, longer striated collar and short neck.

Discussion:

Until now this species has been attributed to LAUBE (1871). In LAUBE's 1869a paper, however, this species is already mentioned, accompanied by a description. Thus, according to the ICZN the year 1869 has to be regarded as the valid date for this species.

MANZONI (1880b: 329) and VINASSA DE REGNY (1897: 144-145) considered *S. ? schwabenaui* as junior synonym of *Dorocidaris papillata* LESKE (a junior synonym of *Cidaris cidaris* LINNÉ, 1758; often confused with *Stylocidaris affinis* (PHILIPPI, 1845); compare MORTENSEN, 1928: 289-298). When the fossil material from Austria is compared in detail, however, differences between *S. ? schwabenaui* and the two extant species reported under the name *"Dorocidaris papillata"* are apparent. *S. ? schwabenaui* has a much wider adradial and admedian zone in the ambital interambulacra than either extant species. Moreover, it shows distinct horizontal grooves in the admedian zone. Thus *S. ? schwabenaui* is considered a distinct species and the opinion of VINASSA DE REGNY is rejected.

Occurrence:

Austria: Badenian (Langhian-Early Serravallian)

Vienna Basin: Kahlenberg, W ([NHMW]); Perchtoldsdorf, NÖ ([NHMW]); Steinebrunn (formerly Steinabrunn), NÖ (LAUBE, 1869a, 1871; MANZONI, 1880b; [NHMW]); Wöllersdorf, NÖ ([NHMW]); Vienna Basin (MANZONI, 1880b)

Eisenstadt-Sopron Basin: E of Großhöflein, near Eisenstadt, Bgld (KAPOUNEK , 1934; [NHMW]); St. Margarethen, Bgld (LAUBE, 1871; MANZONI, 1880b; VADÁSZ, 1915); Walbersdorf, Bgld ([NHMW])

Danube Basin: Mörbisch (formerly Merwisch), near Rust, Bgld (LAUBE, 1871; MANZONI, 1880b); Winden, Bgld (LAUBE, 1871; MANZONI, 1880b; [NHMW])

Paratethys (non-Austrian occurrences): Badenian (Langhian-Early Serravallian)

Vienna Basin: Devínska Kobyla (= Thebener Kogel), near Bratislava (KORNHUBER, 1898); Devínska Nová Ves (= Neudorf an der March, = Dévény-Újfalu), Slovak Republic (LAUBE, 1869a, 1871; MANZONI, 1880b; VADÁSZ, 1915; [MAFI]; [NHMW]); Sandberg, Devínska Nová Ves (= Neudorf an der March), Slovak Republic ([NHMW])

Great Hungarian Basin (Pannonian Basin): Sóskut (or Bia), Hungary (VADÁSZ, 1915)

Transylvanian Basin: ? Mănăstire Polovragi (= Kloster Polowratsch), Romania (REDLICH, 1895)

Mediterranean: unknown

Genus *Eucidaris* POMEL, 1883

Type-species: *Cidarites metularia* LAMARCK, 1816; by subsequent designation (CLARK, 1909: 88, and reply by BATHER, 1909: 88; see also discussions in BATHER, 1908a, b; CLARK, 1908; MORTENSEN, 1909: 39-40)

Diagnosis: The test is flattened, the areoles are well separated, only the proximal 2 to 3 are confluent. The primary tubercles are noncrenulate, except in young stages, where they may be weakly subcrenulate aborally. Madreporite slightly larger than other genital plates. The primary spines typically cylindrical, shaft is abruptly truncated and ends in a crown with central prominence. The ornamentation of the spines consists of low rounded warts disposed in regular, longitudinal series (modified from FELL, 1966a).

Distribution: Late Paleocene (HOLMES, 1999) to Recent – circumtropical and southern temperate (Australasia)

Possibly from the Danian (Paleocene) on (compare the spines figured in as indeterminate cidaroid in JAGT, 2000: pl. 2, figs. 10-12)

Ecology and biogeography: Feeding experiments with *E. tribuloides* showed that this species is an opportunistic omnivore. When starved the animals consumed anything within reach, including wooden and fibre-glass parts of the aquarium. Under normal conditions they showed a preference for rock infested with the endolithic sponge *Cliona lampa*, over the sea-grass *Thalassa testudinum*, rock covered with algae and pieces of fish (MCPHERSON, 1968b). Analysis of the gut content of animals captured in the field showed a significant amount of hard material and it seems likely that clionid sponges and other epi- and endolithic organisms constitute a large amount of the food under natural conditions. The high-feeding rate of *E. tribuloides* observed by MCPHERSON (1968b) indicate that this species is potentially an important bioeroder. Due to the low density in the field (usually <2 indiv./m²), however, this species does not reach the importance of the major bioeroding echinoid of the Caribbean, *Diadema antillarum* (MCPHERSON, 1968b).

Extant species of *Eucidaris* are especially abundant from the intertidal down to 20-30 metres (KIER & GRANT, 1965; MCPHERSON, 1968a; NEBELSICK, 1992; HICKMAN, 1998), but are rarely found beyond the continental shelf (MORTENSEN, 1940; FELL, 1954). *E. tribuloides* is usually observed on sandy or rocky bottoms, within turtle grass or on rocks in crevices or beneath sponges or corals; but not on clean sand, where sea grass or algae are absent (KIER & GRANT, 1965; MCPHERSON, 1968a); its distribution coincides with that of *Tripneustes ventricosus* (KIER & GRANT, 1965). *E. metularia* lives on a variety of hard and soft substrates and is especially common on sand with coral patches, in coral reefs and coral carpets (NEBELSICK, 1992).

E. tribuloides shows relatively slow growth with an increase of test diameter between 1 and 2 mm/month or less (MCPHERSON, 1968a). Based on these rates MCPHERSON (1968a) estimated a minimum age of four to five years for the largest specimens observed in the field. Aquarium kept specimens may live up to seven years. This species is preyed upon by cassid gastropods, crabs and a variety of fishes (MCPHERSON, 1968a)

The spatial distribution of *Eucidaris* was summarised by FELL (1954: 20), who found that it is well correlated with the 15° C winter surface isotherms.

Eucidaris zeamays (SISMONDA, 1842)

(Pl. 2, Figs. 5-6; Pl. 6, Figs. 1-21; Pl. 7, Figs. 1-30)

*	1842	*Cidarites zea-mays* mihi. – SISMONDA: 391
	1846	*[Cidaris] Zea-Mays* E. SISM. – AGASSIZ & DESOR: 336
	1847	*Cidaris Zea-mays*. SISM. – MICHELOTTI: 67
	1847	*Cidaris Zea-mays* E. SISM. – SISMONDA: 8
	1852	*[Cidaris] zea-mays*, E. SISM. – D'ORBIGNY: 142
	1858	*[Cidaris] Zea-mays* E. SISM. – DESOR: 38
.	1901	*Cidaris zeamais* SISM. – AIRAGHI: 167-168; pl. 19, figs. 49-57
#	1901	*Cidaris fragilis* n. f. – AIRAGHI: 165; pl. 19, figs. 63-64 [based on test fragments]
#	1907a	*Cidaris sardica* LAMBERT. – LAMBERT: 14-15; pl. 1, figs. 7-8
	1907	*Cidaris zeamays* SISM. – VADÁSZ: 370, 422
	1910	*P.[legiocidaris] zea-mays* SISMONDA (*Cidaris*). – LAMBERT & THIÉRY: 134

1913a *Plegiocidaris zea-mays* SISM. (*Cidaris*). – COTTREAU: 45
1915 *Cidaris* cf. *zeamays* SISM. – MÁJER: 34
? 1915 *Cidaris melitenni* FORBES.– MÁJER: 35, 88
. 1915 *Cidaris zeamays* SISM. – VADÁSZ: 105; pl. 8 (2), fig. 17
v. 1915 *Cidaris* cfr. *zeamays* SISM. – VADÁSZ: 105; pl. 8 (2), fig. 15
v 1915 *Cidaris melitensis* FORBES. – VADÁSZ: 104; pl. 8 (2), fig. 4
? pp 1915 *Plegiocidaris Peroni* COTT. sp. – VADÁSZ: 104; pl. 8 (2), figs. 18-19
1927b *Cidaris zeamais* SISMONDA, 1842 – LAMBERT: 84-85
1928 *Cidaris Zeamays* SISM. – KUTASSY: 8
? 1928 *Cidaris Desmoulinsi* SISM. – KUTASSY: 8
1934 *Cidaris* cf. *zeamays* SISM. – HORUSITZKY: 324
1966 *Cidaris* cfr. *zeamays* SISM. – KÓKAY: 83
1969 *Plegiocidaris peroni* (COTT.) – GÁBOS & GHIURCA: 82; pl. 1, fig. 1 [misidentified adapical test fragment of *E. zeamays*]
1969 *Cidaris melitensis* FORBES – GÁBOS & GHIURCA: 82-86; pl. 1, fig. 2 [misidentified interambulacral plate of *E. zeamays*]
1969 *Cidaris zeamays* SISM. – GÁBOS & GHIURCA: 86; pl. 1, fig. 3
1974 *Cidaris zeamais* SISMONDA, 1842 – CHAVANON: 51-52; pl. 3, fig. 5a-c
1974 *Cidaris tribuloïdes* (LAMARCK) – PHILIPPE: 7 [misidentification, fide PHILIPPE (1998: 46)]
. 1977 *Cidaris* cf. *desmoulinsi* SISMONDA, 1842 – MĄCZYŃSKA: 194; pl. 1, figs. 7, 13
pp 1977 *Cidaris* sp. div. – MĄCZYŃSKA: 194; pl. 1, figs. 4-6, 9-12, 14-15
non 1977 *Cidaris* sp. div. – MĄCZYŃSKA: 194; pl. 1, fig. 16
? 1981 *C.* cf. *melitensis* FORB. – HALMAI: 106
? 1981 *C.* cf. *zeamays* SISM. – HALMAI: 106
. 1984 *Plegiocidaris zeamaïs* (SISMONDA, 1842) – PHILIPPE: 86; pl. 5, fig. 11
. 1987 *Cidaris zeamais* SISMONDA, 1842 – MĄCZYŃSKA: 146, 148; pl. 1, fig. 1
. 1987 *Cyathocidaris avenionensis* (DESMOULINS, 1837) – MĄCZYŃSKA: 145-146, 148; pl. 1, figs. 4-7; pl. 2, figs. 1a-d
1987 *Cidaris desmoulinsi* SISMONDA, 1842 – MĄCZYŃSKA: 148
. 1988 *Cidaris zeamais* SISMONDA, 1841 – MĄCZYŃSKA: 60; pl. 1, figs. 1-3
. 1989 *Eucidaris zeamais* – PHILIPPE: 27; tab. 1
. 1993 *Cidaris zeamais* SISMONDA, 1842 – MĄCZYŃSKA: 105; pl. 1, fig. 1; pl. 6, figs. 1a, 4
? 1993 *Cidaris* cf. *desmoulinsi* SISMONDA, 1842 – MĄCZYŃSKA: 106; pl. 1, fig. 2
. 1993 *Cyathocidaris avenionensis* (DESMOULINS, 1837) – MĄCZYŃSKA: 106; pl. 1, figs. 3-4; pl. 6, fig. 1c
? 1993 Cidaridae – MĄCZYŃSKA: pl. 1, figs. 8
?pp 1993 *Plegiocidaris peroni* (COTTEAU, 1877) – MĄCZYŃSKA: 106-107; pl. 1, fig. 5
. 1996 *Cidaris zeamais* SISMONDA, 1842 – MĄCZYŃSKA: 40; pl. 1, fig. 1
. 1996 *Cyathocidaris avenionensis* (DESMOULINS, 1837) – MĄCZYŃSKA: 40-41; pl. 1, figs. 2-3
. 1998 *Eucidaris zeamais* (SISMONDA, 1842) – PHILIPPE: 44-46; pl. 4, figs. 8-15
v. 2002b *Eucidaris zeamays* – KROH: 12
v. 2003a *Eucidaris zeamays* (SISMONDA, 1842) – KROH: 158-159; pl. 1, figs. 1-11
2003b *Eucidaris zeamays* (SISMONDA, 1842) – KROH: 248
v. 2003 *Eucidaris zeamays* (SISMONDA) – KROH et al.: 92

Type-material:

Cidaris zeamays SISMONDA, 1842:
Type-material: current whereabouts unknown
Locus typicus: Colli di Torino, Italy
Stratum typicum: "terreno terziario medio"
Age: Early to mid Miocene

Cidaris fragilis AIRAGHI, 1901:
Type-material: figured by AIRAGHI (1901: pl. 19, figs. 63-64); private collection of Mr. ROVESENDA; current whereabouts unknown
Type area: Colli Torinesi (Baldissero, Val Geppi, Pian dei Boschi, Rio Batteria, Grangia, Val Saufrà, S. Antonio), Italy
Age: Elveziano (Early – Mid Miocene)

Cidaris sardica LAMBERT, 1907:
Type-material: figured by LAMBERT (1907a: pl. 1, figs. 7-8); current whereabouts unknown
Locus typicus: Santa Reparata (Cap de la Testa), Sardinia
Stratum typicum: Calcaire à Poissons
Age: "Helvétien", Early to mid Miocene

Material:

Early Badenian (Langhian) – Bad Vöslau (?), NÖ, Austria
NHMW: 11 spines (NHMW 1997z0178/1328, 1328a-d)
Early Badenian (Langhian) – Eisenstadt (Hartl Fm.), Bgld, Austria
NHMW: 36 spines (NHMW 1869.L.802)
Early Badenian (Langhian) – Forchtenau, Bgld, Austria
NHMW: 9 spines (NHMW 2004z0001/0015a)
Early Badenian (Langhian) – Gainfarn, NÖ, Austria
NHMW: 6 spines (NHMW 2004z0076/0007-12, ../0060), 1 test fragment (NHMW 2004z0076/0043)
WANZENBÖCK coll.: 58 primary spine fragments; 2 test fragments
NEITZ coll.: 16 primary spine fragments
Early Badenian (Langhian) – Grund Formation, Grund, NÖ, Austria
NHMW: 1 spine (NHMW 2002z0086/0001)
Early Badenian (Langhian) – Niederleis, NÖ, Austria
NHMW: 10 interambulacral plates (NHMW 2002z0087/0048-49, 2002z0087/00052), 166 primary spines (NHMW 2002z0087/0040-47, 2002z0087/0050, 2002z0088/0002, 2002z0089/0006) and 1 genital plate (NHMW 2002z0087/0051), 5 spines (NHMW 2004z0001/0016a)
Early ? Badenian (Langhian) – Rauchstallbrunngraben (bryozoan marl), near Baden, NÖ
NHMW: 6 test fragments (NHMW 2004z0121/0001-4, ../0009-10), 7 primary spines (NHMW 2004z0121/0009-17), 2 genital plates (NHMW 2004z0121/0005-06), 2 ocular plates (NHMW 2004z0121/0007-08), 2 demipyramids (NHMW 2004z0121/0018-19), 2 epiphyses (NHMW 2004z0121/0020-21), 4 rotulae (NHMW 2004z0121/0022-25), 1 tooth (NHMW 2004z0121/0026), 4 secondary spines (NHMW 2004z0121/0027-30)
Early Badenian (Langhian) – Retznei [Weissenegg Fm., Lafarge quarry (formerly Perlmoser)], Styria, Austria
NHMW: 1 spine (NHMW 2004z0098/0003)
Early Badenian (Langhian) – Stotzing (sandpit Mayer), Bgld, Austria
NHMW: 60 spines (NHMW 2004z0050/0001, ../0027-30), and 20 test fragments (NHMW 2004z0050/0019, .../0020-25)
Early Badenian (Langhian) – Wagna (brickyard Wagna), Styria, Austria
NHMW: 1 spine (NHMW 2004z0001/0040)
Badenian (Langhian-Early Serravallian) – Bruck an der Leitha, NÖ, Austria
NHMW: 14 spine fragments (NHMW 2004z0001/0021a)

Badenian (Langhian-Early Serravallian) – old sandpit between Großhöflein and Kleinhöflein, near Eisenstadt, Bgld
NHMW: 348 spines (NHMW 2003z0081/0003, ../0046-0050), 105 test fragments (NHMW 2003z0081/0004) and 20 demipyramids (NHMW 2003z0081/0005)
Badenian (Langhian-Early Serravallian) – Pöls, near Wildon, Styria, Austria
NHMW: 3 spines (NHMW 1862.XXXIII.93)
Badenian (Langhian-Early Serravallian) – Steinebrunn (formerly Steinabrunn), NÖ, Austria
NHMW: 1 spine (NHMW 1846.37.954a), 9 spines (NHMW 1859.XLV.634B.a), 40 spine fragments (NHMW 1865.I.612a), 2 spine fragments (NHMW 2004z0001/0020b), 1 spine fragment (NHMW 2004z0001/0024a)
Badenian (Langhian-Early Serravallian) –Teischl quarry, (? Hartl Fm., Eisenstadt, Bgld), Austria
NHMW: 4 test fragments (NHMW 2004z0001/0008a)
Late Badenian (Early Serravallian) – Winden (quarry north of the village), Bgld, Austria
NHMW: 1 primary spine (NHMW 2003z0082/0034)

Foreign material for comparison:
Early Badenian (Langhian) – Coşteiu de Sus (= Kostej), Romania
NHMW: 5 spines (NHMW 2004z0001/0003a), 1 epiphysis (NHMW 2003z0084/0003)
Early Badenian (Langhian) – Kemence, Hungary
MAFI: 3 vials with several hundred spines and spine fragments [MAFI Ech 442 (including material figured as *Cidaris* cfr. *zeamays* Sism. by Vadász, 1915: 105; pl. 8, fig. 15; additionally to the *E. zeamays* spines there are numerous other cidaroid and diadematid spine fragments, as well as annelids of the genus *Ditrupa* in these vials)], 13 test fragments [MAFI Ech 443 (the vial includes also an ambulacral plate of a diadematoid)]
Early Badenian (Langhian) – Lăpugiu des Sus (= Lapugy), Romania
NHMW: 2 spines + 1 interambulacral plate (NHMW 1859.XXXVII.59a), 24 spines, 4 interambulacral plates + 1 demipyramid (NHMW 2004z0001/0004a), 68 spines + 1 interambulacral plate (NHMW 1859.XLV.617B.a), 1 test fragment + 4 spines (NHMW 2004z0001/0027a)
Badenian (Langhian-Early Serravallian) – Lysice (= Lissitz), Moravia, Czech Republic
NHMW: 1 spine (NHMW NHMW 1865.XV.Anhang.a)
Badenian (Langhian-Early Serravallian) – Nový Rybik (= Portzteich), near Sedlec, W Lednice, Czech Republic
NHMW: 22 spine fragments (NHMW 2004z0001/0009)
Late Badenian (Early Serravallian) – Budapest-Budaörs, Hungary
MAFI: 2 test fragments [MAFI Ech 1397 (material figured as *Cidaris melitensis* Forbes by Vadász, 1915: 104; pl. 8, fig. 4)]
Late Badenian (Early Serravallian) – Buituri (= Bujtur), Romania
NHMW: 3 spine fragments + 3 interambulacral plates (NHMW 2004z0001/0005a), 21 spine fragments (NHMW 1852.II.1603a)

Description:
Size and shape: The corona, as could be reconstructed from isolated test segments, is small, its diameter rarely exceeding 2 cm. It is flattened orally and aborally and has a circular outline. The peristomal opening and the apical disc are of about equal size.
Apical disc: The genital plates associated with this species have a near hexagonal shape, wider at their adapical border than on their abapical margin. They are ornamented by large rounded granules along the margin of the plates and smaller ones in the centre of the plates. The genital pores are large rounded pores, one on each plate, about halfway between the centre and abapical margin. The madreporite bears many small madreporic pores which are closely spaced between the rounded granules in the centre of the plate. The ocular plates are distinctly smaller and have a triangular shape.The long side (= the abapical side) is strongly indented. The ocular pores are small and rounded and lie at the tip of the indentation of the abapical side.
Ambulacra: The ambulacra are slightly sinuous and consist of small plates with C2 isopores orally and C1 isopores adapically. The transition of the two pore types is gradually. The pores in each pair are oval and subequal in size and shape. The neural canals are small and are positioned adorally on the perradial pores. Adjacent pore pairs are well separated by a broad wall. Each ambulacral plate bears one large marginal tubercle with prominent boss. Larger specimens have also one inner tubercle per ambulacral plate, except in the most adoral plates. In smaller specimens the inner tubercles are only present at the ambital ambulacral plates. The poriferous zones are slightly wider than the interporiferous zones. Each interambulacral plate is bordered by 4 to 8 ambulacral plates.
Interambulacra: The interambulacra are up to four times the width of the ambulacra. Each column contains 4 to 6 plates. One large perforate, noncrenulate primary tubercle with undercut mamelon is present on each plate. The platform is small and the parapet weakly defined in most specimens. The areoles are well defined and surrounded by a ring of scrobicular tubercles, which are slightly larger than the secondary tubercles. In the ambital plates of large specimens the areoles may be confluent (e.g. Kroh, 2003a: pl. 1, fig. 10) or nearly so with very narrow ridges separating the areoles. The admedian zone is wider than the adradial zone.
Primary spines: The spines are short, relatively stout and ornamented by longitudinal rows of large granules. These granules are enlarged in more or the less regular intervals and may form whorls along the shaft. Towards the tip, the size of the granules decreases and they grade into longitudinal ridges, which form a crown at the tip of the spine. The acetabulum has a noncrenulate margin. The milled ring is prominent and indistinctly striated. The collar is short and distinctly striated. Spine cross-section is consistent with that of extant *Eucidaris*.

In some outcrops (e.g. Niederleis, Austria or Lapugy, Romania) the spines still show traces of their original coloration. They are banded with broad dark (dark pink to brownish) and thin light bands, the latter coinciding with the whorls of the ornamentation. Very similar coloration is found in extant *E. metularia* (Lamarck) and *E. tribuloides* (Lamarck).

In bulk samples isolated spine bases and regenerating spines (Pl. 6, Fig. 17) are commonly found. The former result from breakage of the spine through the proximal part of the collar, a process possibly associated with autotomy (Cutress, 1965).
Secondary spines: The secondary spines are small, flattened and have a rectangular outline.
Demipyramids: The demipyramids are typical cidaroid demipyramids, with a very shallow foramen magnum.
Rotulae: cidaroid type
Epiphyses: cidaroid type
Teeth: The teeth are grooved longitudinally and have a pointed tip.

Differential diagnosis:
E. zeamays differs from *E. metularia* (Lamarck), an extant species of the Red Sea by its more coarsely ornamented spines (no whorls of coarse thorns are present in *E. metularia*) and more slender spines (compare Plate 6). Likewise *E. tribuloides* (Lamarck), an Pliocene to extant species of the Caribbean has much less strongly ornamented and, additionally, much larger spines than *E. zeamays*.

E. desmoulinsi (Sismonda, 1842) differs from *E. zeamays* by its larger spines (usually 15 to 25 mm in length), which show a more or less regular ornamentation consisting of 15 to 18 longitudinal rows of small granulae. There are no whorls or other protrusions in the spines of *E. desmoulinsi*. The areoles are well separated by a row of scrobicular tubercles on each interambulacral plate, whereas in *E. zeamays* they are less well separated and may be confluent occasionally. The median zone of the interambulacra is wider in *E. desmoulinsi* than in *E. zeamays*.

Discussion:
The description is based on disarticulated material, since no whole coronas could be obtained from the various outcrops studied. A number of outcrops e.g. the bryozoan marls of Rauchstallbrunngraben near Baden in NÖ, however, yielded only this cidaroid species and thus it was possible to relate all cidaroid skeletal elements found in samples from these outcrops to this species. Even skeletal elements of the Aristotle's lantern and the apical disc, which bear only few or no features diagnostic on species level and which are thus normally not determinable could be associated with this species. This offers the unique chance to study the morphology of nearly all skeletal elements, except the pedicellariae, of a fossil *Eucidaris* species

Primary spine ornamentation varies from relatively slender spines with several (usually between one and three) whorls of coarse thorns to stout spines without whorls (see Plate 6). Spines similar to the latter have been attributed to *Eucidaris* aff. *tribuloides* by PHILIPPE (1998). In the bulk samples examined for the present paper, however, all intermediate morphs between the morph with whorls and that without could be recovered from single samples in several outcrops (e.g. Niederleis, Lapugy, Forchtenau,...). This variation in spine ornamentation is thus interpreted as function of the position on the test and/or an environmental signal (although all intermediates are present in nearly all samples, different morphs are most common in different samples/settings).

MĄCZYŃSKA's (1987: pl. 1, figs. 4-7 and pl. 2, figs. 1a-d; 1993: pl. 1, figs. 3-4; pl. 6, fig. 1c) erroneously determined test fragments of this species as *Cyathocidaris avenionensis*. The record from the Aquitanian/Burdigalian-transition in Gozo, Maltese Islands is based on a single spine found in the C_1-Phosphorite bed by the author.

Occurrence:

Austria: Early to Late Badenian (Langhian-Early Serravallian)

Molasse Zone: Grund Fm., Grund, NÖ (KROH, 2003a; [NHMW])

Vienna Basin: Bad Vöslau, NÖ ([NHMW]); Bruck an der Leitha, NÖ ([NHMW]); Gainfarn, NÖ (KROH, 2002b; [NHMW]); Kaisersteinbruch, Bgld ([NHMW]); Müllendorf (Mühlendorfer Kreide AG quarry), Bgld ([NHMW]); Niederleis, NÖ (KROH, 2003a; [NHMW]); Rauchstallbrunngraben near Baden (bryozoan marls), NÖ ([NHMW]); Steinebrunn, NÖ ([NHMW]); Stotzing (sandpit Mayer), Bgld ([NHMW])

Eisenstadt-Sopron Basin: Eisenstadt (Hartl Fm.), Bgld (KROH et al., 2003; [NHMW]); Forchtenau, Bgld ([NHMW]); between Großhöflein and Kleinhöflein, near Eisenstadt, Bgld ([NHMW])

Danube Basin: Winden, Bgld ([NHMW])

Oberpullendorf Bay: Wiesfleck, Bgld ([NHMW])

Styrian Basin: Pöls, near Wildon, Styria ([NHMW]); Wagna (brickyard Wagna), Styria ([NHMW]); Retznei (Weissenegg Fm., Lafarge quarry), Styria ([NHMW])

Paratethys (non-Austrian occurrences): Karpatian (Late Burdigalian) to Late Badenian (Early Serravallian)

Molasse Zone: Lysice, Moravia, Czech Republic ([NHMW]); Nový Rybik (= Portzteich), near Sedlec, W Lednice, Czech Republic ([NHMW]);

Great Hungarian Basin (Pannonian Basin): Budaörs, Pest, Hungary (VADÁSZ, 1915; [MAFI]); Csomád (Magashegy), Pest, Hungary (HORUSITZKY, 1934; HALMAI, 1981; KROH, 2003b); Herend-Márkó, Veszprém, Hungary (KÓKAY, 1966); Kemence, Pest, Hungary (MÁJER, 1915; VADÁSZ, 1915; [MAFI]); Királd, Hungray (KUTASSY, 1928); Kisalag, Hungary (HALMAI, 1981); Sámsonháza, Nógrád, Hungary ([NHMW]); Szatina, near Kishajmás, Baranya, Hungary (VADÁSZ, 1915)

Fore-Carpathian Basin: Busko, Central Poland (MĄCZYŃSKA, 1993); Korytnica Clays, Korytnica, Poland (MĄCZYŃSKA, 1977, 1987); Niechobrz, near Rzeszów, Southern Poland (MĄCZYŃSKA, 1996); Pińczów, Central Poland (MĄCZYŃSKA, 1993); Rybnica, Southern Poland (MĄCZYŃSKA, 1988); Skowronno, Central Poland (MĄCZYŃSKA, 1993); Szczaworyż, Central Poland (MĄCZYŃSKA, 1993); Żerniki, Central Poland (MĄCZYŃSKA, 1993)

Transylvanian Basin: Berchezoaia, Romania (GÁBOS & GHIURCA, 1969); Buituri (= Bujtur), Hunedoara, Romania (VADÁSZ, 1915; [NHMW]); Coştei (= Kostéj), Timiş, Romania ([NHMW]); Lăpugiu des Sus (= Lapugy), Hunedoara, Romania (VADÁSZ, 1915; GÁBOS & GHIURCA, 1969; [NHMW]); Livezile-Aiud, Romania (GÁBOS & GHIURCA, 1969); Ribicioara (= Ribice), near Brad, Hunedoara, Romania (VADÁSZ, 1907, 1915); Sîg, Romania (GÁBOS & GHIURCA, 1969); Valea Satului, Romania (GÁBOS & GHIURCA, 1969)

Mediterranean: Aquitanian/Burdigalian-transition to Serravallian

Western Mediterranean: Sardinia, Italy: Santa Reparata (Cap de la Testa) (LAMBERT, 1907a)

Central Mediterranean: Maltese Islands: Aquitanian/Burdigalian-transition (dating of REHFELD & JANSSEN, 1995): C_1-Phosphorite Layer between the Lower and Middle *Globigerina* Limestone at Dwejra Point, Gozo ([NHMW]); Piemont Basin, Italy: Colli di Torino (SISMONDA, 1842; AGASSIZ & DESOR, 1846; MICHELOTTI, 1847; SISMONDA, 1847; D'ORBIGNY, 1852; DESOR, 1858; AIRAGHI, 1901; COTTREAU, 1913a)

Rhône Basin: Burdigalian: Secteur des étangs (Fos-sur-Mer, Istres) (PHILIPPE, 1998), Bassin d'Avignon (Les Angles, Sernhac, Théziers, Vedène) (PHILIPPE, 1998); Late Burdigalian (faciès "Marnes Bleues"): Bassin d'Avignon (Caumont-sur-Durance) (PHILIPPE, 1998), Bassin de Faucon-Mollans-Malaucène (Entrechaux, Mollans-sur-Ouvèze) (PHILIPPE, 1974, 1984, 1998), Bassin d'Uzès et de la Tave (Vers) (PHILIPPE, 1998)

Atlantic Ocean: Burdigalian to Langhian

Aquitaine Basin: "falun supérieur" of Mérignac, France (LAMBERT, 1927b; CHAVANON, 1974); Peloua, near Saucats, Bordelais, France (LAMBERT, 1927b; CHAVANON, 1974); Pessac, France (LAMBERT, 1927b; CHAVANON, 1974)

Genus *Plegiocidaris* ? POMEL, 1883

Type-species: *Echinus coronatus* VON SCHLOTHEIM, 1820; by subsequent designation (LAMBERT & THIÉRY, 1910: 135).

Diagnosis: Ambulacra simple throughout the corona, but pluriserially on the peristomal membrane; primary tubercles crenulate; areoles round, well separated; scrobicular tubercles larger than the other secondary tubercles; primary spines thick, with long cylindrical collar and relatively short shaft (only one to three times the length of the collar); shaft diameter twice the collar diameter; transition between shaft and collar oblique, with abrupt change in diameter (modified from FELL, 1966a).

Distribution: Late Triassic to the Late Jurassic; ? Miocene – Europe (FELL, 1966a)

Remarks: Attribution of the present species (*peroni* COTTEAU, 1877), although widely accepted in Neogene echinoderm-literature, is problematic. See discussion for details.

Plegiocidaris ? peroni (COTTEAU, 1877)

(Pl. 2, Fig. 9; Pl. 4, Fig. 12-18)

*	1877	*Cidaris Peroni*, COTTEAU, 1877 – COTTEAU in LOCARD: 231-232 ; pl. 8, figs. 8-12 [based on disarticulated primary spines]
	1895	*Cidaris Munsteri*. SISMONDA, 1844. – COTTEAU: 8-9 [fide AIRAGHI, 1901 and PHILIPPE, 1998]
pp	1895	*Rhabdocidaris compressa*, COTTEAU, 1893. – COTTEAU: 10; pl. 1, fig. 7 [fide PHILIPPE, 1998]
.	1897	*Cidaris Munsteri* E. SISMONDA. – DE LORIOL: 115, 119-121; pl. 4, figs. 3-6
?	1901	*Cidaris Peroni* COTT. – AIRAGHI: 170-171; pl. 19 (1), figs. 23-25 [very poorly preserved]
	1907a	*Plegiocidaris peroni* COTTEAU (*Cidaris*). 1877. – LAMBERT: 9-10; pl. 1, figs. 9-10
	1910a	*Plegiocidaris Peroni* COTTEAU (*Cidaris*), 1877. – LAMBERT: 7-9; pl. 1, figs. 5-17
#	1910a	*Plegiocidaris Cureti* LAMBERT. – LAMBERT: 9; pl. 1, figs. 18-19 [fide PHILIPPE, 1998]
#	1910a	*Dorocidaris Deydieri* LAMBERT. – LAMBERT: 16; pl. 3, figs. 4-6 [fide PHILIPPE, 1998]
? v	1915	*Plegiocidaris Peroni* COTT. sp. – VADÁSZ: 104; pl. 8 (2), fig. 13 [most probably the figured specimen (MAFI Ech 1398) is a *S.? schwabenaui* spine]
non	1915	*Plegiocidaris Peroni* COTT. sp. – VADÁSZ: 104; pl. 8 (2), figs. 18-19 [misidentified *E. zeamays* coronal plates]
	1926	*Plegiocidaris peroni* COTT. – STRAUSZ: 213, 368
?	1930	*Cidaris peroni* COTT. – VENDL: 51
non	1969	*Plegiocidaris peroni* (COTT.) – GÁBOS & GHIURCA: 82; pl. 1, fig. 1 [misidentified adapical test fragment of *E. zeamays*]
	1974	*Plegiocidaris peroni* (COTT.) – ROMAN: 332, 333, 336
?	1981	*Plegiocidaris cureti* LAMB. – HALMAI: 106
	1981	*P. peroni* COTT. – HALMAI: 106
	1984	*Plegiocidaris peroni* (COTTEAU, 1877) – PHILIPPE: 86; pl. 5, figs. 7-10
non	1987	*Plegiocidaris peroni* (COTTEAU, 1877) – MĄCZYŃSKA: 146, 148; pl. 1, figs. 2 [misidentified diadematoid interambulacral plate]
.	1989	*Plegiocidaris peroni* (COTTEAU) – PHILIPPE: 27; tab. 1
non	1993	*Plegiocidaris peroni* (COTTEAU, 1877) – MĄCZYŃSKA: 106-107; pl. 1, fig. 5 [misidentified *E. zeamays* coronal plates]
non	1993	*Plegiocidaris peroni* (COTTEAU, 1877) – MĄCZYŃSKA: 106-107; pl. 6, fig. 1b [misidentified diadematoid interambulacral plate]
.	1998	*Plegiocidaris peroni* (COTTEAU, 1877) – PHILIPPE: 48-50; pl. 3, figs. 20-31
v.	1999	Cidariden-Stachel – HARZHAUSER & KROH: 221
v.	1999	Cidaridae indet. – KROH & HARZHAUSER: 154-155; pl. 7, figs. 1-4
	2003b	*Plegiocidaris ? peroni* (COTTEAU, 1877) – KROH: 249

Type-material:

Type-material: disarticulated spines figured by COTTEAU in LOCARD (1877: pl. 8, figs. 8-12); current whereabouts unknown

Locus typicus: near Bonifacio, Sardinia

Stratum typicum: "couches inférieures"

Age: ? Burdigalian, Early Miocene

Material:

Late Eggenburgian (Early Burdigalian) – Eggenburg (brick-factory Stransky), NÖ, Austria

NHMW: 12 spine fragments (NHMW 2003z0078/0001)

Late Eggenburgian (Early Burdigalian) – Eggendorf, NÖ, Austria

NHMW: 1 spine (NHMW 1998z0048/0094)

Late Eggenburgian (Early Burdigalian) – Grübern (Zogelsdorf Fm.), near Maissau, NÖ, Austria

NHMW: 3 spine fragments (NHMW 1914.VII.23), 1 interambulacral plate and 48 spine fragments (NHMW 2003z0075/0009, 2003z0075/0012, ../12a-b)

Late Eggenburgian (Early Burdigalian) – Klein-Meiseldorf, NÖ, Austria

NHMW: 2 spine fragments (NHMW 1981/56/1)

Late Eggenburgian (Early Burdigalian) – Rohrendorf, E Pulkau, NÖ, Austria

NHMW: 4 spines (NHMW 2003z0077/0001-3)

Late Eggenburgian (Early Burdigalian) – Sonndorf, S Eggenburg, NÖ, Austria

NHMW: slab with 4 spines and 6 spine fragments (NHMW 2003z0076/0001)

Late Eggenburgian (Early Burdigalian) – Unternalb (Retz Fm.), SE Retz, NÖ

NHMW: 4 spine fragments (NHMW 1999z0051/0015)

Although this cannot be demonstrated beyond doubt, these ossicles are considered to be conspecific here.

Description:

Ambulacral plates: not preserved

Interambulacral plates: Only a single isolated, strongly abraded interambulacral plate is preserved. It bears a single large perforate tubercle. The areole is subcircular and a full scrobicular ring is developed. Adradial and interradial zones are subequal and not very wide. Scrobicular and secondary tuberculation obscured.

Primary spines [based mainly on the Sonndorf specimens (NHMW 2003z0076/0001) since these spines are the only ones with well preserved base and ornamentation. The remaining material is abraded, but is considered to be conspecific with the Sonndorf material based on the closely similar morphology (as far as preserved)]: Largest spine fragment 54 mm in length, with a proximal width of 3.7 mm. The spines are straight and slightly tapering distally. They are ornamented by 15 to 20 well separated, longitudinal rows formed by small thorns. The thorns in each row are connected towards each other via shallow ridges. In the distal part of the spines and in some fragments from other outcrops (e.g. Eggendorf) the thorns are less well developed and the interconnecting ridges more prominent, forming an ornamentation very similar to those of the Ottnangian species described above. The space between the longitudinal rows of thorns and the thorns themselves are covered by small granulae and longitudinal striae. The shape of the distal end is unknown. The spine bases are only slightly widened and may have a slightly larger diameter than the thickest part of the shaft. The collar is rather long, being longer than the distance acetabulum margin – milled ring. The neck is moderately well developed and lacks ornamentation. The acetabulum has a crenulate margin (Pl. 4, Fig. 12).

Cross-section (Pl. 2, Fig. 9): The medulla has a diameter of about 30 % of the corresponding spine diameter and consists of an irregular mesh. In the centre of the medulla, the holes of the mesh are larger than at the margin. The lamellae, which are connected to each other by trabeculae, radiate from this central part. The outermost layer of the spine is formed by moderately thick cortex. All three zones are clearly separated.

Differential diagnosis:

This species is easily recognised by its crenulate acetabulum and well developed neck lacking ornamentation. All other species recorded from Austria have noncrenulate tubercles/acetabula. Only in *S. ? polyacantha* a crenulation may occur in aboral tubercles, but they are only subcrenulate, i.e. the crenulation is developed only weakly and only on one side of the tubercle (compare Pl. 3, Fig. 2)

Discussion:
The well preserved material from Sonndorf can be confidently referred to *Plegiocidaris* ? *peroni* (Cotteau, 1877), based on the characteristic crenulation, rather long collar, well developed neck and ornamentation on the shaft. *P. peroni* has recently been re-described by Philippe (1998: 48-50) and the reader is referred to that work for a comprehensive synonymy and description of better preserved coronal material. Albeit rather similar to the primary spines occurring in the Ottnangian of OÖ, the Eggenburgian material is clearly not conspecific.

The generic placement of *peroni* necessitates some remarks: Originally it was attributed to the genus *Cidaris*, like many fossil cidaroids. Lambert (1907a: 9-10) transferred it to the genus *Plegiocidaris*, which has been accepted by most subsequent authors. *Plegiocidaris* Pomel, 1883 (type-species: *Echinus coronatus* von Schlotheim, 1820), however, is characterised by its short spine with long collar and abrupt change in diameter from collar to shaft. The spines are thus a little bit club-shaped and a neck is not developed (compare Fell, 1966a: U337; Hess, 1975: 86, pl. 27, figs. 1-7). According to Fell (1966a) *Plegiocidaris* ranges from the Late Triassic to the Late Jurassic. Albeit there are some similarities (crenulate tubercles, ornamentation of the interambulacral plates) the attribution of *peroni* to the genus *Plegiocidaris* seems doubtful. Unfortunately the available material is not sufficient to revise this taxon and it is thus named in open nomenclature.

Although well figured by Cotteau (1877) this species has often been confused with other cidaroid taxa (e.g. Vadász, 1915 wrongly attributed coronal plates of *Eucidaris zeamays* to *P.* ? *peroni*). Even diadematoid interambulacral plates have been associated with it (Mączyńska, 1987: pl. 1, fig. 2; 1993: pl. 6, fig. 1b).

Occurrence:

Austria: Late Eggenburgian (Early Burdigalian)
Molasse Zone: Eggenburg (brick-factory Stransky), NÖ ([NHMW]); Eggendorf, NÖ ([NHMW]); Grübern (Zogelsdorf Fm.), near Maissau, NÖ ([NHMW]); Klein-Meiseldorf, NÖ ([NHMW]); Rohrendorf, E Pulkau, NÖ ([NHMW]); Sonndorf, S Eggenburg, NÖ ([NHMW]); Retz Fm., Unternalb, NÖ (Harzhauser & Kroh, 1999; Kroh & Harzhauser, 1999; [NHMW])

Paratethys (non-Austrian occurrences): Karpatian (Late Burdigalian), ? Badenian (Langhian-Early Serravallian)
Great Hungarian Basin (Pannonian Basin): ? Bélapátfalva, Heves, Hungary (Vadász, 1915); ? Cinkota, Pest, Hungary (Vadász, 1915); Fót (Somlyó Mt.), Pest, Hungary (Strausz, 1926; Kroh, 2003a); ? Györ (brickyard Schwarz), Györ-Monson-Sopron, Hungary (Vendl, 1930); ? Kisalag, Hungary (Halmai, 1981)

Mediterranean: Burdiaglian to Serravallian
Western Mediterranean: Piemont Basin, Italy: ? Serravalle Scrivia (Airaghi, 1901)
Central Mediterranean: Corsica, France: Candelabra, près Bonifacio, (Cotteau in Locard, 1877; Cotteau, 1895); Sardinia, Italy: Bingia Fargeri (Fangario) (Cotteau, 1895), Capo Sant'Elia, near Cagliari (Cotteau, 1895; Lambert, 1907a), Mont San Michele (Cotteau, 1895), Saint-Advendrace (Cotteau, 1895), Scala Chilivri (Orosei) (Lambert, 1907a)
Rhône Basin: Burdigalian: Littoral de Nerthe (Carro près Carry) (Lambert, 1910a; Roman, 1974), Bassin d'Avignon-Carpentras (Beaucaire, Les Angles, Aramon, Avignon, Barbentane, Saint-Hippolyte-le-Graveyron, Saint-Rémy-de-Provence, sernhac, Théziers, Vedène, Villeneuve-lès-Avignon, Notre-Dame-du-Château) (de Loriol, 1897; Lambert, 1910a; Philippe, 1998), Bassin de Valréas-Visan (Montbrison-sur-le-Lez, Saint-Restitut, Taulignan) (Lambert, 1915a; Philippe, 1998); Late Burdigalian (faciès "Marnes Bleues"): Bassin d'Avignon-Carpentras (le Barroux, Caumont-sur-Durance) (Roman, 1974; Philippe, 1998), Bassin de Faucon-Mollans-Malaucène (Entrechaux, St. Marcellin, St. Romain, Faucon, Mollans-sur-Ouvèze) (Philippe, 1984, 1998); Serravallian: Bassin d'Avignon (Beaumes-de-Venise, Courthézon, Entraigues) (Philippe, 1998), Bassin d'Apt, Reillanne, Forcalquier (Reillanne) (Lambert, 1910a; Philippe, 1998), Bassin de Faucon-Mollans-Malaucène (Entrechaux) (Philippe, 1998), Lyonnais [Lyon (Saint-Paul et le jardin des Plantes), Saint-Fons] (Roman, 1974; Philippe, 1998)

Cidarinae sp. 1

(Pl. 1, Figs. 6-10; Pl. 2, Figs. 7-8; Pl. 5, Figs. 1-26)

	1875a	*Cidaris* sp. – Hoernes: 387
	1875b	unbestimmbare *Cidaris*-Stacheln – Hoernes: 211
	1877	*Cidaris* sp. indet. – Fuchs: 663
?	1926	Cidarisstacheln – Stadler: 82
	1956	*Cidaris* sp. – Sieber: 317
v.	1965	*Cidaris* spec. – Schultz: 283-284
	1969	Stachel regulärer Seeigeln – Steininger: 49; pl. 14, fig. 10

Material:
Ottnangian (Late Burdigalian) – Allerding, OÖ, Austria
Danninger coll.: 1 coronal segment (no inventory no.)
Ottnangian (Late Burdigalian) – between Edermaning and Höbmannsbach, OÖ, Austria
NHMW: 3 primary spine fragments (NHMW 2004z0002/0004)
Ottnangian (Late Burdigalian) – Höbmannsbach, Taufkirchen Bay, ESE Schärding, OÖ, Austria
NHMW: 330 primary spine fragments (NHMW 1978/1966/12a, 12j-n, 12s-t), 54 oral primary spines (NHMW 1978/1966/12b, 12o-r), 13 secondary spines (NHMW 1978/1966/12c, 12u, 12w-y), 36 scrobicular spines (NHMW 1978/1966/12d), 8 isolated spine bases (NHMW 1978/1966/12e)
Ottnangian (Late Burdigalian) – Höbmannsbach (playing ground), OÖ, Austria
NHMW: 2 primary spine fragments (NHMW 2004z0002/0005)
Ottnangian (Late Burdigalian) – Kletzenmarkt, NNW Bad Schallerbach, OÖ, Austria
NHMW: 4 primary spine fragments (NHMW 1978/1966/15a), 1 oral primary spine (NHMW 1978/1966/15b)
Ottnangian (Late Burdigalian) – Mitterndorf (sandpit Denk), 3.2 km NNE Sigharting, NNE Andorf, OÖ, Austria
NHMW: 57 primary spine fragments (NHMW 1986/102/37), 10 primary spine fragments (NHMW 1989/58/24)
Ottnangian (Late Burdigalian) – Offenhausen, S Grieskirchen, OÖ, Austria
NHMW: 17 primary spine fragments (NHMW 1978/1966/14a)
Ottnangian (Late Burdigalian) – Plesching (Alte Königsstrasse), OÖ, Austria
NHMW: 177 primary spine fragments and 3 oral primary spines (NHMW 2004z0002/0001a-g)
Ottnangian (Late Burdigalian) – Rainbach (large sandpit), near Schärding, OÖ, Austria
NHMW: 8 primary spine fragments (NHMW 2004z0002/0002a-f)
Ottnangian (Late Burdigalian) – Rainbach (small sandpit), near Schärding, OÖ, Austria

NHMW: 3 primary spine fragments (NHMW 2004z0002/0003a-b)

Foreign material for comparison:

Ottnangian (Late Burdigalian) – Höch (gravel-pit), SW Passau, Bavaria, Germany

NHMW: 257 primary spine fragments (NHMW 1990/1487/0225a-b), 1 test fragment consisting of two interambulacral plates (NHMW 1990/1487/0225b), 83 primary spine fragments (NHMW 1990/1487/0227a), 73 primary spine fragments (NHMW 1990/1487/0228), 96 primary spine fragments (NHMW 1990/1487/0229); 287 spine fragments (NHMW 2003z0090/0007)

Schultz coll.: 5 interambulacral plates (no inv. no.)

Ottnangian (Late Burdigalian) – Holzbach, near Fürstenzell, SW Passau, Bavaria, Germany

NHMW: 17 primary spine fragments (NHMW 1990/1489/0120a)

Ottnangian (Late Burdigalian) – Kälberbach, near Höch, SW Passau, Bavaria, Germany

NHMW: 1 oral primary spine (NHMW 1990/1488/0006), 70 primary spine fragments (NHMW 1990/1488/0007a)

Although this cannot be demonstrated beyond doubt, these ossicles are considered to be conspecific here.

Description:

Ambulacral plates: The ambulacra are very slightly sinuous. Each plate bears a single large, partitioned isopore (P1 to P2 type) and two marginal tubercles, the perradial of which is distinctly smaller. Both marginal tubercles and pore pairs take up the full height of the ambulacral plates. There is no naked area along the perradial suture. Each interambulacral plate is bordered by 12 ambulacral plates at the ambitus.

Interambulacral plates: Each interambulacral plate bears one large perforate noncrenulate primary tubercle. The areoles are not confluent, but separated by a narrow ridge with smaller scrobicular tubercles ambitally. The areoles are near circular except at the ambitus where they are slightly oval. The scrobicular tubercles are distinctly larger than the other secondary tubercles. The adradial zone is not very wide bearing up to two rows of secondary tubercles additionally to the scrobicular tubercles. The interradial zone is only slightly wider and bears 2-3 irregular rows of secondary tubercles additionally to the scrobicular tubercles. The interradial suture is slightly sunken.

Primary spines (Pl. 5, Figs. 1-11, 16-22): Largest complete spines up to 45 mm in length, individual fragments suggest even longer spines. The spines are straight and slightly tapering distally. Primary spine diameter ranges from 1 to 4 mm in the proximal part.

They are ornamented by 13 to 21 well separated, longitudinal ridges. At one side of each spine the ridges are usually straight and of more or less constant width. On the opposing side the width of the ridges undulates, giving them a beaded appearance. In the proximal part of larger spines the beaded ridges may grade into longitudinal rows of large, lentil-shaped knobs which are connected by shallow ridges. Rarely also spine with short, blunt thorns instead knobs are observed. The space between the ridges, the ridges themselves and the base of the thorns are covered by small granulae. The spines end bluntly or, more rarely, terminate in a small crown. The spine bases are not widened and have about the same diameter as the proximal part of the shaft. The neck is poorly developed and may be slightly constricted only. Ornamentation starts immediately above the finely striated collar. The acetabulum has a noncrenulate margin. Some spines have a triangular to T-shaped cross-section in their distal part.

In cross-section the spines show a microstructure similar to that of *Stylocidaris*-spines, with non-specialised lamellae, large medulla and a thin to moderately thick cortex. Most spines are, however, distinctly oval in diameter and have an oval medulla.

Oral primary spines (Pl. 5, Figs. 23-26): Associated with the "normal" primary spines strongly serrated, flattened oral primaries have been observed. They are rarely more than 10 mm in length and are distinctly curved. Apart from the serrated sides they lack any ornamentation and have a smooth, glossy surface. Only close to the tip up to two longitudinal keels may be developed at the inwards bent side. The serration consist of large diametrically opposed thorns. The association with the above described "normal" primary spines is based on their cooccurrence in all localities where the serrated type was recovered. Additionally some spines with intermediate morphology (poorly developed lateral thorns in the proximal part, many longitudinal ridges in the distal part) are preserved.

Secondary spines (Pl. 5, Figs. 12-15): The scrobicular secondary spines are up to 8 mm in length and have a similar ornamentation as the "normal" primary spines. They are, however, slightly bent, flattened and slightly serrated lateral edges. Nonscrobicular secondary spines were rarely recovered and are very similar to the primary spines but much smaller.

Differential diagnosis:

Spines of the Badenian species *Stylocidaris ? polyacantha* (Reuss, 1860) differ by their ornamentation (longitudinal ridges with well developed thorns), their well developed porcelaineous neck and widened base. Moreover, no serrated oral primary spines have been found in association with the *S. ? polyacantha* species.

Spines of the Badenian *Stylocidaris ? schwabenaui* Laube, 1869 are generally very similar to the Ottnangian spines but differ by their ornamentation (and microstructure). The longitudinal ridges in *S. ? schwabenaui* are usually more numerous and more closely spaced. Moreover, the ridges are made up of large, closely spaced short thorns or nodules, which are better described as beaded longitudinal rows than as ridges (compare Pl. 4, Fig. 10). Additionally, they have a widened base and a well developed neck devoid of any ornamentation. No serrated oral primaries have been found in association with *S. ? schwabenaui*.

Plegiocidaris ? peroni (Cotteau, 1877) differs by its strongly crenulate acetabula/tubercles and characteristic long collar.

Discussion:

Although complete specimens are wanting and preservation of most ossicles is poor the abundance of the material allows a good characterisation of the species. Based on primary spine and coronal plate morphology this species is clearly distinct from the Badenian *S. ? polyacantha* and *S. ? schwabenaui* (see above). Differentiation from the Eggenburgian cidaroid material is more difficult, although the well developed neck and crenulate acetabulum margin of the latter suggest that they are not conspecific.

Generic and specific determination of the Ottnangian cidaroids is very difficult. As stated above cidaroid taxonomy strongly relies on pedicellariae morphology, although Mooi et al. (2000) made a strong case for the importance of primary spine morphology in cidaroid systematics. The horizontal, nonconjugate ambulacral pores and lack of sutural pits or grooves suggest an attribution to the subfamily Cidarinae (compare Fell, 1966a: U330). Extant members of the genera *Cidaris* and *Stylocidaris* are rather similar to the present species. Strongly serrated oral primary spines, however, have been reported in neither. Moreover, primary spines of these genera usually have more strongly developed bases and neck.

Similar shaped oral primary spines are present in the Histocidarinae (genera *Histocidaris* and *Poriocidaris*) and the Ctenocidarinae (genera *Ctenocidaris* and *Aporocidaris*). The former, however, have strong crenulate tubercles, not present in the Ottnangian material and occur in depths below 200 m (the Ottnangian material predominantly comes from shallow-water settings). The Ctenocidarinae are restricted to the sub-Antarctic region and most species are deep-water specialists

(> 2.500 m). For the time being the Ottnangian material is left in open nomenclature until more coronal material becomes available and more extant material is examined.

Most of the Ottnangian cidaroid material comes from higher-energy near-shore settings in the OÖ and Bavarian part of the Molasse Zone. The spines are most frequent in poorly sorted coarse sandy environments. Nearly all specimens were subject to heavy abrasion obscuring ornamentation detail. Coronal plates, even disarticulated interambulacral plates, are extremely rare in most outcrops and seriously underrepresented compared with the number of spine fragments. The single coronal segment available comes from Allerding, OÖ, a breaker zone setting and must be considered as exceptional preservation. A possible explanation would be trapping and rapid burial in e.g. a crustacean burrow.

Occurrence:

Austria: Ottnangian (Late Burdigalian)

Molasse Zone: Allerding, OÖ (Danninger coll.); between Edermaning and Höbmannsbach, OÖ ([NHMW]); Haselbach, near Schärding, OÖ (Schultz, 1965); Höbmannsbach, Taufkirchen Bay, ESE Schärding, OÖ (Schultz, 1965; [NHMW]); Kletzenmarkt, NNW Bad Schallerbach, OÖ ([NHMW]); region of Linz (phosphoritic sands), OÖ (Steininger, 1969); Mitterndorf, NNE Andorf, OÖ ([NHMW]); Plesching, OÖ ([NHMW]); Offenhausen, S Grieskirchen, OÖ ([NHMW]); Ottnang, OÖ (Hoernes, 1875a, b; Fuchs, 1877); Rainbach, near Schärding, OÖ (Schultz, 1965; [NHMW])

Paratethys (non-Austrian occurrences): Ottnangian (Late Burdigalian)

Bavarian Molasse Basin: Höch (gravel-pit), SW Passau, Bavaria, Germany ([NHMW]); Holzbach, near Fürstenzell, SW Passau, Bavaria ([NHMW]); Kälberbach, near Höch, SW Passau, Bavaria (? Stadler, 1926; Sieber, 1956; [NHMW])

Cidaroida indet.

mentioned as Cidaroid spines:

1867 Cidaritenstachel – Karrer: 335, 339-342
1868 Cidaritenstachel – Karrer: 576, 581
1869 Cidaritenstacheln – Bunzel: 81
1877 Cidariten-Stacheln – Karrer: 79, 112, 116, 126, 132, 146-147, 151, 153-56, 159-60, 163, 166-67, 169, 171-73, 181, 187, 188, 195, 201, 222, 223, 225-9, 232-5, 258-9, 266, 277, 282, 285, 297-8, 305-6
1877 Cidariten-Stachel. – Hilber: 261
? 1893 Ocellarplättchen des Scheitelschildes (*Salenia* ?). – Toula: 288 [genital plates; probably from co-occurring cidaroids]
1898 Cidariten-Stachel – Kornhuber: 31
1905 *Cidaris*, sp. – Gaál: 357, 362
1906 Cidariten – Schaffer: 65
1927 riesige Cidaritenstachel – Küpper & Bobies: 3
1928 Große Cidaritenstacheln. – Bobies: 45
1930 *Cidaris*-Stachel. – Vendl: 50
1942 *Cidaris*-Stacheln. – Winkler: 106
1955 Cidarisstachel, Großform. – Tollmann: Tab. 5b
1968 Cidarisstacheln, Großform – Schmid: Anhang

mentioned as *Cidaris* sp.

1853 Stacheln von *Cidaris* – Melion: 707
1856 *Cidaris* – Rolle: 592
1877 *Cidaris* Stachel – Karrer: 104, 193
1882 *Cidaris* sp. – Hilber: 236
1892 *Cidaris* – Hilber: 1032
1893 *Cidaris* n. sp. – Toula: 288
1894 *Cidaris* sp. – Lörenthey: 67
1897 *Cidaris* spec. ind. – Schaffer: 535, 547
1906 *Cidaris* sp. – Vadász: 329 [spine fragments]
1907 *Cidaris* tüskék – Vogl: 245, 306 [= *Cidaris*-Stachel]
1915 *Cidaris* sp. – Májer: 29, 82
1915 *Cidaris* spec. – Toula: 672
1926 *Cidars* sp. – Strausz: 213, 368
1927 *Cidaris* tük. – Horusitzky: 25, 165
1928 *Cidaris* sp. – Bobies: 48
1930 *Cidaris* sp. – Bobies: 26
1953b *Cidaris* sp. – Sieber: 194
1955 *Cidaris* sp. – Tollmann: Tab. 5b
1955 *Cidaris* sp. – Tollmann: tab. 1
1964 *Cidaris* – Kollmann: 540
1965 *Cidaris* spec. – Schultz: 283-284
1968 *Cidaris* sp.– Stancu & Andreescu: 466, tab.
1974 *Cidaris* sp. – Vávra: 345
2000 *Cidaris* – Piller: 87
2002 *Cidaris* sp. – Kazár: 153; fig. 1

mentioned as Cidaridae or Cidaroida indet.:

v. 2002b Cidaridae indet. – Kroh: 12
v. 2003a Cidaroida indet. – Kroh: 159; pl. 2, figs. 6-7
2003b Cidaridae indet. – Kroh: 249

Material:

Ottnangian (Late Burdigalian) – Höbmannsbach, Taufkirchen Bay, ESE Schärding, OÖ, Austria

NHMW: 7 smooth primary spines (NHMW 1978/1966/12f)

Early Badenian (Langhian) – Gainfarn, NÖ, Austria

NHMW: 1 primary spine fragment (NHMW 2004z0076/0020)

Early Badenian (Langhian) – Niederleis, NÖ, Austria

NHMW: 4 fragmented primary spines (NHMW 2002z0087/0006-8)

Early Badenian (Langhian) – Stotzing (sandpit Mayer), Bgld, Austria

NHMW: 1 spine fragment (NHMW 2004z0050/0026), 1 tooth and 1 demipyramid (NHMW2004z0050/0006)

Badenian (Langhian-Early Serravallian) – Forchtenau, NÖ, Austria

NHMW: 1 spine fragment (NHMW 2004z0001/0015e)

Badenian (Langhian-Early Serravallian) – Grinzing, Vienna, Austria

NHMW: 1 spine fragment and 1 demipyramid (NHMW 1846.37.953a)

Badenian (Langhian-Early Serravallian) – old sandpit between Großhöflein and Kleinhöflein, Bgld, Austria

NHMW: 1 scrobicular spine (NHMW 2003z0081/0039)

Late Badenian (Early Serravallian) – Winden [old quarry, N of the Ludlloch (cave), N of the village], Bgld, Austria

NHMW: 1 rotula (NHMW 2003z0082/0042)

Foreign material:

Badenian (Langhian-Early Serravallian) – Nový Rybik (= Portzteich), near Sedlec, W Lednice, Czech Republic

NHMW: 1 genital plate (NHMW 1858.XLVII.78a)

The material discussed under this heading is identifiable to order rank only due to the nature of the studied specimens. The unified discussion under the heading "Cidaroida indet." does not imply that they are conspecific or even congeneric.

Discussion:

Many literature records of cidaroid spines could not be revised since the material on which the records are based is lost or its whereabouts unknown. Additionally, some of the disarticulated cidaroid remains present in the Austrian collections could not be attached to the groups outlined above and are listed here under the heading "Cidaroida indet.".

Occurrence:

Austria: Late Eggenburgian (Early Burdigalian) to Late Badenian (Early Serravallian)

Molasse Zone: Enzersdorf, near Staatz, NÖ (KARRER, 1867); Grübern, NÖ (KARRER, 1867); Höbmannsbach, Taufkirchen Bay, ESE Schärding, OÖ ([NHMW]); Laa an der Thaya, NÖ (KARRER, 1867)

Vienna Basin: Bad Vöslau, NÖ (KARRER, 1877); Baden, NÖ (KARRER, 1877); Brunn am Gebirge, NÖ (KARRER, 1877); Gainfarn, NÖ (KARRER, 1877; KROH, 2002b; [NHMW]); Grinzing, W ([NHMW]); Gumpoldskirchen, NÖ (KARRER, 1877); Heferlberg, near Gumpoldskirchen, NÖ (KÜPPER & BOBIES, 1927); Leobersdorf, NÖ (KARRER, 1877); Maria-Enzersdorf, NÖ (KARRER, 1877); Mödling, NÖ (KARRER, 1877); Müllendorf (Fenk quarry), Bgld (TOLLMANN, 1955); Müllendorf (Mühlendorfer Kreide AG quarry), Bgld (TOLLMANN, 1955); Niederleis, NÖ (KROH, 2003a; [NHMW]); Perchtoldsdorf, NÖ (KARRER, 1868, 1877; WINKLER, 1942); Pötzleinsdorf, W (SIEBER, 1953b); Rauchstallbrunngraben, near Baden, NÖ (KARRER, 1877; BOBIES, 1928, 1930; VÁVRA, 1974); Sooss, near Baden, NÖ (KARRER, 1877); Stotzing (sandpit Mayer), Bgld (KAZÁR, 2002); Wien XVI (SCHAFFER, 1906); Wöllersdorf, NÖ (KARRER, 1877); Vienna Basin (MELION, 1853)

Gaaden Bay: Siegenfeld, NÖ (BOBIES, 1928)

Eisenstadt-Sopron Basin: Eisenstadt (Hartl Fm., Johannesgrotte), Bgld (PILLER, 2000); Eisenstadt (Schloßpark), Bgld (TOLLMANN, 1955); Forchtenau, Bgld ([NHMW]); Großhöflein, Bgld (TOLLMANN, 1955); sandpit between Groß- and Kleinhöflein, near Eisenstadt, Bgld (TOLLMANN, 1955; [NHMW]); Hader Berg, Bgld (SCHMID, 1968); St. Georgen, Bgld (SCHMID, 1968)

Danube Basin: Breitenbrunn, Bgld (SCHMID, 1968); Donnerskirchen, Bgld (SCHMID, 1968)

Styrian Basin: Grubthal, near Gamlitz, Styria (HILBER, 1877); Muggenau, Styria (ROLLE, 1856); Neurath, Styria (ROLLE, 1856); Tafern quarry, Styria (KOLLMANN, 1964)

Paratethys (non-Austrian occurrences): Eggenburgian (Early Burdigalian), Karpatian (Late Burdigalian) to Late Badenian (Early Serravallian)

Molasse Zone: Jaklovec (= Jakloretz), near Ostrava, Czech Republic (KARRER, 1867); Knínice (= Kinitz), Czech Republic ([NHMW]); Kralice (= Kralitz) nad Oslavou, Czech Republic (TOULA, 1893); Nový Rybik (= Portzteich), near Sedlec, W Lednice, Czech Republic (BUNZEL, 1869; [NHMW]); Orlová (= Orlau), near Karvina, Czech Republic (KARRER, 1867)

Vienna Basin: Devínska Kobyla (= Thebener Kogel), near Bratislava (KORNHUBER, 1898); Devínska Nová Ves (= Neudorf an der March), Slovak Republic (TOULA, 1915); Devínska Nová Ves (brickyard), Slovak Republic (SCHAFFER, 1897)

Great Hungarian Basin (Pannonian Basin): Budapest-Rákos, Pest, Hungary (VADÁSZ, 1906); Fót, Pest, Hungary (VOGL, 1907); Fót (Somlyó Mt.), Pest, Hungary (STRAUSZ, 1926; KROH, 2003b); Györ (brickyard Schwarz), Györ-Monson-Sopron, Hungary (VENDL, 1930); Hont, Börzsöny, Hungary (MÁJER, 1915); Mogyoród, near Fót, Pest, Hungary (HORUSITZKY, 1927); Nógrádszakál (= Szakall), Nógrád, Hungary (GAÁL, 1905)

Fore-Carpathian Basin: Mykolaiv (= Mikołajów), S of Lwów, western Ukraine (HILBER, 1882)

Transylvanian Basin: Buituri (= Bujtur), Hunedoara, Romania (LÖRENTHEY, 1894); Rugi-Delineşti, Romania (STANCU & ANDREESCU, 1968)

Zala, Sáva and Dráva Basins: Bartolomae (Pereiraïa beds), Slovenia (HILBER, 1892)

Additional cidaroid taxa reported from the Central Paratethys

***Prionocidaris avenionensis* (DES MOULINS, 1837)**

v 1915 *Cidaris (Cyathocidaris) avenionensis* DESM. sp. – VADÁSZ: 105; pl. 8 (2), figs. 15, 17
1915 *C. (Cyathocidaris) avenionensis* DERM. – MÁJER: 35, 88
1926 *Cidaris avenionensis* DESM. – STRAUSZ: 213, 368
1934 *Cidaris avenionensis* DESM. – HORUSITZKY: 324
1960 *Cyathocidaris avenionensis* (DESMOULINS 1837) – KOJUMDGIEVA & STRACHIMIROV: 228-229; pl. 8, figs. 6-8
1968 *Cyathocidaris avenionensis* DESM. – STANCU & ANDREESCU: 466, tab.
1981 *Cidaris avenionensis* DESM. (= *Cyathocidaris avenionensis*) – HALMAI: 105
non 1987 *Cyathocidaris avenionensis* (DESMOULINS, 1837) – MĄCZYŃSKA: 145-146, 148; pl. 1, figs. 4-7; pl. 2, figs. 1a-d [misidentified *Eucidaris zeamays* plates]
non 1993 *Cyathocidaris avenionensis* (DESMOULINS, 1837) – MĄCZYŃSKA: 106; pl. 1, figs. 3-4; pl. 6, fig. 1c [misidentified *Eucidaris zeamays* plates]
non 1996 *Cyathocidaris avenionensis* (DESMOULINS, 1837) – MĄCZYŃSKA: 40-41; pl. 1, figs. 2-3 [misidentified *Eucidaris zeamays* plates]
2003b *Prionocidaris avenionensis* (DES MOULINS, 1837) – KROH: 248 [literature review]

Reported occurrence: Ottnangian: Cinkota and Püspökhatvan, Hungary (VADÁSZ, 1915); Karpatian: Cinkota (VADÁSZ, 1915), Csomád (Magas-hegy) (HORUSITZKY, 1934; HALMAI, 1981), Fót (VADÁSZ 1915), Fót (Somlyó Mt.) (STRAUSZ, 1926; HALMAI, 1981) all in Hungary; Badenian: Kemence, Hungary (VADÁSZ, 1915); Kisalag, Hungary (HALMAI, 1981); Ribicioara (= Ribice), Romania (VADÁSZ, 1915); Rugi-Delineşti, Romania (STANCU & ANDREESCU, 1968), Tarnene and Opansko bardo, Bulgaria (KOJUMDGIEVA & STRACHIMIROV, 1960)

Remarks: Concerning the generic attribution see above under *P. sismondai*. Some records of *P. avenionensis* species are based on misidentification of other species (e.g. MĄCZYŃSKA, 1987, 1993 and 1996).

The available material (exclusively spines) is not sufficient to judge whether a single species or two (*P. avenionensis* and *P. sismondai*) were present in the Badenian of the Central Paratethys. Ottnangian and Karpatian material preserved at the Hungarian Geological Survey is closely similar to *P. avenionensis* reported from the Rhône Basin by PHILIPPE (1998). Yet, Badenian material from Austria (here referred to *P. sismondai*) clearly falls outside the variation documented by PHILIPPE (1998) for *P. avenionensis*. More specimens, especially coronal plates, are needed to solve this puzzle and to get a clear idea on the spatial and temporal distribution of these forms.

Misidentifications or dubious records of cidaroids

Apart from the taxa discussed above a number of cidaroids have been mentioned from Austrian and nearby outcrops. Reference material of some of these records could be traced and referred to the taxa above. Part of these records, however, remain dubious since reference material is lost or of unknown repository and descriptions/illustrations if present at all are insufficient.

***Dorocidaris papillata* (LESKE, 1778)**

1893 *Dorocidaris papillata* LESKE. – TOULA: 288
1939 *Dorocidaris papillata* LESKE – KAPOUNEK: 74

1958a *Cidaris (Dorocidaris) papillata* Leske – Sieber: 152
1969 *Cidaris (Dorocidaris) papillata* Leske – Gábos & Ghiurca: 86; pl. 1, fig. 5 [? misidentified *S. schwabenaui* or *S. polyacantha* spines]
1981 *Dorocidaris papillata* Leske – Halmai: 106

Reported occurrence: Leitha Limestone between Groß- and Kleinhöflein, Bgld (Kapounek, 1939); Steinebrunn, NÖ (Sieber, 1958a); Kisalag, Hungary (Halmai, 1981); Kralice nad Oslavou (= Kralitz), Moravia, Czech Republic (Toula, 1893); Valea Satului and Berchezoaia, Baia Mare Basin, Romania (Gábos & Ghiurca, 1969)

Remarks: *Dorocidaris papillata* Leske is a junior synonym of *Cidaris cidaris* Linné, 1758 according to Mortensen (1928), but has been often confused with *Stylocidaris affinis* (Philippi). Both are extant species living in the Mediterranean Sea today. Fossil records of this species from Austria are most probable misidentified *Stylocidaris* ? *schwabenaui* or *S.* ? *polyacantha* spines (see above under these species).

Cidaris pseudoserrata Cotteau, 1863

1893 *Cidaris pseudoserrata* Cott. – Toula: 288

Reported occurrence: Kralice nad Oslavou (= Kralitz), Moravia, Czech Republic (Toula, 1893)

Remarks: This species was established for disarticulated material (mainly spines) from the Middle to Late Eocene of south-western France. The spines are distinctly flattened with serrate edges and have strongly crenulate acetabula (compare Cotteau, 1863: 75, pl. 2, figs. 17-22; Cotteau, 1892: 474-478, pl. 311, figs. 1-23). There is no similar material in the Badenian of the Central Paratethys and it remains unclear what species Toula's record refers to actually.

Cidaris subularis d'Archiac in Agassiz & Desor, 1846

1893 *Cidaris subularis* d'Arch. var. – Toula: 288

Reported occurrence: Kralice nad Oslavou (= Kralitz), Moravia, Czech Republic (Toula, 1893)

Remarks: This species was based on disarticulated spines from the Eocene of Biarritz, France (Cotteau, 1892: 422-426; pl. 304, figs. 1-16). It is slightly similar to *Stylocidaris* ? *schwabenaui* spines and Badenian Central Paratethyan records may be misidentifications of such spines.

Cidaris hirta Sismonda, 1842

1857 *Cidaris hirta* Sismonda – Peters: 322

Reported occurrence: Baden, NÖ (Peters, 1857)

Remarks: This species was established for isolated spines from the Pliocene of Asti, Italy and is a junior synonym of *Histocidaris rosaria* (Bronn, 1831) according to Borghi (1999: 112). Austrian material referred to his species could probably be misidentified *Stylocidaris* ? *polyacantha* spines, but differences are strong.

Histocidaris rosaria (Bronn, 1831)

1981 *C. rosaria* Bronn – Halmai: 106

Reported occurrence: Csomád, Oldalhegy, Hungary (Halmai, 1981)

Remarks: *Histocidaris rosaria* (Bronn, 1831) is a species occurring in the Pliocene of Northern Italy. Halmai's (1981) record is based on an unpublished Ph.D. thesis and could not be verified.

Subclass Euechinoidea Bronn, 1860
Infraclass Echinothurioidea Claus, 1880
Order Echinothurioida Claus, 1880
Family Echinothuriidae Thomson, 1872
Subfamily Echinothuriinae Thomson, 1872

Fell (1966b) states that the tubercles in the order Echinothurioida are noncrenulate. In extant *Asthenosoma ijimai* Yoshiwara (USNM E10669), however, both noncrenulate and crenulate primary tubercles were observed in the interambulacra, albeit the former were more common. Smith & Wright (1990: 102) report crenulate primary tubercles in the genus *Kamptosoma*.

Genus *Retzneiosoma* Kroh, nov. gen.

Type-species: *Retzneiosoma jaseneki* Kroh, nov. sp.

Derivatio nominis: Named after the locality where the material was found, and adding the echinothuriid suffix *-soma*.

Type horizon and locality: Lower Badenian (Langhian, Middle Miocene) marls overlying the corallinacean limestone of the Weissenegg Formation, outcropping in the quarries of the Lafarge cement company in Retznei, Styria (Austria).
Diagnosis: Echinothuriid with trigeminate ambulacra with single large primary plate reaching from adradial to perradial suture and smaller, ovate demiplates positioned along the horizontal sutures; demiplates occluded from both adradial and perradial sutures; pores arranged in three vertical rows; primary tubercles crenulate and perforate; tubercles arranged in distinct adradial columns in the aboral interambulacra; membranous gap in aboral interambulacra; hollow, smooth primary spines terminating in a tapering, rounded tip.
Remarks: This new genus differs distinctly from all known Echinothurioida. To investigate the placement of this new taxon a cladistic analysis carried out with the program PARS from the PHYLIP Phylogeny Inference Package [Version 3.6 (alpha3), July, 2002, by Joseph Felsenstein] was done. Using the characters and data matrix provided by Smith & Wright (1990: 101-104; Fig. 19) the analysis resulted in a single most parsimonious tree placing *Retzneiosoma* within the clade Echinothuriinae as defined by Smith & Wright (1990) but failed to fully resolve its position within the clade (see Fig. 11.1). Running the same analysis with the different character set and data matrix from Smith (*"The Echinoid Directory"*, 12. 11. 2003), *Retzneiosoma* is again placed into clade Echinothuriinae but its position is not resolved either. The analysis yielded 5 most parsimonious trees which differ only in the position of *Retzneiosoma* in relation to the other echinothuriinids (Fig. 11.2a-e). A slightly extended set of characters (Tabs. 7-8) yielded a single most parsimonious tree and places *Retzneiosoma* as sister-group to all living echinothuriinids (Fig. 11.3). *Echinothuria* is the sister-group of all living echinothuriinids plus *Retzneiosoma*. This result is supported by the stratigraphic distribution of the taxa in question (*Echinothuria* – Late Cretaceous, *Retzneiosoma* – Miocene).

Extant and fossil echinothuriinids differ markedly from *Retzneiosoma*: *Araeosoma* has much more prominent membranous gaps and a much sparser tuberculation; *Asthenosoma* lacks primary tubercles on its aboral surface; *Calveriosoma* lacks membranous gaps on the aboral surface and bears primary tubercles only on each second or third primary ambulacral plate; *Echinothuria* has only very few primary tubercles on the aboral side and lacks membranous gaps in the interambulacra; *Hapalosoma* similarly has much less prominent and common primary tubercles and demiplates which reach the adradial suture adapically. Moreover, all of these taxa have non-crenulate primary tubercles and primary spines terminating in a hyaline hoof.

Apical disc

1. Ocular and genital plates form a contiguous ring in adults: no (0); yes (1).
2. Genital plates subdivided into a mass of small platelets: no (0); yes (1).
3. Ocular plates very elongate and extended: no (0); yes (1).
4. Gonopores open within genital plate (0); open in distal membranous gap (1).
5. periproctal plates: contiguous pavement (0); separated and embedded in membrane (1).

Ambulacra

7. Smaller elements reduced to minute, ovate accessory plates: no (0); yes one is (1); yes both are (2).
8. On oral surface smaller ambulacral plates occluded from adradial suture: no (0); inner plate occluded (1); both elements occluded (2).
9. On aboral surface primary element occluded from the adradial suture: no (0); yes (1).
10. On oral surface triads heterogeneous and tending to form multicomponent elements: no (0); yes (1).
11. Primary element divided medially into two pieces: no (0); yes (1).
12. On oral surface, pore-pairs marginal (0); forming a band removed from the adradial suture (1).
13. Pore-pairs adorally forming a broad marginal phyllode; no (0); yes (1).
14. Only one accessory plate on ambital and adoral elements: no (0); yes (1).
15. All pores/tube-feet equally developed (0); some pore-pairs reduced (1).
16. Pore-pairs in three discrete columns at ambitus and aborally: no (0); yes (1).

Peristome

17. Buccal plates present: no (0); yes (1).
18. Plates arranged uniserially (0); biserially (1) in each half ambulacrum.
19. Buccal notches: absent (0); feeble (1); well developed (2).

Interambulacra

20. Membranous gaps present orally in interambulacral zones: no (0); yes (1).
21. Membranous gaps present adapically in interambulacral zones: no (0); yes (1).

Tuberculation

22. Primary tubercles: non-crenulate (0); crenulate (1).
23. Areole not sunken (0); sunken (1).
24. Oral tubercles much larger than aboral: no (0); yes (1).
25. Multiple equal tubercles on interambulacral plates: no (0); yes (1).
26. Distinct adradial column of tubercles: no (0); yes (1).

Spines and pedicellariae

27. Oral spines end: in rounded point with fleshy sac (0); in hyaline hoof (1); in flattened blade (2).
28. Primary spines central lumen: empty (0); filled with coarse mesh (1).
29. Sphaeridia: on both accessory plates (0); on inner plate only (1).
30. Tridentate pedicellariae: open blade-like valves with contiguous edges (0); with involute neck and wide lateral gap (1).
31. Ophicephalous pedicellariae: absent (0); present (1).
32. Dactylous pedicellariae: absent (0); present (1).

Teeth

33. CLNP/LLNP system differentiated: no (0); yes (1).
34. CLNP/LLNP system of *Calverisoma*-type (see JENSEN, 1982): no (0); yes (1).
35. Teeth, shape of tip: acuminate (0); rounded (1).

Table 7: Extended set of characters used in the analysis shown in Fig. 1.2 (no. 3) (modified from SMITH "*The Echinoid Directory*", 12.11.2003).

	12345	1 67890	11111 12345	11112 67890	22222 12345	22223 67890	33333 12345
Pelanechinus	10000	01101	00100	00110	00000	00??0	00?0?
Kamptosoma	0?101	12000	00010	01000	01010	12000	0100?
Hemiphormosoma	01001	02200	01000	00010	00111	00000	00?00
Phormosoma	01001	02000	00000	00010	00111	00000	00000
Paraphormosoma	00000	02110	00000	00010	00000	00010	00?00
Hygrosoma	00010	02000	00000	00010	00111	01010	10101
Sperosoma	01010	11100	11001	00010	00001	01010	10101
Tromikosoma	01011	11100	01001	00000	00001	01010	10?01
Echinothuria	10000	02200	01000	10010	00010	11??0	?????
Hapalosoma	10000	02100	01100	10011	10010	11110	01111
Asthenosoma	11000	02200	01100	10010	10011	01110	00111
Araeosoma	10010	02200	01100	10111	10001	11111	01111
Calveriosoma	10010	02200	01100	10111	10011	11?11	00111
Retzneiosoma	?????	?2200	01?00	1????	110?1	100??	????0

Table 8: Character matrix for the analysis shown in Fig. 11 (no. 3) (data from MORTENSEN, 1935; SMITH & WRIGHT, 1990; and SMITH, "*The Echinoid Directory*", 12.11.2003).

Retzneiosoma jaseneki Kroh, nov. gen. et nov. sp.

(Fig. 10; Pl. 9, Figs. 1-12; Pl. 10, Figs. 1-8)

v. 2004a echinothurioid – Kroh: 231
(Early Badenian: Lafarge quarry, Retznei, Stmk)

Type-material:
Holotype: NHMW 2003z0071/0001
Locus typicus: Retznei [Lafarge quarry (formerly Perlmoser)], Styria, Austria.
Stratum typicum: grey marls overlying the Weissenegg Fm.
Age: Early Badenian (Langhian, Middle Miocene)
Derivatio nominis: Named after Manfred Jasenek (Vienna), who collected and donated the holotype.

Diagnosis:
Echinothuriid with trigeminate ambulacra with single large primary plate reaching from adradial to perradial suture and smaller, ovate demiplates positioned along the horizontal sutures; demiplates occluded from both adradial and perradial sutures; pores arranged in three vertical rows; primary tubercles crenulate and perforate; tubercles arranged in distinct adradial columns in the aboral interambulacra; in ambital ambulacral plates with one and two primary tubercles alternate, adapical ambulacral plates bear only a single primary tubercle; membranous gap in aboral interambulacra; hollow, smooth primary spines terminating in a tapering, rounded tip.

Material:
Early Badenian (Langhian) – Retznei [Lafarge quarry (formerly Perlmoser)], Styria, Austria
NHMW: 2 specimens and several disarticulated ossicles [NHMW 2003z0071/0001 (holotype) and 2003z0071/0002 (fragmented specimen), 2003z0071/0003 (isolated spines and coronal plates)]

Dimensions (in mm):

Inv. no.	TL	TW	TH
NHMW 2003z0071/0001	78	60	-

Description:
Size and shape: Test fragile, with imbricated coronal plates. Large flanges are developed internally on both ambulacral and interambulacral plates. The single complete specimen has a largest test diameter of about 78 mm. Both outline and profile are unknown.
Apical disc: unknown
Ambulacral: Ambulacra moderately wide reaching about 50 % of interambulacral width at the ambitus. Ambulacral plates low and horizontally elongated. The arrangement of the ambulacral plates is that of pseudocompound triads. In each triad there is one large demiplate (further called primary plate) extending from perradial to adradial suture and two small, subequal demiplates situated obliquely along the horizontal sutures, about one third the length of the primary plate from the adradial suture. These smaller demiplates have no contact to the adradial or perradial sutures (Fig. 10.A, 10.B). Each demiplate bears a single well developed pore pair (the preservation is not good enough to identify the type). The pore pair on the primary plate lays adradially of the smaller demiplates. The pores form three vertical series.

Tuberculation of the ambulacra is denser than on the interambulacra. Primary plates with one and two large primary tubercles alternate on the aboral side (Fig. 10.A; oral side unknown). Secondary tubercles and miliaries obscured.
Interambulacra: The interambulacra are wide, about two times as wide as the ambulacra at the ambitus. The interambulacral plates are low, and up to ten times wider than high. Between individual interambulacral plates of the aboral side there are small but distinct gaps interpreted as membranous gaps as oc-

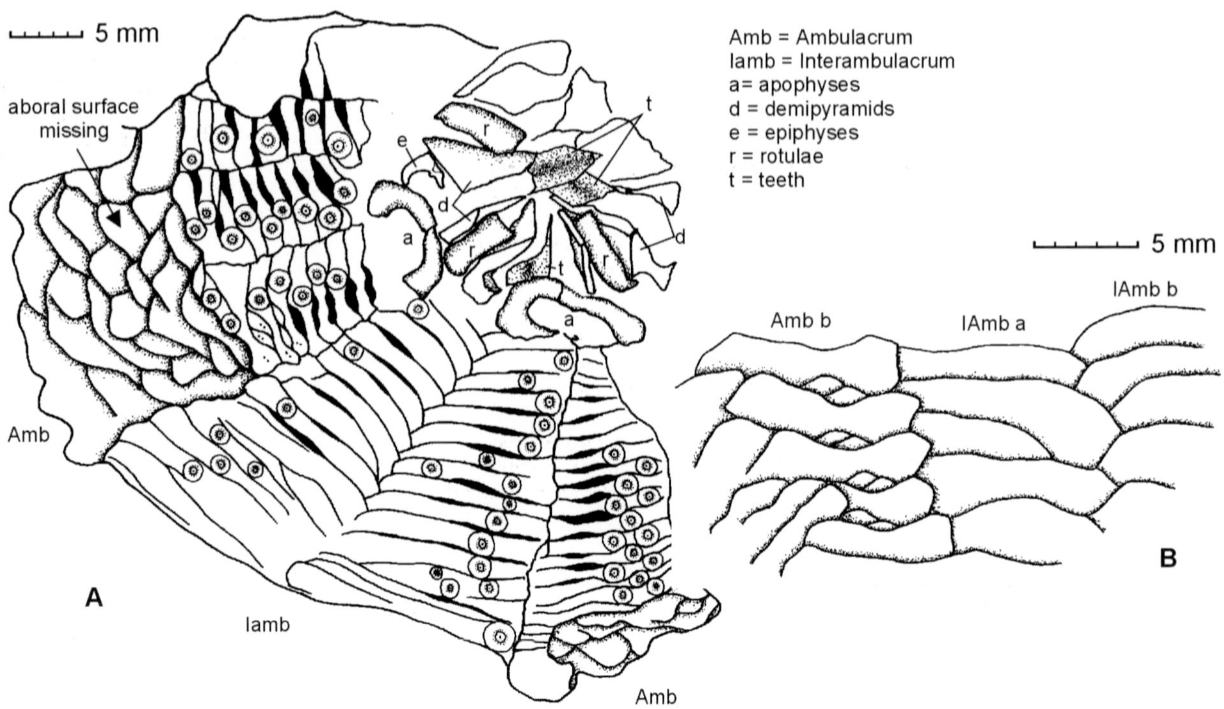

Figure 10: *Retzneiosoma jaseneki* Kroh, nov. sp.: aboral view showing imbricate plates, "membranous" gaps, and aboral tuberculation (A). Notice that large parts of the aboral surface are missing exposing the lantern and the inside of the oral surface (B; showing ambulacral plating).

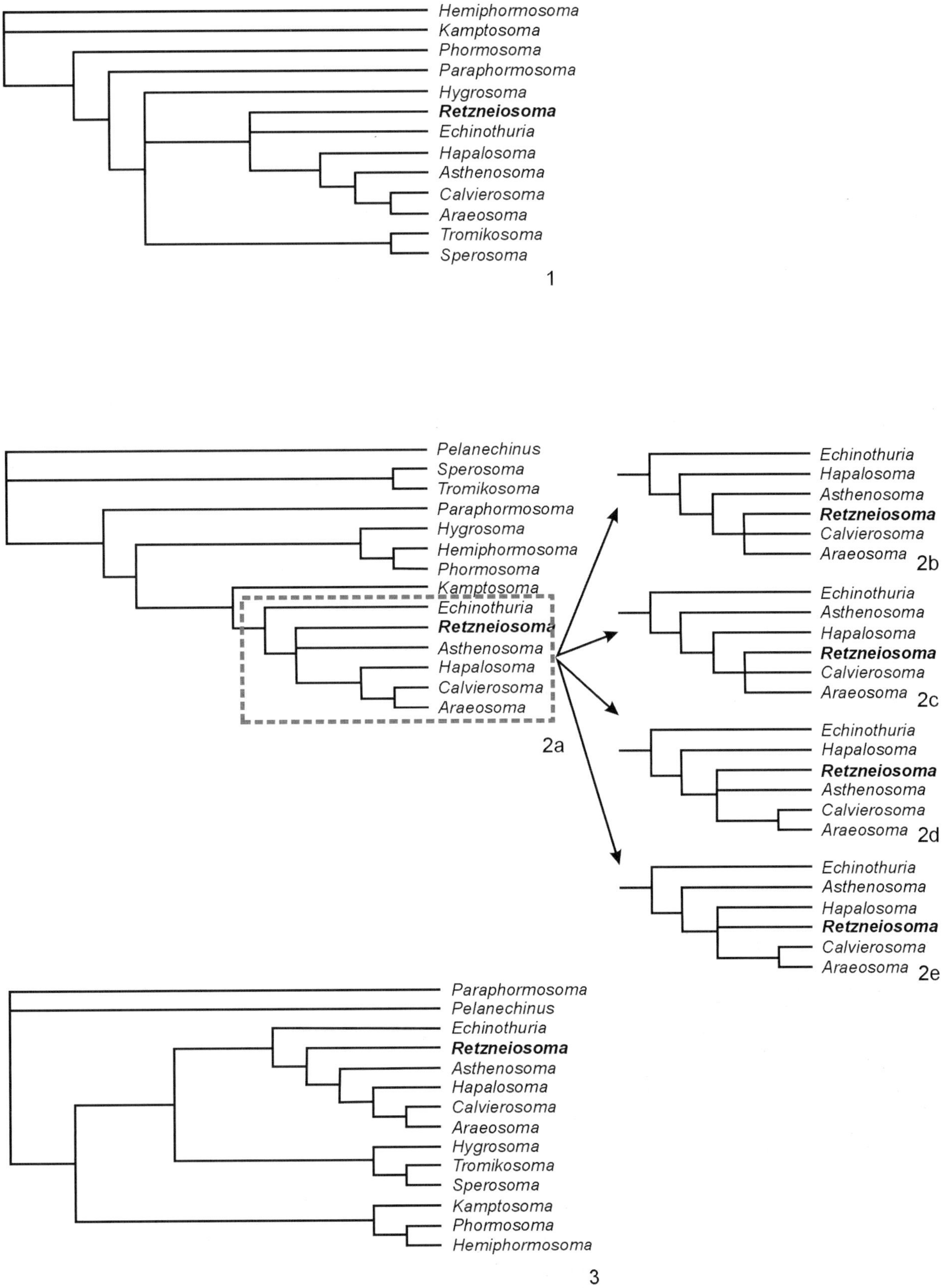

Figure 11: Cladistic analysis of *Retzneiosoma* utilising published characters and data matrices for the order Echinothurioida (Smith & Wright, 1990; Smith, "*The Echinoid Directory*", 12.11.2003). The placement of the new genus *Retzneiosoma* on the echinothurioid tree proposed by Smith & Wright (1990) (no. 1, single most parsimonious tree), Smith ("*The Echinoid Directory*", 12.11.2003) (nos. 2a-e, five equally parsimonious trees), and a single most parsimonious tree (no. 3) resulting from a modified set of characters (tabs. 6, 7).

curring in most members of the subfamily Echinothuriinae. The tuberculation is very spares. Each plate bears only a single primary tubercle, which takes up the full height of the plate. The tubercles are situated close to the adradial suture and form a more or less well defined vertical row (Fig. 10.A). Closer to the ambitus some of the primary tubercles may be situated outside this vertical row, lying more interradially. In some instances there may be more than one primary tubercle per plate. Oral tuberculation unknown.

Lantern and perignatic girdle: The lantern belongs to the aulodont type, with broad, longitudinally grooved teeth, and moderately open foramen magnum. The teeth have acuminate distal tips. The auricles have the form of broad inverted U's.

Primary spines: Primary spines attached to the single complete specimen are poorly preserved and their microstructure is obliterated by pyritization. In no instance a terminal hoof could be observed. Bulk samples from the same level from which the unique complete specimen was recovered yielded several cm long well preserved spines that are tentatively assigned to the same taxon. Additionally fragmented coronal plates with large crenulate perforate primary tubercles similar to those of the complete specimen were found in these bulk samples.

The spines have crenulate bases which fit in size very well to the tubercles observed in the complete specimen. The bases are moderately large, stout and have the shape of a small bucket. The shaft of the spines is very smooth with extremely narrow longitudinal slits which are barely visible, except in high magnification. These slits correspond to the borders between individual wedges of which the spine is construed as seen in cross-section (Pl. 9, Figs. 5-6). There are about 33 to 35 of these wedges which are attached to a hollow cylinder. The cylinder shows regularly arranged perforations forming a reticulate pattern (Pl. 9, Figs. 8-9). The central lumen is empty.

Discussion:

The systematic placement of the new genus has been discussed above. Extant species of the Echinothuriinae occur in the tropical and temperate regions all around the world. Most species are confined to deep sea habitats, although some species of *Asthenosoma* (e.g. *A. ijimai*) are also found in shallow depth. The echinothuriids are epibenthic scavengers, which seem to ingest mainly macroplant debris (Mortensen, 1935).

Apart from *Echinothuria floris* Woodward from the Santonian of England of which 8 specimens are known (Smith & Wright, 1990) *Retzneiosoma jaseneki* is the only other fossil echinothurioid known from "complete" tests. Disarticulated material, mainly spines, have been reported from the Santonian of France (*"Phormosoma" homoei* Lambert, 1907a), the Maastrichtian and Danian of Denmark (*"Araeosoma" mortenseni* Ravn, 1928, *"A." brunnichi* Ravn, 1928 and *"Asthenosoma" striatissimum* Ravn, 1928) and the Netherlands (*Echinothuria* ? sp., Jagt, 2000; Smith & Jeffery, 2000), the Miocene of Sardinia (*"Phormosoma" lovisatoi* Lambert, 1907a), and the Pliocene of New Zealand (*Araeosoma* aff. *thetidis* (Clark), Fell, 1966b; spine and test fragment).

The holotype of *Retzneiosoma jaseneki* is a largely complete, albeit crushed, corona with few spines and the lantern still associated. The specimen is pyritized and aboral and oral plates lie directly on top of each other. Since these animals are highly fragile and prone to rapid disintegration burial must have been fast or even take place while the animal was still alive. The oral side of the holotype is the side still attached to the sediment. It could not be examined without destroying the specimen.

Occurrence:

Austria: Early Badenian (Langhian)
Styrian Basin: Retznei (Lafarge quarry), Styria (Kroh, 2004b; [NHMW])

Infraclass Acroechinoidea Smith, 1981
Order Diadematoida Duncan, 1889
Family Diadematidae Peters, 1853

Diadematidae indet.

("*Diadema desori* Reuss, 1860")
(Pl. 8, Figs. 1-25)

mentioned as Diadematidae indet.:
? 1866a *Diadema* oder *Echinus* – Suess: 99
v. 1999 Diadematidae indet. – Kroh & Harzhauser: 155-156; pl. 8, figs. 1-7
v. 2002b Diadematidae indet. – Kroh: 12
v. 2003a Diadematidae indet. – Kroh: 160; pl. 2, figs. 1-5

mentioned as *Diadema desori* Reuss, 1860:
* 1860 ? *Diadema desori* m. n. sp. – Reuss: 222-223; pl. 3, figs. 3, 3a
1880 *Diadema desori* Rss. – Rzehak: 302
1892b *Diadema Desori* Rss. – Procházka: 342
1893 *Diadema desori* Reuss. – Toula: 289
1900 *Diadema Desori*, Rss. – Procházka: 21, 74
1900 ? *Diadema Desori*, Reuss. – Procházka: 146
1942 Stachelformen die *Diadema desorii* Reuss nahestehen – Toth: 525

mentioned as *Centrostephanus calarensis* Cotteau, 1905:
v. 1915 *Centrostephanus calarensis* Cott. Sp. – Vadász: 106; pl. 8 (2), fig. 11
1915 *Centrostephanus calariensis* Cott. sp. – Májer: 29, 82
1915 *Centrostephanus calariense* Cott. sp.– Májer: 35, 88
1969 *Centrostephanus calarensis* Cott. – Gábos & Ghiurca: 86-87; pl. 1, fig. 7 [undeterminable diadematid spine]
1977 *Centrostephanus calarensis* Cotteau, 1905 – Mączyńska: 195; pl. 1, figs. 8
1987 *Centrostephanus calarensis* Cotteau, 1905 – Mączyńska: 148
1988 *Centrostephanus calarensis* Cotteau, 1905 – Mączyńska: 60; pl. 1, fig. 4
1993 *Centrostephanus calarensis* Cotteau, 1905 – Mączyńska: 107; pl. 1, figs. 6-7

mentioned as *Centrostephanus rhodanicus* (Mayer-Eymar, 1910):
? 1987 *Centrostephanus rhodanicus* (Mayer-Eymar, 1910) – Mączyńska: 148

mentioned as *Centrostephanus airaghii* Lambert, 1907a:
v. 1915 *Centrostephanus Airaghii* Lamb. – Vadász: 106; pl. 8 (2), fig. 12
1915 *C.[entrostephanus] Airaghii* Lamk. – Májer: 35, 88
1928 *Centrostephanus Airaghii* Lamb. – Kutassy: 8
1969 *Centrostephanus airaghii* Lambert – Gábos & Ghiurca: 87; pl. 1, fig. 6 [undeterminable diadematid spine]

mentioned as *Centrostephanus* cf. *longispinus* Philippi, 1845:
1915 *Centrostephanus* cfr. *longispinus* Phil. sp.– Vadász: 106

mentioned as *Centrostephanus* sp.:
1955 *Centrostephanus* sp., regelmäßig skulpturiert – Tollmann: tab. 1
1955 *Centrostephanus* sp., unregelmäßig skulpturiert – Tollmann: tab. 1

misidentifications:
? 1977 *Cidaris* sp. div. – Mączyńska: 194; pl. 1, fig. 16

? 1987 *Plegiocidaris peroni* (COTTEAU, 1877) – MĄCZYŃSKA: 146, 148; pl. 1, figs. 2

? pp 1993 *Plegiocidaris peroni* (COTTEAU, 1877) – MĄCZYŃSKA: 106-107; pl. 6, fig. 1b

Remark: references from the outside the Paratethys are omitted.

Type-material:
Syntypes: 6 spine fragments (NHMW 1859.X.136); Naturhistorisches Museum Wien
Locus typicus: Rudolticé (= Rudelsdorf), Czech Republic
Stratum typicum: "Rudoltitzer Tegel"
Age: Badenian (Langhian-Early Serravallian), Middle Miocene

Material:
Late Eggenburgian (Early Burdigalian) – Eggenburg, NÖ, Austria
NHMW: 9 spine fragments (NHMW 2004z0110/0002, 0005-0007)
Late Eggenburgian (Early Burdigalian) – Grübern (Zogelsdorf Fm.), NÖ, Austria
NHMW: 225 spine fragments (NHMW 2003z0075/0001, 2003z0075/0004) and 9 interambulacral plates (NHMW 2003z0075/0010)
Late Eggenburgian (Early Burdigalian) – Limberg (Zogelsdorf Fm., Hengl quarry), NÖ, Austria
NHMW: 82 spine fragments (NHMW 2003z0002/0007)
Late Eggenburgian (Early Burdigalian) – Maissau (Zogelsdorf Fm.), NÖ, Austria
NHMW: 33 spine fragments (NHMW 1860.L.191b)
Late Eggenburgian (Early Burdigalian) – Unternalb (Retz Fm.), SE Retz, NÖ, Austria
NHMW: ~100 spine fragments and 5 coronal plates (NHMW 1999z0051/0019-22)
Ottnangian (Late Burdigalian) – Höbmannsbach, Taufkirchen Bay, ESE Schärding, OÖ, Austria
NHMW: 4 spine fragments (NHMW 1978/1966/12g, 12z)
Early Badenian (Langhian) – Bad Vöslau (?), NÖ, Austria
NHMW: 1 spine fragment (NHMW 1997z0178/1327)
Early Badenian (Langhian) – Eisenstadt (Hartl Fm.), Bgld, Austria
NHMW: 17 spine fragments (NHMW 1859.L.803), 4 spine fragments (NHMW 1869.L.803)
Early Badenian (Langhian) – Gainfarn, NÖ, Austria
NHMW: 1 interambulacral plate (NHMW 2004z0076/0021), 1 ambulacral plate (NHMW 2004z0076/0022), 5 spine fragments (NHMW 2004z0076/0023-26, ../0029)
Early Badenian (Langhian) – Niederleis, NÖ, Austria
NHMW: 3 interambulacral plates (NHMW 2002z0087/0021, 2002z0087/00023), 1 ambulacral plate (NHMW 2002z0089/0002), 12 spine fragments (NHMW 2002z0087/0018-20, 2002z0087/0022), 1 spine fragment (NHMW 2004z0001/0016b)
Early Badenian (Langhian) – Stotzing (sandpit Mayer), Bgld, Austria
NHMW: 2 spine fragments (NHMW 2004z0050/0003)
Badenian (Langhian-Early Serravallian) – Bruck an der Leitha, NÖ, Austria
NHMW: 26 spine fragments (NHMW 2004z0001/0021b)
Badenian (Langhian-Early Serravallian) – Enzesfeld, NÖ, Austria
NHMW: 1 spine fragment (NHMW 1981/56/7b)
Badenian (Langhian-Early Serravallian) – Grinzing, Vienna, Austria
NHMW: 6 spine fragments + 1 interambulacral plate (NHMW 1846.37.953b)
Badenian (Langhian-Early Serravallian) – old sandpit between Großhöflein and Kleinhöflein, near Eisenstadt, Bgld, Austria
NHMW: 57 spine fragments (NHMW 2003z0081/0001, ../0023-25), 1 epiphysis (NHMW 2003z0081/0038), and 11 test fragments (NHMW 2003z0081/0002)
Badenian (Langhian-Early Serravallian) – Mattersburg, Bgld, Austria
NHMW: 1 spine fragment (NHMW 1981/56/11b)
Badenian (Langhian-Early Serravallian) – Nussdorf, Wien, Austria
NHMW: 3 spine fragments (NHMW 2004z0001/0026)
Badenian (Langhian-Early Serravallian) – Oslip (sandpit E of the village), Bgld, Austria
NHMW: 1 spine fragment (NHMW 2003z0083/0001)
Badenian (Langhian-Early Serravallian) – Spielfeld, Styria, Austria
NHMW: 1 spine fragment (NHMW 2004z0001/0007)
Badenian (Langhian-Early Serravallian) – Steinebrunn (formerly Steinabrunn), NÖ, Austria
NHMW: 36 spine fragments (NHMW 1859.XLV.634), 23 spine fragments (NHMW 1865.I.612b), 33 spine fragments (NHMW 2004z0001/0024b)
Badenian (Langhian-Early Serravallian) –Teischl quarry, (? Hartl Fm., Eisenstadt, Bgld), Austria
NHMW: 1 interambulacral plate (NHMW 2004z0001/0008b)
Late Badenian (Early Serravallian) – Müllendorf (Mühlendorfer Kreide AG quarry), Bgld, Austria
NHMW: 2 spine fragment and 2 intermabulacral plates (NHMW 2003z0083/0001)

Foreign material:
Early Badenian (Langhian) – Coşteiu de Sus (= Kostej), Romania
NHMW: 2 spine fragments (NHMW 2004z0001/0003b)
Early Badenian (Langhian) – Kemence, Hungary
MAFI: *c.* 40 spine fragments [MAFI Ech 439 (material referred to as *Centrostephanus calarensis* COTTEAU by VADÁSZ, 1915)], 1 spine fragment [MAFI Ech 450 (material referred to as *Centrostephanus airaghii* LAMBERT by VADÁSZ, 1915)]
Early Badenian (Langhian) – Lăpugiu des Sus (= Lapugy), Romania
NHMW: 2 spine fragments + 1 interambulacral plate (NHMW 2004z0001/0004b), 5 spine fragments (NHMW 1859.XLV.617B.b), 7 spine fragments (NHMW 2004z0001/0027b), 1 spine (NHMW 2004z0001/0028b)
Badenian (Langhian-Early Serravallian) – Nový Rybik (=Portzteich), near Sedlec, W Lednice, Czech Republic
NHMW: 1 ambulacral + 1 interambulacra plate (NHMW 1858.XLVII.78b)
Badenian (Langhian-Early Serravallian) – Rudolticé (= Rudelsdorf), Czech Republic
NHMW: 6 spine fragments [NHMW 1859.X.136 (**type material** of *Diadema desori* REUSS, 1860)]
Late Badenian (Early Serravallian) – Buituri (= Bujtur), Romania
NHMW: 1 spine fragment (NHMW 2004z0001/0005b), 20 spine fragments + 1 interambulacral plate (NHMW 1852.II.1603b)

Remarks: It is highly probable that several species/genera are present in the material considered here. The unified discussion under the heading "Diadematidae indet." does not imply that they are conspecific. The problems outlined below, however, precludes separation into homogenous units currently.

Description:
Ambulacral plates: Each plate bears one to two large perforate, crenulate marginal tubercles in the middle of the plate. Inner tubercles are usually absent. The plates are trigeminate and show diadematoid compounding.
Interambulacral plates: Most of the plates bear a single, large, perforate, strongly crenulate primary tubercle. Rarely two primary tubercles are developed. The remaining surface of the

plates is usually smooth and lacks morphological details. In some cases few, small secondary tubercles are developed at the margin.

Spines: The spines are very fragile and invariably preserved as fragments only. Spine diameter varies from 0.5 to 2 mm, the fragments are usually between 1 and 10 mm in length. In cross-section they show the typical diadematoid structure: wedge-shaped radial septa attached to a perforated tube with an empty central lumen (Pl. 8, Fig. 25). In some spines, however, the central lumen is crossed by trabeculae (Pl. 8, Fig. 24) [similar to the irregular stereom meshwork reprted from spines of extant diadematids (e.g. *Astropyga*)]. The surface ornamentation of the spines consists of spinous processes arranged in spirals, producing a verticillate pattern. The bases of the spines are distinctly separated from the shaft by a prominent, often oblique crenulate ring.

Discussion:

Spine fragments and coronal plates of the family Diadematidae can be easily identified by their typical verticillate ornamentation and characteristic cross-section (spines, Pl. 8, Figs. 21-25), their large, in most cases strongly crenulate, perforate tubercles on large plates lacking any other ornamentation (interambulacral plates, Pl. 8, Figs. 8, 11, 21) and diadematoid compounding (ambulacral plates, Pl. 8, Figs. 7, 10, 22).

A more refined identification of these disarticulated remains is difficult. Donovan et al. (2001: 3-4), however, claim that diadematoid spines can be identified to genus or even species level. Some authors (e.g. Kutscher, 1985) have even established new species based on disarticulated diadematoid ossicles. Comparison of spine morphology of extant genera (*Diadema*, *Centrostephanus* and *Echinothrix*; compare Fig. 12), however, showed that ornamentation in these three genera is very similar and individual variation along the spine and between spines is strong. Confident identification of disarticulated spines is therefore extremely difficult, if possible at all. In any case the disarticulated material would have to be very well preserved and features seen in cross-section would have to be considered too. Thus, specific and generic attribution might be possible in Pleistocene settings (where comparison can often be restricted to a smaller range of taxa living in that particular region today) but could not be realised for the Miocene of the Central Paratethys, where a large number of extant genera could potentially occur. Moreover, variation of external spine morphology observed in individual samples is large, suggesting either the presence of several diadematoid taxa, or more likely variation along the spines.

In the palaeontological literature several diadematoid species based on isolated spine fragments are recorded from the Paratethys and the Mediterranean [*Diadema desori* Reuss, 1860, *Centrostephanus airaghii* Lambert, 1907a, *C. calarensis* Cotteau, 1905 (The original paper of Cotteau could not be located. Moreover, Cotteau was already deceased in 1905. This taxon is obviously first mentioned by Lambert, 1907a), *C. rhodanicus* (Mayer-Eymar, 1910) (Correct authorship is Lambert, 1910a; the name *rhodanicus* Mayer is a collection name and thus not valid according to the ICZN), *C.* cf. *longispinus* (Philippi, 1845), *Centrostephanus* sp.]. Differential diagnoses of these taxa are, however, insufficient and it is impossible at present to confidently relate the Austrian material to any of these taxa.

Disarticulated diadematoid interambulacral plates were sometimes misidentified as cidaroids [e.g. by Mączyńska (1977: pl. 1, fig. 16) as *Cidaris* sp. and later (1987: pl. 1, fig. 2; 1993: pl. 6, fig. 1b) as *Plegiocidaris peroni*].

Occurrence:

Austria: Late Eggenburgian (Early Burdigalian), Ottnangian (Late Burdigalian), Early to Late Badenian (Langhian-Early Serravallian)

Molasse Zone: Eggenburg, NÖ ([NHMW]); ? Gaudern-dorf, NÖ (Suess, 1866a); Grübern (Zogelsdorf Fm.), NÖ ([NHMW]); Höbmannsbach, ESE Schärding, OÖ ([NHMW]); Maissau (Zogelsdorf Fm.), NÖ ([NHMW]); Unternalb, near Retz, NÖ (Kroh & Harzhauser, 1999; [NHMW])

Vienna Basin: ? Bad Vöslau, NÖ ([NHMW]); Baden, NÖ (Reuss, 1860); Bruck an der Leitha, NÖ ([NHMW]); Enzesfeld, NÖ ([NHMW]); Gainfarn, NÖ (Kroh, 2002b; [NHMW]); Grinzing, W (Reuss, 1860; [NHMW]); Müllendorf (Mühlendorfer Kreide AG quarry), Bgld ([NHMW]); Niederleis, NÖ (Kroh, 2003a; [NHMW]); Nussdorf, W ([NHMW]); Steinabrunn, NÖ (Reuss, 1860; [NHMW]); Stotzing (sandpit Mayer), Bgld ([NHMW])

Gaaden Bay: roadcuts between Sittendorf-Sparbach-Weissenbach, NÖ (Toth, 1942)

Eisenstadt-Sopron Basin: Eisenstadt, Bgld (Tollmann, 1955); Eisenstadt (Hartl Fm.), Bgld ([NHMW]); Großhöflein, Bgld (Tollmann, 1955); Mattersburg, Bgld ([NHMW]); Oslip, Bgld ([NHMW])

Styrian Basin: Spielfeld, Styria ([NHMW])

Paratethys (non-Austrian occurrences): Ottnangian (Late Burdigalian) to Late Badenian (Early Serravallian)

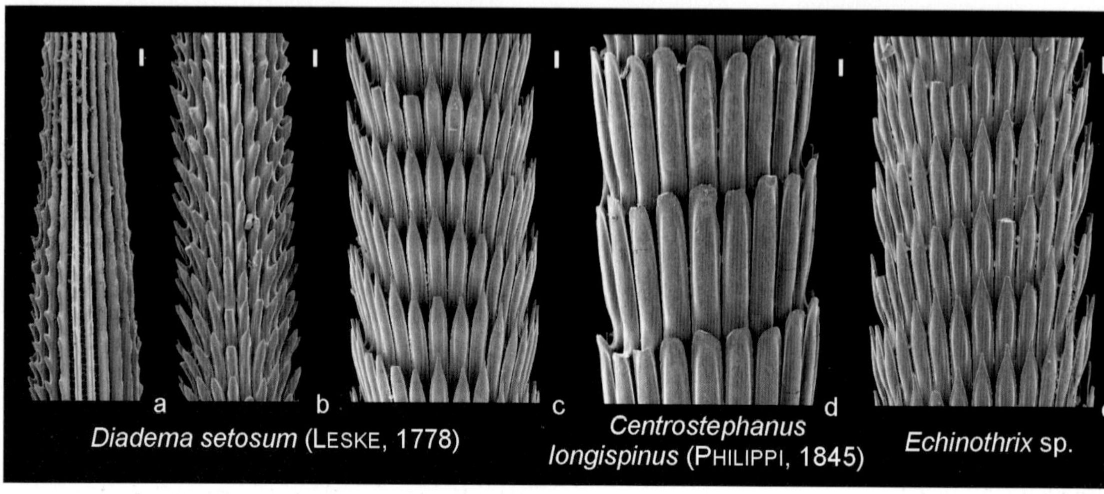

Figure 12: Spine ornamentation of extant Diadematidae [a-c: *Diadema setosum* (Leske, 1778), distal (a, b) and proximal part of the shaft (c); d: *Centrostephanus longispinus* (Philipi, 1845), proximal part; e: *Echinothrix* sp., proximal part]. Scale bars equal 100 µm.

Molasse Zone: Kralice nad Oslavou (= Kralitz), Czech Republic (Toula, 1893); Nový Rybik (= Portzteich), near Sedlec, W Lednice, Czech Republic ([NHMW]); Rudolticé (= Rudelsdorf), Czech Republic (Reuss, 1860; Procházka, 1900; [NHMW]); Židlochovice (= Groß-Seelowitz) (*Leda* beds), Moravia, Czech Republic (Rzehak, 1880)

Great Hungarian Basin (Pannonian Basin): Hont, Börzsöny, Hungary (Májer, 1915; Vadász, 1915); Kemence, Pest, Hungary (Májer, 1915; Vadász, 1915; [MAFI]); Királd, Borsod-Abaúj-Zemplén, Hungary (Kutassy, 1928)

Fore-Carpathian Basin: Kików, Central Poland (Mączyńska, 1993); Korytnica Clays, Korytnica, Poland (Mączyńska, 1977, 1987); Pińczów, Central Poland (Mączyńska, 1993); Rybnica, Southern Poland (Mączyńska, 1988); Skowronno, Central Poland (Mączyńska, 1993); Szczaworyż, Central Poland (Mączyńska, 1993); Wieliczka, Poland (Reuss, 1860); Żerniki, Central Poland (Mączyńska, 1993)

Transylvanian Basin: Berchezoaia, Maramureş, Romania (Gábos & Ghiurca, 1969); Buituri (= Bujtur), Hunedoara, Romania (Vadász, 1915; [NHMW]); Ciceu-Hăşmaş, Sălaj, Romania (Gábos & Ghiurca, 1969); Coştei de Sus (= Kostéj), Timiş, Romania (Gábos & Ghiurca, 1969; [NHMW]); Lăpugiu des Sus (= Lapugy), Hunedoara, Romania (Vadász, 1915; Gábos & Ghiurca, 1969; [NHMW]); Livezile-Aiud, Hunedoara, Romania (Gábos & Ghiurca, 1969); Valea Satului, Romania (Gábos & Ghiurca, 1969)

Cohort Echinacea Claus, 1876

Identification of small regular echinoids of the cohort Echinacea is notoriously difficult. Most outcrops surveyed during the present study yielded mainly poorly preserved, disarticulated material or surface details were obscured by syntaxial rim cement. Earlier workers used mainly size and the number of vertical rows of the interambulacra and ambulacra (e.g. Laube, 1871). The latter feature, however, is strongly size dependant, as the number of large, subequal tubercles increases during growth, as can be seen in growth series of e.g. extant *Sphaerechinus granularis* or *Tripneustes ventricosus*. Thus very small specimens were usually classified as *Arbacina*, those of medium size (~1.5 – 4 cm test diameter) as *Psammechinus* and larger ones as *Echinus*. Although, this method resulted in more or less reproducible determinations, it is highly artificial and does not result in natural groupings. Specific determinations of echinaceans in Laube (1871) are often very arbitrary and could not be reproduced in all instances, even in cases, where the original material could be located.

Owing to these difficulties a very different approach was followed during the present study. Scanning electron microscopic (SEM in the following) images of fossil material from Austria was compared with SEM images of (related) extant taxa (e.g. *Psammechinus microtuberculatus*, *P. miliaris*, *Sphaerechinus granularis*, *Paracentrotus lividus*, *Arbacia lixula*, *Echinometra mathaei*, *Tripneustes gratilla*, *T. ventricosus*, *Genocidaris maculata*, ...) as well as fossil species from the Mediterranean region. Special emphasis was placed on the arrangement of the primary and secondary tubercles, the number and arrangement of the ambulacral pores and (if preserved) the nature of the apical disc.

Superorder Camarodonta Jackson, 1912
Order Temnopleuroida Mortensen, 19412
Family Temnopleuridae Agassiz, 1872
Genus *Arbacina* Pomel, 1869

Type-species: *Echinus monilis* Desmarest, 1816; by original designation (Pomel, 1869: XLI)
Diagnosis: Small hemispherical or subconical corona; no angular pits or pores; tubercles circular, imperforate; sutural depressions in the horizontal suture (see remarks); dense secondary tuberculation (modified from Mortensen, 1943a and Fell & Pawson, 1966)
Distribution: Early Miocene to Pliocene – circum-Mediterranean, Paratethys and western Africa
Remarks: One of the characteristic features of *Arbacina* according to Mortensen (1943a) are sutural depressions along the horizontal sutures of the interambulacra. The term "sutural depressions" is, however, somewhat misleading in this context. In fact these are not true depressions in the plate surface as in e.g. *Temnopleurus*. Instead these "sutural depressions" are areas along the horizontal sutures which are free of secondary tuberculation. Since the secondary tuberculation is very dense everywhere else, with the areoles of the tubercles often touching each other, these areas look like depressions, but in fact are not. This is especially evident in large specimens of *Arbacina* from the Pliocene of Rhodes Island, which has numerous subequal secondary tubercles.

Arbacina catenata (Desor in Agassiz & Desor, 1846)

(Pl. 11, Figs. 1-2)

*	1846	[*Echinus* (*Psammechinus*)] *catenatus* Desor. – Desor in Agassiz & Desor: 369, No. T 69
	1858	[*Psammechinus*] *catenatus* – Desor: 122
	1910a	*Arbacina catenata* Desor (*Psammechinus*), 1846. – Lambert: 27-28; pl. 1, figs. 52-58
	1910a	*Arbacina tenera* de Loriol, 1902 – Lambert: 28-29; pl. 1, figs. 59-62
#	1910a	*Arbacina Savini* Lambert. – Lambert: 30; pl. 1, figs. 71-73 [fide Philippe 1998]
? v	1915	*Arbacina tenera* Lor. – Vadász: 108-109
?	1930	*Arbacina tenera* Lor. – Vendl: 77
	1943a	*Arbacina catenata* (Desor) – Mortensen: 366; fig. 224a
?	1987	*Arbacina catenata* (Desor, 1847) – Mączyńska: 146, 148; pl. 2, figs. 3a-c
?	1993	*Arbacina catenata* (Desor, 1847) – Mączyńska: 108; pl. 2, fig. 1a-c
.	1998	*Arbacina catenata* (Desor, 1846) – Philippe: 55-58; pl. 7, figs. 1-8; pl. 8, figs. 1-2
v.	1999	*Arbacina* – Harzhauser & Kroh: 221
v.	1999	*Arbacina* sp. – Kroh & Harzhauser: 156-158; fig. 4; pl. 7, figs. 5-7
non	2003a	*Genocidaris catenata* (Desor, 1846) – Kroh: 161-162; figs. 2c-d; pl. 3, figs. 1-6 [misidentification]

Type-material:
Provenience and current whereabouts unknown. The only information available is found in Desor (1858: 122): "T. 69. (Original de l'espèce.) Tertiaire (Molasse?) du Midi. Très rare. Mus. d'Avignon.". Philippe (1998: 55), however, was not able to trace the type specimen in French collections. According to him it seems probable that the plaster cast T69 in the collection of Neuchâtel was made from a specimen coming from the Burdigalian of Avignon, France.

Material:
Late Eggenburgian (Early Burdigalian) – Gauderndorf (Himmelreich), NÖ, Austria
NHMW: 1 corona (NHMW 1869.III.41)
Late Eggenburgian (Early Burdigalian) – Limberg (Zogelsdorf Fm., Hengl quarry), near Maissau, NÖ, Austria
NHMW: 1 corona (NHMW 1998z0048/0048)
Late Eggenburgian (Early Burdigalian) – Roggendorf [Zogelsdorf Fm., Teufelslucke (cave)], NÖ, Austria
NHMW: 1 corona (NHMW 2004z0001/0018)
Late Eggenburgian (Early Burdigalian) – Unternalb (Retz Fm.), SE Retz, NÖ, Austria

NHMW: 22 coronas and about one hundred test fragments (NHMW 1999z0051/0016 – 18)

Dimensions:
Largest whole specimen: diameter: 7.1 mm; height: 4.2 mm
estimate average diameter: 3.6 mm; estimate average height: 2.1 mm

Description:
Size and shape: The test is very small, hemispherical with circular outline. In profile the test is domed, with a tumid ambitus. The oral surface is flattened.
Apical system: The apical system is lacking in all specimens. The apical area is about one quarter of the test diameter wide.
Ambulacra: The ambulacra are about half to two third as wide as the interambulacra. The ambulacral plates are trigeminate and belong to the echinoid compound type. The pores are partitioned isopores (type P2 of SMITH, 1978), which are arranged in very slightly undulating vertical series. Each plate bears one large imperforate, noncrenulate (?) marginal tubercle with a distinct boss and a globular mamelon with undercut neck. The bases of the marginal tubercles are indented. Along the adapical and perradial border of each ambulacral plate, several imperforate, noncrenulate inner tubercles are found. Two to three of these situated at the edges of the plates are distinctly enlarged, the remaining ones are small.
Interambulacra: Each interambulacral plate bears one large imperforate, noncrenulate (?) primary tubercle with a distinct boss and a large globular mamelon with an undercut neck. The bases of the primary tubercles are indented. The primary tubercles lie approximately halfway along the adoral suture of each plate. Secondary tuberculation is very dense, leaving little or no space between the areoles of the secondary tubercles.
Peristome: Situated centrally on the ventral surface; circular in outline, with very shallow gill slits. The peristome is about half as wide as the test diameter in specimens of 3 to 6 mm TD.

Differential diagnosis:
According to PHILIPPE (1998: 58) *Arbacina piae* LOVISATO in COTTEAU, 1895 differs from *A. catenata* by its less globular shape, less homogenous secondary tuberculation and circular arrangement of secondary tubercles around the primary tubercles. Illustrations in COTTEAU (1895: pl. 3, figs. 1-6) are, however, rather poor and re-description/new illustration of *A. piae* are needed for comparison with *A. catenata*. Material from the Messinian Upper Coralline Limestone Formation of the Maltese Islands referred to *A. piae* by CHALLIS (1980) is a different species with differing tubercle arrangement.

A. mutellensis DE LORIOL, 1896 differs by its more numerous and stronger projecting tubercles (PHILIPPE 1998).

A. jacquemeti LAMBERT, 1910a differs from *A. catenata* by its more numerous tubercles in ambital coronal plates and different tubercle arrangement (compare PHILIPPE, 1998: pl. 8, figs. 3-4).

Discussion:
The present material is highly similar to material attributed to *A. catenata* by PHILIPPE (1998). Only the weak crenulation of the marginal and primary tubercles seen in pl. 8, fig. 1b-c in PHILIPPE (1998) could not be observed in the Austrian material. This may, however, be related to the poor preservation of the Austrian specimens.

Sutural depressions along the horizontal sutures of the interambulacra, which MORTENSEN (1943a) regarded as critical feature of this genus, could not be observed. Nevertheless, the specimens are placed into the genus *Arbacina*. As PHILIPPE (1998: 57) showed recently, this feature is size-dependant and does not occur in smaller specimens. This could also be confirmed in Pliocene *Arbacina* specimens from Rhodes Island (Greece) in the collection of the NHMW, where the so called "sutural depressions" (see above under the remarks to the genus *Arbacina*) are only developed in larger specimens (*c.* 10 mm test diameter). According to MORTENSEN (1943a) indentations in the margin of the primary and marginal spine bases are not present in *Arbacina*. Yet such indentations, albeit often faint, can be observed in the present specimens and in that of the Rhône Basin (compare PHILIPPE, 1998: pl. 8, figs. 1b-c, 2b-c). This suggests that this feature is not well suited as a sole distinction between these two genera.

In a recent revision of the echinoids of the Rhône Basin, PHILIPPE (1998) synonymised *Arbacina tenera* DE LORIOL, 1902 and *A. savini* LAMBERT, 1910a with this species, *A. tenera* representing juvenile specimens and *A. savini* representing very large ones.

KROH (2003a: 161-162) referred this species to the genus *Genocidaris* based on disarticulated material from the Early Badenian of the northern Vienna Basin. This is, however, resulted from a misidentification of these specimens (see below under *Genocidaris* sp.). Additional material and comparison with supplementary fossil and extant material from various localities suggests that while the Niederleis material does belong to the genus *Genocidaris*, it is not conspecific with *A. catenata*.

Occurrence:

Although repeatedly reported in the literature no Badenian specimens of *A. catenata* could be located. It is thus unclear if this species is really present in the Badenian of the Paratethys or if these records are misidentifications (as e.g. the record of KROH, 2003a is).

Austria: Late Eggenburgian (Early Burdigalian), ? Badenian (Langhian to Early Serravallian)
Molasse Zone: Gauderndorf (Himmelreich), NÖ ([NHMW]); Limberg (Zogelsdorf Fm., Hengl quarry), near Maissau, NÖ ([NHMW]); Teufelslucke (Zogelsdorf Fm.), near Roggendorf, NÖ ([NHMW]); Unternalb (Retz Fm.), near Retz, NÖ (HARZHAUSER & KROH, 1999; KROH & HARZHAUSER, 1999)

Paratethys (non-Austrian occurrences): Eggenburgian (Early Burdigalian), ? Badenian (Langhian-Early Serravallian)
Eisenstadt-Sopron Basin: ? Fertörákos, near Sopron, Györ-Monson-Sopron, Hungary (VENDL, 1930); Sopron-Rákos, Györ-Monson-Sopron, Hungary (VADÁSZ, 1915)
Great Hungarian Basin (Pannonian Basin): ? Budapest-Rákos, Pest, Hungary (VADÁSZ, 1915); ? Mátraverebély, Nógrád, Hungary (VADÁSZ, 1915); ? Sámsonháza, Nógrád, Hungary (VADÁSZ, 1915)
Fore-Carpathian Basin: ? Korytnica Clays, Korytnica, Poland (MĄCZYŃSKA, 1987); ? Pińczów, Central Poland (MĄCZYŃSKA, 1993)

Mediterranean: Burdigalian to Early Langhian
Western Mediterranean: Vence, France: Ste-Colombe and Gattières near Vence (LAMBERT, 1910a)
Rhône Basin: Burdigalian: Secteur des étangs (Fos-sur-Mer, Istres, Saint-Mitre-les-Remparts) (PHILIPPE, 1998), Littoral de Nerthe (Plan d'Arren) (LAMBERT, 1910a), Bassin de Mus-Sommières (Coustourelle, near Sommières) (PHILIPPE, 1998), Secteur de Montpellier (Juvignac) (PHILIPPE, 1998), Bassin d'Avignon (Les Angles, Avignon, Beaucaire, Vedène, Villeneuve-lès-Avignon) (LAMBERT, 1910a; PHILIPPE, 1998), Bassin d'Apt-Reillanne-Forcalquier (Bonnieux, Lacoste, Viens) (PHILIPPE, 1998), Bassin de Faucon-Mollans-Malaucéne (Vaison-la-Romaine) (PHILIPPE, 1998), Bassin de Valréas-Visan (Grignan, Montségur-sur-Lauzon, Nyons, Solérieux, Taulignan) (LAMBERT, 1910a; PHILIPPE, 1998), Dauphiné (Sainte-Marie-d'Alvey) (LAMBERT, 1910a; PHILIPPE, 1998), Late Burdigalian (faciès "Marnes Bleues"): Bassin de Faucon-Mollans-Malaucène (Entrechaux) (PHILIPPE, 1998)

Arbacina cf. *macrophyma* Lambert, 1910
(Pl. 11, Figs. 3-4; Pl. 12, Figs. 1-6)

cf. 1910a *Arbacina macrophyma* Lambert. – Lambert: 31; pl. 2, figs. 2-5

? 1950 *Arbacina monilis* Desmanest. – Szörényi: 140; pl. 1, figs. 1, 1a-b

Material:
Late Badenian (Early Serravallian) – Müllendorf (Mühlendorfer Kreide AG quarry), Bgld, Austria
NHMW: 79 juvenile coronas (NHMW 2004z0006/0001-0003, 0005-0012, 0018-0020, 2004z0045/0004)
GBA: 82 coronas (GBA 2004/1/3)
Late Badenian (Early Serravallian) – Winden (old quarry, N of the Ludlloch (cave), N of the village), Bgld, Austria
NHMW: 1 corona (NHMW 2004z0001/0025b)

Dimensions:
Test diameter ranging from 4.2 to 8.3 mm, most common between 5.5 and 7.7 mm
Test height ranging from 2.1 to 3.9 mm, most common between 2.9 to 3.7 mm

Description:
Size and shape: The test rarely exceeds 7 mm in diameter and is hemispherical with circular outline. In profile the test is domed, with a tumid ambitus. The oral surface is slightly flattened.
Apical system: The apical system is dicyclic with the oculars widely exsert. The madreporite bears only few madreporic pores most of which are situated near the suture to ocular III and interambulacrum 2. The apical area is about one quarter of the test diameter wide. Each genital plate bears a single large tubercle, on the madreporite (genital plate 2) there may also be two. In one specimen (Pl. 11, Fig. 4a) the genital plates are ornamented by ridges.
Ambulacra: The ambulacra are about two third as wide as the interambulacra. The ambulacral plates are trigeminate and belong to the echinoid compound type. The pores are partitioned isopores (type P2 of Smith, 1978), which are arranged in arcs of three. Each plate bears one large imperforate, noncrenulate marginal tubercle. The bases of the marginal tubercles are not indented. Secondary tuberculation moderately dense, consisting of rather large imperforate, noncrenulate inner tubercles.
Interambulacra: Each interambulacral plate bears one large imperforate, noncrenulate primary tubercle with a large globular mamelon with undercut neck. The bases of the primary tubercles are not indented. The primary tubercles lie approximately halfway along the adoral suture of each plate. Secondary tuberculation is moderately dense. Secondary tubercles along the adapical suture may be distinctly enlarged.
Peristome: The peristome is about half as wide as the test diameter, but ratio decreases as the animals grow. It has a circular outline. Gill slits are not discernible.

Discussion:
The studied specimens are tentatively assigned to *Arbacina macrophyma* Lambert, 1910a. Unfortunately, no re-description and new illustrations of this species exist. All subsequent authors simply refer to Lambert's original description and illustrations. The present specimens from the Badenian of Müllendorf differ from *A. catenata* reported here from the Eggenburgian by its different tuberculation. In the Badenian species the primary tubercles are larger, show no indented bases and less dense secondary tuberculation (compare Pl. 11, Figs. 1d and 2 with Pl. 11, Fig. 3).

This species is easily confused with co-occurring juvenile *Schizechinus* sp. The latter may be recognised by its hemicyclic apical disc, less dense secondary tuberculation with large areas free of tubercles in the interambulacra. These areas are covered by small granulae not present in *Arbacina* (compare Pl. 12, Fig 7). Additionally the arrangement of the ambulacral pores in arcs of three is much more strongly developed in *Schizechinus* (compare Pl. 12, Fig. 7).

Occurrence:
Austria: Late Badenian (Early Serravallian)
Vienna Basin: Müllendorf (Mühlendorfer Kreide AG quarry), Bgld ([GBA]; [NHMW])
Danube Basin: Winden, Bgld ([NHMW])

Paratethys (non-Austrian occurrences): ? Badenian (Langhian-Early Serravallian)
Great Hungarian Basin (Pannonian Basin): ? Magyaregregy, Baranya, Hungary (Szörényi, 1950)

Genus *Genocidaris* Agassiz, 1869

Type-species: *Genocidaris maculata* Agassiz, 1869; by original designation (Agassiz, 1869: 262).
Diagnosis: Small hemispherical corona; noncrenulate, imperforate primary tubercles with indented bases; radiating sculpture (ridges) around primary and marginal tubercles; apical disc dicyclic with widely exsert oculars; single plate covering periproct (modified from Mortensen, 1943a and Fell & Pawson, 1966)
Distribution: Miocene to Recent – Caribbean Sea, North Atlantic Ocean, Mediterranean Sea, Central Paratethys
Ecology and biogeography: *G. maculata*, the only extant species of this genus, is found on sands with coralline algae and muddy sands with *Peyssonnelia* between 12 to 500 metres water depth in the Mediterranean (Riedl, 1983). Mortensen (1943a: 263) found mainly bryozoan fragments in the intestines of this species.

Genocidaris sp.
(Pl. 13, Figs. 1-12)

v. 2001 *Brochopleurus* sp. – Schmid et al.: 13 [misidentification]

v. 2003a *Genocidaris catenata* (Desor, 1846) – Kroh: 161-162; figs. 2c-d; pl. 3, figs. 1-6 [misidentification]

Material:
Early ? Badenian (Langhian) – Rauchstallbrunngraben (bryozoan marl), near Baden, NÖ, Austria
NHMW: 9 test fragments (NHMW 2004z0121/31-39)
Early Badenian (Langhian) – Niederleis, NÖ, Austria
NHMW: 6 test fragments (NHMW 2002z0087/0053, 2002z0089/0053, 2002z0090/0003-6)
Badenian (Langhian-Early Serravallian) – old sandpit between Großhöflein and Kleinhöflein, Bgld, Austria
NHMW: 2 test fragments (NHMW 2003z0081/0013-14)
Late Badenian (Early Serravallian) – St. Margarethen (Kummer quarry), Bgld, Austria
NHMW: 2 specimens (NHMW 1986/98, 1998/138/128)

Dimensions (in mm):

Inv. no.	test ø	apic. area ø	ps ø
NHMW 1986/98	8.6	-	4.1
NHMW 1998/138/128	9.3	2.4	-

Description:
Size and shape: The test is very small, hemispherical with a circular outline. In profile, the test is domed, with a tumid ambitus.

<u>Apical disc</u>: Apical disc not preserved; diameter of apical area about 25 % of the test diameter.
<u>Ambulacra</u>: The ambulacra are about two-third to half the width of the interambulacra. The ambulacral plates are trigeminate with pores arranged in indistinct arcs of three and belong to the echinoid compound type. The pores are oblique P1 to P2 isopores (compare Smith, 1978). Individual pore pairs are separated by granulated ridges. Each ambulacral plate bears one large, imperforate marginal tubercle, which lies slightly adorally of the centre of each plate. The bases of the marginal tubercles are distinctly indented (see Pl. 13, Fig. 1). Along the adapical and perradial border of each ambulacral plate, several imperforate, noncrenulate inner tubercles are seen, two to four of which are enlarged at the ambitus. The base of the marginal tubercles is connected to the bases of the inner tubercles by low ridges, producing a radiating pattern.
<u>Interambulacra</u>: Each interambulacral plate bears one large, imperforate primary tubercle. The bases of the primary tubercles are distinctly indented (see Pl. 13, Figs. 2, 5, 7). Similar to the ambulacra, the bases of the primary tubercles are connected to the bases of the secondary tubercles by low ridges, producing a radiating pattern. The secondary and miliary tuberculation is dense, consisting of small noncrenulate, imperforate tubercles.
<u>Peristome</u>: The peristome lies centrally on the oral side, is large, about 50 % of the test diameter, and circular. The gill slits are shallow and indistinct.

Discussion:
Apart from two whole specimens from the Kummer quarry in St. Margarethen only fragmented and/or disarticulated material is available. Despite this unfavourable preservation even individual plates are usually readily recognisable based on the radiating pattern produced by the low ridges connecting the primary/marginal tubercle with the secondary/inner tubercles. Based on this radiating pattern, the indented tubercles and the lack of sutural pits the material is placed into the genus *Genocidaris*. Species of the genus *Arbacina* may have a very similar tubercle arrangement but lack the distinct radiating pattern and the strong indentations in the bases of the primary respectively marginal tubercles (there may be faint indentations in *Arbacina*, contrary to the statement of Mortensen, 1943a). Two other extant genera show a similar ornamentation: *Lamprechinus* and *Trigonocidaris*. The latter, however, has crenulate tubercles (at least some). Judging from the illustrations in Mortensen (1943a) *Lamprechinus* has less secondary tubercles and longer ridges (no material of these two genera was available for SEM analysis).

In most specimens the primary and marginal tubercles are weakly crenulate. Some, however, are obviously noncrenulate (Pl. 13, Fig. 3b) and others show the crenulation very well (Pl. 13, Fig. 5b). A similar situation is present the Pliocene *Genocidaris maculata* from Northern Italy, which usually has noncrenulate tubercles, but commonly shows crenulation on some or all tubercles in some specimens (pers. comm. E. Borghi, March 2004; see Fig. 13).

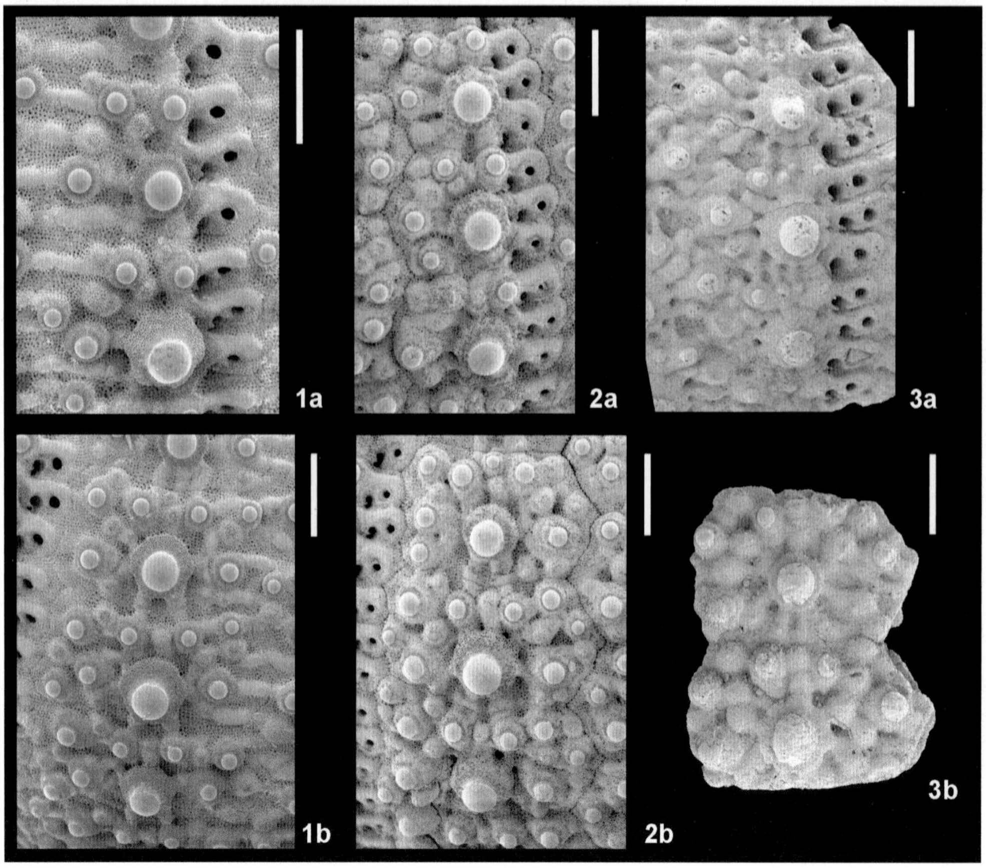

Figure 13: Comparison of the ornamentation in *Genocidaris maculata* (Agassiz, 1869) (1a-b; recent, off Florida Keys, Gulf of Mexico, USNM E12092), *Genocidaris maculata pliorecens* Borghi, 1995 (2a-b; Pliocene of Castell'Arquato, Northern Italy; NHMW coll.), and *Genocidaris* sp. (3a-b; Badenian of Rauchstallbrunngraben, near Baden, NÖ; NHMW 2004z0121/0036 and ../0034). Figures "a" show the ambulacra, "b" the interambulacra; scale bars equal 0.5 mm.

Kroh (2003a: 161-162) determined fragmented material of this species as *Arbacina catenata* (Desor, 1846) and transferred that species the genus *Genocidaris*. This, however, was an error as subsequent investigations of new material from additional localities showed. While it was correct to place the specimens into the genus *Genocidaris*, they are not conspecific with the species of Desor (1846).

Up to now this genus included a single extant species: *G. maculata* Agassiz, 1869, restricted to the Atlantic Ocean and the Mediterranean Sea. For fossil specimens from the Pliocene of Castell'Arquato, Italy the subspecies *G. maculata pliorecens* Borghi, 1995 was established (Borghi, 1995).

Occurrence:

Austria: Early to Late Badenian (Langhian-Early Serravallian)
Vienna Basin: Niederleis, NÖ (Kroh, 2003a; [NHMW]); Rauchstallbrunngraben, near Baden, NÖ ([NHMW])
Eisenstadt-Sopron Basin: between Großhöflein and Kleinhöflein, near Eisenstadt, Bgld ([NHMW]); St. Margarethen (Kummer quarry), Bgld (Schmid et al., 2001; [NHMW])

Order Echinoida Claus, 1876
Family Echinidae Gray, 1825
Genus *Psammechinus* Agassiz & Desor, 1846

Type-species: *Echinus miliaris* Müller; by subsequent designation (Lambert & Thiéry, 1910: 239)
Diagnosis: Ambitus is circular; ambulacral plates trigeminate with primary tubercles on each; buccal membrane densely plated, with contiguous or even imbricated plates; secondary spines numerous and smooth; apical disc dicyclic (Fell & Pawson, 1966).
Distribution: Miocene to Recent – North Atlantic and Mediterranean
Ecology and biogeography: The two extant species of Psammechinus are restricted to the Mediterranean (*P. microtuberculatus*) and the eastern Atlantic Ocean (*P. miliaris*). They occur most commonly between 4 and 100 m water depth. *P. miliaris* may also inhabit intertidal settings and *P. microtuberculatus* is most common in the *Zostera* zone. Both feed on algae and small invertebrates (Mortensen, 1943b: 134-148).

Psammechinus dubius dubius (Agassiz, 1840)
(Pl. 14, Figs. 1-4)

*	1840a	*Echinus dubius* Ag. – Agassiz: 84-85; pl. 22, figs. 4-6
	1840b	[*Echinometra*] *margaritifera* Nic. – Agassiz: 12 [fide Lambert & Jeannet, 1928: 147, 165-166]
	1840b	[*Echinus*] *dubius* Ag. – Agassiz: 12
	1875	*Psammechinus dubius*, Agassiz. – de Loriol: 29-30; pl. 2, figs. 6, 6a-c, 7, 7a-b
#	1912a	*Psammechinus extraalpinus* Schaff. – Schaffer: 187; pl. 59, figs. 7-10
	1913a	*Psammechinus extraalpinus* Schff. –Schaffer: 10, 84-85, 154-155; pl. 10, figs. 7-8
v pp	1915	*Psammechinus dubius* Ag. – Vadász: 110
	1971a	*Psammechinus* (?) *extraalpinus* Schaffer, 1912 – Steininger: 593-594; pl. 1, figs. 3-4
	1971b	*Psammechinus* (?) *extraalpinus* Schaffer – Steininger: 118
	1998	*Psammechinus dubius* (Agassiz, 1840) – Philippe: 65-67; pl. 8, figs. 5-8 [cum syn.]
	2003b	*Psammechinus dubius* (L. Agassiz, 1840) – Kroh: 249

Type Material:
Echinus dubius Agassiz, 1840:
Syntypes: figured in Agassiz (1840a: pl. 22, figs. 4-6); current whereabouts unknown
Locus typicus: Chaux-de-Fonds, Switzerland
Stratum typicum: ? Upper marine Molasse (= Obere Meeresmolasse)
Age: Middle Burdigalian

Psammechinus extraalpinus Schaffer, 1912:
Syntypes: figured in Schaffer (1912a: 187; pl. 59, figs. 7-10); Krahuletz Museum, Eggenburg, NÖ
Locus typicus: Reinprechtspölla, NÖ
Age: Late Eggenburgian (Early Burdigalian)

Material:
Late Eggenburgian (Early Burdigalian) – Reinprechtspölla, NÖ, Austria
NHMW: 16 specimens (NHMW 1997z0178/1832a-d, 1914.VII.55a-j, 2002z0181/0003)
IPUW: 3 specimens (no inventory no.)
KM: 4 specimens [KM22-24, 65 (**syntypes** of *Psammechinus extraalpinus* Schaffer, 1912a)]
Late Eggenburgian (Early Burdigalian) – Zogelsdorf (Zogelsdorf Fm., Johannisbruch), NÖ, Austria
NHMW: 1 specimen (NHMW 2002z0182/0001)

Foreign material for comparison:
Obere Meeresmolasse (Middle Burdigalian) – Hohburggraben, Belpberg, Switzerland
NMB: 16 specimens (NMBE 5011426-5011440, 5011442)
Obere Meeresmolasse (Middle Burdigalian) – between Seitenberg and Studweidholz, N Bütschelegg, Switzerland
NMB: 1 specimen (NMBE 5011444)
Obere Meeresmolasse (Middle Burdigalian) – Seitenberg, near Niedermuhlern, Switzerland
NMB: 1 specimen (NMBE 5011443)
Karpatian (Late Burdigalian) – Drégely, Nógrád, Hungary
MAFI: 2 specimens (MAFI Ech 447)

Dimensions:

Inv. No.	test ø	TH
NHMW 1997z0178/1832a	28.9	18.2
NHMW 1997z0178/1832b	22.5	13.5
NHMW 1997z0178/1832c	~21.0	13.8
NHMW 1997z0178/1832d	20.5	10.4
NHMW 1914.VII.55a	21.6	12.3
NHMW 1914.VII.55b	26.3	14.7
NHMW 1914.VII.55c	19.1	9.8
NHMW 1914.VII.55d	18.0	9.6
NHMW 1914.VII.55e	24.9	-
NHMW 1914.VII.55f	20.0	9.7
NHMW 1914.VII.55g	26.5	15.8
NHMW 1914.VII.55h	24.6	~13.6
NHMW 1914.VII.55i	18.3	-
NHMW 1914.VII.55j	26.6	~ 15

Description:
Size and shape: Test of small to medium size, the test diameter ranging from 18.0 to 28.9 mm in the investigated specimens. The outline of the corona is circular. The aboral surface is domed, the oral surface flattened and depressed around the peristome. The ambitus is tumid.
Apical disc: The apical disc is not preserved in any specimen. The apical area is small, occupying no more than 20 % of the test diameter.
Ambulacra: The ambulacra consist of trigeminate plates of the echinoid compound type. The pores are partitioned isopores of the P2 type and are arranged in arcs of three. Each plate bears one large noncrenulate, imperforate marginal tubercle and several smaller inner tubercles. The most perradial inner tu-

bercle is distinctly enlarged. The interporiferous zone of a single ambulacral column is about one and a half to two times as wide as the poriferous zone at the ambitus.
Interambulacra: The interambulacra are about one and a half times as wide as the ambulacra at the ambitus. Each plate bears one noncrenulate, imperforate primary tubercle, which lies slightly below the centre of the plate. The secondary tuberculation is dense. Two to four (at the ambitus) secondary tubercles are distinctly enlarged on each side of the primary tubercle. These enlarged secondary tubercles and the primary tubercle are arranged in a more or less horizontal row. Along the interradial suture of the interambulacra a narrow "naked zone" is present in some specimens.
Peristome: The peristome is large, about 40-45 % of the test diameter and has a circular outline. The gill slits are very shallow.
Periproct: not preserved.
Primary spines: The primary spines have a smooth base with a granulated ring. The acetabulum has a noncrenulate margin and the shaft is striated longitudinally.

Discussion:
The studied specimens clearly belong to *Psammechinus dubius dubius* (Agassiz, 1840a). All features correspond well to the original description (Agassiz, 1840a) and the re-description of Philippe (1998). A throughout revision and complete synonymy of closely related taxa was recently published by Philippe (1998) and the reader is referred to that paper.

The type material of *Psammechinus extraalpinus* Schaffer, 1912a is indistinguishable from the material referred to *P. dubius dubius* and is therefore placed into the synonymy of the latter.

Fell & Pawson (1966: U433) stated that the placement of Miocene species referred to *Psammechinus* is not supported by evidence of buccal plates, a feature they regarded as critical for this genus. It is true that no buccal plates are preserved in any specimen known from a Miocene *Psammechinus* species. Nevertheless, placement of *P. dubius* and its subspecies into this genus is well supported by other features. Comparison with extant *P. microtuberculatus* and *P. miliaris* from the Mediterranean and the North Atlantic respectively, reveals high similarities in tuberculation patterns, ambulacral pores structure and arrangement, overall shape and type of apical disc [dicyclic with clearly exsert ocular plates; Agassiz (1840a) and pers. observ. on Swiss Material (specimen NMBE 5011427)].

Vadász (1915) recorded this species from several localities in Hungary. Only the material from the Karpatian of Drégely, Nógrád (MAFI Ech 447), however, can be identified to species level. The remaining specimens [MAFI Ech 300 (Fót, Pest), MAFI Ech 453 (Piliny, Nógrád)] are indeterminable internal casts or strongly abraded coronas.

Although this (sub-)species was reported in several papers from Badenian deposits of the Central Paratethys (e.g. Mączyńska, 1979, 1987, 1988; Kazár, 2002) this could not be substantiated (see below under *Psammechinus dubius* ?). Most records from the Badenian of Austria and all specimens examined belong to the subspecies discussed below.

Occurrence:

Austria: Early ? to Late Eggenburgian (Early Burdigalian), ? Badenian (Langhian-Early Serravallian)
Molasse Zone: Achberg, near Maria Dreieichen (Scutellensande), NÖ (Steininger, 1971a, b); Fels am Wagram, NÖ {Steininger, 1971a [record questionable; in an earlier, extensive monograph on this locality (Steininger, 1963), this species is not mentioned]}; Maissau, NÖ (Schaffer, 1912a); Reinprechtspölla, NÖ (Schaffer, 1912a, 1913a; Steininger, 1971a); Zogelsdorf (Johannisbruch), NÖ ([NHMW])

Paratethys (non-Austrian occurrences): "Obere Meeres-molasse" (Middle Burdigalian) to Karpatian (Late Burdigalian)
Swiss Molasse: Belpberg, Switzerland ([NMB]); Chaux-de-Fonds, Switzerland (Agassiz, 1840a, b; de Loriol, 1875); between Seitemberg and Studweidholz, N Bütschelegg, Switzerland ([NMB]); Sidenberg, near Niedermuhlern, Switzerland ([NMB]); Ste-Croix, Vaud, Switzerland (de Loriol, 1875); Verrières, Neuchâtel, Switzerland (de Loriol, 1875)
Great Hungarian Basin (Pannonian Basin): Drégely, Nógrád, Hungary (Vadász, 1915; Mihály, 1985; Kroh, 2003b [MAFI])

Mediterranean: Aquitanian to Serravallian
Rhône Basin: numerous localities (see Philippe, 1998: 65)

Psammechinus cf. *dubius gauthieri* Cotteau, 1885

(Figs. 14-15; Pl. 15, Figs. 1-3)

v pp 1869a *Psammechinus mirabilis* Nicolet sp. – Laube: 182
v 1869a *Psammechinus monilis* Desmar. sp. – Laube: 182
v pp 1870 *Psammechinus mirabilis* Nic. – Laube: 314
v 1870 *Psammechinus monilis* Derm. – Laube: 314
v pp 1871 *Psammechinus mirabilis* Nicolet sp. – Laube: 59
v pp 1871 *Psammechinus monilis* Desmarest sp. – Laube: 59
? 1958a *Psammechinus dubius* Ag. – Sieber: 152

Material:
Badenian (Langhian-Early Serravallian) – Steinebrunn (formerly Steinabrunn), NÖ, Austria
NHMW: 1 corona (NHMW 1846.37.954a), 2 coronas [NHMW 9023; These specimens are most probably the reference specimens of Laube (1869a, 1870, 1871)], 3 coronas on a slab (NHMW 2004z0001/0029a-c), 1 corona with partially preserved apical disc (NHMW 2004z0001/0030)

Foreign material for comparison:
Badenian (Langhian-Early Serravallian) – Úvaly (= Garschenthal or Garschönthal), near Valtice (= Feldsberg), Czech Republic
NHMW: 9 coronas (NHMW 1859.XLV.630), 6 coronas + 1 fragment (NHMW 1859.XLV.633, 633a), 1 corona with preserved apical disc (NHMW 1872.V.42a), 1 corona (NHMW 1872.V.42b), 3 partially fragmented coronas (NHMW 1875.XXV.44), 4 coronas (NHMW 1904.VIII.93)

Dimensions (in mm):

Inv. No.	test ø	TH	ps ø	pc ø
NHMW 9023a	18.5	13.3	7.9	-
NHMW 9023b	16.7	11.8	6.9	2.0
NHMW 1859.XLV.633a	13.0	7.5	5.5	-
NHMW 1859.XLV.633b	15.5	9.9	6.3	-
NHMW 1859.XLV.633c	15.6	9.3	-	-
NHMW 1859.XLV.633d	16.6	10.3	7.2	-
NHMW 1859.XLV.633e	17.1	9.9	7.3	-
NHMW 1859.XLV.633f	18.3	11.1	-	-

Description:
Size and shape: Test small, with circular outline rarely exceeding 20 mm test diameter. In profile the test is high, with a tumid ambitus.
Apical disc: Apical disc dicyclic with oculars widely exsert (Fig. 14). Both genital and ocular plates densely tuberculated. Genital plate 2, the madreporite, strongly inflated and covered by numerous moderately large madreporic pores. All 5 gonopores opened. In a single specimen (NHMW 9023b, Fig. 14.B) a

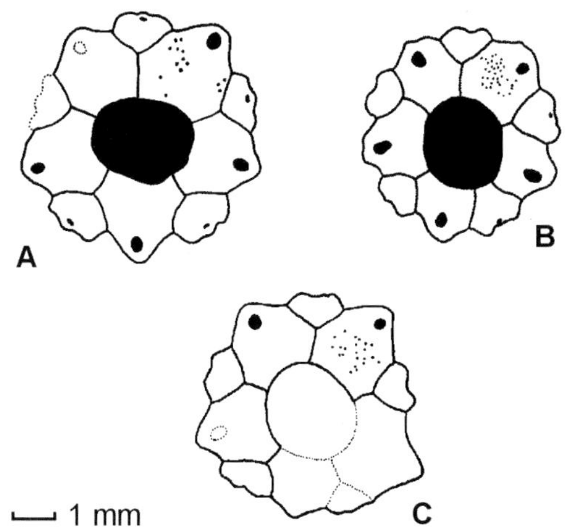

Figure 14: *Psammechinus* cf. *dubius gauthieri* COTTEAU, 1885: apical disc [A: NHMW 1872.V.42 from Úvaly (= Garschenthal), Czech Republic; B: NHMW 8023b and C: NHMW 2004z0001/0030 from Steinebrunn, NÖ].

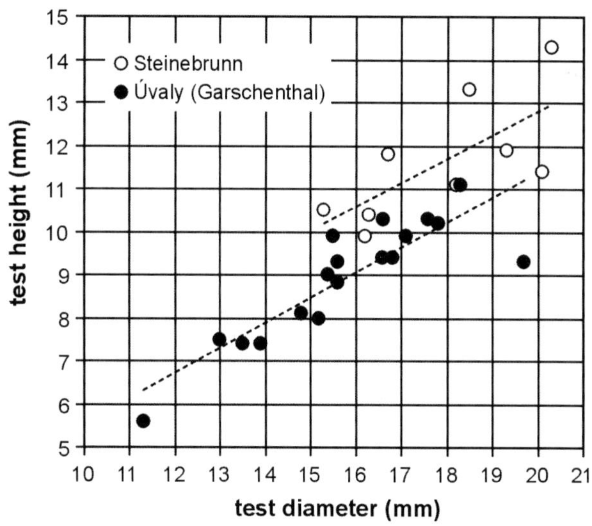

Figure 15: *Psammechinus* cf. *dubius gauthieri* COTTEAU, 1885: variation of test height (plotted against test diameter) in material from Steinabrunn (NÖ, Austria) and Úvaly (Czech Republic).

hemicyclic apical disc with ocular I insert and ocular II exsert was observed. As all other features are identical with the remaining material this specimen is considered as freak.

Ambulacra: The ambulacra consist of trigeminate plates of the echinoid compound type. The pores are partitioned isopores of the P2 type and are arranged in arcs of three. Each plate bears one large noncrenulate, imperforate marginal tubercle and several smaller inner tubercles. In ambital plates the most perradially situated of the inner tubercles is distinctly enlarged. The ambulacra are up to two third as wide as the interambulacra at the ambitus.

Interambulacra: Each interambulacral plate bears one moderately large noncrenulate, imperforate primary tubercle, which lies halfway along the adoral margin of the plate. The secondary tuberculation is dense. Ambitally there are 1-2 distinctly enlarged secondary tubercles on each side of the primary tubercle. Together with the latter they form a more ore less horizontal row on each plate.

Peristome: The peristome is large, about 43 % of the test diameter and has a subcircular outline. Gill slits are present, but developed as faint notches.

Periproct: The periproct is small, about 12 % of TD in diameter and lies centrally within the apical disc. It is very slightly oval, elongated along the I-3 axis.

Discussion:

The specimens come from a biodetritic packstone to grainstone consisting mainly of fragmented corallinaceans, bryozoans and echinoderm debris. Most outcrops are situated at the Austria-Czech border on a small hill called Veitsberg, which lies approximately halfway between Steinebrunn and the former Garschenthal (today called Úvaly). In Steinebrunn a second outcrop was situated, which yielded a high-diversity mollusc fauna and (among else) disarticulated echinoderm material from a pelitic sand facies. This locality is sometimes referred to as "Steinabrunn-Milchhaus" and is not to be confused with the limestone-facies locality discussed here.

Apart from the hemicyclic apical disc in a single specimen all features are typical for the genus *Psammechinus*. Especially tuberculation pattern and ambulacral pore structure and arrangement are very similar to *P. microtuberculatus* or *P. dubius dubius*. *Arbacina* specimens of comparable test diameter have a much smaller peristome with only *c.* 33 % TD in diameter. *Schizechinus* specimens of similar size have distinct, already well developed gill slits and differ strongly in tuberculation patterns.

The specimens differ only slightly from typical *P. dubius dubius*. The major differences are their more numerous (in specimens of comparable size) and more tightly arranged secondary tubercles. Moreover, some of the specimens are rather high (with test height ranging up to *c.* 70 %), compared to *P. dubius dubius* (height ranging from 48.5.6 to 63.0 % diameter, with a mean of 55.5 %). Similar specimens occur in the Burdigalian and Langhian of the Rhône Basin. There they are attributed to *Psammechinus gauthieri* COTTEAU, 1885 (COTTEAU, 1885: 65-66; pl. 8, figs. 15-17; no. 45), which is currently considered as subspecies of *P. dubius* (PHILIPPE, 1998: 72-73; pl. 9, figs. 7-9) (The arrangement of the ambulacral pore in COTTEAU's figure is incorrectly illustrated. In echinoid-type compound plates it is always the middle pore pair on each plate the lies most adradially, not the lowermost).

Similarly strong variation in test height as observed in this species was documented for several extant echinaceans in the Mediterranean (ERNST, 1973a). Especially apparent is that different populations from (slightly) different environments show strong differences in coronal height. This is paralleled by the "populations" from the two outcrops Steinebrunn and Úvaly (= Garschenthal) (Fig. 15) which are located very close to each other (several 100 meters). ERNST (1973a) related such differences in coronal height to differences in wave exposure and food availability.

Occurrence:

Austria: Badenian (Langhian-Early Serravallian)

Vienna Basin: Nussdorf, W (LAUBE, 1869a, 1871); Steinebrunn (formerly Steinabrunn), NÖ (LAUBE, 1869a, 1871; SIEBER, 1958a; [NHMW])

Paratethys (non-Austrian occurrences): Badenian (Langhian-Early Serravallian)

Vienna Basin: Úvaly (= Garschenthal or Garschönthal), near Valtice (= Feldsberg), Czech Republic (LAUBE, 1869a, 1871; [NHMW])

Psammechinus dubius ? (Agassiz, 1840)

? 1882 *Psammechinus monilis* Desm. – Hilber: 236
? v pp 1915 *Psammechinus dubius* Ag. – Vadász: 110 [indeterminable echinaceans]
1938 *Psammechinus extraalpinus* Schaffer 1912 – Poljak: 174-175; pl. 2, figs. 2, 2a
? 1953 *Psammechinus dubius* (Agassiz), 1840. – Szörényi: 11, 58
? 1969 *Psammechinus dubius* (Agassiz) – Gábos & Ghiurca: 88; pl. 2, fig. 2 [illustration & description not sufficient for revision]
1970 schizechinid (?) echinoids – Radwański: 382
1973 *Psammechinus* sp. – Radwański: pl. 7
? 1979 *Psammechinus dubius* (Agassiz, 1846) – Mączyńska: 31, pl. 1, figs. 1-2 [very small coronas, poorly preserved, probably better referred to as Echinacea indet.]
? 1981 *Psammechinus dubius* Ag. – Halmai: 106
? 1984 *Psammechinus dubius* (Ag.) – Kókay et al.: 290
? 1985 *Psammechinus dubius* (Agassiz, 1846) – Mihály: 238; pl. 1, figs.3-5 [indeterminable juvenile Echinacean]
? 1987 *Psammechinus dubius* (L. Agassiz, 1840) – Mączyńska: 146, 148; pl. 2, figs. 6-7; pl. 3, figs. 1-7 [juvenile echinaceans, precise determination difficult]
1988 *Psammechinus dubius* (Agassiz, 1840) – Mączyńska: 61; pl. 1, figs. 7-9; pl. 3, fig. 11; pl. 5, fig. 4; pl. 6, fig. 1
? 1988 *Psammechinus dubius laqueatus* Lambert – Mączyńska: pl. 3, figs. 1-7 [juvenile echinaceans, precise determination difficult]
? 1993 *Psammechinus dubius* (L. Agassiz, 1840) – Mączyńska: 108-109; pl. 2, figs. 4-5 [small juveniles, identification doubtful]
1996 *Psammechinus dubius* – Machalski: 26; unnumbered fig. on p. 25 (small specimens)
? 2002 *Psammechinus dubius* – Kazár: 153; fig. 1
2004 *Psammechinus dubius* (L. agassiz, 1840) – Radwański & Wysocka: 384, 385-386; pl. 2, figs. 3, 3a; pl. 3, figs. 1-2; pl. 4-5; pl. 6, figs. 2b (small specimens)

Discussion:
Most of the records of *Psammechinus dubius* from the Badenian of the Central Paratethys are based on either poorly preserved or juvenile specimens (often both). These specimens would probably be better referred to as Echinacea indet. More convincing specimens of this species have been reported from the Early Badenian of the Carpathian Foredeep (Mączyńska, 1988; Machalski, 1996; Radwański & Wysocka, 2004). They come from a mass occurrences at Świniary in Southern Poland, which have been interpreted as tempestites (Radwański & Wysocka, 2004). As most specimens from these mass-occurrences retain at least parts of their spine cover or are obscured by syntaxial rim cement or adhering sediment their determination is difficult. Only a single illustration of their tuberculation pattern is available in the literature (Mączyńska, 1988: pl. 5, fig. 4). Unfortunately this is from the adapical region of the corona and therefore poorly suited for interspecific comparison. Convincing evidence for the attribution to *Psammechinus dubius* has yet to be published. Based on the available information the attribution can only be regarded as tentative at best.

Occurrence:

Austria: ? Early Badenian (Langhian)
Vienna Basin: ? Stotzing (sandpit Mayer), Bgld, Austria (Kazár, 2002)

Paratethys (non-Austrian occurrences): Karpatian (Late Burdigalian) Eggenburgian (Early Burdigalian) to Early Badenian (Langhian)
Great Hungarian Basin (Pannonian Basin): ? Budafolk, Hungary (Vadász, 1915); ? Budapest, Hungary (Kókay et al., 1984; Mihály, 1985); ? Fót, Hungary (Vadász, 1915; Halmai, 1981); ? Piliny, Hungary (Vadász, 1915)
Fore-Carpathian Basin: Huta Różaniecka, southeastern Poland (Mączyńska, 1979); ? Korytnica Clays, Korytnica, Poland (Mączyńska, 1987); ? Mikołajów, S of Lwów, western Ukraine (Hilber, 1882; Szörényi, 1953); Nawodzice, Southern Poland (Mączyńska, 1988); ? Rybnice, southern Poland (Mączyńska, 1988); ? Pińczów, Central Poland (Mączyńska, 1993); Świniary, southern Poland (Mączyńska, 1988)
Transylvanian Basin: Berchezoaia, Bucium, Ciolt, and Valea Satului, Romania (Gábos & Ghiurca, 1969)
Zala, Sáva and Dráva Basins: Jelav Potok, near Lonđice, Croatia (Poljak, 1938)

Psammechinus sp.

(Fig. 16; Pl. 15, Figs. 4-7)

? 1882 *Psammechinus* sp. – Hilber: 236
? 1893 *Psammechinus* sp. – Toula: 289
? 1958b *Psammechinus* – Sieber: 297
? 1961 *Psamechinus* – Schaffer: 149
? 1987 *Psammechinus* sp. – Mączyńska: 146; pl. 2, figs. 5a-c
v. 2002b *Psammechinus* sp. – Kroh: 12

Material:
Ottnangian (Late Burdigalian) – Allerding, OÖ
NHMW: 3 test fragments (NHMW 2004z0005/0002)
Danninger coll.: 2 specimens (no reference no.; a cast of the figured specimen is stored at the NHMW under 2004z0005/0001), 3 test fragments (NHMW 2004z0005/0002)
Early Badenian (Langhian) – Gainfarn, NÖ, Austria
NHMW: 3 test fragments (NHMW 2004z0076/0001-3)
Badenian (Langhian-Early Serravallian) – Oslip, Bgld, Austria
Fürnkranz coll.: 1 corona (no inventory no.)

Dimensions (in mm):

Inv. No.	test ø	TH
Danninger coll. spec. 1	11.8	~ 5
Danninger coll. spec. 2	20.8	-

Description:
Size and shape: Test small, with circular outline. The aboral surface is arched; ambitus tumid.
Apical disc: not preserved
Ambulacra: The ambulacra consist of trigeminate plates of the echinoid compound type. The pores are partitioned isopores of the P2 type and are arranged in arcs of three. Each plate bears one large noncrenulate, imperforate marginal tubercle and few smaller inner tubercles. The ambulacra are about two third as wide as the interambulacra at the ambitus.
Interambulacra: Each interambulacral plate bears one moderately large noncrenulate, imperforate primary tubercle, which lies halfway along the adoral margin of the plate. The secondary tuberculation is dense, several secondary tubercles are distinctly enlarge especially those adradially of the primary tubercle. No "naked zone" along the interradial suture.
Peristome: The peristome is large, about 45 % of the test diameter and has a circular outline. Shallow gill slits are present.

Discussion:
The investigated material, 2 subadult specimens and some disarticulated coronal plates from the Ottnangian, can be at-

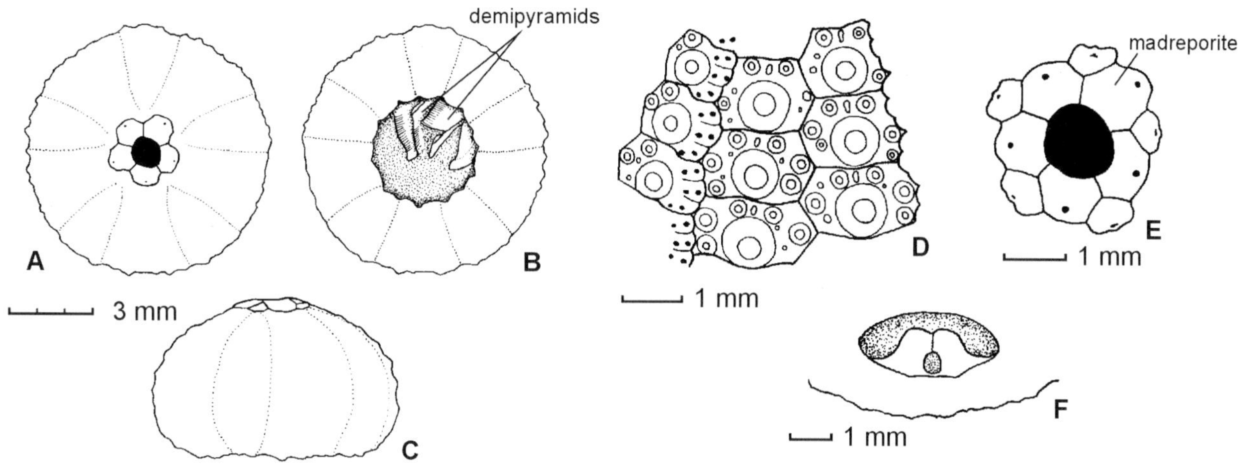

Figure 16: *Psammechinus* sp. juv.: aboral (A), oral (B) and lateral view (C), detail of supra-ambital tuberculation (D), apical disc (E) and apophyses (F) (FÜRNKRANZ coll.; Oslip, Bgld).

tributed to the genus *Psammechinus* on base of the tuberculation pattern. Despite the good preservation of the individual ossicles (no rim cement) a specific determination of such small (young) specimens is difficult. The tuberculation patterns are similar to those of the Early Miocene species *Psammechinus dubius dubius* (see above), but not identical. Until more material becomes available this species is left in open nomenclature.

Additionally, disarticulated coronal plates from the Badenian of Gainfarn are attributed to the genus *Psammechinus*. Poor preservation, lack of whole coronas and the scarcity of the material preclude a more refined determination at the time being.

The material discussed under this heading is identifiable to genus rank only. Its unified discussion under the heading "*Psammechinus* sp." does not imply that they are conspecific.

Psammechinus sp. is mentioned several times in the geological literature of Austria. These records could not be revised, since the material on which they are based could not be located. Some of them might actually be based on misidentified *Schizechinus* or *Arbacina* species.

Juvenile *Psammechinus* specimens (Figs. 16.A-F) can be distinguished from juvenile *Arbacina* specimens by their less dense secondary tuberculation and from *Schizechinus* by the lack of finely granular surface between the tubercles.

Occurrence:

Austria: Ottnangian (Late Burdigalian), Badenian (Langhian-Early Serravallian)

Molasse Zone: Allerding, OÖ ([DANNINGER coll.])

Vienna Basin: Gainfarn, NÖ (KROH, 2002b [NHMW]); ? Müllendorf (Mühlendorfer Kreide AG quarry), Bgld (SCHAFFER, 1961); ? Steinebrunn (= Steinabrunn), NÖ (SIEBER, 1958b)

Paratethys (non-Austrian occurrences): Badenian (Langhian-Early Serravallian)

Molasse Zone: Kralice nad Oslavou (= Kralitz), Moravia, Czech Republic (TOULA, 1893);

Fore-Carpathian Basin: ? Korytnica Clays, Korytnica, Poland (MĄCZYŃSKA, 1987); ? Mykolaiv (= Mikołajów), S of Lwów, western Ukraine (HILBER, 1882)

Family Echinometridae GRAY, 1825
Genus *Echinometra* GRAY, 1825

Type-species: *Echinus lucunter* LINNÉ, 1758, p. 665; by original designation (GRAY, 1825: 426).

Diagnosis: Test oval in outline, with long axis passing through ocular I and genital plate 3. Apical disc dicyclic to hemicyclic (ocular V commonly exsert). Ambulacra polygeminate with 4-10 pore-pairs. Pore pairs arranged in arcs in the type species; arcs separated by adradial secondary tubercles. One large imperforate, noncrenulate primary tubercle on each ambulacral and interambulacral plate. Peristome subcircular to slightly oval, with shallow gill slits. No phyllodes. Spines stout and pointed; usually a little less than the test diameter long (modified from FELL & PAWSON 1966 and SMITH "*The Echinoid Directory*", 26.01.04)

Distribution: Early Miocene to Recent – circumtropical

Echinometra mathaei (DE BLAINVILLE, 1825)

(Pl. 16, Figs. 1a-f)

*	1825	*Echinus mathaei* – DE BLAINVILLE: tome 37, p. 94 [not seen; fide PHILIPPE (1998)]
*	1902	*Echinometra miocenica* – DE LORIOL: 12-13; pl. 3, figs. 4-5 [not seen; fide PHILIPPE (1998)]
	1906	*Echinometra lucunter* (LESKE). – GREGORY: 253 [misidentified *E. mathaei* fide PHILIPPE (1998)]
	1910a	*Ellipsechinus miocenicus* DE LORIOL (*Echinometra*), 1902. – LAMBERT: 47-48; pl. 3, figs. 21-23
	1913a	*Echinometra miocenica* DE LORIOL. – COTTREAU: 85-86; text-fig. 12
	1931	*Echinometra mathaei* (de BLAINVILLE) – BRIGHTON: 324
	1943b	*Echinometra Mathaei* (BALINVILLE). – MORTENSEN: 381-393; figs. 155-194; pl. 42, figs. 1-10; pl. 47, figs. 1-4; pl. 65, figs. 16-26 [cum. syn.]
	1960	*Echinometra miocenica* LORIOL 1902 – KOJUMDGIEVA & STRACHIMIROV: 229; pl. 9, fig. 2a-c
?	1968	*Echinometra miocenica* LOR. – STANCU & ANDREESCU: 466, 468, tab.; fig. 79 [very small juvenile, ~2-3 mm TL]
	1981	*Echinometra mathaei* (BLAINVILLE, 1825) BLAINVILLE, 1830 – DOLLFUSS & ROMAN: 7274; pl. 18, figs. 3-6; pl. 19, figs. 5-7
	1985	*Echinometra mathaei* (BLAINVILLE) – ALI: 285; fig. 3
.	1990	*Echinometra mathaei* (BLAINVILLE) – NEGRETTI et al.: 446-450; pl. 1, figs. 1-4; pl. 2, figs. 1-4
.	1998	*Echinometra mathaei* (DE BLAINVILLE, 1825) – PHILIPPE: 78-81; pl. 6, figs. 1-4
v.	2004a	*Echinometra mathaei* – KROH: 229

Material:
Badenian (Langhian-Early Serravallian) – Hundsheim, near Deutsch Altenburg, NÖ, Austria
NHMW: 1 specimen (NHMW 1939 No.2)

Dimensions (in mm):

Inv. No.	TL	TW	TH	ps ø
NHMW 1939 No.2	38.9	32.8	18.6	17.5

Description:
Size and shape: Test of medium size with oval outline. In profile test cushion-shaped with tumid ambitus. Oral surface slightly concave in lateral view.
Apical disc: not preserved
Ambulacra: The ambulacra consist of tetrageminate plates of the echinoid compound type. In some of the ambulacra there few quinquegeminate plates in the aboral part of the ambulacra {IIb: 3rd plate, IIIb: 4th plate, IVa: 3rd and 5th plates, IVb: 3rd plate, Va: 6th and 7th plates, Vb: 6th plate [counted from the (missing) apical disc towards the ambitus; numbering of ambulacra based on the assumption that the long axis passes through ocular I and genital plate 3 as in the extant specimens]}. The pores are partitioned isopores and belong to the P1 type aborally, respectively P2 to P3 type adorally. The pore pairs are arranged in arcs, separated by adradial secondary tubercles. Each plate bears one large noncrenulate, imperforate marginal tubercle and few small inner tubercles. The ambulacra are nearly half as wide as the interambulacra at the ambitus.
Interambulacra: Each interambulacral plate bears one large noncrenulate, imperforate primary tubercle, situated halfway along the adoral margin of the plate. Secondary tuberculation consists of a vertical adradial row of tubercles in each half-interambulacrum and a zigzagging row of secondary tubercles along the interradial suture. Remaining secondary/miliary tubercles small and inconspicuous.
Peristome: The peristome is large and has a subcircular outline. Shallow gill slits are present.
Periproct: not preserved

Discussion:
Comparison with extant specimens of *Echinometra mathaei* from the Philippines revealed no significant differences, besides a slightly smaller peristome (relative to overall size) and slightly smaller primary tubercles. These two features vary, however, also within the extant material at hand and are considered irrelevant. Like in the fossil specimens, some of the extant specimens also have few quinquegeminate plates in their aboral ambulacra. For a throughout synonymy, biogeographical data of the extant animals and differential diagnoses for this and the other living congeners see MORTENSEN (1943b).

Echinometra miocenica DE LORIOL, 1902 from the Early Miocene of Southern France was placed in the synonymy of *E. mathaei* by NEGRETTI et al. (1990) and recently re-described by PHILIPPE (1998).

This record is just the third record of *Echinometra* from the Paratethys and the first one from Austria. The first record is from Bulgaria and is based on two specimens, one of them well preserved. (KOJUMDGIEVA & STRACHIMIROV, 1960). The second record, from Romania (STANCU & ANDREESCU, 1968), is dubious as it is based on an extremely small juvenile (TL ~ 2-3 mm). Although highly abundant in extant settings (see below) *Echinometra* is obviously extremely rare in the sediments of the Mediterranean and the Paratethys. The reason for the rarity of this genus is connected to the poor preservation potential in its preferred habitat (KIER, 1977: 171; GREENSTEIN, 1993: 590-591, appendix; compare also DONOVAN & GORDON, 1993) and the poor preservation potential of these rocky intertidal and extremely shallow subtidal environments, where erosion prevails (KIER, 1977: 171; SCHOPF, 1978: 267). The situation is similar in the Caribbean, where two extant species of *Echinometra* occur abundantly in the shallow water habitats, yet are only rarely documented in the fossils record (GORDON, 1991; GORDON & DONOVAN, 1992; DONOVAN & GORDON, 1993; DONOVAN & COLLINS, 1997).

Palaeoecology:
Today *Echinometra mathaei* lives in the rocky intertidal and sublittoral down to approximately 30 metres. The population density can be quite high (up to 100 animals/m^2) and thus this species is an important bioeroder and sediment producer. *E. mathaei*, like its congeners, is restricted to the tropical climate zone. Successful reproduction takes place in a narrow temperature interval between 28 to 36° C, although normal development only occurs at temperatures below 34° C (RUPP, 1973). The animals are epibenthic grazers and feed mainly upon epi- and endolithic algae, as well as on algal drift (MCCLANAHAN & MUTHIGA, 2001). *Echinometra* forms boreholes in the rocky intertidal by repeatedly removal of the uppermost millimetres of rock with its lantern. These boreholes are a common feature of the upper margin of rocky intertidal and reef flat zones, where they form a distinct zone at mean low water line (SCHOPPE & WERDINGS, 1996).

The single specimen from Hundsheim lacks ambital tuberculation in the columns 3b, IVa, IVb, 4a and 4b. There the test surface is scratched by irregularly arranged grooves (Pl. 16, Fig. 1d). Some of these grooves are arranged in an inconspicuous stellate pattern. These traces are interpreted as gnawing marks/grazing traces produced by regular sea urchins (probably by this species [not specimen] itself). Such gnawing marks have been introduced as *Gnathichnus pentax* BROMLEY, 1975 into the trace fossil literature (see BROMLEY, 1975). Later several other ichnospecies have been established for similar traces (MICHALIK, 1977; BRETON et al., 1992), but their validity has been questioned (RADWAŃSKA, 1999). Such traces have been found on a variety of substrates including other echinoids, crinoids, asteroids, brachiopods, oysters, and others (BROMLEY, 1975; RADWAŃSKA, 1999; NEUMANN, 2000; CARRASCO, 2003 and references therein). Some patches of epibiont overgrowth of corallinaceans and (subsidiary) bryozoans is still preserved on the studied specimen.

Occurrence:

Austria: Badenian (Langhian-Early Serravallian)
Vienna Basin: Hundsheim, near Deutsch Altenburg, NÖ (KROH, 2004a; [NHMW])

Paratethys (non-Austrian occurrences): Badenian (Langhian-Early Serravallian)
Transylvanian Basin: Rugi-Delineşti, Romania (STANCU & ANDREESCU, 1968)
Lom Basin (Bulgaria): Ochrid, Palilula, and Urovene, Northern Bulgaria (KOJUMDGIEVA & STRACHIMIROV, 1960)

Mediterranean: Late Burdigalian or Tortonian
Western Mediterranean: Florinas, Sardinia (NEGRETTI et al., 1990)
Rhône Basin: Sériège, Hérault, France (DE LORIOL, 1902; LAMBERT, 1910a; COTTREAU, 1913a; NEGRETTI et al., 1990; PHILIPPE, 1998)

Indian Ocean and Red Sea: Early Pliocene to Recent (see MORTENSEN, 1943b; DOLLFUSS & ROMAN, 1981; ALI, 1985)

Family Toxopneustidae TROSCHEL, 1872
Genus *Schizechinus* POMEL, 1869

Type-species: see remarks
Diagnosis: Corona large, more or less hemispherical; ambulacral plates trigeminate with a single marginal tubercle on each; inner tubercles form regular series parallel to the marginal series; interambulacra with up to 6 rows of subequal tubercles

(secondary tubercles nearly as large as primary tubercle) in each half ambulacrum; apical disc hemicyclic with oculars I and V insert; peristome with distinct, moderately deep gill slits (strongly modified from Fell & Pawson, 1966).

Distribution: Miocene to Pliocene – circum-Mediterranean

Remarks: Lambert & Thiéry (1914: 247) designated *Anapesus tuberculatus* Pomel, 1887 as type-species of this genus. As this species was not included into original paper of Pomel (1869: XLII) this cannot be a valid designation (ICZN, 4th ed.). Three of the four species mentioned by Pomel (1869) are extant forms that are not at all related with what is considered as *Schizechinus* today. As the fourth species (*Echinus serresii* des Moulins, 1837) is a hybrid species (of a Late Cretaceous taxon and an Early Miocene one) a proposal to the ICZN was submitted to suppress *E. serressii* des Moulins, 1837, to validate *Echinus serresii* Agassiz & Desor, 1846 and designate the latter as type-species of *Schizechinus* (decision pending).

Palaeoecology and biogeography: As the genus *Schizechinus* is only known from fossils little is known on its biology. It occurs commonly in carbonatic and less commonly siliciclastic shallow water sediments throughout the Mediterranean and the Central Paratethys. It is common in Northern Africa (from Morocco in the west to Egypt in the East), Saudi Arabian and reaches a far north as Poland and the western Ukraine. Recently it has also been reported from the Miocene of Barrow Island in north-western Australia (McNamara & Kendrick, 1994) and the Pliocene of Papua New Guinea (Lindley, 2003).

Schizechinus seems to be closely related to the extant *Sphaerechinus*, a monotypic genus of the Mediterranean and Eastern Atlantic Ocean. In the Mediterranean *S. granularis* occurs between 1 to 120 m depth, but is commonest between 8-12 metres, on coarse sandy to gravel bottoms with sea grass (Riedl, 1983; Ernst, 1973a). According to Ernst (1973a), however, water depth is not relevant for the distribution of this echinoid, instead it is controlled by the presence of suitable substrates and most important the presence of macroalgae on which the urchin feeds.

Schizechinus hungaricus (Laube, 1869)

(Figs. 17-19; Pl. 17, Figs. 1-6; Pl. 18, Figs. 1-19)

1869a *Psammechinus Serresii* Desmou. sp. – Laube: 182
pp 1869a *Psammechinus mirabilis* Nicolet sp. – Laube: 182
1869a *Psammechinus Duciei* Wright. – Laube: 182
* v. 1869a *Echinus hungaricus* Laube. – Laube: 182
# v. 1869a *Echinus dux* Laube. – Laube: 182
1870 *Psammechinus Serresii* Derm. – Laube: 314
pp 1870 *Psammechinus mirabilis* Nic. – Laube: 314
1870 *Psammechinus Duciei* Wright. – Laube: 314
v. 1870 *Echinus dux* Laube. – Laube: 314
v. 1870 *Echinus hungaricus* Laube. – Laube: 314
1871 *Psammechinus Serresii* Desmoulins sp. – Laube: 58-59
pp 1871 *Psammechinus mirabilis* Nicolet sp. – Laube: 59
1871 *Psammechinus Duciei* Wright. – Laube: 59
v. 1871 *Echinus hungaricus* Laube. – Laube: 60-61; pl. 16, figs. 3, 3a
v. 1871 *Echinus dux* Laube. – Laube: 60; pl. 16, figs. 2, 2a
non 1880b *Echinus hungaricus*, Laube. – Manzoni: 330-331 [= *Schizechinus serialis* Pomel, 1887, fide Borghi (1993: 9)]
pp 1891 *Echinus duciei*, Wright, 1855. – Gregory: 590
pp 1891 *Echinus hungaricus*, Laube, 1871. – Gregory: 592
1892a *Echinus* cf. *Hungaricus* Laube. – Procházka: 734, 743
non 1897 *Echinus hungaricus* Laube. – Vinassa de Regny: 147 [= *Schizechinus serialis* Pomel, 1887, fide Borghi (1993: 9)]
non 1898 *Echinus aff. hungaricus* Laub. – Airaghi: 364; pl. 1, fig. 4 [= *Schizechinus serialis* Pomel, 1887, fide Borghi (1993)]
non 1901 *Anapesus hungaricus* (Lbe.) Lamb. – Airaghi: 174; pl. 19 (1), fig. 3
non 1908 *Echinus hungaricus* Lbe. – Airaghi: 255, 256-257, 259 [= *Schizechinus serialis* Pomel, 1887, fide Borghi (1993: 9)]
1930 *Schizechinus hungaricus* Laube. – Vendl: 76
1953 *Schizechinus duciei* (Wright), 1855 – Szörényi: 12, 58-59; pl. 8, figs. 1, 1a, 2
1953 *Schizechinus hungaricus* (Laube), 1871 – Szörényi: 12, 59-60; pl. 1, fig. 1
? 1958a *Schizechinus duciei* (Wright) – Sieber: 152
? 1979 *Schizechinus duciei* (Wright, 1855) – Mączyńska: 30, pl. 1, figs. 3-5
1979 *Schizechinus dux* (Laube, 1871) – Mączyńska: 30-31, pl. 1, fig. 6; pl. 2, fig. 1
1988 *Schizechinus dux* (Laube, 1871) – Mączyńska: 60; pl. 1, fig. 10-11
1988 *Schizechinus* cf. *hungaricus* (Laube, 1871) – Mączyńska: 60; pl. 2, fig. 2
? pp 1988 *Arbacina* sp. – Mączyńska: 60; pl. 1, fig. 5 [misidentification]
v. 1998 *Psammechinus mirabilis* Nicolet – Schultz: 116; pl. 52b, figs. 6
v. 2001 *Schizechinus dux* (Laube) – Schmid et al.: 13; pl. 2; pl. 3, fig. 1
v. 2002 *Schizechinus* – Nebelsick & Kroh: 381

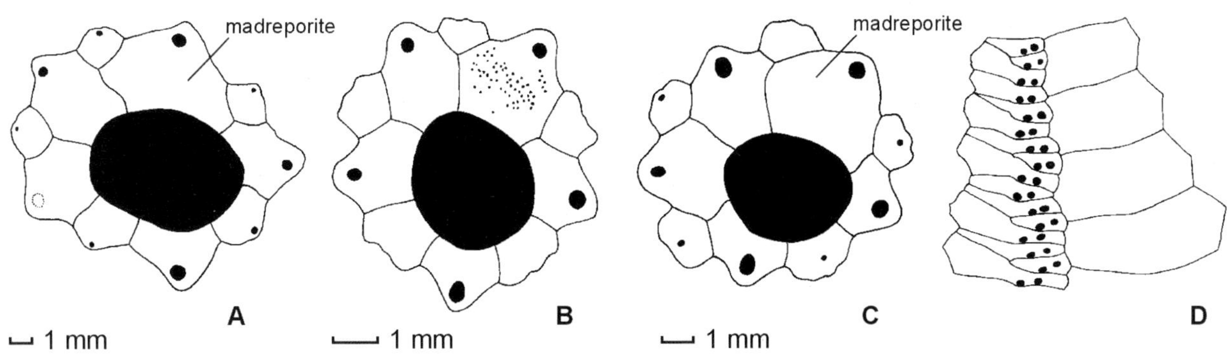

Figure 17: *Schizechinus hungaricus* (Laube, 1869): apical disc (A-C) and ambulacral plate structure (D) (A: St. Margarethen, Bgld, NHMW 1986/118/1; B: Winden, Bgld, Wanzenböck coll.; C: Oslip, Bgld, Wanzenböck coll.; D: Winden, Bgld, NHMW 1992/0084). In the great majority of the specimens oculars I and V are insert, in a single, abnorm specimen (C) only ocular I is insert. Figure A was drawn from the inside of the test and is mirrored for easier comparison.

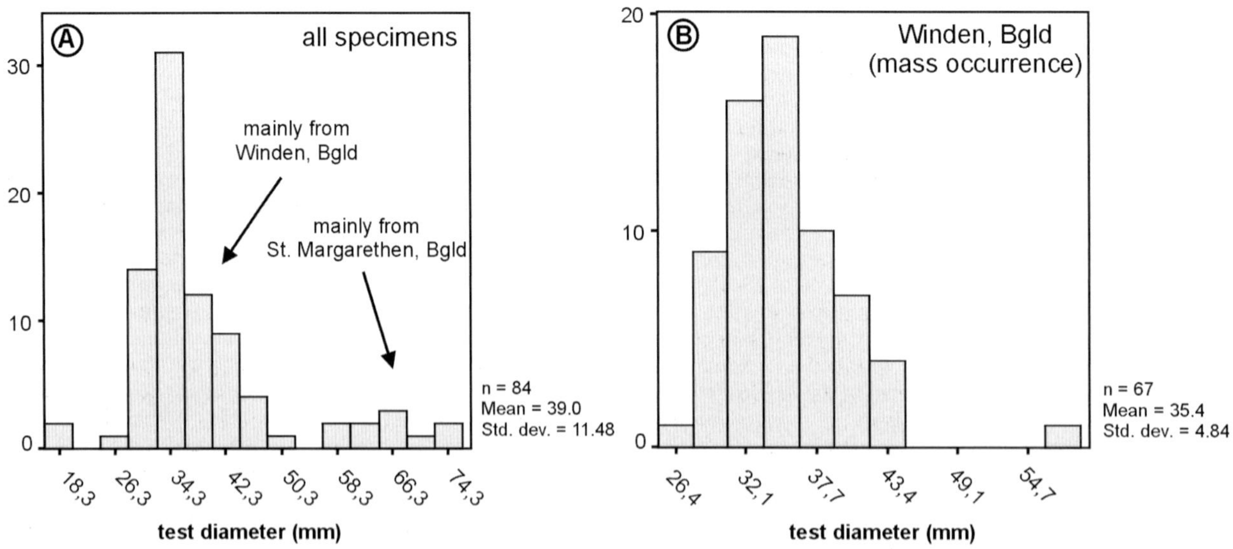

Figure 18: *Schizechinus hungaricus* (Laube, 1869): size frequency diagrams of all specimens (A) and the subset from the mass occurrence at Winden, Bgld (B).

Type-material:

Echinus hungaricus Laube, 1869:

Holotype: specimen MAFI Ech 227 from Bia, Hungary, figured by Laube (1871: pl. 16, figs. 3, 3a); located in the collection of the Geological Museum at the Geological Survey of Hungary in Budapest.

Locus typicus: Biatorbágy (= Bia), Hungary

Stratum typicum: Leitha limestone

Age: Late Badenian (Early Serravallian), Middle Miocene

Echinus dux Laube, 1869:

Lectotype (designated herein; Pl. 17, Fig. 4): specimen NHMW 1859.XLV.632 from Steinebrunn, figured by Laube (1871: pl. 16, figs. 2, 2a); located in the Naturhistorisches Museum Wien.

Locus typicus: Steinebrunn (formerly Steinabrunn), NÖ

Age: Badenian (Langhian-Early Serravallian), Middle Miocene

Remarks: The second specimen mentioned by Laube (1871: 60) from Valtice (= Feldsberg), Czech Republic could not be located in the Austrian and Hungarian collections.

Material:

Early Badenian (Langhian) – Wiesfleck (sandpit Hohlreich), Bgld, Austria
NHMW: 1 specimen (NHMW 2004z0001/0031)

Badenian (Langhian-Early Serravallian) – Purbach, Bgld, Austria
NHMW: 2 specimens (NHMW 2003z0033/0002-3)

Badenian (Langhian-Early Serravallian) – Ritzing, Bgld, Austria
NHMW: 1 specimen (NHMW 1848.III.61)

Badenian (Langhian-Early Serravallian) – Steinebrunn (formerly Steinabrunn), NÖ, Austria
NHMW: 1 corona (**lectotype** of *E. dux* Laube, 1869; NHMW 1859.XLV.632), 1 test fragment (NHMW 1861.XXXV.138), 2 coronas (NHMW A1050, 1997z0178/0771)

Late Badenian (Early Serravallian) – Biatorbágy (= Bia), Hungary
MAFI: 1 specimen [MAFI Ech 227 (**holotype** of *Echinus hungaricus* Vadász, 1915)]

Late Badenian (Early Serravallian) – St. Margarethen (Kummer quarry), Bgld, Austria
NHMW: 1 test fragment (NHMW 1976/1844/53)
Wanzenböck coll.: slab with three specimens preserving spines and lantern (no inv. no.)

Late Badenian (Early Serravallian) – Winden (Nirgl quarry, N of the village), Bgld, Austria
NHMW: 4 test fragments, 1 demipyramid (NHMW 1859.L.831a)

Late Badenian (Early Serravallian) – Winden [old quarry, N of the Ludlloch (cave), N of the village], Bgld, Austria
NHMW: 62 specimens (NHMW 1992/0084), 80 partially fragmented coronas (2004z0003/0001), 1 partial corona (NHMW 2004z0001/0025a), 182 test fragments (NHMW 2003z0082/0001, ../0025, ../0029-30), 5 genital plates (NHMW 2003z0082/0002, ../0026-27), 113 demipyramids (NHMW 2003z0082/0003, ../0010-12), 11 epiphyses (NHMW 2003z0082/0004, ../0020-21), 42 rotulae (NHMW 2003z0082/0005, ../0014-17), 11 teeth (NHMW 2003z0082/0006, ../0013), 112 spines (NHMW 2003z0082/0007, ../0022-24, ../0028, ../0043-45), 2 compass elements (NHMW 2003z0082/00018-19)

Dimensions (in mm):

Inv. No.	test ø	TH	ps ø	pc ø
NHMW 2004z0001/0031	16.8	8.8	~ 7.8	-
NHMW 1848.III.61	~28	13.8	-	-
NHMW A1050	33.8	17.3	-	2.7
NHMW 1859.XLV.632	~ 47	23.7	-	-

Description:

Size and shape: The test is of medium to large size, ranging from 15 to 75 mm test diameter in the material studied. The outline of the test is circular to very slightly pentagonal. In profile the test is moderately high and arched to high and domed. The ambitus is tumid.

Apical disc: The apical disc lies centrally on the aboral side, it is relatively small, measuring between 15 to 17 % TL along the III-5 axis. It is hemicyclic, oculars I and V being insert, all others exsert (Figs. 17.A-B). The only exception is a single specimen from Oslip, Bgld (coll. Wanzenböck) in which only ocular I is insert (Figs. 17.C). The genital plates are large and perforated by a single rounded gonopore. The madreporite is the largest of the genital plates and bears many small madreporic pores. The ocular plates are much smaller than the genital plates and bear a small ocular pore each.

Ambulacra: The ambulacra consist of trigeminate plates compound in the echinoid type (Fig. 17.D). The pores are partitioned isopores of the P2 type and are arranged in arches of three. Adjacent pore pairs are separated by broad, granulate walls. Each plate bears one large noncrenulate, imperforate marginal tubercle with undercut mamelon and several smaller inner tubercles. The one lying perradially to the marginal tubercle is distinctly enlarged and nearly as large as the marginal tubercle in ambital ambulacral plates. This feature forms a pattern of four parallel series of tubercles per ambulacrum. The interporiferous zone of a single ambulacral column is about one and a half times as wide as the poriferous zone at the ambitus.

Interambulacra: The interambulacra are about twice as wide as the ambulacra at the ambitus. Each plate bears one large noncrenulate, imperforate primary tubercle with undercut mamelon. The secondary tubercles besides the primary tubercles are distinctly enlarged and nearly as large at the ambitus. They form regular series parallel to the series of primary tubercles on the ambulacra. The number of enlarged secondary tubercles per plate increases from the apex, where they are not present, or only slightly enlarged, to the ambitus. In large specimens there may be as many as six parallel rows in each half interambulacrum. On each second plate the number of enlarged secondary tubercles adradial to the primary tubercle is doubled vertically (compare Pl. 17, Fig. 6). "Normal" secondary tubercles and miliary tubercles are spread loosely among the larger tubercles, but are most common along the adapical border of each plate.

Peristome: Peristome circular with moderately deep gill slits. The diameter of the peristome varies between 30 and 33 % test diameter. Auricles are present in the perignathic girdle.

Periproct: The periproct is irregularly oval and elongated roughly along the 1-IV or I-3 axis. Its long axis is about 50 to 55 % of the length of the apical disc respectively 7.3 – 9.5 % of TL.

Spines: The spines are short, with a smooth base, a granulated ring and noncrenulate acetabulum. The shaft is longitudinally striated.

Lantern: The lantern is camarodont, the foramen magnum is closed by the epiphyses. The demipyramids are camarodont demipyramids, with a striated interpyramidal face and a well developed styloid process. The shape of the rotulae and compass elements can be seen on Plate 18 (rotulae: Figs. 5-8; compass elements: Figs. 9-10). The teeth are keeled and have a bluntly pointed tip.

Differential diagnosis:

S. duciei (Wright, 1855), a species from the Messinian of Malta, differs from the species considered here by its consistently flatter profile and the presence of zigzagging walls between the pore pairs (Wright, 1855; Cottreau, 1913a; Challis, 1980; but see discussion). *S. saheliensis* (Pomel) was considered a synonym of the former species by Cottreau (1913a). *S. pentagonalis* (Kier, 1972), from the Miocene of Saudi Arabia and Iran (NHMW coll.), differs by its pentagonal outline and less prominent tuberculation. *S. angulatus* (Pomel, 1887) has also a markedly pentagonal outline. *S. tuberculatus* (Pomel, 1887) has much larger gill slits, judging by Pomel's figure, but is otherwise very similar. *S. chateleti* Lambert, 1910 is based on a single specimen of dubious provenience and is probably synonymous to *S. duciei* (Philippe, 1998). *S. serialis* Pomel, 1887 from the Pliocene of Algeria and Northern Italy differs by its more prominent gill slits, higher ambital interambulacral plates (usually 25% of plate width; vs. ~ 20 % in *S. hungaricus*) and smaller tubercles (diameter usually ~ 50% of corresponding plate height vs. ~ 70% in *S. hungaricus*) (compare Fig. 19, and description and illustrations in Borghi, 1993).

Discussion:

Until now this species has been attributed to Laube (1871). In Laube's 1869a paper, however, this species is already mentioned, accompanied by a differential diagnosis. Thus, according to the ICZN the year 1869 has to be regarded as the valid date for this species.

Although, at a first glance it may seem possible to distinguish two separate species within the material studied. More detailed investigations reveal that only overall size and height of the test separate these two morphotypes and that all other features i.e. tuberculation are identical. Statistical analysis of the complete data set (Kolmogorov-Smirnov test on normal distribution) reveals that the test diameter is not normally distributed (p=0.002; Fig. 18.A). Most specimens belong to the smaller morphotype and were found in a mass occurrence preserved in

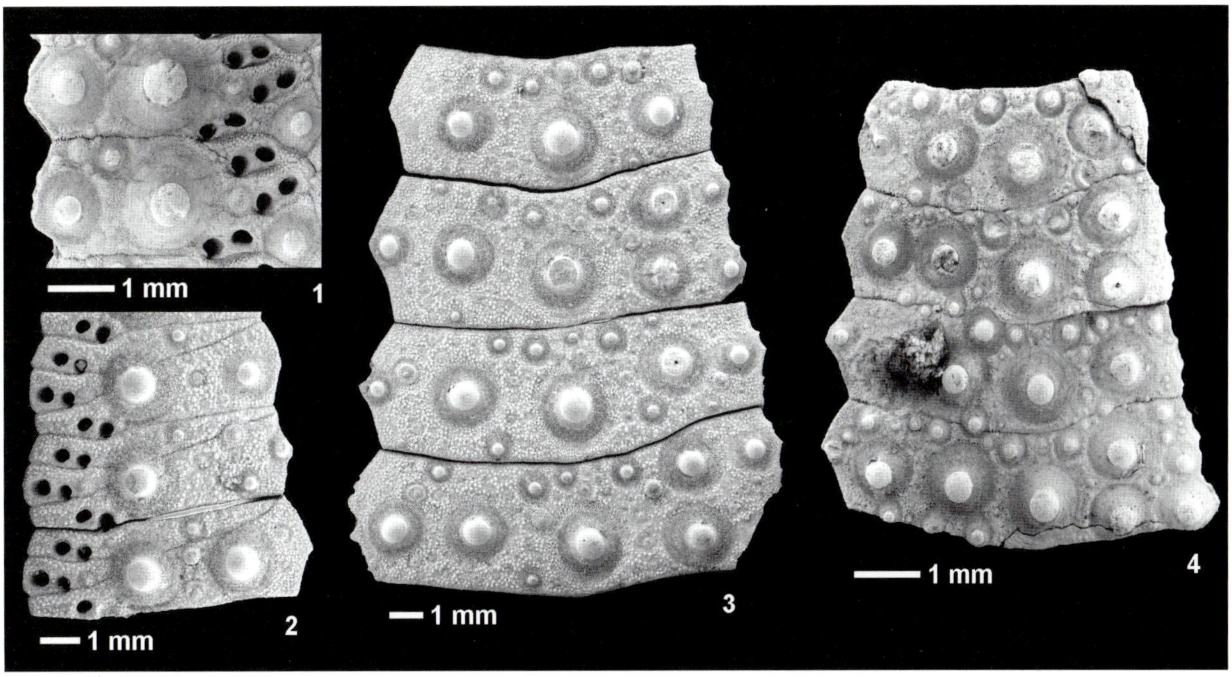

Figure 19: Comparison between *Schizechinus hungaricus* (Laube, 1869) [1, 4: Badenian (Middle Miocene) of Winden, Bgld; NHMW 2003z0082/0029-30] and *Schizaster serialis* (Pomel, 1887) [2, 3: Late Pliocene (Gelasian) of the Stirone Rive (Salsomaggiore-Parma), Northern Italy; NHMW coll., don. E. Borghi]

a single bed at the locality Winden (Bgld). This occurrence is interpreted as representing a local mass-mortality event (Kroh et al., 2002). Albeit such mass mortality events do usually not contribute significantly to the fossil record (see discussion of the *Diadema antillarum* mass mortality event in 1983-84 in Lessios, 1988 and Greenstein, 1989) in this case the skeletal material was preserved due to the special sedimentary setting. Size-frequency diagrams (Fig. 18.B) of the test diameter of the specimens from this deposit show that only a narrow size range is preserved and that the data set is normally distributed (Kolmogorov-Smirnov test, p=0.291). Most probably only a single year-class is present in this material. Thus, the smaller morphotype is over-represented in the fossil record and is considered here as young adult of the larger morphotype.

Other authors (e.g. Laube, 1871; Mączyńska, 1979) placed the smaller morphotype into the species *Schizechinus duciei* (Wright), a species from the Messinian of Malta. According to Mączyńska (1979) *S. hungaricus* differs from *S. duciei* by the larger size of the test, flatter base and lower profile [although she states that the test height of her two specimens (of *S. dux*) "displays a considerable variability" (Mączyńska, 1979: 31)].

Here only a single species of *Schizechinus* is recognised in the Badenian (Langhian-Serravallian) strata of the Central Paratethys. The oldest available name for which is *S. hungaricus* (Laube, 1869). The relation to the Messinian *S. duciei* (Wright, 1855) from the Maltese islands remains to be investigated, but the two may well turn out to be conspecific as suggested already by Gregory (1891: 592).

The names *E. dux* and *E. hungaricus* were simultaneously published. Vinassa de Regny (1897: 147-148), following a suggestion of Manzoni (1880b: 331), placed *E. dux* in the synonymy of *E. hungaricus*. According to the ICZN rule of "the First Reviser" (4th ed., Article 24.2.) the name *E. hungaricus* thus has priority over *E. dux* (it is irrelevant in this context that the material of both Manzoni and Vinasse de Regny was not conspecific with *E. hungaricus/dux*).

According to Ernst (1973a) length/height ratio in extant regular sea urchins is very variable and clearly influenced by environmental parameters. Thus, care should be taken to overemphasis differences in the height of fossil regular echinoids, especially when the specimens come from the same or adjacent outcrops or related contemporaneous strata.

Part of the types and reference material of Laube (1869a, 1870 and 1871) could be traced in the collections of the Naturhistorisches Museum Wien (NHMW), the Geological Survey of Austria (GBA) and the Museum of the Geological Survey of Hungary (MAFI). Material attributed to the taxa *Psammechinus duciei* (Wright), *Psammechinus serresii* (des Moulins), *Psammechinus monilis* (Desmarest), *Psammechinus mirabilis* (Nicolet), *Echinus dux* Laube and *Echinus hungaricus* Laube represent growth stages of a single taxon, namely *Schizechinus hungaricus* (Laube). During the growth of *Schizechinus* (but also of many other toxopneustids, like e.g. *Sphaerechinus granularis, Toxopneustes pileolus, Tripneustes ventricosus*) the number of subequal tubercles on each interambulacral plate (and to a lesser extent on the ambulacral plates) increases. Thus a specimen with 16.8 mm TD (NHMW 2004z0001/0031, Wiesfleck, Bgld) has one large primary tubercle roughly in the centre of each plate. Ambitally one secondary tubercle on each side of the primary tubercle is distinctly enlarged in that specimen (Pl. 17, Fig. 1a-b). In another specimen (NHMW 1848.III.61, Ritzing, Bgld; referred to *P. serresii* by Laube, 1871) these two secondary tubercles are as large as the primary tubercle ambitally (Pl. 17, figs. 2a-b). In a 33.5 mm specimen (NHMW A1050, Steinabrunn, NÖ; referred to *P. mirabilis* by Laube, 1871) a third enlarged secondary tubercle appears interradially at the ambitus (Pl. 17, figs. 3a-b). In the lectotype of *E. dux* with 46.7 mm largest test diameter (NHMW 1859.XLV.632) there are already five subequal tubercles in each interambulacral column ambitally (Pl. 17, Fig. 4). Finally, in the type of *E. hungaricus* with *c.* 75 mm TD (MAFI Ech 227) there are up to six subequal tubercles on each interambulacral plate ambitally. Therefore, it is impossible to use the number of subequal tubercles in the interambulacra and ambulacra to differentiate between species in the way Laube (1871) did.

Lambert (1897: 120; in a literature review) remarked that *Echinus dux* and *E. hungaricus* are closely related to *Anapesus* Holmes, an opinion that was followed by Airaghi (1901). The later, however, is today considered as junior synonym of *Arbacia* (Fell & Pawson, 1966: U409). There is no doubt that *E. dux* and *E. hungaricus* belong to the genus *Schizechinus*, based on their multiple subequal primary tubercles, hemicyclic apical disc with oculars I and V insert and the deep gill slits.

Fragmented material of *Schizechinus* from the Badenian of Rybnica, Poland was attributed to the genus *Arbacina* (Mączyńska, 1988). This, however, can be ruled out based on tuberculation pattern and size of the fragments. This material is tentatively assigned to this species based on the co-occurrence with whole coronas at the same locality.

Occurrence:

Austria: Early to Late Badenian (Langhian-Early Serravallian)
- Vienna Basin: Steinebrunn (formerly Steinabrunn), NÖ (Laube, 1869a, 1871; ? Sieber, 1958a; Schultz, 1998; [NHMW])
- Eisenstadt-Sopron Basin: St. Margarethen (Kummer quarry), Bgld (Schmid et al., 2001; [NHMW]); Walbersdorf, Bgld (Procházka, 1892a)
- Danube Basin: Purbach, Bgld ([NHMW]); Winden, Bgld ([NHMW])
- Oberpullendorf Bay: Ritzing, Bgld (Laube, 1869a; [NHMW]); Wiesfleck, Bgld ([NHMW])

Paratethys (non-Austrian occurrences): Early to Late Badenian (Langhian-Early Serravallian)
- Vienna Basin: Úvaly (= Garschenthal), near Valtice (=Feldsberg), Czech Republic (Laube, 1869a, 1871; Gregory, 1891)
- Eisenstadt-Sopron Basin: Fertörákos, Györ-Monson-Sopron, Hungary (Vendl, 1930)
- Great Hungarian Basin (Pannonian Basin): Biatorbágy (= Bia or Bid), Pest, Hungary (Laube, 1869a, 1871; Gregory, 1891; [MAFI])
- Fore-Carpathian Basin: Huta Lubycka, southeastern Poland (Mączyńska, 1979); Monastyrz, Ukraine (Mączyńska, 1979); Potylicz, near Rawa Ruska, western Ukraine (Szörényi, 1953); Rybnica, Southern Poland (Mączyńska, 1988); Szuszkowce, near Białozurka, , Ukraine (Szörényi, 1953);? Trzęsiny, southeastern Poland (Mączyńska, 1979); Wybranówska, western Ukraine (Szörényi, 1953)

Mediterranean: unknown (but may be represented by *S. duciei*, see discussion)

Schizechinus sp.

(Pl. 12, Figs. 7-8)

v. 2002b *Schizechinus* sp. – Kroh: 12
v. 2003a *Schizechinus* sp. – Kroh: 162-163; pl. 3, figs. 7-10

Material:

Early Badenian (Langhian) – Eisenstadt (Hartl Fm., Hartl hill), Bgld, Austria
Wanzenböck coll.: 1 test fragment (W30)
Early Badenian (Langhian) – Gainfarn, NÖ, Austria
NHMW: 1 test fragment (NHMW 2004z0076/0006), 1 rotula (NHMW 2004z0076/0017)
Early Badenian (Langhian) – Niederleis, NÖ, Austria

NHMW: 15 test fragments (NHMW 2002z0087/0057-62, 2002z0089/0008, 2002z0090/0007-8)
Badenian (Langhian-Early Serravallian) – Grinzing, Vienna, Austria
NHMW: 2 test fragments (NHMW 1846.37.953d)
Badenian (Langhian-Early Serravallian) – Steinebrunn (formerly Steinabrunn), NÖ, Austria
NHMW: 2 test fragments with preserved apical disc (NHMW 1846.37.955)
Late Badenian (Early Serravallian) – Müllendorf (Mühlendorfer Kreide AG quarry), Bgld, Austria
NHMW: 23 juvenile coronas (NHMW 2004z0006/0004, 0021, 2004z0045/0005)
GBA: 30 juvenile coronas (GBA 2004/1/2)
Late Badenian (Early Serravallian) – St. Margarethen (Kummer quarry), Bgld, Austria
NHMW: 1 natural cast (NHMW 2004z0001/0042), 1 fragment (NHMW 2004z0001/0043)

Dimensions (juveniles):
Test diameter ranging from 4.4 to 15.8 mm, most common between 7.1 and 11.8 mm
Test height ranging from 2.1 to 6.3 mm, most common between 3.4 to 5.9 mm

Description:
Ambulacra: The ambulacra consist of trigeminate plates of the echinoid compound type. The pores are partitioned isopores and are arranged in arcs of three, forming a zigzag pattern. Each plate bears one large noncrenulate, imperforate marginal tubercle with undercut mamelon and several smaller inner tubercles.
Interambulacra: Each interambulacral plate bears one large noncrenulate, imperforate primary tubercle with undercut mamelon. The secondary tubercles besides the primary tubercles are distinctly enlarged, forming a horizontal row. The number of these subequal tubercles increases with test size and position on the corona. On many plates (usually every second plate, at the ambitus) there are two enlarged secondary tubercles arranged in a vertical row adradially of the primary tubercle. "Normal" secondary tubercles and miliary tubercles are spread loosely among the larger tubercles, but are, commonest along the adapical border of each plate.
Peristome: The peristome shows moderately deep gill slits.

Discussion:
Although the material is highly fragmentary it can be assigned to the genus *Schizechinus* without much doubt. Based on a comparison with specimens of *Schizechinus* from Winden and St. Margarethen (Bgld, Austria), where this genus is fairly common and where specimens preserving apical disc, associated spines and jaws are known (compare e.g. Schmid et al., 2001: pl. 2; pl. 3, fig. 1). Diagnostic features are the tuberculation pattern, the arrangement of the ambulacral pores and the presence of moderately deep gill slits. Although most of the remains will probably belong to *S. hungaricus* (Laube, 1869) (see above), this cannot be proven as such fragmentary and disarticulated material lacks enough diagnostic features for identification to species level.

Juvenile of *Schizechinus* (e.g. Pl. 12, Figs. 7-8) are easily confused with juveniles and subadult specimens of the genera *Arbacina* and *Psammechinus*. In *Schizechinus*, however, secondary tuberculation is always less dense and the primary tubercles are larger than in the other two genera. Moreover, the remaining surface of the coronal plates (i.e. that not occupied by tubercles) is always distinctly granular, a feature not present in the other two genera.

Occurrence:
Austria: Early to Late Badenian (Langhian Early Serravallian)
Vienna Basin: Gainfarn, NÖ (Kroh, 2002b; [NHMW]); Grinzing, W ([NHMW]); Müllendorf (Mühlendorfer Kreide AG quarry), Bgld ([GBA]; [NHMW]); Niederleis, NÖ (Kroh, 2003a; [NHMW]); Steinebrunn (formerly Steinabrunn), NÖ ([NHMW])
Eisenstadt-Sopron Basin: Eisenstadt (Hartl Fm., Hartl hill), Bgld, ([NHMW]); St. Margarethen (Kummer quarry), Bgld ([NHMW])

Genus *Tripneustes* Agassiz, 1841b

Type-species: *Echinus ventricosus* Lamarck, 1816, by original designation (Agassiz, 1841b: 7).
Diagnosis: Corona large, hemispherical to sub-spherical; ambulacra trigeminate with primary tubercles on each 3rd or 4th plate; ambulacral pores arranged in three vertical series in each half ambulacrum; conspicuous naked median space in both ambulacra and interambulacra; apical disc with oculars I and V insert (modified from Fell & Pawson, 1966).
Distribution: Early Miocene to Recent – circumtropical
Ecology and biogeography: Extant *Tripneustes ventricosus* lives in subtidal environments between 0 to 55 metres (Hendler et al., 1995), where it inhabits a variety of habitats ranging from coral reefs to sea grass beds on sand bottoms (compare data summarised in Lawrence & Agatsuma, 2001: tab. 1). Juveniles may also be common in the intertidal, but adults are generally restricted to the subtidal (Moore et al., 1963). Most records are from *Thalassa testudinum* sea grass beds, where it reaches densities of up to 2.5 individuals/m² (Lawrence & Agatsuma, 2001: tab. 1, 2). Data reported for *T. gratilla* are comparable, although densities up to 20 individuals/m² are reported (during spawning periods, Régis & Thomassin 1982) and depth range of this species is larger (0-75 metres, Mortensen, 1943a). Extant *Tripneustes* feed mainly on sea grass and (to a lesser extent) macro algae, although coralline algae and other endolithic microalgae are also reported. Variation of diet seems to depend on availability and habitat (Lawrence & Agatsuma, 2001). Growth in extant *Tripneustes* is very rapid cage experiments by Lewis (1960) showed that the urchins reach test diameters of about 80 to 90 mm in the first year and had a life span of about two years.

Today *Tripneustes* is limited to the tropical climate zone, its poleward distribution being limited by temperature. Successful reproduction and normal embryonic development is limited to a narrow temperature range. The critical temperature for spawning lies around 22 °C in *T. ventricosus* (Moore et al., 1963), below it spawning does not occur. Plutei mortality at 20 and 30 °C is *c.* 100 % in contrast to 10 % at 25 °C (Cameron et al., 1985). Individual animals can withstand winter temperatures below 15 °C (*T. gratilla*, Dafni, 1992), but mass mortalities have been reported to occur when the temperature drops below 20 °C (*T. gratilla*: Tokioka, 1963, 1966; *T. ventricosus*: Moore et al.,1963). The highest latitude of *T. gratilla* is in the south-eastern Pacific Easter Islands (Fell, 1974), for which a temperature range of 17.5 to 24 °C is reported (DeSalvo et al., 1988).

Tripneustes planus Agassiz in Agassiz & Desor, 1846

(Pl. 19, Figs. 1-4)

	1840b	*Echinus planus* Ag. – Agassiz: 12 [*nomen nudum*]
*	1846	[*Tripneustes*] *planus* Agass. – Agassiz in Agassiz & Desor: 364, no. X 63
#	1846	[*Tripneustes*] *Parkinsoni* Agass. – Agassiz in Agassiz & Desor: 364, no. S 60
	1858	*Tripneustes Parkinsoni* Agass. – Desor: 132; pl. 18, fig.9
	1858	[*Tripneustes*] *planus* Agass. – Desor: 132-133
	1877	*Hipponoe parkinsoni* (Agassiz), Cotteau, 1877 – Cotteau in Locard: 233-235; pl. 8, figs. 13-16

1895 *Hipponoë Parkinsoni* (Agassiz), Cotteau. – Cotteau: 10

1910a *Tripneustes Parkinsoni* Agassiz, 1846. – Lambert: 44-46; pl. 2, figs. 21-24

1910a *Tripneustes planus* Agassiz (*Echinus*), 1840. – Lambert: 46-47; pl. 2, figs. 1

1913a *Tripneustes Parkinsoni* Agassiz. – Cottreau: 51, 84-85; pl.1, fig. 17

1928 *Tripneustes planus* Agassiz (*Echinus*) – Lambert & Jeannet: 135, no. X.63

1928 *Tripneustes Parkinsoni* Agassiz – Lambert & Jeannet: 179, no. S.60

1931 *Tripneustes Parkinsoni* Agassiz. – Lambert: 45, 208

1973 *Tripneustes parkinsoni* Agassiz, 1846 – Philippe: 183-184

1974 *Tripneustes parkinsoni* Ag. – Roman: 332, 333, 334

1974 *Tripneustes planus* (Ag.) – Roman: 334

. 1984 *Tripneustes parkinsoni* Agassiz, 1846 – Negretti: 89-90: pl. 1, fig. 6

. 1989 *Tripneustes planus* (Agassiz) – Philippe: 28, tab. 1

. 1990 *Tripneustes planus* (Agassiz, 1840) – Philippe et al.: 240

. 1998 *Tripneustes planus* (Agassiz, 1840) – Philippe: 61-64; pl. 4, figs. 23-24; pl. 5, figs. 1-2 [cum. syn.]

Type-material:

Tripneustes planus Agassiz in Agassiz & Desor, 1846:
Holotype: specimen figured by Lambert (1910a: pl. 2, fig. 1); housed at the Ecole nationale superieur des Mines, Lyon, France (Philippe, 1998: 64).
Locus typicus: Villeneuve, lès-Avignon, Gard, France
Age: Burdigalian, Early Miocene

Tripneustes parkinsoni Agassiz in Agassiz & Desor, 1846:
Holotype: specimen MGS AI-2 44, figured by Desor (1858: pl. 18, fig. 9 [contrary to Desor's figure only a small part of the aboral surface is preserved according to Philippe, 1998: 64]); housed at the Muséum d'Histoire naturelle de Genève (Philippe, 1998: 280)
Locus typicus: Lavalduc, France (Philippe, 1998: 280)
Age: Burdigalian, Early Miocene

Material:

Late Eggenburgian (Early Burdigalian) – Limberg (Zogelsdorf Fm., Hengl quarry), NÖ, Austria
NHMW: 36 test fragments (NHMW 2003z0002/0001-5)
Late Eggenburgian (Early Burdigalian) – Roggendorf [Zogelsdorf Fm., Teufelslucke (cave)], NÖ, Austria
NHMW: 1 test fragment (NHMW 2003z0001/0001)

Description:

Size and shape: The size, thickness and curvature of the studied test fragments indicate a test diameter of up to 70 mm and a depressed profile. The test thickness ranges from 1.0 to 2.8 mm in ambital test fragments.
Ambulacra: The ambulacra are trigeminate with pores arranged in three vertical series in each half ambulacrum. The ad- and perradial series are more or less straight; the middle series forms a zigzag pattern. Primary tubercles are found on each third plate. Enlarged secondary tubercles are found perradially of the primary tubercle and between the vertical pore series. The development and distribution of the secondary tubercles is highly variable, depending on the position of the fragment on the test. In subambital test fragments, not every ambulacral plate reaches the perradial suture, approximately every third plate (the ones adorally of these, which bear the primary tubercles) terminates just short of the perradial suture.
Interambulacra: The interambulacra bear multiple subequal nonperforate, noncrenulate tubercles on each plate. "Normal" secondary tubercles are sparse. They are found mainly along the adapical and adradial sutures.

Differential diagnosis:

According to Philippe (1998: 64) *T. planus* differs from the extant representatives of the genus by its more depressed shape and more numerous, but less well developed tubercles.

For the differences to *T.* cf. *ventricosus* recorded from the Badenian of the Central Paratethys see below under that species.

Discussion:

When Philippe (1998) synonymised *T. planus* and *T. parkinsoni* in his recent revision of the Miocene echinoids of the Rhône Basin, he failed to realise that *planus* Agassiz, 1840b is in fact a *nomen nudum* according to ICZN, since it is not accompanied by either description or illustration. The designation of a holotype alone is not enough to establish a valid species (ICZN 4th ed., 2000, Article 12.1.). Therefore Agassiz in Agassiz & Desor, 1846 is the correct authorship.

The description given above is based on the abraded and fragmentary material preserved in the Austrian collections. An extensive description of this species can be found in Philippe (1998), who described whole coronas with preserved apical disc and partly attached spines from the Burdigalian of the Rhône Basin. He also demonstrated that only a single species was present in the Rhône Basin during the Burdigalian and showed that *T. parkinsoni* Agassiz, 1846 and *T. planus* Agassiz, 1846 are conspecific. As proposed already by Cottreau (1913a: 51, 85).

Occurrence:

Austria: Late Eggenburgian (Early Burdigalian)
Molasse Zone: Limberg (Zogelsdorf Fm., Hengl quarry), NÖ ([NHMW]); Teufelslucke (Zogelsdorf Fm.), near Roggendorf, NÖ ([NHMW])

Mediterranean: Burdigalian
Western Mediterranean: Corisca, France: Santa-Manza (Cotteau in Locard, 1877); Morroco: Melilla, (Lambert, 1931); Sardinia, Italy: Capo Sant'Elia, near Cagliari (Cotteau, 1895), Fontanazzo (Cotteau, 1895)
Eastern Mediterranean: Gebel Geneffé, Egypt (Lambert, 1932)
Rhône Basin: Burdigalian: Littoral de la Nerthe (Beaumaderie, St-Paul près St-Rémy) (Lambert, 1910a; Roman, 1974; Negretti, 1984; Philippe et al., 1990), Secteur des étangs (Fos-sur-Mer, Istres, Port-de-Bouc, Saint-Mitre-les-remparts) (Agassiz in Agassiz & Desor, 1846; Desor, 1858; Lambert, 1910a; Cottreau, 1913a; Lambert & Jeannet, 1928; Philippe, 1998), Bassin d'Avignon-Carpentras (Aramon, Saint-Etienne-du-Grès, Théziers, Vaison-la-Romaine, Vedène, Villeneuve-les-Avignon) (Agassiz in Agassiz & Desor, 1846; Desor, 1858; Lambert, 1910a; Lambert & Jeannet, 1928; Philippe, 1998), Bassin d'Apt-Reillanne-Forcalquier (Apt, Auribeau, Buoux, Lacoste, Ménerbes, Reillanne, Rocsalières, Viens) (Lambert, 1910a; Roman, 1974; Philippe, 1998); Late Burdigalian (faciès "Marnes Bleues"): Bassin de Faucon-Mollans-Malaucène (Entrechaux) (Philippe, 1998)

Tripneustes cf. *ventricosus* (Lamarck, 1816)

(Fig. 20; Pl. 19, Figs. 5-13; Pl. 20, Figs. 1-10)

cf 1943a *Tripneustes ventricosus* (Lamarck). – Mortensen: 490-498, figs. 299-304; pl. 33, fig. 4; pl. 36, figs. 1-4; pl. 37, figs. 3, 11-12; pl. 38, figs. 5-8; pl. 61, figs. 3, 6, 7, 9, 15-17

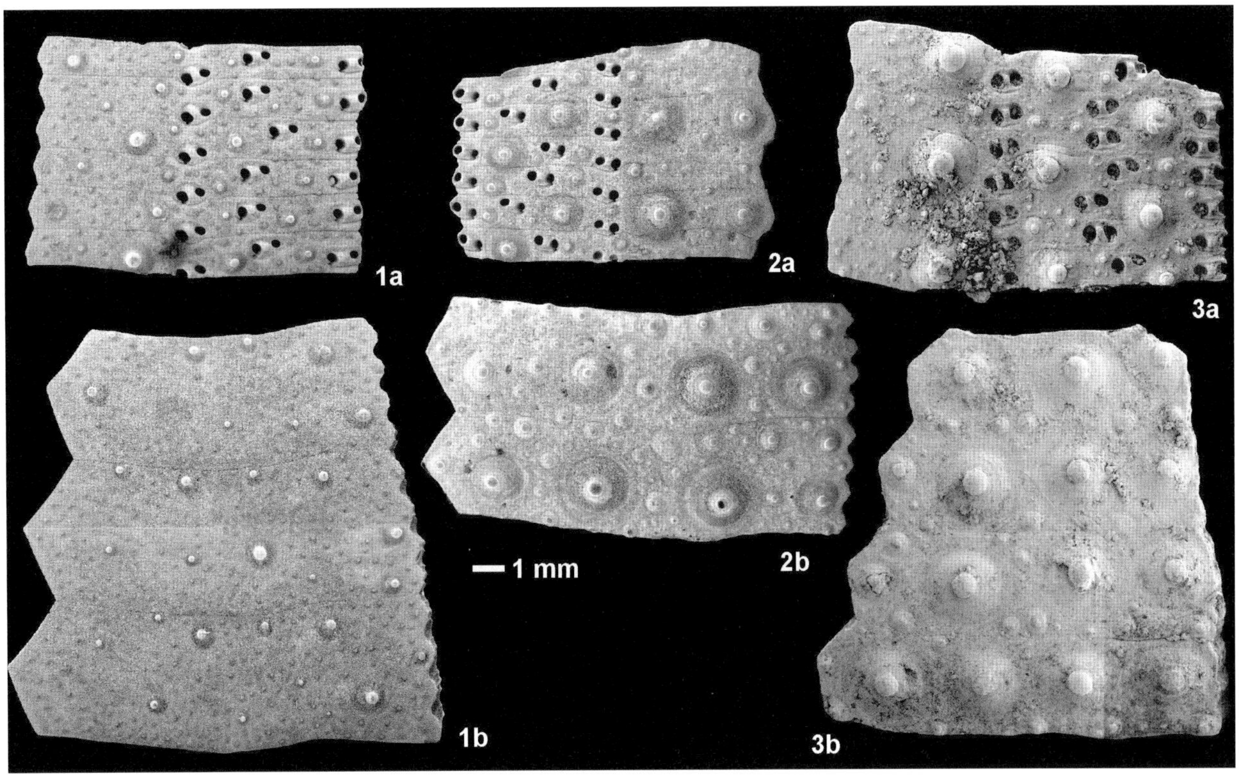

Figure 20: Comparison between extant and fossil species of *Tripneustes*: 1a-b: *T. gratilla* (LINNÉ, 1758), recent, Safaga, Red Sea; 2a-b: *T. ventricosus* (LAMARCK, 1816), recent, aquarium reared; 3a-b: *T.* cf. *ventricosus* (LAMARCK, 1816), Badenian (Middle Miocene) of Gainfarn, NÖ (3a: NHMW 2003z0003/0004; 3b: NHMW 2003z0003/0001)

\# 1951 *Tripneustes ventricosus austriacus* nov. ssp. – TAUBER: 312-318; figs. 2, 3, 4b; pl. 1, figs. a-b

1978 *T.[ripneustes] ventricosus* (LAMARCK) *austriacus* TAUBER, 1951 – KIER & LAWSON: 44

v. 2002b *Tripneustes* cf. *ventricosus* – KROH: 12

Type-material:

Tripneustes ventricosus austriacus TAUBER, 1951:

Holotype: 1 subambital test fragment figured by TAUBER (1951: figs. 1a-b); originally in the private collection of TAUBER; presumed to be lost (see discussion)

Locus typicus: Kalksburg

Stratum typicum: "Grobsand unter der fossilreichen Kalkbank"

Age: Badenian (Langhian-Early Serravallian), Middle Miocene

Material:

Early Badenian (Langhian) – Gainfarn, NÖ, Austria

NHMW: 10 test fragments (NHMW 2003z0003/0001-9, 2004z0076/0069), 9 spine fragments (NHMW 2003z0003/0011-19), 1 rotula (NHMW 2003z0003/00010), 9 spines (NHMW 2003z0003/0011-19)

WANZENBÖCK coll.: 128 test fragments, 5 rotulae, 127 spine fragments

NEITZ coll.: 13 test fragments, 33 primary spine fragments

Early Badenian (Langhian) – Hartl Fm., Hartl hill, Eisenstadt, Bgld, Austria

WANZENBÖCK coll.: 1 test fragment (W29)

Early Badenian (Langhian) – Stotzing (sandpit Mayer), Bgld, Austria

WANZENBÖCK coll.: 1 test fragment (no inv. no.)

Description:

Size and shape: The size, thickness and curvature of the studied test fragments indicate a test diameter of more than 80 mm and a high, domed profile. The test thickness ranges from 1.5 to 2.3 mm in ambital test fragments.

Ambulacra (Pl. 19, Figs. 9-10; Pl. 20, Figs. 1-2): The ambulacra are trigeminate with pores arranged in three vertical series in each half ambulacrum. The ad- and perradial series are more or less straight, the middle series forms a slightly undulating pattern. Primary tubercles are found on each third plate. Enlarged secondary tubercles are found perradially of the primary tubercle and between the vertical pore series. The development and distribution of the secondary tubercles is highly variable, depending on the position of the fragment on the test. In subambital test fragments, not every ambulacral plate reaches the perradial suture, approximately every third plate (the ones adorally of these, which bear the primary tubercles) terminates just short of the perradial suture. Adapically there is a conspicuous naked median area in the ambulacra.

Interambulacra (Pl. 19, Figs. 11, 13; Pl. 20, Figs. 3, 9): The interambulacra bear multiple subequal nonperforate, noncrenulate tubercles on each plate. "Normal" secondary tubercles are sparse. They are found mainly along the adapical, adradial and interradial sutures. Adapically there is a conspicuous naked median area in the interambulacra.

Spines (Pl. 19, Figs. 12a-b; Pl. 20, Figs. 4-7): The spines are rather short and stout with a bluntly pointed tip. The shaft is striate, with small granules on the ridges. The bases of the spines are about as wide as the basal part of the shaft, only the ring is wider.

Rotulae (Pl. 20, Fig. 8): Rotulae found in association with the coronal fragments described above are indistinguishable from rotulae of extant *T. ventricosus*.

Differential diagnosis:

The present material differs from the extant *T. ventricosus* (LAMARCK, 1816) (Fig. 20.B) from the Caribbean Sea by its larger primary tubercles in the interambulacra, which are about 75 to

80 % of the plate height in diameter (vs. 50-60 % in *T. ventricosus*). Moreover, the mamelon of these tubercles is much larger in *T.* cf. *ventricosus* than in *T. ventricosus* (30-35 % of corresponding tubercle diameter vs. 10-15 % in *T. ventricosus*). Additionally, the small secondary and miliary tubercles are less numerous in the interambulacra of *T.* cf. *ventricosus* than in *T. ventricosus*.

In the extant *T. gratilla* (LINNÉ, 1758) (Fig. 20.A) from the Red Sea there is only a single primary tubercle on each ambital interambulacral plate, not multiple subequal tubercles as in *T.* cf. *ventricosus*. Moreover, the middle pore row of each poriferous zone is more or less straight in *T. gratilla*, whereas it is undulating in *T.* cf. *ventricosus*.

T. planus AGASSIZ in AGASSIZ & DESOR, 1846 from the Burdigalian of the Rhône Basin and Austria differs from this species by its low profile, thinner test, and lack of a naked median zone in the interambulacra (compare PHILIPPE, 1988: pl. 5, figs. 1-2).

Discussion:
The studied test fragments are very similar to extant *Tripneustes ventricosus*, differences are small (see above). TAUBER (1951) established the subspecies *T. ventricosus austriacus* based on a single subambital test fragment from the Badenian of Kalksburg, Vienna. Unfortunately the specimen, which was located in the private collection of TAUBER could not be found. After his death in 1984 part of the collection of TAUBER was given to the Naturhistorisches Museum Wien, but the type specimen of TAUBER's subspecies was not among that material (pers. comm. H. SUMMESBERGER and O. SCHULTZ, 2000). Since the specimen could not be traced in the NHMW collection or in any other public and private collection available to the author, it is presumed to be lost.

Since the material available for examination consists of small test fragments only, the specimens are named in open nomenclature. The question if the present material is really conspecific with the extant Caribbean and West African species *T. ventricosus* and the validity of TAUBER's subspecies must wait until more, better preserved material becomes available.

Occurrence:

Austria: Badenian (Langhian-Early Serravallian)
Vienna Basin: Gainfarn, NÖ (KROH, 2002b; [NHMW]); Kalksburg, W (TAUBER, 1951); Stotzing (sandpit Mayer), Bgld ([WANZENBÖCK coll.])
Eisenstadt-Sopron Basin: Eisenstadt (Hartl Fm., Hartl hill), Bgld ([WANZENBÖCK coll.])

Tripneustes sp.

(pl. 21, Figs. 9-12)

	1915	*Tripneustes* sp. – VADÁSZ: 112
	1955	*Tripneustes* ? sp. – TOLLMANN: tab. 1
?	1981	*Tripneustes* sp. – HALMAI: 106
v.	2003a	*Tripneustes* sp. – KROH: 163-164; figs. 3a, c; pl. 2, figs. 8-10

Material:
Late Eggenburgian (Early Burdigalian) – Grübern (Zogelsdorf Fm.), NÖ, Austria
NHMW: 6 spine fragments (NHMW 2003z0075/0002)
Late Eggenburgian (Early Burdigalian) – Limberg (Zogelsdorf Fm., Hengl quarry), NÖ, Austria
NHMW: 103 spine fragments (NHMW 2003z0002/0006)
Early Badenian (Langhian) – Bad Vöslau (?), NÖ, Austria
NHMW: 3 spine fragments (NHMW 1997z0178/1326a-c)
Early Badenian (Langhian) – Niederleis, NÖ, Austria
NHMW: 3 spine fragments (NHMW 2002z0087/0066-68)
Early Badenian (Langhian) – Stotzing (sandpit Mayer), Bgld, Austria
NHMW: 4 spine fragments and 1 test fragment (NHMW 2004z0050/0004-5)
Badenian (Langhian-Early Serravallian) – Forchtenau, Bgld, Austria
NHMW: 4 spine fragments (NHMW 2004z0001/0006, 2004z0001/0015b)
Badenian (Langhian-Early Serravallian) – old sandpit between Großhöflein and Kleinhöflein, near Eisenstadt, Bgld, Austria
NHMW: 7 test fragments (NHMW 2003z0081/0006, ../0040), 18 spines (NHMW 2003z0081/0007, ../0027-28)
Badenian (Langhian-Early Serravallian) – Steinebrunn (formerly Steinabrunn), NÖ, Austria
NHMW: 2 spine fragments (NHMW 1859.XLV.631b), 22 spine fragments (NHMW 1865.I.612d, 612m)

Foreign material for comparison:
Badenian (Langhian-Early Serravallian) – Nový Rybik (= Portzteich), near Sedlec, W Lednice, Czech Republic
NHMW: 1 spine fragment and 1 test fragments (NHMW 1858.XLVII.78c)

Description:
Primary spines: The examined spine fragments are striate, with small granules on the ridges. The base of the spine is about as wide as the basal part of the shaft, only the ring is wider. Where the base is corroded, a stacked pattern of tubes is revealed. The cross-sections of the spines is highly similar to that seen in extant *Tripneustes ventricosus* and to that of spines found in association with *T.* cf. *ventricosus* in the Badenian of Gainfarn.

Discussion:
The material discussed under this heading is identifiable to genus rank. The unified discussion under the heading "*Tripneustes* sp." does not imply all these specimens are conspecific. It is very likely that the Eggenburgian ones belong to *T. planus* and the Badenian ones to *T.* cf. *ventricosus*.

Spines of *Tripneustes* are clearly identifiable by their external (ornamentation) and internal (cross-section) morphology. Comparison with spines of extant *Tripneustes* species (KROH, 2003a: fig. 3) shows that they are closely similar. A specific determination of isolated spines, however, is impossible (at least at the current state of knowledge).

Occurrence:

Austria: Late Eggenburgian (Early Burdigalian), Early and Late Badenian (Langhian-Early Serravallian)
Molasse Zone: Grübern (Zogelsdorf Fm.), NÖ ([NHMW]); Limberg (Zogelsdorf Fm., Hengl quarry), NÖ ([NHMW])
Vienna Basin: ? Bad Vöslau, NÖ ([NHMW]); Müllendorf (Mühlendorfer Kreide AG quarry), Bgld (TOLLMANN, 1955); Niederleis, NÖ (KROH, 2003a; [NHMW]); Steinebrunn (formerly Steinabrunn), NÖ ([NHMW]); Stotzing (sandpit Mayer), Bgld ([NHMW])
Eisenstadt-Sopron Basin: Eisenstadt, Bgld (TOLLMANN, 1955); Eisenstadt (Hartl Fm., Hartl hill), Bgld (TOLLMANN, 1955; [NHMW]); Forchtenau, Bgld ([NHMW]); Großhöflein, Bgld (TOLLMANN, 1955); between Großhöflein and Kleinhöflein (old sandpit), near Eisenstadt, Bgld ([NHMW]); Kleinhöflein, Bgld (TOLLMANN, 1955)

Paratethys (non-Austrian occurrences): ? Karpatian (Late Burdigalian), Badenian (Langhian-Early Serravallian)
Vienna Basin: Nový Rybik (= Portzteich), near Sedlec, W Lednice, Czech Republic ([NHMW])
Great Hungarian Basin (Pannonian Basin): ? Kisalag, Hungary (HALMAI, 1981)
Transylvanian Basin: Lăpugiu des Sus (= Lapugy), Hunedoara, Romania (VADÁSZ, 1915)

Echinacea indet.

(Pl. 21, Figs. 1-8)

? 1902 Echinusstachel – RZEHAK: 48, 60
pp 1952 „Seeigelstachel, Echinidenkiefer, Pedicellarien" – KÜHN: 124
pp 1955 Echiniden: Platten, Zähne, Kieferteile, Echiniden-Stachel indet. – TOLLMANN: tab. 1
1968 Piese din lanterna lui Aristotel – STANCU & ANDREESCU: 468, figs. 80-81 [demipyramids]
1977 Elements of the Aristotle's lanterns – MĄCZYŃSKA: pl. 2, figs. 1-6
1987 Isolated lantern parts – MĄCZYŃSKA: 147; pl. 7, fig. 2; pl. 8, figs. 1-2
1988 Aristotle's lantern – MĄCZYŃSKA: pl. 2, figs. 3a-b
pp ? 1992 Seeigelstachel – SAUER & al.: fig. 76H
1993 Loose elements of the Aristotle's Lantern – MĄCZYŃSKA: pl. 6, figs. 1i-h, 2a-c, 3
1996 Elements of Aristotle's Lanterns – MĄCZYŃSKA: pl. 1, figs. 11a-b
v pp 1999 Undetermined echinoid ossicles – KROH & HARZHAUSER: 164; pl. 9, fig. 1-7
v. 2003a Echinacea indet. – KROH: 164-165; pl. 4, figs. 1-14
pp v 2003 *Arbacina* sp. – KROH et al.: 92

Material:

Late Eggenburgian (Early Burdigalian) – Eggenburg, NÖ, Austria
NHMW: 150 spine fragments, demipyramids and rotulae (NHMW 2004z0110/0001, 0008-0012)
Late Eggenburgian (Early Burdigalian) – Grübern (Zogelsdorf Fm.), near Maissau, NÖ, Austria
NHMW: 10 coronal fragments and 6 spines (NHMW 2003z0075/0003, 2003z0075/0011)
Late Eggenburgian (Early Burdigalian) – Limberg (Zogelsdorf Fm., Hengl quarry), NÖ, Austria
NHMW: 2 spine fragments (NHMW 2003z0002/0008)
Late Eggenburgian (Early Burdigalian) – Maissau (Zogelsdorf Fm.), NÖ, Austria
NHMW: 7 spine fragments (NHMW 1860.L.191a), 1 juvenile corona (NHMW 1869.V)
Late Eggenburgian (Early Burdigalian) – Unternalb (Retz Fm.), SE Retz, NÖ, Austria
NHMW: numerous spine fragments, demipyramids and rotulae (NHMW 1999z0051/0023-24, 27)
Ottnangian (Late Burdigalian) – Höbmannsbach, Taufkirchen Bay, ESE Schärding, OÖ, Austria
NHMW: 53 spines (NHMW 1978/1966/12h, 12v, 12h1-12h4)
Ottnangian (Late Burdigalian) – Mitterndorf (sandpit Denk), 3.2 km NNE Sigharting, NNE Andorf, OÖ, Austria
NHMW: 2 spines (NHMW 1989/58/25a)
Ottnangian (Late Burdigalian) – Offenhausen, S Grieskirchen, OÖ, Austria
NHMW: 17 primary spine fragments (NHMW 1978/1966/14b)
Early Badenian (Langhian) – Eisenstadt (Hartl Fm.), Bgld, Austria
NHMW: 19 coronas and test fragments (NHMW 1659.L.801)
Early Badenian (Langhian) – Gainfarn, NÖ, Austria
NHMW: 7 demipyramids (NHMW 2004z0076/0013-16, ../0018-19, ../0038), 4 test fragments (NHMW 2004z0076/0031-33, ../0044)
Early Badenian (Langhian) – Niederleis, NÖ, Austria
NHMW: 77 rotulae type 1 (tentatively referred to *Schizechinus* sp. by KROH, 2003a; NHMW 2002z0087/0031-36, NHMW 2002z0089/0005), 1 rotula type 2 (tentatively referred to *Genocidaris* by KROH, 2003a;NHMW 2002z0087/0026), 4 primary spines type 1 (NHMW 2002z0087/0027-30), 1 primary spine type 2 (NHMW 2002z0087/0024), 12 demipyramids (NHMW 2002z0087/0025, 2002z0089/0003-4, 2002z0090/0002).
Early ? Badenian (Langhian) – Rauchstallbrunngraben (bryozoan marl), near Baden, NÖ
NHMW: 11 test fragments (NHMW 2004z0121/0040-50)
Early Badenian (Langhian) – Stotzing (sandpit Mayer), Bgld, Austria
NHMW: 2 spines (NHMW 2004z0050/0007), and 8 test fragments (NHMW 2004z0050/0008)
Early Badenian (Langhian) – Wiesfleck (sandpit Hohlreich), Bgld, Austria
NHMW: 1 specimen (NHMW 2004z0113/0003)
Badenian (Langhian-Early Serravallian) – Bruck an der Leitha, NÖ, Austria
NHMW: 27 spine fragments (NHMW 2004z0001/0021c)
Badenian (Langhian-Early Serravallian) – Brunn am Steinfeld, NÖ, Austria
NHMW: 6 rotulae (NHMW 1997z0178/1586), 34 demipyramids and 3 epiphyses (NHMW 1997z0178/1587)
Badenian (Langhian-Early Serravallian) – Enzesfeld, NÖ, Austria
NHMW: 8 spine fragments (NHMW 1981/56/7a)
Badenian (Langhian-Early Serravallian) – Grinzing, Vienna, Austria
NHMW: 2 demipyramids (NHMW 1846.37.953c)
Badenian (Langhian-Early Serravallian) – old sandpit between Großhöflein and Kleinhöflein, near Eisenstadt, Bgld, Austria
NHMW: 139 test fragments (NHMW 2003z0081/0008, ../0019-22), 3 juvenile coronas (NHMW 2003z0081/0016-18), 77 spines (NHMW 2003z0081/0009, ../0026), 3 rotulae (NHMW 2003z0081/0033-35), 2 epiphyses (NHMW 2003z0081/0036-37), and 78 demipyramids (NHMW 2003z0081/00010)
Badenian (Langhian-Early Serravallian) – Mattersburg, Bgld, Austria
NHMW: 1 spine fragment (NHMW 1981/56/11a)
Badenian (Langhian-Early Serravallian) – Nussdorf, Wien, Austria
NHMW: 1 abraded corona (NHMW 2004z0001/0019), 14 spines (NHMW 2004z0001/0022), 2 rotulae (NHMW 2004z0001/0013), 3 demipyramids (NHMW 2004z0001/0023)
Badenian (Langhian-Early Serravallian) – Oslip (sandpit E of the village), Bgld, Austria
NHMW: 5 demipyramids (NHMW 2004z0083/0002), and 1 rotula (NHMW 2004z0083/0003)
Badenian (Langhian-Early Serravallian) – Perchtoldsdorf (Brunnergasse, highway-ramp), NÖ, Austria
NHMW: 1 spine + 1 rotula (NHMW 1981/56/9b)
Badenian (Langhian-Early Serravallian) – Steinebrunn (formerly Steinabrunn), NÖ, Austria
NHMW: 3 abraded coronas (NHMW 1857.XIV.55), 1 poorly preserved corona (NHMW 1857.XIV.56), 123 spines (NHMW 1859.XLV.631a), 323 spines (NHMW 1865.I.612c, 612g-612l), 8 spine fragments (NHMW 1865.I.612f), 34 rotulae ("*Schizechinus*-type", NHMW 2004z0001/0020a)
Badenian (Langhian-Early Serravallian) –Teischl quarry, (? Hartl Fm., Eisenstadt), Bgld, Austria
NHMW: 18 spine and 3 test fragments (NHMW 2004z0001/0008c)
Late Badenian (Early Serravallian) – Müllendorf (Mühlendorfer Kreide AG quarry), Bgld, Austria
NHMW: 69 abraded coronas (NHMW 2004z0006/0013-0017, 2004z0045/0001), 3 demipyramids (NHMW 2004 z0045/0002), and 1 rotula (NHMW 2004z0045/0003)
Late Badenian (Early Serravallian) –Winden (Nirgl quarry, N of the village), Bgld, Austria
NHMW: 43 spines (NHMW 1859.L.832b)

Foreign material for comparison:
Early Badenian (Langhian) – Coşteiu de Sus (= Kostej), Romania
NHMW: 1 test fragment (NHMW 2003z0084/0005)
Early Badenian (Langhian) – Lăpugiu des Sus (= Lapugy), Romania
NHMW: 5 juvenile coronas (NHMW 1859.XXXVII.59b), 2 coronal plates, 1 demipyramid, 1 juvenile corona + 1 spine fragment (NHMW 2004z0001/0004c), 1 spine (NHMW 1859.XLV.617B.c), 3 test fragments, 1 juvenile corona, 4 spine fragments (NHMW 2004z0001/0027c), 1 spine (NHMW 2004z0001/0028a)
Late Badenian (Early Serravallian) – Buituri (= Bujtur), Romania
NHMW: 1 juvenile corona (NHMW 1852.II.1603c)
Late Badenian (Early Serravallian) – Devínska Nová Ves (= Neudorf an der March), Slovak Republic
NHMW: 15 spines (NHMW 1859.XXXVIII.88)

Remark: The material discussed under this heading is identifiable to superorder rank only due to the nature of the studied specimens. The unified discussion under the heading "Echinacea indet." does not imply that they are conspecific or even congeneric.

Discussion:
Disarticulated coronal plates as well as primary spines and lantern elements (mostly rotulae and demipyramids, more rarely epiphyses) of echinaceans are quite common in bulk samples from shallow water environments from the Eggenburgian, Ottnangian and Badenian of Austria. In most cases a more refined identification is very difficult or not possible at all (depending on the preservation). Based on comparison with fossil [*Schizechinus hungaricus* (Laube, 1869)] and extant (*Genocidaris maculata*) material Kroh (2003a) could tentatively assign rotulae from the Niederleis section to coronal material of co-occurring echinacean taxa. In the scope of this study it was not possible to follow a similarly refined approach and all this material is listed under Echinacea indet. In case of the primary spines, however, short remarks are given in the plate description, as spine ornamentation is very characteristic (see Plate 21).

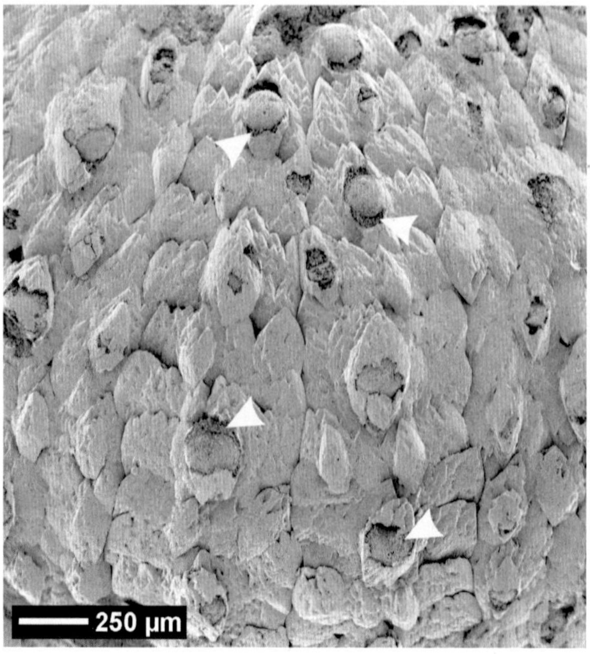

Figure 21: Echinacea indet.: Syntaxial rim cement on the primary tubercles of a juvenile echinacean producing a spindle-shaped appearance. Note the micrite trapped between cement and original shell (arrows).

Additional to the fragmented and disarticulated ossicles juvenile echinacean coronas that could not be referred to specific taxa are discussed under this heading. In many cases the coronas are too small to study tuberculation patterns and ambulacral pores. Especially the presence of syntaxial rim cement, a common diagenetic feature, hampers detailed study of the surface morphology. In many small coronas, presumably belonging to the genera *Arbacina*, *Genocidaris* or *Psammechinus*, the cement crystals strongly modify the shape of the tubercles. Instead of being more or less circular as in unaltered specimens, they are often vertically elongated becoming spindle-shaped. As this pattern is very consistent and very similar in all these specimens one could think that the vertically elongated tubercles are an original feature and not a diagenetic alteration. Albeit there is a genus which is reported to have oval, vertically elongated tubercles (*Leiocyphus* Cotteau, 1866: p. 760-764, pl. 1185, figs. 8-14; compare Fell & Pawson, 1966: U423), this is clearly a diagenetic modification in the present material. Specimens from the Hartl Formation clearly show that the tubercles are circular and the spindle-shape is caused by syntaxial rim cement (micritic material trapped between the original skeletal material and the cement produced a discontinuity surface where part of the cement crystals fell of during cleaning in the ultra sonic bath, thereby revealing the shape of the original test; see Fig. 21).

Occurrence:

Austria: Late Eggenburgian (Early Burdigalian), Ottnangian (Late Burdigalian), Early to Late Badenian (Langhian-Early Serravallian)

Molasse Zone: Eggenburg, NÖ ([NHMW]); Grübern (Zogelsdorf Fm.), near Maissau, NÖ ([NHMW]); Höbmannsbach, Taufkirchen Bay, ESE Schärding, OÖ ([NHMW]); Limberg (Zogelsdorf Fm., Hengl quarry), NÖ ([NHMW]); Maissau (Zogelsdorf Fm.), NÖ ([NHMW]); Mitterndorf (sandpit Denk), NNE Andorf, OÖ ([NHMW]); Offenhausen, S Grieskirchen, OÖ ([NHMW]); Unternalb (Retz Fm.), near Retz, NÖ (Kroh & Harzhauser, 1999; [NHMW])

Vienna Basin: Bruck an der Leitha, NÖ ([NHMW]); Brunn am Steinfeld, NÖ (Kühn, 1952; [NHMW]); Enzesfeld, NÖ ([NHMW]); Grinzing, W ([NHMW]); Müllendorf (Mühlendorfer Kreide AG quarry), Bgld (Tollmann, 1955; [NHMW]); Niederleis, NÖ (Kroh, 2003a; [NHMW]); Nussdorf, W ([NHMW]); Perchtoldsdorf, NÖ ([NHMW]); Rauchstallbrunngraben, near Baden, NÖ (Kühn, 1952; [NHMW]); Steinebrunn (formerly Steinabrunn), NÖ ([NHMW]); Stotzing (sandpit Mayer), Bgld ([NHMW])

Eisenstadt-Sopron Basin: Eisenstadt, Bgld (Tollmann, 1955); Eisenstadt (Hartl Fm., Hartl hill), Bgld (Tollmann, 1955; Kroh et al., 2003 [NHMW]); Großhöflein, Bgld (Tollmann, 1955); between Großhöflein and Kleinhöflein (old sandpit), Bgld ([NHMW]); Kleinhöflein, Bgld (Tollmann, 1955); Mattersburg, Bgld ([NHMW]); Oslip, Bgld ([NHMW]); ? St. Georgen (sandpit), Bgld (Sauer & al., 1992)

Danube Basin: Winden (Nirgl quarry), Bgld ([NHMW])

Paratethys (non-Austrian occurrences): Eggenburgian (Early Burdigalian) – Late Badenian (Early Serravallian)

Vienna Basin: Březí (= Bratelsbrunn), Czech Republic (Rzehak, 1902); Devínska Nová Ves (= Neudorf an der March), Slovak Republic ([NHMW]); Mikulov (= Nikolsburg), Czech Republic (Rzehak, 1902)

Fore-Carpathian Basin: Busko, Central Poland (Mączyńska, 1993); Korytnica Clays, Korytnica, Poland (Mączyńska, 1977, 1987); Niechobrz near Rzeszów, Southern Poland (Mączyńska, 1996); Pińczów, Central Poland (Mączyńska, 1993); Skowronno, Central Poland (Mączyńska, 1993); Świniary, Southern Poland (Mączyńska, 1988)

Transylvanian Basin: Buituri (= Bujtur), Hunedoara, Romania ([NHMW]); Coşteiu de Sus (= Kostej), Romania ([NHMW]); Lăpugiu des Sus (= Lapugy), Hunedoara, Romania ([NHMW]); Rugi-Delineşti, Romania (STANCU & ANDREESCU, 1968)

Additional echinaceans reported from the Central Paratethys

***Psammechinus* ? sp.**

1988 *Schizechinus chateleti* LAMBERT, 1910 – MĄCZYŃSKA: 60-61; pl. 2, fig. 1
1996 *Schizechinus chateleti* – MACHALSKI: 26; journal cover, unnumbered fig. on p. 25 (large specimen)
2004 "Giant *Psammechinus* sp." – RADWAŃSKI & WYSOCKA: 386; pl. 6, figs. 2a-b (large specimen)

Remarks: This species was erroneously attributed to *Schizechinus chateleti* by MĄCZYŃSKA (1988). Aboral tuberculation, however, clearly rules out a attribution to the genus *Schizechinus*, which is (among else) characterised by serially arranged subequal secondary tubercles (see under *S. hungaricus* for details on morphology and more information on *S. chateleti*). This was correctly recognised by RADWAŃSKI & WYSOCKA (2004: 386) who tentatively referred this species to the genus *Psammechinus*, naming it "Giant *Psammechinus* sp.".

This species is characterised by trigeminate (?) ambulacra with a marginal tubercle on each ambulacral plate, interambulacra with one moderately large primary tubercle on each plate and numerous irregularly arranged secondary tubercles. On the oral surface the secondary tubercles seem to be larger. The peristome is very small and has obviously poorly developed gill slits (as far as can be judged from the photograph in MĄCZYŃSKA, 1988).

Two specimens of this species are known, one in the collection of the Museum of the Earth, Polish Academy of Sciences, Warsaw (MĄCZYŃSKA's specimen) and one in the private collection of Marcin MACHALSKI (illustrated by MACHALSKI, 1996 and RADWAŃSKI & WYSOCKA, 2004). Neither of the specimens was examined by the present author. Albeit it seems clear that the species is distinctly different form any echinacean reported from the Central Paratethys so far it is very difficult to evaluate its status. Therefore the tentative attribution of RADWAŃSKI & WYSOCKA (2004) is accepted here.

Paratethys (non-Austrian occurrences): Early Badenian (Langhian)
Fore-Carpathian Basin: Świniary, Southern Poland (MĄCZYŃSKA, 1988; MACHALSKI, 1996; RADWAŃSKI & WYSOCKA, 2004)

Misidentifications or dubious records of echinaceans

Apart from the taxa discussed above a number of other echinacean taxa have been mentioned from Austrian and nearby outcrops. Reference material of some of these records could be traced and referred to the taxa above (e.g. "*Echinus*" *hungaricus*, "*Psammechinus*" *mirablis* and *serresii*). Part of these records, however, remain dubious since reference material is lost or of unknown repository and descriptions/illustrations if present at all are insufficient.

***Arbacina monilis* (DESMAREST in DEFRANCE, 1816)**

1893 *Psammechinus monilis* DSM.– TOULA: 288
1894 *Psammechinus* cf. *monilis* DESMAREST – MÁRTONFI: 153
v 1915 *Arbacina monilis* DESM. sp. – VADÁSZ: 109
1953 *Arbacina monilis* (DESMAREST), 1822. – SZÖRÉNYI: 11, 58
1958a *Arbacina monilis* (DESM.) – SIEBER: 152
1969 *Arbacina monilis* (DESMAREST) – GÁBOS & GHIURCA: 88; pl. 2, figs. 4-5 [illustration & description not sufficient for revision]
1977 *Arbacina monilis* DESMARET – MĄCZYŃSKA: 195; pl. 3, figs. 6-9
1979 *Arbacina* cf. *monilis* DESMARET – MĄCZYŃSKA: 30
1984 *Arbacina monilis* (DESMAR.) – KÓKAY et al.: 290
1985 *Arbacina monilis* (DESMAREST, 1858) – MIHÁLY: 237-238; pl. 1, figs.1-2 [indeterminable juvenile echinaceans]
1987 *Arbacina monilis* (DESMAREST, 1822) – MĄCZYŃSKA: 146, 148

Reported occurrence: Badenian (Langhian-Early Serravallian): Budapest, Hungary (KÓKAY et al., 1984; MIHÁLY, 1985); Buituri (= Bujtur), Hunedoara, Romania (MÁRTONFI, 1894); Garáb, Hungary (VADÁSZ, 1915; MIHÁLY, 1985); Huta Różaniecka, southeastern Poland (MĄCZYŃSKA, 1979); Korytnica Clays, Korytnica, Poland (MĄCZYŃSKA, 1977, 1987); Kralice nad Oslavou (= Kralitz), Moravia, Czech Republic (TOULA , 1893); Mátraverebély (Meszestetö hill), Hungary (VADÁSZ, 1915; MIHÁLY, 1985); Mayaregregy, Hungary (MIHÁLY, 1985); Mykolaiv (= Mikołajów) and Radziechów (= Radziejów), S of Lwów, western Ukraine (SZÖRÉNYI, 1953); Minişu des Sus (= Felménes), Arad, Romania (LÓCZY, 1877); Steinebrunn near Drasenhofen, NÖ, Austria (SIEBER, 1958a); Vienna Basin, Austria (LÓCZY, 1877); Valea Satului, Berchezoaia, Ciolt and Buciumi, Baia Mare Basin, Romania (GÁBOS & GHIURCA, 1969).

Remarks: This species occurs in the Middle Miocene of Martignas, Western France (see CHAVANON, 1974: 74 for a synonymy list and additional information). A modern re-description and new illustrations are needed. Most probably Central Paratethyan records are misidentifications. Many of the specimens referred to this species are in fact indeterminable echinaceans [e.g. VADÁSZ, 1915 (MAFI Ech 314 and 454); MIHÁLY, 1985: pl. 1, figs.1-2].

***Arbacina tenera* DE LORIOL, 1902**

? v 1915 *Arbacina tenera* LOR. – VADÁSZ: 108 [indeterminable echinaceans]
? 1969 *Arbacina tenera* LORIOL – GÁBOS & GHIURCA: 87-88; pl. 1, fig. 4 [illustration & description not sufficient for revision]

Reported occurrence: Badenian of the Baia Mare and Transylvanian Basins, Romania (GÁBOS & GHIURCA, 1969); Badenian of Budapest-Rákos, Sámsonháza (Mogyorós-puszta), Sopron-Rákos, and ? Mátraverebély (Meszestetö hill, southern slope), in Hungary (VADÁSZ, 1915).

Remarks: *Arbacina tenera* DE LORIOL was synonymised with *A. catenata* (DESOR) by PHILIPPE (1998: 55-58, see above). During this study none of the reported Badenian occurrences of *A. catenata* could be confirmed. Most records were either misidentifications or based on poorly preserved, indeterminable material. The illustration and description in GÁBOS & GHIURCA (1969) is not sufficient for a revised identification, the identity of their material remaining unclear.

VADÁSZ's (1915) records are based on small, poorly preserved specimens. One of them (MAFI Ech 299 from the Early Badenian of Sámsonháza) could be examined by the present author. This specimen is indeterminable at species level, the whereabouts of the other specimens is unknown.

***Arbacina* sp.**

1915 *Arbacina* sp.– Májer: 35, 88
1915 *Arbacina* sp. – Vadász: 109-110; pl. 8 (2), fig. 5 [isolated spines]
1953 *Arbacina* ? n. sp. – Szörényi: 11, 58; pl. 6, fig. 7
1955 *Arbacina* sp. – Tollmann: tab. 1
1981 *Arbacina* sp. – Halmai: 105
1987 *Arbacina* sp. – Mączyńska: 146, 148; pl. 2, fig. 4
pp 1988 *Arbacina* sp. – Mączyńska: 60; pl. 1, fig. 6
1993 *Arbacina* sp. – Mączyńska: 108; pl. 2, figs. 2-3
1996 *Arbacina* sp. – Mączyńska: 41; pl. 1, fig. 4
2000 *Arbacina* – Piller: 87

Reported occurrence: Badenian (Langhian-Early Serravallian): Csomád, Hungary (Halmai, 1981); Großhöflein, Bgld, Austria (Tollmann, 1955); Hartl, Eisenstadt, Bgld, Austria (Tollmann, 1955); Kleinhöflein, Bgld, Austria (Tollmann, 1955); Kemence, Pest, Hungary (Májer, 1915); Kików, Central Poland (Mączyńska, 1993); Korytnica Clays, Korytnica, Poland (Mączyńska, 1987); Lăpugiu des Sus (= Lapugy), Hunedoara, Romania (Vadász, 1915); Müllendorf, Bgld, Austria (Tollmann, 1955); Pińczów, Central Poland (Mączyńska, 1993); Rybnica, southern Poland (Mączyńska, 1988); Szczaworyż, Central Poland (Mączyńska, 1993); Szuszkowce, near Białozurka, Volhynia, Ukraine (Szörényi, 1953); Żerniki, Central Poland (Mączyńska, 1993).

Remarks: In most cases these records are based on juvenile or heavily altered echinaceans and echinacean spines. The generic attribution to *Arbacina*, at least in some of these records, seems doubtful.

The specimen figured by Mączyńska (1988: pl. 1, fig. 5) is a fragment of *Schizechinus hungaricus*.

***Echinus* sp.**

1877 *Echinus* – Fuchs: 669
1958b *Echinus* – Sieber: 297
1981 *Echinus* sp. – Halmai: 106

Reported occurrence: Karpatian of Csomád and Kisalag, Hungary (Halmai, 1981); Badenian (Langhian-Early Serravallian): Austria (Fuchs, 1877); Steinebrunn near Drasenhofen, NÖ, Austria (Sieber, 1958b).

Remarks: Non of the museum collections or the numerous bulk samples investigated during this study yielded material assignable to the genus *Echinus*. The literature records of *Echinus* from the Central Paratethys probably refer to *Psammechinus* or, even more likely, *Schizechinus*.

"*Echinus*" *cucurbites* Mercati

1807 *Echinus Cucurbites* Mercati Linn. – Stütz: 43
1822 *Echinus cucurbites* Mercati – Rasoumovsky: 6

Reported occurrence: Belvédère, Vienna, Austria (Stütz, 1807; Rasoumovsky, 1822)

Remarks: Stütz (1807) refers to the german translation of Linné's "*Regnum Lampideum*" by Gmelin where the name *Echinus cucurbites* Mercati is obviously mentioned. Being pre-Linnéan (Mercati, 1717-19) the name is not valid. There is a chance that it has been validly established by a later author (e.g. Gmelin), but the present author was not able to find any reference to such a taxon. It is thus unclear what species this record refers to actually, moreover, as no echinoids have been reported from this locality in the last 150 years.

"*Echinus*" *floridus* Mercati

1807 *Echinus floridus* [...] Mercati – Stütz: 75
? 1807 See-Igel der dem des Mercati gleicht – Stütz: 111

Reported occurrence: Calvariberg, near Baden (Stütz, 1807); Sooss. near Baden, NÖ, Austria (? Stütz, 1807; Rasoumovsky, 1822)

Remarks: see remarks under "*Echinus*" *cucurbites* Mercarti.

***Psammechinus michelotti* Desor, 1858**

1894 *Psammechinus Michelotti* Des. sp. – Mártonfi: 153
1906 *Psammechinus Michelotti* Desor. – Vadász: 329-330; pl. 10, figs. 2a-c

Reported occurrence: Late Badenian (Early Serravallian) of Budapest-Rákos, Pest, Hungary (Vadász, 1906) and Buituri (= Bujtur), Hunedoara, Romania (Mártonfi, 1894).

Remarks: The specimens reported as *P. michelotti* by Vadász (1906) were later (1915: 108) attributed to *A. tenera* by him (see above under that species).

P. michelottii was established by Desor (1858: 454) for a specimen from the Late Miocene of Serravalle de Servivia. Desor gives only a very brief description but no illustrations and it is impossible to judge whether Mártonfi's of Vadász's records are conspecific with Desor's specimen.

"*Prionechinus*" *felmenesensis* Lambert & Thiéry, 1911

1877 *Echinus* cf. *dux* Laube. – Lóczy: 63; pl. 5, figs. 3a-d
1911 *P.*[*rionechinus*] *felmenesensis* Lambert et Thiéry. – Lambert & Thiéry: 230
1915 *Prionechinus felménesensis* Lamb. & Thiéry. – Vadász: 107; pl. 9 (3), figs. 1-4
1953 *Prionechinus felménesensis* Lambert & Thiéry, 1911. – Szörényi: 10, 57

Reported occurrence: Late Badenian (Early Serravallian) of Minişu des Sus (= Felménes), Arad, Romania (Lóczy, 1877); Early Badenian (Langhian) of Szuszkowce, near Białozurka, Ukraine (Szörényi, 1953); Badenian (Langhian-Early Serravallian) of the Vienna Basin (Lóczy, 1877)

Remarks: It is difficult comment on this species without having examined Lóczy's material. The descriptions of Lóczy (1877) and Vadász (1915) are ambiguous. The former illustrates plates with moderately dense, small tubercles, the latter plates crowed with numerous subequal secondary tubercles (Remark: both figure the same, single specimen of this species). No similar specimens have been encountered during the present study (the only taxon that has a somewhat similar ornamentation is *Arbacina* cf. *macrophyma*). Szörényi's (1953) record added no additional information on the classification of this species.

"*Prionechinus*" *loczyi* Lambert & Thiéry, 1911

1877 *Psammechinus* cf. *monilis* Desmar. – Lóczy: 62; pl. 5, figs. 4a-d
1911 *P.*[*rionechinus*] *Loczyi* Lambert et Thiéry. – Lambert & Thiéry: 230
1915 *Prionechinus Lóczyi* Lamb. & Thiéry. – Vadász: 108
1953 *Prionechinus lóczyi* Lambert & Thiéry, 1911. – Szörényi: 11, 57

Reported occurrence: Late Badenian (Early Serravallian) of Minişu des Sus (= Felménes), Arad, Romania (Lóczy, 1877); Early Badenian (Langhian) of Mykolaiv (= Mikołajów) and Radziechów (= Radziejów), S of Lwów, western Ukraine (Szörényi, 1953); Badenian (Langhian-Early Serravallian) of the Vienna Basin (Lóczy, 1877)

Remarks: According to Vadász (1915: 108) the only known specimen of this species is lost. Thus the illustrations by Lóczy (1877) are the only information available (the description is very brief and lacks details). Unfortunately the Lóczy's drawing fits with a large number of juvenile echinaceans and the present author is neither able to associate *P. loczyi* with any other species, nor to confirm the its distinctiveness.

Schizechinus delphinus (Lambert, 1906)

1887b *Psammechinus Duciei*, Wright. – Koch: 261
1915 *Psammechinus delphinus* Defr. sp. – Vadász: 110 [based on Koch's specimen]

Reported occurrence: Late Badenian (Early Serravallian) of Gârbova de Sus (= Gîrbova de Sus, = Felsö-Orbó), Romania (Koch, 1887b; Vadász, 1915)

Remarks: This species is usually attributed to Defrance, but was never published and clearly is a collection name and thus not valid according to the ICZN. It has recently been re-described by Philippe (1998) under the name *Psammechinus delphinus* (Defrance), but clearly belongs to the genus *Schizechinus* based on its tuberculation pattern (numerous subequal tubercles on the ambital interambulacra, arrangement of the small secondary tubercles).

Koch's records is based on a single worn fragment, which also forms the base of Vadász's record. At present it is unclear to whether the determination as *S. duciei* or as *S. delphinus* is correct, or whether it belongs to the contemporaneous *S. hungaricus*. The situation is complicated by the lack of a throughout revision of the genus *Schizechinus* involving all Mediterranean and Paratethyan species.

Parasalenia fontannesi (Cotteau, 1888)

? 1977 *Parasalenia fontannesi* Cotteau, 1888 – Mączyńska: 195-196; pl. 3, figs. 1-5
non 1987 *Parasalenia fontannesi* Cotteau, 1888 – Mączyńska: 146, 148; pl. 2, fig.2 [misidentified diadematoid ambulacral plate]
? 1993 *Parasalenia fontannesi* Cotteau, 1888 – Mączyńska: 109; pl. 3, figs.1-3; pl. 6, figs. 1d-1f
? 1996 *Parasalenia fontannesi* Cotteau, 1888 – Mączyńska: 41; pl. 1, figs.8a-b, 9a-b

Reported occurrence: Badenian: Korytnica Clays, Poland (Mączyńska, 1977); Busko, Kików, Skowronno, Szczaworyż, and Żerniki, Central Poland (Mączyńska, 1993); Niechobrz near Rzeszów, Southern Poland (Mączyńska, 1996)

Remarks: The attribution of the material figured by Mączyńska (1977, 1993, 1996) to *Parasalenia fontannesi* is very doubtful. As no description is given in any of the cited papers, the records can only be judged by the illustration. The only features that the illustrated specimens have in common are the lateral deformation (clearly post-mortem) and their being juvenile echinaceans. What little can be made out of the tuberculation does not look like *Parasalenia* at all. *P. fontannesi* was recently re-described and very well illustrated by Philippe (1998). His figures clearly show that *P. fontannesi*, as extant *Parasalenia* species, is characterised by a large, imperforate primary tubercles with a huge mamelon on each plate and few, tiny secondary tubercles (see Philippe, 1998: pl. 6, figs. 7b-c). The Polish material doesn't even come close and has to be considered as misidentified. Besides, apart from the records of Mączyńska P. fontannesi was up till now only undoubtedly recorded from Aquitanian strata, while the Polish specimens are of Badenian (Langhian-Early Serravallian) age.

The ambulacral plate figured by Mączyńska (1987: pl. 2, fig. 2) does definitely not belong to *Parasalenia*, yet not even to an echinacean. The perforated, strongly crenulate primary tubercle in combination with the shape of the plate and the ambulacral pores is characteristic of a diadematoid.

P. fontannesi thus has to be considered absent from the Central Paratethys, until more convincing specimens are described.

Parasalenia sp.

? 1993 *Parasalenia* sp. – Mączyńska: 109; pl. 3, figs.4-5
? 2004 *Parasalenia* sp. – Radwański & Wysocka: 385

Reported occurrence: Pińczów, Central Poland (Mączyńska, 1993); Świniary, Poland (Radwański & Wysocka, 2004)

Remarks: The ambulacral plate figured by Mączyńska (1993: pl. 3, fig. 5) again seems to belong to a diadematoid (tough not so clear as in the pervious case due to the poor preservation; see above under *P. fontannesi*). The worn corona figured by Mączyńska (1993: pl. 3, fig. 4a-b), could be a juvenile *Schizechinus* or a *Psammechinus*, but is certainly no *Parasalenia* (based on the tuberculation of the ambital interambulacra which never shows multiple subequal tubercles/enlarged secondary tubercles in *Parasalenia*).

"*Cyphosoma*" sp.

? 1893 *Cyphosoma*. – Toula: 289
? 1974 *Cyphosoma* Kalabis: 315

Reported occurrence: Kralice nad Oslavou (= Kralitz), Czech Republic (Toula, 1893)

Remarks: *Cyphsoma* is a junior synonym of *Phymosoma*, a genus ranging from the Late Jurassic to the Eocene. Toula's record is most probably based on a misidentified Miocene echinacean (e.g. *Schizechinus*).

Salenia ? sp.

1893 Ocellarplättchen des Scheitelschildes (*Salenia* ?). – Toula: 288

Reported occurrence: Kralice nad Oslavou (= Kralitz), Czech Republic (Toula, 1893)

Remarks: *Salenia is a* Cretaceous genus, Toula's record probably refers to misidentified genital plates of cidaroids.

Cohort Irregularia Latreille, 1825
Order Echinoneoida Jensen, 1982
Suborder Echinoneina Clark, 1925
Family Echinoneidae Agassiz & Desor, 1847
Genus *Echinoneus* Leske, 1778

Type-species: *Echinoneus cyclostomus* Leske, 1778; by subsequent designation (Clark, 1917: 101).
Diagnosis: Test ovoid; ambulacra in groups of 3; pore pairs uniserially adapically, in groups of 3 adorally. Poriferous zones slightly sunken. Apical disc tetrabasal. Peristome oblique, irregular. Buccal membrane with small plates. Periproct infra-

marginal. Tubercles imperforate or perforate and noncrenulate (modified from Wagner & Durham, 1966; Donovan, 1993).
Distribution: Oligocene to Recent – circumtropical (from Wagner & Durham, 1966b and Donovan, 1993)

Echinoneus aff. *abnormalis* de Loriol, 1883

(Fig. 23; Pl. 48, Figs. 4-5)

v. 2004a *Echinoneus* – Kroh: 229
(Late Badenian: Müllendorf, Kreide AG quarry, Bgld)

Material:
Late Badenian (Early Serravallian) – Müllendorf (Mühlendorfer Kreide AG quarry), Bgld, Austria
NHMW: 1 specimen (NHMW 2002z0172/0001)
Wanzenböck coll.: 2 specimens (W23, W24)
Temel coll.: 1 specimen (casts of this specimen are housed at the NHMW 2002z0172/0002)

Dimensions (in mm):

Inv. No.	TL	TW	TH
NHMW 2002z0172/0001	41.8	32.8	23.3
Temel coll.	21.0	15.7	9.9
Wanzenböck coll. W23	7.7	5.7	3.5
Wanzenböck coll. W24	18.9	15.0	9.3

Description:
Size and shape: Test ovoid, of small to medium size, ranging from 21 to 42 mm test length in the studied specimens. The maximum width lies centrally in large specimens. In small specimens it lies posteriorly, coinciding with line 1a-4b. Maximum height lies subcentrally. Test width ranging from 74.8 to 78.5 % TL in the studied material and test height from 47.1 to 55.7 %TL.
Apical disc: Apical disc situated slightly anteriorly, about 45 % TL away from the anterior margin. It is holectypoid with genital plate 5 still present (not reduced as in the tetrabasal apical disc of *E. abnormalis* or *E. cyclostomus*), similar to the condition in *Holectypus* (see Fig. 22.D and compare Melville & Durham, 1966: U228, fig. 170.6). The madreporite (G2) is strongly enlarged and bears numerous small madreporic pores. The other genital plates are rather small (see Fig. 22.C). Four small, circular genital pores are present.
Ambulacra: The ambulacra are nonpetaloid with small, closely spaced, slightly sunken, partitioned isopores adapically. Adorally the pores are less closely spaced, strongly oblique and arranged in weak arches of three (giving the poriferous zone a slightly undulating appearance). Around the peristome the pores are indistinct. The poriferous zones are slightly depressed and very narrow. The interporiferous zones are distinctly inflated adapically and are up to ten times as wide as a single poriferous zone. The tuberculation of the interporiferous zones is similar to that of the interambulacra.
Interambulacra: The interambulacra are densely covered by small perforate, noncrenulate primary tubercles (Fig. 23). Miliary and secondary tubercles are found on the ridges between the slightly sunken areoles of the primary tubercles. They are imperforate and noncrenulate. On the oral side the primary tubercles are up to twice as large as on the aboral surface.
Peristome: The peristome lies centrally on the oral side. It is irregularly oval to triangular and strongly oblique, elongated along the 2-V axis. It shows allometric growth and measures 4.5 x 2.4 mm in the 21.0 mm TL specimen and 6.4 x 4.5 mm in the 41.8 mm TL specimen.
Periproct: The periproct lies inframarginally in interambulacrum 5, just behind the peristome. It is teardrop-shaped and elongated antero-posteriorly. It is larger than the peristome, measuring 5.2 x 3.5 mm in the 21.0 mm TL specimen and ~9 x ~5.5 mm in the 41.8 mm TL specimen.

Discussion:
The studied specimens are closely similar to the extant species *E. abnormalis* (see de Loriol, 1883: 41; pl. 5, fig. 2, 2a-d; Mortensen, 1948a: 80-81). They differ, however, by the holectypoid nature of the apical disc and more numerous secondary and/or glassy tubercles (difficult to distinguish in the fossil specimens). The scarcity and moderate preservation of the material precludes both statistical and detailed morphological comparisons (e.g. plate patterns) of extant and fossil material.

Based on the perforated primary tubercles Lambert & Thiéry (1921: 331) established the genus *Koehleraster* for this species. Mortensen (1948a: 74, 81), however, was uncertain if this feature alone justified the generic distinction and tentatively placed the genus into the synonymy of *Echinoneus*.

E. abnormalis differs from *E. cyclostomus* Leske, 1778 by the presence of perforate tubercles instead of imperforate and by its less well developed glassy tubercles (Clark, 1925: 176; Mortensen, 1948a: 80-81).

This is the first record of the genus *Echinoneus* from the Miocene of the Central Paratethys. It is also one of the first fossil echinoneids with perforate tubercles. Those fossil species for which data on the tubercles are available [e.g. *E. melitensis*

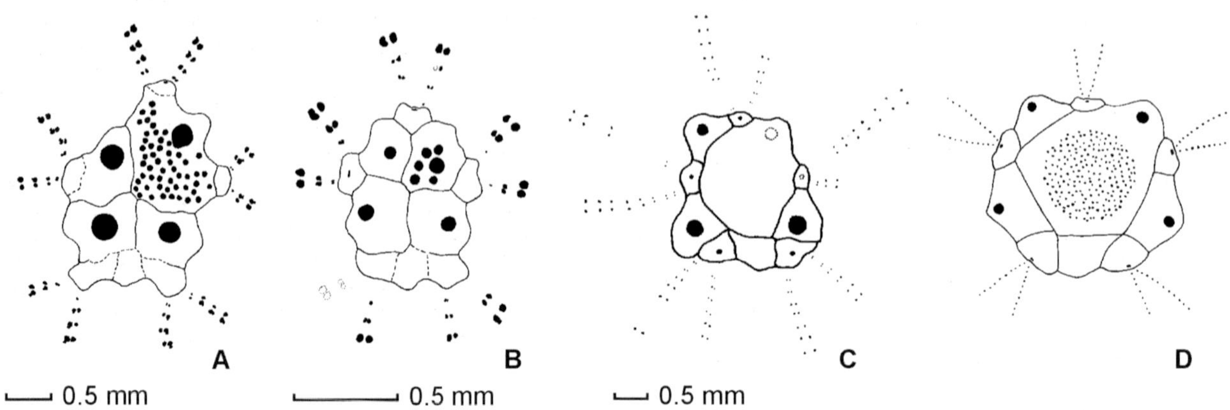

Figure 22: Comparison of the apical discs of *Echinoneus cyclostomus* Leske, 1778 (A: recent, Mauritius, NHMW coll.; drawn from SEM photograph), *Echinoneus abnormalis* de Loriol, 1883 (B: recent, Aldabra, Indian Ocean, BMNH-Zool. 1972.1.12.14; drawn from SEM photograph), *Echinoneus* aff. *abnormalis* de Loriol, 1883 [C: Late Badenian (Early Serravallian), Müllendorf (Kreide AG quarry), Bgld; NHMW 2002z0172/0001; camera lucida drawing], and *Holectypus depressus* (Leske, 1778) (D: redrawn from Wagner & Durham, 1966: U441, fig. 329/1c; scale unknown). Poorly visible sutures and pores stippled.

Figure 23: *Echinoneus* aff. *abnormalis* de LORIOL, 1883: white arrows indicate perforate primary tubercles [Müllendorf (Kreide AG quarry), Bgld; NHMW 2002z0172/0001].

(WRIGHT, 1864), see CHALLIS (1980: 102-105) and PHILIPPE (1998: 84-86); *E. thomasi* PERON & GAUTHIER in COTTEAU et al., 1891, *E. artini* GAUTHIER in FOURTAU, 1900 see BRIGHTON (1929: 89)] have imperforate tubercles and were associated with *E. cyclostomus* recently (CHALLIS, 1980; PHILIPPE, 1998; NÉRAUDEAU et al., 1999; NÉRAUDEAU et al., 2001). Most fossil records of *Echinoneus* are, however, based on poorly preserved material and differential diagnoses provided by the authors are usually insufficient. LAMBERT (1928: 21) stated that: "La distinction des diverses espèces d'*Echinoneus* est souvent assez délicate, mais ce n'est pas une raison pour confondre avec l'espèce vivante, *E. cyclostomus* LESKE, les diverses espèces du Miocène et l'Oligocène des Antilles." MORTENSEN (1948a) in contrast suggested that most fossil species might fall into the synonymy of *E. cyclostomus* because none of the "diagnostic" features allowed a differentiation without knowing the provenience of the material. Indeed material form the Oligocene to Pleistocene of the Caribbean studied by DONOVAN recently is reported to be morphologically indistinguishable from extant *E. cyclostomus* (DONOVAN, 1993: 382; DONOVAN & VEALE, 1996: 633-635; DIXON & DONOVAN, 1998: 104; DONOVAN, 2001: 183, SIMPSON, 2001: 34).

Ecology:

Nearly nothing is known on the ecology of extant *E. abnormalis.* Most authors (e.g. MORTENSEN, 1948a) assumed a similar habitat as for *E. cyclostomus*. Ecological data on *E. cyclostomus* can be found in MORTENSEN (1948a: 75-80), FONTAINE (1953), KIER & GRANT (1965: 11-12, 26), ROSE (1978: 299-306) and HENDLER et al. (1995: 227-228). Generally, *E. cyclostomus* seems to be a cryptic, if not always really burrowing species. It is found in the shallow sublittoral of the tropical region (circumtropical) except at the west coasts of America, Africa and Australia. Usually it is associated with rocks or reef debris of coarse sand to gravel size and is often found attached to the underside of coral slabs. It can also be associated with coralline algal debris (MORTENSEN, 1948a). According to HENDLER et al. (1995: 227-228) *E. cyclostomus* is common on reef flats and in other reef habitats. FONTAINE (1953) and ROSE (1978: 300) suggested that this species indicates the proximity of reefs and might, therefore, be useful in palaeoecological reconstruction. SLOAN et al. (1979: 119) reported *E. abnormalis* from sand under coral heads and boulders in the shallow sublittoral of the lagoon of the Aldabra Atoll.

Both in extant environments (KIER & GRANT, 1965: 12) and the fossil record (ZAMMIT-MAEMPEL, 1979: 9; CHALLIS, 1980: 325; BOGGILD & ROSE, 1984: fig. 2; DONOVAN & VEALE, 1996: 632-635) the co-occurrence of *Echinoneus* with *Brissus* in biodetritic sediments close to reef-like structures was recorded. In the fossil Austrian occurrence (Müllendorf, Bgld) *Echinoneus* is also associated with a *Brissus* species and found within a bioclastic coralline algal wackestone.

Occurrence:

Austria: Late Badenian (Early Serravallian)
Vienna Basin: Müllendorf (Mühlendorfer Kreide AG quarry), Bgld (KROH, 2004a; [NHMW])

Paratethys (non-Austrian occurrences): not recorded

Mediterranean: not recorded

Superorder Microstomata SMITH, 1984
Series Neognathostomata SMITH, 1981
Order Clypeasteroida AGASSIZ, 1872
Suborder Clypeasterina AGASSIZ, 1872
Family Clypeasteridae AGASSIZ, 1835
Genus *Clypeaster* LAMARCK, 1801

Type species: *Echinus rosaceus* LINNÉ, 1758, p. 665; by subsequent designation (DES MOULINS, 1835: 183).
Diagnosis: Medium-sized to large, test flattened to highly campanulate, margin rounded to flattened and inflated; peristome usually in deep infundibulum; oral surface flat to concave; petals variable, closed and rounded to open or sublyrate, with outer pores elongate, inner ones rounded, commonly connected by grooves; periproct usually inframarginal, rarely marginal, situated between 3rd and 4th, or 4th and 5th pair of coronal plates; buccal membrane naked, with imbedded irregular spicules; internal supports variable in abundance, consisting of thin laminae and pillars; wall of test sometimes double, separated by pillars (DURHAM, 1966).
Distribution: Late Eocene to Recent – circumtropical (DURHAM, 1966; ALI, 1983)
Remarks: *Clypeaster* is possibly the most diverse echinoid genus in the Cenozoic. In total more than 480 nominal species and/or subspecies have been established (LAMBERT & THIÉRY, 1909-1925; KIER & LAWSON, 1978). Several attempts were made to divide *Clypeaster* in subgenera or sections, resulting in 38 genus/sub-genus rank taxa. Variation in external test morphology and shape of the petals is very great, but no systematic basis for subgeneric grouping can be recognised according to DURHAM (1955: 118; 1966: U243). Thus these taxa were considered synonyms of *Clypeaster* by him.

The (palaeo-)biogeography of the genus *Clypeaster* was studied by ALI (1983), who postulated a Middle Eocene origin in south-western Europe, based on the report of two species from the Late Lutetian of Catalonia, Spain by VIA & PADRENY (1970). While being largely confined to the circum-Mediterranean Region in the Late Eocene, the Oligocene was a time of dispersal for this group and the genus *Clypeaster* reached Central America in the West and the Indo-Malayan region in the East. In the Miocene *Clypeaster* had its largest geographical distribution, finally reaching Australia in the Middle Miocene (MCNAMARA & KENDRICK, 1994). During the late Pliocene to Pleistocene the diversity of the genus *Clypeaster* was greatly reduced and although *Clypeaster* managed to gain a foothold again in the Mediterranean after the Messinian Salinity Crisis (NÉRAUDEAU et al., 1999; ROSE & WOOD, 1999), it vanished from this region during the Late Pliocene (NÉRAUDEAU et al., 2001). Today *Clypeaster* is found in the shallow sublittoral of the tropics, all around the world (for details see MORTENSEN, 1948b). In some regions the genus still reaches high diversities, as e.g. in the Caribbean, where ten (possibly eleven) species are reported, several of them co-occurring at single localities (HOPKINS, 1988). The co-occurrence of several different clypeasteroid species at single localities may be explained by resource partitioning (*Clypeaster rosaceous*, *C. subdepressus*, *Encope michelini* and *Leodia sexiesperforata* all facilitate different size fractions of the sediment; see Telford & Mooi, 1986, and Telford et al. 1987). Additionally, offset in spawning time might be a mechanism leading to reproductive isolation, but needs to be demonstrated for the Caribbean *Clypeaster* species. The distribution of the Caribbean *Clypeaster* species is at least partially controlled by temperature, only two species range beyond 30° N latitude (Hopkins, 1988: 337, fig. 1). The biogeography of the extant clypeasteroids was discussed by Ghiold & Hoffman (1984, 1985, 1986).

As already pointed out by numerous authors (e.g. Kalabis, 1949: 81) many nominally described species of *Clypeaster* are in fact junior synonyms, owing to the practice of describing new species based on single, often poorly preserved specimens (e.g. Vadász, 1915; Lovisato, 1909, 1910, 1911a, b, 1912a, b, 1913, 1914a, b), and incomplete knowledge of the literature and stratigraphy. The genus *Clypeaster* with its more than 480 nominal species (most of which were described from the Miocene of the Mediterranean region) is thus one of the best examples of taxonomic over-completeness.

One of the major problems in the systematics of this genus is the exceptional high intraspecific variability displayed by (at least some) *Clypeaster* species. Profile shape, outline, thickness of the test, tubercle densities, thickness of the margin, etc. can vary considerably within a single species, especially between populations from different habitats/sedimentary environments. Even the width and depth of the infundibulum and the (relative) length of the petals show a distinct degree of intraspecific variation. Yet, it is these features which were and are primarily used to distinguish the various species (also in the systematics of extant forms) in the genus *Clypeaster*. Attempts to employ the internal support system for classificatory purposes (Checchia-Rispoli, 1920, 1923; Kalabis 1935, 1937c; Roman, 1952) did until now not result in features that can be confidently employed for *Clypeaster* systematics. Although the internal structures can in some cases be revealed by x-ray images evaluation of the differences is very difficult. In the majority of the specimens, however, x-ray images are extremely difficult to prepare (due to low contrast between skeletal material and sedimentary infilling, differing test thickness,...) and thus variation of the internal support system within individual species is poorly known. While it might be possible to employ additional features (e.g. pedicellariae, ...) for the definition of extant species, this is impossible for fossil forms. Therefore special care has to be taken when working with fossil species of this genus and apart from morphology temporal and spatial distribution of the taxa should also be taken into account.

Another potential problem in *Clypeaster* systematics is the ability of echinoids to form hybrids, even across genus boundaries (Mortensen, 1912; Onoda, 1943). Although not demonstrated to occur in the genus *Clypeaster* until now, hybridisation could be one possible explanation for the occurrence of intermediate forms between otherwise well distinguished species.

A further problem in *Clypeaster* systematics is our often inadequate knowledge of the type material of many species, its whereabouts, provenance and morphology. As stated above in the introduction, the author re-examined the material of species recorded from the Austrian Neogene, whenever possible. Special emphasis was also placed on the search for the material of the Naturhistorisches Museum Wien examined by Michelin during 1856 to 1857. This material comprises a large part of the reference specimens of Michelin's *Clypeaster* monograph. About 80 percent of the material could be located in the collections of the Naturhistorisches Museum Wien, including several types and figured specimens.

The synonymy lists include primarily Austrian and other Paratethyal references because these could either be evaluated by the original material or topotypic specimens. It is very difficult to evaluate the extremely high number of *Clypeaster* records in the geological and palaeontological literature of the Mediterranean region without seeing the material the records refer to. Similarly differential diagnoses are restricted to cooccurring taxa in the Paratethys and a few well-known Mediterranean forms. It would be impossible and also out of scope of this paper to give differential diagnoses to all the *Clypeaster* species reported from the Mediterranean Neogene (more than 250 species).

Notes to morphometry in the genus *Clypeaster*: A large number of measurements were made to improve and substantiate the morphological analysis of the Austrian *Clypeaster* species. During this a number of problems were encountered: a) measurements were often difficult or impossible due to inadequate preservation of many specimens. This hampered multivariate approaches, since most analyses exclude incomplete data sets; b) many potential informative features are very difficult to measure using simple distance measurement (e.g. inflation of the interporiferous zone, width of the infundibulum at the transition to the "normal" flattened oral surface, profile outline,...). A schematic drawing (Fig. 24) shows which variables were measured. The raw data can be found in Table 9.

Notes to clypeasteroid nomenclature: The term "infundibulum" is applied in at least two different ways in the literature. While some authors apply it only when the peristome lies in a concave, funnel-like, depression at the apex of the concavity near the centre of the oral surface (e.g. in *Clypeaster rosaceus*), others use it in all *Clypeaster* species, even when oral surface is flattened and horizontal all the way towards the edge of the peristome (e.g. *Clypeaster subdepressus*). Here it is used in the second sense (e.g. stating in the descriptions that there is a very narrow and shallow infundibulum in species like *C. campanulatus*, rather than saying there is none).

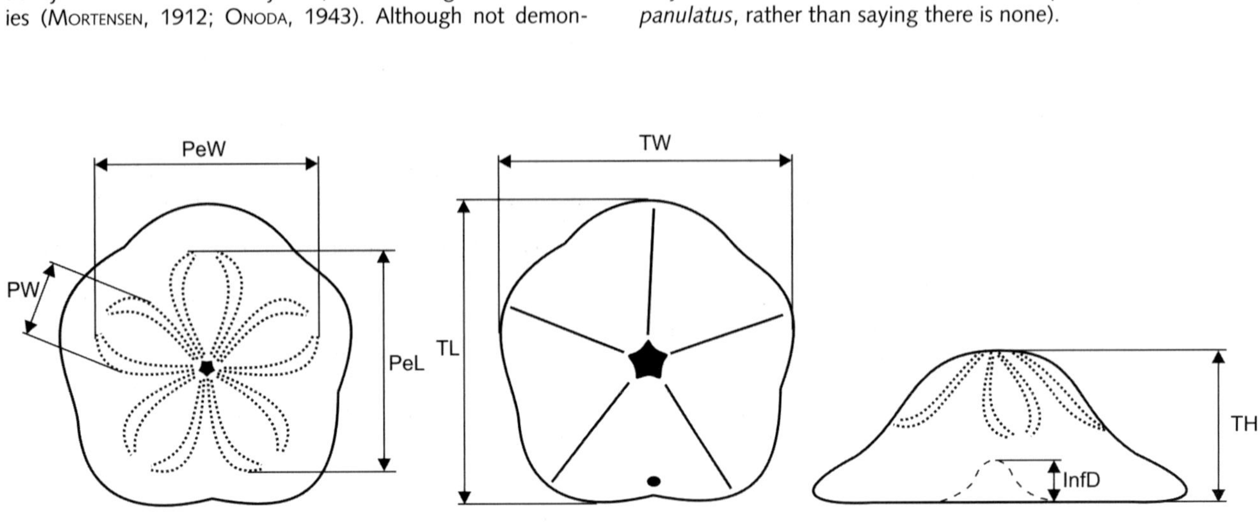

Figure 24: Schematic drawing of *Clypeaster* showing the variables measured.

Clypeaster calabrus Seguenza, 1880

(Pl. 31, Figs. 1-3; Pl. 32, Figs. 1-3)

1868 *C. intermedius* – Karrer: 570 (footnote)
1869 *Clypeaster intermedius* Dues. – Fuchs: 194
1869a *Clyp. intermedius* Desmoulins. – Laube: 183
1870 *Clypeaster intermedius* Mich. – Laube: 314
pp 1871 *Clypeaster intermedius* Desmoulins. – Laube: 64-65 [Austrian material]
1877 *Clypeaster intermedius* Desm. – Hilber: 261, 262
1877 *Clypeaster intermedius* Desm. – Karrer: 170, 312
1877 *Clypeaster intermedius* Desm. – Lóczy: 63
1878 *Clypeaster intermedius* Desm. – Hilber: 553, 560, 563, 575
* 1880 *C.[lypeaster] intermedius* des Moulins, Var. *Calabra*. n. – Seguenza: 54; pl. 5, figs. 3, 3a-b
1907 *[Clypeaster] intermedius* Desm. – Schaffer: 28
1907 *Clypeaster intermedius* Desm. – Schaffer: 35
1915 *Cl. excentricus* Vad. – Májer: 35, 88
1915 *Cl. danubicus* Vad. – Májer: 35, 88
v. 1915 *Clypeaster Scillae* Desm. var. *alienus* Vad. – Vadász: 129-131; fig. 23
v pp 1915 *Clypeaster crassus* Ag. – Vadász: 131-132; figs. 24
v. 1915 *Clypeaster excentricus* n. sp. – Vadász: 133; fig. 26; pl. 11 (5), fig. 9
v. 1915 *Clypeaster danubicus* n. sp. – Vadász: 134; fig. 27
v. 1915 *Clypeaster acclivis* Pomel. – Vadász: 136-137; fig. 29
pp 1925 *Clypeaster calabrus* – Checchia-Rispoli: 18, pl. 7, figs. 1, 1a-b, 2, 2a-b, 3; pl. 9, fig. 2; pl. 10, figs. 1, 1a-b, 2, 2a-b, 3; pl. 13, figs. 1, 1a-b, 2 [the specimen on pl. 9, fig. 2 was placed into *C. scillae* by Imbesi Smedile, 1959]
1928 *Clypeaster intermedius* Desm. – Bobies: 48
? 1936 *Clypeaster intermedius* Désm. – Szörényi: 300-301; figs. 55-56 [abnorm 4-rayed specimen]
1938a *Clypeaster calabrus* Segu. – Kalabis: 2, 5, 7, 10
1938b Clypeaster scillae aff. var. *calabra* Segu. – Kalabis: 4-5, 8
1938 *Clypeaster gregoryi* Lambert 1913 – Poljak: 181-182; pl. 5, fig. 1
1939 *Clypeaster intermedius*. – Kapounek: 72
1941 *Clypeaster acclivis* Pomel. – Meznerics: 87; pl. 1, figs. 1-2
1941 *Clypeaster epianthus* n. sp. – Meznerics: 87-88; pl. 2, fig. 1; pl. 3, fig. 2
1941 *Clypeaster kemencensis* n. sp. – Meznerics: 88; pl. 2, fig. 3; pl. 3, fig. 6
1942 *[Clypeaster] intermedius* Desm. – Schaffer: 130
1942 *Clypeaster intermedius* Desm. – Schaffer: 94
1949 *Clypeaster intermedius* Desmoulins – Schouppé: 143
? 1958 *Clypeaster calabrus* Seguenza – Imbesi Smedile: 17-18; pl. 2, figs. 1, 1a-b, 2, 2a; pl. 3, fig. 1
non 1967 *Clypeaster calabrus* (Seguenza) – Marcopoulou-Diacantoni: 365-367; pl. 16, fig. 1; pl. 18, figs. 1-2
? 1968 *Clypeaster intermedius* Desm. – Flügel & Heritsch: 44; pl. 4, fig. 2
1969 *Clypeaster grandiflorus* Bronn – Mitrović-Petrović: 128; pl. 10, fig. 2; pl. 11, figs. 1, 1a
1975 *Clypeaster calabrus* (Seguenza, 1880). – Kalabis: 176
v. 1982 *Clypeaster scillae* Desmarest – Rögl et al.: 65; un-numbered fig. [misidentification]
v. 1998 *Clypeaster intermedius* Michl.– Schultz: 116; pl. 52b, figs. 7
2001 *Clypeaster calabrus* Seguenza 1880 (Lambert, 1906) – Ventura & Zanfrà: 66

Type-material:

Clypeaster calabrus Seguenza, 1880:
Syntypes: Two syntypes, one of which is figured by Seguenza (1880: pl. 5, figs. 3, 3a-b); current whereabouts unknown
Locus typicus: Stilo, Calabria
Age: Early to ?Middle Miocene [Aquitanian according to Seguenza (1880: 54), "Elveziano" according to Checchia-Rispoli (1925: 18)]

Clypeaster danubicus Vadász, 1915:
Holotype: MAFI Ech 230, figured by Vadász (1915: fig. 27); housed at the Museum of the Hungarian Geological Survey
Locus typicus: Kemence, Pest, Hungary
Stratum typicum: Leitha limestone
Age: Late Badenian (Early Serravallian), Middle Miocene

Clypeaster excentricus Vadász, 1915:
Lectotype (designated by Boda, 1964: 205): MAFI Ech 126; housed at the Museum of the Hungarian Geological Survey
Paralectotype: MAFI Ech 435 from the Early Badenian of Gombhegy, Kemence, Hungary; and another specimen from Letkés, Hungary (not located)
Locus typicus: Kemence, Pest, Hungary
Stratum typicum: Leitha limestone
Age: Late Badenian (Early Serravallian), Middle Miocene

Clypeaster scillae alienus Vadász, 1915:
Syntype: MAFI Ech 123, figured by Vadász (1915: fig. 23); housed at the Museum of the Hungarian Geological Survey; second specimen mentioned by Vadász not located
Locus typicus: Gârbova de Sus (= Gîrbova de Sus, = Felsö-Orbó), Romania
Age: Late Badenian (Early Serravallian), Middle Miocene

Clypeaster epianthus Meznerics, 1941:
Holotype: specimen figured by Meznerics (1941: pl. 2, fig. 1; pl. 3, fig. 2); private collection of R. Streda; current whereabouts unknown
Locus typicus: Mátraszölös (= Mátraszöllös), Hungary
Age: Badenian (Langhian-Early Serravallian)

Clypeaster kemencensis Meznerics, 1941:
Holotype: specimen figured by Meznerics (1941: pl. 2, fig. 3; pl. 3, fig. 6); private collection of R. Streda; current whereabouts unknown
Locus typicus: Kemence, Hungary
Stratum typicum: Leitha limestone
Age: Late Badenian (Early Serravallian), Middle Miocene

Material:

Early ? Badenian (Langhian) – Rauchstallbrunngraben, near Baden, NÖ, Austria
NHMW: 3 specimens (NHMW 1978/2040/2, 1978/2040/6, 1978/2040/14)
Early Badenian (Langhian) – Stotzing (sandpit Mayer), Bgld, Austria
NHMW: 1 specimen (NHMW 2004z0093/0022)
Badenian (Langhian-Early Serravallian) – Brunn am Steinfeld, NÖ, Austria
NHMW: 2 specimens (NHMW 1859.LIV.1, 2003z0051/0001)
Badenian (Langhian-Early Serravallian) – Kalksburg, Vienna, Austria
NHMW: 21 specimens (NHMW 1866.XLVII.36, 1890.X.3-6,1890.X.61903.III.7, 1904.VIII.73-79, 1904.VIII.81, 2003z0049/0001 to 0006, 2003z0049/0030)
Badenian (Langhian-Early Serravallian) – Petersdorf, Styria, Austria
NHMW: 1 specimen (NHMW 1868.VIII.60)
Badenian (Langhian-Early Serravallian) – unknown locality, Austria

Species	Inventory No.	Location	TL	TW	TH	PeL	PeW	APP_W	Loc.[1]	InfD
C. calabrus	NHMW 2003z0049/0001	Kalksburg. W	140	124	44	96	85	26.9	2	-
C. calabrus	NHMW 2003z0049/0002	Kalksburg. W	156	138	50	-	-	-	-	18.0
C. calabrus	NHMW 2003z0049/0003	Kalksburg. W	137	117	47	95	81[2]	25.5	2	18.9
C. calabrus	NHMW 1904.VIII.76	Kalksburg. W	131	113	41	86	-	21.4	2	15.0
C. calabrus	NHMW 1890.X.5	Kalksburg. W	106	93	36	69	68	20.5	4	-
C. calabrus	NHMW 1904.VIII.73	Kalksburg. W	123	110	45	81	-	21.4	2	-
C. calabrus	NHMW 1904.VIII.75	Kalksburg. W	138	129	43	96[2]	89	25.9	2	-
C. calabrus	NHMW 1904.VIII.78	Kalksburg. W	107	93	35	-	-	18.8	4	15.3
C. calabrus	NHMW 1904.VIII.79	Kalksburg. W	101	92	-	-	-	-	-	-
C. calabrus	NHMW 2003z0049/0004	Kalksburg. W	135	115	47	-	-	-	-	20.3
C. calabrus	NHMW 1903.III.7	Kalksburg. W	129	109	31	-	-	-	-	-
C. calabrus	NHMW 1890.X.3	Kalksburg. W	131	121	46	84	78	23	2	-
C. calabrus	NHMW 2003z0049/0006	Kalksburg. W	131	108	44	-	75	25.8	4	17.6
C. calabrus	NHMW 1904.VIII.77	Kalksburg. W	130	120	45	80	73	20.8	2	-
C. calabrus	NHMW 2004z0112/0003	Müllendorf. Bgld	98	84	30	63	55	15.5	2	-
C. calabrus	NHMW 1997z0178/1753a	Müllendorf. Bgld	123	116	37	83	75	22.5	2	14.6
C. calabrus	NHMW 1997z0178/1753b	Müllendorf. Bgld	84	74	30	59	51	15.7	2	12.7
C. calabrus	NHMW 1868.VIII.60	Petersdorf. Bgld	123	107	37	-	-	27.1	4	15.7
C. calabrus	NHMW 1978/2040/6	Baden (RBG). NÖ	102	86	32	68	58	-	-	-
C. calabrus	NHMW 1978/2040/2	Baden (RBG). NÖ	116	97	38	71	60	19.6	4	-
C. campanulatus f. *acuminatus*	NHMW 1904.VIII.85	Kalksburg. W	147	133	62	104	96	30.7	4	12.1
C. campanulatus f. *acuminatus*	NHMW 1904.VIII.91	Perchtoldsdorf. NÖ	150	133	63	108	105	32.6	2	16.1
C. campanulatus f. *acuminatus*	NHMW A2736	Perchtoldsdorf. NÖ	148	133	66	115	108	36.3	4	-
C. campanulatus f. *aff. gibbosus*	NHMW 1857.38.23	Kalksburg. W	170	144	52	117	107	38.8	1	-
C. campanulatus f. *aff. reidii*	NHMW 1904.VIII.87	Kalksburg. W	152	138	63	113	-	32.8	2	15.9
C. campanulatus f. *aff. reidii*	NHMW 1890.X.8	Perchtoldsdorf. NÖ	134	126	60	103	98	28.8	2	-
C. campanulatus f. *campanulatus*	NHMW 1867.XXII.1	Hainburg. NÖ	-	132	60	99	97[2]	33.9	4	-
C. campanulatus f. *campanulatus*	NHMW 2003z0049/0008	Kalksburg. W	177	166	95	120	-	-	-	-
C. campanulatus f. *campanulatus*	NHMW 2003z0049/0007	Kalksburg. W	150	137	59	105	96	27.8	2	18.0
C. campanulatus f. *campanulatus*	NHMW 1978/2040/3	Baden (RBG). NÖ	154	134	60	99	93	30.6	2	-
C. campanulatus f. *campanulatus*	NHMW 1978/2040/4	Baden (RBG). NÖ	149	137	63	-	97	31.2	2	17.2
C. campanulatus f. *campanulatus*	NHMW 1976/1843/17	Baden (RBG). NÖ	140	122	60	98	90	28.8	2	17.9
C. campanulatus f. *campanulatus*	NHMW 1976/1843/16	Baden (RBG). NÖ	147	134	86[2]	107	100	31.8	4	16.5
C. campanulatus f. *partschi*	NHMW 1904.VIII.63	Kalksburg. W	147	140	52	102	97	27.6	2	-
C. campanulatus f. *partschi*	NHMW 1857.38.35	Kalksburg. W	151	140	50	94	88	25.9	2	-
C. campanulatus f. *partschi*	NHMW 2003z0049/0011	Kalksburg. W	147	139	51	101	101	28.8	4	14.6
C. campanulatus f. *partschi*	NHMW 1866.I.1274	Kalksburg. W	157	137	55	107	97	31.7	2	-
C. campanulatus f. *partschi*	NHMW 2003z0049/0012	Kalksburg. W	198	152	62[2]	128	108	34.5	2	-
C. campanulatus f. *partschi*	NHMW 2003z0049/0013	Kalksburg. W	186	165	64	-	-	31	2	-
C. campanulatus f. *partschi*	NHMW 2003z0049/0014	Kalksburg. W	161	146	59	-	-	-	-	17.1
C. campanulatus f. *partschi*	NHMW 2003z0049/0015	Kalksburg. W	153	143	54	96[2]	89	30.1	4	15.1
C. campanulatus f. *partschi*	NHMW A2740	Kalksburg. W	167	155	62	-	-	-	-	14.9
C. campanulatus f. *partschi*	NHMW A2744	Kalksburg. W	162	144	51	114	104	33	2	14.5
C. campanulatus f. *partschi*	NHMW 1857.38.31	Kalksburg. W	169	161	59	116	108	33.1	2	-
C. campanulatus f. *partschi*	NHMW 1904.VIII.66	Kalksburg. W	172	159	64	108	103	30	2	-
C. campanulatus f. *partschi*	NHMW 2003z0049/0009	Kalksburg. W	149	138	49	-	-	-	-	18.3
C. campanulatus f. *partschi*	NHMW 1904.VIII.69	Kalksburg. W	162	151	52	102	95	27.9	2	-
C. campanulatus f. *partschi*	NHMW 1904.VIII.68	Kalksburg. W	144	135	48	-	-	-	-	-
C. campanulatus f. *partschi*	NHMW 1858.XXV.39	Kalksburg. W	143	132	42	88	85	25	2	-
C. campanulatus f. *partschi*	NHMW 1858.III.132	Kalksburg. W	152	-	47	102	-	27.6	2	14.6
C. campanulatus f. *partschi*	NHMW 2003z0049/0010	Kalksburg. W	145	133	49	-	-	-	-	-
C. campanulatus f. *partschi*	NHMW 1858.III.18	Kalksburg. W	186	172	53	114	113	29.2	2	14.5
C. campanulatus f. *partschi*	NHMW 1857.38.29	Kalksburg. W	158	145	50	100	95	-	-	-
C. campanulatus f. *partschi*	NHMW 1858.III.19	Kalksburg. W	162	154	50	96	92	23.8	4	18.3
C. campanulatus f. *partschi*	NHMW 1857.38.30	Kalksburg. W	176	159	62	118	113	32.9	4	14.8
C. campanulatus f. *partschi*	NHMW 1857.38.26	Kalksburg. W	159	142	78	-	-	26.2	4	-
C. campanulatus f. *partschi*	NHMW 1904.VIII.88	Kalksburg. W	182	161	61	117	102	33.4	4	-
C. campanulatus f. *partschi*	NHMW 1857.38.32	Kalksburg. W	152	139	59	101	97	30	4	14.7
C. campanulatus f. *partschi*	NHMW 1904.VIII.71	Perchtoldsdorf. NÖ	150	134	52	102	92	31.8	4	-
C. campanulatus f. *partschi*	NHMW A2737	Kalksburg. W	162	147	52	98	91	24.3	4	-
C. campanulatus f. *partschi*	NHMW 2003z0049/0028	Kalksburg. W	162	149	60	-	-	29.4	2	-
C. campanulatus f. *portentosus*	NHMW 1978/2040/1	Baden (RBG). NÖ	148	141	64	100	103	28.2	4	-
C. campanulatus f. *portentosus*	NHMW 2003z0055/0002	Baden (RBG). NÖ	164	152	83	-	-	31.4	2	-

Table 9: Dimensions of the *Clypeaster* specimens used in the present study.

Species	Inventory No.	Location	TL	TW	TH	PeL	PeW	APP_W	Loc.[1]	InfD
C. campanulatus f. *pyramidalis*	NHMW 1905.VII.10	Leibnitz. Stmk	155	145	85	-	-	31.2	2	-
C. campanulatus f. *pyramidalis*	NHMW 1843.LIII.1	Baden (M). NÖ	167	153	84	121	119	34.8	2	-
C. campanulatus f. *pyramidalis*	NHMW 1904.VIII.83	Baden (RBG). NÖ	151	137	86	105	100	32.9	2	-
C. campanulatus f. *pyramidalis*	NHMW 1904.VIII.89	Baden (RBG). NÖ	148	134	70	-	-	29.5	4	-
C. scillae	NHMW 1978/2022/3	Aflenz. Stmk	87	73	33	57	51	14.3	4	-
C. scillae	NHMW 1978/2022/2	Aflenz. Stmk	85	70	35	59[2]	48	15.1	2	-
C. scillae	NHMW 1978/2022/1	Aflenz. Stmk	91	66	31	-	-	14	4	16.2
C. scillae	NHMW 1858.XII.8	Brunn/Steinfeld. NÖ	89	85	31	65	58	16.2	2	-
C. scillae	NHMW 2003z0056/0001	Eisenstadt. Bgld	92	73	35	60	50	14.6	4	13.7
C. scillae	NHMW 1869.I.664	Großhöflein. Bgld	98	82	-	69	55	16.7	2	-
C. scillae	NHMW 1867.XXII.2	Hainburg. NÖ	85	72	30	57	49	14.5	2	11.6
C. scillae	NHMW 1858.XV.56	Höflein. Bgld	92	72	35	-	-	-	-	-
C. scillae	NHMW 1868.VIII.42	Höflein. Bgld	91	74	42	-	-	-	-	-
C. scillae	NHMW 1852.II.1547	Šahy. SK	99	81	37	74	62	21.1	4	18.7
C. scillae	NHMW 1852.II.1548	Šahy. SK	92	77	36	-	-	-	-	-
C. scillae	NHMW 1904.VIII.74	Kalksburg. W	136	110	49	96	76	-	-	22.6
C. scillae	NHMW 1857.38.28	Kalksburg. W	98	77	40	66	54	-	-	-
C. scillae	NHMW 1865.XXXV.22	Kemence. HU	116	94	41	76	64	20.3	2	-
C. scillae	NHMW 1865.XXXV.16	Kemence. HU	55	44	15	33	28	-	-	-
C. scillae	NHMW 2003z0057/0001	Kemence. HU	86	67	35	-	-	14.3	4	15.0
C. scillae	NHMW 1865.XXXV.19	Kemence. HU	92	71	32	60	49	16.3	4	15.2
C. scillae	NHMW 2003z0057/0002	Kemence. HU	95	76	36	63	51	15.9	4	-
C. scillae	NHMW 1865.XXXV.21	Kemence. HU	78	62	28	52	42	12.8	2	15.6
C. scillae	NHMW 1865.XXXV.20	Kemence. HU	86	70	28	-	-	-	-	-
C. scillae	NHMW 1858.XXXIX.17	Kemence. HU	75	59	29	50	41	13.4	2	-
C. scillae	NHMW 1858.XXXIX.14	Kemence. HU	106	82	43	75	62	17.5	2	-
C. scillae	NHMW 1858.XXXIX.15	Kemence. HU	72	61	30	-	-	11.6	2	-
C. scillae	NHMW 1858.XXXIX.16	Kemence. HU	86	68	35	58	47	13.5	2	15.7
C. scillae	NHMW 1852.II.1546	Kemence. HU	88	75	30	-	-	-	-	-
C. scillae	NHMW 1885.XXXV.18	Kemence. HU	92	74	35	-	-	15	4	-
C. scillae	NHMW 1865.XXXV.17	Kemence. HU	47	38	15	29	24	7.8	2	7.6
C. scillae	NHMW 2003z0058/0001	Leibnitz. Stmk	97	78	28	63	50	14.5	2	-
C. scillae	NHMW 1978/2020/1	Mannersdorf. NÖ	106	88	40	73	57	21.2	2	>20
C. scillae	NHMW 1978/2020/3	Mannersdorf. NÖ	103	81	38	-	-	-	-	>16
C. scillae	NHMW 1978/2020/6	Mannersdorf. NÖ	112	87	35	78	-	18.5	4	-
C. scillae	NHMW 1978/2020/4	Mannersdorf. NÖ	108	85	40	73	56	17.1	4	-
C. scillae	NHMW 1978/2020/5	Mannersdorf. NÖ	116	88	42	81	-	17.9	2	18.7
C. scillae	NHMW 1978/2020/17	Mannersdorf. NÖ	96	76	32	63	52	15.6	2	-
C. scillae	NHMW 1978/2020/16	Mannersdorf. NÖ	106	89	36	72	61	16.8	4	-
C. scillae	NHMW 1978/2020/7	Mannersdorf. NÖ	112	86	39	77	-	17	2	17.7
C. scillae	NHMW 1978/2020/13	Mannersdorf. NÖ	78	63	31	52	42	13.5	4	15.9
C. scillae	NHMW 1978/2020/14	Mannersdorf. NÖ	81	63	34	-	-	-	-	-
C. scillae	NHMW 1978/2020/11	Mannersdorf. NÖ	97	74	34	65	51	16.2	4	16.2
C. scillae	NHMW 1978/2020/18	Mannersdorf. NÖ	83	65	27	56	44	14.7	2	15.7
C. scillae	NHMW 1978/2020/8	Mannersdorf. NÖ	94	73	66	-	-	15.5	4	-
C. scillae	NHMW 1978/2020/20	Mannersdorf. NÖ	91	74	24	57	47	15.1	4	14.5
C. scillae	NHMW 1978/2020/10	Mannersdorf. NÖ	101	78	36	-	-	19[2]	2	19.8
C. scillae	NHMW 1978/2020/12	Mannersdorf. NÖ	86	66	34	55	44	12.8	2	18.6
C. scillae	NHMW 1978/2020/9	Mannersdorf. NÖ	97	75	-	-	-	-	-	-
C. scillae	NHMW 1978/2020/19	Mannersdorf. NÖ	104	84	31	66	54	16.3	2	16.2
C. scillae	NHMW 1978/2020/15	Mannersdorf. NÖ	106	81	47	-	-	-	-	-
C. scillae	NHMW 2004z0072/0003	Mannersdorf. NÖ	82	67	30	53	44	13.4	4	16.6
C. scillae	NHMW A2328	Müllendorf. Bgld	99	79	34	67	55	18.5	2	15.0
C. scillae	NHMW 2003z0031/0006	Müllendorf. Bgld	66	55	19	-	-	12.8	4	9.2
C. scillae	NHMW 2004z0112/0007	Müllendorf. Bgld	85	70	31	59	50	15.6	2	-
C. scillae	NHMW 1997z0178/1745	Müllendorf. Bgld	111	89	39	78	62	20	2	14.9
C. scillae	NHMW 1976/1843/8	Baden (RBG). NÖ	108	89	33	70	61	16.6	4	17.4
C. scillae	NHMW 2004z0098/0006	Retznei. Stmk	57	48	23	39	33	10.7	2	10.7
C. scillae	NHMW 1978/2021/1	St. Nikolai. Stmk	93	78	35	-	-	-	-	-
C. scillae	NHMW 1978/2021/2	St. Nikolai. Stmk	98	77	29	-	-	15.9	2	-
C. scillae	NHMW 1868.I.Anhang	Gamlitz. Stmk	112	91	38	74	59	19.4	4	-

Abbreviations for variables: see Fig. 24; SK = Slovak Republic, HU = Hungary, RGB = Rauchstallbrunngraben, M = Mitterberg
Footnotes: 1 = Location of measurement; 2 = approximation

Table 9: Dimensions of the *Clypeaster* specimens used in the present study.

NHMW: 1 specimen (NHMW 2003z0050/0001, 2003z0050/0003 [specimen figured in Rögl et. al., 1982])

Late Badenian (Early Serravallian) – Müllendorf (Mühlendorfer Kreide AG quarry), Bgld, Austria

NHMW: 9 specimens (NHMW 1997z0178/1741, ../ 1746a-c, ../ 1753a-b, ../1755, 2004z0112/0003-4)

GBA: 3 specimens (no inventory numbers)

Foreign material for comparison:

Early Badenian (Langhian) – Kemence (Gombhegy), Pest, Hungary

MAFI: 2 specimens [MAFI Ech 126 and Ech 435 (**syntypes** of *Clypeaster excentricus* Vadász, 1915: 133, fig. 26, pl. 11 (5), fig. 9)]

Late Badenian (Early Serravallian) – Gârbova de Sus (= Gîrbova de Sus, = Felsö-Orbó), Romania

MAFI: 1 specimen [MAFI Ech 123 (**syntype** of *Clypeaster scillae alienus* Vadász, 1915: 129-131, fig. 23), Ech 134 (specimen figured as *Clypeaster acclivis* Pomel by Vadász, 1915: 136, fig. 29)]

Late Badenian (Early Serravallian) – Kemence, Pest, Hungary

MAFI: 2 specimens [MAFI Ech 230 (**holotype** of *Clypeaster danubicus* Vadász, 1915: 134, fig. 27), Ech 239 (specimen figured as *Clypeaster crassus* Ag. by Vadász, 1915: 131-132, fig. 24)]

Dimensions: see Tab. 9

Description:

Size and shape: The test is of medium to large size, elongated antero-posteriorly, with a pentagonal outline. The anterior margin is bluntly pointed, the posterior margin transversely truncated, distinctly indented and less thick than the anterior margin. The ambitus is moderately thick and may be tumid in some specimens, but is distinctly thinner than in *C. scillae*. The lateral margins are distinctly indented in interambulacra 1 and 4. There may be also marginal indentations in interambulacra 2 and 3 but they are usually less well developed. The maximum width lies anterior of the apical disc, where ambulacral columns IIa and IIIb reach the ambitus. In profile, the test has a trapezoid shape with a moderately domed petaloid area. The maximum height lies around the apical disc on the raised adapical parts of the interporiferous zones of the ambulacra. The oral surface is distinctly flattened, with a broad, moderately deep infundibulum.

Apical disc: The apical disc lies subcentrally and is slightly depressed. It belongs to the monobasal type with a large, central, subpentagonal madreporite and 5 circular gonopores.

Ambulacra: All five ambulacra are petaloid, straight and closing distally. The frontal petal is longest, the anterior paired petals shortest. The length of the petals ranges from 57 to 66 % of the corresponding test radius. The poriferous zones are slightly depressed and relatively wide. The pore pairs are conjugated anisopores. Adjacent pore pairs are separated by a narrow ridge with a single row of up to 8 primary tubercles. The interporiferous zones are strongly inflated and usually 2.4 to 2.7 times as wide as single poriferous zone at the widest point of the petals. They are crowded with primary tubercles similar to those on the interambulacra. On the oral surface shallow, simple unbranched food grooves are present in the axis of the ambulacra.

Interambulacra: The interambulacra are distinctly depressed adapically between the petals. They are crowded with perforate crenulate primary tubercles in sunken areoles. Between the primary tubercles, dense miliary tuberculation is present. Tubercle size and density increases towards the margin. Between the petals the tubercle density is usually low and the tubercles are slightly larger. On the oral surface tubercles are larger than on the aboral side and more closely spaced, their areoles nearly touching each other.

Peristome: The peristome is subcircular to very slightly oval and lies subcentrally on the oral side of the test in a moderately deep, broad infundibulum with gently sloping walls.

Periproct: The periproct is subcircular and lies inframarginally, about 3 to 4 mm away from the posterior margin.

Internal support system: unknown

Differential diagnosis:

For the difference to *C. barcinensis* ? Lambert, 1906, see above under that species.

C. campanulatus (Schlotheim, 1820) (and its phenotypes), a species occurring in the Badenian (Langhian-Early Serravallian) of the Paratethys, differs by its larger test height, broader petals, more strongly raised petaloid area, and thicker margin.

C. folium Agassiz in Agassiz & Desor, 1847, a species restricted to the Badenian (Langhian-Early Serravallian) in the Paratethys, is distinguished from this species by its lower test height, more elongated outline with deeper marginal indentations in the interambulacra, smaller and less raised petaloid area, and deeper food grooves.

C. intermedius des Moulins, 1837, an Early Miocene species of the Rhône Basin, differs from *C. calabrus* by its broader petals, less strongly inflated interporiferous zones, but more inflated interambulacra between the petals, deeper, more steep-walled infundibulum and less strong marginal indentations.

C. latirostris Michelin, 1861, a species occurring in the Late Eggenburgian (Early Burdigalian) of the Molasse Zone, is distinguished from this species by its subequal test length and width, lower test height, thin margin, slightly inflated interporiferous zones, and the broad, very shallow infundibulum.

For the difference to *C. neudorfensis* Lambert, 1927, see below under that species.

C. scillae des Moulins, 1837, a co-occurring species, differs from *C. calabrus* by its (usually) higher test, more elongated outline (see Fig. 25.A), more strongly raised petaloid area, thicker and more tumid margin, deeper, steep-walled infundibulum (Fig. 25.B) and less strong marginal indentations. High morphotypes of *C. calabrus* differ from *C. scillae* by their different profile: in *C. scillae* the whole aboral surface is inflated and the margin is thick; in *C. calabrus*, in contrast, only the petaloid area is inflated and the margin is relatively thin [compare e.g. the figure of "*C. danubicus*" in Vadász (1915: 134, fig. 7), which represents such a high morphotype of *C. calabrus*]. Additionally, *C. scillae* has an inflated oral surface with cushion-shaped interambulacra, whereas *C. calabrus* has a distinctly flattened surface.

Discussion:

The studied specimens are placed in the species *Clypeaster calabrus* Seguenza, 1880. The same is true for most of the specimens of the Badenian deposits of Austria formerly referred to *C. intermedius*, as already observed by Kalabis (1937a: 2, 7-8; 1949: 93-96). *C. intermedius* has recently been re-described and illustrated by Philippe (1998: 112-118) and it seems clear now that records from the Middle Miocene of the Paratethys are misidentifications.

Lambert (1906b) raised *C. intermedius calabra* Seguenza, 1880 to species rank and also associated *C. petaliferus* Seguenza, 1880 with that species. Cottreau (1913a: 58) rejected this and placed both *C. intermedius calabra* and *C. petaliferus* into the synonymy of *C. intermedius* des Moulins, 1837. Checchia-Rispoli (1925: 20) opposed Cottreau and insisted that both *C. calabrus* and *C. petaliferus* were distinct species. Lambert (1927a: 14) commented on Cottreau's action in the following way: «M. Cottreau réunit ces deux formes au *C. intermedius*; cette manière de voir peut être fondée, en ce qui concerne *C. petaliferus*, mais en ce qui concerne *C. calabrus* je ne saurais m'y rallier; sa forme est trop différente de l'espèce se rapprocherait plutôt *C. crassicostatus*.» ("Mr. Cottreau united these two forms with *C. intermedius*; while this opinion can be accepted, with regard to *C. petaliferus*, but with regard to *C. calabrus* I cannot endorse this; its form is too different from

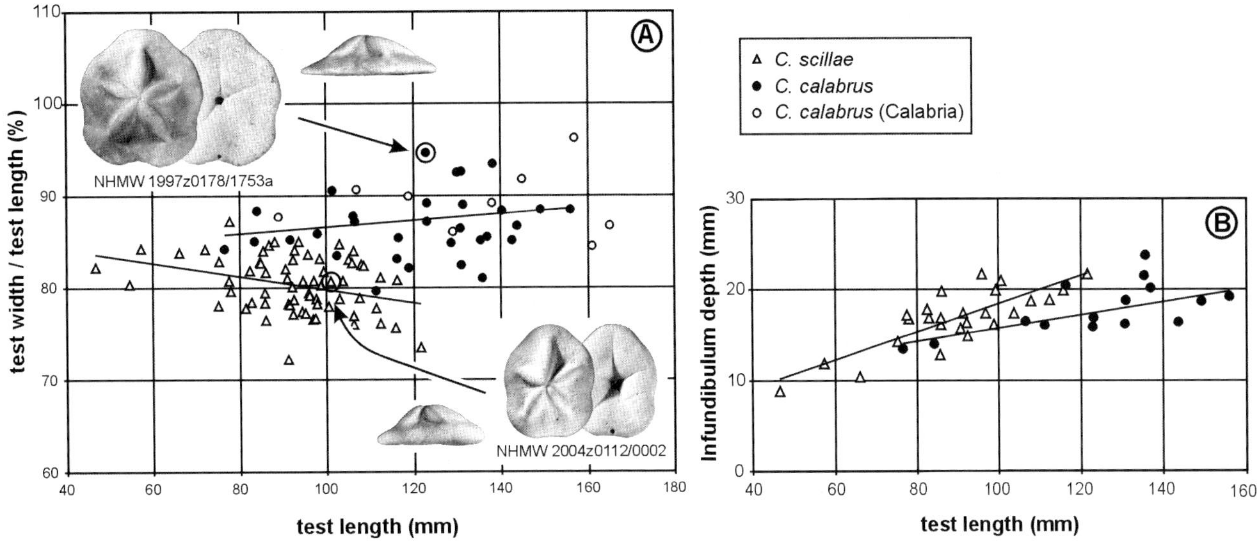

Figure 25: Morphometric comparison between *C. calabrus* SEGUENZA, 1880 and *C. scillae* DES MOULINS, 1837: Variation of relative test width (A) and infundibulum depth (B). Data of *C. calabrus* specimens from the type-area have been included in diagram A for comparison (from SEGUENZA, 1880 and VENTURA & ZANFRÀ, 2001).

that species and would rather approach *C. crassicostatus.*"). KALABIS (1949: 94, 97) discussed the different opinions expressed by former authors and concluded that since *C. intermedius* and *C. calabrus* did not coexist in any locality they represent an example of geographic speciation and have thus to be considered subspecies. I do, however, not support his view and, like LAMBERT, regard *C. calabrus* as distinct species for the following reasons: a) the two species are clearly separable and to my knowledge no intermediate forms with *C. intermedius* exist; b) the re-description of the type material of *C. intermedius* and additional material from the type region by PHILIPPE (1998) give us a much better knowledge of the variation in *C. intermedius* than ever before and although the two species seem to be clearly related, they are well differentiated; c) the two species are, most probably, not contemporaneous (Middle Miocene records of *C. intermedius* are dubious, as are Early Miocene records of *C. calabrus*).

Another question is the relation between *C. calabrus* and *C. scillae*. Like KALABIS (1937b: 2, 1938b: 3-5 but see also the discussion in 1949: 96) I initially thought that *C. calabrus* and *C. scillae* could be the extreme morphotypes of a single species with strong morphological variation, probably related to the sediment grain size or other environmental factors. This observation was based on a large series of specimens from the Vienna Basin (Mannersdorf and Kalksburg, Austria) and the Pannonian Basin (Kemence, Hungary). Research on specimens from further localities and different palaeoenvironmental settings, however, suggests that this is not the case. Although, there is an environmental signal in test morphology, it is less strong than initially thought and seems to concern mostly test height and to a certain extent thickness of the margin. Moreover, the two species co-occur in several localities (e.g. Müllendorf, Austria), within the same sedimentary units, where they are well differentiated. I thus consider *C. calabrus* and *C. scillae* distinct species with slightly overlapping morpho-space (when all individuals of several time horizons and palaeoenvironmental settings are examined together; compare Fig. 25.A). Another, although, probably less likely possibility is that these two species have the ability to interbreed. Hybridisation is known to occur in many groups of the echinoids (MORTENSEN, 1912; ONODA, 1943; LESSIOS & PEARSE, 1996; LESSIOS et al., 2001a, b; RAHMAN et al., 2001), probably even across genus boundaries (MORTENSEN, 1912) and the hybrids usually share features of both parent taxa.

The two species *Clypeaster epianthus* MEZNERICS, 1941 and *C. kemencensis* MEZNERICS, 1941 which have been established each on a single specimen of Mátraszölös respectively Kemence (both in Hungary) are placed into the synonymy of *C. calabrus* here. The type specimens were housed in a private collection and could not be located. Nevertheless it is clear from the descriptions and illustrations that they fall well within the variability of the species discussed here. MEZNERICS (1941) was correct in arguing that the two species cannot be placed into *C. scillae* or synonymous (*C. grandiflorus*, *C. crassus*, ...) species but failed to realise that SEGUENZA had already proposed a name for such forms. There are no features which would allow to differentiate these two species from *C. calabrus*.

Occurrence:

Austria: Badenian (Langhian-Early Serravallian)

Vienna Basin: Brunn am Steinfeld (KALABIS, 1938a; [NHMW]); Hainburg, NÖ (KALABIS, 1938b); Kalksburg, Vienna (KARRER, 1868; FUCHS, 1869; LAUBE, 1869a; KARRER, 1877; SCHAFFER, 1907; KALABIS, 1938a+b; SCHAFFER, 1942; KALABIS, 1975; SCHULTZ, 1998; [NHMW]); Müllendorf (Mühlendorfer Kreide AG quarry), Bgld (KAPOUNEK, 1939; RÖGL et al., 1982; [NHMW]); Perchtoldsdorf, NÖ (KALABIS, 1938a+b); Rauchstallbrunngraben, Baden, NÖ (LAUBE, 1869a, 1871; KARRER, 1877; SCHAFFER, 1907; BOBIES, 1928; SCHAFFER, 1942; KALABIS, 1975; [NHMW]); Rodaun, Vienna (KALABIS, 1938b); Vienna Basin (LÓCZY, 1877)

Styrian Basin: Gamlitz, Styria (HILBER, 1877); Kainberg, Styria (HILBER, 1878); Petersdorf, Styria [NHMW]; Retznei near Ehrenhausen, Styria (SCHOUPPÉ, 1949); St. Nikolai im Sausal, Styria (HILBER, 1878); Wildon, Styria (HILBER, 1878; ? FLÜGEL & HERITSCH, 1968)

Zala, Sáva and Dráva Basins: Kaznica, Croatia (POLJAK, 1938); Vujići, Bosnia & Herzegovina (MITROVIĆ-PETROVIĆ, 1969)

Paratethys (non-Austrian occurrences): Badenian (Langhian-Early Serravallian)

Molasse Zone: Podivín, Czech Republic (KALABIS, 1975)

Great Hungarian Basin (Pannonian Basin): Kemence (Gombhegy), Pest, Hungary (MÁJER, 1915; VADÁSZ, 1915; [MAFI]); Kemence, Pest, Hungary (VADÁSZ, 1915; MEZNE-

Rics, 1941; [MAFI]); Mátraszölös, Hungary (? Szörényi, 1936; Meznerics, 1941; Kalabis, 1975)

Transylvanian Basin: Gârbova de Sus (= Gîrbova de Sus, = Felsö-Orbó), Romania (Vadász, 1915; [MAFI]); Minişu des Sus (= Felménes), Arad, Romania (Lóczy, 1877)

Mediterranean: Early to Middle Miocene

Central Mediterranean: Belcastro, Calabria, Italy (pp Checchia-Rispoli, 1925; Ventura & Zanfrà, 2001); Stilo, Calabria, Italy (Seguenza, 1880; ? Imbesi Smedile, 1958)

Clypeaster campanulatus (Schlotheim, 1820)

(Pl. 22 to 27)

Material: see below under the respective formae

Dimensions: see Tab. 9

Description:

Size and shape: The test is of large size, with a subpentagonal outline. Test width varies between 85 to 95 % TL with very few exceptions (see Fig. 26.A). The margin of the test is moderately thick and rounded but not tumid (like e.g. in *C. scillae*). In aboral view the anterior margin is bluntly pointed and the posterior margin transversely truncated with only weak interambulacral indentations (if present at all). The maximum width lies anterior of the centre, where ambulacral columns IIa and IVb reach the ambitus. Profile outline and inflation of the petaloid area are extremely variable (see Plates 22 to 27 and Fig. 26.B). Consequently, the profile may be low and rounded to very high and pyramidal or even nearly hemispherical. The apex is usually somewhat flattened and coincides with the apical disc. The oral surface is distinctly flattened with a very narrow infundibulum.

Apical disc: The apical disc lies subcentrally and belongs to the monobasal type with a large, central, subpentagonal madreporite and 5 circular gonopores.

Ambulacra: Petals straight and open distally (IPZ width usually >7 mm distally). The frontal petal is usually the longest; the paired petals are more or less subequal in length, the posterior ones being a little bit longer than the anterior ones in some specimens. The length of the petals is quite variable and ranging from 63 to 75 % of the corresponding test radius. The poriferous zones are depressed and relatively wide. They are widest near the distal end of the petals. The pore pairs are conjugated anisopores. Adjacent pore pairs are separated by a narrow ridge with a single row of 2 to 15 primary tubercles (depending on position within the petal, size of the specimen and possibly also environmental parameters such as sediment grain size). The interporiferous zones are inflated, medially flattened and usually 2.5 to 3.5 times as wide as a single poriferous zone at the widest point of the petals. They are crowded with primary tubercles similar to those on the interambulacra. Outside the petals only minute microunipores are present, which are observable only in exceptionally preserved specimens/fragments. On the oral surface simple unbranched food grooves are present in the axes of the ambulacra.

Interambulacra: Adapically the interambulacra are slightly inflated between the petals, forming weak keels between them. They are crowded with perforate, crenulate primary tubercles

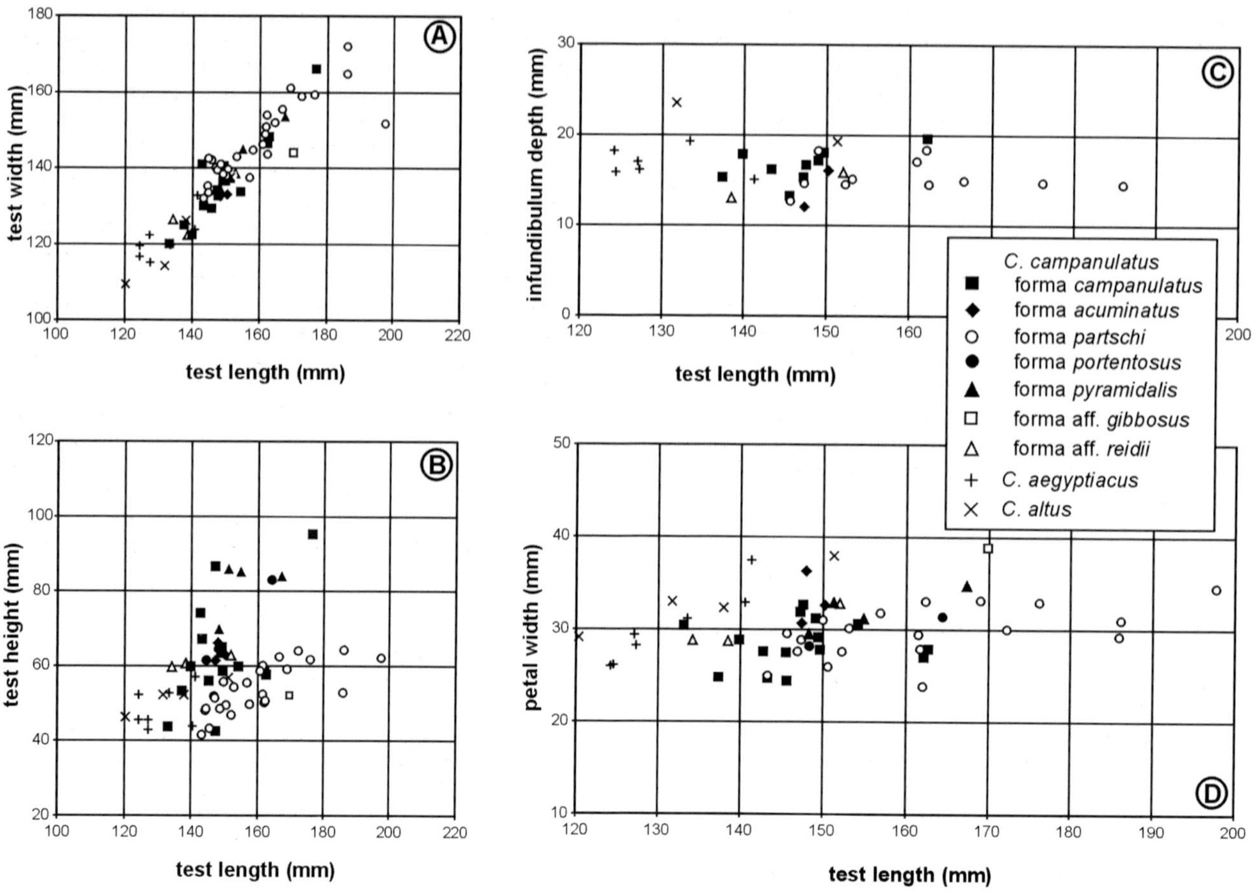

Figure 26: Morphological variation in the *Clypeaster campanulatus* group: test width (A), test height (B), infundibulum depth (C) and maximum petal width (D; measured in petals II or IV). Additionally, data of the nominal species *C. altus* from the Messinian of the Maltese Islands and *C. aegyptiacus* from the Pliocene of Egypt are included for comparison.

in sunken areoles. Between the primary tubercles, dense miliary tuberculation is present. Tubercle density and size is more or less homogenous on the aboral surface. On the oral surface tubercles are larger, more closely spaced and tubercle size increases towards the peristome. Adorally the interambulacra are flattened, producing an even surface, which is not inclined towards the peristome.

Peristome: The peristome is circular and lies centrally on the oral side of the test in a narrow and moderately deep infundibulum (see Fig. 26.C).

Periproct: The periproct is subcircular to very slightly oval, transversely elongated. Its diameter is about half or less that of the peristome. It lies inframarginally, usually about 5 to 6 mm away from the posterior margin.

Internal support system: The test of this species is double-walled in the marginal areas and reinforced by a moderately dense internal pillar system. The pillars are most abundant in the margin, in the central region around the lantern and in the interambulacra (see Pl. 26, Fig. 3).

Differential diagnosis:

C. altus Leske, 1778, from the Messinian of Malta differs by its longer and broader petals, thicker margin and broad infundibulum. Similarities in overall shape and profile occur but are considered to represent parallel adaptations to environmental conditions here.

For the difference to *C. barcinensis* ? Lambert, 1906 see above under that species.

C. calabrus Seguenza, 1880, a co-occurring species, differs by its very different profile, concave oral surface with broad infundibulum, narrower petals, much more inflated interporiferous zone, and deep marginal indentations.

C. folium Agassiz in Agassiz & Desor, 1847, from the Badenian (Langhian-Early Serravallian) of the Paratethys, is distinguished from this species by its flattened test with deep marginal indentations in the interambulacra, thinner margin, smaller petaloid area, broader infundibulum, and deeper food grooves.

C. intermedius des Moulins, 1837, an Early Miocene species of the Rhône Basin, differs from *C. campanulatus* (and its phenotypes) by its strongly different profile, lower test height, narrower petals, less strongly raised petaloid area, larger infundibulum, and thinner margin.

C. latirostris Michelin, 1861, a species occurring in the Eggenburgian (Early Burdigalian) of the Molasse Zone, is distinguished from this species by its subequal test length and width, lower test height, thin margin, only slightly inflated interporiferous zones, and the broad, very shallow infundibulum.

For the difference to *C. neudorfensis* Lambert, 1927, see below under that species.

C. scillae des Moulins, 1837, a co-occurring species, differs by its strongly different profile, concave oral surface with broad infundibulum, narrower petals, much more inflated interporiferous zone.

Discussion:

As outlined above in the introduction to the genus *Clypeaster*, this group is an outstanding example of taxonomic over-completeness. High intraspecific morphological variability and the use typological species concepts by earlier palaeontologists led to the recognition of an enormous number of species. In the Paratethys more than 50 nominal species were described (Michelin, 1861; Vadász, 1915; ...). Although the genus *Clypeaster* can show high diversity in the modern tropics (e.g. the Caribbean, see Hopkins, 1988), 50 synchronous nominal species in an area slightly larger than the modern Mediterranean are unparalleled in modern times.

It is difficult to accept all these as biological species having evolved in different parts of the Paratethys as reproductively isolated groups. Furthermore, these groups are interconnected by transitional morphs which co-occur at single localities. Hence they cannot be interpreted as chronosubspecies (*sensu* Willmann, 1987) or as biological subspecies (as done by Kalabis, 1949) since at least some are clearly sympatric. Hybridization which is known to occur in a variety of echinoid groups (Mortensen, 1912; Onoda, 1943; Lessios & Pearse, 1996) may also be a possible factor.

As these numerous species can be relatively well characterised and differentiated from other *Clypeaster* species in the study area I decided to treat the *Clypeaster campanulatus* types as extreme morphologies of a single species. I do, however, not renounce describing the characteristic formae. Although being aware that these taxa do not accurately fit to the biological species concept, ecological information might be lost if these were treated as a single superspecies.

These formae are named according to the junior synonym of *C. campanulatus*, whose type material is closest to that particular phenotype. Care has to be taken, however, that only names of species synonymous of *C. campanulatus* are used. Other names like *reidii* Wright, 1855 are not available, since they probably represent other, distinct species are synonyms of other, distinct species [like here in the case of *reidii* Wright, 1855 which was shown to be a junior synonym of *C. altus* Leske, 1778 by Challis, 1980 (A comprehensive revision of the Maltese clypeasterids is still missing. It is possible that Challis (1980) overestimated the variability of *C. altus* and that *C. campanulatus* co-occurs with the former species there)]. It is interesting that clearly distinct species like *C. campanulatus* and *C. altus* both exhibit a large intraspecific variation and may indeed form very similar phenotypes [concerning profile outline and overall shape; Marcopoulou-Diacantoni (1967) for example, described a large number of species using various names from different regions and stratigraphic horizons. Many of the specimens on which her identifications are based, however, could also be interpreted as phenotypes of *C. altus*, the intraspecific variability of which was documented by Challis (1980)]. Addressing the subject of parallel evolution and development of similar phenotypes in different species of *Clypeaster* Rose & Poddubiuk (1987: 464) suggested that: "Individual species morphology may thus be of more obvious palaeoenvironmental than taxonomic or stratigraphic significance."

The records of *C. campanulatus* from the Burdigalian of the Rhône Basin by Lambert (1913a) and Philippe (1998) are unconvincing. Both records lack an illustration or description. Moreover, there are also no other undoubted Early Miocene records of this species. Records of *C. campanulatus* or its synonyms and formae from sediments of the Mediterranean area are difficult to evaluate, in many cases no other information besides the name of the taxon is given [e.g. *C. alticostatus* and *C. portentosus* in the Miocene of Cyprus (Cowper Reed, 1932, 1933; Henson et al., 1949)]. But also in better documented records it is often difficult to doubtlessly confirm the specific identity with the Paratethyan species *C. campanulatus*. Anyhow, the present work focuses on the records from the Paratethys where the reference specimens or at least material from the respective localities were seen by the author. Records of *C. campanulatus* from the Mediterranean area are omitted from the synonymy lists in most cases.

forma *campanulatus* (Schlotheim, 1820)
(Pl. 22; Figs. 1a-c)

	1767	unbekannte und bisher noch nicht beschriebene Versteinerung – Anonymus: 702-706 [first mention of the two *Clypeaster* specimens described and illustrated by Knorr & Walch]
.	1771	*Scutum angulare altum* – Walch: 215-217
.	1771	[*Scutum angulare altum*] – Knorr: pl. IX.d., fig. 1.
*	1820	*Echinites campanulatus*. – Schlotheim: 323
?	1868	*Cl. altus* – Karrer: 570 (footnote)
?	1869	*Clypeaster altus* Lam. – Fuchs: 194

	1874	*Clypeaster altus conicus* – QUENSTEDT: pl. 82, fig. 13
pp	1875	*Scutum altum* – QUENSTEDT: 533-534
	1875	*Scutum altum conicum* – QUENSTEDT: 534-538
pp	1875	*Scutum altum campanulatum* – QUENSTEDT: 538-539
?	1877	*Clypeaster altus* LAM. – KARRER: 312
pp	1891	*Clypeaster altus*, (LESKE), 1778. – GREGORY: 593-596
?	1899	*Clypeaster altus* LMK. – ROTH VON TELEGD: 95
?	1907	*Clypeaster altus* LAM. – SCHAFFER: 35
# ?	1915	*Clypeaster hungaricus* n. sp. – VADÁSZ: 155-156, figs. 47-48 [type material lost]
?	1915	*Clypeaster agassizi* SISM. – VADÁSZ: 162-165, figs. 54-55 [reference material lost]
# v.	1915	*Clypeaster periplanus* n. sp. – VADÁSZ: 193-194, figs. 83-84
# ?	1915	*Clypeaster periplanus* var. *subtilis* – VADÁSZ: 194 [type material lost]
?	1937a	*Clypeaster campanulatus* SCHLOTH. – KALABIS: 46
	1938b	*Clypeaster campanulatus* (SCHLOTHEIM) – KALABIS: 5-7, 8-10
	1938	*Clypeaster campanulatus* SCHLOTHEIM 1920 – POLJAK: 186-187; pl. 6, fig. 1
?	1942	*Clypeaster altus* LAM. – SCHAFFER: 94
	1949	*Clypeaster campanulatus* (SCHLOTHEIM, 1820) – KALABIS: 42-54; 97-111
	1949	*Clypeaster campanulatus campanulatus* (SCHLOTHEIM, 1820) – KALABIS: 42-46; 97-101; pl. 3, figs. 1-3
	1949	*Clypeaster altus* LAMBERT – SCHOUPPÉ: 143
?	1967	*Clypeaster campanulatus* SCHLOTHEIM, 1820 – MARCOPOULOU-DIACANTONI: 367-368; pl. 12, fig. 1
	1969	*Clypeaster suboblongus* POMEL – MITROVIĆ-PETROVIĆ: 127; pl. 7, fig. 2; pl. 8, fig. 1
	1969	*Clypeaster sequenzai* VAD. – MITROVIĆ-PETROVIĆ: 133; pl. 19, figs. 1, 1a; pl. 20, fig. 1
	1981	*Clypeaster reidii* WRIGHT– MITROVIĆ-PETROVIĆ: 179; figs. 9-10
	1984	*Clypeaster sequenzai* VADASZ – MITROVIĆ-PETROVIĆ: 233; pl. 13, figs. 2, 2a-b
#	1990	*Clypeaster gombosae* n. sp. – MIHÁLY: 238-239, 240-241; pl. 2, fig.2; pl. 3, fig. 1; pl. 4, fig. 1
v.	2004	*Clypeaster* – HARZHAUSER et al.: 82, fig.

Type-material:

Clypeaster campanulatus (SCHLOTHEIM, 1820):
Syntypes: specimen figured by KNORR (1771: pl. IX.d., fig. 1) and an additional specimen mentioned by WALCH (1771); collection of Mister GEINITZ (donated by R. von MOLL, Vienna); current whereabouts unknown [see also ANONYMOUS (1767)]
Locus typicus: Baden, NÖ, Austria (probably from the quarries in the valley called Rauchstallbrunngraben)
Stratum typicum: "gelber Sandstein"
Age: Badenian (Langhian-Early Serravallian), Middle Miocene

Clypeaster periplanus VADÁSZ, 1915:
Holotype: MAFI Ech 146 (VADÁSZ, 1915: figs. 83-84); Museum of the Hungarian Geological Survey, Budapest
Locus typicus: Gârbova de Sus (= Felsö-Orbó), Romania
Age: Late Badenian (Early Serravallian), Middle Miocene

Clypeaster gombosae MIHÁLY, 1990:
Holotype: MAFI Coll. Ech. 340; Museum of the Hungarian Geological Survey, Budapest
Locus typicus: Gomb-Hill, Kemence, Hungary
Stratum typicum: Sámsonháza Fm.
Age: Early Badenian (Langhian), Middle Miocene

Material:

Early ? Badenian (Langhian) – Rauchstallbrunngraben, near Baden, NÖ, Austria
NHMW: 7 specimens (NHMW 1952/14, 1976/1843/14 to 1976/1843/17, 1978/2040/3, 1978/2040/4), 7 casts ["Prägesteinkerne"] (NHMW 1836.XXIII, 1978/2040/8, 1978/2040/9, 1978/2040/12, 1978/2040/13, 1978/2040/15, 1978/2040/16)
Badenian (Langhian-Early Serravallian) – Brunn am Steinfeld, NÖ, Austria
NHMW: 1 specimen (NHMW 2003z0051/0002)
Badenian (Langhian-Early Serravallian) – Hainburg (*Halitherium*-locality), NÖ, Austria
NHMW: 1 specimen (NHMW 1867.XXII.1)
Badenian (Langhian-Early Serravallian) – Kalksburg, Vienna, Austria
NHMW: 2 specimens (NHMW 2003z0049/0007, 2003z0049/0008)
Badenian (Langhian-Early Serravallian) – Lugoj (surroundings of), Romania
NHMW: 1 specimen (NHMW 1873.XIV.41)
Late Badenian (Early Serravallian) – Gârbova de Sus (= Felsö-Orbó), Romania
MAFI: 1 specimen [MAFI Ech 146 (**holotype** of *Clypeaster periplanus* VADÁSZ, 1915: 193-194, figs. 83-84)]
Late Badenian (Early Serravallian) – Müllendorf (quarry of the Mühlendorfer Kreidewerke AG), Bgld, Austria
NHMW: 6 specimens (NHMW 1997z0178/1747-1748, ../1750-1751, 2004z0112/0001, ../0006)

Transitional specimen *campanulatus* (SCHLOTHEIM, 1820) – *acuminatus* DESOR in AGASSIZ & DESOR, 1847:
Badenian (Langhian-Early Serravallian) – Rauchstall-brunngraben, near Baden, NÖ, Austria
NHMW: 1 specimen (NHMW 1978/2040/5)

Transitional specimens *campanulatus* (SCHLOTHEIM, 1820) – *partschii* MICHELIN, 1861:
Badenian (Langhian-Early Serravallian) – Kalksburg, Vienna, Austria
NHMW: 3 specimens (NHMW A2737, 2003z0049/0028 to 0029)

Description:

Morphotype with a strongly raised rounded petaloid area, the sides of which form a distinct angle with the marginal area outside the petals (in profile view). The area around the apical disc is somewhat flattened. The petals are narrower than in morphotypes where the whole aboral surface is inflated. Test height between 40 to 60 % TL.

Remarks:

This phenotype can easily be recognised by its distinct hat-like test profile. It is most common in the localities Rauchstallbrunngraben near Baden (where the type material of *campanulatus* comes from) and Müllendorf. Intermediate morphs connecting this form to *partschii* MICHELIN, 1861, *pyramidalis* MICHELIN, 1861 and *acuminatus* DESOR in AGASSIZ & DESOR, 1847 exist.

Some specimens of this phenotype from the type locality Rauchstallbrunngraben near Baden show an extremely thin and unusually sharp margin instead of the normally 5 to 10 mm thick margin. A section through such a specimen revealed that this was caused by postsedimentary compaction and leaching of the test (i.e. the specimens are "Prägesteinkerne"; see Fig. 27). The skeletal material is completely dissolved and partly replaced by an extremely thin layer of newly formed calcite (i.e. a pseudomorphosis). Therefore the extremely thin margin of these specimens does not have any value as specific character, it is caused postmortally by diagenesis. KALABIS (1938b: 6) reached a very similar explanation for this phenomenon, based on a large series of specimens from this locality.

Based on VADÁSZ's description and illustrations *C. hungaricus* and *C. agassizi* are very similar to *C. campanulatus*. They fit

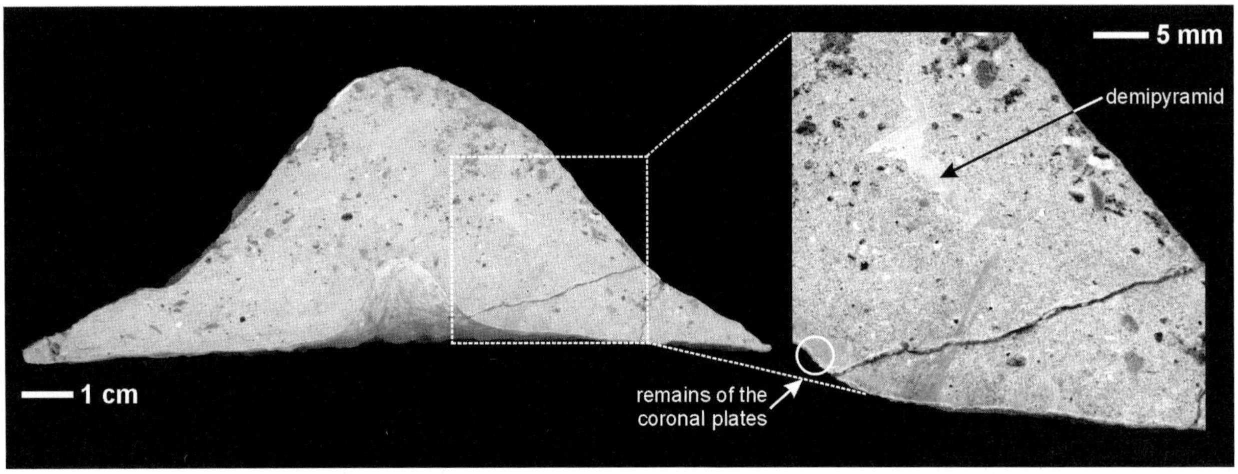

Figure 27: *Clypeaster campanulatus* (SCHLOTHEIM, 1820): vertical cross section of a "Prägesteinkern" from Rauchstallbrunngraben, near Baden, NÖ (NHMW 1978/2040/16; the original test is nearly completely leached, only traces remain)

best with material recovered from the locality Rauchstallbrunngraben, near Baden, the type locality of *campanulatus*. The type material of *hungaricus* and *agassizi* could not be located at the Museum of the Hungarian Geological Survey (where all other types of Vadász are kept) and is considered lost.

forma *partschii* MICHELIN, 1861
(Pl. 23; Figs. 1-2; Pl. 26, Fig. 1a-c)

* v. 1861 *Clypeaster Partschii*, MICHELIN – MICHELIN: 127-128; pl. 17, figs. 3; pl. 30, figs. a-f
1867 *Clyp. Partschi* MICH. – STACHE: 143
v. 1868 *Clypeaster Partschi* MICH. – KARRER: 570 (footnote)
v. 1869 *Clypeaster Partschi* MICHEL. – FUCHS: 194
v. 1869a *Clyp. Partschi* MICHELIN. – LAUBE: 183
v. 1870 *Clypeaster Partschi* MICH. – LAUBE: 314
v. 1871 *Clypeaster Partschi* MICHELIN. – LAUBE: 64
v. 1877 *Clypeaster Partschi* MICH. – KARRER: 312
? 1885 *Clypeaster confusus* (*Partschii* antea, non MICHELIN) – POMEL: pl. B22, figs. 1-5 [fide LAMBERT, 1927a]
? 1885 *Clypeaster paratinus* – POMEL: pl. B27, figs. 1-6 [fide LAMBERT, 1927a]
1906 *Clypeaster Partschi*, MICHELIN. – VADÁSZ: 331
1907 *Clypeaster Partschi* MICH. – SCHAFFER: 35
v. 1915 *Clypeaster sardiniensis* COTTEAU. – VADÁSZ: 137-138; figs. 30
\# v. 1915 *Clypeaster sardiniensis* COTT. var. *ellipticus* VAD. – VADÁSZ: 137-138; figs. 31
v. 1915 *Clypeaster crassicostatus* SISM. – VADÁSZ: 139-140; figs. 32-33
\# ? v 1915 *Clypeaster inflatus* n.sp. – VADÁSZ: 142; figs. 35
pp 1915 *Clypeaster campanulatus* SCHLOTH. sp. – VADÁSZ: 144-145
\# 1915 *Clypeaster campanulatus* SCHL. var. *declinatus* VAD. – VADÁSZ: 144-145, fig. 37a [type material lost]
\# 1915 *Clypeaster campanulatus* SCHL. var. *rotundus* VAD. – VADÁSZ: 144-145, fig. 37b [type material lost]
\# 1915 *Clypeaster campanulatus* SCHL. var. *sphaericus* VAD. – VADÁSZ: 144-145 [type material lost]
\# 1915 *Clypeaster depressus* n. sp. – VADÁSZ: 150-152, figs. 42-43 [type material lost]
\# v. 1915 *Clypeaster Lambert* LOV. var. *extensus* n. var. – VADÁSZ: 172-173, figs. 64-65
1915 *Clypeaster suboblongus* POM. – VADÁSZ: 183-184, fig. 74
\# v. 1915 *Clypeaster felménesensis* n. sp. – VADÁSZ: 183-184, fig. 74 [subadult specimen]
v. 1915 *Clypeaster partschi* MICH. – VADÁSZ: 186-187, figs. 76-77
? 1915 *Clypeaster petaliferus* SEGU. – VADÁSZ: 190-191, fig. 80
? # 1915 *Clypeaster gracilis* n. sp. – VADÁSZ: 191-192, figs. 81-82 [type material lost]
v. 1924b *Clypeaster Partschi* Mich. – SCHAFFER: 478; fig. 594
? 1931 *Clypeaster Partschi* MICHELIN. – LAMBERT: 87
1936 *Clypeaster partschi* MICHELIN – PAUCĂ: 143, 194-195; pl. 1, figs. 1-2
1937a *Clypeaster partschii* MICH. – KALABIS: 45, fig. 2
1938a *Clypeaster partschi* MICHELIN – KALABIS: 3, 6, 9, 11
1938 *Clypeaster partschi* MICHELIN 1863 – POLJAK: 189-190
1939 *Clypeaster Partschi* MICH. – KAPOUNEK: 72
1942 *Clypeaster Partschi* MICH. – SCHAFFER: 94
1949 *Clypeaster campanulatus partschi* MICHELIN, 1861 – KALABIS: 46-48; 101-104; pl. 4, figs. 1-2
1953 *Clypeaster partschi* MICHELIN, 1861. – SZÖRÉNYI: 68-69
1954 *Clypeaster campanulatus partschi* MICHEL. – SIEBER: pl. 10, fig. 4
? 1969 *Clypeaster sardiniensis* COTTEAU – MIHÁLY: 256
1969 *Clypeaster sardiniensis* COTT. – MITROVIĆ-PETROVIĆ: 125; pl. 3, fig. 3; pl. 4, figs. 1, 1a
1969 *Clypeaster olisiponensis* MICH. – MITROVIĆ-PETROVIĆ: 130-131; pl. 14, fig. 2; pl. 15; figs. 1, 1a
1969 *Clypeaster gibbosus* (RISSO) M. DE SERR. – MITROVIĆ-PETROVIĆ: 133-134; pl. 20, fig. 2; pl. 21; figs. 1, 1a
non 1970 *Clypeaster* cf. *partschi* Michelin, 1861. – MARCOPOULOU-DIACANTONI: 252; pl. 23, fig. 3
1984 *Clypeaster sardiniensis* COTT.– KÓKAY et al.: 288
1984 *Clypeaster reidii* WRIGHT – MITROVIĆ-PETROVIĆ: 222-223; pl. 2, fig. 2; pl. 3, figs. 1, 1a
1984 *Clypeaster sardiniensis* COTTEAU – MITROVIĆ-PETROVIĆ: 230; pl. 10, figs. 1, 1a-b
v. 2002b *Clypeaster* gr. *altus* (LAMARCK) – KROH: 11

Type-material:

Clypeaster partschii MICHELIN, 1861:

Syntypes: sixteen specimens in the collection of the Naturhistorsiches Museum Wien [NHMW 1857.38.24, 1857.38.26, 1857.38.29, 1857.38.30 (probably the specimen figured by MICHELIN, 1861), 1857.38.31-35, 1858.III.16-19, 1858.III.132, 2003z0049/0007, 2003z0049/0010]

Locus typicus: Kalksburg, Vienna

Age: Badenian (Langhian-Early Serravallian), Middle Miocene

Remarks: All specimens are accompanied by hand-written labels of MICHELIN. Both the labels and the specimens bear corresponding numbers. There is, therefore, little doubt that these specimens represent MICHELIN's reference specimens.

Clypeaster felménesensis VADÁSZ, 1915:

Holotype: Ech 149 (VADÁSZ, 1915: fig. 74); Museum of the Hungarian Geological Survey, Budapest

Locus typicus: Minişu des Sus (= Felménes), Arad, Romania

Age: Late Badenian (Early Serravallian), Middle Miocene

Clypeaster inflatus VADÁSZ, 1915:

Holotype: Ech 148 (VADÁSZ, 1915: fig. 35); Museum of the Hungarian Geological Survey, Budapest

Locus typicus: Slatina-Timiş (= Temes-Szlatina), Romania

Age: Late Badenian (Early Serravallian), Middle Miocene

Clypeaster lamberti extensus VADÁSZ, 1915:

Holotype: Ech 232 (VADÁSZ, 1915: figs. 64-65); Museum of the Hungarian Geological Survey, Budapest

Locus typicus: Alsótold, Nógrád, Hungary

Age: Late Badenian (Early Serravallian), Middle Miocene

Clypeaster sardiniensis ellipticus VADÁSZ, 1915:

Holotype: MAFI Ech 155 (VADÁSZ, 1915: fig. 31); Museum of the Hungarian Geological Survey, Budapest

Locus typicus: Armeniş (= Örményes), Romania

Age: Late Badenian (Early Serravallian), Middle Miocene

Material:

Early Badenian (Langhian) – Gainfarn, NÖ, Austria
  NHMW: 1 specimen (NHMW 2003z0003/20)
Early ? Badenian (Langhian) – Rauchstallbrunngraben, near Baden, NÖ, Austria
  NHMW: 3 specimens (NHMW 1976/1843/22, 1978/2040/11, 1978/2040/7)
Badenian (Langhian-Early Serravallian) – Kalksburg, Vienna, Austria
  NHMW: 44 specimens, among them the **syntypes** of *C. partschi* MICHELIN (see above under Type-material) (NHMW 1857.38.24, 1857.38.25, 1857.38.29 to 1857.38.31, 1857.38.33 to 1857.38.35, 1858.III.16 to 1858.III.19, 1858.III.132, 1858.XXV.39, 1866.I.1274, 1890.X.1, 1904.VIII.63, 1904.VIII.65 to 1904.VIII.69, 1904.VIII.86, A2738, A2740, A2744, 1997z0178/1672, 2003z0049/0009 to 0026)
Badenian (Langhian-Early Serravallian) – Perchtolsdorf, NÖ, Austria
  NHMW: 3 specimens (NHMW 1904.VIII.70, A 2735, 2003z0052/0001)
Badenian (Langhian-Early Serravallian) – Steinebrunn, NÖ, Austria
  NHMW: 1 specimen (NHMW 1859.XLV.628)
Badenian (Langhian-Early Serravallian) – Vöslau, NÖ, Austria
  NHMW: 1 specimen (NHMW 2003z0053/0001)
Badenian (Langhian-Early Serravallian) – Kemence, Hungary
  NHMW: 1 specimen (NHMW 1858.XXXIX.13)
  MAFI: 1 specimen [MAFI Ech 132 (referred to as *Clypeaster crassicostatus* SISM. by VADÁSZ, 1915: 139-140, figs. 32-33)]
Late Badenian (Early Serravallian) – Alsótold, Nógrád, Hungary
  MAFI: 1 specimen [MAFI Ech 232 (**holotype** of *Clypeaster lamberti extensus* VADÁSZ, 1915: 172-73, figs. 64-65)]
Late Badenian (Early Serravallian) – Armeniş (= Örményes), Romania
  MAFI: 1 specimen [MAFI Ech 155 (**holotype** of *Clypeaster sardiniensis ellipticus* VADÁSZ, 1915: 138, fig. 31)]
Late Badenian (Early Serravallian) – Biatorbágy (= Bia), Pest, Hungary
  MAFI: 1 specimen [MAFI Ech 154 (referred to as *Clypeaster sardiniensis* COTT. by VADÁSZ, 1915: 137-138)]
Late Badenian (Early Serravallian) – Gârbova de Sus (= Felsö-Orbó), Romania
  MAFI: 1 specimen [MAFI Ech 262 (figured as *Clypeaster partschi* MICH. by VADÁSZ, 1915: 186-187, figs. 76-77)]
Late Badenian (Early Serravallian) – Globu Craiovei (= Globukrajova), Romania
  MAFI: 3 specimen [MAFI Ech 271 (**syntype** of *Clypeaster acuminatus robustus* VADÁSZ, 1915: 149-150), Ech 272 and 278 (referred to as *Clypeaster sardiniensis* COTT. by VADÁSZ, 1915: 137-138)]
Late Badenian (Early Serravallian) – Luncaviţa (= Lunkavica), Romania
  MAFI: 1 specimen [MAFI Ech 277 (referred to as *Clypeaster sardiniensis* COTT. by VADÁSZ, 1915: 137-138)]
Late Badenian (Early Serravallian) – Minişu des Sus (= Felménes), Arad, Romania
  MAFI: 1 specimen [MAFI Ech 149 (**holotype** of *Clypeaster felménesensis* VADÁSZ, 1915: 183-184, fig. 74)]
Late Badenian (Early Serravallian) – Müllendorf (quarry of the Mühlendorfer Kreidewerke AG), Bgld, Austria
  NHMW: 2 specimens (NHMW 1997z0178/1752, 2004z0112/0005)

Transitional specimens *partschii* MICHELIN, 1861 – *acuminatus* DESOR in AGASSIZ & DESOR, 1847:

Badenian (Langhian-Early Serravallian) – Kalksburg, Vienna, Austria
  NHMW: 2 specimens (NHMW 1857.38.26, 1857.38.27)
Late Badenian (Early Serravallian) – Minişu des Sus (= Felménes), Arad, Romania
  MAFI: 1 specimen [MAFI Ech 280 (**syntype** of *Clypeaster acuminatus robustus* VADÁSZ, 1915: 149-150)]
Late Badenian (Early Serravallian) – Slatina-Timiş (= Temes-Szlatina), Romania
  MAFI: 1 specimen [MAFI Ech 147 (**syntype** of *Clypeaster acuminatus robustus* VADÁSZ, 1915: 149-150, fig. 41)]

Transitional specimens *partschii* MICHELIN, 1861 – aff. *gibbosus* DE SERRES, 1829:

Early Badenian (Langhian) – Gamlitz (Unterer Gnaser Bruch), near Ehrenhausen, Styria, Austria
  NHMW: 1 specimen (NHMW 1868.I.Anhang.2)
Badenian (Langhian-Early Serravallian) – Kalksburg, Vienna, Austria
  NHMW: 3 specimens (NHMW 1904.VIII.84, 1904.VIII.92, 2003z0049/0027)

Transitional specimens *partschii* MICHELIN, 1861 – aff. *reidii* WRIGHT, 1855:

Badenian (Langhian-Early Serravallian) – Kalksburg, Vienna, Austria
  NHMW: 1 specimen (NHMW 1857.38.32)
Badenian (Langhian-Early Serravallian) – Perchtoldsdorf, NÖ, Austria
  NHMW: 2 specimens (NHMW 1890.X.9, 1904.VIII.71)

Description:

This morphotype has a low test height and a gently arched profile. The petaloid area is moderately inflated and its sides are either flush with the surface outside the petalodium or they form a very weak angle. Test height between 29 to 37 % TL.

Remarks:

This phenotype is very common at the locality Kalksburg, Vienna, where it occurs in a poorly consolidated conglomerate with a grain size usually ranging from 2 to 5 mm. Intermediate morphs connecting this form with *campanulatus* (Schlotheim, 1820), *acuminatus* Desor in Agassiz & Desor, 1847, *portentosus* Michelin, 1861 and to forms determined by earlier workers as *gibbosus* de Serres, 1829 and *reidii* Wright, 1855 exist.

Lambert (1931) was one of the first to recognise the large variability in this group and synonymised the species *C. confusus* Pomel, 1887, *C. paratinus* Pomel, 1887, *C. bassanii* Lovisato, 1911, *C. subconoideus* Vadász, 1915, *C. convexus* Vadász, 1915, *C. gracilis* Vadász, 1915 and *C. periplanus* Vadász, 1915 with *C. partschii* Michelin, 1861.

In most papers this species is referred to as *partschi*, in contrast to the original spelling *partschii* by Michelin (1861). According to the ICZN (4th ed., Articles 33.4. and 33.5.) the original spelling is to be used and *partschi* has to be regarded as "incorrect subsequent spelling" (not as emendation) regardless whether the change from –ii to –i was intentional or not.

The record of *C. campanulatus* forma *partschi* from the Ukrainian part of the Carpathian Foredeep by Szörényi (1953, as *C. partschi*) is peculiar. *C. campanulatus* was never before or afterwards reported from that region (neither from Poland, nor from the northern Czech Republic). Until confirmed by additional finds this record should be treated with caution (Szörényi's material consisted entirely of Museum specimens, it is easily possible that the *C. campanulatus* specimen was accompanied by an erroneous label). The only other species of *Clypeaster* occurring in the Carpathian Foredeep is *C. scillae* which was reported by Mączyńska (1993, 1996) from Pińczów and Niechobrz in Poland.

forma *acuminatus* Desor in Agassiz & Desor, 1847
(Fig. 28; Pl. 27; Figs. 1a-c)

* v 1847a [*Clypeaster*] *acuminatus* Desor. – Desor in Agassiz & Desor: 130 ["moulage en plâtre R.63"]
v. 1858 [*Clypeaster*] *acuminatus* Desor. – Desor: 242
v. 1861 *Clypeaster acuminatus*, Desor – Michelin: 119-120; pl. 21, figs. a-e
v 1867 *Clypeaster acuminatus* Des. – Stache: 143
1869 *Clypeaster acuminatus* Desor. – Fuchs: 194
v. 1869a *Clyp. acuminatus* Desor. – Laube: 183
v. 1870 *Clypeaster acuminatus* Desm. – Laube: 314
v. 1871 *Clypeaster acuminatus* Desor. – Laube: 63
1877 *Clypeaster acuminatus* Defr. – Karrer: 170
v. 1877 *Clypeaster acuminatus* Des. – Karrer: 312
1878 *Clypeaster acuminatus* Desor. – Hilber: 563, 575
? 1880 C.[*lypeaster*] *acuminatus* Desor. Var. *devians* – Seguenza: 54; pl. 5, figs. 2, 2a-b
1887b *Clypeaster acuminatus*, Desor. – Koch: 264
1894 *Clypeaster acuminatus*. Desor. – Lörenthey: 59
1907 *Clypeaster acuminatus* Defr – Schaffer: 28
v. 1915 *Clypeaster acuminatus* desor. – Vadász: 149-150
v. 1915 *Clypeaster acuminatus* des. var. *robustus* vad. – Vadász: 149-150, fig. 41 [transitional to forma *partschi*]
1915 *Clypeaster Gauthieri* Lov. – Vadász: 153-155, figs. 45-46
? 1915 *Clypeaster subacutus* Pomel. – Vadász: 161-162
v. 1915 *Clypeaster subacutus* Pomel var. *depressus* Vad. – Vadász: 161-162, fig. 53
v. 1915 *Clypeaster subconoideus* n. sp. – Vadász: 165-166, figs. 56-57
v. 1915 *Clypeaster dacicus* n. sp. – Vadász: 167-168, figs. 58-59
1915 *Clypeaster transylvanicus* n. sp. – Vadász: 170-171, figs. 62-63 [type material lost]
1915 *Clypeaster convexus* n. sp. – Vadász: 178-179, figs. 70-71 [type material lost]
1915 *Clypeaster subaltus* n. sp. – Vadász: 181-182, figs. 72-73 [type material lost]
1928 *Clypeaster acuminatus* Defr. – Bobies: 48
v. 1928 *Clypeaster acuminatus* Desor – Lambert & Jeannet: 189
1938 *Clypeaster altus* Klein (*Scutum angulare*) 1734, 1778 – Poljak: 183-184; pl. 6, fig. 2
1941 *Clypeaster acuminatus* Desor. – Meznerics: 89-90; pl. 2, fig. 5
1942 *Clypeaster acuminatus* Defr – Schaffer: 130
. 1949 *Clypeaster campanulatus acuminatus* Desor, 1847. – Kalabis: 49-51, 104-106; pl. 4, figs. 3-5
? 1969 *Clypeaster* cf. *acuminatus* Cotteau – Mihály: 256
1969 *Clypeaster reidii* Wright – Mitrović-Petrović: 127; pl. 6, figs. 1, 1a; pl. 7, fig. 1
1984 *Clypeaster* cf. *acuminatus* Desor – Kókay et al.: 288

Type-material:

Clypeaster acuminatus Desor in Agassiz & Desor, 1847:
Holotype: collection of Michelin, current whereabouts unknown; plaster casts (moulage en plâtre R.63 of Agassiz) of the holotype are kept at the Musée d'Histoire Naturelle de Neuchâtel and at the Naturhistorisches Museum Wien (NHMW 1857.Gipsabguss.5)
Locus typicus: Šahy (= Ipoly-Shag), Slovak Republic
Age: Badenian (Langhian-Early Serravallian), Middle Miocene

Clypeaster acuminatus robustus Vadász, 1915:
Syntypes: three specimens (MAFI Ech 147, 271, 280; Vadász, 1915: 149-150, fig. 41); Museum of the Hungarian Geological Survey, Budapest
Type area: Globu Craiovei (= Globukrajova), Minişu des Sus (= Felménes), and Slatina-Timiş (= Temes-Szlatina), Romania
Age: Late Badenian (Early Serravallian), Middle Miocene

Clypeaster dacicus Vadász, 1915:
Holotype: MAFI Ech 118 (Vadász, 1915: figs. 58-59); Museum of the Hungarian Geological Survey, Budapest
Locus typicus: Gârbova de Sus (= Felsö-Orbó), Romania
Age: Late Badenian (Early Serravallian), Middle Miocene

Clypeaster subacutus depressus Vadász, 1915:
Holotype: MAFI Ech 122 (Vadász, 1915: fig. 53); Museum of the Hungarian Geological Survey, Budapest
Locus typicus: Gârbova de Sus (= Felsö-Orbó), Romania
Age: Late Badenian (Early Serravallian), Middle Miocene

Clypeaster subconoideus Vadász, 1915:
Holotype: MAFI Ech 120 (Vadász, 1915: fig. 56-57); Museum of the Hungarian Geological Survey, Budapest
Locus typicus: Gârbova de Sus (= Felsö-Orbó), Romania
Age: Late Badenian (Early Serravallian), Middle Miocene

Material:

Early Badenian (Langhian) – Letkés, Pest, Hungary
MAFI: 1 specimen [MAFI Ech 131 (referred to *Clypeaster acuminatus* Desor by Vadász, 1915: 149-150)]
Badenian (Langhian-Early Serravallian) – Kalksburg, Vienna, Austria
NHMW: 1 specimen (NHMW 1904.VIII.85)
Badenian (Langhian-Early Serravallian) – Kemence, Pest, Hungary
NHMW: 1 specimen (NHMW 1865.XXXV.15)
Badenian (Langhian-Early Serravallian) – Perchtolsdorf, NÖ, Austria
NHMW: 2 specimens (NHMW 1904.VIII.91, A2736)
Badenian (Langhian-Early Serravallian) – Šahy (= Ipoly-Shag), Slovak Republic

Figure 28: *Clypeaster campanulatus* forma *acuminatus* DESOR in AGASSIZ & DESOR, 1847: plaster cast of the holotype from Kemence, Hungary ("moulage R.63"; NHMW 1857.Gipsabguss.5)

NHMW: 1 plaster cast of the **holotype** [moulage en plâtre R.63] of *Clypeaster acuminatus* Desor in AGASSIZ & DESOR, 1847 (NHMW 1857.Gipsabguss.5)

Late Badenian (Early Serravallian) – Gârbova de Sus (= Felsö-Orbó), Romania

MAFI: 3 specimens [MAFI Ech 118 (**holotype** of *Clypeaster dacicus* VADÁSZ, 1915: : 167-168, figs. 58-59), Ech 120 (**holotype** of *Clypeaster subconoideus* VADÁSZ, 1915: 165-166, figs. 56-57), Ech 122 (**holotype** of *Clypeaster subacutus depressus* VADÁSZ, 1915: 162, fig. 53)]

Transitional specimen *acuminatus* DESOR in AGASSIZ & DESOR, 1847 – aff. *reidii* WRIGHT, 1855:

Early Badenian (Langhian) – Letkés, Pest, Hungary

MAFI: 1 specimen [MAFI Ech 233 (**holotype** of *Clypeaster angulatus oblonus* VADÁSZ, 1915: 175-177, fig. 69)]

Description:

Characterised by its low-conical profile and its long and broad petals (see Fig. 28). The surface of the petaloid area is more or less flush with the surface outside the petalodium. Test height is moderate and ranges from between 40 to 48 % TL.

Remarks:

This phenotype is never very common. A complete series of intermediate morphs connect this form with *partschii* MICHELIN, 1861. Some of these intermediates have been named *C. acuminatus robustus* by VADÁSZ (1915). There are, however, no clear-cut boundaries between this forma, *partschii* (low test height end) and *pyramidalis* (high test height end). Specimens with increasingly flattened adapical area connect this type with aff. *gibbosus* and aff. *reidii*. As in the former, it is impossible to really separate any well defined units from these series. Except forma *partschii*, which is very common (probably resulting from over-complete sampling of the locality Kalksburg), most of the phenotypes and their intermediates are more or less equally abundant. Therefore it is thought rather unlikely that they represent a complex of cryptic sympatric or hybridising species [as has been suggested for extant species of *Echinocardium* with high haplotype diversity and overlapping morphospace (compare results and hypotheses presented in: LAURIN et al., 1994; WÖRHEIDE, 1995; FÉRAL et al., 1998; CHENULI & FÉRAL, 2003)]. These morphological changes observed in *C. campanulatus* are thus interpreted as environmental signal.

In another group of species (*C. calabrus* – *C. scillae*; see above, respectively below) a similar problem of overlapping morphospace exists. In that group, however, intermediates are rather rare and do not form a continuous series connecting two extremes. Furthermore, individual groups are rather well defined and identification of most specimens is rather straight forward. In that particular case, therefore, an interbreeding species pair seems much more likely.

forma aff. *reidii* WRIGHT, 1855

(Pl. 25; Figs. 2a-c)

1915 *Clypeaster Reidii* WRIGHT. – VADÁSZ: 145
1915 *Cl. Almerai* LAMB. – MÁJER: 35, 88
1915 *Cl. angulatus* VAD. – MÁJER: 35, 88
1915 *Clypeaster hemisphaericus* n. sp. – VADÁSZ: 168-169, figs. 60-61 [type material lost]

v ? 1915 *Clypeaster Almerai* Lamb. – Vadász: 173-175, fig. 66
v. 1915 *Clypeaster angulatus* n.sp. – Vadász: 175-177, figs. 67-68 [transitional to forma *acuminatus*]
v. 1915 *Clypeaster angulatus* Vad. var. *oblongus* Vad. – Vadász: 175-177, fig. 69 [transitional to forma *acuminatus*]
1949 *Clypeaster campanulatus reidii* Wright, 1855 – Kalabis: 51-52; 106-109; pl. 5, figs. 1-4
1955 *Clypeaster angulatus oblongus* Vadas – Tollmann: Tab. 5b
1996 *Clypeaster hemisphaericus* Vadasz – Ghiurca: 190, unnumbered fig.

Type-material:
Clypeaster angulatus Vadász, 1915:
Lectotype (designated herein): MAFI Ech 121 (Vadász, 1915: figs. 67-68); Museum of the Hungarian Geological Survey, Budapest
Locus typicus: unclear [label gives no clue; Vadász (1915: 177) lists three localities: Gârbova de Sus (= Felsö-Orbó), Romania; Kemence and Letkés, Hungary]
Age: Badenian (Langhian-Early Serravallian), Middle Miocene

Clypeaster angulatus oblonus Vadász, 1915:
Holotype: MAFI Ech 233 (Vadász, 1915: fig. 69); Museum of the Hungarian Geological Survey, Budapest
Locus typicus: Letkés, Pest, Hungary
Age: Early Badenian (Langhian), Middle Miocene

Material:
Early Badenian (Langhian) – Gombhegy, Kemence, Pest, Hungary
MAFI: 1 specimen [MAFI Ech 237 (specimen strongly damaged; referred to *Clypeaster almerai* Lamb. by Vadász, 1915: 173-175, fig. 66)]
Badenian (Langhian-Early Serravallian) – Gârbova de Sus (= Felsö-Orbó) or Kemence or Letkés (Hungary or Romania)
MAFI: 1 specimen [MAFI Ech 121 (**syntype** of *Clypeaster angulatus* Vadász, 1915: 175-177, fig. 67-68)]
Badenian (Langhian-Early Serravallian) – Kalksburg, Vienna, Austria
NHMW: 2 specimens (NHMW 1890.X.7, 1904.VIII.87)
Badenian (Langhian-Early Serravallian) – Perchtoldsdorf, NÖ, Austria
NHMW: 1 specimen (NHMW 1890.X.8)
Late Badenian (Early Serravallian) – Müllendorf (quarry of the Mühlendorfer Kreidewerke AG), Bgld, Austria
NHMW: 1 specimens (NHMW 1997z0178/1754)

Description:
This morphotype is characterised by its high, near hemispherical profile and antero-posteriorly elongated outline. The petaloid area is strongly inflated and slightly larger than in other morphotypes. The surface of the petaloid area is more or less flush with the surface outside the petalodium or forms a very weak angel with it. Test height is moderate and ranges from between 40 to 45 % TL.

Remarks:
This phenotype of *C. campanulatus* is very similar (especially in profile view) to *C. reidii* Wright, 1855 from the Messinian of the Maltese Islands. Yet as demonstrated by Challis (1980) *C. reidii* is a junior synonym of *C. altus* Leske, 1778, a species considered to be different from *C. campanulatus* here. The astonishing similarities in overall test shape, profile and intraspecific variability exhibited could reflect similar responses to environmental factors.

This phenotype is quite rare. Transitional forms connecting this morphotype with *partschii* Michelin, 1861, *campanulatus* (Schlotheim, 1820) and forms identified as *gibbosus* de Serres, 1829 by former authors exist.

forma aff. *gibbosus* de Serres, 1829
(Pl. 25; Figs. 1a-c)

v pp 1861 *Clypeaster gibbosus*, Marcel de Serres – Michelin: 120-121; pl. 22, figs. a-g; pl. 23, figs. 1a-c
v pp 1861 *Clypeaster michelotti*, Agassiz – Michelin: 132 [specimen from Kalksburg]
1866b *Clypeaster gibbosus* M. de Serres – Suess: 251
1868 *Cl. gibbosus* – Karrer: 570 (footnote)
1869 *Clypeaster gibbosus* Serr. – Fuchs: 194
1869 *Clypeaster Michelotti* Agass. – Fuchs: 194
1869a *Clyp. gibbosus* Risso. – Laube: 183
1870 *Clypeaster gibbosus* Mich. – Laube: 314
1871 *Clypeaster gibbosus* Risso sp. – Laube: 63-64
1877 *Clypeaster gibbosus* Risso sp. – Hilber: 262
1877 *Clypeaster gibbosus* Risso. – Karrer: 170
1877 *Clypeaster gibbosus* Serr. – Karrer: 312
1877 *Clypeaster Michelotti* Agass. – Karrer: 312
1887b *Clypeaster* cfr. *gibbosus*, Risso sp. – Koch: 265
1894 *Clypeaster* cfr. *gibbosus*. Risso – Lörenthey: 59
1907 [*Clypeaster*] *gibbosus* Risso – Schaffer: 28
v pp 1915 *Clypeaster campanulatus* Schloth. sp. – Vadász: 144-145 [MAFI Ech 236]
v 1915 *Clypeaster delagadoi* Lor. – Vadász: 146-147; figs. 38-39
1928 *Clypeaster gibbosus* Risso. – Bobies: 48
1942 [*Clypeaster*] *gibbosus* Risso – Schaffer: 130

Material:
Early Badenian (Langhian) – Gamlitz (Unterer Gnaser Bruch), near Ehrenhausen, Styria, Austria
NHMW: 1 specimen (NHMW 1868.I.Anhang.1)
Badenian (Langhian-Early Serravallian) – Kalksburg, Vienna, Austria
NHMW: 3 specimens (NHMW 1857.38.23, 1904.VIII.72, 1904.VIII.88)
Late Badenian (Early Serravallian) – Gârbova de Sus (= Felsö-Orbó), Romania
MAFI: 1 specimen [MAFI Ech 236 (referred to *Clypeaster campanulatus* Schloth. by Vadász, 1915: 144-145)]
Late Badenian (Early Serravallian) – Kishajmás, Baranya, Hungary
MAFI: 1 specimen [MAFI Ech 235 (referred to *Clypeaster delagadoi* Lor. by Vadász, 1915: 146-147; figs. 38-39)]

Description:
This morphotype is very similar to *C. campanulatus* forma aff. *reidii* Wright, 1855 and is also characterised by a strongly inflated petaloid area. In aff. *gibbosus*, however, the test is flattened adapically and somewhat lower. The test is trapezoidal in profile. Test height is moderate and ranges from between 30 to 35 % TL.

Remarks:
C. gibbosus de Serres, 1829 is rather poorly known, the original description is insufficient (as stated already by Desor, 1858). The re-description and excellent illustrations by Michelin (1861; partly based on the specimens considered here), which at least in part, are based on the type material of *C. gibbosus* have been questioned by subsequent authors. Lambert (1912: 99; 1913a: 117) pointed out that the descriptions of de Serres and Michelin contradict each other and suggested that there was some confusion of the type material. Unfortunately he was unable to trace the type material. Thus the question whether Michelin's description corresponds to the species of de Serres or not remains open.

forma *pyramidalis* Michelin, 1861
(Pl. 24; Figs. 1-2)

* v. 1861 *Clypeaster pyramidalis*, Michelin – Michelin: 124-125; pl. 27, figs. a-e
v. 1869 *Clypeaster pyramidalis* Michel. – Fuchs: 194
v. 1869a *Clyp. pyramidalis* Michelin. – Laube: 183
v. 1870 *Clypeaster pyramidalis* Mich. – Laube: 314
v. 1871 *Clypeaster pyramidalis* Michelin. – Laube: 64
v. 1877 *Clypeaster pyramidalis* Mich. – Karrer: 312
? 1880 *C.[lypeaster] pyramidalis* Michelin. – Seguenza: 54
1880 *C.[lypeaster] pyramidalis* Michelin – Seguenza: 86; pl. 9, figs. 1, 1a
1880 *C.[lypeaster] pyramidalis* Michelin Var. *brevior* – Seguenza: 86; pl. 9, figs. 1b-c
1880 *C.[lypeaster] portentosus* des Moulins Var. *elatior* – Seguenza: 86; pl. 9, fig. 3
1887b *Clypeaster pyramidalis*, Michelin. – Koch: 265
1894 *Clypeaster pyramidalis* Michelin. – Hilber: 394
1915 *Cl. hungaricus* Vad. var. *dispar* Vad. – Májer: 35, 88
? 1915 *Clypeaster hungaricus* n. sp. – Vadász: 155-157; fig. 47-48
? v. 1915 *Clypeaster hungaricus* var. *dispar*. – Vadász: 155-156
? 1915 *Clypeaster Seguenzai* n. nom. – Vadász: 158-159, figs. 49-50 [reference material lost]
1915 *Clypeaster magnus* n. sp. – Vadász: 160-161, figs. 51-52 [type material lost]
? 1931 *Clypeaster pyramidalis* Michelin. – Lambert: 45-46
1938a *Clypeaster portentosus* Desm. var. *pyramidalis* Michelin – Kalabis: 3, 6, 9, 11
1938 *Clypeaster pyramidalis* Michelin 1861 – Poljak: 187; pl. 7, figs. 1-2
1941 *Clypeaster pyramidalis* Michelin. – Meznerics: 89; pl. 2, fig. 8; pl. 3, fig. 5 [transitional between forma *acuminatus* and forma *pyramidalis*]
? 1955 *Clypeaster* aff. *seguenzai* Vadas – Tollmann: Tab. 5b
1969 *Clypeaster turritus* Phil. – Mitrović-Petrović: 132-133; pl. 17, fig. 2; pl. 19; figs. 1, 1a
1984 *Clypeaster pyramidalis* Michelin – Mitrović-Petrović: 226-227; pl. 6, figs. 1, 1a; pl. 7, fig. 1
1984 *Clypeaster Lamberti* Lovisato – Mitrović-Petrović: 227-228; pl. 7, figs. 2, 2a; pl. 8, fig. 1
1984 *Clypeaster malladai* Lambert – Mitrović-Petrović: 228-229; pl. 8, fig. 2; pl. 9, fig. 1, 1a
1984 *Clypeaster gustavi* Lovisato – Mitrović-Petrović: 231; pl. 11, figs. 1, 1a-b
1984 *Clypeaster subacutus depressus* Vadasz – Mitrović-Petrović: 232; pl. 12, figs. 2, 2a

Type-material:
Clypeaster pyramidalis Michelin, 1861:
Holotype: NMHW 1843.LIII.1 (Michelin, 1861: pl. 27, figs. a-e); Naturhistorisches Museum Wien
Locus typicus: Mitterberg near Baden, NÖ, Austria
Age: Badenian (Langhian-Early Serravallian), Middle Miocene

Clypeaster hungaricus dispar Vadász, 1915:
Holotype: MAFI Ech 145 (Vadász, 1915: 156); Museum of the Hungarian Geological Survey, Budapest
Locus typicus: Gombhegy, Kemence, Pest, Hungary
Age: Early Badenian (Langhian), Middle Miocene

Material:
Early Badenian (Langhian) – Leibnitz, Styria, Austria
NHMW: 1 specimen (NHMW 1905.VII.10)
Early Badenian (Langhian) – Rauchstallbrunngraben, near Baden, NÖ, Austria
NHMW: 2 specimens (NHMW 1904.VIII.83, 1904.VIII.89)
? Early Badenian (Langhian) – Kemence (Gombhegy), Pest, Hungary
MAFI: 1 specimen {MAFI Ech 145 [**holotype** of *Clypeaster hungaricus dispar* Vadász, 1915: 156; specimen extremely similar to these classified as *pyramidalis* and *portentosus* by Michelin (1861)]}
Badenian (Langhian-Serravallian) – Mitterberg, near Baden, NÖ, Austria
NHMW: 1 specimen [NHMW 1843.LIII.1 (**holotype** of *C. pyramidalis* Michelin, 1861)]

Description:
Characterised by its high, conical profile. The petaloid area, like in aff. *reidii*, is slightly larger than in other phenotypes, but otherwise there are few differences. Test height is high and ranges from between 47 to 57 % TL.

Remarks:
The original description was based on a single specimen which is still housed at the Naturhistorisches Museum Wien. In the meantime a few additional specimens have been collected. These show that test height (the sole feature distinguishing this form from forma *acuminatus* and forma *portentosus*) is quite variable. This form is connected by transitional morphs with *acuminatus* and *portentosus* and via these formae with the remaining formae of *C. campanulatus*.

forma *portentosus* Michelin, 1861
(Pl. 22; Figs. 2-3)

? 1829 *Clypeaster altus*. Lamarck. C. I. – de Serres: 157 [fide des Moulins, 1837; *no description or illustration*]
? 1837 *C.[lypeaster] portentosus*. Nob. – des Moulins: 64 [*nomen nudum*]
*v pp 1861 *Clypeaster portentosus*, des Moulins – Michelin: 125-126; pl. 28, figs. a-e
v pp 1861 *Clypeaster alticostatus*, Michelin – Michelin: 126-127; pl. 29, figs. a, b
v. 1869a *Clyp. portentosus* Desmoulins. – Laube: 183
? 1870 *Clypeaster alticostatus* Mich. – Laube: 314
v. 1871 *Clypeaster portentosus* Desmoulins. – Laube: 64
? 1871 *Clypeaster alticostatus* Michelin. – Laube: 64
1877 *Clypeaster alticostatus* Mich. – Karrer: 170
? 1879 *Clypeaster portentosus*, Desmoulin. – Hermite: 239
? 1880 *C.[lypeaster] alticostatus* Michelin – Seguenza: 87
? 1891 *Clypeaster alticostatus*, Michelin, 1861 ? – Péron & Gauthier in Cotteau, Péron & Gauthier: 210-213
? 1906a *Clypeaster alticostatus* Michelin, 1861 – Lambert: 88
1907 *[Clypeaster] alticostatus* Mich. – Schaffer: 28
1928 *Clypeaster alticostatus* Micht. – Bobies: 48
1936 *Clypeaster alticostatus* Mich. – Paucă: 143
1938a *Clypeaster portentosus* Desm. var. *alticostatus* Michelin – Kalabis: 3, 5, 8-9, 11
? 1938 *Clypeaster alticostatus* Michelin 1861 – Poljak: 184-185
1938 *Clypeaster portentosus* Desmoulins 1837 – Poljak: 187-188
1942 *[Clypeaster] alticostatus* Mich. – Schaffer: 130
? 1970 *Clypeaster portentosus* – Marcopoulou-Diacantoni: 261; pl. 24, figs. 1, 4

Type-material:

Clypeaster alticostatus MICHELIN, 1861:
Syntype: NHMW 2003z0055/0001 from the Early Badenian (Langhian) of Rauchstallbrunngraben, near Baden (NÖ, Austria), figured by MICHELIN (1861: pl. 29, figs. a, b); Naturhistorisches Museum Wien
Remarks: Albeit the present specimen is the figured specimen of MICHELIN it is not chosen as lectotype since it is very poorly preserved (MICHELIN's figure is strongly reconstructed). MICHELIN mentioned additional syntypes from Santa Manza in Corsica, which are probably better suited for a lectotype designation, but could not be located.

Clypeaster portentosus MICHELIN, 1861:
Syntype: NHMW 1846.37.960; Naturhistorisches Museum Wien
Locus typicus: Eisenstadt, Bgld, Austria
Age: Badenian (Langhian-Early Serravallian), Middle Miocene
Remarks: Specimen accompanied by a hand-written label of MICHELIN, proving that this species was a syntype. The figured specimen could not be located.

Material:

Early ? Badenian (Langhian) – Rauchstallbrunngraben, near Baden, NÖ, Austria
NHMW: 6 specimens [NHMW 1904.VIII.90, 1978/2040/1, 1978/2040/10, 2003z0055/0001 (**syntype** of *C. alticostatus* MICHELIN, 1861), 2003z0055/0002, 1890.X.2 (specimen labelled as Kalksburg but preservation and attached sediment suggest the locality Rauchstallbrunngraben)]
Badenian (Langhian-Early Serravallian) – Baden, NÖ, Austria
NHMW: 1 specimen (NHMW 2003z0054/0001)
Badenian (Langhian-Early Serravallian) – Eisenstadt, Bgld, Austria
NHMW: 1 specimen (NHMW 1846.37.960 [**sytype** of *C. portentosus* MICHELIN, 1861])

Description:

Characterised by its highly raised petaloid area and conical to bell-shaped profile, intermediate in shape between *pyramidalis* and *campanulatus*. Test height is about 50 % of TL.

Remarks:

C. portentosus DES MOULINS, 1837 was established by indication to *C. altus* in DE SERRES (1829). In DE SERRES's work, however, there is no description or illustration. Thus the species cannot be regarded as validly established by indication. Therefore the name *C. portentosus* DES MOULINS, 1837 is a *nomen nudum*. MICHELIN (1861) is the next author who gives a description. Therefore according to the ICZN the species has to be attributed to MICHELIN. AGASSIZ & DESOR (1847a: 130) mention this species in the synonymy of *C. altus* LESKE, but give no description and DESOR (1858: 240) regards it as probable synonym of *C. turritus* AGASSIZ. Thus both cannot be considered as valid authors for this species.

Part of the type material (including the figured specimen) of *Clypeaster alticostatus* MICHELIN, 1861 is kept at the Naturhistorisches Museum Wien. The figured specimen is very poorly preserved (see Pl. 22, Figs. 3a-b) and belongs to the forma discussed here (*portentosus*). Thus, until the remaining type material (originally in the MICHELIN collection) of *C. alticostatus* is located and described *C. alticostatus* has to be considered a *nomen dubium*.

Occurrence:

Austria: Early to Late Badenian (Langhian-Early Serravallian)
Vienna Basin: Bad Fischau, NÖ (LAUBE, 1869a, 1871; KALABIS, 1938a); Bad Vöslau (MICHELIN, 1861; LAUBE, 1869a, 1871; [NHMW]); Baden (WALCH, 1771; KNORR, 1771; QUENSTEDT, 1875; KARRER, 1868); Brunn am Steinfeld, NÖ (LAUBE, 1869a, 1871; [NHMW]); Gainfarn, NÖ (KROH, 2002b; [NHMW]); Hundsheim Hill, Hainburg an der Donau, NÖ (STACHE, 1867; [NHMW]); Kalksburg, Vienna (MICHELIN, 1861; SUESS, 1866b; KARRER, 1868; FUCHS, 1869; LAUBE, 1869a, 1871; KARRER, 1877; SCHAFFER, 1907; KALABIS, 1937a, 1938a; SCHAFFER, 1942; [NHMW]); Mannersdorf (KALABIS, 1938a); Mitterberg (= Mittenberg) near Baden, NÖ (MICHELIN, 1861; LAUBE, 1869a, 1871; [NHMW]); Müllendorf (Mühlendorfer Kreide AG quarry), Bgld (KAPOUNEK, 1929; TOLLMANN, 1955); Perchtoldsdorf, NÖ (KALABIS, 1938a; [NHMW]); Rauchstallbrunngraben, near Baden, NÖ (MICHELIN, 1861; SUESS, 1866b; LAUBE, 1869a, 1871; KARRER, 1877; pp GREGORY, 1891; SCHAFFER, 1907; BOBIES, 1928; KALABIS, 1938a; SCHAFFER, 1942; [NHMW]); Rodaun, Vienna (KARRER, 1868); Steinebrunn (formerly Steinabrunn), NÖ (KALABIS, 1938a; [NHMW]); Vienna (QUENSTEDT, 1875); Wallersdopff (probably misspelling of Wöllersdorf), near Baden, NÖ (MICHELIN, 1861); Wiener Neustadt, NÖ (MICHELIN, 1861); Wöllersdorf, NÖ (SUESS, 1866b; KARRER, 1868; LAUBE, 1869a, 1871)
Eisenstadt-Sopron Basin: Eisenstadt, Bgld (MICHELIN, 1861; LAUBE, 1871; pp GREGORY, 1891; [NHMW])
Styrian Basin: Gamlitz, Styria ([NHMW]); Kainberg, Styria (HILBER, 1878); Leibnitz, Styria ([NHMW]); Retznei, near Ehrenhausen, Styria (SCHOUPPÉ, 1949); Weinleiten, near Gamlitz, Styria (HILBER, 1877); between Wiesfleck and Schreibersdorf, Bgld (HILBER, 1894)

Paratethys (non-Austrian occurrences): Early to Late Badenian (Langhian-Early Serravallian)
Molasse Zone: Mušlov, Czech Republic (KALABIS, 1938a); Židlochovice (= Selowitz), Moravia, Czech Republic (QUENSTEDT, 1875); Znojmo, Czech Republic (KALABIS, 1938a)
Vienna Basin: Devínska Nová Ves, Slovak Republic (KALABIS, 1938b, 1949); Kinberk (= Kienberg), near Mikulov, Czech Republic (KALABIS, 1938a, 1949); Mikulov, Czech Republic (KALABIS, 1937a, 1938a, 1949; HARZHAUSER et al., 2004)
Great Hungarian Basin (Pannonian Basin): Alsótold (=Alsó-Told), Nógrád, Hungary (VADÁSZ, 1915; [MAFI]); Biatorbágy (= Bia), Pest, Hungary (VADÁSZ, 1915; [MAFI]); Budapest, Pest, Hungary (KÓKAY et al., 1984); Budapest Rákos, Pest, Hungary (VADÁSZ, 1906); Budapest-Budatétény (= Tétény), Hungary (VADÁSZ, 1915); Hidasd, Baranya, Hungary (VADÁSZ, 1915); Kemence, Pest, Hungary (MICHELIN, 1861; LAUBE, 1869a, 1871; MÁJER, 1915; VADÁSZ, 1915; [MAFI]; [NHMW]); Kemence (Sámsonháza Fm., Gomb-Hill), Pest, Hungary (MIHÁLY, 1990; [MAFI]); Kishajmás, Baranya, Hungary (VADÁSZ, 1915; [MAFI]); Kovácsszénnája (= Kovacéna), Baranya, Hungary (VADÁSZ, 1915); Mátraverebély, Nógrád, Hungary (MEZNERICS, 1941); Mátraszölös, Nógrád, Hungary (MEZNERICS, 1941); Letkés, Pest, Hungary (VADÁSZ, 1915; [MAFI]); Lunkavica, Romania (VADÁSZ, 1915; [MAFI]); Nagypall, Baranya, Hungary (VADÁSZ, 1915); Pécs, Baranya, Hungary (VADÁSZ, 1915); Šahy (= Ipoly-Shag), Slovak Republic (DESOR in AGASSIZ & DESOR, 1847a; DESOR, 1858; LAMBERT & JEANNET, 1928; [NHMW]); Sámsonháza, Nógrád, Hungary (VADÁSZ, 1915); Várpalota (Bánta-puszta), Veszprém, Hungary (VADÁSZ, 1915);
Fore-Carpathian Basin: Krekhiv (= Krehov, = Krechów), near Rawa Ruska, western Ukraine (SZÖRÉNYI, 1953; record based on a single specimen, never reported again from the Carathian Foredeep)
Transylvanian Basin: Armeniş (= Örményes), Romania (VADÁSZ, 1915; [MAFI]); Buciumi, Romania (GHIURCA, 1996); Cacovalerii (= Cacova, = Kákova), Romania (KOCH, 1887b); Cricău (= Krakkó), Romania (KOCH, 1887b); Gârbova (= Szász-Orbó),, Romania (VADÁSZ, 1915); Gârbova de Sus (= Felsö-Orbó), Romania (KOCH, 1887b; VADÁSZ, 1915;

[MAFI]); Globu Craiovei (= Globukrajova), Romania (VADÁSZ, 1915; [MAFI]); Livezile (= Vládháza), Romania (VADÁSZ, 1915); Lugoj, Romania [NHMW]; Minişu des Sus (= Felménes), Arad, Romania (VADÁSZ, 1915; [MAFI]); Pietroasa (= Csegez), Romania (ROTH VON TELEGD, 1899; VADÁSZ, 1915); Rachiş (= Oláh-Rákos), Romania (KOCH, 1887b; LŐRENTHEY, 1894; VADÁSZ, 1915); Slatina-Timiş (= Temes-Szlatina), Romania (VADÁSZ, 1915; [MAFI]); region E of Taşad, Romania (PAUCĂ, 1936)

Zala, Sáva and Dráva Basins: Baćin Dol, Cepeliš Brijeg, Giznik, Glogovica, ? Ostrožin, Samobor, Utinje Potok and Zrinj, Croatia (POLJAK, 1938); Derventa, Gradačac, Kozare, Maglajc-Poljane and Ugljevik, Bosnia & Herzegovina (MITROVIĆ-PETROVIĆ, 1969)

Mediterranean*: Middle (?) Miocene

Aegean Sea: ? region Ano Assites Malévysiou, Hiraklion, Creta (MARCOPOULOU-DIACANTONI, 1970)

* This species is probably much more widespread in the Mediterranean. However, the large number of nominal *Clypeaster* species and the taxonomic mess in this genus prevented me from getting a clear picture on the distribution of this species in the Mediterranean without re-examining the Mediterranean specimens.

Clypeaster folium AGASSIZ in AGASSIZ & DESOR, 1847

(Fig. 29; Pl. 27, Figs. 2a-c)

?	1829	*Clypeaster scutellatus* Nobis. – DE SERRES: 157 (fide LAMBERT & JEANNET, 1928: 179)
*	1847a	[*Clypeaster*] *Folium* AGASS. – AGASSIZ & DESOR: 131, S61.
pp	1858	[*Clypeaster*] *Folium* AGASS. – DESOR: 243
.	1861	*Clypeaster folium*, AGASSIZ – MICHELIN: 139; pl. 20, figs. 2a-d
#	1887	*Clypeaster subfolium* – POMEL: 184 (fide LAMBERT & THIÉRY, 1914: 309)
	1887b	*Clypeaster* cfr. *folium*, AGASSIZ. – KOCH: 265-266; pl. 5, figs. 2a-c, 3a-b, 4
#	1887b	*Clypeaster Herepeyi*, nov. sp. – KOCH: 266-268; pl. 5, figs. 5a-c
?	1891	*Clypeaster folium*, AGASSIZ, 1847. – COTTEAU et al.: 164-165
?	1891	*Clypeaster subfolium*, POMEL, 1887. – COTTEAU et al.: 165-167
#	1895	*Clypeaster Lovisatoi*, COTTEAU, 1893 – COTTEAU: 26, pl. 4, figs. 1-2 (fide LAMBERT, 1907a: 48)
	1910	*Clypeaster Lovisatoi* – LOVISATO: 137-140; pl. 18 (3), fig. 3
.	1915	*Clypeaster folium* AG. – VADÁSZ: 195-196; fig. 85
v.	1915	*Clypeaster subfolium* POM. – VADÁSZ: 196-197; fig. 86-88
#	1915	*Clypeaster sublaganoides* n.sp. – VADÁSZ: 198; fig. 89; pl. 5, fig. 5 [type material lost]
?	1915	*Clypeaster folium* AG. – VADÁSZ: 198-199; fig. 90
	1935	*Clypeaster subfolium* POM. – KALABIS: 276; figs. 2a-b [x-ray image]
? non	1937c	*Clypeaster subfolium* POM. – KALABIS: 200; fig. 1 [x-ray]
	1937c	*Clypeaster subfolium* POM. – KALABIS: 200; fig. 2 [x-ray]
?	1938	*Clypeaster beaumonti* SISMONDA 1841 – POLJAK: 191; pl. 5, fig. 2
?	1958	*Clypeaster folium* AGASSIZ – IMBESI SMEDILE: 32-33; pl. 11, figs. 3, 3a-b
	1960	*Clypeaster folium* AG. nov. ssp. – SOMOS & KÓKAY: 342-343, 346-347; pl. 17, fig. 2.
	2001	*Clypeaster marginatus* LAMARCK, 1816 – MIKUŽ & MITROVIĆ-PETROVIĆ: 54-55; pl. 3, figs. 1, 1a; pl. 4, figs. 3, 3a; pl. 5, figs. 1, 1a
	2001	*Clypeaster folium* AGASSIZ, 1847 – MIKUŽ & MITROVIĆ-PETROVIĆ: 55-56; pl. 3, figs. 2, 2a-b
	2001	*Clypeaster* cf. *laganoides* AGASSIZ, 1847 – MIKUŽ & MITROVIĆ-PETROVIĆ: 56-57; pl. 4, figs. 1, 1a
	2001	*Clypeaster* cf. *scutellatus* SERRES – MIKUŽ & MITROVIĆ-PETROVIĆ: 57-58; pl. 5, figs. 2, 2a
	2001	*Clypeaster balillai* (LOVISATO, 1911) – MIKUŽ & MITROVIĆ-PETROVIĆ: 58-60; pl. 4, fig. 2; pl. 6, figs. 2, 2a-b
	2001	*Clypeaster* cf. *balillai* (LOVISATO, 1911) – MIKUŽ & MITROVIĆ-PETROVIĆ: 60-61; pl. 6, figs. 1, 1a
	2002	*Clypeaster subfolium* – KAZÁR: 153; fig. 1

Type-material:
Holotype: the specimen figured by MICHELIN (1861: 139; pl. 20, figs. 2a-d); Museum of Genève, Deluc collection (according to LAMBERT, 1912: 121 and LAMBERT & JEANNET, 1928)
Locus typicus: Palermo, Sicily, Italy
Age: Miocene

Material:
Early Badenian (Langhian) – Stotzing (sandpit Mayer), Bgld, Austria

NHMW: 4 specimens (NHMW 2004z0093/0019-21, 2005z0007/0001)

Wanzenböck coll.: numerous specimens (others without inventory numbers)

Foreign material for comparison:
Early Badenian (Langhian) – Mátraverebély, Nográd, Hungary

MAFI: 1 specimen [MAFI Ech 434 (material referred to as *Clypeaster subfolium* POM. by VADÁSZ, 1915: 196-197)]

Late Badenian (Early Serravallian) – Gârbova de Sus (= Felsö-Orbó), Romania

MAFI: 4 specimens [MAFI Ech 317, 318, 319, 320 (material referred to as *Clypeaster subfolium* POM. by VADÁSZ, 1915: 196-197; fig. 86-88)]

Dimensions (in mm):

Inv. no.	TL	TW	TH	remarks
W22	99.6	91.0	19.4	
NHMW 2004z0093/0020	97.1	92.6	18.3	
NHMW 2004z0093/0019	53.5	49.3	10.8	deformed
NHMW 2004z0093/0021	94.5	80.3	-	deformed

Description

Size and shape: The test is of medium to large size with antero-posteriorly elongated, irregularly subpentagonal outline. There are characteristic marginal indentations in the interambulacra, especially in interambulacra 1, 4 and 5. These indentations can be very strongly developed (e.g. Pl. 27, Figs. 2a-b) but may also be less prominent. The margin of the test is thin, with a rounded profile anterior and a sharp profile posterior. The maximum width lies anterior of the centre, where ambulacral columns IIa and IVb reach the ambitus. In profile, the test is very low with a moderately raised petaloid area. The apex coincides with the apical disc. The oral surface is flattened with deep food grooves and a moderately deep, broad infundibulum.

Apical disc: The apical disc lies subcentrally, varying from a slightly posterior to a slightly anteriorly displaced position (rarely exceeding 3% TL displacement). It is monobasal with a large central, subpentagonal madreporite and 5 circular gonopores.

Ambulacra: All five ambulacra are distinctly petaloid. The petals are straight, broad and relatively short, extending about 44 to 48 % of the corresponding test radius. The frontal petal is the longest; the posterior paired petals are shortest (but difference between APP and PPP is usually slight). The poriferous

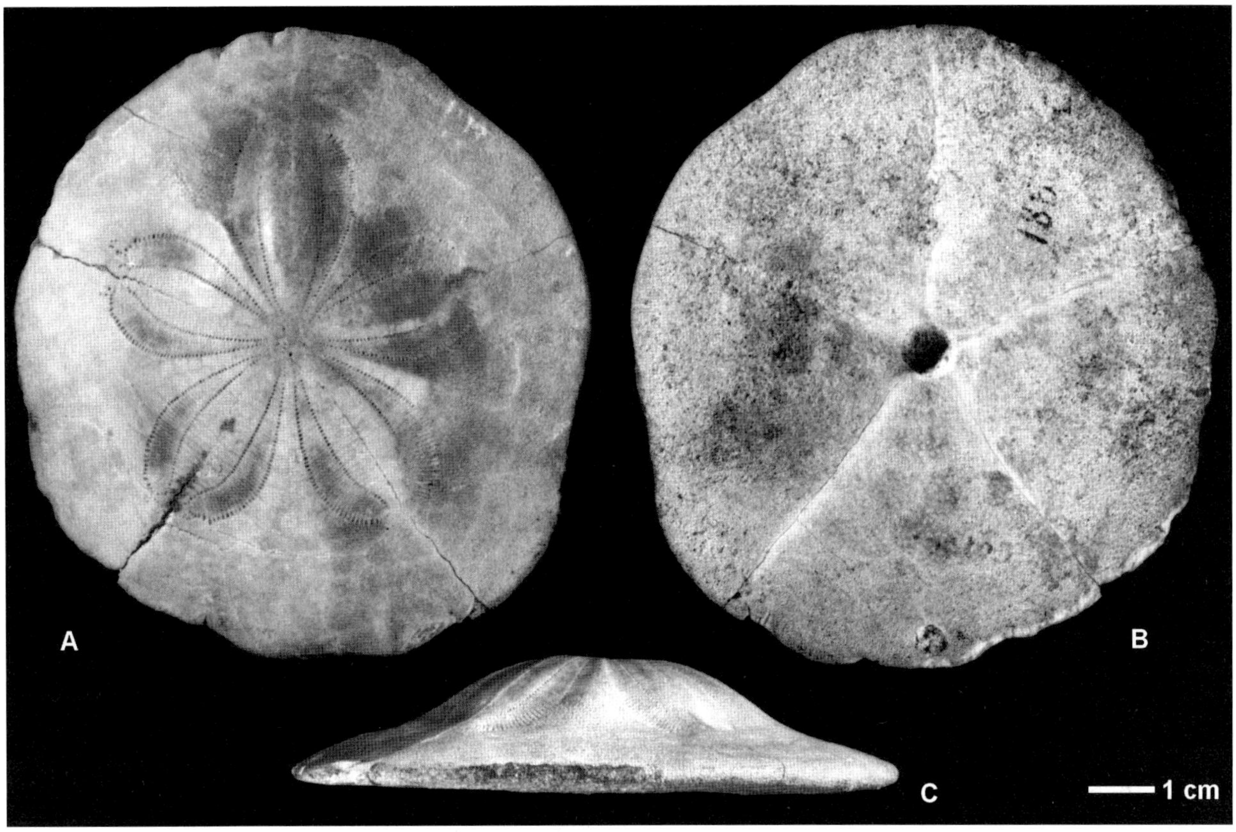

Figure 29: *Clypeaster latirostris* MICHELIN, 1861: lectotype (Burdigalian ? of Corsica; NHMW 1860.XIX.6).

zones are slightly depressed and moderately wide. The pore pairs are conjugated anisopores. Adjacent pore pairs are separated by a narrow ridge with a single row of up to 6 primary tubercles. The interporiferous zones are moderately inflated and usually 2.6 to 2.8 times as wide as a single poriferous zone at the widest point of the petals. They are crowded with primary tubercles similar to those on the interambulacra. On the oral surface deep, simple unbranched food grooves are present in the axis of the ambulacra.
Interambulacra: The interambulacra are slightly inflated adapically between the petals. They are crowded with very small perforate crenulate primary tubercles in sunken areoles. Between the primary tubercles, dense miliary tuberculation is present. Tubercle density and size is rather homogenous on the aboral surface. On the oral surface tubercles are much larger (about twice as large). The interambulacra are slightly flattened on the oral surface, producing an even surface, except adorally, where they are inclined towards the peristome.
Peristome: The peristome is circular and lies centrally on the oral side of the test in a wide, moderately deep infundibulum with gently sloping walls.
Periproct: The periproct is subcircular and lies inframarginally, about 3 mm away from the posterior margin.
Internal support system: The test of this species is single-walled but reinforced by a dense internal pillar system especially in the marginal, thinner parts of the corona.

Differential diagnosis:
For the difference to *C. barcinensis* ? LAMBERT, 1906 see above under that species.

C. calabrus SEGUENZA, 1880, a species occurring in the Badenian (Langhian – Early Serravallian) in the Paratethys, differs from *C. folium* by its higher test, with more strongly raised petaloid area, thicker margin, deeper infundibulum and more strongly inflated interporiferous zones.

C. campanulatus (SCHLOTHEIM, 1820) (and its phenotypes), a species occurring in the Badenian (Langhian – Early Serravallian) of the Paratethys, differs by its strongly different profile, larger test height, broader and longer petals, more strongly raised petaloid area and thicker margin.

C. intermedius DES MOULINS, 1837, an Early Miocene species of the Rhône Basin, differs from *C. folium* by its higher test, lack of prominent marginal indentations in the interambulacra, much larger and more raised petaloid area, and its deeper infundibulum.

For the difference to *C. latirostris* MICHELIN, 1861 see below under that species.

For the difference to *C. neudorfensis* LAMBERT, 1927 see below under that species.

For the difference to *C. scillae* DES MOULINS, 1837 see below under that species.

C. marginatus LESKE, 1778, from the Early Miocene of western France differs by its larger petalodium, more strongly inflate interporiferous zones, more elongated petals and narrower infundibulum. The Late Miocene Maltese species, which has often been confused with *C. marginatus* [compare LAMBERT (1927b: 98-101)] differs by its pentagonal outline, much narrower posterior end, thinner margin, narrower infundibulum (interambulacra not gently sloping to the peristome), smaller petalodium and more strongly raised petaloid area, and large dish-like marginal area.

Discussion:
According to LAMBERT & JEANNET (1928: 179) this species is synonymous to *C. scutellatus* DE SERRES. Furthermore, they place *C. lovisatoi* COTTEAU (non SEGUENZA) into the synonymy of *C. scutellatus* [not, however, *C. martini*, tentatively placed into the synonymy of *C. folium* by DESOR (1858: 243)]. On basis of the description in DE SERRES (1829: 157) I am, however, not able to validate this statement and therefore continue to use the name

C. *folium*, the type specimen of which was figured by MICHELIN (1861: pl. 20, figs. 2a-d). Moreover, specimen attributed to *C. scutellatus* (e.g. MICHELIN, 1861: 131-132; pl. 23, figs. 2a-f) are very different from *C. folium*. To the author's knowledge, the holotype of *C. scutellatus* was never figured.

Juvenile and subadult specimens of *C. folium* differ slightly from adult specimens in outline and shape of the oral side. While the marginal indentations are often very distinct and characteristic in adults, they are less well developed in juveniles and subadults of up to approximately 50 mm test length. The same is true for the oral surface, while it is distinctly flattened in adults, it is slightly concave in juveniles and subadults (this is caused by the infundibulum, which needs proportionally more space in smaller specimens than in larger ones). These differences led VADÁSZ (1915) to propose two new species for such subadult specimens: *C. sublaganoides* and *C.* n. sp.

LAMBERT & THIÉRY (1914: 309) stated that *C. subfolium* POMEL, 1887 is a synonym of *C. folium* AGASSIZ in AGASSIZ & DESOR, 1847 and I agree with them. In fact the two species were nearly always reported from the same localities (e.g. COTTEAU et al., 1891; VADÁSZ, 1915;...) and the differential diagnostic features reported can in part be explained by the differences between juvenile/subadult and fully adult specimens.

Clypeaster herepeyi KOCH, 1887 was synonymised with *C. folium* by VADÁSZ (1915: 195) based on a re-examination of the holotype. Although I had no opportunity to examine KOCH's (1887b) type specimen, I fully agree with VADÁSZ based on KOCH's description.

Occurrence:

Austria: Early Badenian (Langhian)

Vienna Basin: Stotzing (sandpit Mayer), Bgld (KAZÁR, 2002; [NHMW])

Paratethys (non-Austrian occurrences): Early to Late Badenian (Langhian-Early Serravallian)

Styrian Basin: Šentilj, Slovenske gorice, NE Slovenia (MIKUŽ & MITROVIĆ-PETROVIĆ, 2001)

Great Hungarian Basin (Pannonian Basin): Hird, Hungary (SOMOS & KÓKAY, 1960); Hor. Strháre (= Felsö-Esztergály), Slovak Republic (VADÁSZ, 1915); Márkháza, Nógrád, Hungary (VADÁSZ, 1915); Matraszölös, Nógrád, Hungary (KALABIS, 1937c); Mátraverebély, Nógrád, Hungary (VADÁSZ, 1915; [MAFI]); Sámsonháza near Matraszölös, Nógrád, Hungary (KALABIS, 1937c)

Transylvanian Basin: Gârbova de Sus (= Felsö-Orbó), Romania (KOCH, 1887b; VADÁSZ, 1915; KALABIS, 1937c; [MAFI])

Zala, Sáva and Dráva Basins: ? Zaprešić Breg, Croatia (POLJAK, 1938)

Mediterranean: Middle (?) Miocene

Western Mediterranean: Capo Sant'Elia, Sardinia (COTTEAU, 1895); Coroneddu between Tresnuraghes and Bosa, Sardinia (COTTEAU, 1895); Monte Alvu and Canales near Modulo (Planargia), Sardinia (COTTEAU, 1895); ? Oued Sebt, E of Tizi-Ouzou, Kabylie, Algeria (COTTEAU et al., 1891); Palermo, Sicily, Italy (AGASSIZ & DESOR, 1847a; DESOR, 1858; MICHELIN, 1861); south of Piano, Sardinia (COTTEAU, 1895); near Pozzomaggiore, Sardinia (COTTEAU, 1895)

Central Mediterranean: ? Tropea-Brattirò, Calabria (IMBESI SMEDILE, 1958)

Clypeaster intermedius DES MOULINS, 1837

(Pl. 28, Fig. 1-2)

	1768	*Scuta angularia* [...] KLEIN – WALCH: 182-183
	1768	[*Scuta angularia* KLEIN] – KNORR: pl. E.V., figs. 1-2
*	1837	*C.*[*lypeaster*] *intermedius*. Nob. – DES MOULINS: 65-65, no. 15
	1840b	*Clypeaster crassicostatus* AG. – AGASSIZ: 6, Q.12 [*nomen nudum*] [fide PHILIPPE, 1998: 112]
pp	1847a	[*Clypeaster*] *scutellatus* M. de S. – AGASSIZ & DESOR: 131
pp	1858	[*Clypeaster*] *Scillae* – DESOR: 241
?	1858	*Clypeaster grandiflorus* BRONN. – DESOR: pl. 29, figs. 1, 2, 2a, 3, 3a [fide PHILIPPE, 1998: 112]
pp	1861	*Clypeaster intermedius*, DES MOULINS – MICHELIN: 128-129
non	1861	*Clypeaster intermedius*, DES MOULINS – MICHELIN: pl. 31, figs. 1a-g [= *C. scillae*, fide Lambert, 1905: 103]
non	1868	*C. intermedius* – KARRER: 570 (footnote)
non	1869	*Clypeaster intermedius* DUES. – FUCHS: 194
non	1869a	*Clyp. intermedius* DESMOULINS. – LAUBE: 183
non	1870	*Clypeaster intermedius* MICH. – LAUBE: 314
non	1871	*Clypeaster intermedius* DESMOULINS. – LAUBE: 64-65
non	1877	*Clypeaster intermedius* DESM. – HILBER: 261, 262
non	1877	*Clypeaster intermedius* DESM. – KARRER: 170, 312
? pp	1877	*Clypeaster intermedius* DESM. – LÓCZY: 3
non	1877	*Clypeaster intermedius* DESM. – LÓCZY: 3
non	1878	*Clypeaster intermedius* DESM. – HILBER: 553, 560, 563, 575
	1906a	*Clypeaster intermedius* DES MOULINS, 1837 – LAMBERT: 85-88
non	1907	[*Clypeaster*] *intermedius* DESM. – SCHAFFER: 28
non	1907	*Clypeaster intermedius* DESM. – SCHAFFER: 35
#	1912a	*Clypeaster sub-Partschi* SCHFF. – SCHAFFER: 188; text-fig. 1; pl. 60, figs. 3
	1913a	*Clypeaster intermedius* DESMOULINS, 1837. – LAMBERT: 105-107
	1913a	*Clypeaster crassicostatus* SISMONDA, 1841 – LAMBERT: 107-108; pl. 8, fig. 4
? #	1913a	*Clypeaster Guebhardi* LAMBERT. – LAMBERT: 108-110; pl. 8, figs. 1-3
#	1913a	*Clypeaster coronalis* LAMBERT. – LAMBERT: 120; pl. 10, figs. 4-6
	1920	*Clypeaster intermedius*, DESMOULINS, 1837 – FOURTAU: 50
	1927a	*Clypeaster intermedius* DESMOULINS, 1837 – LAMBERT: 13-14
non	1928	*Clypeaster intermedius* DESM. – BOBIES: 48
?	1937a	*Clypeaster* sp. – KALABIS: 44, fig. 1
?	1937b	*Clypeaster* – KALABIS: 109, 111
?	1938a	*Clypeaster* sp. – KALABIS: 5, 11
	1938a	*Clypeaster sub-partschi* SCHAFFER – KALABIS: 6, 9, 11
non	1939	*Clypeaster intermedius*. – KAPOUNEK: 72
non	1942	[*Clypeaster*] *intermedius* DESM. – SCHAFFER: 130
non	1942	*Clypeaster intermedius* DESM. – SCHAFFER: 94
pp	1949	*Clypeaster intermedius intermedius* DESMOULINS, 1837. – KALABIS: 38-42, 91-97; pl. 2, figs. 1-3
non	1949	*Clypeaster intermedius* DESMOULINS – SCHOUPPÉ: 143
non	1968	*Clypeaster intermedius* DESM. – FLÜGEL & HERITSCH: 44; pl. 4, fig. 2
	1971a	*Clypeaster subpartschi* SCHAFFER, 1912 – STEININGER: 594; pl. 3, fig.2
	1971c	*Clypeaster subpartschi* SCHAFFER– STEININGER: 129
pp	1975	*C. subpartschi* SCHAFFER – KALABIS: 315
	1984	*Clypeaster intermedius* DESMOULINS, 1837 – NEGRETTI: 99-100; pl. 3, figs. 1-6; pl. 4, figs. 2-3
	1998	*Clypeaster intermedius* DESMOULINS, 1837 et ses morphes. – PHILIPPE: 112-119; pl. 11, figs. 4a-c, 5a-b, 6a-b; pl. 12, figs. 1, 2a-c, 3a-c, 4a-c
non	1998	*Clypeaster intermedius* MICHL.– SCHULTZ: 116; pl. 52b, figs. 7

Type-material:

Clypeaster intermedius DES MOULINS, 1837:

Holotype: the specimen figured by PHILIPPE (1998: pl. 11, figs. 4a-b); specimen no. CR 5 837 (new number MR 3 001 901); Muséum Requien, Avignon, France

Locus typicus: unknown

Age: Burdigalian, Early Miocene

Remarks: More information on the type material, its history and a detailed description of the holotype can be found in PHILIPPE (1998: 112-118).

Clypeaster subpartschi SCHAFFER, 1912:

Holotype (Pl. 28, Fig. 1a-c): specimen figured by SCHAFFER (1912a: 188; text-fig. 1; pl. 60, figs. 3); housed at the Krahuletz-Museum, Eggenburg, NÖ, Austria

Locus typicus: Schloßthal, Roggendorf, NÖ, Austria

Stratum typicum: Zogelsdorf Fm.

Age: Late Eggenburgian (Early Burdigalian), Early Miocene

Material:

Late Eggenburgian (Burdigalian) – Roggendorf (Zogelsdorf Fm., Schloßthal), NÖ, Austria

KM: 1 specimen (holotype; no inventory number)

Late Eggenburgian (Burdigalian) – Grübern (Zogelsdorf Fm.), near Maissau, NÖ, Austria

HM: 1 specimen (HM 644)

Dimensions (in mm):

	TL	TW	TH
Holotype (KM)	135	131	43
HM 644	99	92	24

Description:

Size and shape: The test is of large size, slightly antero-posteriorly elongated, with a pentagonal outline. The margin is rounded and moderately thick. The lateral margins are very slightly indented in interambulacra 1, 4 and 5. The maximum width lies anterior of the apical disc, where ambulacral columns IIa and IIIb reach the ambitus. The profile is low arched, with a slightly elevated petaloid area. The maximum height lies subcentrally around the apical disc. The oral surface is flattened, with a broad but shallow infundibulum.

Apical disc: The apical disc lies subcentrally and is not well preserved in the studied specimen.

Ambulacra: All five ambulacra are petaloid, straight and slightly closing distally. The petals are more or less subequal in length and extend about 70 % of the corresponding test radius. The poriferous zones are slightly depressed and moderately wide. The pore pairs are conjugated anisopores. Adjacent pore pairs are separated by narrow ridges with a single row of up to 9 primary tubercles. The interporiferous zones are moderately inflated, flattened and up to four times as wide as a single poriferous zone at the widest point of the petals. They are crowded with primary tubercles similar to those on the interambulacra. On the oral surface shallow, simple unbranched food grooves are present in the axis of the ambulacra.

Interambulacra: Adapically the interambulacra are slightly depressed between the petals. They are crowded with perforate crenulate primary tubercles in sunken areoles. Between the primary tubercles, dense miliary tuberculation is present. Aboral tuberculation is rather homogenous. On the oral surface tubercles are larger and more closely spaced than on the aboral side.

Peristome: The peristome lies subcentrally on the oral side of the test in a moderately deep depression with gently sloping walls.

Periproct: The periproct is subcircular and lies inframarginally.

Internal support system: unknown

Differential diagnosis:

C. calabrus SEGUENZA, 1880, a Middle Miocene species, differs from *C. intermedius* by its higher test, more strongly raised petaloid area, deeper, steep-walled infundibulum and stronger marginal indentations.

C. campanulatus (SCHLOTHEIM, 1820) (and its phenotypes), a species occurring in the Badenian (Langhian-Early Serravallian) in the Paratethys, differs by its strongly different profile, larger test height, broader petals, more strongly raised petaloid area and thicker margin.

C. folium AGASSIZ IN AGASSIZ & DESOR, 1847, a species restricted to the Early Badenian (Langhian) in the Paratethys, is distinguished from this species by its lower test height, more elongated outline with much deeper marginal indentations in the interambulacra, smaller and less raised petaloid area, and deeper food grooves.

C. latirostris MICHELIN, 1861, a co-occurring species, is distinguished from this species by its subequal test length and width, lower test height, with less raised petaloid area, thinner margin, more flattened oral surface and broad, very shallow infundibulum.

For the difference to *C. neudorfensis* LAMBERT, 1927 see above under that species.

C. scillae DES MOULINS, 1837, an Early to Middle Miocene species, differs from *C. intermedius* by its higher test, more elongated outline, more strongly raised petaloid area, thicker and more tumid margin, deeper, steep-walled infundibulum and less strong marginal indentations (see also PHILIPPE, 1998).

Discussion:

Clypeaster intermedius DES MOULINS, 1837 is a common and rather well defined species, occurring abundantly in the Aquitanian to Burdigalian (Middle Miocene records need to be confirmed) of the Mediterranean. The type material of this species has recently been re-described and illustrated by PHILIPPE (1998: 112-118). The incorrect range extension to the Late Miocene of the Central Paratethys by PHILIPPE (1998) are based on the erroneous dating of Paratethyan sediments as Tortonian, which are in fact Middle Miocene in age (Langhian-Early Serravallian, the Badenian stage in the Paratethys). Still, as discussed below, even the Middle Miocene occurrence off this species remains to be substantiated. The Middle Miocene occurrence in Eggenburg listed by PHILIPPE (1998: 118) in fact refers to SCHAFFER's specimens which are Late Eggenburgian (Early Burdigalian) in age.

PHILIPPE (1998) placed a number of other species into the synonymy of *C. intermedius*: *C. crassicostatus* AGASSIZ, 1840; *C. diversicostatus* ABICH, 1857; *C. latirostris ventiensis* LAMBERT, 1906 non TOURNOUER, 1897; *C. pillai* LOVISATO, 1909; *C. lamarmorai* LOVISATO, 1910; *C. subpartschi* SCHAFFER, 1912; and *C. guebhardi* LAMBERT, 1913a.

Clypeaster subpartschi SCHAFFER, 1912 is based on a single specimen from the Eggenburgian (Early Burdigalian) of the Molasse Zone. Although this area and time interval has been in the focus of intense research during the last twenty years only a single additional specimens was recovered (HM collection; Pl. 28, Figs. 2a-c). The holotype of *C. subpartschi* is rather similar to some of the *C. intermedius* specimens from the Rhône Basin (e.g. PHILIPPE, 1998: pl. 12, figs. 2a-c). KALABIS (1949: 95) regarded *C. subpartschi* conspecific with *C. intermedius*, based on the type specimen and additional material from Služín (Czech Republic) (but see below). Later, however, KALABIS (1974: 315) renounced his view and stated that *C. intermedius* is not present in the Paratethys at all. Indeed, all Badenian (Langhian to Early Serravallian, Middle Miocene) records of *C. intermedius* species that could be evaluated in the course of this study turned out to be erroneous [most records refer to specimens which were misidentified and belong to *C. scillae* or *C. calabrus* instead; the reason for the common confusion of *C. intermedius* and *C. scillae* resulted from an error of MICHELIN (1861: pl. 31, fig. 1a-g) who depicted a *C. scillae* specimen as *C. intermedius* (see also LAMBERT, 1913a: 103; and VADÁSZ, 1915: 131); MICHELIN (1861) himself, however, never announced *C. intermedius* from the Paratethys Region, neither did Vadász (1915); the records of *C. intermedius* by Szörényi (1936) (an abnormal 4-rayed specimen probably belonging to *C. calabrus*) and Poljak (1938) are highly questionable and could not be confirmed].

Here *C. subpartschi* is considered as junior synonym of *C. intermedius* following Kalabis (1949) and Philippe (1998) as no features were found which would allow to separate the two confidently.

Kalabis (1949: 38-42, 91-97; pl. 2, figs. 1-3) recorded *C. intermedius intermedius* from the locality Služín in the Czech Republic. Later he referred the same specimens into (Kalabis, 1974: 315) to *C. subpartschi.* Yet, they are not very well preserved show some differences to the specimens of *C. intermedius* from the Rhône Basin (see Philippe, 1998) and the Eggenburgian of Austria. Thus, their attribution to *C. intermedius* can be considered as tentative only.

Unfortunately, the age of the locality Služín is unclear. Although Kalabis (1949: 96) referred it to the "Tortonien" [= Badenian (at that time incorrectly correlated with the Late Miocene Tortonian stage of the Mediterranean)] published faunal lists (Kalabis, 1937b) are inconclusive. No younger paper are available and research by Czech colleague in the in the last decades yielded no additional data (written comm. Kamil Zágoršek, July, 2003).

Occurrence:

Austria: Late Eggenburgian (Early Burdigalian)
Molasse Zone: ? Eggenburg (Brunnstube), NÖ (Steininger, 1971a, c; reference material not located); Grübern (Zogelsdorf Fm.), NÖ ([HM]); Roggendorf, NÖ (Schaffer, 1912a; Kalabis, 1938a; Steininger, 1971a)

Paratethys (non-Austrian occurrences): Miocene
Molasse Zone: ? Služín, Czech Republic (Kalabis, 1937a+b, 1938a, 1949, 1975)

Mediterranean: Aquitanian to Burdigalian
Rhône Basin: Aquitanian: Littoral de la Nerthe (Martigues, Sausset-les-Pins), France (Philippe, 1998)
Burdigalian: Littoral de la Nerthe (Lavalduc, Martigues, Sausset-les-Pins), France (Lambert, 1913a; Philippe, 1998); France (Philippe, 1998); Basse-Provence (Port-de-Bouc), France (Philippe, 1998); Languedoc [Montpellier (le Boutonner)], France (Lambert, 1913a; Philippe, 1998); Bassin de Valréas-Visan (Bollène, Chamaret, Montségur-su-Lauzon, Saint-Paul-Trois-Châteaux, Taulignan), France (Lambert, 1913a; Philippe, 1998); Sillon périalpine (Oriol-en-Royans), France (Lambert, 1913a; Philippe, 1998)
Western Mediterranean: Alpes-Maritimes (Tourettes-sur-Loup), France (Lambert, 1913a); Catalonia (St. Pau d'Ordal, St. Sadurní de Noya, Olérdola, and Viladellops), Spain (Lambert, 1927a)
Eastern Mediterranean: Gebel Geneifa, Egypt (Fourtau, 1920)

Clypeaster latirostris Michelin, 1861

(Fig. 29; Pl. 29, Figs. 1-4)

non 1840b *Clypeaster latirostris* Ag. – Agassiz: 6 [fide Lambert & Jeannet, 1928: 125] [*nomen nudum*]
non 1847a [*Clypeaster*] *Scillæ* Desml. – Agassiz & Desor: 131 [placed *C. latirostris* Ag. into the synonymy of *C. scillae*]
non 1858 [*Clypeaster*] *scutellatus* Marcel de Serres – Desor: 242 [placed *C. latirostris* Ag. into the synonymy of *C. scutellatus*]
* v. 1861 *Clypeaster latirostris*, Agassiz – Michelin: 137-138; pl. 15, figs. 2a-d; pl. 36, fig. 2
1869a *Clyp. latirostris* Ag. – Laube: 183
1870 *Clypeaster latirostris* Ag. – Laube: 314
1871 *Clypeaster latirostris* Agassiz. – Laube: 65
? 1879 *Clypeaster latirostris*, Agass. – Hermite: 253
1912 *Clypeaster ventiencis vindobonensis* – Lambert: 97
? 1912 *Clypeaster ventiencis* Tournouer, 1879. – Lambert: 97-100
1912a *Clypeaster latirostris* Ag. – Schaffer: 187-188; pl. 60, figs. 1-2
1913a *Clypeaster latirostris* Ag. – Schaffer: 56; 157-158; pl. 10, figs. 10
v non 1915 *Clypeaster latirostris* Ag. var. *vindobonensis* Lamb. – Vadász: 186
non 1928 *Clypeaster latirostris* Agassiz – Lambert & Jeannet: 125
1938a *Clypeaster latirostris* Ag. – Kalabis: 4, 5, 9,10
1943 *Clypeaster latirostris* Ag. – Schaffer: 521
? 1949 *C. ventiensis* sensu latissimo – Kalabis: 113
1951 *Clypeaster latirostris* Ag. – Schaffer & Grill: 709
? 1956 *Clypeaster latirostris* Ag. var. *laganoides* Agassiz – Socin: 7
? 1958 *Clypeaster latirostris* Agassiz – Imbesi Smedile: 16-17; pl. 1, figs. 4, 4a-b, 5, 5a
1962 *Clypeaster latirostris* – Papp & Thenius in Kühn: 118
? 1967 *Clypeaster latirostris* Agassiz – Menesini: 149-151; pl. 51 (6), figs. 1, 1a
non 1970 *Clypeaster latirostris* Agassiz, 1840. – Marcopoulou-Diacantoni: 249-250; pl. 22, figs. 3a-b [misidentified "*C. marginatus*" (Maltese form)]
1971a *Clypeaster latirostris* Agassiz, 1840 – Steininger: 594; pl. 3, fig.1
1971c *Clypeaster latirostris* Agassiz – Steininger: 129
1974 *Clypeaster latirostris* Agassiz – Thenius: 47
? 1975 *Clypeaster ventiensis vindobonensis* Lambert, 1912. – Kalabis: 175

Type-material:
Lectotype (designated herein; Fig. 29): the specimen figured by Michelin (1861: pl. 15, figs. 2a-d); housed at the Naturhistorisches Museum Wien (Geologische Abteilung; inventory number NHMW 1860.XIX.6).
Paralectotypes: According to Michelin (1861: 138) his material came from the Naturhistorisches Museum Wien and from his own collection, the latter is housed at the École des Mines, Paris today. None of the specimens from Gauderndorf, Austria is accompanied by a label indicating that they were studied by Michelin. Yet at least two of the specimens at hand (namely specimens 1848.III.67 and 1860.L.443) had already been in the collection of the NHMW at the time of Michelin's visit and have to be considered as potential type material. The same can be assumed for the two specimens from Santa Manza, Corsica (1857.XV.89 and 1857.XV.90)
Locus typicus: Corsica
Age: ? Burdigalian, Early Miocene

Material:
Late Eggenburgian (Early Burdigalian) – Eggenburg, NÖ, Austria
NHMW: 1 specimen (NHMW 1904.VIII.25)
Late Eggenburgian (Early Burdigalian) – Eggenburg area (Zogelsdorf Fm.), NÖ, Austria
NHMW: 1 specimen (NHMW 1848.III.67)
Late Eggenburgian (Early Burdigalian) – Gauderndorf (Himmelreich), NÖ, Austria
NHMW: 4 specimens (NHMW 1860.L.443 (**paralectotype** of *C. latirostris* Michelin, 1861), 1866.XX.20 [=1869.III.37], 1869.III.36, 2004z0042/0001-2)

Late Eggenburgian (Early Burdigalian) – Gauderndorf, NÖ, Austria
NHMW: 2 specimens [NHMW 1848.III.67 (**paralectotype** of *C. latirostris* Michelin, 1861), 1850.IX.84, 1997z0178/0510]
GBA: 1 specimen (GBA 2004/1/19)
Late Eggenburgian (Early Burdigalian) – Grübern, near Maissau, NÖ, Austria
NHMW: 1 specimen (NHMW1997z0178/0526)
Late Eggenburgian (Early Burdigalian) – Limberg (Zogelsdorf Fm., Hengl quarry), NÖ, Austria
NHMW: 1 specimen (NHMW 2004z0042/0003)
Late Eggenburgian (Early Burdigalian) – Pulkau (Retz Fm.), NÖ, Austria
NHMW: 3 specimens (NHMW 2004z0042/0004-5, ../0008)
Late Eggenburgian (Early Burdigalian) – Straning, N of Limberg, NÖ, Austria
IPUW: 2 specimens (no inventory no.)

Foreign material for comparison:
Burdigalian ? – Cannetto, Corsica
NHMW: 1 specimen (NHMW 1976/1843/6)
Burdigalian ? – Corsica
NHMW: 1 specimen [NHMW 1860.XIX.6 (**lectotype** of *C. latirostris* Michelin, 1861: pl. 15, figs. 2a-d)]
Burdigalian ? – Santa Manza, Corsica
NHMW: 2 specimens [NHMW 1857.XV.89, 1857.XV.90 (**paralectotypes** of *C. latirostris* Michelin, 1861)]

Dimensions (in mm):

Inv. No.	TL	TW	TH
NHMW 1848.III.67	137.6	130.9	34.8
NHMW 1866.xx.20	129.8	125.3	27.2
NHMW 1997z0178/0526	111.2	111.0	23.7
NHMW 2004z0042/00003	104.1	97.3	~ 20

Description
Size and shape: The test is of large size, with a pentagonal outline and subequal test length and width. The margin of the test is thin, with a rounded profile anterior and a sharp profile posterior. In aboral view the anterior margin is bluntly pointed and the posterior margin transversely truncated. All interambulacra except interambulacrum 5 show slight marginal indentations. The maximum width lies anterior of the centre, where ambulacral columns IIa and IVb reach the ambitus. In profile, the test is very low with a moderately raised petaloid area. The apex is somewhat flattened and coincides with the apical disc. The oral surface is flattened with a broad, very shallow infundibulum.
Apical disc: The apical disc lies subcentrally and belongs to the monobasal type with a large central, subpentagonal madreporite and 5 circular gonopores.
Ambulacra: All five ambulacra are distinctly petaloid, straight and moderately closing distally. The frontal petal is the longest; the paired petals are subequal in length, the posterior ones being a little bit longer than the anterior ones in some specimens. The length of the petals is about 58 to 64 % of the corresponding test radius. The poriferous zones are slightly depressed and relatively wide and are widest near the distal end of the petals. The pore pairs are conjugated anisopores. Adjacent pore pairs are separated by narrow ridges with a single row of 7 to 12 primary tubercles. The interporiferous zone are strongly inflated and usually 2.1 to 2.7 times as wide as single poriferous zone at the widest point of the petals. They are crowded with primary tubercles similar to those on the interambulacra. Outside the petals only minute microunipores are present, which are observable only in exceptionally preserved specimens/fragments. On the oral surface simple unbranched food grooves are present in the axis of the ambulacra.
Interambulacra: The interambulacra are moderately inflated adapically between the petals. They are crowded with perforate crenulate primary tubercles in sunken areoles. Between the primary tubercles, dense miliary tuberculation is present. Tubercle density and size is rather homogenous on the aboral surface. On the oral surface tubercles are slightly larger and more closely spaced. On the oral surface the interambulacra are slightly flattened, producing an even surface, which is not inclined towards the peristome.
Peristome: The peristome is circular and lies centrally on the oral side of the test in a broad, very shallow infundibulum.
Periproct: The periproct is subcircular to very slightly oval, transversely elongated. It lies inframarginally, about 2.2 to 4 mm away from the posterior margin.
Internal support system: The test of this species is single-walled but reinforced by a dense internal pillar system. The pillars are most abundant close to the margin and along the midlines of ambulacra and interambulacra.

Differential diagnosis:
For the difference to *C. barcinensis* ? Lambert, 1906 see above under that species.
For the difference to *C. calabrus* Seguenza, 1880 see above under that species.
C. campanulatus (Schlotheim, 1820) (and its phenotypes), a species occurring in the Badenian (Langhian – Early Serravallian) in the Paratethys, differs by its strongly different profile, larger test height, broader petals, more strongly raised petaloid area and thicker margin.
C. folium Agassiz in Agassiz & Desor, 1847, a species restricted to the Early Badenian (Langhian) in the Paratethys, is distinguished from this species by its more elongated outline with deep marginal indentations in the interambulacra, smaller petaloid area, broader infundibulum and deeper food grooves.
For the difference to *C. scillae* des Moulins, 1837 see below under that species.
C. intermedius des Moulins, 1837, an Early Miocene species of the Rhône Basin, is distinguished from this species by its more elongated outline, higher test, more raised petaloid area, thicker margin, less flattened oral surface and deeper infundibulum.
For the difference to *C. neudorfensis* Lambert, 1927 see below under that species.
For the difference to *C. scillae* des Moulins, 1837 see below under that species.

Discussion:
Clypeaster latirostris Agassiz (1840b, p. 4) is a *nomen nudum* according to the ICZN rules 4th ed., 2000, Article 12.1. since it is neither figured nor described in Agassiz (1840b). Agassiz & Desor (1847a: 131) and Desor (1858: 242) are not available as authors because the name *latirostris* is just listed as junior synonym of *C. scillae* and *C. scutellatus* respectively. The next author who published a valid diagnosis for this name is Michelin (1861: 137-138; pl. 15, figs. 2a-d; pl. 34, fig. 2), which is thus the valid author for *C. latirostris* (ICZN 4th ed., Articles 11 and 12). According to Lambert & Thiéry (1928) the material ("moulage" 59 and 60 of the Agassiz collection) listed under that name in Agassiz (1840b: 6) does not correspond to the material described and figured by Michelin (1861). This, however, is irrelevant in this matter as *C. latirostris* Agassiz, 1840 is a *nomen nudum*.
Lambert (1912: 97) established the name *Clypeaster ventiencis* [*sic*] *vindobonensis* for the Austrian material described by Michelin (1861). {*C. latirostris vintiensis* [*sic*] Tournouër, 1877 is actually a *nomen nudum* as the original reference lacks any description, illustration or other indication (see Tournouër, 1877: 844). Lambert (1912: 97) elevated this subspecies to species rank, but misspelled the species name}. Curiously the figure (Michelin, 1861: pl. 15, fig. 2) Lambert (1912: 97) refers to depicts a specimen (NHMW 1860.XIX.6) coming from Corsica, not from Austria. This specimen is the one selected as lectotype

for *C. latirostris* MICHELIN. Therefore the name *Clypeaster vintiensis vindobonensis* LAMBERT, 1912 is superfluous.

Part of the type material of *C. latirostris* MICHELIN, 1861 is housed at the Naturhistorisches Museum Wien, among them the specimen figured on plate 15, figures 2-d of MICHELIN's monograph. The latter specimen is chosen as lectotype. It is a specimen coming from Corsica, most probably from the Burdigalian part of the succession, although the exact provenience of the specimens is unknown. The Austria specimens are rather similar to that of Corsica, albeit larger. In the lectotype, the petaloid area is slightly more inflated than in the Austrian specimens and the test outline a little bit more elongated. These differences, however, are here considered subordinate and attributed to individual variation.

The specimen referred to *Clypeaster latirostris* AG. var. *vindobonensis* LAMB. in VADÁSZ (1915: 186) is a strongly damaged fragment (housed in the collection of the Hungarian Geological Survey, Budapest as MAFI Ech 279). It is very different from *C. latirostris* and probably related to *C. campanulatus* but better referred to as *Clypeaster* sp. on account of its poor preservation.

Some of the specimens from the Burdigalian of the Rhône Basin attributed to *C. intermedius* by PHILIPPE (1998) may actually belong to *C. latirostris* (e.g. the specimen figured on pl. 12, figs. 2a-c)

Occurrence:

Austria: Late Eggenburgian (Early Burdigalian)

Molasse Zone: Eggenburg, NÖ (SCHAFFER, 1912; KALABIS, 1938a ;SCHAFFER, 1943; SCHAFFER & GRILL, 1951); Eggenburg (Brunnstube), NÖ (STEININGER, 1971a, c); Eggenburg (Zogelsdorf Fm), NÖ ([NHMW]); "Eggenburger Schichten" (PAPP & THENIUS in KÜHN, 1962; THENIUS, 1974; ? KALABIS, 1975); Gauderndorf (and Gauderndorf, Himmelreich), NÖ (MICHELIN, 1861; LAUBE, 1869a, 1871; SCHAFFER, 1912; KALABIS, 1938a; [NHMW]); Grübern, near Maissau, NÖ ([NHMW]); Klein-Meiseldorf, NÖ (SCHAFFER, 1912a, 1913a); Limberg (Zogelsdorf Fm., Hengl quarry), NÖ [NHMW]; Pulkau (Retz Fm.), NÖ ([NHMW]); Straning, NÖ ([IPUW])

Paratethys (non-Austrian occurrences): Eggenburgian (Early Burdigalian)

Molasse Zone: ? Slatinky near Prostějov, Czech Republic (KALABIS, 1949)

Mediterranean: Aquitanian (?) – Burdigalian

Western Mediterranean: Cannetto, Corsica, France [NHMW]; Corsica (MICHELIN, 1861; LAUBE, 1871); Le Palendurier, Alpes-Maritimes, France (LAMBERT, 1912); ? Posano, Volterra, Prov. Pisa, Italy (MENESINI, 1967); Saint-Jannet, Alpes-Maritimes, France (LAMBERT, 1912); ? Santa Cristobal, Menorca, Baleares, Spain (HERMITE, 1879); Santa Manza, Corsica [NHMW]; Ste-Colombe, Alpes-Maritimes, France (LAMBERT, 1912); Tourrettes-sur-Loup, Alpes-Maritimes, France (LAMBERT, 1912); Vence, Alpes-Maritimes, France (LAMBERT, 1912)

Central Mediterranean: ? Stilo, Calabria (IMBESI SMEDILE, 1958)

Indian Ocean: ? Miocene

East African coast: ? Uanane and Hafun, Somalia (SOCIN, 1956)

Clypeaster neudorfensis LAMBERT, 1927

(Pl. 30, Figs. a-c)

v pp 1861 *Clypeaster melitensis*, MICHELIN – MICHELIN: 129-130; pl. 32, figs. a-c; pl. 33, figs. a
non 1861 *Clypeaster melitensis*, MICHELIN – MICHELIN: 129-130; pl. 32, figs. d; pl. 33, figs. b-c
v. 1870 *Clypeaster Melitensis* MICH. – LAUBE: 314
v. pp 1871 *Clypeaster Melitensis* MICHELIN. – LAUBE: 65
1898 *Clypeaster melitensis* MICHELIN – KORNHUBER: 31
pp 1913a *Clypeaster Guebhardi* LAMBERT. – LAMBERT: 109
? 1915 *Clypeaster tenuipetalus* SEGU. – VADÁSZ: 188-189, figs. 78-79
? 1917 *Clypeaster Carapezzai* CHECCHIA-RISPOLI. – CHECCHIA-RISPOLI, 1917: 61-62; fig. 3; pl. 8(13), fig. 2; pl. 10(15), fig. 8
* v. 1927b *C.[lypeaster] neudorfensis* – LAMBERT: 101
v. pp 1949 *Clypeaster campanulatus carapezzai* CHECCHIA-RISPOLI, 1917 – KALABIS: 52-54, 109-111; pl. 7, fig. 1
pp 1931 *Clypeaster Guebhardi* LAMBERT. – LAMBERT: 49
1938a *Clypeaster* [*partschi* var.] *neudorfensis* LAMBERT-KALABIS: 6, 12
1938b *Clypeaster* [*campanulatus* var.] *neudorfensis* LAMB. – KALABIS: 6-7, 9
? pp 1949 *Clypeaster campanulatus carapezzi* CHECCHIA-RISPOLI, 1917 – KALABIS: 52-54, 109-111; pl. 6, figs. 1-3

Type Material:

Holotype: specimen figured by MICHELIN (1861: pl. 32, figs. a-c; pl. 33, figs. a); specimen no. 34 of MICHELIN; housed at the Naturhistorisches Museum Wien (Inv. No. NHMW 1851. VI.119)

Locus typicus: Devínska Nová Ves (= Neudorf an der March), Slovak Republic

Age: Late Badenian (Early Serravallian), Middle Miocene

Material:

Late Badenian (Early Serravallian) – Devínska Nová Ves, Slovak Republic

NHMW: 1 specimen [NHMW 1851.VI.119 (**holotype**; specimen no. 34 of MICHELIN)]

? Badenian (Langhian-Early Serravallian) – Rauchstallbrunngraben near Baden, NÖ, Austria

NHMW: 3 specimens [NHMW 1846.37.959, 2003z0055/0003 to 0004 (specimens nos. 20 to 22 of MICHELIN)]

Dimensisons (in mm):

Inv. No.	TL	TW	TH
NHMW 1851.VI.119	> 160	147.3	42.1

Description:

Size and shape: The test is of large size, with an antero-posteriorly elongated, subpentagonal outline. Test width is approximately 86 % of TL (assuming a TL of *c.* 170 mm). The margin of the test is relatively thin and only poorly rounded. No marginal indentations are present. The maximum width lies anterior of the centre, where ambulacral columns IIa and IVb reach the ambitus. In profile test low, wedge shaped with gently sloping anterior and more steeply loping posterior part. The apex lies at the posterior margin of the apical disc. The oral surface is distinctly flattened with a very narrow infundibulum.

Apical disc: The apical disc lies slightly posterior of the centre, approximately 52 % TL (assuming a TL of ~170 mm) away from the anterior margin. It is slightly inclined anteriorly and is poorly preserved, but seems to be monobasal.

Ambulacra: The petals are lanceolate in shape, straight and moderately closed distally (width of the IPZ 4.8 to 7 mm distally). The frontal petal is longest, the posterior petals shortest. The length of the petals is about 70 % of the corresponding test radius. The poriferous zones are depressed and relatively wide. The pores are conjugated anisopores. The interporiferous zones are inflated, medially flattened and range from 1.8 to 2.0 times a single poriferous in width. On the oral surface shallow, simple, unbranched food grooves are present in the axis of the ambulacra.

Interambulacra: Adapically the interambulacra are slightly inflated between the petals, forming weak keels between them. The surface is poorly preserved, but the tuberculation of the aboral surface seems to have been fairly homogenous. Adorally all surface details except food grooves and peristome are obscured.

Peristome: The peristome lies subcentrally to slightly posterior of the centre on the oral surface. It is covered by sediment in the holotype. The infundibulum is very narrow.

Periproct: not preserved

Internal support system: unknown

Differential diagnosis:

C. altus LESKE, 1778 from the Messinian of Malta differs by its higher test, different profile shape, broader petals, thicker margin and broad infundibulum.

For the difference to *C. barcinensis*? LAMBERT, 1906 see above under that species.

C. calabrus SEGUENZA, 1880, a co-occurring species, differs by its very different profile, thicker margin, concave oral surface with broad infundibulum, narrower petals, much more inflated interporiferous zone and deep marginal indentations.

For the difference to *C. campanulatus* (SCHLOTHEIM, 1820) see below under "Discussion".

C. folium AGASSIZ in AGASSIZ & DESOR, 1847, a species occurring in the Badenian in the Paratethys, is distinguished from this species by its more flattened test with deep marginal indentations in the interambulacra, smaller petaloid area, broader infundibulum and deeper food grooves.

C. intermedius DES MOULINS, 1837, a species occurring in the Burdigalian of the Rhône Basin and the Paratethys, differs by its narrower and more strongly inflated petals, and different profile.

C. latirostris MICHELIN, 1861, a species occurring in the Eggenburgian (Early Burdigalian) of the Molasse Zone, is distinguished from this species by its subequal test length and width, lower test height, and the broad, very shallow infundibulum.

C. marginatus LESKE, 1778 from the Early Miocene of western France differs by its lower profile, strong marginal indentations, small petaloid area and large dish-like marginal area.

The Late Miocene Maltese species, which has often been confused with *C. marginatus* [compare LAMBERT (1927b: 98-101)] differs by its very different shape with a rather small, strongly raised petaloid area and large dish-like marginal area.

C. scillae DES MOULINS, 1837, a co-occurring species, differs by its strongly different profile, thick margin, concave oral surface with broad infundibulum, narrower petals, much more inflated interporiferous zone.

Discussion:

The type material of *Clypeaster melitensis* MICHELIN, 1861 comprised two specimens: an internal cast from the Late Miocene of Malta and a well preserved specimen from the Middle Miocene of the Vienna Basin. Apparently those two specimens are not conspecific and consequently LAMBERT (1927b: 100-101) proposed the name *C. neudorfensis* for the specimen from the Vienna Basin. According to him *C. melitensis* should be restricted to the Maltese species. He included specimens from the Late Miocene of Malta previously erroneously identified as *C. marginatus*. Although, I agree with LAMBERT that the name *C. marginatus* is used for two different species (an Early Miocene Atlantic taxon and the Late Miocene Maltese one) I do not subscribe to his idea to use *C. melitensis* for the latter. The type of *C. melitensis* is an internal cast and it is impossible to determine it to species level or to securely associate other specimens with this species. I would propose to consider *C. melitensis* as *nomen dubium* and to establish a new species name for the Maltese species should no other name be available.

Earlier (LAMBERT, 1913a: 109) and also later (LAMBERT, 1931: 49), however, LAMBERT associated the specimen from Neudorf with *C. guebhardi* LAMBERT, 1913. This species is a junior synonym of the Early Miocene *C. intermedius* according to PHILIPPE (1998) and is clearly a different species (compare PHILIPPE, 1998: 112-119; pl. 11, figs. 4a-c, 5a-b, 6a-b; pl. 12, figs. 1, 2a-c, 3a-c, 4a-c).

KALABIS (1949: 52-54, 109-111) agreed with LAMBERT (1927b), but considered *C. neudorfensis* a junior synonym of *C. campanulatus carapezzai* CHECCHIA-RISPOLI, 1917. The type material of *C. carapezzai*, however, is very poorly preserved [see CHECCHIA-RISPOLI, 1917: pl. 8(13), fig. 2; pl. 10(15), fig. 8] and it is nearly impossible to decide whether the material is conspecific with the Vienna Basin material. Therefore the name *C. neudorfensis* is retained and *C. carapezzai* included into the synonymy with a question mark only.

Concerning the question whether *C. neudorfensis* could be a phenotype of *C. campanulatus* (i.e. range within the morphological variability of *C. campanulatus*; see also KALABIS, 1938a, b) or not the following can be stated: Only a single specimen, the holotype, can be attributed to this species without doubt (all the references in the synonymy are based on the holotype). All other specimens which were placed in this species are usually poorly preserved and their determination is unclear. The type specimen of *C. neudorfensis* differs from *C. campanulatus* (including its phenotypes) by its lanceolate, long, more closed and relatively narrow petals, its low test height (<25 % TL; whereas even the lowest *C. campanulatus* forma *partschii* usually have a TH of usually have a TH of >30% TL), slightly thinner margin, more elongated outline and conspicuous profile shape. Yet we know nothing about the variability of these features in *C. neudorfensis* and it is easily possible that the type of *C. neudorfensis* is yet another extreme phenotype of *C. campanulatus*. For the time being, however, *C. neudorfensis* is kept as separate species until more material becomes available to investigate this matter at a broader scale.

Three additional specimens in the collection of the NHMW which were labelled as "*C. tarabellianus*" by MICHELIN initially (later he crossed that species name out on the labels and replaced it with "*altus*? LMK.") and later as "*C. campanulatus* aff. var. *neudorfensis*" by KALABIS are tentatively placed into *C. neudorfensis*. All three come from grey sandstone of Rauchstallbrunngraben near Baden, NÖ (a facies not longer exposed) and share the low test height and thin margin with the *C. neudorfensis* holotype. Their petals, however, are broader and slightly shorter.

Occurrence:

Austria: Badenian (Langhian-Early Serravallian)

Vienna Basin: ? Rauchstallbrunngraben, near Baden, NÖ ([NHMW])

Paratethys (non-Austrian occurrences): Late Badenian (Early Serravallian)

Vienna Basin: Devínska Kobyla, near Bratislava, Slovak Republic (KORNHUBER, 1898); Devínska Nová Ves, Slovak Republic (pp MICHELIN, 1861; pp LAUBE, 1871; pp LAMBERT, 1913a; LAMBERT, 1982; pp LAMBERT, 1931; KALABIS, 1938a, b, 1949; [NHMW])

Transylvanian Basin: ? Gârbova de Sus (= Felsö-Orbó), Romania (VADÁSZ, 1915)

Mediterranean: ? Middle Miocene

Central Mediterranean: ? Canicattini, province Siracusa, south-central Sicily, Italy (CHECCHIA-RISPOLI, 1917)

Clypeaster scillae DES MOULINS, 1837

(Pl. 33; Figs. 1-3; Pl. 34, Figs. 1-8; Pl. 35, Figs. 1-2)

* 1837 *C.[lypeaster] scillæ*. Nob. – DES MOULINS: 65-65, no. 13
1837b *Clypeaster grandiflorus* – BRONN: 14; pl. xxxvi, fig. 9a-d [*nomen nudum*]
pp 1838 *Clypeaster grandiflorus* nob. α. *humilis* (fig. nostr.) – BRONN: 903-905
1840b *Clypeaster crassus* AG. – AGASSIZ: 6 (moulage en plâtre 55) [*nomen nudum*]
1840b *Clypeaster crassicostatus* AG. – AGASSIZ: 6 (moulage en plâtre Q 12) [*nomen nudum*]
pp 1847a *[Clypeaster] Scillæ* DESML. – AGASSIZ & DESOR: 131 [placed *C. latirostris* AG. into the synonymy of *C. scillae*]
1852 *Clypeaster grandiflorus* LAM. – ČŽJŽEK: 51
1855 *Clypeaster crassicostatus* AG. – ANDRAE: 302
1856a *Clypeaster grandiflorus* (α, 903, pars). – BRONN in BRONN & ROEMER: 324-325
1856b *Clypeaster grandiflorus* BR. – BRONN & ROEMER: 83; pl. 36, fig. 9a-d
1856 *Clypeaster crassicostatus* AG. – ROLLE: 591
1856 *Clypeaster grandiflorus* BRONN excl. synon. (*C. crassus* Ag.) – ROLLE: 591
1857 *[Clypeaster] Scillæ*, id. – PICTET: 220
pp 1858 *[Clypeaster] Scillæ* DESMOUL. – DESOR: 241; pl. 29, figs. 1-3
1859 *Clypeaster crassicostatus* – ZOLLIKOFER: 217
1861 *Clypeaster Scillæ*, DES MOULINS – MICHELIN: 114; pl. 16, figs. 1a-f
1861 *Clypeaster crassicostatus*, AGASSIZ – MICHELIN: 115; pl. 17, figs. 1a-f
1867 *Clyp. crassicostatus* AG. – STACHE: 143
1868 *Cl. Scillae* – KARRER: 570 (footnote)
1869 *Clypeaster Scillae* DESM. – FUCHS: 194
1869a *Clypeaster crassicostatus* AGASSIZ. – LAUBE: 183
1869a *Clypeaster Scillae* DESMOULINS. – LAUBE: 183
1870 *Clypeaster crassicostatus* DESM. – LAUBE: 314
1870 *Clypeaster Scillae* DESM. – LAUBE: 314
1871 *Clypeaster crassicostatus* AGASSIZ. – LAUBE: 63
1871 *Clypeaster Scillae* DESMOULINS. – LAUBE: 63
1871 *Clypeaster grandiflorus* BRONN – STUR: 590
? 1871 *Clypeaster crassicostatus* AG. – STUR: 590
1874 *Clypeaster grandiflorus* – QUENSTEDT: pl. 82, fig. 12, 18
1875 *Clypeaster grandiflorus* – QUENSTEDT: 539-541
1877 *Clypeaster Scillae* DESM. – KARRER: 312
1878 *Clypeaster crassicostatus* AG. – HILBER: 560, 575
1879 *Clypeaster* aff. *Clyp. crassicostatus*, DESOR. – HERMITE: 242
1879 *Clypeaster crassicostatus*, AGASS. – HERMITE: 253
1885 *Clypeaster crassicostatus, Agass.* – POMEL: pl. B25, figs. 1-8
1885 *Clypeaster Scillee*? DESM.- POMEL: pl. B26, figs. 1-7
1887b *Clypeaster crassicostatus*, AGASSIZ. – KOCH: 263-264
1887 *Clypeaster bunopetalus* – POMEL: p. 204 [refers to the specimen figured as *C. scillee*? in 1885]
1891 *Clypeaster pentadactylus*, PERON et GAUTHIER, 1891. – PÉRON & GAUTHIER in COTTEAU, PÉRON & GAUTHIER: 183-187; pl. 6, figs. 4-5
1894 *Clypeaster crassicostatus*. AGASS. – LÖRENTHEY: 61
1898 *Clypeaster crassicostatus* AG. – KORNHUBER: 31
. 1900 *Clypeaster Geneffensis* GAUTHIER 1898 – GAUTHIER in FOURTAU: 705-707; pl. 3, figs. 4-6
. 1900 *Clypeaster pentadactylus* PERON et GAUTHIER 1891 – FOURTAU: 708-709; pl. 4, figs. 5-9
1902 *Clypeaster Vasseli* GAUTHIER, 1901. – GAUTHIER in FOURTAU: 97; pl. 4, figs. 1-3 [fide FOURTAU, 1920, who placed it in the synonymy of *C. crassus*]
1905 *Clypeaster crassicostatus*, AG. – GAÁL: 362
1906b *Clypeaster crassus* – LAMBERT: 27; pl. 7, figs. 1-2 [not seen, fide PHILIPPE, 1998]
1906b *Clypeaster grandiflorus* – LAMBERT: 23-26 [not seen, fide PHILIPPE, 1998]
1912 *Clypeaster Scillæ* DESMOULINS, 1837. – LAMBERT: 93-94
1912 *Clypeaster crassus* AGASSIZ, 1840. – LAMBERT: 94-95; pl. 7, fig. 6
1912 *Clypeaster grandiflorus* BRONN, 1838. – LAMBERT: 95-96
1912 *Clypeaster pentadactylus* PERN et GAUTHIER, 1891. – LAMBERT: 96-97
1913a *Clypeaster grandiflorus* – COTTREAU: 103; pl. 10, fig. 2
. 1913a *Clypeaster pentadactylus* PERON et GAUTHIER – COTTREAU: 102-104; pl. 10, figs. 1, 1a; pl. 13, figs. 1, 1a
1914b *Clypeaster Manini* LOV. –LOVISATO: 557-561; pl. 9, figs. 1a-d
1915 *Clypeaster Scillæ* DESMOULINS. – LAMBERT: 223
1915 *Clypeaster grandiflorus* BRONN. – MÁJER: 35, 88
1915 *Cl. Scillae* DERM. – MÁJER: 35, 88
? 1915 *Cl. Crassis* AG. – MÁJER: 35, 88
? 1915 *Cl. crassicostatus* SISM. – MÁJER: 35, 88
v. 1915 *Clypeaster grandiflorus* BRONN. – VADÁSZ: 126-128; pl. 12 (6), fig. 4
v. 1915 *Clypeaster grandiflorus* BRONN. var. *anteacutus* VAD. – VADÁSZ: 126-128; figs. 19, 21
v. 1915 *Clypeaster grandiflorus* BRONN. var. *rhabdopetalus* POM. – VADÁSZ: 126-128; figs. 20, 22
1915 *Clypeaster grandiflorus* BRONN. var. *basiconcava* – VADÁSZ: 128 [type material lost]
1915 *Clypeaster grandiflorus* BRONN. var. *altipetalus* – VADÁSZ: 128 [type material lost]
v. 1915 *Clypeaster Scillae* DESMOULINS. – VADÁSZ: 129-131
v ? 1915 *Clypeaster digitalis* n. sp. – VADÁSZ: 135-136; fig. 28
v. 1915 *Clypeaster altus* KLEIN. – VADÁSZ: 152-153, figs. 44
v pp 1915 *Clypeaster crassus* AG. Junges Exemplar. – VADÁSZ: pl. 6, fig. 2
1920 *Clypeaster crassus*, AGASSIZ 1840 – FOURTAU: 47-48
1920 *Clypeaster geneffensis*, GAUTHIER 1898 – FOURTAU: 49-50
1920 *Clypeaster pentadactylus*, PERON et GAUTHIER 1891 – FOURTAU: 53-55
1920 *Clypeaster Scillæ*, DESMOULINS 1837 – FOURTAU: 56
1927b *Clypeaster Zanoni* CHECCHIA-RISPOLI – CHECCHIA-RISPOLI: 12-15; pl. 3, figs. 1, 1a; pl. 4, figs. 1-2
1927a *Clypeaster crassicostatus* SISMONDA, 1841 – LAMBERT: 14-15
1927a *Clypeaster Scillae* DESMOULINS, 1837 – LAMBERT: 15
1927b *Clypeaster Scillae* DESMOULINS, 1837. – LAMBERT: 94
1931 *Clypeaster Scillæ* DESMOULINS. – LAMBERT: 45
1931 *Clypeaster Scillæ* DESMOULINS variété *geneffensis* GAUTHIER – LAMBERT: 212; pl. 8, figs. 1-3
1938a *Clypeaster scillae* DESM. – KALABIS: 2-3, 5, 8, 10
1938a *Clypeaster scillae* DESM. var. *grandiflorus* BRONN – KALABIS: 3, 5, 8, 10
1938b *Clypeaster scillae* DESM. – KALABIS: 3-5, 8-9
1938 *Clypeaster pentadactylus* PERON & GAUTHIER 1891 – POLJAK: 177-178; pl. 3, figs. 1, 1a
1938 *Clypeaster airaghi* LAMBERT 1913 – POLJAK: 178; pl. 3, fig. 2

1938 *Clypeaster scillae* Desmoulins 1837 – Poljak: 175-176; pl. 2, fig. 3
? 1938 *Clypeaster crassus* Agassiz 1840 – Poljak: 176-177
? 1938 *Clypeaster scutiformis* Lamarck 1824 – Poljak: 182-183; pl. 2, fig. 4 [very probably a juvenile *C. scillae*]
1939 *Clypeaster Scillae* Dreg. – Kapounek: 67
1939 *Clypeaster grandiflorus* Bronn. – Kapounek: 72
1939 *Clypeaster Scillae* Desm. – Kapounek: 72, 74
1949 *Clypeaster scillae scillae* Desmoulins, 1837. – Kalabis: 33-35, 86-89; pl. 1, figs. 1-2
1949 *Clypeaster scillae crassus* L. Agassiz, 1840. – Kalabis: 35-37, 89-91; pl. 1, figs. 3-5
1949 *Clypeaster crassicostatus* Sismonda – Schouppé: 143
1955 *Clypeaster crassicostatus* Sism. – Tollmann: Tab. 5b
1955 *Clypeaster grandiflorus* Bronn – Tollmann: Tab. 5b
1955 *Clypeaster scillae* Desm. – Tollmann: Tab. 5b
non 1958 *Clypeaster Scillae* Desmoulins – Imbesi Smedile: 20-21, 35-36; pl. 3, figs. 3, 3a-b; pl. 14, figs. 2, 2a; pl. 15, fig. 1
non 1958 *Clypeaster grandiflorus* Bronn – Imbesi Smedile: pl. 16, figs. 4, 4a-b; pl. 18, figs. 1, 1a-b [probably a juvenile *C. altus*]
1960 *Clypeaster* (*Pliophyma*) *altus* (Klein 1734, in Leske 1778) – Kojumdgieva & Strachimirov: 230; pl. 7, fig. 9; pl. 8, fig. 16-8 [misidentified *C. scillae*]
1962a *Clypeaster scillae* – Thenius: 27; pl. 6, fig. 33
1963 *Clypeaster Zanoni* Checchia Rispoli – Comaschi Caria: 22-23; pl. 4, figs. 1-2
v. 1964 *Clypeaster campanulatus partschi* Michel. – Ertl: 253-255; figs. 1-2 [misidentification]
1968 *Clypeaster scillae* – Küpper: pl. 4, fig. 32
1968 *Clypeaster grandiflorus* Bronn. – Schmid: Anhang
1969 *Clypeaster scillae alienus* Vad. – Mitrović-Petrović: 125-126; pl. 4, fig. 2; pl. 5, figs. 1, 1a
1969 *Clypeaster altus* Klein – Mitrović-Petrović: 126; pl. 5, figs. 2, 2a
1969 *Clypeaster grandiflorus anteacutus* Vad. – Mitrović-Petrović: 129; pl. 12, figs. 1, 1a-b
1969 *Clypeaster manini* Lov. – Mitrović-Petrović: 129-130; pl. 13, figs. 1, 1a-b
1969 *Clypeaster chiassi* Lov. – Mitrović-Petrović: 130; pl. 14, figs. 1, 1a-b
1969 *Clypeaster danubicus* Vad. – Mitrović-Petrović: 131; pl. 15, fig. 2; pl. 16; figs. 1, 1a
1969 *Clypeaster majocchii* Lov. – Mitrović-Petrović: 131-132; pl. 16, figs. 2, 2a-b
1969 *Clypeaster mostoi* Lov. – Mitrović-Petrović: 132; pl. 17, figs. 1, 1a-b
1969 *Clypeaster zanoni* Ch-Risp. – Mitrović-Petrović: 134; pl. 22, figs. 1, 1a-b
pp 1969 *Clypeaster acclivis* Pom. – Mitrović-Petrović: 135; pl. 23, figs. 1, 1a; pl. 24, fig. 1 [pl. 23, figs. 2a-b may be a juvenile *C. calabrus*]
1970 *Clypeaster scillae* – Thenius: p. 211; pl. 3, fig. 32
1972 *Clypeaster crassicostatus* Ag. – Mitrović-Petrović: 71, 74; figs. 5, 5a
1984 *Clypeaster tauricus* Desor – Mitrović-Petrović: 221-222; pl. 1, figs. 1, 1a; pl. 2, fig. 1
1984 *Clypeaster chiassi* Lovisato – Mitrović-Petrović: 225; pl. 5, figs. 1, 1a-b [juvenile specimen]
1984 *Clypeaster grandiflorus anteacutus* Vadasz – Mitrović-Petrović: 225; pl. 5, figs. 2, 2a-b
1984 *Clypeaster scillae* Desmoulins – Mitrović-Petrović: 232; pl. 12, figs. 1, 1a-b
? 1984 *Clypeaster acclivis* Pomel – Mitrović-Petrović: 233; pl. 13, figs. 1, 1a-b [juvenile specimen]
1993 *Clypeaster scillae* Desmoulins, 1837 – Mączyńska: 110; pl. 4, figs. 1a-1d; pl. 5, fig. 1
1996 *Clypeaster scillae* Desmoulins, 1837 – Mączyńska: 41; pl. 2, figs. a-e
1996 *Clypeaster altus* Klein – Ghiurca: 190, unnumbered fig.
1997 *Clypeaster scillae alienus* Vadász – Majcen et al.: 106; pl. 6, fig. 1
v. 1998 *Clypeaster* sp. – Hiden: pl. 1, figs. 1-5
1998 *Clypeaster scillae* Desmoulins, 1837 – Philippe: 119-122; pl. 17, figs. 1-2
1999 *Clypeaster altus* – Piller & Harzhauser: 224; fig. 25
2000 *Clypeaster* sp. – Zorn: 19; fig. 8
v. 2001 *Clypeaster scillae* (Michelin) – Schmid et al.: 13, 23
. 2001 *Clypeaster scillae* Desmoulins, 1837 – Mikuž & Mitrović-Petrović: 50-52; pl. 1, figs. 1, 1a, 2
. 2001 *Clypeaster scillae* cf. *alienus* Vadász, 1915 – Mikuž & Mitrović-Petrović: 52-53; pl. 1, figs. 3, 3a-b; pl. 2, figs. 2, 2a-b
v. 2002b *C.*[*lypeaster*] *scillae* des Moulins – Kroh: 11
v. 2002 *Clypeaster* sp. – Plöchinger & Karanitsch: 165, fig. 354
2002 *Clypeaster* – Plöchinger & Karanitsch: 168, fig. 364
2004 *Clypeaster scillae* – Harzhauser et al.: 82, fig.

Type-material:
Clypeaster scillae des Moulins, 1837:
Holotype: current whereabouts unknown
Locus typicus: "Villeneuve-lès-Avignon"; according to Philippe (1998: 120) there is no doubt that this means the quarries of Montagnet, near Angles, Gard, France
Age: Burdigalian, Early Miocene

Clypeaster grandiflorus Bronn, 1838:
Holotype: specimen figured by Bronn (1837: 14; pl. xxxvi, fig. 9a-d); collection of Bronn; current whereabouts unknown
Locus typicus: probably Kemence, Pest, Hungary (see below)
Stratum typicum: Leitha limestone
Age: Badenian (Langhian-Early Serravallian), Middle Miocene
Remarks: Bronn mentioned the species for the first time in 1837 without description or locality information, but provided a figure (Bronn, 1837: 14; pl. xxxvi, fig. 9a-d). A figure alone is not sufficient for the establishment of a species according to the ICZN. Thus *grandiflorus* Bronn, 1837 is a *nomen nudum*. In 1838 referring to his figure from 1837 he mentioned *grandiflorus* again, but this time accompanied by a description, numerous localities and a synonymy list. The correct, valid, authorship for *grandiflorus*, therefore, is Bronn, 1838.

As Bronn included many records in his synonymy list that are not conspecific and mentioned several questionable localities the question which specimens can be regarded as types arises. Since Bronn (1838) explicitly refers to the figure of his 1837 paper, that would be a logical choice. The problem is that we do not know where that specimen came from. In a later, revised edition of his "*Lethaea geognostica*" (Bronn & Roemer, 1856) the specimen is again figured and this time only the locality Kemence, Hungary is mentioned. It thus seems likely that this is where the specimen came from. Furthermore, the list of localities in Bronn (1838) gives us further hints: several of the localities are marked by an exclamation mark. This sign means that Bronn had specimens from the so marked localities in his collection. Under "*Clypeaster grandiflorus* nob. α. humilis (fig. nostr.)" in Bronn (1838: 903-905) there are just two localities marked this way "in der Tegel-Formation in ! Steyermark, zu ! Kemenze im Honther Comitat in Ungarn". As Kemence thus occurs again it is proposed to consider that locality as type locality of *C. grandiflorus*.

There are no features which would allow to confidently separate *C. grandiflorus* and *C. scillae*, therefore *C. grandiflorus* is placed into the synonymy of the latter. As many other, probably not conspecific taxa are mentioned in BRONN's synonymy the record is added to the synonymy list with the critical sign "pp" (meaning that part of the material included by BRONN in *C. grandiflorus* might not be conspecific with *C. scillae*).

Clypeaster bunopetalus POMEL, 1887:
Holotype: specimen figured by POMEL (1885: pl. B26, figs. 1-7); current whereabouts unknown
Locus typicus: Djebel Mouzaia, Algeria
Stratum typicum: "terrain cartennien"
Age: Middle ? Miocene

Clypeaster pentadactylus PERON & GAUTHIER in COTTEAU et al., 1891:
Holotype: specimen figured by COTTEAU et al. (1891: pl. 6, figs. 4-5); current whereabouts unknown
Locus typicus: El Hammam, Valley of Oed Abdi, Aurès, Algeria
Age: Langhian, Middle Miocene

Clypeaster vasseli GAUTHIER in FOURTAU, 1902:
Holotype: specimen figured by FOURTAU (1902: pl. 4, figs. 1-3); Muséum de Paris
Locus typicus: Gebel Géneffé, opposite the railroad station, Eastern Desert, Egypt
Age: "Vindobonien", Middle Miocene

Clypeaster manini LOVISATO, 1914:
Syntypes: originally in the collection of LOVISATO; current whereabouts unknown
Type region: numerous localities in Sardinia, Italy

Clypeaster grandiflorus anteacutus VADÁSZ, 1915:
Holotype: MAFI Ech 125, figured by VADÁSZ (1915: figs. 19, 21); housed at the Museum of the Hungarian Geological Survey
Locus typicus: Gârbova de Sus (= Felsö-Orbó), Romania
Age: Late Badenian (Early Serravallian), Middle Miocene

Clypeaster grandiflorus altipetalus VADÁSZ, 1915:
Holotype: single specimen mentioned by VADÁSZ (1915: 128); not preserved at the Museum of the Hungarian Geological Survey, where the VADÁSZ-material is housed
Locus typicus: Kemence, Pest, Hungary
Stratum typicum: Leitha limestone
Age: Badenian (Langhian-Early Serravallian), Middle Miocene

Clypeaster grandiflorus basiconcava VADÁSZ, 1915:
Syntypes: several specimens mentioned by VADÁSZ (1915: 128); not preserved at the Museum of the Hungarian Geological Survey, where the VADÁSZ-material is housed
Locus typicus: Gârbova de Sus (= Felsö-Orbó), Romania
Age: Late Badenian (Early Serravallian), Middle Miocene

? *Clypeaster digitalis* VADÁSZ, 1915:
Holotype: MAFI Ech 150, figured by VADÁSZ (1915: fig. 28); housed at the Museum of the Hungarian Geological Survey
Locus typicus: Livezile (= Ùrháza, = Vládháza), Romania
Age: Late Badenian (Early Serravallian), Middle Miocene
Remarks: Specimen damaged (posterior end missing), very probably synonymous with *C. scillae*

Clypeaster zanoni CHECCHIA-RISPOLI, 1927:
Syntypes: current whereabouts unknown
Locus typicus: Faehiat, Cyrenaica, Libya
Age: Middle Miocene

Material:
Early Badenian (Langhian) – Aflenz, Styria, Austria
NHMW: 4 specimens (NHMW 1978/2022/1, 1978/2022/2, 1978/2022/3, 2004z0072/0004)
Early Badenian (Langhian) – Gamlitz (Unterer Gnaser Bruch), near Ehrenhausen, Styria, Austria
NHMW: 1 specimen (NHMW 1868.I.Anhang)
Early Badenian (Langhian) – Leibnitz, Styria, Austria
NHMW: 1 specimen (NHMW 2003z0058/0001)
Early Badenian (Langhian) – St. Nikolai, Styria, Austria
NHMW: 2 specimens (NHMW 1978/2021/1, 1978/2021/2)
Early ? Badenian (Langhian) – Rauchstallbrunngraben, near Baden, NÖ, Austria
NHMW: 2 specimens (NHMW 1976/1843/8, 1997z0178/1680)
Early Badenian (Langhian) – Retznei [Weissenegg Fm., Lafarge quarry (formerly Perlmoser)], Styria, Austria
NHMW: 3 specimens (NHMW 1997z0178/1831, 2004z0098/0006-7)
Early Badenian (Langhian) – Wiesfleck (sandpit Hohlreich), Bgld, Austria
NHMW: 2 specimen (NHMW 2004z0113/0001-2)
Badenian (Langhian-Early Serravallian) – Brunn am Steinfeld, NÖ, Austria
NHMW: 1 specimen [NHMW 1858.XII.8 (abnormal specimen, probably deformed due to recovery after sublethal predation)]
Badenian (Langhian-Early Serravallian) – Eisenstadt, Bgld, Austria
NHMW: 1 specimen (NHMW 2003z0056/0001)
Badenian (Langhian-Early Serravallian) – Großhöflein, near Eisenstadt, Bgld, Austria
NHMW: 4 specimens (NHMW 1858.XV.56, 1869.I.663, 1869.I.664, 1868.VIII.42)
Badenian (Langhian-Early Serravallian) – Hainburg (*Halitherium*-locality of STACHE, 1867), NÖ, Austria
NHMW: 1 specimen (NHMW 1867.XXII.2)
Badenian (Langhian-Early Serravallian) – Kalksburg, Vienna, Austria
IPUW: 1 specimen (no inventory no.)
Badenian (Langhian-Early Serravallian) – north-western Leitha Mountains, Bgld-NÖ boundary region, Austria
NHMW: 1 specimen (NHMW 1976/1825 [specimen figured by ERTL, 1964])
Badenian (Langhian-Early Serravallian) – Mannersdorf, Bgld, Austria
NHMW: 21 specimens (NHMW 1978/2020/1-20, 2004z0072/0003)
Badenian (Langhian-Early Serravallian) – Neudorf near Sopron (the locality could not be traced, either in Austria or in Hungary)
NHMW: 1 specimen (NHMW 2003z0059/0001)
Late Badenian (Early Serravallian) – Hornstein (= Szarvkö), Bgld, Austria
MAFI: 1 specimen [MAFI Ech 275 (referred to *C. scillae* DESMOULINS by VADÁSZ, 1915: 129-131)]
Late Badenian (Early Serravallian) – Müllendorf (quarry of the Mühlendorfer Kreidewerke AG), Bgld, Austria
NHMW: 4 specimens (NHMW A2328, 1997z0178/1745, 2003z0031/0006, 2004z0112/0002, ../0007)
Late Badenian (Early Serravallian) – St. Margarethen (Kummer quarry), Bgld, Austria
NHMW: 2 specimens (NHMW 1976/1844/57, 2004z0001/0041)

Foreign material for comparison purposes:
Middle Miocene – Ragusa, Sicily, Italy
NHMW: 1 specimen (NHMW 2003z0060/0001)
Middle Miocene – Una river, Bosanski Novi (= Bosnisch Novi), Bosnia and Herzegovina
NHMW: 1 specimen (NHMW 1888.93)

Early Badenian (Langhian) – Gombhegy, Kemence, Pest, Hungary

MAFI: 5 specimens [MAFI Ech 156 (material referred to *C. grandiflorus* BRONN by VADÁSZ, 1915: 126-128); Ech 321 and Ech 430 (specimens referred to *C. grandiflorus* BRONN. var. *rhabdopetalus* POM. by VADÁSZ, 1915: 126-128); Ech 324 and Ech 325 (specimens referred to *C. scillae* DESMOULINS by VADÁSZ, 1915: 129-131)]

Early Badenian (Langhian) – Mátraverbély (Meszetetö), Nográd, Hungary

MAFI: 2 specimens [MAFI Ech 322 and Ech 323 (referred to *C. scillae* DESMOULINS by VADÁSZ, 1915: 129-131)]

Badenian (Langhian-Early Serravallian) – Šahy (= Ipoly-Shag), Slovak Republic

NHMW: 2 specimens (NHMW 1852.II.1547, 1852.II.1548)

Late Badenian (Early Serravallian) – Gârbova de Sus (= Felsö-Orbó), Romania

MAFI: 5 specimens [MAFI Ech 125 (**holotype** of *C. grandiflorus anteacutus* VADÁSZ, 1915: 126-128; figs. 19, 21), Ech 135 (material referred to *C. grandiflorus* BRONN. var. *rhabdopetalus* POM. by VADÁSZ, 1915: 126-128; figs. 20, 22), Ech 136 (material referred to *C. altus* KLEIN by VADÁSZ, 1915: 152-153, figs. 44), Ech 259 & 316 (material referred to *C. grandiflorus* BRONN by VADÁSZ, 1915: 126-128), Ech 260 (referred to *C. scillae* DESMOULINS by VADÁSZ, 1915: 129-131)]

Late Badenian (Early Serravallian) – Kemence, Pest, Hungary

NHMW: 14 specimens (NHMW 1852.II.1546, 1858.XXXIX.14, 1858.XXXIX.15, 1858.XXXIX.16, 1858.XXXIX.17, 1865.XXXV.16 (9047), 1865.XXXV.17 (9046), 1865.XXXV.19, 1865.XXXV.20, 1865.XXXV.21, 1865.XXXV.22, 1885.XXXV.18 (A1056), 2003z0057/0001 to 0002)

MAFI: 1 specimen [MAFI Ech 138 (referred to *C. crassus* AG. by VADÁSZ, 1915: pl. 6, fig. 2)]

? Late Badenian (Early Serravallian) – Livezile (= Vládháza), Romania

MAFI: 1 specimen [MAFI Ech 150 (holotype of *Clypeaster digitalis* VADÁSZ, 1915: 135-136; fig. 28)]

Dimensions: see Tab. 9

Description:

Size and shape: The test is of medium to large size, elongated antero-posteriorly, with a subpentagonal outline. The anterior margin is bluntly pointed, very thick and tumid. The posterior margin is transversely truncated, slightly indented and less thick than the anterior margin, but still tumid. The lateral margins may be slightly indented in interambulacra 1 and 4. The maximum width lies both anterior and posterior of the apical disc, where ambulacra II and IV respectively I and V reach the ambitus. Test width usually varies between 75 and 85% TL and shows allometric growth, larger specimens showing a more elongated outline. In profile, the test has a trapezoid shape with a domed petaloid area. The maximum height lies around the apical disc on the raised adapical parts of the ambulacral interporiferous zones. The margin is of varying thickness and tumidity; but in all investigated specimens it is very thick and tumid. The oral surface is inclined towards the peristome, which lies in a deep infundibulum.

Apical disc: The apical disc lies subcentrally and is slightly depressed. It belongs to the monobasal type with a large central, subpentagonal madreporite and 5 circular gonopores.

Ambulacra: All five ambulacra are petaloid, straight and distally closing. The frontal petal is longest, the paired petals are subequal in length, the posterior ones sometimes being a little bit longer than the anterior ones. The petals extend roughly two third of the corresponding test radius. The poriferous zones are slightly depressed and relatively wide, being widest near the distal end of the petals. The pores are conjugated anisopores. Adjacent pore pairs are separated by narrow ridges with a single row of up to 5 primary tubercles. The interporiferous zones are strongly inflated and usually 2.2 to 3.5 times as wide as a single poriferous zone at the widest point of the petals. They are crowded by primary tubercles similar to those on the interambulacra. On the oral surface simple unbranched food grooves are present in the axis of the ambulacra.

Interambulacra: The interambulacra are distinctly depressed adapically between the petals. They are crowded with perforate crenulate primary tubercles in sunken areoles. Between the primary tubercles, dense miliary tuberculation is present. Between the petals the tubercle density is usually very low and the tubercles large and widely spaced. Tubercles size and density increases towards the margin. On the oral surface tubercles are closely spaced, their areoles nearly touching each other. On the oral surface the interambulacra are slightly inflated between the ambulacra, giving the a "swollen" appearance.

Peristome: The peristome is oval, transversely elongated and lies subcentrally on the oral side of the test in a deep, broad infundibulum with sloping walls.

Periproct: The periproct is subcircular to very slightly oval, transversely elongated. It lies inframarginally, about 3 to 4.5 mm away from the posterior margin.

Internal support system: Contrary to other species of *Clypeaster*, *C. scillae* has a rather weakly developed internal support system. Though the test is double walled and relatively thick, internal pillars are less common and present in the central part around the lantern only. There is a large cavity for the intestines running around the central part which is free from pillars. There are only two perforated septa one anterior and one next to the periproct (compare Pl. 34, Fig. 1). A dense marginal framework as in *C. barcinensis*? for example is completely missing.

Differential diagnosis:

For the difference to *C. barcinensis* ? LAMBERT, 1906 see above under that species.

For the difference to *C. calabrus* SEGUENZA, 1880 see above under that species.

C. campanulatus (SCHLOTHEIM, 1820) (and its phenotypes), a co-occurring species, differs by its strongly different profile, flattened oral surface, broader petals, which are more closed distally and have a much less inflated interporiferous zone and steep-walled infundibulum.

C. folium AGASSIZ in AGASSIZ & DESOR, 1847, a co-occurring species, is distinguished from this species by its lower test height, thin and rather sharp margin, distinct marginal indentation in all interambulacra, short petals, shallow infundibulum and deep food grooves.

C. intermedius DES MOULINS, 1837, an Early Miocene species of the Rhône Basin, differs from *C. scillae* by its different profile, less elongated outline, less strongly raised petaloid area, thinner and less tumid margin, and broader petals with less inflated interporiferous zones (see also PHILIPPE, 1998).

C. latirostris MICHELIN, 1861, a species occurring in the Eggenburgian (Early Burdigalian) of the Molasse Zone, is distinguished from this species by its subequal test length and width, lower test height, thin margin, slightly inflated interporiferous zones, flattened oral surface and the broad, very shallow infundibulum.

Discussion:

Clypeaster scillae DES MOULINS, 1835 is a common and rather well known species, occurring abundantly in the Burdigalian to Serravallian of the Mediterranean and the Badenian of the Paratethys. In the latter it is one of the most wide-spread *Clypeaster* species, occurring as far north as Poland and the Ukraine where most other species are missing. It is characterised by its pentagonal outline, with sub-parallel sides, strongly inflated interporiferous zones, very thick, tumid margin and broad infundibulum with sloping walls.

The following species were considered junior synonyms of *C. scillae* by PHILIPPE (1998: 119) and other authors: *Clypeaster grandiflorus* BRONN, 1838, *C. crassus* AGASSIZ, 1840 [*nomen nudum*], *C. crassicostatus* AGASSIZ, 1840 [*nomen nudum*], *C. bunopetalus* POMEL, 1887, *C. pentadactylus* PERON & GAUTHIER in COTTEAU et al., 1891, *C. vasseli* GAUTHIER in FOURTAU, 1902, *C. manini* LOVISATO, 1914 and *C. zanoni* CHECCHIA-RISPOLI, 1927. Based on the descriptions and illustrations of these species in the original publications, but without having seen the type material I can only agree with PHILIPPE that these species really seem to fall within the morphological variation exhibited by *C. scillae*. Several of these names have also been used for Austrian material both in publications and in collections. The records/specimens of these which I could examine belong indeed all to *C. scillae*. It is interesting to note that some authors recognised the over-splitting in this species very early (e.g. VADÁSZ, 1915) but still kept them as different species, although noting that these species were often impossible to distinguish.

PHILIPPI (1842, 1851) used the position of the gonopores as one of the major features to distinguish between the species *C. scillae*, *C. altus* and *C. turritus*. It is true that the position of the gonopores varies in relation to the madreporite (sometimes the gonopores lie directly at the apices of the madreporite, in other cases they are removed from the madreporite). This variation, however, occurs within single species and cannot be used as diagnostic feature. A very drastic example of this is the well known, extant *Clypeaster rosaceous* where the position of the gonopores varies very strongly. The function of this variation is unclear.

Occurrence:

Austria: Early to Late Badenian (Langhian-Early Serravallian)

Vienna Basin: Brunn am Gebirge, NÖ (MICHELIN, 1861; LAUBE, 1869a, 1871); Brunn am Steinfeld, NÖ (KALABIS, 1938a; [NHMW]); Gainfarn, NÖ (KROH, 2002b; PLÖCHINGER & KARANITSCH, 2002); Hainburg an der Donau, NÖ (LAUBE, 1869a, 1871; KALABIS, 1938b); Hundsheim hills, Hainburg an der Donau, NÖ (STACHE, 1867; [NHMW]); Kalksburg, Vienna (FUCHS, 1869; QUENSTEDT, 1874, 1875; KARRER, 1877; KALABIS, 1938b; THENIUS, 1962a, 1970; [NHMW]); Mannersdorf, NÖ ([NHMW]); Müllendorf (Mühlendorfer Kreide AG quarry), Bgld (KALABIS, 1938a+b; KAPOUNEK, 1939; TOLLMANN, 1955; HIDEN, 1998; ? PILLER & HARZHAUSER, 1999; ZORN, 2000; PLÖCHINGER & KARANITSCH, 2002; HARZHAUSER et al., 2004; [NHMW]); north-western Leitha Mts., Bgld/NÖ border region (ERTL, 1964; [NHMW]); Rauchstallbrunngraben, near Baden, NÖ (KALABIS, 1938a+b; [NHMW]); Rodaun, Vienna (KARRER, 1868); Hornstein (= Szarvkö), Bgld (VADÁSZ, 1915; [MAFI]); Vienna Basin (KÜPPER, 1968)

Eisenstadt-Sopron Basin: Eisenstadt, Bgld (LAUBE, 1869a, 1871; [NHMW]); Großhöflein, Bgld (CŽJŽEK, 1852; LAUBE, 1869a, 1871; KALABIS, 1938a; [NHMW]); St. Georgen (Hader hill), Bgld (SCHMID, 1968); Eisenstadt (Hartl Fm., Hartl hill), Bgld (KAPOUNEK, 1939); Höflein, near Eisenstadt, Bgld (MICHELIN, 1861; LAUBE, 1869a; KALABIS, 1938b); Kleinhöflein, near Eisenstadt, Bgld (CŽJŽEK, 1852); between Großhöflein and Kleinhöflein, near Eisenstadt, Bgld (KAPOUNEK, 1939; TOLLMANN, 1955); Neudorf near Sopron, Austria/Hungary border region ([NHMW]); St. Margarethen (Kummer quarry), Bgld (SCHMID et al., 2001; [NHMW])

Oberpullendorf Bay: Ritzing, Bgld (QUENSTEDT, 1874, 1875)

Styrian Basin: Aflenz, Styria ([NHMW]); Ehrenhausen, Styria (LAUBE, 1869a, 1871); Eichberg, near Zirknitz (LAUBE, 1869a, 1871); Gamlitz, near Ehrenhausen, Styria ([NHMW]); Leibnitz, Styria ([NHMW]); Nikolai-hill, St. Nikolai im Sausal, Styria (ROLLE, 1856; STUR, 1871; HILBER, 1878; [NHMW]); Ober-Mureck, Styria (ANDRAE, 1855); Retznei (Weissenegg Fm., Lafarge quarry), Styria (SCHOUPPÉ, 1949; HIDEN, 1998; [NHMW])

Paratethys (non-Austrian occurrences): Early to Late Badenian (Langhian-Early Serravallian)

Fore-Carpathian Basin: Niechobrz near Rzeszów, Southern Poland (MĄCZYŃSKA, 1996); Pińczów, Central Poland (MĄCZYŃSKA, 1993)

Molasse Zone: Židlochovice, Czech Republic (KALABIS, 1938a, 1949); Podivín, Czech Republic (KALABIS, 1949);

Vienna Basin: Devínska Kobyla, near Bratislava, Slovak Republic (KORNHUBER, 1898); Devínska Nová Ves, Slovak Republic (LAUBE, 1869a, 1871)

Great Hungarian Basin (Pannonian Basin): Hor. Strháre (= Felsö-Esztergály), Slovak Republic (GAÁL, 1905); Kemence, Pest, Hungary (BRONN, 1838; BRONN in BRONN & ROEMER, 1856; MICHELIN, 1861; KARRER, 1868; LAUBE, 1869a, 1871; COTTREAU, 1913a; VADÁSZ, 1915; KALABIS, 1938b; [MAFI]); Kemence (Gombhegy), Pest, Hungary (MÁJER, 1915; [MAFI]); Letkés, Pest, Hungary (VADÁSZ, 1915); Mátraverebély, Nógrád, Hungary (VADÁSZ, 1915; [MAFI]); Nagymaros, Pest, Hungary (VADÁSZ, 1915); Šahy (= Ipoly-Shag), Slovak Republic ([MAFI])

Transylvanian Basin: Buciumi, Romania (GHIURCA, 1996); Gârbova de Sus (= Felsö-Orbó), Romania (KOCH, 1887b; LÖRENTHEY, 1894; GAÁL, 1905; VADÁSZ, 1915; [MAFI]); Livezile (= Ùrháza, = Vládháza), Romania (VADÁSZ, 1915; [MAFI]); Moldoveneşti (= Várfalva), Cluj, Romania (KOCH, 1887b)

Lom Basin: Opanez and Opansko bardo, Northern Bulgaria (KOJUMDGIEVA & STRACHIMIROV, 1960)

Styrian Basin: Poljance (= Pöltschach, train station), Slovenia (ZOLLIKOFER, 1859); Šentilj in Slovenske gorice; NE Slovenia (MIKUŽ & MITROVIĆ-PETROVIĆ, 2001)

Sáva Basin: Una river, Bosanski Novi (Bosnisch Novi), Bosnia and Herzegovina ([NHMW])

Zala, Sáva and Dráva Basins: Slovenia (MAJCEN et al., 1997); Hrastovica, Kaznica Potok, ? Repušnica Potok, Sutjeska, Utinje Potok and Zrinj, Croatia (POLJAK, 1938); Bosanska Kostajnica, Gradačac and Ugljevik, Bosnia & Herzegovina (MITROVIĆ-PETROVIĆ, 1969)

Mediterranean: Burdigalian to Serravallian

Rhône Basin: Burdigalian: ? Aramon, Bassin d'Avignon, France (PHILIPPE, 1998); Beaucaire; Bassin d'Avignon, France (PHILIPPE, 1998); Buoux, Bassin d'Apt-Reillanne-Forcalquier, France (PHILIPPE, 1998); Couronne, Bouches-du-Rhône, France (LAMBERT, 1912); Istres, Secteur des étang, Bouches-du-Rhône, France (LAMBERT, 1912; COTTREAU, 1913a; PHILIPPE, 1998); Lacoste, Bassin d'Apt-Reillanne-Forcalquier, France (PHILIPPE, 1998); Martigues, Littoral de la Nerthe, France (PHILIPPE, 1998); Martigues, Secteur des étang, France (PHILIPPE, 1998); Ménerbes, Bassin d'Apt-Reillanne-Forcalquier, France (PHILIPPE, 1998); Montagnet, near Angles, Gard, France (DES MOULINS, 1837; LAMBERT, 1912; PHILIPPE, 1998); St. Mite-les-Remparts, Secteur des étang, France (PHILIPPE, 1998); Taillades, Vaucluse, France (LAMBERT, 1912)

Langhian: not recorded

Serravallian: Cucuron, Bassin Sud-Luberon, France (PHILIPPE, 1998)

Western Mediterranean: Catalonia, Spain (LAMBERT, 1927a); Calada Bianaca, Sardinia, Italy (COMASCHI CARIA, 1963); Chabbat-el-Kotta, Algeria (POMEL, 1885, 1887); Corsica, France (AGASSIZ, 1840b; PICTET, 1857; MICHELIN, 1861; LAUBE, 1871); Deya, Mallorca, Baleares, Spain (HERMITE, 1879); Djebel Mouzaia, Algeria (POMEL, 1885, 1887); El Hammam, valley of Oed Abdi, Aurès, Algeria (COTTEAU, PÉRON & GAUTHIER, 1891); La Superga, near Torino, Italy (MICHELIN, 1861; LAUBE, 1871); Monte Alvu near Bosa (Planargia), Sardinia, Italy (LOVISATO, 1914b); "regione Pilaghe (d)e Badde di Pozzomaggiore", Sardinia, Italy (LOVISATO, 1914b); Poetto al Capo S. Elia di Cagliari, Sardinia, Italy (LOVISATO, 1914b); Portotorres, Sardinia, Italy (LOVISATO,

1914b); south of Rif, Leben valley, Morocco (LAMBERT, 1931); Santa Christobal, Menrca, Baleares, Spain (HERMITE, 1879); Santa Manza, Corsica, France (MICHELIN, 1861; LAUBE, 1871); Santa-Maria near Ronca, Italy (MICHELIN, 1861; LAUBE, 1871); Sedini e Castelsardo, Sardinia, Italy (LOVISATO, 1914b); St. Pau d'Ordal, Catalonia, Spain (LAMBERT, 1927a); Vence, Alpes-Maritimes, France (LAMBERT, 1912);

Central Mediterranean: Faehiat, Cyrenaica, Libya (CHECCHIA-RISPOLI, 1927b); Ragusa, Sicily, Italy ([NHMW])

Eastern Mediterranean: Gebel Geneffé, Eastern Desert, Egypt (FOURTAU, 1901, 1902, 1920; LAMBERT, 1931); Geneifa (= Geneffé) massif, Eastern Desert, Egypt (FOURTAU, 1920); Morea (= Morée), Greece (PICTET, 1857)

Atlantic Ocean: Early-Middle Miocene

Aquitaine Basin: Caupian and Léognan, France (LAMBERT, 1927b; CHAVANON, 1974)

Clypeaster sp. indet.

? 1807 *Echinus rosaceus* – STÜTZ: 75-76
1831 *Clypeaster* (*Echinanthus marginatus* (?) of LESKE) – SEDGWICK & MURCHISON: 391, 419
1856 *Clypeaster* – ROLLE: 583, 592
1867 *Clypeaster* – MÜRLE: 333
1868 *Clypeaster* – FUCHS: 597
1875 *Clypeaster* – FUCHS: 59; No. 112
1877 *Clypeaster* div. sp. – FUCHS: 667
1877 *Clypeaster* – HILBER: 265
1877 *Clypeaster* sp. – KARRER: 78, 168, 229, 242, 262
1877 *Clypeaster* div. sp. – KARRER: 303
1882 *Clypeaster* sp. – HILBER: 236
1893 *Clypeaster* sp. – TOULA: 288
1894 Többféle *Clypeaster* töredéke. – LÖRENTHEY: 61
1900 *Clypeaster* – FUCHS: 882
1902 *Clypeaster* – FUCHS: 351, 355
1903a *Clypeaster* – FUCHS: 7
1903 Clypeastern – HOERNES: 955
1903 *Clypeaster* – HOERNES: 957, 964, 965, 1095
1905 *Clypeaster* – FABIAN: 20
1906 *Clypeaster* – SCHAFFER: 78, 90
1907 *Clypeaster* – SCHAFFER: 25, 26
1908 *Clypeaster* sp. – SCHAFFER: 46
1908 *Clypeaster* sp. – SCHAFFER: 85
1912b *Clypeaster* – SCHAFFER: 45, 47, 53
1915 *Clypeaster* sp. – MÁJER: 36, 89
1924 *Clypeaster* – ABEL: 60
1924a *Clypeaster* – SCHAFFER: 43, 46, 51
1924b *Clypeaster* – SCHAFFER: 485
1927a *Clypeaster* – SCHAFFER: 78
1928 *Clypeaster* – SCHAFFER: 202
1930 *Clypeaster* sp. – BOBIES: 26
1930 *Clypeaster* sp. – VENDL: 77
1938 *Clypeaster* sp. – POLJAK: 191-192
1942 *Clypeaster* – SCHAFFER: 128, 142
1943 *Clypeaster* – JANOSCHEK in SCHAFFER: 441, 443, 445, 447
1951 *Clypeaster* – JANOSCHEK in SCHAFFER: 559, 561, 563, 565
1953b *Clypeaster* sp. – SIEBER: 194
1955 *Clypeaster* – THENIUS: 110-111; pl. 13, fig. 33
1955 *Clypeaster* sp. – TOLLMANN: Tab. 5b
1958a *Clypeaster* sp. – SIEBER: 152
1961 *Clypeaster* – SCHAFFER: 149
1961 *Clypeaster* – WESSELY: 319
1962 *Clypeaster* – PAPP & THENIUS in KÜHN: 145
1962 *Clypeaster* – THENIUS: 108-109; pl. 8, fig. 33
1965 *Clypeaster* sp. (flache Art) – KOLLMANN: 541
1965 *Clypeaster* – KOLLMANN: 549
1965 *Clypeaster* – FUCHS: 169
1968 *Clypeaster* – FLÜGEL & HERITSCH: 35, 99
1968 *Clypeaster* sp. – SCHMID: Anhang
1969 *Clypeaster exscentricus* VAD. – MITROVIĆ-PETROVIĆ: 136; pl. 25, figs. 1, 1a-b
1970 *Clypeaster* – THENIUS: p. 215
1970 *Clypeaster* – THENIUS: p. 216
1974 *Clypeaster* sp.– THENIUS: 70-71; fig. 13/24
1974 *Clypeaster* sp. – VÁVRA: 345
1977 *Clypeaster* sp. – EBNER & GRÄF: 161
1978 *Clypeaster* sp. – KOLLMANN & RÖGL: 165
1978b *Clypeaster* – PAPP et al.: 33
1979 *Clypeaster* sp. – THENIUS: 52; pl. 3, fig. 24
1980 *Clypeaster* sp. – BRIX: 44; pl. 6, fig. 23
1982 *Clypeaster* – FENNINGER & WASSERMANN: 50
1983 *Clypeaster* – DULLO: 19
1983 *Clypeaster* sp. – THENIUS: 119; pl. 7, fig. 24
1984 *Clypeaster* sp. – KÓKAY et al.: 290
1985 *Clypeaster* sp. indet.– MIHÁLY: 238
1985 *Clypeaster* – TOLLMANN: 501
1985 *Clypeaster* – TOLLMANN: 564
1987 *Clypeaster* sp. – MĄCZYŃSKA: 146, 148; pl. 4, figs. 6a-b
1988 *Clypeaster* sp. – BRIX: 35
1989 *Clypeaster* sp. – NEBELSICK: 15
1990 *Clypeaster* sp. – FRIEBE: 245, 254
1991a *Clypeaster* sp. – NEBELSICK et al.: 88
1991b *Clypeaster* – NEBELSICK et al.: 119
? 1993 *Echinolampas* sp. – MĄCZYŃSKA: 111-112; pl. 2, figs. 6, 6a-b
1997 *Clypeaster* sp. – MAJCEN et al.: 106; pl. 6, fig. 2
1998 *Clypeaster* sp. – HIDEN: pl. 1, figs. 6
1999 *Clypeaster* sp. – ROHATSCH: 23; fig. 6 [probably a *C. campanulatus*]
2000 *Clypeaster* – PILLER: 87
2001 *Clypeaster* sp. – MIKUŽ & MITROVIĆ-PETROVIĆ: 53-54; pl. 2, figs. 1, 1a, 3, 3a
2002 *Clypeaster* sp. – KAZÁR: 153; fig. 1
2002 *Clypeaster* – WEIDERT: 141, fig.
v. 2002b *Clypeaster* sp. – KROH: 12
v. 2002 *Clypeaster* sp. – PLÖCHINGER & KARANITSCH: 161, fig. 332
v. 2002 *Clypeaster* sp. – PLÖCHINGER & KARANITSCH: 165, fig. 353

Remarks: In the case of *Clypeaster* sp. international references are omitted from the synonymy. In some cases non-Austrian Paratethyan references, if accompanied by a description or illustration are included.

Material:

Late Eggenburgian (Early Burdigalian) – Eggenburg (train station, Zogelsdorf Fm.), NÖ, Austria
NHMW: 2 specimens (NHMW 1981/55/1)

Late Eggenburgian (Early Burdigalian) – Limberg (Zogelsdorf Fm., Hengl quarry), NÖ, Austria
NHMW: 2 specimens (NHMW 2004z0042/0006-7)

Early Badenian (Langhian) – Gainfarn, NÖ, Austria
NHMW: 1 fragment (NHMW 2004z0076/0068)

Early Badenian (Langhian) – Stotzing (sandpit Mayer), Bgld, Austria
NHMW: 1 specimen (NHMW 2004z0093/0024)

Early ? Badenian (Langhian) – Rauchstallbrunngraben, near Baden, NÖ, Austria
NHMW: 2 specimens (NHMW 1846.37.962, 1846.37.963)

Badenian (Langhian-Early Serravallian) – Baden, NÖ, Austria
NHMW: 1 specimen (NHMW 2003z0054/0002 to 0003)

Badenian (Langhian-Early Serravallian) – Bad Fischau, NÖ, Austria
NHMW: 1 specimen (NHMW 1860.XIX.4/1861.I.527 [two inv. nos., one specimen])

Badenian (Langhian-Early Serravallian) – Mannersdorf, Bgld, Austria
NHMW: 1 specimen (NHMW A3158)
Badenian (Langhian-Early Serravallian) – Devínska Nová Ves, Slovak Republic
NHMW: 1 specimen (NHMW 1852.II.1552)
Badenian (Langhian-Early Serravallian) – Oslip, Bgld, Austria
NHMW: numerous fragments (NHMW 2003z0083/0004)
Badenian (Langhian-Early Serravallian) – Perchtoldsdorf, NÖ, Austria
NHMW: 2 specimens (NHMW 2003z0052/0002-0003)
Badenian (Langhian-Early Serravallian) – Sooss, near Baden, NÖ, Austria
NHMW: 1 specimen (NHMW 1855.XLV.991)
Badenian (Langhian-Early Serravallian) – Steinebrunn (formerly Steinabrunn), NÖ, Austria
NHMW: 8 specimens (NHMW 1846.37.956, 1846.37.956.b [*Clypeaster*-demipyramids])
Badenian (Langhian-Early Serravallian) – Wöllersdorf, NÖ, Austria
NHMW: 1 specimen (NHMW 2003z0061/0001)
Late Badenian (Early Serravallian) – Müllendorf (quarry of the Mühlendorfer Kreidewerke AG), Bgld, Austria
NHMW: 5 fragments (NHMW 2004z0112/0008)
Late Badenian (Early Serravallian) – Stiefingbach (old mill, Pesendorf 16), Styria, Austria
NHMW: 1 specimen (NHMW 2004z0114/0001)

Discussion:

Being easily recognisable, the genus *Clypeaster* has often been reported in the geological/palaeontological literature of Austria. Most of the records do not state a specific species and come without description and or illustration. Thus they cannot safely be referred to any of the species discussed above and are therefore listed here as *Clypeaster* sp. indet. to provide complimentary information on the distribution of this genus in the Neogene of Austria. Additionally specifically undeterminable *Clypeaster* specimens in the collection of the Naturhistorisches Museum Wien are listed, which document the occurrence of this genus in previously not mentioned localities. Most of these specimens certainly belong to the species discussed above. Some of them, however, like the specimen from Stiefingbach might represent other species, not reported from Austria up till now. The poor preservation of the material, however, does not warrant discussing them separately.

The fragments figured as *Echinolampas* sp. by Mączyńska (1993: pl. 2, figs. 6, 6a-b) belong to the genus *Clypeaster*, this is indicated by the size-relation of interambulacral and ambulacral plates, as well as by the presence of an internal buttress system.

The record of *Echinus rosaceus* by Stütz (1807: 75-76) from the Calvariberg near Baden (NÖ) most probably refers to a *Clypeaster* species.

Occurrence:

Austria: Late Eggenburgian (Early Burdigalian), Badenian (Langhian-Early Serravallian)

Molasse Zone: Eggenburg (Brunnstube), NÖ (Nebelsick et al., 1991a); Kühnring valley, near Eggenburg (Fuchs, 1868); Limberg (Hengl quarry), NÖ (Nebelsick et al., 1991b); Schmidabach, near Eggenburg (Fuchs, 1900); Zogelsdorf Fm., region of Eggenburg, NÖ (Nebelsick, 1989)

Vienna Basin: Baden, NÖ ([NHMW]); Baden (Calvariberg), NÖ (Stütz, 1807); Bad Fischau, NÖ (Rohatsch, 1999; [NHMW]); Brunn am Steinfeld, NÖ (Karrer, 1877); Gainfarn (Kroh, 2002b; Plöchinger & Karanitsch, 2002); Grinzing (Villa Schöller), Vienna (Fuchs, 1875; Schaffer, 1906); Hainburg hills, NÖ (Wessely, 1961); Hainburg an der Donau, NÖ (Mürle, 1867); Hasenberg, NÖ (Karrer, 1877); Hornstein, Bgld (Tollmann, 1955); Jadlkogel between Baden and Gumpoldskirchen, NÖ (Karrer, 1877); Kaisersteinbruch (Leitha Limestone), Bgld (Schaffer, 1908); Kaisersteinbruch (as reworked clasts in Sarmatian sediments), Bgld (Fuchs, 1902); Kalksburg, Vienna (Hoernes, 1903; Janoschek in Schaffer, 1943, 1951; Thenius, 1955, 1962, 1974, 1979; 1983); Leitha Mountains (Janoschek in Schaffer, 1943, 1951; Schmid, 1968; Tollmann, 1985); Lindabrunn, NÖ (Schaffer, 1942; Brix, 1980); Mannersdorf, NÖ ([NHMW]); Mödling, NÖ (Karrer, 1877); Möllersdorf, NÖ (Karrer, 1877); Müllendorf (Mühlendorfer Kreide AG quarry), Bgld (Schaffer, 1908; Schaffer, 1961; Dullo, 1983); Müllendorf (Fenk quarry), Bgld (Tollmann, 1955); Nußdorf (Grünes Kreuz), Vienna (Schaffer, 1906); Perchtoldsdorf, NÖ (Karrer, 1877); Pötzleinsdorf, Vienna (Sieber, 1953b; Thenius, 1970); Rauchstallbrunngraben, near Baden (Fuchs, 1903a; Schaffer, 1907, 1912b; Abel, 1924; Schaffer, 1924a; Bobies, 1930; Schaffer, 1942; Vávra, 1974; Brix, 1988; Plöchinger & Karanitsch, 2002; [NHMW]); Sooss, near Baden, NÖ (Karrer, 1877; [NHMW]); Steinebrunn, NÖ (Sieber, 1958a; [NHMW]); Stotzing (sandpit Mayer), Bgld (Kazár, 2002); Vienna (Thenius, 1970); Vienna Basin (Hoernes, 1903; Schaffer, 1924b, 1927a, 1928), Wöllersdorf, NÖ ([NHMW])

Eisenstadt-Sopron Basin: Burgenland (Weidert, 2002); Eisenstadt (Gloriette), Bgld (Tollmann, 1955); Eisenstadt (Johannesgrotte), Bgld (Piller, 2000); between Großhöflein and Kleinhöflein, Bgld (Tollmann, 1955); Oslip, Bgld (Fuchs, 1965; Dullo, 1983)

Styrian Basin: Altenberg, Styria (Hiden, 1998); Eckberg, near Gamlitz, Styria (Hilber, 1877); mid-styrian ridge, between Wildon and Ehrenhausen, Styria (Flügel & Heritsch, 1968; Tollmann, 1985); hills of Neurath and Muggenau, Styria (Rolle, 1856); Pinkafeld, Bgld (Hoernes, 1903); Retznei [Lafarge quarry (formerly Perlmoser)], Styria (Flügel & Heritsch, 1968; Friebe, 1990); between St. Georgen and Pichla, Styria (Kollmann, 1964); Stiefingbach, Pesendorf, near Wildon, Styria (Kollmann & Rögl, 1978; [NHMW]); Sukdull, Styria (Fabian, 1905); Weitendorf, Styria (Ebner & Gräf, 1977; Fenninger & Wassermann, 1982); Wiesfleck, Styria (Kollmann, 1964); Wildon-Weißenegg, Styria (Friebe, 1990); Zirknitz valley, Styria (Sedgwick & Murchison, 1831)

Paratethys (non-Austrian occurrences): Badenian (Langhian-Early Serravallian)

Fore-Carpathian Basin: Korytnica Clays, Korytnica, Central Poland (Mączyńska, 1987); Lackie małe, [Zloczow (= Zolochiv) county], Ukraine (Hilber, 1882); Mykolaiv (= Mikołajów), S of Lwów, Ukraine (Hilber, 1882); Pińczów, Central Poland (Mączyńska, 1993)

Molasse Zone: Kralice nad Oslavou (= Kralitz), Moravia, Czech Republic (Toula, 1893)

Vienna Basin: Devínska Nová Ves, Slovak Republic ([NHMW])

Eisenstadt-Sopron Basin: Fertörákos, near Sopron, Hungary (Vendl, 1930)

Great Hungarian Basin (Pannonian Basin): Budapest, Hungary (Kókay et al., 1984; Mihály, 1985); Kemence (Gombhegy), Pest, Hungary (Májer, 1915)

Styrian Basin: Šentilj in Slovenske gorice; NE Slovenia (Mikuž & Mitrović-Petrović, 2001)

Transylvanian Basin: Gârbova de Sus (= Felsö-Orbó), Romania (Lörenthey, 1894)

Zala, Sáva and Dráva Basins: Zrinj, Croatia (Poljak, 1938); Bosanska Kostajnica, Bosnia & Herzegovina (Mitrović-Petrović, 1969)

Clypeasteridae indet.

1978b Clypeasteridae – Papp et al.: 31
1978 Clypeasteriden – Papp & Steininger: 140
1978 Clypeasteriden – Steininger & Papp: 199

Discussion:
In most cases these records will probably be based on fragments of *Clypeaster*, the most common Miocene echinoid in Austria.

Occurrence:

Austria: Badenian (Langhian-Early Serravallian)
Vienna Basin: Müllendorf (Fenk quarry), Bgld (Steininger & Papp, 1978); Rauchstallbrunngraben, near Baden (Papp & Steininger, 1978)

Suborder Laganina Mortensen, 1948b
Family Fibulariidae Gray, 1855
Genus *Echinocyamus* van Phelsum, 1774

Type-species: *Spatangus pusillus* Müller, 1776, by subsequent designation (ICZN 1954a, opinion 207, p. 350, 8th March 1954)
Diagnosis: Test moderately flattened; monobasal apical disc with a single hydropore, which is not located in a groove; periproct round or transversely elongated, situated between 1st and 2nd pair of coronal plates; development of petals variable ranging from poorly developed with few pore pairs to well developed with 7 or more pore pairs in each column; pore pairs usually strongly oblique; no spicules in the tube feet; accessory podia arranged in long lines along circumferential sutures; 5 pairs of internal radial partitions (one pair in each interambulacrum); periproct transversely oval; few small plates in periproctal membrane; plates in periproctal membrane without spine tubercles; some species show sexual dimorphism in genital pore size and position (compare Kier, 1967, 1969). (modified from Durham, 1966 and Mooi, 1989)
Distribution: Eocene to Recent – cosmopolitan (except polar regions; see Mironov & Sagidachny, 1984: figs. 9, 10)
Remarks: In 1891 Lambert (1891: 749) proposed to interchange the names *Echinocyamus* and *Fibularia*, completely reversing the usage from that of all former authors. The echinoderm research community split into two fractions on this issue. While a large number of authors (e.g. Capeder, Cotteau, Cottreau, de Loriol, ...), among them most of those working with extant echinoid rejected Lambert's proposal, many palaeontologists followed him (e.g. Airaghi, Checchia-Rispoli, Jeannet, Vadász, Szörényi, ...). Despite the clarifying statements of Mortensen (1948b: 172) and the "*Treatise of Invertebrate Paleontology*" (Durham, in Moore, 1966), and the ICZN decision (opinion 207, 1954) there were still authors who follow Lambert's opinion. Until now no true *Fibularia* has been recorded from the Central Paratethys, all reported and newly collected specimens belong into the genus *Echinocyamus* (*sensu* Mortensen, 1948b).

More than 110 nominal species, 11 of which are extant, of *Echinocyamus* have been described until now (compare Lambert & Thiéry, 1909-1925 [under *Fibularia*, see above]; Kier & Lawson, 1978; Mironov & Sagaidachny, 1984). Most of these taxa have not been described and illustrated in a modern way, only very few were examined using SEM techniques. Furthermore, the morphological plasticity of extant fibulariid taxa remains jet to be investigated and consequently very different features were regarded as diagnostic at species level [e.g. Capeder (1906) considered the arrangement of genital and ocular pores as very significant, while Mortensen (1948b: 160), an authority on extant echinoids, disregarded it completely]. Therefore, interspecific comparison is very difficult and differential diagnoses are restricted to taxa occurring in (or reported from) the Central Paratethys in the present paper. Although often applied to material from the Central Paratethys, the species established by Capeder (1906) seem not to be conspecific with the Paratethyan material. Because of the problems outlined above and the fact that a world-wide revision on fossil and extant *Echinocyamus* and *Fibularia* is currently being prepared by Rich Mooi (pers. comm. 20. March 2001) the investigations on the fossil *Echinocyamus* species of Austria are mainly restricted to re-describing Laube's *Echinocyamus transylvanicus* and selecting a lectotype for that species. Based on the re-examination of the type material, it seems possible to attach all known specimens from the Badenian of Austria to *E. transylvanicus* and a second species. Additionally, a third *Echinocyamus* species from the Eggenburgian is described in open nomenclature and some remarks on other species reported from the Central Paratethys are given.

Echinocyamus transylvanicus Laube, 1869

(Figs. 30, 31.B, 32, 34.1; Pl. 36, Figs. 1-5; Pl. 37, Figs. 1-4)

*v.	1869a	*Echinocyamus Transylvanicus* Laube. – Laube: 182
v.	1870	*Echinocyamus transsylvanicus* Lbe. – Laube: 314
v.	1871	*Echinocyamus transsylvanicus* Laube. – Laube: 61; pl. 16, fig. 4
.	1877	*Echinocyamus transsylvaniae* Lbe. – Karrer: 181
	1887b	*Echinocyamus transylvanicus*, Laube. – Koch: 261
non	1897	*Echinocyamus pusillus* Muell. – Vinassa de Regny: 149
?	1905	*Echinocyamus transylvanicus*, Lbe. ? – Gaál: 357, 362
non	1915	*Fibularia pusilla* Müll. sp. – Vadász: 113-114; pl. 9 (3), figs. 7-9
	1915	*Fibularia stellata* Cap. – Vadász: 114; pl. 9 (3), figs. 5-6
	1960	*Echinocyamus marioi* Lovisato in Cotteau 1895 – Kojumdgieva & Strachimirov: 231; pl. 9, figs. 3a-c

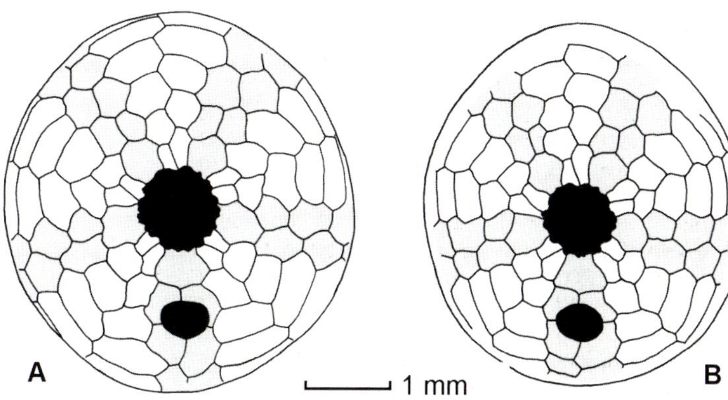

Figure 30: *Echinocyamus transylvanicus* Laube, 1869: oral plating [A: Lapugy, Romania (lectotypus, NHMW 1860.XL.529a); B: Baden, NÖ (NHMW 1852.II.1612a)].

?	1969	*Echinocyamus pusillus* (Müller) – Gábos & Ghiurca: 88; pl. 2, figs. 3, 6 [? misidentified *E. transylvanicus*]
?	1969	*Echinocyamus* cfr. *stellatus* Capeder – Gábos & Ghiurca: 88-89; pl. 2, fig. 1 [? misidentified *E. transylvanicus*]
?	1977	*Echinocyamus circularis* Capeder, 1906 – Mączyńska: 197; pl. 6, figs. 4-5
?	1987	*Echinocyamus circularis* Capeder, 1906 – Mączyńska: 148
?	1988	*Echinocyamus circularis* Capeder, 1906 – Mączyńska: 61; pl. 4, fig. 1-2

Type-material:
Although no repository for Laube's types is given in any of his works (1869a, 1870, 1871) the majority of other specimens he investigated came either from the collection of the k. & k. Hofmuseum (now Naturhistorisches Museum Wien) or from the Geologische Reichsanstalt (now Geological Survey of Austria). Only in the former collection *Echinocyamus* specimens from the localities mentioned by Laube [Lăpugiu des Sus (= Lapugy), and Coşteiu de Sus (= Kostej), both in Romania] could be located. Based on the acquisition numbers (acquired in the years 1859 and 1860) of the specimens we know that they were already in the collection at the time of Laube's investigations. These specimens, therefore, are regarded as syntypes.
Lectotype (designated herein): NHMW 1860.XL.529a (see Pl. 36, Figs. 1a-c)
Paralectotypes: NHMW 1859.XLV.615 (see Pl. 36, Figs. 3a-c), 1860.XL.529b (see Pl. 36, Figs. 2a-c)
Locus typicus: Lăpugiu des Sus (= Lapugy), Romania
Age: Early Badenian (Langhian, Middle Miocene)

Material:
Early Badenian (Langhian) – Baden Tegel, Baden, NÖ, Austria
NHMW: 2 whole coronas and several fragments [NHMW 1852.II.1612a-d (2 whole coronas, 1 fragment, 1 pyramid with attached tooth),1 juvenile specimen (NHMW 2003z0079/0001)

Foreign material for comparison:
Early Badenian (Langhian) – Coşteiu de Sus (= Kostej), Romania
NHMW: 19 whole coronas and several fragments (NHMW 2003z0084/0001a-t)
Early Badenian (Langhian) – Lăpugiu des Sus (= Lapugy), Romania
NHMW: 4 whole coronas and several fragments [NHMW 1859.XLV.615, 1860.XL.529a-d (2 whole coronas, 1 pyramid with attached tooth and several fragments), 2003z0073/0001]

Dimensions (in mm):

Inv. No.	TL	TW	TH
Baden, Austria			
NHMW 1852.II.1612b	3.98	3.56	2.63
NHMW 1852.II.1612c	2.70	2.34	1.54
Coşteiu de Sus (= Kostej), Romania			
NHMW 2003z0084/0001a	4.11	3.67	2.33
NHMW 2003z0084/0001b	4.28	3.50	2.22
NHMW 2003z0084/0001c	4.22	3.50	2.44
NHMW 2003z0084/0001d	3.33	2.78	1.78
NHMW 2003z0084/0001e	2.00	1.72	1.11
NHMW 2003z0084/0001f	3.17	2.83	1.72
NHMW 2003z0084/0001g	2.06	1.78	1.17
NHMW 2003z0084/0001h	4.50	3.61	2.67
NHMW 2003z0084/0001i	4.44	3.78	2.33
NHMW 2003z0084/0001j	4.28	3.50	2.17
NHMW 2003z0084/0001k	3.72	3.28	2.06
NHMW 2003z0084/0001l	3.39	2.83	2.00
NHMW 2003z0084/0001m	3.11	2.61	1.94
NHMW 2003z0084/0001n	3.28	2.83	1.72
NHMW 2003z0084/0001o	2.83	2.50	1.72
NHMW 2003z0084/0001p	2.78	2.44	1.44
NHMW 2003z0084/0001q	2.78	2.28	1.44
NHMW 2003z0084/0001r	2.22	1.89	1.22
NHMW 2003z0084/0001s	1.89	1.61	1.06
Lăpugiu des Sus (= Lapugy), Romania			
NHMW 1859.XLV.615	2.02	1.79	1.29
NHMW 1860.XL.529a	4.33	4.12	2.75
NHMW 1860.XL.529c	3.98	3.57	2.08
NHMW 2003z0073/0001	3.79	3.42	2.25

Description:
Size and shape: Test very small, usually less than 6 mm in length. Outline oval (antero-posteriorly elongated) to subcircular (test width ranges from 80.2 to 95.2 % TL, mean: 86.3 %). Maximum width slightly posterior of centre. In profile test is high arched to slightly conical with a tumid ambitus (test height ranging from 50.6 to 66.0 % TL, mean: 56.6 %). The maximum height coincides with the position of apical disc. The oral surface is flattened, without infundibulum.
Apical disc: The apical disc lies slightly anterior of the centre (~ 47 % of TL from the anterior margin) at the highest point of the test. Four moderately large subcircular genital pores are present. Gonopore size varies, probably due to sexual dimorphism as documented already for a number of extant and fossil species (Mortensen, 1948b; Kier, 1969). The five ocular pores are small and the anterior ones lie slightly outside the polygon formed by the genital pores. Only a single circular hydropore is present, it lies at the point where the imaginary lines drawn between genital pore 1 and 3, respectively 2 and 4 intersect (see Pl. 36, Fig. 5).
Ambulacra: Adapically more or less well developed petals are present. They consist of 2 to 6 strongly oblique partitioned isopores in each column. The petals are straight, open distally, with parallel poriferous. The frontal petal is longest, the paired petals are subequal in length, *c.* 90 % of the frontal one. They extend about 50 to 55 % of the corresponding test radius (measured in planar aboral view). The interporiferous zones are as wide or slightly wider than a single poriferous zone. Outside the petals, distinct arcs of accessory pores (micro-unipores) which lie transversely to the axis of the ambulacra are found. The size of these accessory pores varies, but they are always distinctly smaller than the respiratory pores in the petals.
Interambulacra: The interambulacra are not inflated or depressed between the ambulacra. They are covered by small perforate tubercles, between which many militaries are found.
Peristome: The peristome is subcircular and rather large (~20 % of TL in diameter in adult specimens). It lies subcentrally on the oral surface.
Periproct: About half the size of the peristome (~10 % of TL wide), of oval size, transversely elongated. It lies halfway between the peristome and the posterior margin or slightly posterior. The position of the periproct coincides with the border between the first and second pairs of postbasicoronal plates (Fig. 30).
Internal support structures: Five pairs of internal radial partitions are present in the ambulacra (see Pl. 37, Fig. 3), proving the attribution to the genus *Echinocyamus*.

Differential diagnosis:
The Aquitanian to Langhian Maltese *Echinocyamus*, which has been referred to *E. stellatus* Capeder, 1906 by various authors (e.g. Challis, 1980) differs from *E. transylvanicus* by its larger petaloid area, its raised apical disc, and its slightly inflated, cushion-shaped coronal plates (see Figs. 34.6a-c).

Echinocyamus ovatus (Münster in Goldfuss, 1829) from the Late Oligocene of Germany differs by its rostrate posterior end, smaller petalodium with fewer ambulacral pores (in specimens of comparable size) and more anteriorly eccentric apical disc (see Figs. 34.4a-c).

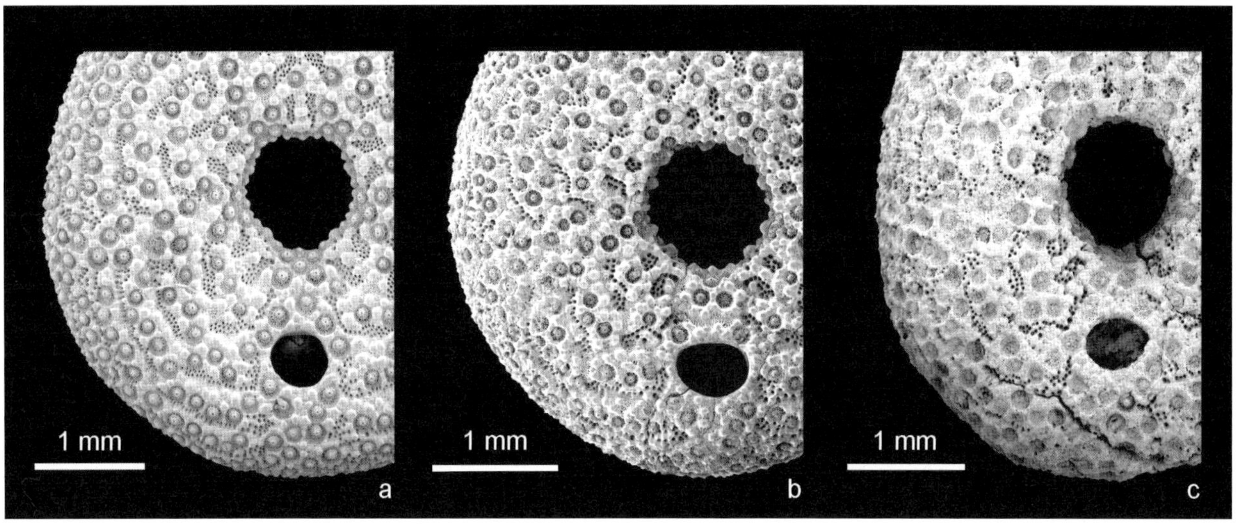

Figure 31: *Echinocyamus*, comparison of the distribution of the accessory pores [A: *E. pusillus* (Müller, 1776), recent, Mediterranean Sea, Croatia (NHMW coll.); B: *E. transylvanicus* Laube, 1869, Badenian of Lapugy, Romania (NHMW 1860.XL.529a); C: *E. pseudopusillus* Cotteau, 1895, Badenian of Gainfarn, NÖ (mirrored for easier comparison, NHMW 2004z0076/0045)].

The contemporaneous *Echinocyamus* species described under the name *Fibularia calariensis* from Hungary by Vadász (1915) and as *Echinocyamus pseudopusillus* Cotteau from Poland by Mączyńska (1977) differs by its strongly elongated outline, "pointed" anterior margin and more flattened profile.

The common Polish species *Echinocyamus* sp. B [illustrated as *E. linearis* Capeder, 1906 by Mączyńska (1977)] differs strongly by its elongated, often subpentagonal outline and the submarginal position of its periproct (see Figs. 34.5a-c).

Discussion:

Albeit few specimens of this species are known, they are rather well preserved and allow a good characterisation of this species. Vinassa de Regny (1897: 149) synonymised *E. transylvanicus* with the extant *E. pusillus*, an opinion that had been followed by Vadász (1915: 113-114) and Mortensen (1948b: 182). Based on direct comparison (employing SEM techniques) of the type material of *E. transylvanicus* with extant specimens, the former seems to be well differentiated from the Pliocene to extant species *E. pusillus* (found throughout the Lusitanian bioprovince and the North Sea). Apart from the obvious differences in outline and profile, *E. transylvanicus* has a much smaller number of accessory pores per plate (compare Fig. 31) and fewer pore pairs in the petals (Figs. 34.3a-b).

Material from the Badenian of Poland ascribed to *Echinocyamus circularis* Capeder, 1906 by Mączyńska (1977, 1987 and 1988) are very similar to *E. transylvanicus* and might be conspecific.

Occurrence:

Austria: Early Badenian (Langhian)

Vienna Basin: Baden Tegel, Baden, NÖ (Laube, 1871; Karrer, 1877; Koch, 1887b; [NHMW])

Paratethys (non-Austrian occurrences): Early Badenian (Langhian)

Great Hungarian Basin (Pannonian Basin): Kishajmás (= Háromház), Baranya, Hungary (Vadász, 1915); ? Nógrádszakál (= Szakall), Nógrád, Hungary (Gaál, 1905)

Fore-Carpathian Basin: ? marly sands overlying the Korytnica Clays, Korytnica, Poland (Mączyńska, 1977, 1987); ? Rybnice, Southern Poland (Mączyńska, 1988)

Transylvanian Basin: ? Berchezoaia, Romania (Gábos & Ghiurca, 1969); ? Ciceu-Hăşmaş, Romania (Gábos & Ghiurca, 1969); ? Ciolt, Romania (Gábos & Ghiurca, 1969); Coşteiu de Sus (= Kostéj), Romania (Laube, 1869a, 1871; Koch, 1887b; ? Gábos & Ghiurca, 1969; [NHMW]); ? Gîrbova de Sus, Romania (Gábos & Ghiurca, 1969); Lăpugiu des Sus (= Lapugy), Romania (Laube, 1869a, 1871; Koch, 1887b; ? Gaál, 1905; Vadász, 1915; [NHMW]); ? Valea Satului, Romania (Gábos & Ghiurca, 1969); ? Văleni, Romania (Gábos & Ghiurca, 1969)

Lom Basin: Opansko bardo Northern Bulgaria (Kojumdgieva & Strachimirov, 1960)

Echinocyamus pseudopusillus Cotteau, 1895

(Figs. 31.C, 32, 33, 34.2, Pl. 37, Figs. 5-6; Pl. 38, Figs. 1, 3)

pp 1915 *Fibularia pusilla* Müll. sp. – Vadász: 113-114; pl. 9 (3), figs. 7-9

? 1927 *Fibularia* aff. *pusilla* Müll. – Horusitzky: 25, 165

? 1950 *Fibularia pusilla* Müller, 1776. – Szörényi: 141-142; pl. 1, figs. 2, 2a, 3, 3a

1950 *Fibularia pseudopusilla* Cotteau, 1895. – Szörényi: 142; pl. 1, figs. 4, 4a

1950 *Fibularia lecointreae* Lambert, 1908. – Szörényi: 143; pl. 1, figs. 5, 5a

? 1950 *Fibularia stellata* Capider, 1906. – Szörényi: 143; pl. 2, figs. 2, 2a

1953 *Fibularia lecointreae* Lambert, 1907. – Szörényi: 60

? 1953 *Fibularia pseudopusilla* [Cotteau], 1895. – Szörényi: 60

1960 *Echinocyamus stellatus* Capeder 1906 – Kojumdgieva & Strachimirov: 231; pl. 8, figs. 10a-c

? 1968 *Echinocyamus stellatus* Caper. – Stancu & Andreescu: 466, 468, tab.; figs. 78a-b

? 1977 *Echinocyamus pusillus* (O. F. Müller, 1776) – Mączyńska: 196; pl. 5, figs. 1-4

non 1977 *Echinocyamus pseudopusillus* Cotteau, 1895 – Mączyńska: 196; pl. 6, figs. 1-3

? 1979 *Echinocyamus pusillus* (O. F. Müller, 1776) – Mączyńska: 31, pl. 2, figs. 2-4

non 1987 *Echinocyamus pseudopusillus* Cotteau, 1895 – Mączyńska: 148

? 1987 *Echinocyamus pusillus* (O. F. Müller, 1776) – Mączyńska: 146, 148; pl. 4, figs. 1-4

? 1988 *Fibularia pusilla* Müll. – Kókay: 264

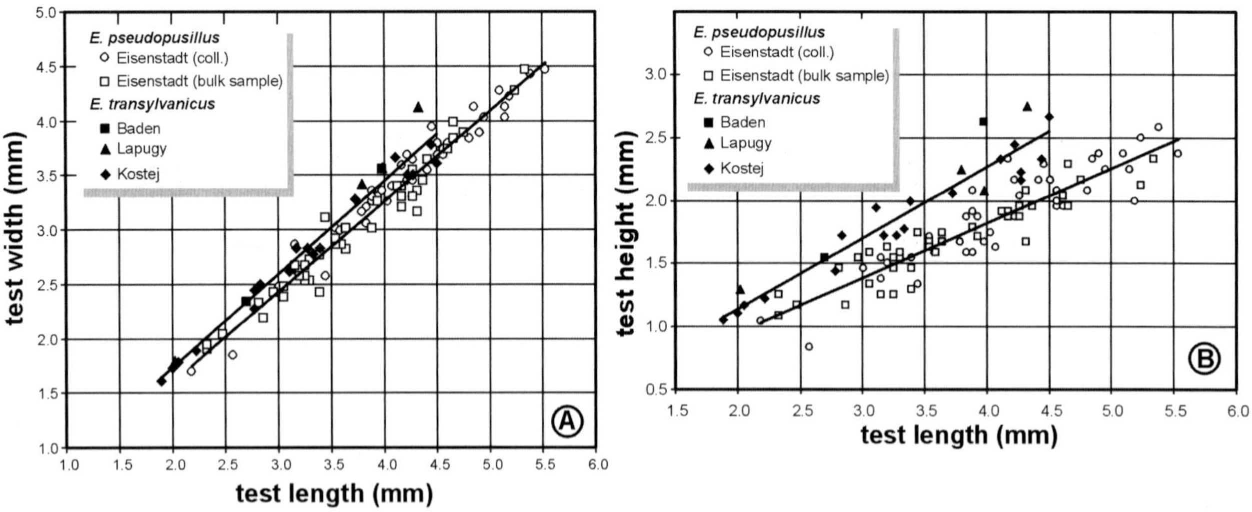

Figure 32: Morphometric comparison between *Echinocyamus transylvanicus* Laube, 1869 and *E. pseudopusillus* Cotteau, 1895 (A: test length vs. test width; B: test length vs. test height).

? 1988 *Echinocyamus pusillus* (O. F. Müller, 1776) – Mączyńska: 61; pl. 3, figs. 8a-c, 9a-c, 10a-c
1988 *Echinocyamus pseudopusillus* Cotteau, 1895 – Mączyńska: 61; pl. 4, fig. 3a-c
? 1989 *Echinocyamus pseudopusillus* Cotteau, 1895 – Schmid: 7
1993 *Echinocyamus pusillus* (O. F. Müller, 1776) – Mączyńska: 110-111; pl. 3, figs. 6-7
? 1996 *Echinocyamus pusillus* (O. F. Müller, 1776) – Mączyńska: 41; pl. 1, figs. 5-7 [crushed specimens]
v. 2002b *Echinocyamus* sp. – Kroh: 12
v. 2003 *Echinocyamus stellatus* Capeder – Kroh et al.: 92
? 2004 *Echinocyamus pusillus* (O.F. Müller, 1776) – Radwański & Wysocka: 385

Material:
Early Badenian (Langhian) – Gainfarn, NÖ, Austria
NHMW: 2 specimens (NHMW 2004z0076/0045-46)
Wanzenböck coll.: 30 specimens
Early Badenian (Langhian) – Eisenstadt (Hartl Fm.), Bgld, Austria
NHMW: 47 whole coronas and several fragments [NHMW 1859.L.800 (45 whole coronas and several fragments), 1859.L.800a-b (2 whole coronas)]
Early Badenian (Langhian) – Eisenstadt (Hartl Fm., Hartllucke), Bgld, Austria
NHMW: 45 whole coronas and several fragments (NHMW 2003z0072/0001/1-45)
Early ? Badenian (Langhian) – Rauchstallbrunngraben (bryozoan marl), near Baden, NÖ
NHMW: 1 specimen (NHMW 2004z0121/0051)
Early Badenian (Langhian) – Stotzing (sandpit Mayer), Bgld, Austria
NHMW: 1 specimen (NHMW 2004z0050/0030)
Badenian (Langhian-Early Serravallian) – Brunn am Steinfeld, NÖ, Austria
NHMW: 3 specimens (NHMW 1997z0178/1585)
Badenian (Langhian-Early Serravallian) – old sandpit between Großhöflein and Kleinhöflein, Bgld, Austria
NHMW: 3 specimens (NHMW 2003z0081/0015, ../0043-44)

Foreign material for comparison:
Early Badenian (Langhian) – Buituri (= Bujtur), Romania
NHMW: 2 specimens (NHMW 1852.II.1603g)

Dimensions (in mm):

Inv. No.	TL	TW	TH
Hartl. Fm., Hartllucke, Eisenstadt, Austria			
NHMW 2003z0072/0001/1	4.76	3.88	2.17
NHMW 2003z0072/0001/2	4.66	3.98	2.29
NHMW 2003z0072/0001/3	5.34	4.47	2.33
NHMW 2003z0072/0001/4	5.24	4.27	2.13
NHMW 2003z0072/0001/5	4.61	3.74	2.04
NHMW 2003z0072/0001/6	4.17	3.40	1.92
NHMW 2003z0072/0001/7	4.37	3.45	1.96
NHMW 2003z0072/0001/8	4.27	3.54	1.96
NHMW 2003z0072/0001/9	3.64	2.82	1.67
NHMW 2003z0072/0001/10	4.66	3.83	1.96
NHMW 2003z0072/0001/11	4.17	3.30	1.92
NHMW 2003z0072/0001/13	4.32	3.16	2.08
NHMW 2003z0072/0001/14	3.93	3.25	1.71
NHMW 2003z0072/0001/15	4.17	3.20	1.88
NHMW 2003z0072/0001/16	4.32	3.35	1.67
NHMW 2003z0072/0001/17	4.27	3.30	1.88
NHMW 2003z0072/0001/18	2.86	2.18	1.17
NHMW 2003z0072/0001/19	2.33	1.89	1.08
NHMW 2003z0072/0001/21	3.88	3.01	1.79
NHMW 2003z0072/0001/22	3.45	3.11	1.75
NHMW 2003z0072/0001/23	4.13	3.40	1.92
NHMW 2003z0072/0001/24	3.64	3.01	1.75
NHMW 2003z0072/0001/25	3.64	3.01	1.67
NHMW 2003z0072/0001/26	3.64	2.82	1.67
NHMW 2003z0072/0001/27	3.59	2.86	1.58
NHMW 2003z0072/0001/28	3.54	2.86	1.67
NHMW 2003z0072/0001/29	3.40	2.77	1.46
NHMW 2003z0072/0001/30	3.25	2.52	1.46
NHMW 2003z0072/0001/31	3.40	2.43	1.29
NHMW 2003z0072/0001/32	3.30	2.72	1.54
NHMW 2003z0072/0001/33	3.06	2.48	1.58
NHMW 2003z0072/0001/34	3.16	2.67	1.25
NHMW 2003z0072/0001/35	3.54	2.91	1.63
NHMW 2003z0072/0001/36	3.25	2.67	1.46
NHMW 2003z0072/0001/37	3.30	2.52	1.58
NHMW 2003z0072/0001/38	3.25	2.52	1.25
NHMW 2003z0072/0001/39	3.06	2.38	1.33
NHMW 2003z0072/0001/40	3.25	2.57	1.54
NHMW 2003z0072/0001/41	3.20	2.57	1.63
NHMW 2003z0072/0001/42	2.96	2.43	1.54
NHMW 2003z0072/0001/43	2.33	1.94	1.25
NHMW 2003z0072/0001/44	2.48	2.04	1.17
NHMW 2003z0072/0001/45	2.82	2.33	1.46

Description:

Size and shape: Test very small, usually less than 6 mm in length. Outline oval (antero-posteriorly elongated) to egg-shaped (test width ranges from 71.4 to 90.8 % TL, mean: 81.3 %). Maximum width lying slightly posterior of centre. In profile test is low arched with a tumid ambitus (test height usually ranging from 32.4 to 55.9 % TL, mean: 45.7 %). The maximum height coincides with the position of apical disc. The oral surface is flattened, without infundibulum.

Apical disc: The apical disc lies slightly anterior of the centre (~ 47 % of TL from the anterior margin) at the highest point of the test. Four moderately large subcircular genital pores are present. As in the former species gonopore size varies, probably due to sexual dimorphism. The five ocular pores are small and vary in their position regarding the polygon formed by the genital pores. Only a single circular hydropore is present, it lies at the point where the imaginary lines drawn between genital pore 1 and 3, respectively 2 and 4 intersect (Pl. 38, Fig. 1e).

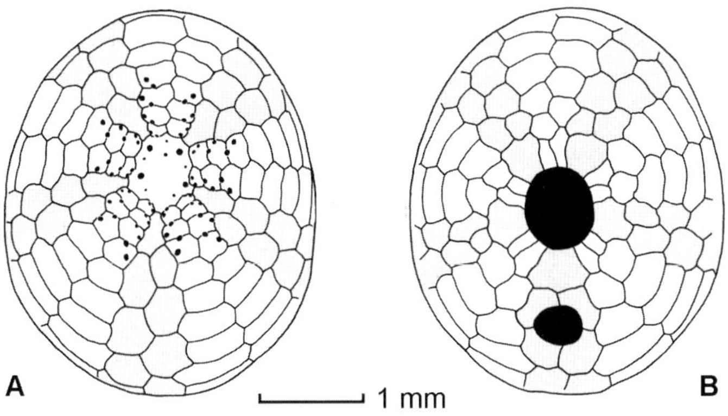

Figure 33: *Echinocyamus pseudopusillus* Cotteau, 1895: aboral (A) and oral (B) plating (Hartl Fm., Eisenstadt, Bgld; NHMW 1859.L.800b).

Ambulacra: Adapically more or less well developed petals are present. They consist of 4 to 7 strongly oblique pore pairs in each column. The petals are straight, open distally, with poriferous zones that converge towards each other at the tip of each petal. The frontal petal is longest, the paired petals are subequal in length, approximately 80 to 85 % of the frontal one. They extend about 50 to 55 % of the corresponding test radius. The interporiferous zones are as wide as or slightly wider than a single poriferous zone (Figs. 33.A). Outside the petals, distinct arcs of accessory pores (micro-unipores) which lie transversely to the axis of the ambulacra are found. The size of these accessory pores varies, but they are always distinctly smaller than the respiratory pores in the petals.

Interambulacra: The interambulacra are not inflated or depressed between the ambulacra. They are covered by small perforate tubercles, between which many miliary tubercles are found.

Peristome: The peristome is subcircular and rather large (~20 % of TL in diameter), lying subcentrally on the oral surface. The posterior margin of the peristome is usually depressed, sometimes forming a shallow groove connecting the peristome to the periproct. In some specimens the oral surface is slightly depressed around the peristome but not so much as to form a true infundibulum.

Periproct: About half the size of the peristome (~10 % of TL wide), of oval size, transversely elongated. It lies halfway between the peristome and the posterior margin or slightly posterior (Fig. 33.B).

Internal support structures: Five pairs of internal radial partitions are present in the ambulacra (see Pl. 38, Fig. 3a), substantiating the attribution to the genus *Echinocyamus*.

Differential diagnosis:

The Aquitanian to Langhian Maltese *Echinocyamus*, which has been referred to *E. stellatus* Capeder, 1906 by various authors (e.g. Challis, 1980) differs from this species by its more rounded outline, usually higher test, fewer respiratory pores (even in large specimens there are usually not more then 5 pore pairs in each ambulacral column), its raised apical disc, and its slightly inflated, cushion-shaped coronal plates (see Figs. 34.6a-c).

Echinocyamus ovatus (Münster in Goldfuss, 1829) from the Late Oligocene of Germany differs by its more rounded outline, rostrate posterior end, much smaller petalodium with fewer respiratory pores and more anteriorly eccentric apical disc (see Figs. 34.4a-c).

The contemporaneous *Echinocyamus calariensis* (Lambert, 1907) differs by its strongly elongated outline, "pointed" anterior margin and more flattened profile.

E. transylvanicus Laube, 1869, a contemporaneous species, differs from this species by its more rounded outline, higher test, fewer respiratory pores and not depressed posterior margin of the peristome (see Figs. 32.A-B, Figs. 34.1a-c).

Echinocyamus sp. B from the Badenian of Poland differs strongly by its submarginal periproct, fewer respiratory pores (in specimens of comparable size) and slightly subpentagonal outline (see Figs. 34.5a-c).

Discussion:

Although common, this species is rarely well preserved in the Badenian sediments of the Central Paratethys. No less than eight different species have been described from this age and region. Most of those are extremely similar to each other and vary only by slight differences in test outline and/or elevation. Szörényi (1950), for example, described four species of *Echinocyamus* (under the genus *Fibularia*) from the Badenian of the Mecsek Mts. (Hungary). Her differential diagnoses, descriptions and illustrations, however, do not allow a confident separation of these forms and seem to fall well within the variation displayed by *E. pseudopusillus*.

Most specimens of *E. pseudopusillus* come from psammitic sediments, often in association with rich bryozoan faunas, whereas the other Badenian species discussed here (*E. transylvanicus*) usually occurs in pelitic sediments. Since the two species differ mainly in shape characters rather than structural differences they could represent end-members of an ecological gradient. Currently, however, not enough data are available to prove or to reject this hypothesis and thus the question is set aside for the time being. Moreover, most of the 11 extant species likewise differ mainly in shape characteristics when only the corona is considered (compare the data presented by Mironov & Sagidachny, 1984). In the Upper Oligocene of Landes (Aquitaine Basin, France; Cahuzac & Roman, 1994) there are also two species [attributed to *E.* cf. *bouillei* (Lambert in Castex & Lambert) and *E. pusillus* (Müller) by the authors] differing mainly in corona outline.

Contemporaneous material of *E. stellatus* Capeder, 1906 from the Langhian of Malta (> 1000 specimens from a single horizon/locality), however, shows that intraspecific variation in outline (ranging from nearly circular to oval [most common] to slightly subpentagonal [rare]) and elevation of the test (strong variation, ranging from low arched to highly domed) can be quite severe. To further explore this situation more material, especially of *E. transylvanicus*, is needed.

Up to 5 species co-occur in extant setting of comparable size as the Central Paratethys [e.g. the south-eastern Australian part of the Tasman Sea or the Hawaiian Islands; compare Mironov & Sagidachny (1984)].

E. pseudopusillus is very similar to the Pliocene to extant species *E. pusillus* (Müller) found throughout the Lusitanian bioprovince and the North Sea. Indeed some fossil specimens from the Hungarian and Polish Middle Miocene have been attributed to the extant species (see Vadász, 1915; Mączyńska, 1977). Whether *E. pseudopusillus* is a junior synonym of *E. pusillus* or not cannot be decided here because more material has to be investigated (possible with the aid of morphometric methods). The extant form seems to be differentiated from this species by its non convergent, straight poriferous zones in the petals, a larger number of accessory pores per plate (compare Fig. 31), and a more "pointed" outline at least (see Figs. 34.3a-c).

Occurrence:

Austria: Early to Late Badenian (Langhian-Early Serravallian)

Vienna Basin: Brunn am Steinfeld, NÖ ([NHMW]); Gainfarn, NÖ (Kroh, 2002b; [NHMW]); Nußdorf, Vienna (Schmid, 1989); Rauchstallbrunngraben (bryozoan marl), near Baden, NÖ [NHMW]; Stotzing (sandpit Mayer), Bgld [NMHW]

Eisenstadt-Sopron Basin: Eisenstadt (Hartl Fm.), Bgld (Kroh et al., 2003 [NHMW]); old sandpit between Großhöflein and Kleinhöflein, Bgld ([NHMW])

Paratethys (non-Austrian occurrences): ? Karpatian, Early to Late Badenian (Langhian-Early Serravallian)

Great Hungarian Basin (Pannonian Basin): Acsa, Pest, Hungary (Vadász, 1915); ? borehole Alcsútdoboz, SE Budapest (Kókay, 1988); Garáb, Nógrád, Hungary (Vadász, 1915); Háromház, Baranya, (Vadász, 1915); Magyaregregy, Csigadulö, Mecsek Mts., Hungary (Szörényi, 1950); Magyaregregy, Leányköi árok, Mecsek Mts., Hungary (Szörényi, 1950); Mátraverebély (Meszestetö), Nógrád, Hungary (Vadász, 1915); ? Mogyoród, near Fót, NW Budapest, Hungary (Horusitzky, 1927); Piliny, Nógrád, Hungary (Vadász, 1915); Nógrádszakál (= Szakall), Nógrád, Hungary (Vadász, 1915)

Fore-Carpathian Basin: Busko, Poland (Mączyńska, 1993); ? Chomentów, Poland (Mączyńska, 1987); ? Huta Różaniecka and Huta Lubycka, Poland (Mączyńska, 1979); ? Karsy, Poland (Mączyńska, 1987); ? marly sands overlying the Korytnica Clays, Poland (pp Mączyńska, 1977); Mykolaiv (= Mikołajów), Ukraine (Szörényi, 1953); ? Niechobrz near Rzeszów, Poland (Mączyńska, 1996); Pińczów, Central Poland (Mączyńska, 1993); Radziechów (= Radziejów), Ukraine (Szörényi, 1953); Rybnice, Poland (Mączyńska, 1988); ? Świniary, Poland (Radwański & Wysocka, 2004); ? Zhukov (= Zukowce), Ukraine (Szörényi, 1953)

Transylvanian Basin: Buituri (= Bujtur), Romania ([NHMW]);? Coşteiu de Sus (= Kostej), Romania (Vadász, 1915); Gârbova de Sus (= Felsö-Orbó), Romania (Vadász, 1915); ? Lăpugiu des Sus (= Lapugy), Romania (Vadász, 1915); Rugi-Delineşti, Romania (Stancu & Andreescu, 1968)

Lom Basin: Bivolare, Opanez and Opansko bardo Northern Bulgaria (Kojumdgieva & Strachimirov, 1960)

Echinocyamus sp. A

(Fig. 35.A-C)

Material:

Early Eggenburgian (Early Burdigalian) – Gösing (Fels Fm.), NÖ, Austria

NHMW: 5 specimens (NHMW 1998z0048/0064, NHMW 2003z0074/0001-2)

Dimensions (in mm):

Inv. No.	TL	TH	TH*
NHMW 1998z0048/0064a	7.8	7.0	> 2.7
NHMW 1998z0048/0064b	7.0	5.9	> 2.4
NHMW 1998z0048/0064c	6.4	5.4	> 1.9
NHMW 2003z0074/0001	6.5	5.3	> 1.8
NHMW 2003z0074/0002	6.0	5.5	> 1.4

* All specimens crushed during diagenesis by sediment compaction.

Description:

Size and shape: Test small, TL ranging from 6 to 7.8 mm in the studied specimens. Outline conspicuously egg-shaped, anteroposteriorly elongated, the posterior end being much wider than anterior one. Maximum width far posterior, about halfway between centre and posterior margin. Profile shape not discernible in the present material, all specimens crushed. The oral surface is flattened, with a wide depression around the peristome.

Apical disc: The apical disc lies anterior of the centre, about one third of TL from the anterior margin. Four moderately large subcircular genital pores are present. Only a single circular hydropore is present, lying slightly anteriorly eccentric. Ocular pores poorly visible due to syntaxial rim cement formation.

Ambulacra: Adapically well developed petals are present. The petals are formed by 7 to 10 strongly oblique partitioned isopores in each column. The petals are straight and open distally. The poriferous zones of the paired petals are straight, that of the frontal petal are diverging distally. The frontal petal is longest, the paired petals are subequal in length. The frontal petal extends about two third of the corresponding test radius, whereas the posterior paired petals extend only half of the corresponding test radius. The anterior paired petals are intermediate. The interporiferous zones are about one to one and a half times as wide as a single poriferous zone. Outside the petals, distinct arcs of accessory pores (micro-unipores) which lie transversely to the axis of the ambulacra are found.

Interambulacra: The interambulacra are not inflated or depressed between the ambulacra. They are covered by small perforate tubercles, between which many miliary tubercles are found.

Peristome: The peristome is subcircular and rather large (nearly 20 % of TL in diameter). It lies subcentrally on the oral surface in a wide depression.

Periproct: The periproct lies at the margin of the circum-oral depression, slightly closer to the margin than to the peristome. It is irregularly oval and about half the size of the peristome.

Internal support structures: unknown

Discussion:

Although the presence of internal partitions could not be shown in the present material due to the poor preservation, all other features support an attribution to the genus *Echinocyamus*. The specimens are clearly not conspecific with *E. transylvanicus*, *E.* sp. B or *E. pseudopusillus* from the Badenian (Langhian-Early Serravallian). The Eggenburgian species differs from the Badenian one by its larger size, pronounced egg-shaped outline, more respiratory pores in the petals, straight or even diverging poriferous zones. Due to the limited amount of material and the poor state of preservation (all specimens crushed and most overgrown by syntaxial rim cement) the specimens could not be securely referred to any known species and have to be named in open nomenclature.

Occurrence:

Austria: Early Eggenburgian (Early Burdigalian)

Molasse Zone: Gösing (Fels Fm.), NÖ ([NHMW])

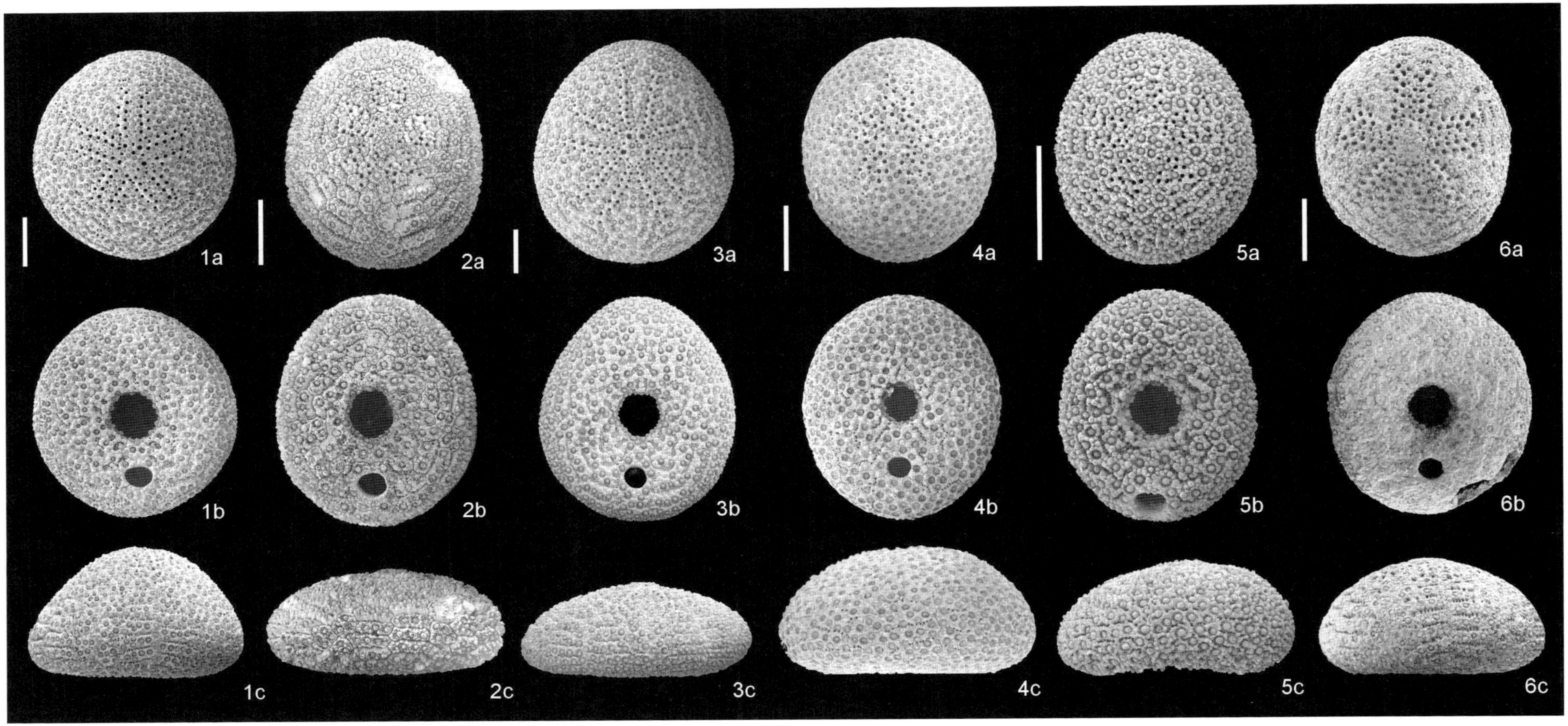

Figure 34: Comparison between selected European species of *Echinocyamus* from the Late Oligocene to Recent.
1: *Echinocyamus transylvanicus* Laube, 1869: Early Badenian (Langhian), Lapugy, Romania (NHMW.1860.XL.529a); note high profile
2: *Echinocyamus pseudopusillus* Cotteau, 1895: Early Badenian (Langhian), basal Hartl Fm., Eisenstadt, Bgld (NHMW 1859.L.800b); note low profile and small petalodium
3: *Echinocyamus pusillus* (Müller, 1776): recent, Mediterranean Sea, coast of Croatia (NHMW coll.); note egg-shaped outline, large petalodium and well developed petals
4: *Echinocyamus ovatus* (Münster in Goldfuss, 1829): Late Oligocene of Astrup, Germany (Univ. Münster coll. AS1-60); note peculiar outline and small petalodium
5: *Echinocyamus* sp. B: Early Badenian (Langhian) of Korytnica, Central Poland (Nebelsick coll., Tübingen); note submarginal position of periproct (same position in large specimens)
6: *Echinocyamus stellatus* Capeder, 1906: Langhian, Xwieni Bay Mb, Upper Globigerina Limestone Fm., Gozo, Malta (NHMW coll.); note large petaloid area

Scale bars equal 1 mm each; a: aboral view, b: oral view, c: right lateral view.

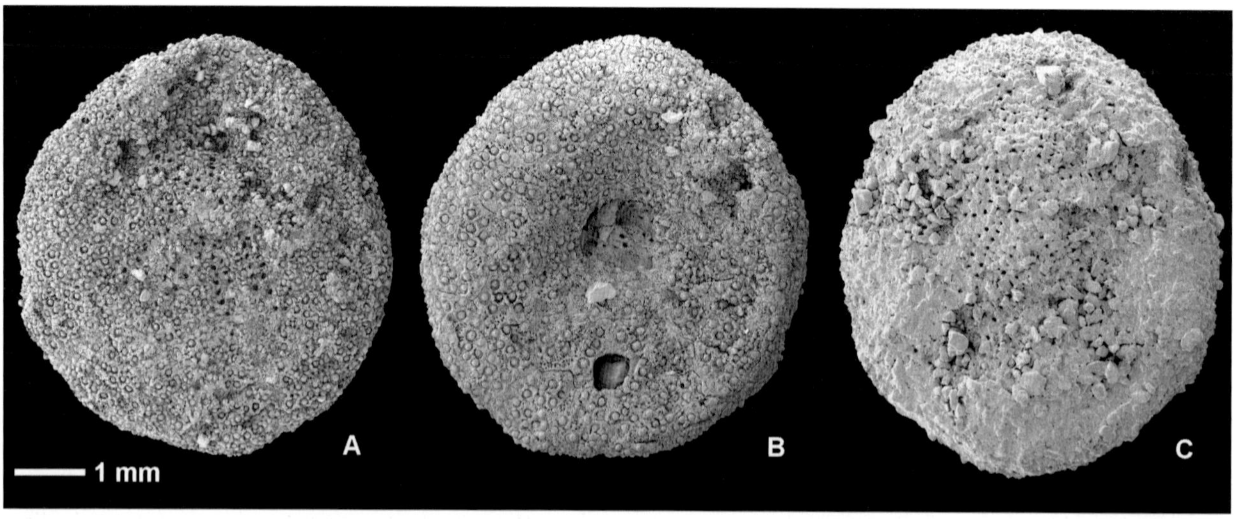

Figure 35. *Echinocyamus* sp. A: aboral (A, C) and oral view (B) (Eggenburgian of Gösing, NÖ; A-B: NHMW 2003z0074/0001, C: NHMW 1998z0048/0064c)

Other *Echinocyamus* species reported from the Central Paratethys:

Apart from the species discussed above (and records of species placed into their synonymy) three other species of *Echinocyamus* have been reported from the Central Paratethys. Two of these (*E. calariensis* and *E.* sp. B) are sufficiently well known and are considered to be distinct from those discussed above. A third (*E. sandalinus*) is based on a single specimen, which has not been re-examined, its status is unclear.

Echinocyamus calariensis **(Lambert, 1907)**

v 1915 *Fibularia calariensis* Lamb. – Vadász: 114; pl. 8 (2), figs. 22
1977 *Echinocyamus pseudopusillus* Cotteau, 1895 – Mączyńska: 196; pl. 6, figs. 1-3
1987 *Echinocyamus pseudopusillus* Cotteau, 1895 – Mączyńska: 148

Reported occurrence: Badenian (Langhian-Early Serravallian) of Hungary (locality Acsa, Pest, Vadász, 1915), Romania [locality Rachiş (= Oláh-Rákos), Vadász, 1915], and Early Badenian (Langhian) of Poland (marly sands overlying the Korytnica Clays; under the name *E. pseudopusillus*, Mączyńska, 1977).

Remarks: This species is characterised by its very elongated outline, pointed anterior margin and low, flattened profile. It was reported under the name *Fibularia calariensis* Lambert by Vadász (1915: 114; pl. 2, fig. 22) and as *Echinocyamus pseudopusillus* Cotteau by Mączyńska (1977: 196; pl. 6, figs. 1-3 and 1987: 148, not 1988: 61; pl. 4, fig. 3). This form does not fall within the morphological range of the species discussed above. After examining the material of Vadász housed at the Hungarian Geological Survey (MAFI Ech 248, 303) I would indeed regard these specimens as specifically different from the other species discussed here (at least MAFI Ech 248; the three specimens of Ech 303 could also be deformed *E. pseudopusillus*). This form was not found in Austrian localities until now, only a single, poorly preserved specimen (NHMW 2004z0076/0047) from the Early Badenian of Gainfarn might be tentatively assigned to this species (Pl. 38, Figs. 2a-c).

Echinocyamus **sp. B**

non 1906 *Echinocyamus linearis* n. f. – Capeder: 517; figs. 12a-c
1977 *Echinocyamus linearis* Capeder, 1906 – Mączyńska: 197; pl. 7, figs. 1-6
1987 *Echinocyamus linearis* Capeder, 1906 – Mączyńska: 148
1993 *Echinocyamus linearis* Capeder, 1906 – Mączyńska: 111; pl. 3, fig. 8; pl. 6, fig. 1g
2003 *Echinocyamus linearis* Capeder, 1906 – Ceranka & Złotnik: 491ff.; figs. 1A-E

Reported occurrence: Early Badenian of Poland [marly sands overlying the Korytnica Clays (Mączyńska, 1977, 1987; Ceranka & Złotnik, 2003) and the localities Pińczów and Skowronno in Central Poland (Mączyńska, 1993)].

Remarks: This species is characterised by its submarginal position of its periproct, its elongated outline, subpentagonal shape and low number of pore pairs in the petals. It is clearly different from the type material of *E. linearis* figured by Capeder (1906: 517, figs. 12a-c), which has a periproct lying halfway between peristome and posterior margin. Among the numerous nominal species of the genus *Echinocyamus* there are few with a similar posterior position of the periproct. Among the extant species this is only the Australian endemic *Echinocyamus platytatus* Clark, 1914, which occurs in fine sands between 9 to 365 m all along the southern coast of Australia from Port Stephens (New South Wales) to Point d'Entrecasteaux (Western Australia) (Miskelly, 2002). Additionally three fossil species have a similar periproct position: *E. lecointreae* (Lambert, 1907a) from the "Helvetian" of the "Faluns Touraine" of Western France (Indre & Loire), *E. studeri* (Sismonda, 1841) from the Miocene of the "Colli Torinesi" and Monte Gargano (d'Alessandro et al., 1979) in Italy, and *E. lebesconti* Bazin, 1884 from the Middle Miocene of the Bretagne (Western France). Unfortunately, the available descriptions are insufficient to decide whether the Polish material is conspecific with any of them or not.

Echinocyamus ? sandalinus **(Szörényi, 1953)**

1953 *Fibularia sandalina* n.sp. – Szörényi: 60-61; pl. 5, figs. 6, 6a

Reported occurrence: Early Badenian (Langhian) of Zhukov (= Zukowce), Ukraine Szörényi, 1953).

Remarks: Only the short description and poor illustrations of Szörényi (1953) are available of this species. The single described and figured specimen is slightly elongated antero-posteriorly, has a pointed anterior and a transversely truncated posterior margin, a low, wedge-shaped profile (thicker anteriorly) and its periproct lies roughly halfway between the peristome and the posterior margin. The development of the petaloid area and other details are unknown. Based on the available information alone it is impossible to confidently relate this species to any other. A re-examination of the type-material is necessary.

Szörényi (1953) followed Lambert's (1891) use of the names *Fibularia* and *Echinocyamus*. According to the current usage (Mortensen, 1948b; Durham, 1966) this species has to be transferred to *Echinocyamus*, although the presence of internal partitions remains to be shown.

***Echinocyamus* ? sp.**

	1893	*Echinocyamus* sp. – Toula: 288
	1927	*Fibularia* sp. – Horusitzky: 27, 167
	1894	*Echinocyamus ovatus*. Ag. – Lörenthey: 59
	1981	*Echinocyamus studeri* Sism. – Halmai: 106
pp	2003b	*Echinocyamus* sp. – Kroh: 249

Reported occurrence: Karpatian (Late Burdigalian) of Kisalag (Halmai, 1981) and Mogyoród, NW Budapest, Hungary (Horusitzky, 1927; pp Kroh, 2003b); Badenian (Langhian-Early Serravallian) of Kralice nad Oslavou (= Kralitz), Moravia, Czech Republic (Toula, 1893), and Rachiş (= Oláh-Rákos), Alba, Romania (Lörenthey, 1894)

Remarks: The present records lack sufficient information for revision and the reference material could not be traced. It is likely that they refer to any of the species described above. The record of *Fibularia* sp. by (Horusitzky, 1927) most probably also refers to an *Echinocyamus*, as convincing records of *Fibularia* from the Miocene of the Central Paratethys are missing. The species *Echinocyamus ovatus* Agassiz recorded by (Lörenthey, 1894) is unknown to the present author. It is possible that Lörenthey meant *E. ovatus* (Münster in Goldfuss, 1829), a species from the Late Oligocene of Germany. Halmai's (1981) record lacks any decription or illustration and is based on an unpublished Ph.D. thesis. It remains to be shown if either species occurs in the Miocene of the Central Paratethys.

Suborder Scutellina Haeckel, 1896
Family Scutellidae Gray, 1825
Genus *Parascutella* Durham, 1953

Type-species: *Scutella leognanensis* Lambert, 1903; by original designation (Durham, 1953: 173)
Diagnosis: Test large and thin; petals extending about 65-70 % of the corresponding test radius. Anterior petals shorter than posterior ones. Five to six ambulacral and four to five interambulacral coronal plates on oral surface. Interambulacra about as wide as ambulacra at the ambitus. Periproct submarginal, between the 3rd pair of postbasicoronal plates (modified from Durham, 1953, 1955, 1966).
Distribution: Miocene – Europe (Durham, 1966)
Remarks: When revising the Clypeasteroida Durham (1953, 1955) found that two very different groups were attributed to the genus *Scutella*, partly explaining the confusion in the literature. Based on a re-examination of the type-species Durham provided an emended diagnosis for the genus *Scutella* and established a new genus, *Parascutella*, for a group of species erroneously attributed to the former. The two genera differ strongly in the position of their periproct in relation to the coronal plates. In *Scutella* the periproct lies between the 1st pair of postbasicoronal plates, about halfway between the peristome and the posterior margin. In *Parascutella* it lies submarginally, between the 3rd pair of postbasicoronal plates. Additionally the two genera differ by the length and structure of the petals. In the former the petals are usually half as long as the corresponding test radius and closed, while in the latter they are two third as long and distally open. The frontal petal is the longest of the petals in *Scutella* and the shortest in *Parascutella*.

Although this separation was very clearly documented and published three times (Durham, 1953, 1955, 1966) it was largely ignored by the palaeontologists working in the Paratethys region (e.g. Schaffer, Mihály, Mączyńska). All Paratethyan specimens attributed to the genus *Scutella* previously have to be transferred to the genus *Parascutella*. Until now no "true" *Scutella* species have been found in the Neogene of the Paratethys (see Ali, 1998 and Kroh, 2002a).

Parascutella gibbercula (de Serres, 1829)

(Fig. 36, 37; Pl. 39, Figs.1a-b; Pl. 40, Figs. 1-2; Pl. 41, Figs. 1-3; Pl. 42, Figs. 1a-c)

*	1829	*Scutella gibercula.* Nobis. – de Serres: 156
?	1830	*Scutella subrotunda*, Lam. – Eichwald: 195-196
	1835	*Sc.[utella] gibberula* M. de S. – Agassiz: 188
	1837	*S.[cutella] gibercula*. M. de Serres – des Moulins: 80
?	1837	*Scutella subrotunda* Lam. – Pusch: 181, tab.
	1841a	*Scutella gibbercula* de Marcel de Serres – Agassiz: 86
pp	1847a	*Scutella subrotunda* Lamk. – Agassiz & Desor: 134
?	1852	*Scutella subrotunda* Leske. – Eichwald: pl. 3, figs. 1a-c
?	1853	*Scutella subrotunda* Leske. – Eichwald: 47
pp	1858	[*Scutella*] *Faujasii* Defr. – Desor: 233
?	1860	*Scutella Faujasii* – Foetterle: 76
	1869	*Scutella Faujasii* – Fuchs: 194
	1869	*Scutella Vindobonensis* Laube sp. ined. – Fuchs: 194
v. #	1869a	*Scutella vindobonensis* Laube. – Laube: 183
	1870	*Scutella Faujasii* Defr. – Wolf: 38
v.	1870	*Scutella Vindobonensis* Lbe. – Laube: 314
v.	1871	*Scutella Vindobonensis* Laube. – Laube: 62; pl. 17, fig. 1
	1873	*Scutella vindobonensis* Laube. – Stur: 91
#	1874	*Scutella kalksburgensis* – Wiesbaur: 164-165.
	1874	*Scutella Vindobonensis* – Quenstedt: pl. 82, figs. 19-21; pl. 83, fig. 1
	1875	*Scutella Vindobonensis* – Quenstedt: 542-544
v.	1877	*Scutella vindobonensis* Laube. – Karrer: 78, 303, 312
	1877	*Scutella vindobonensis* Laube. – Lóczy: 63
	1881	*Scutella Vindobonensis* Laube. – Franzenau: 33, 86
	1887b	*Scutella Vindobonensis*, Laube. – Koch: 261-262
? #	1887b	*Scutella pygmea*, nov. sp. – Koch: 262-263; pl. 5, figs. 1a-c
	1888b	*Scutella Vindobonensis*, Laube – Nemes: 22, 32
?	1888b	*Scutella pygmaea*, Koch, nov. sp. – Nemes: 22, 32
	1894	*Scutella vindobonensis*. – Lörenthey: 67
	1894	*Scutella vindobonensis*, Laube. – Mártonfi: 153
	1905	*Scutella vindobonensis*, Lbe. – Gaál: 344, 362
	1906	*Scutella Vindobonensis* Lbe. – Schaffer: 69
	1906	*Scutella vindobonensis* Lbe. – Vadász: 329
	1907	*Scutella Vindobonensis* Laube – Schaffer: 35
.	1912	*Scutella Gibbercula* M. de Serres, 1829. – Lambert: 73-74; pl. 5, figs. 1-2
	1913a	*S.[cutella] gibbercula* M. de Serres. – Cottreau: 53
v ? #	1915	*Scutellina hungarica* nov. sp. – Vadász: 112-113; pl. 10 (4), figs. 2-3
	1915	*Scutella vindobonensis* Lbe. – Májer: 35, 88
v.	1915	*Scutella vindobonensis* Lbe. – Vadász: 115-117; figs. 12-13

v. 1915 *Scutella gibbercula* M. de Serr. – Vadász: 117-118, fig. 14
v ? 1915 *Scutella leognanensis* Lambert. – Vadász: 119-120; figs. 15-16
v. 1915 *Scutella paulensis* Ag. – Vadász: 121, fig. 17
? 1915 *Scutella pygmea* Koch. – Vadász: 122; pl. 9 (3), figs. 10-11
v. 1922 *Scutella vindobonensis*, Laube – Abel: 243, fig. 202
1928 *Scutella gibbercula* [...] M. de Serres – Lambert & Jeannet: 206, no. V.56
1930 *Scutella vindobonensis* Laube. – Vendl: 76
1930 *Scutella leognanensis* Lamb. – Vendl: 77
1931 *Scutella vindobonensis* Laube – Janoschek: 68, 78, 79, 81, 83, 84, 85, 86, 87 (fide Schaffer, 1962)
1936 *Scutella vindobonensis* Laube. – Paucă: 143
1939 *Scutella vindobonensis* Lbe. – Kapounek: 72
1942 *Scutella Vindobonensis* Laube –Schaffer: 94
1953b *Scutella vindobonensis* Laube – Sieber: 194
. 1953 *Scutella vindobonensis* Laube, 1871. – Szörényi: 62-64; pl. 1, figs. 4, 4a-b
? 1953 *Scutella cf. leoganensis* Lambert, 1903. – Szörényi: 64-65; pl. 1, fig. 7
? 1953 *Scutella paulensis* Agassiz, 1841. – Szörényi: 65-66; pl. 1, figs. 1-3, 3a-b; pl. 2, fig. 2
? # 1953 *Scutella almerai parva* n. ssp. – Szörényi: 67; pl. 1, figs. 6, 6a-b
? # 1953 *Scutella eichwaldi* n. sp. – Szörényi: 67-68; pl. 1, figs. 5, 5b
1958a *Scutella* aff. *vindobonensis* Laube – Sieber: 152
v. 1959 *Scutella* C. – Schaffer: 255
v. 1959 *Scutella* D. – Schaffer: 255
v. 1959 *Scutella* E. – Schaffer: 255
1959 *Scutella* F. – Schaffer: 255
1959 *Scutella* G. – Schaffer: 255
1960 *Scutella* aff. *gibbercula* M. de Serres – Roman: 89-90; pl. 8, fig. 3
#. 1960 *Scutella vindobonensis planata* Kókay nov. ssp. – Somos & Kókay: 341-342, 346; 3 figs.; pl. 16, figs. 1-2; pl. 17, fig. 1.
? 1960 *Scutella* sp. – Somos & Kókay: 342
1962 *Scutella styriaca* n. sp. – Schaffer: 153-154; fig. 9; pl. 19, fig. 1
v. # 1962 *Scutella multiconcava* n. sp. – Schaffer: 154-156; fig. 10; pl. 17, figs. 4-5; pl. 18, figs. 1-5
v. 1962 *Scutella vindobonensis vindobonensis* Lbe. – Schaffer: 156-157; fig. 1; pl. 16, figs. 1, 6; pl. 19, fig. 3
v. # 1962 *Scutella vindobonensis secunda* n. ssp. – Schaffer: 157-159; fig. 11; pl. 19, figs. 2, 4
1962a *Scutella vindobonensis* – Thenius: 27
v. 1963 *Scutella vindobonensis* Laube – Müller, 521; figs. 683B a-c
1965 *Scutella styriaca* Schaffer – Kollmann: 541
1969 *Scutella vindobonensis* Laube – Mihály: 254-255
#. 1969 *Scutella szoerenyiae* nov. sp. – Mihály: 255; pl. 1, figs. 1-2; pl. 2, fig. 1
. 1969 *Scutellina hungarica* Vadász – Mihály: 255; pl. 2, figs. 2-3
1969 *Scutella gibbercula* Serr. – Mitrović-Petrović: 137-138; pl. 27, figs. 1, 1a
1969 *Scutella vindobonensis* Lbe. – Mitrović-Petrović: 138; pl. 27, fig. 2; pl. 28, figs. 1, 1a
1969 *Scutella multiconcava* Serr. – Mitrović-Petrović: 138-139; pl. 29, fig. 1; pl. 30, fig. 1
1969 *Scutella styrica* Schaf. – Mitrović-Petrović: 139; pl. 29, figs. 2; pl. 30, fig. 2
1970 *Scutella vindobonensis* – Thenius: p. 216
1974 *Scutella gibbercula* Ag. – Roman: 334-338
1974 *Scutella vindobonensis* – Thenius: 27
1975 *Scutella vindobonensis* Laube, 1871 – Kalabis: 176
1978 *S.[cutella] szoerenyiae* Sándor, 1969 – Kier & Lawson: 67 [correct author is Mihály, Sándor is his first name]
v. 1978 *S.[cutella] multiconcava* Schaffer, 1962 – Kier & Lawson: 67
. 1978 *S.[cutella] styriaca* Schaffer, 1962 – Kier & Lawson: 67
. 1978 *S.[cutella] vindobonenis* Laube *planata* Kókay in Somos and Kókay, 1960 – Kier & Lawson: 67
v. 1978 *S.[cutella] vindobonenis* Laube *secunda* Schaffer, 1962 – Kier & Lawson: 67
1978 *Scutella styriaca* Schaffer – Kollmann & Rögl: 165
v. 1978 *Scutella vindobonensis* Laube – Müller, 566; figs. 695B a-c
. 1981 *Scutella multiconcava* Schaffer, 1962. – Codrea: 141-145; figs. 3-5
. 1984 *Scutella vindobonensis* Laube – Kókay et al.: 288, 290
. 1984 *Scutella ("Scutellina") hungarica* (Vadász) – Kókay et al.: 288
. 1984 *Scutella szoerenyiae* Mihály – Kókay et al.: 288
. 1984 *Scutella* n. sp. 1– Kókay et al.: 288
. 1984 *Scutella* n. sp. 2 – Kókay et al.: 288
. 1985 *Scutella szoerenyiae* Mihály, 1969 – Mihály: 238-239, 258-259; pl. 2, figs.1-3
. 1985 *Scutella hungarica* (Vadász, 1914) nov. comb. – Mihály: 239-240, 259-260; pl. 1, figs.6-10
. 1985 *Scutella pygmea* Koch, 1887 – Mihály: 240; pl. 3, figs.1-4
#. 1985 *Scutella romani* n. sp. – Mihály: 240-241, 260; pl. 2, figs.4-6
#. 1985 *Scutella muelleri* n. sp. – Mihály: 241, 260-261; pl. 3, figs.5-6; pl. 4, fig. 1
. 1985 *Scutella vindobonensis* Laube, 1871 – Mihály: 241-242
1988 *Scutella gibbercula* N. de Serres, 1829 – Saraiman: 37, pl. 1, figs. 1-2
1988 *Scutella vindobonensis* Laube, 1871 – Saraiman: 37, pl. 1, figs. 3-4
#. 1990 *Scutella vindobonensis altus* n. ssp. – Mihály: 237-238, 240; pl. 1, figs.1-2; pl. 2, fig. 1
1996 *Scutella subtrigona* Laube – Ghiurca: 189, unnumbered fig. [misidentification]
. 1998 *Parascutella gibbercula* (de Serres, 1829) – Philippe: 150-151; pl. 15, figs. 7 a-d
non 1998 *Parascutella striatula* (de Serres, 1829) – Philippe: 143-145; pl. 15, figs. 1-3
v. 2002 *Parascutella vindobonensis* – Plöchinger & Karanitsch: 161, fig. 336

Type-material:

Scutella gibbercula de Serres, 1829:
Holotype: according to Philippe (1998: 150, 315) the specimen figured by Lambert (1912: 73-74) could probably be the holotype of de Serres' species [see also comment of Cottreau (1913b) and reply of Lambert (1913b)]; a cast of that specimen is housed in the collection Lambert at the Muséum national Paris under the number L 18 399
Locus typicus: Cadenet, Vaucluse, France
Age: Tortonian (according to Philippe, 1998: 150)

Scutella vindobonensis Laube, 1869:
Lectotype (Pl. 39, Figs. 1a-b; Pl. 40, Fig. 2): NHMW 1904.VIII.59, figured by Laube (1871: 62; pl. 17, fig. 1), designated by Schaffer (1962: 156); originally in the collection of Karrer, since 1903 at the Naturhistorisches Museum Wien, Geologische Abteilung
Paralectotypes: most of the *Ps. gibbercula* specimens from the

NHMW collection registered before 1869 have to be considered paralectotypes (see below under "Material").
Locus typicus: Kalksburg near Vienna, Austria
Age: Badenian (Langhian-Early Serravallian), Middle Miocene

Scutella kalksburgensis Wiesbaur, 1874:
Holotype: Naturhistorisches Museum Wien, Geologische Abteilung (no located)
Locus typicus: Kalksburg near Vienna, Austria
Age: Badenian (Langhian-Early Serravallian), Middle Miocene
Remarks: None of the *Ps. gibbercula* species from Kalksburg in the collection of the Naturhistorisches Museum Wien is accompanied by a label or any other indication that it might represent the type material of this species. Unfortunately, the type material was never figured, so there is no possibility to identify the type material of this species. However, since all the scutellids from the locality Kalksburg examined during the present study are *Ps. gibbercula*, it is only valid to assume that *S. kalksburgensis* is a junior synonym of that species. Moreover, the description and diagnostic features given by Wiesbaur (1874) fall well within the variation of *Ps. gibbercula*.

? *Scutella pygmea* Koch, 1887b:
Holotype: the specimen figured by Koch (1887b: pl. 5, figs. 1a-c) and Vadász [1915: pl. 9 (3), figs. 10-11 (seems more accurate and fits better with the descriptions than Koch's figures]; current whereabouts unknown
Locus typicus: Buituri (= Bujtur), 20 km S of Deva, Hunedoara, Romania
Age: Late Badenian (Early Serravallian)
Remarks: The holotype is a poorly preserved juvenile scutellid. Vadász (1915:122) argued that the slightly thickened, rounded margin indicates that the specimen is no juvenile, which according to his opinion should have a thin, sharp margin. In *Ps. gibbercula*, however, a moderately thick, rounded margin is developed throughout the ontogeny and can already be observed in specimens between 15 and 45 mm TL, which still lack gonopores. *S. pygmea* is here tentatively placed into the synonymy of the co-occurring *Ps. gibbercula* on base of the high similarity to juveniles of that species.

? *Scutellina hungarica* Vadász, 1915:
Holotype: MAFI Ech 228, figured by Vadász (1915: pl. 10 (4), figs. 2-3); Geological Survey of Hungary, Budapest
Locus typicus: Biatorbágy (= Bia), Hungary
Age: Late Badenian (Early Serravallian), Middle Miocene
Remarks: The holotype of this species is a poorly preserved juvenile scutellid. It is very similar to juveniles of *Ps. gibbercula*, the only other scutellid present in the Late Badenian of the Central Paratethys and is tentatively placed into the synonymy of the latter. Material attributed to *S. hungarica* by Mihály (1969 and 1985) is clearly conspecific with *Ps. gibbercula*.

? *Scutella almerai parva* Szörényi, 1953:
Holotype: specimen no. 29 figured by Szörényi (1953: pl. 1, figs. 6, 6a-b); University of Lwów, Ukraine
Locus typicus: Velyki Birky (= Borki-Wielkie), near Tarnopol, Ukraine
Age: Badenian (Langhian-Early Serravallian), Middle Miocene

? *Scutella eichwaldi* Szörényi, 1953:
Holotype: specimen no. 15 figured by Szörényi (1953: pl. 1, figs. 5, 5b); University of Lwów, Ukraine
Locus typicus: Zalesce (= Zalezce), Ukraine
Age: Badenian (Langhian-Early Serravallian), Middle Miocene

Scutella vindobonensis planata Kókay, 1960 *in* Somos & Kókay, 1960:
Holotype: MAFI Ech 638, figured by Somos & Kókay (1960: 341-342, 346; 1 fig.; pl. 16, fig.2); Geological Survey of Hungary, Budapest
Locus typicus: Railroad cut near Hird, Hungary
Age: Badenian (Langhian-Early Serravallian), Middle Miocene

Scutella multiconcava Schaffer, 1962:
Holotype: NHMW 390/1961 figured by Schaffer (1962: pl. 18, figs. 1-3); Naturhistorisches Museum Wien, Geologische Abteilung
Figured specimen: NHMW 1964/590 figured by Schaffer (1962: pl. 17, figs. 4) and NMW 391/1961 figured by Schaffer (1962: pl. 15, figs. 3); Naturhistorisches Museum Wien, Geologische Abteilung
Locus typicus: Neckenmarkt, Bgld, Austria
Age: Badenian (Langhian-Early Serravallian), Middle Miocene
Remarks: The holotype (see Pl. 41, Fig. 1a-c) is a pathological specimen. Sometime during its ontogeny the specimen seems to have been injured at the apical disc. This resulted in a posterior and left lateral displacement of the ocular plates, which do not border the madreporite. As a whole this also accounts for the smaller TL/TW ratio than in other specimens from the type locality.

Scutella styriaca Schaffer, 1962:
Holotype: IPUW 1559 figured by Schaffer (1962: pl. 19, fig. 1); Institute of Palaeontology, University of Vienna (not located)
Locus typicus: quarry near "Tittenberg" (probably identically with Seggauberg), south-west Leibnitz, Styria, Austria
Age: Badenian (Langhian-Early Serravallian), Middle Miocene
Remarks: Although this species is registered in the inventory of the Institute of Palaeontology at the University of Vienna it could not be located in the collections.

Scutella vindobonensis secunda Schaffer, 1962:
Holotype (Fig. 36.B, Pl. 40, Figs. 1a-c): NHMW 1964/589, figured by Schaffer (1962: pl. 19, fig. 4); Naturhistorisches Museum Wien, Geologische Abteilung
Locus typicus: Sandberg, near Devínska Nová Ves (= Neudorf an der March), Slovak Republic
Age: Badenian (Langhian-Early Serravallian), Middle Miocene
Remarks: The holotype of this species has been registered at two institutions the Naturhistorisches Museum Wien (as NHMW 1964/589), where it is still located and erroneously at the Institute of Palaeontology, University of Vienna (as IPUW 1560).

Scutella szoerenyiae Mihály, 1969:
Holotype: MAFI Ech 3, figured by Mihály (1969: pl. 1, figs.1-2; pl. 2, fig. 1); Geological Survey of Hungary, Budapest
Locus typicus: Budapest-Rákos, Hungary
Age: Late Badenian (Early Serravallian), Middle Miocene

Scutella romani Mihály, 1985:
Holotype: MAFI Ech 1, figured by Mihály (1985: pl. 2, fig. 4); Geological Survey of Hungary, Budapest
Paratype: MAFI Ech 2, figured by Mihály (1985: pl. 2, fig. 5); Geological Survey of Hungary, Budapest
Locus typicus: Gyakorló út (street), Budapest, Xth district, Hungary
Age: Late Badenian (Early Serravallian), Middle Miocene

Scutella muelleri Mihály, 1985:
Holotype: MAFI Ech 358, figured by Mihály (1985: pl. 3, figs. 5-6, pl. 4, fig. 1); Geological Survey of Hungary, Budapest
Locus typicus: Gyakorló út (street), Budapest, Xth district, Hungary
Age: Late Badenian (Early Serravallian), Middle Miocene

Scutella vindobonensis altus Mihály, 1990:
Holotype: MAFI Ech 400, figured by Mihály (1990: pl. 1,

figs.1-2; pl. 2, fig. 1); Geological Survey of Hungary, Budapest
Locus typicus: Budapest, XIVth district, Hungary
Age: Late Badenian (Early Serravallian), Middle Miocene

Material:
Early Badenian (Langhian-Early Serravallian) – Retznei [Weissenegg Fm., Lafarge quarry (formerly Perlmoser)], Styria, Austria
  WANZENBÖCK coll.: 2 specimens (no inv. nos.)
Badenian (Langhian-Early Serravallian) – Bad Fischau, NÖ, Austria
  NHMW: 1 specimen (NHMW 1904.VIII.62)
Badenian (Langhian-Early Serravallian) – Brunn am Steinfeld, NÖ, Austria
  NHMW: 2 specimens (NHMW 2004z0001/0048, 2004z0001/0051)
Badenian (Langhian-Early Serravallian) – Hainburg (*Halitherium* locality of STACHE, 1867), NÖ, Austria
  NHMW: 1 specimen (NHMW 1867.XXII.3)
Badenian (Langhian-Early Serravallian) – Haschendorf, W Deutschkreutz, Bgld, Austria
  NHMW: 5 specimens (NHMW 1848.III.63-64, 1848.III.66, 2004z0001/0052, 2004z0001/0060)
Badenian (Langhian-Early Serravallian) – Hasenberg, near Brunn am Gebirge, NÖ, Austria
  NHMW: 2 specimens (NHMW 1904.VIII.61, 2004z0001/0047)
Badenian (Langhian-Early Serravallian) – Kalksburg, Vienna, Austria
  NHMW: 12 specimens [NHMW 1857.38.21, 1857.38.22, 1858.III.13-14, 1858.XXV.40, 1904.VIII.56, 1904.VIII.57 (figured specimen of SCHAFFER, 1962: pl. 15, fig. 2) 1904.VIII.58, 1904.VIII.59 (**lectotype** of *Scutella vindobonensis* LAUBE, 1869), 2004z0001/0059, 2004z0072/0002, 2004z0111/0001]
Badenian (Langhian-Early Serravallian) – Neckenmarkt, Bgld, Austria
  NHMW: 6 specimens [NHMW 291/1961, 390/1961 (**holotype** of *Scutella multiconcava* SCHAFFER, 1962: pl. 18, figs. 1-2), 391/1961 (figured specimen of SCHAFFER, 1962: pl. 15, fig.3), 1964/590 (figured specimen of SCHAFFER, 1962: pl. 17, fig. 4), 2004z0001/0055-56, 2004z0001/0058]
Badenian (Langhian-Early Serravallian) – Oslip, Bgld, Austria
  NHMW: 1 specimen (NHMW 1976/1843/18) and numerous test fragments (NHMW 2003z0083/0006)
Badenian (Langhian-Early Serravallian) – Pötzleinsdorf, Vienna, Austria
  NHMW: 1 specimen (NHMW 2004z0072/0001)
Badenian (Langhian-Early Serravallian) – Rauchstallbrunngraben, near Baden, NÖ, Austria
  NHMW: 1 specimen (NHMW 2004z0001/0062)
Badenian (Langhian-Early Serravallian) – Siegenfeld, Gaaden Bay, NÖ, Austria
  NHMW: 1 specimen (NHMW 2004z0111/0002)
Late Badenian (Early Serravallian) – Müllendorf (Mühlendorfer Kreide AG quarry), Bgld, Austria
  NHMW: 19 specimens (NHMW 1981/55/20, 1997z0178/1754, ../1756-1757, 2002z0181/0023-34, 2004z0112/0009-11),
Late Badenian (Early Serravallian) – Unterpetersdorf (= Alsópéterfa), Bgld, Austria
  NHMW: 2 specimens [MAFI Ech-283 (reference material of VADÁSZ, 1915)]

Foreign material for comparison:
Badenian (Langhian-Early Serravallian) – "Bia gravel", Soskút, Hungary
  NHMW: 4 specimens (NHMW 1865.XXX.78-80)
  Remark: According to VADÁSZ (1915: 105, footnote) there are only strata of Sarmatian age outcropping at Soskút and material from the marine Badenian comes from the nearby Leitha Limestone of Bia.
Badenian (Langhian-Early Serravallian) – Kemence, Hungary
  NHMW: 1 specimen (NHMW 2004z0001/0054)
Late Badenian (Early Serravallian) – Armeniş (= Örményes), Romania
  NHMW: 2 specimens and 1 fragment [MAFI Ech-290-292 (reference material of VADÁSZ, 1915)]
Late Badenian (Early Serravallian) – Biatorbágy (= Bia), near Budapest, Hungary
  NHMW: 3 specimens [NHMW 1859.XLIII.26; MAFI Ech-137 (figured specimen of VADÁSZ, 1915: fig. 12), ? MAFI Ech-228 (**holotype** of *Scutellina hungarica* VADÁSZ, 1915: pl. 10 (4), figs. 2-3)]
Late Badenian (Early Serravallian) – Budatétény (= Tétény), Hungary
  NHMW: 1 specimen (NHMW 1997z0178/2090)
Late Badenian (Early Serravallian) – Buituri (= Bujtur), Romania
  MAFI: 2 specimens [MAFI Ech-240 (figured specimen of VADÁSZ, 1915: fig. 14), Ech-264 (reference material of VADÁSZ, 1915)]
Late Badenian (Early Serravallian) – Devínska Nová Ves (= Neudorf an der March), Slovak Republic
  NHMW: 1 fragment (NHMW 1857.XIX.46), 8 specimens [NHMW 1851.VI.120, 1851.VI.120.6, 1866.I.1256, 1929.No.16, 1964/589 (**holotype** of *Scutella vindobonensis secunda* SCHAFFER, 1962: pl. 19, fig. 4), 2004z0001/0046, 2004z0111/0003], 2 partially leached specimens (NHMW 1859.XXXVIII.100)
Late Badenian (Early Serravallian) – Sandberg, Devínska Nová Ves (= Neudorf an der March), Slovak Republic
  NHMW: 3 specimens (NHMW 2004z0001/0050). 2 partially leached and fragmented specimens (NHMW 2004z0001/0049)
? Late Badenian (Early Serravallian) – Gârbova de Sus (= Felsö-Orbó), Romania
  NHMW: 2 specimens [MAFI Ech-130 (figured specimen of VADÁSZ, 1915: fig. 16), Ech-234 (figured specimen of VADÁSZ, 1915: fig. 15)]
Late Badenian (Early Serravallian) – Globu Craiovei (= Globukrajova), Romania
  NHMW: 1 specimen and 1 fragment [MAFI Ech-286, 289 (reference material of VADÁSZ, 1915)]
Late Badenian (Early Serravallian) – Harka, Györ-Moson-Sopron, Hungary
  NHMW: 1 specimen [MAFI Ech-284 (reference material of VADÁSZ, 1915)]
Late Badenian (Early Serravallian) – Iablanitja (= Bélajablánc), Romania
  NHMW: 1 specimen [MAFI Ech-282 (reference material of VADÁSZ, 1915)]
Late Badenian (Early Serravallian) – Minişu des Sus (= Felménes), Romania
  NHMW: 1 specimen and 1 fragment [MAFI Ech-285, 288 (reference material of VADÁSZ, 1915)]
Late Badenian (Early Serravallian) – Paptelek, Hungary
  NHMW: 1 specimen [MAFI Ech-263 (reference material of VADÁSZ, 1915)]
Late Badenian (Early Serravallian) – Sopron (= Ödenburg), Hungary
  NHMW: 1 specimen (NHMW 2004z0001/0053)
Late Badenian (Early Serravallian) – Tusa (= Tusza), Romania
  NHMW: 1 specimen and 2 fragments [MAFI Ech-287 (reference material of VADÁSZ, 1915), Ech-158 (figured specimen of VADÁSZ, 1915: fig. 17)]

Dimensions (in mm):

Inv. No.	TL	TW	TH
NHMW 1997z0178/1756	158.6	172.5	28.9
NHMW 1997z0178/1757	125.5	137.8	20.1

NHMW 1964/589	133.3	~140	21.5
NHMW 390/1961	95.8	99.4	20.0
NHMW 1964/590	92.6	102.3	~17
NHMW 1904.VIII.59	150.3	163.8	30.6

Description:

Size and shape: Test large to very large, ranging from 15 (juvenile specimen with gonopores still closed) to 159 mm TL. In a 44.6 mm TL specimen the gonopores are still closed. The outline of the test is subcircular, slightly transversely elongate (rarely antero-posteriorly elongated individuals are found, most of these, however, show a regenerated marginal area and thus do not show the characteristic outline of this species). The anterior margin is rounded with shallow indentations corresponding to the ad-, inter- and perradial sutures. The posterior margin is transversely truncated, with two strong indentations corresponding to the perradial sutures of the ambulacra I and V. The anal notch is shallow but distinct and rather broad. In juveniles and specimens with thicker coronal plates (i.e. specimens from high-energy settings, e.g. from Neckenmarkt), however, it may be deeper. The maximum width lies slightly posterior of the apical disc, at the line interambulacra 1b-4a. In profile the test is distinctly wedge shaped, with a thicker, rounded anterior margin and a thinner, relatively sharp posterior margin. The maximum height lies 10 to 13 % of the TL posterior of the centre of the apical disc along the central suture of interambulacrum 5. The test width is about 103 to 110 % of the TL and the height about 13-25 % of the TL.

Apical disc: The apical disc lies subcentrally and belongs to the tetrabasal type. The madreporite is pentagonal and crowded with many small madreporic pores. Four gonopores are situated at the apices of the madreporite, except in interambulacrum 5. The ocular pores are small and rounded.

Ambulacra: The ambulacra are petaloid. The petals are straight, moderately closed distally and are usually flush with the interambulacra (rarely they may be slightly depressed). They extend about two third of the corresponding test radius. The frontal petal is slightly shorter than the anterior paired petals, which in turn are slightly short than the posterior paired petals. The ambulacra pores within the petals are septate conjugated isopores. That means that the two pores in each pair are connected by an extremely elongated, narrow groove, which deepens towards the adradial pore. Contrary to the condition in most other clypeasteroid echinoids (Smith, 1980c) (see Fig. 37.A-C), the adradial pair of this groove is subdivided by numerous vertical septae (see Fig. 37.D-E). Outside the petals only minute micro-unipores are found. Each poriferous zone is about 1.4 to 1.9 times as wide as the interporiferous zone.

Interambulacra: The interambulacra are covered by small, crenulate, perforate tubercles, which are slightly larger adorally than adapically.

Food grooves: The food grooves are restricted to the oral surface, they are simple in smaller and more complex in larger specimens, where many secondary grooves branch from the main grooves. They bifurcate at about 17 to 20 % of the corresponding test radius from the peristome. From this point on they run along the midline of each ambulacral plate column. Just before they reach the ambitus, they bend in direction of the interambulacra.

Peristome: The peristome is situated subcentrally on the oral surface and has a circular outline. It is about 4.5 mm in diameter in a 125 mm specimen.

Periproct: The periproct is small, rounded to oval (transversely elongated) with a diameter of 2 to 2.5 mm. It lies marginally along the interradial suture of interambulacrum 5 between the 3rd pair of postbasicoronal plates (Fig. 36), usually around 6 % TL from the posterior margin. A very shallow, broad groove connects it with the posterior test margin. In juvenile specimens (up to *c.* 25-30 mm) the periproct may lie marginally within the anal notch.

Differential diagnosis:

The type species of *Parascutella*, *Ps. leognanensis* (Lambert, 1903) from the Early Miocene of Western France, differs from *Ps. gibbercula* by its flatter profile, without conspicuous elevation in adapical interambulacrum 5, its broader interporiferous zones and more closed petals [compare Agassiz, 1841a (under the name *Scutella subrotunda*) and Durham, 1953].

Parascutella paulensis (Agassiz, 1841), from the Burdigalian of the Mediterranean and Eggenburgian to Karpatian of the Central Paratethys, likewise differs from *Ps. gibbercula* by its less strongly developed marginal indentations, its rounded posterior margin without or with faint anal notch, and flattened profile with central maximum height.

Parmulechinus hoebarthi (Kühn, 1936) from the Early Eggenburgian (Early Burdigalian) differs from *Parascutella gibbercula* by it shorter petals (*c.* half the length of the corresponding test radius), its antero-posteriorly elongated outline, and the different position of its peristome (between the 4th or 5th pair of postbasicoronal plates).

For the difference between *Ps. gibbercula* and *Ps. jaquemeti* (de Loriol, 1902), *Ps. striatula* (de Serres, 1829), *Ps. michaleti* (Lambert, 1906), *Ps. depereti* (Lambert, 1912), and *Ps. savini* (Lambert, 1915) the reader is referred to Philippe (1998).

Discussion:

Originally the species was named *Scutella gibercula* by de Serres (1829: 156). Agassiz (1841a: 86) changed the name to *S. gibbercula* (there is no indication whether the change was intentional or unintentional). According to the ICZN rules this represents an unjustified emendation or incorrect subsequent spelling. The name *S. gibbercula* is in prevailing usage during the last 150 years and is attributed to the original author and date of the original spelling. According to the ICZN it is therefore deemed to be a justified emendation, respectively a correct original spelling (in case of an unintentional change) (ICZN 4th ed., 2000, Article 33.2.3.1, respectively Article 33.3.1).

Because of the marginal position of the periproct between the 3rd pair of postbasicoronal plates and the long petals (two third of the corresponding test radius), this species is placed into the genus *Parascutella* Durham, 1953, as proposed by Philippe (1998) already.

Specimens of this species found in the Miocene of Austria and its neighbouring countries were long known under the name *Scutella vindobonensis* Laube, 1871. However, apart from the fact that the name was already established in 1869, *S. vindobonensis* is a junior synonym of *S. gibbercula* de Serres, 1829. Both the original diagnosis of de Serres (1829: 156) and subsequent descriptions and illustrations by Lambert (1912: 73-74; pl. 5, figs. 1-2) and Philippe (1998: 150-51; pl. 15, figs. 7a-d) correspond very well to the present material including Laube's holotype. This was already suggested by Vadász (1915: 117-118), who considered *S. gibbercula* as local variety of *S. vindobonensis*. Szörényi (1953: 64), however, rejected this and stated that those two are separate species based on the following features: 1) *S. gibbercula* has no deep anal notch only a shallow indentation; 2) a slightly concave oral surface; 3) a posteriorly eccentric apex; 4) a frontal ambulacrum which is shorter than the paired ambulacra; and 5) interporiferous zones which are slightly smaller than a single poriferous zone. Comparing these features with the material at hand it is apparent that all these features are also present in the specimens originally classified as *S. vindobonensis*, even in the holotype. The only differences that can be observed between *S. gibbercula* specimens from France and specimens of the Central Paratethys formerly classified as *S. vindobonensis* are the more strongly inflated interambulacrum 5 and that the thicker ambitus in the latter. There are, however, numerous specimens which do not show these differences, or which are intermediate. For such specimens the nominal subspecies *vindobonensis planata* Kókay and *vindobonensis secunda* Schaffer have been established. The close relation between *gibber-*

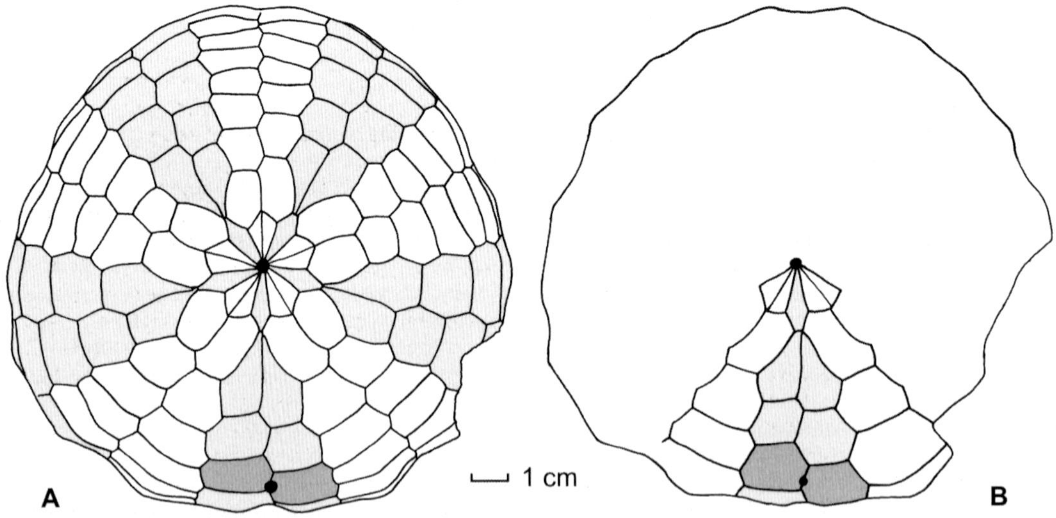

Figure 36: *Parascutella gibbercula* (de Serres, 1829): oral plating [A: Kalksburg, Vienna (NHMW 2004z0111/0001); B: holotype of *Scutella vindobonensis secunda* Schaffer, 1962, Sandberg, near Devínska Nová Ves (Neudorf an der March), Slovak Republic (NHMW 1964/589)]. Interambulacra shaded in light grey; post-basicoronal plate pair enclosing the periproct shaded in dark grey.

cula and *vindobonensis* have already been observed by Lambert (1912: 74) and Philippe (1998: 151). Since all other features are identical and the overlap of the former features is very large a clear distinction on species or even subspecies level is not possible. Moreover, these features seem to be correlated to the substratum on/in which the animals lived, since the conditions typical for the *vindobonensis*-morphology is developed in larger specimens coming from carbonate environments, whereas they are no present in specimens coming from predominantly siliciclastic sediments. Consequently *Ps. vindobonensis* is considered a junior synonym of *Ps. gibbercula* here.

A number of subspecies have been established within *S. vindobonensis*: *S. vindobonensis planata* Kókay, 1960 *in* Somos & Kókay, 1960 and *S. vindobonensis secunda* Schaffer, 1962, for specimens with a lower test profile and thinner margin; and *S. vindobonensis altus* Mihály, 1990 for specimens with especially high test profile. However, since neither of these subspecies is characteristic for a certain time slice or region and they occur erratically, as do other extreme morphotypes, they are poorly justified. Moreover they are connected by numerous intermediate morphologies and any separation would be highly artificial.

Scutella kalksburgensis Wiesbaur, 1974 a species based on a single specimen from the type locality of *S. vindobonensis*, is also considered a junior synonym of *S. gibbercula* de Serres, 1829 (see above under type-material).

In 1962 Schaffer established a number of new "*Scutella*"-species (now referred to *Parascutella*) from the Middle Miocene of Austria. According to Schaffer (1962) his *Scutella styriaca* is characterised by its profile and its constant "Marginopetalabstand" of 28-30 mm (distance between tip of the petals and the margin of the test). The latter feature, albeit repeatedly used by Schaffer (1962) is poorly suited for distinguishing scutellid species. During the growth of the animals this distance increases. If used at all it can only be employed for specimens of similar size or if expressed as relation (e.g. percentage of the TL). In scutellids, however, the relative length of the petals is very constant within genera. In *Parascutella* the petals usually extend about two third and in *Scutella* usually half of the corresponding test radius. The test profile is better suited to distinguish between species, but still can vary strongly within a single species, especially between specimens from different palaeoenvironments. In the case of *styriaca* the profile is nearly identical with that of specimens Schaffer himself placed into another species (*S. vindobonensis secunda*; compare Schaffer, 1962: fig. 9 with fig. 11). Comparing large series of scutellids with TL ranging form *c.* 15 mm to more than 150 mm coming from various localities throughout the Central Paratethys it becomes evident that variation of profile can be quite strong. Within *Ps. gibbercula* the profile ranges from a very flat type to a very high markedly triangular shape. These differences have been used to establish subspecies (*secunda*, *planata* and *altus*) within *Ps. vindobonensis*, a junior synonym of *Ps. gibbercula* (see above). As stated above, these subspecies are, however, highly artificial as they are connected by series of intermediate morphotypes and occur all sympatric and synchronous. Moreover, *Ps. gibbercula* shows allometric growth and relative test height (and triangularity of the profile) decreases with increasing size. A constant feature of *Ps. gibbercula* is the presence of a bulge in adapical interambulacrum 5. The inflation of this bulge varies, but it is always present and coincides with the maximum test height. Schaffer's statement that such a bulge is not present in *S. styriaca* (Schaffer, 1962: 154) is not true as can be seen on his fig. 9 and in the material itself. There are no features which allow confident recognition of this species. Therefore, *S. styriaca* is considered as yet another junior synonym of *Ps. gibbercula*.

Similarly the species *Scutella multiconcava* Schaffer, 1962 cannot be maintained as separate species. The characteristic features provided by Schaffer (1962) include small size, a strong bulge in adapical interambulacrum 5, a constant "Marginopetalabstand" of 17 mm, inframarginal position of the periproct, a semicircular anal notch and strongly branched food grooves. As discussed above the "Marginopetalabstand" is poorly suited to characterise scutellid species. The bulge in adapical interambulacrum 5, albeit certainly very prominent in Schaffer's specimens, is shared by all the nominal species discussed here. The same is valid for the inframarginal position of the peristome and the anal notch. Only the strong branching of the food grooves is peculiar for *S. multiconcava*. Yet, this is also poorly suited as diagnostic feature since the depth and exact arrangement of the food grooves varies strongly between individuals. The smaller size, higher profile and as a whole more "stout" appearance might be an adaptation to the local palaeoenvironment (e.g. higher energy, different sediment,...). *S. multiconcava* is thus considered a local morphotype and junior synonym of *Ps. gibbercula*.

Scutella szoerenyiae MIHÁLY, 1969 is based on a single specimen with wavy margin. The latter and the deep anal notch, as well as the large test height were considered as diagnostic features distinguishing this species from *S. vindobonensis* by MIHÁLY (1969: 255). The wavy margin, however, is a clear sign of regeneration following sub-lethal predation, a feature commonly encountered in fossil and extant sand dollars (e.g. VADÁSZ, 1914, 1915; FRAZER et al., 1991; NEBELSICK, 1999; KOWALEWSKI & NEBELSICK, 2003; LAWRENCE et al., 2004). As regeneration in these echinoids does not reproduce the original outline but rather seals scars and gaps in the skeleton this feature could also be an explanation for the unusual deep anal notch. As other diagnostic features are absent *S. szoerenyiae* is here considered a junior synonym of *Ps. gibbercula*. MIHÁLY (1985: 259) stated that *S. szoerenyiae* is closely related to *Abertella aberti* and "*Scutella*" *floridana* (the latter is a junior synonym of the former according to COOKE, 1959: 44-45). Apart from slight superficial similarities in overall shape, however, there is no relation between these species and MIHÁLY'S species, apart from the fact that they are members of the suborder Scutellina.

The specimens attributed to the species *Scutella hungarica* (VADÁSZ, 1914), *Scutella pygmea* KOCH, 1887b, *Scutella romani* MIHÁLY, 1985 and *Scutella muelleri* MIHÁLY, 1985 are characterised by their small size and the fact that all occur within a single locality and horizon. Based on the comparison with growth series of *Ps. gibbercula* from Austrian localities these specimens are here considered as juvenile and subadult specimens of *Ps. gibbercula* (which, under the name *S. vindobonensis*, was reported from the same locality). The changing position of the periproct from an aboral-supramarginal position to a marginal and finally submarginal position is very characteristic for the ontogeny of scutelline echinoids and was documented for *Dendraster excentricus* by DURHAM (1953: 102) and could also be shown for *Parmulechinus hoebarthi* (this paper, see below under that species). There are no features that would allow any of these "species" to be distinguished from juvenile *Ps. gibbercula* and consequently *S. romani* and *S. muelleri* are placed into the synonymy of the former. Obviously the idea that part of his material might be juvenile specimens never came to MIHÁLY'S mind, as can also bee seen by the fact that he established a new genus and species for juveniles of *Amphiope* in the same paper (see below).

The type material of both *Scutella pygmea* KOCH, 1887b and *Scutellina hungarica* VADÁSZ, 1915 are poorly preserved juvenile scutellids. They are tentatively placed into the synonymy of *Ps. gibbercula* here based on the high similarity to juveniles of that species (see also above under "Type-material") and the co-occurrence of adult specimens of *Ps. gibbercula*.

SZÖRÉNYI (1953) described a large number of different scutellid species, among them a new species and a new subspecies, from the Early Badenian of the western Ukraine. The specimen figured as *S. vindobonensis* by SZÖRÉNYI (1953: pl. 1, figs. 4, 4a) clearly belongs to *S. gibbercula* (based on outline, profile and shape of the petals). The specimens figured as *Scutella paulensis* by her are more difficult to determine. Apparently they have few marginal indentations, a rather low profile and narrow petals. Yet the specimens are uncharacteristic for *Parascutella paulensis* (compare illustrations in PHILIPPE, 1998). They are here tentatively referred to *Ps. gibbercula*, but need to be re-examined to confirm this placement. The strong weathering of the specimens may account for the rather narrow appearance of the petals (contrary to the rather broad petals of typical *Ps. gibbercula*). *Scutella eichwaldi* and *S. almerai parva* are subadult specimens that fall well within the variation observed in *Ps. gibbercula* and are tentatively referred to this species here.

PHILIPPE (1998: 143, 145) referred the material figured as *Scutella leognanensis* by VADÁSZ (1915) to *Ps. striatula* (DE SERRES, 1829). Based on a re-examination of these two specimens they are here tentatively referred to *Ps. gibbercula*. They differ from the typical *Ps. gibbercula* by their less strong marginal indentations and rounded posterior margin (the margin of one specimen is, however, strongly regenerated and that of the other specimen damaged, hampering comparison), the very shallow or lacking anal notch and low profile with the maximum height subcentrally. They could also be related to *Ps. paulensis*, which is known at least until the Karpatian (Late Burdigalian) of the Central Paratethys and may also have been present in the Early Badenian. A relation to the poorly known *Ps. striatula* (DE SERRES, 1829) [see COTTREAU (1913a: 88-92, figs. 15-16), NEGRETTI (1984: 106-107; pl. 5, figs. 4-5, pl. 6, fig. 3) and PHILIPPE (1998: 143-145; pl. 15, figs. 1-3)] seems to be unlikely, as this is a form with very distinct marginal indentations in the ambulacra, a strong anal notch and is until now only known from the Burdigalian of France.

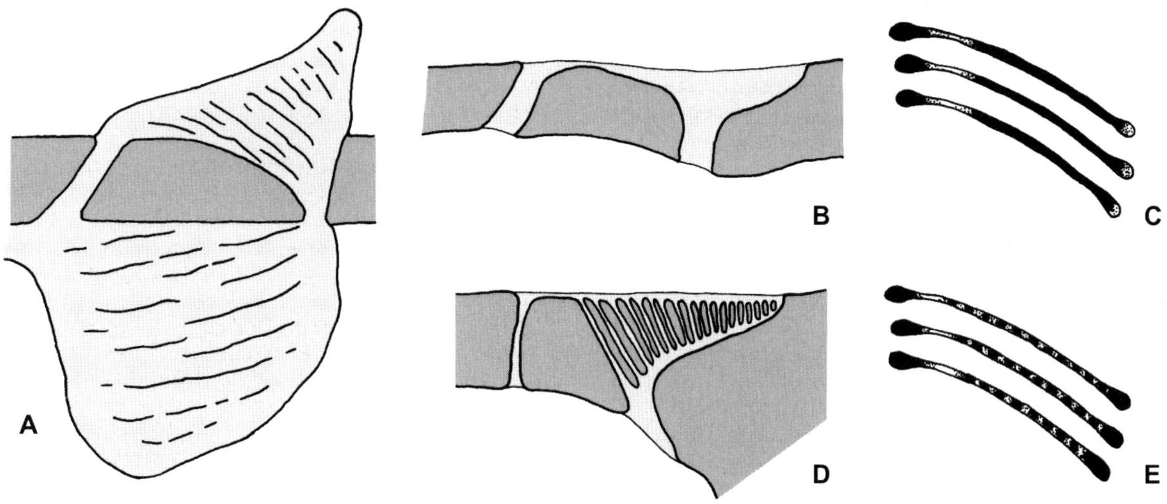

Figure 37: Comparison of respiratory pore structure in *Parascutella* vs. that typical for other scutelline sand dollars. A: section through a respiratory tube foot in extant *Mellita quinquiesperforata* (redrawn from SMITH, 1984); B-C: vertical section (B) and aboral view (C) of a typical elongate isopore in the petals of a scutelline sand dollar (*Mellita quinquiesperforata*); D-E: vertical section (D) and aboral view (E) of a septate pore in the petals of a *Parascutella* (*Ps. gibbercula*).

Occurrence:

Austria: Early to Late Badenian (Langhian-Early Serravallian

Vienna Basin: Brunn am Steinfeld, NÖ (KARRER, 1877; SCHAFFER, 1962; [NHMW]); Bad Fischau, NÖ (PLÖCHINGER & KARANITSCH, 2002; [NHMW]); Dornbach, Vienna (SCHAFFER, 1906); Hainburg, NÖ (LAUBE, 1869a, 1871; [NHMW]); Hasenberg, near Brunn am Gebirge, NÖ (KARRER, 1877; SCHAFFER, 1962; [NHMW]); Kalksburg, Vienna (FUCHS, 1869; LAUBE, 1869a, pp 1871; STUR, 1873; WIESBAUR, 1874; QUENSTEDT, 1874, 1875; KARRER, 1877; SCHAFFER, 1907; ABEL, 1922; SCHAFFER, 1942; SCHAFFER, 1962; THENIUS, 1962a; MÜLLER, 1964, 1978; [NHMW]); Müllendorf (Mühlendorfer Kreide AG quarry), Bgld (KAPOUNEK, 1939; SCHAFFER, 1962; [NHMW]); Perchtoldsdorf, NÖ (KARRER, 1877); Pötzleinsdorf, Vienna (SIEBER, 1953b; THENIUS, 1970; [NHMW]); Rauchstallbrunngraben, near Baden, NÖ (SCHAFFER, 1962; [NHMW]); Steinabrunn, NÖ (SIEBER, 1958a; SCHAFFER, 1962); Steinberg, near Niederkreuzstetten, NÖ (SCHAFFER, 1962); Vienna Basin (THENIUS, 1974)

Eisenstadt-Sopron Basin: Oslip, Bgld ([NHMW])

Oberpullendorf Bay: Haschendorf (= Hasfalva), W Deutschkreutz, Bgld (LAUBE, 1869a, 1871; VADÁSZ, 1915; SCHAFFER, 1962; [NHMW]); Neckenmarkt, Bgld (JANOSCHEK, 1931; SCHAFFER, 1962; [NHMW]); Ritzing, Bgld (QUENSTEDT, 1874, 1875; JANOSCHEK, 1931); Unterpetersdorf (= Alsópéterfa), Bgld, Austria (WOLF, 1870; LAUBE, 1871; VADÁSZ, 1915; [MAFI])

Styrian Basin: Retznei (Weissenegg Fm., Lafarge quarry), Styria ([WANZENBÖCK coll.]); between St. Georgen and Pichla, Styria (KOLLMANN, 1965); Tittenberg (probably identical with Leibnitz-Seggauberg), south-west Leibnitz, Styria (SCHAFFER, 1962); "Grazer Bucht" (SCHAFFER, 1959); Stiefingbach, Pesendorf, near Wildon, Styria (KOLLMANN & RÖGL, 1978; [NHMW])

Paratethys (non-Austrian occurrences): Early to Late Badenian (Langhian-Early Serravallian)

Vienna Basin: Devínska Nová Ves (= Neudorf an der March), Slovak Republic (SCHAFFER, 1962; [NHMW]); Sandberg, Devínska Nová Ves (= Neudorf an der March), Slovak Republic (SCHAFFER, 1962; [NHMW])

Eisenstadt-Sopron Basin: Fertörákos, Györ-Moson-Sopron, Hungary (VENDL, 1930); Harka (= Harkau), Györ-Moson-Sopron, Hungary (WOLF, 1870; VADÁSZ, 1915; MIHÁLY, 1969; [MAFI]); Sopron, Györ-Moson-Sopron, Hungary (? FOETTERLE, 1860; [NHMW])

Great Hungarian Basin (Pannonian Basin): Biatorbágy (= Bia), Pest, Hungary (LÓCZY, 1887; VADÁSZ, 1915; SCHAFFER, 1962; MIHÁLY, 1969, 1985; KÓKAY et al., 1984; [NHMW]; [MAFI]); Bodolyaber (= Egyházbér), Baranya, Hungary (VADÁSZ, 1915; MIHÁLY, 1969); Budapest, Xth district, Hungary (MIHÁLY, 1985); Budapest, XIVth district, Hungary (MIHÁLY, 1990); Budapest-Gyakorló, Hungary (MIHÁLY, 1985); Budapest-Kerepesi, Hungary (MIHÁLY, 1985); Budapest Rákos, Hungary (FRANZENAU, 1881; VADÁSZ, 1906, 1915; MIHÁLY, 1969, 1985); Érd, Pest (VADÁSZ, 1915); Érdliget, Pest, Hungary (MIHÁLY, 1969); Hidas, Hungary (VADÁSZ, 1915); Hird, Hungary (SOMOS & KÓKAY, 1960); Hor. Strháre (= Felsö-Esztergály), Slovak Republic (GAÁL, 1905); Kemence, Nógrád, Hungary (MÁJER, 1915; SCHAFFER, 1962; MIHÁLY, 1969; [NHMW]); Kisgeresd, Baranya, Hungary (VADÁSZ, 1915; MIHÁLY, 1969); Mátraverebély, Nógrád, Hungary (MIHÁLY, 1969); Ösagárd, Nógrád, Hungary (VADÁSZ, 1915; MIHÁLY, 1969); Paptelek, Hungary (VADÁSZ, 1915; [MAFI]); Sámsonháza, Nógrád, Hungary (MIHÁLY, 1969); Soskút, Pest, Hungary (LAUBE, 1869a, 1871; SCHAFFER, 1962; [NHMW]); Szatina, Baranya, Hungary (VADÁSZ, 1915; MIHÁLY, 1969)

Fore-Carpathian Basin: ? Bilka, Ukraine (EICHWALD, 1830; PUSCH, 1837); Velyki Birky (= Borki-Wielkie), near Tarnopol, western Ukraine (SZÖRÉNYI, 1953); Kurzany, near Kovo-Bircza, western Ukraine (SZÖRÉNYI, 1953; SCHAFFER, 1962); ? Salisce (Salisze), Ukraine (EICHWALD, 1830; PUSCH, 1837); Zhukov (= Zukowce, = Shukowze), western Ukraine (? EICHWALD, 1830; ? PUSCH, 1837; ? EICHWALD, 1852, 1853; SZÖRÉNYI, 1953; SCHAFFER, 1962); ? Zalesce (= Zalezce), western Ukraine (SZÖRÉNYI, 1953; SCHAFFER, 1962)

Transylvanian Basin: Armeniş (= Örményes), Romania (VADÁSZ, 1915; [MAFI]); Brăteşti, Beiuş Basin, NW Romania (CODREA, 1981); Buituri (= Bujtur), Romania (KOCH, 1887b; NEMES, 1888b; LÖRENTHEY, 1894; MÁRTONFI, 1894; VADÁSZ, 1915; [MAFI]); Felsö-Orbó, Romania (KOCH, 1887b; GAÁL, 1905; ? VADÁSZ, 1915; [MAFI]); Globu Craiovei (= Globukrajova), Romania (VADÁSZ, 1915; [MAFI]); Iablanitja (= Bélajablánc, Romania) (VADÁSZ, 1915; [MAFI]); Lăpugiu des Sus (= Lapugy), Romania (SCHAFFER, 1962); Minişel (= Kresztaménes), Arad, Romania (VADÁSZ, 1915); Minişu des Sus (= Felménes), Romania (LÓCZY, 1887; VADÁSZ, 1915; [MAFI]); région E de Taşad, Romania (PAUCĂ, 1936); Tusa (= Tusza), Romania (VADÁSZ, 1915; [MAFI])

Zala, Sáva and Dráva Basins: Derventa and Ugljevik, Bosnia & Herzegovina (MITROVIĆ-PETROVIĆ, 1969)

Varna Gulf: Murfatlar, Basarabi, Romania (SARAIMAN, 1988)

Mediterranean: Serravallian to Tortonian

Rhône Basin: Cabrière-d'Aigues, south of the Luberon hills, Vaucluse, France (LAMBERT, 1912; PHILIPPE, 1998); Cadenet, south of the Luberon hills, Vaucluse, France (DES MOULINS, 1837; AGASSIZ, 1841a; LAMBERT, 1912; PHILIPPE, 1998); Cucuron, south of the Luberon hills, Vaucluse, France (LAMBERT, 1912; ROMAN, 1974; PHILIPPE, 1998); Vaugines, south of the Luberon hills, Vaucluse, France (LAMBERT, 1912; PHILIPPE, 1998); Vaucluse, France (COTTREAU, 1913a)

Eastern Mediterranean: Karaman, Turkey (ROMAN, 1960)

Parascutella paulensis (AGASSIZ, 1841)

(Pl. 43, Figs. 1-2; Pl. 44, Figs. 1a-d)

	1841a	*Scutella paulensis* Ag. – AGASSIZ: 83; pl. 19, fig. 8-10
	1858	[*Scutella*] *Paulensis* AGASS. – DESOR: 233
pp	1871	*Scutella Vindobonensis* LAUBE. – LAUBE: 62; pl. 17, fig. 1
?	1892	*Scutella paulensis*, AGASSIZ. – GOURRET: 130
?	1907a	*Scutella paulensis* AGASSIZ, 1841. – LAMBERT: 43; pl. 3, figs. 1-2
	1912	*Scutella paulensis* AGASSIZ, 1841 – LAMBERT: 64-66; pl. 4, fig. 10-13
	1913a	*Scutella paulensis* AGASSIZ. – COTTREAU: 131-135; fig. 40, pl. 3, fig. 1-9; pl. 4, fig. 1-6
	1915a	*Scutella paulensis* AGASSIZ. – LAMBERT: 214
v non	1915	*Scutella paulensis* AG. – VADÁSZ: 121, fig. 17
? non	1953	*Scutella paulensis* AGASSIZ, 1841. – SZÖRÉNYI: 65-66; pl. 1, figs. 1-3, 3a-b; pl. 2, fig. 2 [see above under *P. gibbercula*]
v.	1959	*Scutella* B. – SCHAFFER: 255
#v.	1962	*Scutella media* n. sp. – SCHAFFER: 151-153; fig. 1; pl. 17, figs. 2-3
	1974	*Scutella paulensis* AG. – ROMAN: 331-338
	1978	*S.*[*cutella*] *media* SCHAFFER, 1962 – KIER & LAWSON: 67
	1984	*Scutella paulensis* AGASSIZ, 1841 – NEGRETTI: 103-105, pl. 4, figs. 4-6; pl. 5. figs. 1-3
	1989	*Parascutella paulensis* – PHILIPPE: 29; tab. 1
	1990	*Parascutella paulensis* (AGASSIZ, 1841) – PHILIPPE et al.: 243

pp	1998	*Parascutella paulensis* (Agassiz, 1841) – Phillippe: 129-141; figs. 6-11; pl. 13, figs. 1-8, 11; pl. 14, figs. 1-3
non	1998	*Parascutella paulensis* (Agassiz, 1841) – Phillippe: pl. 13, figs. 9-10
v.	2002a	*Parascutella paulensis* (Agassiz, 1841) – Kroh: 306-308; pl. 1, figs. 1-5
	2003b	*Parascutella paulensis* (Agassiz, 1841) – Kroh: 250

Type-material:

Scutella paulensis Agassiz, 1841:
Holotype: a cast of the holotype is located in the collection of the Muséum d'histoire naturelle de Neuchâtel under the number MNS 281 according to Philippe (1998: 307)
Locus typicus: St. Paul-trois-Châteaux, Rhône Basin, France
Age: Burdigalian, Early Miocene

Scutella media Schaffer, 1962:
Holotype: specimen NHMW 1861.L.137; Naturhistorisches Museum Wien, Geologische Abteilung
Locus typicus: Niederkreuzstetten, NÖ
Stratum typicum: Korneuburg Fm.
Age: Karpatian, Late Burdigalian

Material:

Late Eggenburgian (Early Burdigalian) – Limberg (Zogelsdorf Fm., Hengl quarry), NÖ, Austria
NHMW: 1 specimen (NHMW 1998z0048/0050)
Karpatian (Late Burdigalian) – Neubau, near Niederkreuzstetten, NÖ, Austria
NHMW: 3 specimens [NHMW 1861.L.137 (**holotype** of *Scutella media* Schaffer, 1962), 1999z0097/0001, 2004z0077/0001] and 18 fragments (NHMW 1861.L.138, 1981/55/2)
Karpatian (Late Burdigalian) – Niederkreuzstetten (railway crossing 400 m north of the village), NÖ, Austria
NHMW: 2 specimens (NHMW 1999z0097/0001, 2004z0077/0001)

Dimensions (in mm):

Inv. No.	TL	TW	TH	Remarks
1861.L.137	113.2	129.6	~10-12	deformed
1999z0097/0001	118.7	128.5	>12	collapsed
1998z0048/0050	86.9	> 82	10.7	fragmented
2004z0077/0001	118.8	121.2	~11.5	

Description:

Size and shape: Test large, subcircular, slightly transversely elongated. The anterior margin is rounded with very shallow indentations corresponding to the ad-, inter- and perradial sutures (usually much less distinct than in *Ps. gibbercula*). Posterior margin trapezoidal, rounded to transversely truncated with two strong indentations corresponding to the perradial sutures of ambulacra I and V. Anal notch extremely shallow, if present at all. The maximum width lies posterior of the apical disc. In profile, the test is low, with convex aboral and flat oral surface. The maximum height lies approximately at the posterior end of the apical disc. The test width is about 105 to 115 % of TL, the height about 10 %.
Apical disc: The apical disc lies centrally and is tetrabasal with 4 rounded gonopores. The madreporite is pentagonal and crowded by many small madreporic pores.
Ambulacra: The ambulacra are petaloid. The petals are straight, closing distally and are flush with the interambulacra. They extend about two third of the corresponding test radius. The anterior petals are shorter than the posterior ones. The frontal petal is the shortest and is slightly wider than the paired ones. Within the petals the ambulacral pores are septate conjugated isopores (see above under *Ps. gibbercula*), outside the petals only minute microunipores are found. Each poriferous zone is about one and a half times as wide as the interporiferous zone.
Interambulacra: The interambulacra are covered by small, crenulate, perforate tubercles, which are slightly larger adorally than adapically.
Food grooves: The food grooves are restricted to the oral surface, they are simple and bifurcate at about one fifth of the corresponding test radius from the peristome. From this point on they run along the midline of each ambulacral plate row. Just before they reach the ambitus, they bend in direction of the interambulacra.
Peristome: The peristome is subcircular and about up to 4 mm in diameter. It is situated centrally on the oral surface.
Periproct: The periproct is small, rounded to oval (trans-versely elongated) with a diameter of 2 to 2.5 mm. It lies marginally along the interradial suture of interambulacrum 5 between the 3rd pair of postbasicoronal plates.

Differential diagnosis:

The type species of *Parascutella*, *Ps. leognanensis* (Lambert, 1903) from the Early Miocene of Western France, differs from *Ps. paulensis* by its more strongly developed marginal indentations, a prominent anal notch, and its straight, transversely truncated posterior margin [compare Agassiz, 1841a (under the name *Scutella subrotunda*) and Durham, 1953].

Parascutella gibbercula (de Serres, 1829), from the Late Badenian (Early Serravallian) of the Central Paratethys and the Serravallian to Tortonian of the Mediterranean, likewise differs from *Ps. paulensis* by its more strongly developed marginal indentations, its straight, transversely truncated posterior margin with distinct anal notch, its posteriorly eccentric maximum height, the wedge-shaped profile with strong inflation in adapical interambulacrum 5 and the smaller interporiferous zones [the latter feature, however, may be strongly influenced by the preservation of the material, see Schaffer (1962: 136)].

Parmulechinus hoebarthi (Kühn, 1936) from the Early Eggenburgian (Early Burdigalian) differs from *Parascutella paulensis* by its shorter petals (*c.* half the length of the corresponding test radius), its antero-posteriorly elongated outline, its strong anal notch, and the different position of its peristome (between the 4th or 5th pair of postbasicoronal plates).

For the difference between *Ps. paulensis* and *Ps. jaquemeti* (de Loriol, 1902), *Ps. striatula* (de Serres, 1829), *Ps. michaleti* (Lambert, 1906) [Durham (1955: 153) placed this species in *Parmulechinus*, but Philippe (1998) refers it to the genus *Parascutella*; until the oral plating is investigated, however, this question cannot be answered], *Ps. depereti* (Lambert, 1912), and *Ps. savini* (Lambert, 1915) the reader is referred to Philippe (1998).

Discussion:

Because of the marginal position of the periproct between the 3rd pair of postbasicoronal plates and the long petals (two third of the corresponding test radius), the specimens studied are placed into the genus *Parascutella* Durham, 1953. Based on the high similarities of test features Kroh (2002a: 306-307) placed the species *Scutella media* Schaffer, 1962 into the synonymy of *Ps. paulensis* (Agassiz, 1841). According to Cottreau (1913a: 53) and Philippe (1998: 129-133) *S. lorioli* Lambert 1906 and *S. deydieri* Lambert, 1912 are also junior synonyms of *Ps. paulensis*.

Philippe (1998: 129-141) attributed a number of specimens (e.g. the specimens on pl. 13, figs. 9-10) from the Serravallian of Southern France to *Ps. paulensis* which show a number of differences to the Burdigalian material of *Ps. paulensis* (compare Cottreau, 1913a: pls. 3, figs. 1-9 and pl. 4, figs. 1-6). These specimens are more similar to *Ps. gibbercula* as they show a distinct anal notch, more prominent marginal indentations, and a wedge-shaped profile. The attribution of these specimens to *Ps. paulensis* seems to be questionable.

Occurrence:

Austria: Late Eggenburgian (Early Burdigalian), Karpatian (Late Burdigalian)

Molasse Zone: Limberg (Zogelsdorf Fm., Hengl quarry), NÖ [NHMW]

Vienna Basin: Niederkreuzstetten, NÖ, Austria (pp LAUBE, 1871); Neubau, near Niederkreuzstetten, NÖ, Austria (SCHAFFER, 1959, 1962; KROH, 2002a, 2003b; [NHMW])

Mediterranean: Aquitanian-Burdigalian, ? Langhian, ? Serravallian

Rhône Basin: Aquitanian: Carry-le-Rouet (Nerthe), Sausset-les-Pins (Nerthe); all in France (ROMAN, 1974; NEGRETTI, 1984; PHILIPPE et al., 1990; PHILIPPE, 1998)

Burdigalian: Istres, Manosque, Montpezat, Juvignac, Beaucaire, Buoux, Céreste, Ménerbes, Montjustin, Saigon, Saint-Martin-de-Vastillon, Reillanne, Viens, Bllène, Clansayes, Grignan, Montbrison-sur-le-Lez, Saint-Paul-Trois-Châteaux, Saint-Restitut, Solérieux, Suez-la-Rousse, Taulignan, Autichamp, Crest, Divajeu, Bouvante (Vercors); all in France (LAMBERT, 1912; COTTREAU, 1913a; ROMAN, 1974; PHILIPPE, 1998)

? Langhian: Châteauneuf-Miravail, France (PHILIPPE, 1998)

? Serravallian: Istres; Lambesc, Rognes, Salon-de-Provence, Ansouis, Cadenet, Curcuron, la Motte-d'Aigues, Sannes, Mérindol-les-Oliviers, Puyméras; all in France (PHILIPPE, 1998)

Parascutella ? sp.

	1877	*Scutella* – FUCHS: 667
	1877	*Scutella* – KARRER: 262
	1894	*Scutella* (*subrotunda* LAM?) – LÖRENTHEY: 59
	1908	*Scutella* – SCHAFFER: 36
	1924	*Scutella* – ABEL: 60
	1924b	*Scutella* – SCHAFFER: 485
	1927a	*Scutella* – SCHAFFER: 78
	1943	*Scutella* – JANOSCHEK in SCHAFFER: 445
	1951	*Scutella* – JANOSCHEK in SCHAFFER: 563
	1961	*Scutella* –SCHAFFER: 149
	1961	*Scutella* – WESSELY: 319
	1965	Scutellen – FUCHS: 169
	1968	*Scutella* sp. – SCHMID: Anhang
	1974	*Scutella* – THENIUS: 69
	1978b	*Scutella* – PAPP et al.: 33
	1983	*Scutella* – DULLO: 19
	1985	*Scutella* – TOLLMANN: 501
	1987	*Scutella* (?*Parascutella*) sp. – MĄCZYŃSKA: 146-148; pl. 3, fig. 8
pp	1987	*Echinolampas* sp. – MĄCZYŃSKA: 147, 149; pl. 4, fig. 5
?	1991	*Scutella* sp. – ROETZEL et al.: 48
	1991b	*Parascutella* – NEBELSICK et al.: 119
v.	2002b	*Parascutella* sp. – KROH: 12
	2004	*Scutella* – RADWAŃSKI & WYSOCKA: 384

Material:

Early Badenian (Langhian) – Gainfarn, NÖ, Austria
NHMW: 2 fragments (NHMW 2004z0076/0066-67)
NEITZ coll.: 1 fragment (no inventory number)

Badenian (Langhian-Serravallian) – Neckenmarkt, Burgenland, Austria
NHMW: 5 fragments (NHMW 2004z0001/0057)

Discussion:

A number of literature records of the genus "*Scutella*" from the Neogene of the Central Paratethys could not be referred to any of the species mentioned above. In most cases the records lack any description and/or illustrations or are based on highly fragmentary material precluding identification to species level. The material is here referred to the genus *Parascutella* as the "true" *Scutella* (emend. DURHAM, 1953) was never recorded, nor were any specimens referable to that genus recovered during the preparation of the present paper. As observed already by ALI (1998: 543) it seems highly probable that *Scutella* never reached the Paratethys. It is also lacking from all well documented fauna from the northern coast of the Mediterranean [e.g. the Rhône Basin (see PHILIPPE, 1998)].

The "*Echinolampas* sp." fragment from the Korytnica Clays figured by MĄCZYŃSKA (1987: pl. 4, fig. 5) is most certainly a scutellid fragment. This can be concluded by the presence of elongated conjugate anisopores, which are typical for scutelline echinoids and the small diameter of the tubercles, which is much larger in echinolampadids.

Occurrence:

Austria: ? Egerian (Chattian-Aquitanian), Late Eggenburgian (Early Burdigalian), Badenian (Langhian-Early Serravallian)

Molasse Zone: Limberg (Hengl quarry), NÖ (NEBELSICK et al., 1991b); ? Steyregg (sandpit Treul), OÖ (ROETZEL et al., 1991)

Vienna Basin: Gainfarn, NÖ (KROH, 2002b); Hainburg Mts. (WESSELY, 1961); Mannersdorf, NÖ (SCHAFFER, 1908); Mödling (KARRER, 1877); Müllendorf, Bgld (SCHAFFER, 1961; DULLO, 1983); Rauch-stallbrunngraben, near Baden, NÖ (ABEL, 1924); Vienna Basin (FUCHS, 1877; SCHAFFER, 1924b, 1927a; THENIUS, 1974)

Eisenstadt-Sopron Basin: Leitha Mountains (JANOSCHEK in SCHAFFER, 1943, 1951; TOLLMANN, 1985); Oslip, Bgld (FUCHS, 1965; DULLO, 1983)

Danube Basin: Donnerskirchen, Bgld (SCHMID, 1968)

Oberpullendorf Bay: Neckenmarkt, Bgld ([NHMW])

Paratethys (non-Austrian occurrences): Early Badenian (Langhian)

Fore-Carpathian Basin: Korytnica Clays, Poland (MĄCZYŃSKA, 1987); Świniary, Poland (RADWAŃSKI & WYSOCKA, 2004)

Transylvanian Basin: Rachiş (= Oláh-Rákos), Romania (LÖRENTHEY, 1894)

Genus *Parmulechinus* LAMBERT, 1906d

= *Stenaster* LAMBERT, 1905 non *Stenaster* BILLINGS, 1885 (see HOTCHKISS, 1976 for further information on the latter)

Type-species: *Stenaster labriei* LAMBERT, 1905 = *Scutella striatula* AGASSIZ, 1841 non DE SERRES, 1829; by original designation (LAMBERT, 1910c: 63). *Stenaster labriei* was considered a junior synonym of *Scutella agassizi* OPPENHEIM, 1902 by LAMBERT (1915b: 19-29)

Diagnosis: Medium-sized to large, thin test; petals small, about 50 % of corresponding test radius. Ambitus broadly indented at the ambulacra. Six to seven ambulacral and four to five interambulacral coronal plates on oral surface. Interambulacra about half as wide as the ambulacra at the ambitus. Periproct marginal to submarginal, between the 4th or 5th pair of postbasicoronal plates (modified from DURHAM, 1953, 1955 and 1966).

Distribution: Oligocene to Early Miocene – Europe (DURHAM, 1966)

Parmulechinus hoebarthi (KÜHN, 1936)

(Figs. 38-41; Pl. 44, Fig. 2; Pl. 45, Figs. 1-7; Pl. 46, Figs. 1-4)

*v.	1936	*Scutella hōbarthi* nov. spec. – KÜHN: 40-43; figs. 2-3; pl. 1, figs. 1-3
v.	1936	Jugendform?. – KÜHN: 43; pl. 1, figs. 4
v.	1959	*Scutella* A (=*hōbarthi* Kühn). – SCHAFFER: 255

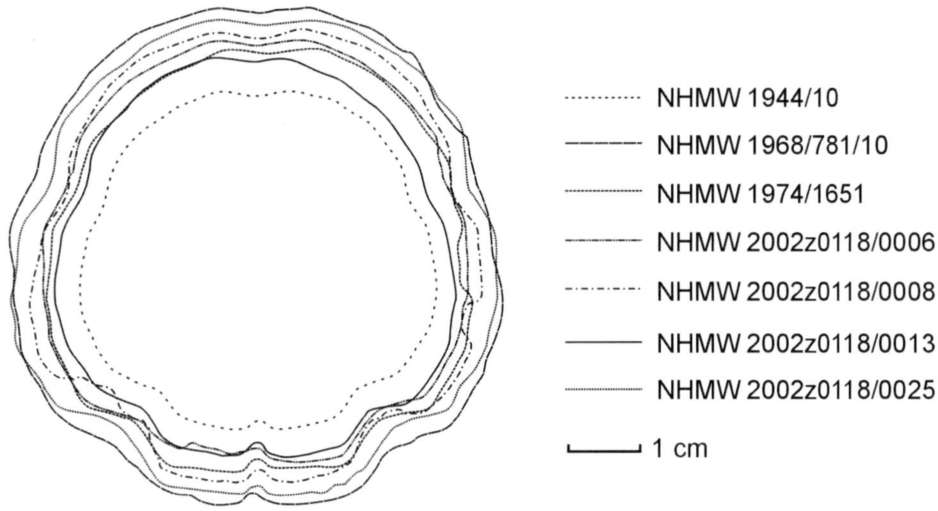

Figure 38: Variation in outline in *Parmulechinus hoebarthi* (KÜHN, 1936).

v. 1962 *Scutella hōbarthi* KUEHN – PAPP & THENIUS in KÜHN: 433
v. 1962 *Scutella hōbarthi* KÜHN – SCHAFFER: 151; pl. 17, fig. 1
v. 1971a *Scutella hōbarthi* KÜHN, 1936 – STEININGER: 595; pl. 2, fig.1-2
. 1971b *Scutella hōbarthi* KÜHN – STEININGER: 119
1971e *Scutella hoebarthi* KÜHN – STEININGER: 167
. 1974 *Scutella höbarthi* KÜHN – THENIUS: 47
v. 1978 *S.[cutella] hobarthi* KÜHN, 1936 – KIER & LAWSON: 67
v. 1991 *Parascutella hōbarthi* – NEBELSICK: 75-79
v. 1998 *"Scutella " hoebarthi* – SCHULTZ: 118; pl. 53, fig. 1
v. 1999 *Parascutella hōbarthi* – NEBELSICK: 351-352; fig. 4B-D, fig. 5B, D
v. 2002 *Parascutella hōbarthi* – NEBELSICK & KROH: 380-381; tab. 2

Type-material:
Holotype: specimen figured by KÜHN,1936: 40-43; pl. 1, fig. 1; housed at the Höbarth-Museum, Horn, NÖ under the inventory number 577 together with two other adult and 4 juvenile specimens of the same species
Locus typicus: Eichberg (= Achberg) east of Horn, near Maria Dreieichen, NÖ
Stratum typicum: "Scutellensand", Loibersdorf Fm.
Age: Early Eggenburgian (Early Burdigalian), Early Miocene

Material:
Early Eggenburgian (Early Burdigalian) – Eichberg near Maria Dreieichen, NÖ, Austria
- NHMW: 72 specimens [NHMW 1321/69/1-5, 134/19551-2, 1935(No.69)/1/94-97 (**figured specimens** of KÜHN, 1936), 1935/80/1-6, 1944/10, 1968/781/1-11, 1972/1546/1-8, 1974/1651, 1997z0178/0492 (5 specimens), 1998z0048/0058, 2002z0118/0001-27]
- Höbarth Museum: 3 adult and 4 juvenile specimens [No. 577 (among those specimens is the **holotype** of *S. hoebarthi*)]
- Krahuletz Museum: 8 specimens (without inventory numbers)
- IPUW: 1 juvenile specimen (IPUW 1992/195b), 26 adult specimens (IPUW no inventory nos.)
- NEBELSICK coll.: 35 specimens (without inventory numbers)

Dimensions (in mm):

Inv. No.	TL	TW	TH
Holotype (HM 577/1)	63.6	69.6	~9.2
HM 577/2	61.9	64.5	~9.8
HM 577/3	59.2	59.6	7.9
NHMW 1944/10	45.0	48.7	6.0
NHMW 1972/1546/7	48.4	50.2	7.5
NHMW 2002z0118/0013	52.9	54.8	7.0
NHMW 1974/1651	56.4	58.5	8.0
NHMW 1968/781/7	56.4	54.9	7.8
NHMW 2002z0118/0008	60.1	61.0	7.3
NHMW 1972/1546/5	62.4	63.3	8.1
NHMW 2002z0118/0006	62.9	67.1	9.4
NHMW 2002z0118/0004	63.1	65.8	6.8
NHMW 1968/781/1	64.5	64.3	8.8
NHMW 1955/1	66.2	66.8	8.5
NHMW 2002z0118/0014	68.9	69.4	9.5

Description:
Size and shape: The test is of medium size, ranging from 43 to 74 mm TL in the studied material (apart from two juvenile specimens measuring 5.2 respectively 8.6 mm). The outline of the test is subcircular to trapezoidal with distinct indentations in the ambulacra. The indentations are strongest in ambulacra I and V (Fig. 38). Additionally, there is a distinct, narrow anal notch. The aboral surface is slightly concave, the oral surface flat. The maximum height coincides with the apical disc. The test width ranges from 92 to 114 % TL (mean = 104 %), the height from 10 to 18 % TL (mean = 14.2 %).
Apical disc: The apical disc is tetrabasal and lies centrally on the aboral side [distance to the anterior margin ranges from 45 to 51 % TL (mean = 48 %)]. The madreporite is pentagonal and crowded with numerous small madreporic pores. Four gonopores lie at the apices of the madreporite, except in interambulacrum 5.
Ambulacra: The ambulacra are petaloid. The petals are more or less of equal length, only in some specimens the posterior paired petals may be up to 10 % longer than the others. The petals extend about 50 % of the corresponding test radius, only the posterior paired petals are slight longer (mean of 53 % TL). The petals are straight, open and have a lanceolate shape. They consist of closely spaced conjugate anisopores. The frontal petal is more widely open than the paired petals. The poriferous zones are very slightly depressed and each about 0.5 to

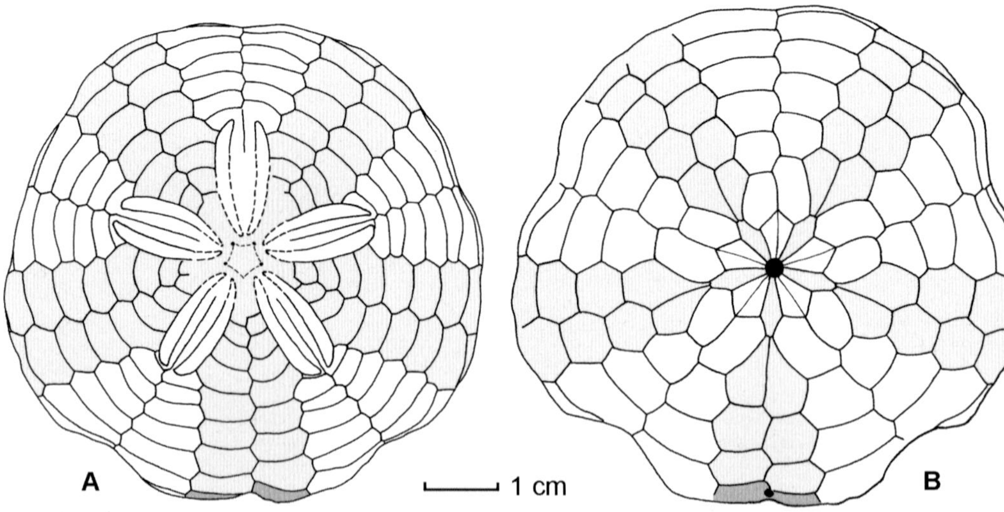

Figure 39: *Parmulechinus hoebarthi* (KÜHN, 1936): aboral (A) and oral (B) plating [Loibersdorf Fm., Eichberg, near Maria Dreieichen, NÖ (A: NHMW 1968/781/10, B: NHMW 2002z0118/0025)]. Interambulacra shaded in light grey; postbasicoronal plate pair enclosing the periproct shaded in dark grey.

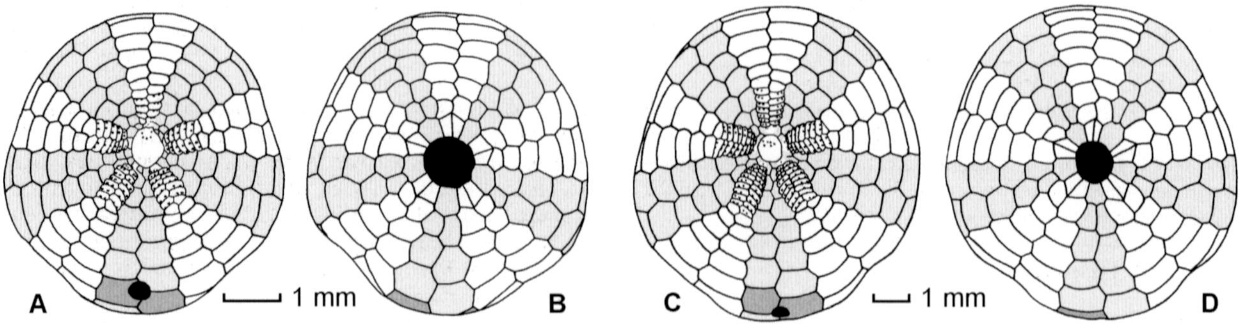

Figure 40: *Parmulechinus hoebarthi* (KÜHN, 1936): aboral (A, C) and oral (B, D) plating in juvenile specimens [Loibersdorf Fm., Eichberg, near Maria Dreieichen, NÖ (A, B: IPUW 1992/195b, B: NHMW 1936(No. 69)/1/95)]. Interambulacra shaded in light grey; postbasicoronal plate pair enclosing the periproct shaded in dark grey.

0.7 times as wide as the interporiferous zone. The shape of the basicoronal plates can be seen in Fig. 39.B.

Interambulacra: The interambulacra are about 0.6 to 0.7 times as wide as the ambulacra at the ambitus. They are densely crowded by small perforate crenulate tubercles.

Food grooves: The food grooves are simple, bifurcating at about 22 % of the corresponding test radius from the peristome. From this point on they run along the midline of each ambulacral plate row. Just before they reach the ambitus, they bend in direction of the interambulacra (Fig. 41).

Peristome: The peristome lies centrally on the oral surface. It is rather small, ranging from 1.0 to 3.3 mm in the studied material (corresponds to 1.7 to 5.2 % TL, with a mean of 3.9 %) and has a circular outline.

Periproct: The periproct very small, subcircular and lies marginally, directly at the posterior margin of the oral side. Its posterior margin is depressed and opens into the anal notch. It is 0.65 to 1.62 mm wide (this corresponds to 1.4 to 2.6 % TL, with a mean of 2.0 %). Concerning the plate sutures, the periproct lies between the 4th pair of postbasicoronal plates in interambulacrum 5 (Figs. 39.B, 40.A-C). In a juvenile specimen [NHMW 1935(No.69).I.95] of 8.6 mm TL it lies between the 5th pair (Figs. 40.C-D).

Differential diagnosis:

Parmulechinus agassizi (OPPENHEIM, 1902) [senior synonym of the type-species *Stenaster labriei* LAMBERT, 1905 (= *Scutella striatula* AGASSIZ, 1841a non DE SERRES, 1829, see LAMBERT, 1910c: 63) according to LAMBERT (1915b: 19-29)] differs from *Pm. hoebarthi* by its less strong marginal indentations (especially in ambulacra II, III and IV), by its shorter and more closed petals (around 44 % of the corresponding test radius), and its narrower interporiferous zones [compare illustrations in AGASSIZ (1841a: pl. 18, figs. 1-5) and CHAVANON (1974)]

Parmulechinus lamberti (AIRAGHI, 1901) from the Oligocene of Italy differs from *Pm. hoebarthi* by its shorter petals (ranging from 43 to 47 % of the corresponding test radius) and its different outline without prominent marginal indentations [based on AIRAGHI (1901), this species was referred to the genus *Parmulechinus* by DURHAM (1955:153)].

Parmulechinus paronai (AIRAGHI, 1901) and *Pm. isseli* (AIRAGHI, 1901), both from the Oligocene of Italy, are insufficiently known for an adequate comparison with the present species and need to be re-described. While the former has an outline similar to *Pm. hoebarthi*, the latter has a distinctly different, oval, antero-posteriorly elongated outline [both species were referred to the genus *Parmulechinus* by DURHAM (1955:153)].

Parmulechinus subtetragonus (GRATELOUP, 1836) from the Oligocene of France differs from *Pm. hoebarthi* by its shorter and lanceolate petals (ranging from 38 to 45 % of the corresponding test radius), its more tetragonal outline with stronger and broader indentations in ambulacra II, III and IV, the different ratio between the ambulacra and interambulacra at the ambitus (*c.* 1:0.5), and differently shaped basicoronal plates

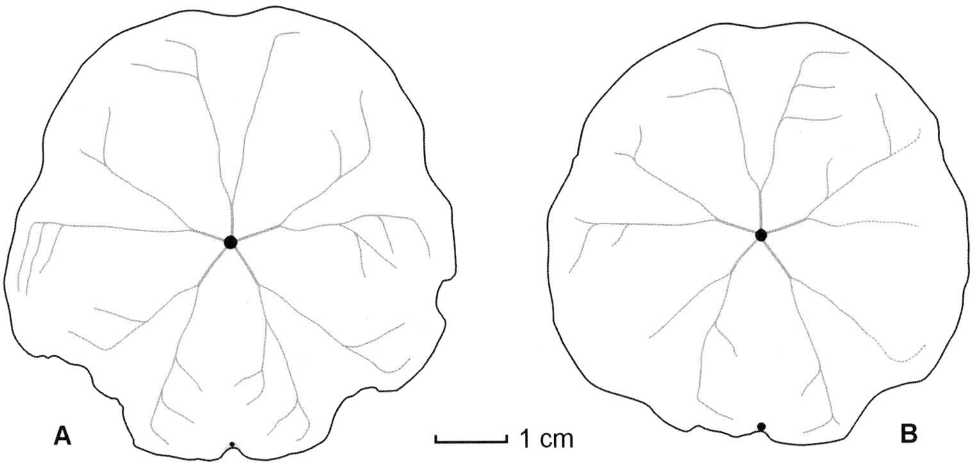

Figure 41: *Parmulechinus hoebarthi* (Kühn, 1936): arrangement of the food grooves [Loibersdorf Fm., Eichberg, near Maria Dreieichen, NÖ (A: NHMW 2002z0118/0008, B: NHMW 2002z0118/0006)].

[compare Grateloup (1836), Cotteau (1889-1894) and the oral plating in Durham (1955: fig. 18e)].

Parmulechinus subtrigonus (Koch, 1887) from the Late Oligocene of Romania differs from *Pm. hoebarthi* by its shorter petals (ranging from 37 to 41 % of the corresponding test radius), its different outline without prominent marginal indentations, the different ratio between the ambulacra and interambulacra at the ambitus (1:0.45), its marginal periproct (in adult specimens), and differently shaped basicoronal plates (interambulacral basicoronal plates about as wide as ambulacral basicoronal plates) [based on Koch (1887a) and material from the type-locality in the collection of the NHMW].

For the differences to the species *Parascutella gibbercula* and *Ps. paulensis* see above under that species.

Discussion:
Based on the inframarginal position of the periproct between the 4th to 5th pair of postbasicoronal plates, the short petals, the strongly indented ambulacra and the relationship between ambulacra and interambulacra at the ambitus, the species is referred to the genus *Parmulechinus*. The species name was modified from *höbarthi* to *hoebarthi* according to the ICZN rules (Article 32.5.2.1).

Occurrence:

Austria: Early Eggenburgian (Early Burdigalian)

Molasse Zone: Achberg (= Eichberg) near Maria Dreieichen (Loibersdorf Fm.), Horn Basin, NÖ (Kühn, 1936; Schaffer, 1959, 1962; Papp & Thenius in Kühn, 1962; Steininger, 1971a, b; Thenius, 1974; Nebelsick, 1991, 1999; Schultz, 1998; Nebelsick & Kroh, 2002; [NHMW]); Obernholz, NÖ (Steininger, 1971e)

Family Astriclypeidae Stefanini, 1911
Genus *Amphiope* Agassiz, 1840

Type-species: *Scutella bioculata* des Moulins, 1837; by subsequent designation Lambert (1907b: 49).
Diagnosis: Sand-dollar shaped test with two subcircular to transversely elongated (except in Oligocene species) lunules in ambulacra I and V; apical disc slightly eccentric anteriorly; four gonopores; well developed petals of subequal length; posterior ambulacra discontinuous adorally, others variable; interambulacra about as wide as ambulacra at the ambitus; primordial interambulacral plates much larger than primordial ambulacra; periproct on oral side, near the margin; food grooves bifurcating just outside primordial plates (modified from Durham, 1966).
Distribution: Late Oligocene to Late Miocene, ?Early Pliocene – Central and Southern Europe, Northern Africa (probably also in the Middle East, India and Angola)
Remarks: Although Mortensen (1948b: 413) and Comaschi Caria (1972: 42) considered *Amphiope* a subgenus of *Echinodiscus* this view found no acceptance among the echinoderm research community (e.g. Durham, 1955, 1966; Philippe, 1998,...). The genus *Kieria* Míhaly, 1985 (type-species *Kieria semseyana* Míhaly, 1985 from the Badenian of Budapest, Hungary) was established for small astriclypeid specimens with notches instead of lunules in the posterior ambulacra. This, however, is a common phenomenon during the ontogeny of astriclypeids and the type material of *Kieria semseyana* is here interpreted as juvenile *Amphiope* sp. as suggested already by Mooi (1989). Thus *Kieria* has to be considered as junior synonym of *Amphiope*.

Amphiope bioculata (des Moulins, 1837)

(Fig. 42.A-B; Pl. 46, Figs. 5a-b; Pl. 47, Figs. 1a-b; Pl. 48, Figs. 1-3)

* 1837 *S.[cutella] bioculata*. Nob. – des Moulins: 72
1841a *Amphiope bioculata* Ag. – Agassiz: 73; pl. 11, fig. 1-5
1846 *A.[mphiope] bioculata* Agass. – Pictet: 155
1847a *[Lobophora (Amphiope)] bioculata* Agass. – Agassiz & Desor: 136
1847a *[Lobophora (Amphiope)] elliptica* Desor – Agassiz & Desor: 136
1848 *Amphiope bioculata* (Ag.) – Gras: 40
1858 *Amphiope bioculata* Agass. – Desor: 236
1858 *[Amphiope] elliptica* Desor – Desor: 236
v. 1869a *Amphiope perspicillata* Ag. – Laube: 182
v. 1869a *Amphiope elliptica* Desor. – Laube: 182
v. 1870 *Amphiope perspicillata* Ag. – Laube: 314
v. 1870 *Amphiope elliptica* Ag. – Laube: 314
v. 1871 *Amphiope perspicillata* Agassiz. – Laube: 61
v. 1871 *Amphiope elliptica* Desor. – Laube: 61-62; pl. 16, fig. 5
v. 1880 *Amphiope* nov. sp. – Hoernes: 194
v. 1883 *Amphiope Styriaca* – Hoernes: 46-50; fig. 1
1892 *Amphiope bioculata*. – Gourret: 130
1892 *Amphiope elliptici*, Desor. – Gourret: 130
1895 *Amphiope Lovisatoi*, Cotteau, 1895. – Cotteau: 16; pl. 3, fig. 15
1907b *Amphiope bioculatas*, Desmoulins (*Scutella*) 1837 – Lambert: 50-53
1907b *Amphiope Lorioli* Lambert – Lambert: 56 (fide Philippe, 1998)

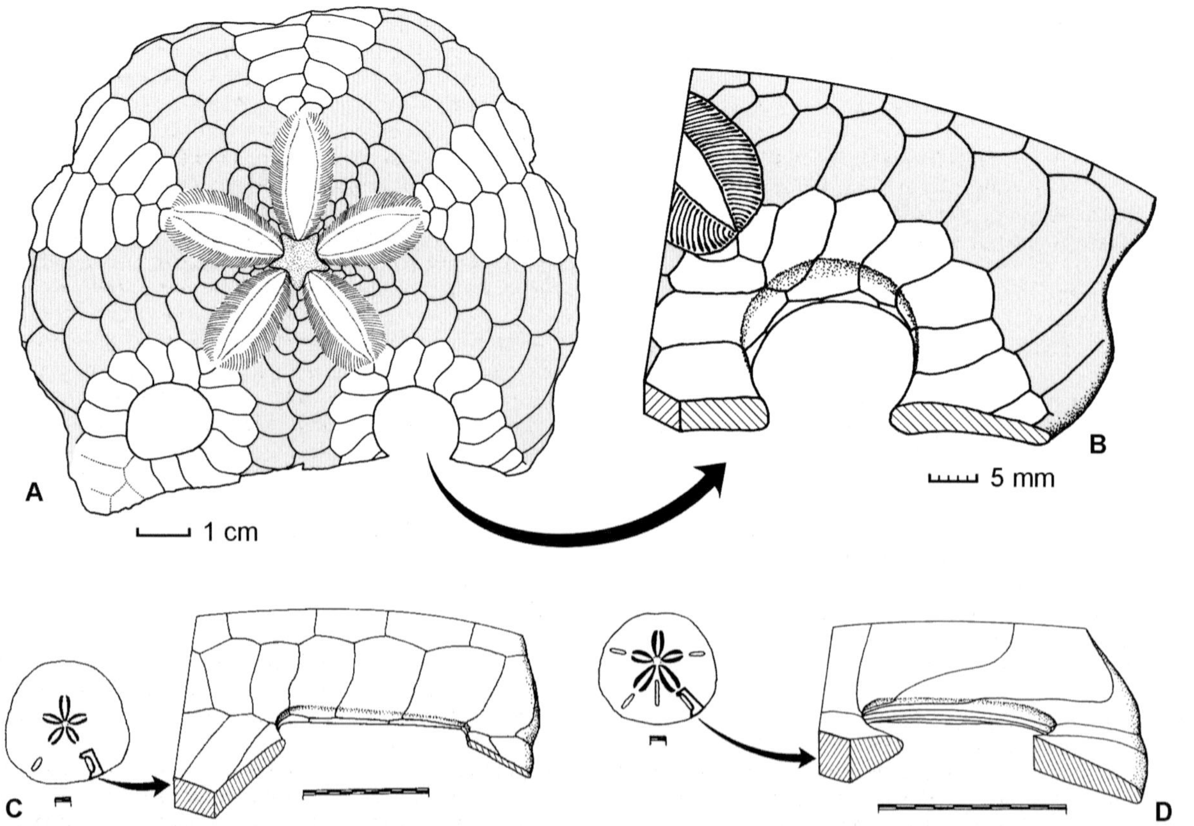

Figure 42: *Amphiope bioculata* (des Moulins, 1837): aboral plating (A) and lunule structure (B; oblique view) [holotype of *A. styriaca* Reuss, 1883 (UGP 1880.XX.5)]; Comparison with the cross-linked lunules of extant astriclypids [C: *Echinodiscus bisperforatus* (Leske), modified from Mooi (1989: fig. 30a); scale bars equal 10 mm] and the festooned lunules of mellitids [D: *Mellita quinquiesperforata* (Leske), modified from Mooi (1989: fig. 30b); scale bars equal 10 mm].

1907b *Amphiope Baquiei* Lambert – Lambert: 56-57; text-fig. (fide Philippe, 1998)
v. 1907b *Amphiope styriaca* Hoernes, 1883. – Lambert: 57-58
1910a *Amphiope elliptica* Desor – Lambert: pl. 3, fig. 24
1911a *Amphiope Montezemoloi* Lov. – Lovisato: 42-47; pl. 6 (3), figs. 1a-b
1912 *Amphiope bioculata* Desmoulins (*Scutella*), 1837. – Lambert: 75-77
1912 *Amphiope elliptica* Desor, 1847. – Lambert: 77-79; pl. 6, fig. 1-3
#v. 1912 *Amphiope Laubei* – Lambert: 79 [nom. nov. for *A. elliptica* Laube, 1871 non Desor, 1858]
1912 *Amphiope Sarasini* Lambert. – Lambert: 80-82; pl. 3, fig. 24 [under the name *A. elliptica*]; pl. 4, fig. 8-10
1912 *Amphiope Baquiei* Lambert, 1907. – Lambert: 83-84; pl. 5, fig. 6-8; pl. 6, fig.4
1912 *Amphiope transversifora* Lambert. – Lambert: 84-85; pl. 7, fig. 3-5
1912 *Amphiope Deydieri* Lambert. – Lambert: 85-86; pl. 6, fig. 5-7
1913a *Amphiope elliptica* Desor – Cottreau: 94-98; figs. 21-22; pl. 8, fig. 1
v. 1913a *Amphiope styriaca* Hoern.– Cottreau: 98, 99; fig. 23/3
1913a *Amphiope bioculata* Desm. (*Scutella*) – Cottreau: 135-139; pl. 5, fig. 1-8; pl. 6, figs. 1-12
1915a *Amphiope elliptica* Desor, 1847. – Lambert: 219-220
1915a *Amphiope transversifora* Lambert. – Lambert: 220
1915a *Amphiope Ludovici* Lambert. – Lambert: 220; pl. 6, figs. 1-3 (under the name *A. elliptica*); pl. 16, figs. 14-15
1915a *Amphiope Baquiei* Lambert, 1907. – Lambert: 221
1915a *Amphiope Sarasini* Lambert. – Lambert: 221
v. 1915 *Amphiope bioculata* Desmoul. sp. – Vadász: 122-124; fig. 18
1955 *Amphiope bioculata,* Desm. var. *montezemoloi* Lov. – Comaschi Caria: 184, pl.14-15
1962 *Amphiope* sp. – Schaffer: 159
v. 1963 *Amphiope elliptica* Desor. – Müller: 522; figs. 685A a-b
v. 1978 *Amphiope elliptica* Desor. – Müller: 569; figs. 697A a-b
1981 *Amphiope bioculata* Desmoulins – Mitrović-Petrović: 180; figs. 11a-c
1984 *Amphiope bioculata* Desmar.– Kókay et al.: 288
? 1984 *Amphiope ludovici* Lamb. – Kókay et al.: 288
? 1984 Echinoidea n. gen. és n. sp. – Kókay et al.: 288 [refers to the specimens named *Kieria semseyana* by Mihály (1985)]
1985 *Amphiope bioculata* (Desmoulins, 1837) – Mihály: 242
? 1985 *Amphiope ludovici* Lambert, 1915 – Mihály: 242-243; pl. 4, figs. 7-8
? 1985 *Kieria semseyana* n. sp. – Mihály: 243, 261; pl. 4, figs. 2-6
1998 *Amphiope bioculata* des Moulins – Cahuzac & Cluzaud: 426; pl. 2, fig. 3

	1998	*Amphiope bioculata* (DESMOULINS, 1837) – PHILIPPE: 152-167; pl. 16, figs. 1-8; text-figs. 12-17 [cum syn]
?	2001	*Amphiope* gr. *bioculata* – NÉRAUDEAU et al.: 52; tab. 1
v.	2002a	*Amphiope elliptica* DESOR, 1847 – KROH: fig. 1 [reproduced from LAUBE, 1871]
v.	2002a	*Amphiope* sp. – KROH: 308-309
v.	2003b	*Amphiope* sp. – KROH: 250

Type-material:
Scutella bioculata DES MOULINS, 1837:
Type-specimens: current whereabouts of the type material unknown.
Type area: "terrain tertiaire, Bordeaux et Sure près Bollène" (DES MOULINS, 1837: 72-73), the latter locality is Suze-la-Rousse, France according to PHILIPPE (1998:153)

Amphiope styriaca HOERNES, 1883:
Holotype: specimen UGP 1880.XX.5; housed in the collection of the Institute of Geology and Palaeontology, Karl-Franzens-University Graz, Austria
Locus typicus: Seggauberg (formerly Sekkauer Berg), near Leibnitz, Styria, Austria
Age: Early ? Badenian (Langhian), Middle Miocene

Amphiope laubei LAMBERT, 1912:
Holotype: specimen NHMW 1849.XXIII.39 figured by LAUBE (1871: pl. 16, fig. 5), preserved in the collection of the Naturhistorisches Museum Wien, Geologische Abteilung
Locus typicus: Niederkreuzstetten, NÖ, Austria
Stratum typicum: Korneuburg Fm.
Age: Karpatian (Late Burdigalian)
Remarks: LAUBE's figure may be a composite image of specimen NHMW 1849.XXIII.39 and 1849.XXIII.40.

Material:
Karpatian (Late Burdigalian) – Niederkreuzstetten, NÖ
NHMW: 3 specimens [NHMW 1849.XXIII.39 (holotype *A. laubei* LAMBERT, 1912 and reference specimen of *A. elliptica* of LAUBE, 1869a, 1870 and 1871), 1849.XXIII.40 (reference specimen to *A. perspicillata* of LAUBE, 1869a, 1870 and 1871), ? 1861.L.136 (poorly preserved specimen, specific identification uncertain)]
Early ? Badenian (Langhian) – Seggauberg, near Leibnitz, Styria, Austria
LMJ (ex. UGP coll.): 1 specimen [UGP 1880.XX.5 (**holotype** of *Amphiope styriaca* HOERNES, 1883)]

Foreign material for comparison:
Badenian (Langhian-Early Serravallian) – Budafok, Hungary
MAFI: 1 specimen [MAFI Ech-312 (figured specimen of VADÁSZ, 1915)]
Late Burdigalian – Gebel Gharra, near Suez, Egypt
NHMW: 6 specimens (NHMW 2001z0043/0024-27, 57-58) and 10 fragments (NHMW 2001z0043/0059, 102-103)

Dimensions (in mm):

Inv. No.	TL	TW	TH
NHMW 1849.XXIII.39	~55	~57	9.8
NHMW 1849.XXIII.40	70.4	>70	~10
NHMW 1861.L.136	>75	>78.5	13.6
UGP 1880.XX.5	> 100	108.7	11.6

Description:
Size and shape: Medium-sized to large, disc-shaped test with irregular circular outline and subequal length and width. Two large and subcircular to slightly triangular (Badenian material) or small and subcircular to elliptical (Karpatian material), lunules are present in ambulacra I and V. Marginal indentations are well developed in ambulacrum III and interambulacra 1 and 4. Faint indentations present in ambulacra II and IV. Maximum width situated subcentrally, slightly posterior of apical disc. In profile test low with the maximum height anterior of centre in the uppermost part of petal III. Ambitus very thin and sharp; oral surface flattened and slightly concave.
Apical disc: Apical disc approximately centrally or slightly anterior of centre, belonging to the monobasal type with 4 gonopores. Madreporite star-shaped.
Ambulacra: Adapically the ambulacra form well developed petals. These are lanceolate in shape, straight and closed distally. Petals similar in length, frontal one longest, anterior paired petals about 90 % and posterior paired petals about 80 % length of the frontal petal. Frontal petal about 50 % of the corresponding test radius. Anterior and posterior paired petals only 45 respectively 35 % of the corresponding test radius. The pores within the petals are typical, closely spaced conjugated anisopores. Apart from trailing podia (*sensu* MOOI, 1989) only minute microunipores can be found outside the petals. Poriferous zones slightly depressed and interporiferous zones slightly raised. Interporiferous zones as wide as single poriferous zone. Tuberculation on the aboral side, as far as preserved, is homogenous and consists of very small perforate, crenulate tubercles which are densely crowded.

On the oral side the ambulacra are depressed along their central sutures. The typical tubercle differentiation in locomotory and geniculate spine fields (compare MOOI, 1989: fig. 33b) can be observed. Generally tubercles are larger on the oral side than aborally.
Interambulacra: Two faint ridges are developed in each interambulacrum aborally. They run along the midline of each interambulacral column. Otherwise the surface is smooth and densely crowded with small crenulate, perforate tubercles, as on the ambulacra.

On the oral surface the interambulacra are slightly inflated, except adorally, where a radially elongated pit is present along their central suture. As stated above the typical sand-dollar tubercle differentiation can be observed.
Food Grooves: Food grooves well developed but restricted to the oral surface. Primary branching occurs close to the peristome, at the edge of the primordial plates, secondary branching near the margin. No food grooves reach the margin. The posterior pair of the food grooves runs around the lunules.
Lunules: In each one of the posterior ambulacra a subcircular to irregularly triangular (Badenian material) to slightly transversely elongate (Karpatian material) lunule is found. They are situated about halfway between the tip of the posterior paired petals and the margin of the test. Lunule structure corresponds to that of *Echinodiscus* [see Fig. 42.C; from MOOI (1989)] with cross linked lunule wall (Fig. 42.B).
Peristome: The peristome is situated subcentrally to slightly anterior of the centre, just below the apical disc. It is small (3.1 mm in diameter) and subcircular to subpentagonal in outline with food grooves leading into it.
Periproct: The anus is preserved in a single specimen only (NHMW 1849.XXIII.40). There it is small and subcircular, lying near the margin in interambulacrum 5, between the first and second pair of postbasicoronal plates.

Differential diagnosis:
Currently there is only a single other species of *Amphiope* that is described and illustrated in a modern way and thus allows comparison with the species investigated here. Most other nominal species are in serious need of re-examination and revision, a task which is out of the scope of the present contribution.

A. boulei COTTREAU, 1913 from the Aquitanian of the Southern Rhône Basin differs from *A. bioculata* by its smaller petalodium (i.e. in *A. bioculata* the tips of the posterior petals nearly reach the lunules, regardless of specimen size, whereas in *A. boulei* there is a large gap between posterior petal tips and margin of the lunules), small and axially elongated lunules (whereas they are large and subcircular or transversely elongated in most *A. bioculata*) and more distinct anal notch but otherwise less indented margin (compare re-description of *A. boulei* by PHILIPPE, 1998).

Discussion:

With approximately 40 nominal species (see e.g. LAMBERT, 1907b; LOVISATO, 1911a, 1914a; COTTREAU, 1913a; LAMBERT & THIÉRY, 1909-1925; COMASCHI CARIA, 1955; KIER & LAWSON, 1978) *Amphiope* is another Miocene genus with poorly resolved taxonomy. Most of the nominal species are in serious need of revision and many may turn out to be synonymous as indicated by investigations on the forms of the Rhône Basin (PHILIPPE, 1998) where just 2 out of 7 described species proved to be valid.

LAMBERT (1907b: 58) remarked to *A. styriaca* that *"et elle paraît distinguer du* A. bioculata *par ses zones interporifères plus effilées."* ("... and it appears to be distinguished from *A. bioculata* by its more frayed/elongate/lanceolate interporiferous zones."). It is not quite clear what LAMBERT really meant with *"plus effilées"*. The present author cannot see any marked differences between the interporiferous zones in the holotype of *A. styriaca* and those of *A. bioculata* (based on material from the Late Burdigalian of Egypt in the NHMW coll., and on the photographs in PHILIPPE, 1998). The only obvious difference in the holotype is that is that its aboral surface is slightly leached due to diagenetic alteration, resulting in slightly enlarged pores and marked visibility of the sutures. In fact, the type specimen of *Amphiope styriaca* falls well within the variation of "population" D of *Amphiope bioculata* from the Serravallian of Beaume, near Rognes (Bouches-du-Rhône) described and illustrated by PHILIPPE (1998: 161-166, fig. 17, pl. 16, figs. 5, 8a-b). It is especially similar to the large forms of this "population" and to the material from the Middle Miocene of Sardinia described as *Amphiope montezemoloi* by LOVISATO (1911a: 42-47). This species is characterised by its large size, irregularly rounded to slightly elliptical outline, large subtrigonal lunules and transversely truncated posterior margin. It thus differs morphologically from typical *A. bioculata*, which has a more regularly shaped outline (with distinct indentations in ambulacra II, III and IV) and slightly oval, laterally elongated lunule. PHILIPPE (1998: 166), however, based on an analysis of a large number of specimens from 4 different "populations" of different provenience and age found that both outline of the test and shape of the lunules are extremely variable even within a single horizon/outcrop. Similarly, features as the position of the periproct and the eccentricity of the apical disc may vary significantly. This could also be confirmed by an investigation of a number of specimens from the Late Burdigalian of Egypt by KROH (unpublished data). Thus *A. styriaca* is considered a junior synonym of *A. bioculata*.

LAUBE (1871: 61) reported two species of the genus *Amphiope* "aus dem Sande in den Leithakalkschichten von Niederkreuzstetten". SCHAFFER (1962), however, stated that LAUBE's material did not come from sands within the [Badenian (Langhian-Early Serravallian)] Leitha Limestone, but from the "Helvetian" [Karpatian (Late Burdigalian)] sands outcropping in the area around Niederkreuzstetten, NÖ. This seems to be correct, as the preservation of the available specimens is highly similar to that of *Parascutella paulensis* collected from the Karpatian sands of Niederkreuzstetten.

Although initial presumed to be lost (KROH, 2002a) Laube's material could be recovered from the NHMW collection since. The whereabouts of SCHAFFER (1962) material is still unknown. The three specimens of LAUBE are rather similar and differ mainly by there different preservation and shape of the lunules. In the specimen (NHMW 1849.XX.III.39) figured as *A. elliptica* by LAUBE (1871) the lunules are very small and almost circular (contrary to his figure), in the two others the lunules are slightly larger and elliptical. LAUBE's attribution to *A. elliptica* (NHMW 1849.XX.III.39) and *A. perspicillata* (NHMW 1849.XX.III.40) was based on determinations made by Hardouin MICHELIN during his visit in Vienna in 1857 or 1858 (MICHELIN, however, had crossed out *perspicillata* on the label of specimen NHMW 1849.XX.III.40 and attributed this specimen to *A. elliptica* too).

For the specimen figured as *Amphiope elliptica* by LAUBE (1871) LAMBERT (1912: 79) established the name *A. laubei* as he considered it specifically distinct from *A. elliptica*. According to him *A. laubei* differs from *A. elliptica* (a synonym of *A. bioculata* according to PHILIPPE, 1998) by *"à pétales saillants et étroites lunules, plutôt ovales qu'elliptiques, nettement transverses et à bords tuméflés, a sa face supérieure plus tourmentée"* (LAMBERT, 1912: 79). Yet when LAUBE's material is compared with the French specimens described in detail by PHILIPPE (1998) it is evident that it falls well within the variability of the French specimens. In particular the material ascribed to population "B" from the Burdigalian of Saint Christobal (PHILIPPE, 1998: 161, fig. 15, pl. 16, fig. 6) is extremely similar. Contrary to LAMBERT's statement the petals and lunule margin of LAUBE's specimen are much less swollen than the figure [LAUBE, 1871: pl. 16 (1), fig. 5] suggests and aboral surface is very similar to that of French specimens of *A. bioculata*. Consequently *A. laubei* is placed into the synonymy of *A. bioculata*.

As mentioned above, *Kieria semseyana* MÍHALY, 1985 is interpreted as juvenile *Amphiope* sp., it is likely that it represents juveniles of *A. bioculata*, since it is morphological very similar to those (compare with PHILIPPE, 1998: figs. 16a-e) and co-occurs with adult specimens of *A. bioculata*. Similarly the specimen determined as *A. ludovici* from the same locality by MIHÁLY (1985) is most probably also a juvenile specimen of *A. bioculata*.

Although SEILACHER (1979: p. 198 and fig. 8) reports the structure of the lunules in *Amphiope* to correspond to the festooned type (see Fig. 42.D) this is not the case in the present species (see Fig. 42.B). MOOI (1989) considered the lunules of *Amphiope* to be of the cross-linked type, as in *Echinodiscus* (see Fig. 42.C), which is supported by the material studied by the present author.

Occurrence:

<u>Austria</u>: Karpatian (Late Burdigalian), Late Badenian (Early Serravallian)

Vienna Basin: Niederkreuzstetten, NÖ, Austria (LAUBE, 1869a, 1871; MÜLLER, 1963, 1978; KROH, 2002a, 2003b; [NHMW]); Neubau, near Niederkreuzstetten, NÖ, Austria (SCHAFFER, 1962)

Styrian Basin: Seggauberg, near Leibnitz, Styria, Austria (HOERNES, 1880, 1883; LAMBERT, 1907b)

<u>Paratethys (non-Austrian occurrences)</u>: Late Badenian (Early Serravallian)

Great Hungarian Basin (Pannonian Basin): Budafok, Hungary (VADÁSZ, 1915); Budapest-Gyakorló, Pest, Hungary (KÓKAY et al., 1984; MIHÁLY, 1985)

Zala, Sáva and Dráva Basins: unnamed locality in former Yugoslavia (MITROVIĆ-PETROVIĆ, 1981)

<u>Mediterranean</u>: Aquitanian – Serravallian

Rhône Basin: Aquitanian ("Massalien"): Littoral de la Nerthe (Carry-le-Rouet, Sausset-les-Pins), France (DESOR, 1858; GOURRET, 1892; LAMBERT, 1912; COTTREAU, 1913a; LAMBERT, 1915a; PHILIPPE,1998)

Burdigalian: Littoral de la Nerthe (Martigues, Sausset-les-Pins), France (PHILIPPE, 1998); Bassins de Apt-Reillanne-Forcalquier (Céreste, Mane, Reillanne, Villemus), France (PHILIPPE, 1998); Bassin de Vairéas-Visan (Bollène, St Paul-Trois-Châteaux), France (AGASSIZ, 1841a; DESOR, 1858; LAMBERT, 1912, 1915a; PHILIPPE, 1998); Sillon pénalpin/ bassin de Crest (Crest), France (PHILIPPE, 1998);

Langhian: Bas-Languedoc occidental (Aspiran, Lespignan,

Nissan-lez-Ensérune), France (Philippe, 1998); Vallée du Jabron (Chateauneuf-Miravai), France (Philippe, 1998)

Serravallian: Secteur des étangs (Istres, St Chamas), France (Philippe, 1998); Bassin de Aix (Lambesc, Rognes, Vernègues), France (Philippe, 1998); Bas-Languedoc (Aspiran, Bassan, Boujan, Cruzy, Lapalme, Nissan-lez-Ensérune, Quarante, St Félix de-Lodez), France (Lambert, 1910a; Philippe, 1998); Bordure Sud-Luberon (Ansouis, Cabnères-de Aigues, Cucuron, la Motte-de Algues, Sannes, Vaugines), France (Lambert, 1912; Philippe, 1998); Bassin de Digne (Tanaron), France (Philippe, 1998); Bassin de Faucon-Mollans-Malaucène (Puyméras, St-Romain-en-Viennois), France (Philippe, 1998); Bassin de Vairéas-Visan (St Paul-Trois-Châteaux, Suze la-Rousse), France (Philippe, 1998)

Western Mediterranean: Cruzy, near Nissan, Herault, France (Lambert , 1912; Cottreau, 1913a; Lambert, 1915a); Ploaghe à Chiaramonti, Sardinia (Cotteau, 1895); S. Giorgio, Sardinia (Lovisato, 1911a; Comaschi Caria, 1955)

Eastern Mediterranean: Gebel Gharra, Eastern Desert, Egypt (Kroh, unpublished data)

Atlantic Ocean: Aquitanian to Burdigalian, Serravallian [some authors, e.g. Chavanon (1974), refer the material from this region to other species of *Amphiope* (i.e. *A. ovalifora* des Moulins, 1837, *A. caupianensis* Chavanon, 1974)]

Aquitaine Basin: Bordeaux, France (Agassiz, 1841a); Bougue, Landes, France (Cahuzac & Cluzaud, 1988); Faluns de Touraine, France (Agassiz, 1841a; Agassiz & Desor, 1847a; Desor, 1858); Gornac, Gironde, France (Agassiz & Desor, 1847a); Saint-Maure, Indre-et-Loir, France (Agassiz & Desor, 1847a; Desor, 1858)

Additional clypeasteroid species reported from the Central Paratethys

Clypeaster barcinensis ? Lambert, 1906

(Pl. 32, Fig. 4)

?* 1906a *Clypeaster barcinensis* – Lambert: 84-85; pl. 6, fig. 7
? 1927a *Clypeaster barcinesis*, Lambert, 1906 – Lambert: 11-12; pl. 5, figs. 1-2
. 1935 *Stolonoclypus* sp. – Kalabis: 275; figs. 1a-b [x-ray]
. 1938a *Stolonoclypus* sp. – Kalabis: 4, 10
. 1938b *Stolonoclypus* sp. – Kalabis: 7, 8, 9-10
. 1949 *Stolonoclypus barcinensis* (J. Lambert, 1906) – Kalabis: 54-56, 111-113; pl. 8, figs. 1-5

Type Material:
Syntypes: specimens figured by Lambert (1906a: pl. 6, fig. 7; 1927a: pl. 5, figs. 1-2); housed at the Museum de Geologia de Barcelona
Locus typicus: Montjuich, Catalonia, Spain
Age: Tortonian, Late Miocene according to Lambert (1927a: 12)

Material:
Late Badenian (Early Serravallian) – Devínska Nová Ves, Slovak Republic
NHMW: 2 fragmentary specimens (NHMW 2003z0048/0001 to 2)

Dimensions (in mm):

Inv. No.	TL	TW	TH
NHMW 2003z0048/0001	> 130	120.4	~ 25
NHMW 2003z0048/0002	> 92	>96	~ 29

Description:
Size and shape: The test is of medium to large size, with an antero-posteriorly elongated sub-pentagonal outline. Test width is approximately 90 % of TL. The margin of the test is thin but rounded and shows shallow but distinct marginal indentations in the interambulacra 1, 4 and 5. The maximum width lies anterior of the centre, where ambulacral columns IIa and IVb reach the ambitus. In profile the test is low with a slightly inflated petaloid area. The apex coincides with the apical disc and lies approximately centrally. The oral surface is distinctly flattened with a shallow and very wide infundibulum.
Apical disc: The apical disc lies subcentrally on the aboral side of the test. It is not preserved in the investigated specimens.
Ambulacra: The petals are ellipsoidal in shape, straight and moderately closed distally (width of the IPZ 4 to 6 mm distally). The anterior paired petals are shortest, the frontal and the posterior ones subequal in length. The length of the petals is about 53 to 58 % of the corresponding test radius. The poriferous zones are depressed and moderately wide. The pore pairs are conjugated anisopores. The interporiferous zones are inflated, medially flattened and range from 2.4 to 2.8 times a single poriferous in width. On the oral surface deep, simple, unbranched food grooves are present in the axes of the ambulacra. The tuberculation of the interporiferous zones is similar to that of the interambulacra, but tubercle density is higher. The tubercles of the interporiferous zones are arranged in series perpendicular to the medial suture of the ambulacra.
Interambulacra: Adapically the interambulacra are slightly inflated between the petals, forming weak keels between them. The tuberculation consists of evenly spaced, small, crenulate, perforate primary tubercles in sunken areoles. Due to the poor preservation of the surface miliary and secondary tubercles are not visible. Likewise no details of the oral tuberculation can be seen on Kalabis' specimens.
Peristome: The peristome lies subcentrally in a shallow but very wide infundibulum. Its shape is not visible in the present specimens.
Periproct: The periproct is poorly preserved in all examined specimens. It is situated inframarginally in interambulacrum 5, about 3 to 4 mm away from the posterior margin.
Internal support system: X-ray images of this species were provided by Kalabis (1935: fig. 1; 1949: pl. 8, fig. 5). These images show an extremely dense pillar and ridge system in the marginal parts of the test, a broad intestine canal and few pillars in the petaloid area.

Differential diagnosis:
C. altus Leske, 1778, from the Messinian of Malta differs by its higher test, different profile shape, broader petals, thicker margin and deeper infundibulum.

C. calabrus Seguenza, 1880, a co-occurring species, differs by its higher profile, thicker margin, concave oral surface with deep infundibulum, narrower petals, more inflated interporiferous zone, and deep marginal indentations.

C. campanulatus (Schlotheim, 1820) (and its phenotypes) differ from this species by its higher profile, longer and broader petals and narrower infundibulum.

C. folium Agassiz in Agassiz & Desor, 1847, a species restricted to the Badenian (Langhian-Early Serravallian) in the Paratethys can, although overally similar to *C. barcinensis* ?, be clearly distinguished from this species. *C. folium* has even shorter petals (44 to 48 % of the corresponding test radius vs. 53 to 58 % in *C. barcinensis* ?), a slightly anteriorly displaced apical disc and petaloid area, and deeper marginal indentations in all interambulacra.

C. intermedius des Moulins, 1837, a species occurring in the Burdigalian of the Rhône Basin and possibly the Paratethys, differs by its narrower and more strongly inflated petals, thicker margin, and higher profile.

C. latirostris Michelin, 1861, a species occurring in the Eggenburgian (Early Burdigalian) of the Molasse Zone, is distinguished from this species by its subequal test length and width, longer and less closed petals, and less inflated interporiferous zones.

C. neudorfensis LAMBERT, 1927, a co-occurring species, differs by its higher profile, lanceolate and longer petals, sharper margin and narrow infundibulum.

C. scillae DES MOULINS, 1837, a co-occurring species, differs by its higher profile, thicker margin, concave oral surface with deep infundibulum, narrower petals and more inflated interporiferous zone.

Discussion:
The types of *C. barcinensis* LAMBERT, 1906, are a fragmented specimen showing only the petaloid area and an internal cast, the oral side of which is still embedded in the matrix (presumably preserving the original shell of the oral surface), the aboral surface, however, is nearly completely lost. To the author's knowledge no additional topotypes have been described in the literature. It is thus very difficult to verify KALABIS' (1949) identification of material from the Vienna Basin with this species. The shape of the petals although overall similar, is slightly different in the Vienna Basin specimens, where the petals are shorter and wider. Additionally, the margin seems to be more rounded and slightly thicker in the latter. The conspicuous oblique arrangement of the tubercles in the interporiferous zones of the petals which LAMBERT (1906a: 84) regarded as characteristic feature of this species is not very evident. Instead an arrangement in series perpendicular to the perradial sutures is more prominent. Currently the identification of the Vienna Basin specimens with the Spanish species cannot be but tentative and has to be critically revised until more and better preserved material of both becomes available.

Occurrence:

Austria: not recorded until now, but occurrence very likely

Paratethys (non-Austrian occurrences): Late Badenian (Early Serravallian)
Vienna Basin: Devínska Nová Ves (= Neudorf an der March), Slovak Republic (KALABIS, 1935, 1938a, b, 1949; [NHMW])

Mediterranean: Tortonian, Late Miocene
Western Mediterranean: Montjuich, Catalonia, Spain (LAMBERT, 1906a, 1927a)

Clypeaster myriophyma ? POMEL, 1887

? 1885 *Clypeaster myriophyma* – POMEL: pl. B44, figs. 1-6
v 1915 *Clypeaster myriophyma* POM. – VADÁSZ: 147-149; fig. 40

Material:
Late Badenian (Early Serravallian) – Gârbova de Sus (= Felsö-Orbó), Romania
MAFI: 1 specimens (MAFI Ech 133)

Discussion:
Record based on a single specimen which is similar to members of the *C. campanulatus* group. It differs, however, by its distinctly oval outline and rounded posterior end (as opposed to a transversely truncated one in *C. campanulatus*). Whether or not it is really conspecific with the Algerian type material can not be decided currently.

Occurrence:

Paratethys: Late Badenian (Early Serravallian)
Transylvanian Basin: Gârbova de Sus (= Gîrbova de Sus, = Felsö-Orbó), Romania (VADÁSZ, 1915)

Mediterranean: Early or Middle Miocene
Western Mediterranean: Beni Chougran, Algeria (POMEL, 1885-87)

Unconfirmed records of clypeasteroids from Austria and the Central Paratethys

Additionally to the species described above several other species are mentioned in the geological literature. Because descriptions and illustrations are either insufficient or completely lacking and the material on which the records are based could not be located it is difficult to evaluate them. Most of these records might be misidentifications, although, it cannot be excluded that some might be valid.

Clypeaster airaghii LAMBERT, 1913

1949 *Clypeaster airaghi* LAMBERT – SCHOUPPÉ: 143

Reported occurrence: quarry in Retznei, near Ehrenhausen, Styria (SCHOUPPÉ, 1949).

Remarks: The name *C. airaghii* was established by LAMBERT (1913a: 108) for a specimen from the Early Miocene of the Colli Torinesi, Italy figured as *C. crassicostatus* Ag. by AIRAGHI [1901: pl. 12 (6), figs. 6, 6a]. The specimen figured by AIRAGHI is extremely similar to *C. scillae* and might well belong to that species. Austrian material referred to *C. airaghii* thus represents most probably *C. scillae*.

Clypeaster angustus POMEL, 1887

1915 *Clypeaster angustus* POM. var. – VADÁSZ: 180

Reported occurrence: Late Badenian (Early Serravallian) of Gârbova de Sus (= Felsö-Orbó), Romania (VADÁSZ, 1915)

Remarks: As for many others no illustration of this species was published by VADÁSZ. The description is not sufficient to confidently refer the record to any of the species discussed above.

Clypeaster coronalis LAMBERT, 1913

1915 *Clypeaster coronalis* LAMB. – VADÁSZ: 143; fig. 36
1969 *Clypeaster coronalis* LAMBERT – MIHÁLY: 256
1984 *Clypeaster coronalis* LAMB. – KÓKAY et al.: 288

Reported occurrence: Badenian (Langhian-Early Serravallian) of Iablanitja (= Bélajablánc, = Jablanica), Caraş-Severin, Romania (VADÁSZ, 1915); Late Badenian (Early Serravallian) of Budapest, Hungary (MIHÁLY, 1969; KÓKAY et al., 1984)

Remarks: It is questionable whether any of the specimens from the Badenian of the Central Paratethys referred to this species really belongs to the Early Miocene *C. coronalis* (now considered to be a junior synonym of *C. intermedius* by PHILIPPE, 1998). Unfortunately, none of the specimen could not be located in the MAFI collection and the specific identifications remain doubtful.

Clypeaster gregoryi LAMBERT, 1913

1949 *Clypeaster gregoryi* LAMBERT – SCHOUPPÉ: 143

Reported occurrence: quarry in Retznei, near Ehrenhausen, Styria (SCHOUPPÉ, 1949).

Remarks: This species was based on a single specimen from the Burdigalian (?) of Tourettes-sur-Loup, Vence, France (LAMBERT,

1913a: 110, pl. 9, figs. 6-9). It is unclear how SCHOUPPÉ's material looked like. If it was similar to the specimen figured by LAMBERT, it might have been a misidentified *C. campanulatus* (a flat phenotype, like e.g. forma *partschi*).

***Clypeaster megastoma mediterraneus* VADÁSZ, 1915**

1915 *Clypeaster megastoma* POM. var. *mediterraneus* n. var. – VADÁSZ: 179-180

Reported occurrence: Late Badenian (Early Serravallian) of Gârbova de Sus (= Felsö-Orbó), Romania (VADÁSZ, 1915)

Remarks: The only specimen available was, unfortunately, not illustrated by VADÁSZ (1915). It is thus unclear whether the name refers to a specimen of the *C. campanulatus* group or rather of *C. scillae* or *C. calabrus*. The mention of a large (22 mm wide), deep peristome indicates that the latter group is more likely. What VADÁSZ observed, however, was not the peristome but rather the infundibulum, which is quite large and deep in *C. scillae*, whereas it is narrow and much less deep in *C. campanulatus*.

***Clypeaster tarabellianus* GRATELOUP, 1836**

1865 *Clypeaster tarbellianus* GRAT. – KORNHUBER: 95

Reported occurrence: Devínska Nová Ves, Slovak Republic (KORNHUBER, 1865).

Remarks: The status of *C. tarabellianus* GRATELOUP, 1836 is unclear, it is only rarely mentioned in the systematic literature (e.g. MICHELIN, 1861) and a modern revision is lacking. GRATELOUP's (1836: pl. 1, figs. 5a-c) figure show a *Clypeaster* with highly vaulted petalodium (reminiscent of *C. campanulatus*) and strong marginal indentations in the interambulacra (similar to *C. folium* or *C. calabrus*). It is uncertain what species KORNHUBER refers to actually.

***Clypeaster* cf. *parvituberculatus* POMEL, 1887**

1915 *Clypeaster* cfr. *parvituberculatus* POM. – VADÁSZ: 152

Reported occurrence: Late Badenian (Early Serravallian) of Gârbova de Sus (= Felsö-Orbó), Romania (VADÁSZ, 1915)

Remarks: Based on a single poorly preserved specimen that was neither described nor illustrated by VADÁSZ. The specimen was not found in the MAFI collection and the record thus remains doubtful

***Clypeaster* cf. *petalodes* POMEL, 1887**

1915 *Clypeaster* cfr. *petalodes* POM.– VADÁSZ: 182-183

Reported occurrence: Badenian (Langhian-Early Serravallian) of Pécsvárad, Baranya, Hungary (VADÁSZ, 1915)

Remarks: Yet again another species that lacks sufficient information for revision and could not be located in the Hungarian collections. The occurrence of this species in the Central Paratethys is thus considered doubtful.

***Parascutella ? guebhardi* (LAMBERT, 1915)**

1988b *Scutella* cfr. *guebharti* – KÓKAY: 3 (unpaginated)
1991 *Scutella guebhardti* LAUBE [sic] – KÓKAY: 105
2003b *Parascutella guebhardi* (LAMBERT, 1915) – KROH: 250

Reported occurrence: Karpatian of Bántapuszta, near Varpalota, Hungary (KÓKAY, 1988, 1991)

Remarks: According to the description of LAMBERT (1915a) the periproct of *P. guebhardi* lies very close to the posterior margin. Albeit no plating pattern of the posterior interambulacrum is available it seems likely that the structural position of the periproct corresponds to that of *Parascutella* an the species is tentatively attributed to this genus here. KÓKAY (1988, 1991) reported this species without any description or illustration, so a literature based revision of this record is impossible. Until KÓKAY's material is re-examined the record must be considered doubtful.

***Parascutella ? lusitanica* (DE LORIOL, 1896)**

1988b *Scutella lusitanica* – KÓKAY: 3 (unpaginated)
1991 *Scutella lusitanica* LOR. – KÓKAY: 105
2003b *Parascutella lusitanica* (LORIOL, 1896) – KROH: 250

Reported occurrence: Karpatian of Bántapuszta, near Varpalota, Hungary (KÓKAY, 1988, 1991)

Remarks: Like the former species this species is also tentatively referred to *Parascutella* on basis of its inframarginal periproct. Like *P. ? guebhardi* this species was recorded without any description or illustration by KÓKAY (1988, 1991) and the record must be considered doubtful until specimen-based revision.

Order Cassiduloida CLAUS, 1880
Family Echinolampadidae GRAY, 1851
Genus *Echinolampas* GRAY, 1825

Type-species: *Echinus oviformis* GMELIN, 1788; by subsequent designation (POMEL, 1883: 62). Currently this species is considered as junior synonym of *Echinanthus ovatus* LESKE, 1778 (compare DÖDERLEIN, 1906; CLARK & ROWE, 1971; DOLLFUS & ROMAN, 1981).
Diagnosis: Medium to large forms, often with high test and elongate to circular outline; apical disc monobasal; moderately developed petals, sometimes lanceolate, open, or closing distally; poriferous zones usually unequal; interporiferous zones wide; single pores in ambulacral plates beyond petals; periproct inframarginal, transverse; peristome transverse, pentagonal; bourrelets well developed; phyllodes single pored, usually moderately developed; phyllodal pores from two to three series in each half-ambulacrum; buccal pores present; often large differences between tubercle density and size between oral and aboral surface; in many extant species a narrow, naked, granular zone is present in oral interambulacrum 5 (modified from KIER, 1962).
Distribution: Eocene to Recent – cosmopolitan (KIER, 1962)
Remarks: *Echinolampas* is one of the most diverse genera in the Cenozoic. According to ROMAN (1965) more than 290 nominal species have been established and since ROMAN's work 18 additional species and/or subspecies have been proposed (KIER & LAWSON, 1978). Several attempts were made to divide *Echinolampas* in subgenera or sections (e.g. LAMBERT & THIÉRY, 1921: 277-384, 1924: 385; MORTENSEN, 1948a: 272 ff.) but these were rejected by KIER (1962: 106-107).

The morphology and evolution of this genus was studied in detail by ROMAN (1965), who provided also an annotated list of the then known species of *Echinolampas*. The biogeography and possible migration pathways have been studied by ROMAN (1977). According to him the genus shows two phases of extreme diversification, in the late Middle Eocene and Late Miocene. The latter was followed by a drastic, rapid decline in the Pliocene, possibly related to climatic deterioration and habitat loss due to the collision between the African and Eurasian plates and resulting geodynamic processes (Roman, 1977). Mooi (1990a) provided an overview of the extant cassiduloids, summarising all available data on the 10 extant species of *Echinolampas*. Cassiduloid phylogeny was studied by Suter (1994a, b) and its relation to the clypeasteroids by Smith (2001), but relationships within this group are still largely unresolved and proposed trees have low boot-strap support.

In the Miocene of Austria seven species of the genus *Echinolampas* are recorded in this study – two in the Eggenburgian (Early Burdigalian), two in the Early Badenian (Langhian) and three in the Late Badenian (Early Serravallian). These species can be distinguished on base of test morphology, including features as relative length of the poriferous zones within the individual petals, relative test height, relative test width, peristome and periproct diameter in relation to test length, tubercle densities, number of tubercles across the interporiferous zones, development of bourrelets and position of the periproct. Similar features were used to distinguish extant species within this genus (e.g. Mortensen, 1948a; Mooi, 1990a). Many of these features show strong correlation which each other and features as tubercle density vary strongly with growth. In the scatter diagrams (e.g. Figs. 44, 46-48) individual groups are, however, well delineated.

Due to the large number of nominal species in this genus and the lack of adequate descriptions, serial measurements and material of most species the differential diagnoses had to be restricted to the species occurring in the Central Paratethys. Moreover, a complete revision of all *Echinolampas* species world-wide or even just the Mediterranean is definitely out of scope of this study. Especially since the subtle differences between species led to both extreme splitting on one hand and extreme lumping on the other. Hence, temporal and spatial distributions of most species are often blurred.

Ecology and biogeography: For a group of the size of *Echinolampas* surprisingly little is known about the biology of the extant species of this genus. Cram (1970) studied the larval development of the South African species *Echinolampas crassa* Bell, which took 41 to 43 days from fertilisation to metamorphosis. Between 30 to 35 days after fertilisation the pluteal swimming habit changes from a near-surface-swimming habit to a bottom-swimming habit (Cram, 1971).

The False Bay (near Cape Town, South Africa) population of *E. crassa* represents the most south and westward known occurrence of this species (Thum & Allen, 1975), no occurrences are known along the Atlantic coast of Southern Africa. Coming from the north the distribution limit of *Echinolampas* ends with the Cape Verde Islands, where *E. rangii* was reported. The False Bay population of *E. crassa* occurs at a depth between 12 and 24 metres (Cram, 1971; Thum & Allen, 1975, 1976). The animals are completely burrowed (3 to 5 cm sediment covering the apex). The substrate is shell gravel consisting of barnacle fragments, mollusc shells and sea-urchin spines. The carbonate content ranges from 75 to 95 %. Due to the prevailing southwesterly sea swell the sediment at False Bay forms macro-ripples with ripple amplitudes between 10 to 15 cm and wavelengths of 90 cm. According to the data obtained by repeated sampling through 13 months in the years 1973-74 by Thum & Allen (1975) *E. crassa* shows a preference for ripple slopes and an avoidance of troughs. Population densities ranged from 0.26 to 1.1 individuals per square metre, with higher densities in autumn and winter. Size-frequency distributions of this population showed a peculiar absence of juveniles and of specimens over 125 mm test length and a considerable variation during the survey interval. Thum & Allen (1975: 373) estimated a growth rate of about 0.5 cm per year for adult specimens. During the survey interval inshore sea temperature (measured at 3 m depth) ranged from 13 to 19° C, sea bottom temperature (measured at 15 m depth) 11 to 17° C. Reproduction (spawning) was correlated with the decrease of temperature immediately after the time of maximum temperature in December and January (Thum & Allen, 1976). Comparison of particle size between substrate (less than 7 % are < 0.5 mm in diameter, ~ 80 % are between 0.5 to 4 mm and usually less than 15 % are > 4 mm) and gut (less than 3 % are < 0.5 mm in diameter, ~ 93 % are between 0.5 to 4 mm and usually less than 3 % > 4 mm) samples showed a significant preference for particles between 1 and 2 mm (~44 to 54 % of the gut content) with no obvious increase of size of particles ingested in larger animals (although the size of the peristome increases significantly during growth; size range studied: 83 to 116 mm test length). Thum & Allen (1976: 31) suggested that the preference for this particular grain size in feeding partly explains the avoidance of ripple troughs.

McNamara & Phillip (1980c: 11-12, fig. 6) suggested that some species of *Echinolampas* might have a similar orientation in the sediment as *Rhyncholampas pacificus* (see Agassiz, 1873: 555), being buried into the sediment only to the tips of the petals, resulting in a tilted position due to the different length of anterior and posterior paired petals (based in part on staining patterns observed on the coronas of dead specimens). According to them this is supported by their observation that during the growth of *Echinolampas* the oral part of the test more or less maintains its profile, while the aboral surface becomes more domed, resulting in a similar depth of the peristome in the sediment through growth.

Depth distribution data of the extant species of *Echinolampas* summarised by Mooi (1990a) show that most species were recorded from depths between 8 and 400 m, with the exception of *E. rangii* from Senegal and the Cape Verde Islands which occurs between 1570 and 1670 m. *E. ovata* (Leske) occurs in the littoral zone of the Red Sea between 9 to 75 m depth (Mortensen, 1948a) and in Australia in the intertidal on fine carbonate sand without mud, where it is completely buried during the day and partly exposed at night (Vail in Mooi, 1990a: 70).

Echinolampas barcinensis Lambert, 1906

(Figs. 43.1, 44; Pl. 49, Figs. 1-2; Pl. 50, Figs. 1a-c)

.	1896	*Echinolampas hemisphaericus*, Agassiz (Lamarck) var. *maxima* – de Loriol: 40; pl. 12, fig. 1
*	1906a	*Echinolampas barcinensis* Lambert – Lambert: 90-93; fig. 2
?	1915	*Echinolampas hemisphaericus* Lam. var. *maxima* Lor. – Vadász: 208-209; fig. 95
v.	1915	*Echinolampas barcinensis* Lamb. – Vadász: 210-211; fig. 96
?	1915	*Echinolampas transylvanicus* n. sp. – Vadász: 214-215; fig. 100; pl. 11 (5), fig. 8
?	1915	*Echinolampas subconoideus* n. sp. – Vadász: 213-214; fig. 99; pl. 12 (6), fig. 7
?	1909	*Echinolampas Lamarmorai* Lovisato – Lambert: 131 (*nomen nudum* according to Roman, 1965: 270)
	1910	*Echinolampas Spanoi* Lov. – Lovisato: 142-144; pl. 18(3), figs. 1a-c [fide Lambert, 1927a and Roman, 1965]
	1912a	*Echinolampas Spanoi*. – Lovisato: 113
.	1927a	*Echinolampas barcinensis* Lambert, 1906 – Lambert: 24
	1936	*Echinolampas barcinensis* Lambert – Paucă: 143, 195; pl. 1, fig. 7
	1939	*Echinolampas laurillardi* Agassiz – Taviani: 38; pl. 2 (4), figs. 14a-b
	1953	*Echinolampas barcinensis* Lambert, 106. – Szörényi: 85
.	1965	[*Echinolampas*] *barcinensis* Lambert 1906c – Roman: 270
?	1969	*Echinolampas hemisphaericus* var. *depressa* Lambert – Mihály: 256
?	1984	*Echinolampas hemisphaeicus* Lamk. var. *depressa* Lamb. – Kókay et al.: 288

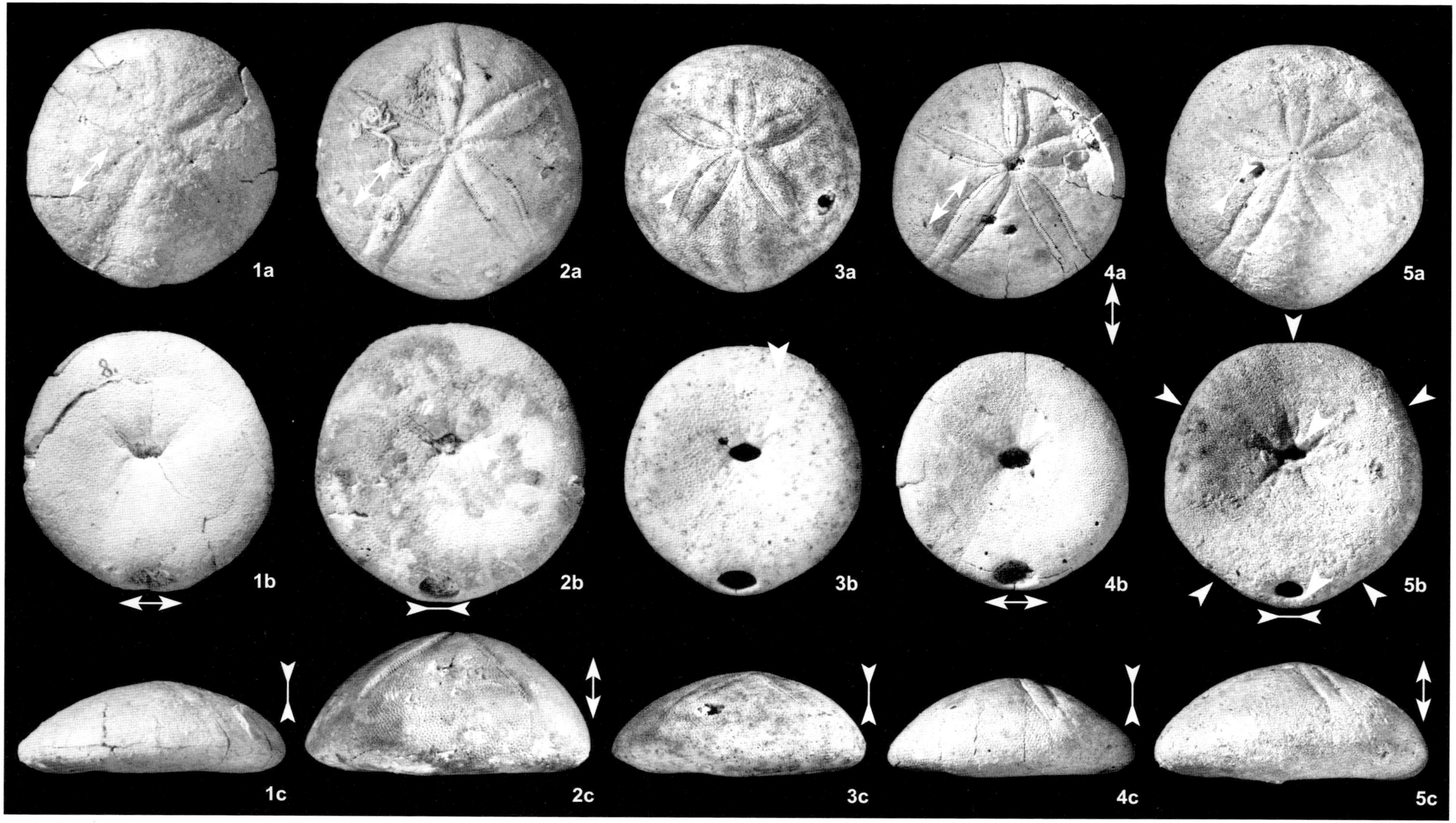

Figure 43: Comparison between the discussed *Echinolampas* species (arrows indicate important features)
1: *Echinolampas barcinensis* Lambert, 1906, Müllendorf, Bgld (NHMW 1997z0178/1743)
2: *Echinolampas hemisphaerica* (Lamarck, 1816), Budatéhény, Hungary (NHMW 1997z0178/2424)
3: *Echinolampas richardi* Desmarest, in Brongniart, 1829: Leognan, western France (NHMW 1914.VI.348b)
4: *Echinolampas sayni* Lambert, 1913, 1a-c: Gauderndorf, NÖ (NHMW 1850.IX.83)
5: *Echinolampas schultzi* Kroh, nom. nov., Gauderndorf, NÖ (NHMW 2003z0036/0001)

a: aboral view, b: oral view, c: lateral view (figures 1c, 4c and 5c mirrored for easier comparison)

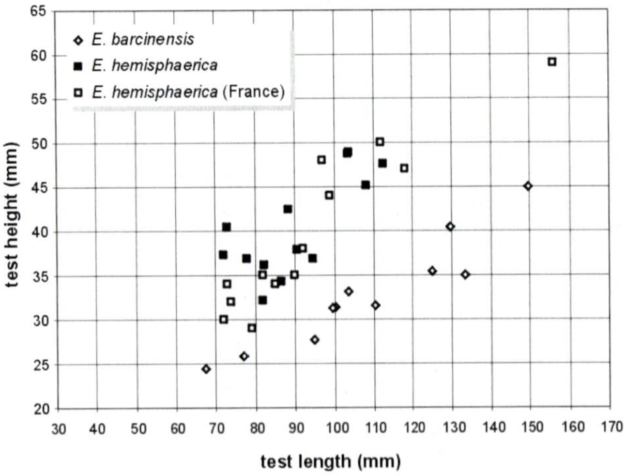

Figure 44: Morphometric comparison (test length vs. test width) between *Echinolampas hemisphaerica* (Lamarck, 1816) and *E. barcinensis* Lambert, 1906. Data of French *E. hemisphaerica* from the literature (Philippe, 1998) is represented by open symbols.

Type-material:
Holotype: the specimen described by Lambert (1906a: 90-93, fig. 2)
Locus typicus: Vilovi, near Panades, Catalonia, Spain
Age: "Helvétien inférieur", Middle ? Miocene

Material:
Early Badenian (Langhian) – Retznei [Weissenegg Fm., Lafarge quarry (formerly Perlmoser)], Styria, Austria
IPUW: 1 specimen (no inventory no.)
coll. Meindl: 1 specimen (no inventory no.)
Late Badenian (Early Serravallian) – Hof, Bgld, Austria
NHMW: 1 specimen (NHMW 2002z0181/0014)
Late Badenian (Early Serravallian) – Hornstein (= Szarvkö), Bgld, Austria
MAFI: 1 specimen [MAFI Ech 281 (reference material of *E. barcinensis* in Vadász, 1915)]
Late Badenian (Early Serravallian) – Mannersdorf, NÖ, Austria
NHMW: 4 specimens (NHMW 1978/2020/21-22, 2002z0181/0006-7)
Late Badenian (Early Serravallian) – Müllendorf (Mühlendorfer Kreide AG quarry), Bgld, Austria
NHMW: 4 specimens (NHMW 1997z0178/1743, 2003z0031/0001-4)
Late Badenian (Early Serravallian) – Schrattenberg, NÖ, Austria
NHMW: 1 specimen (NHMW 2002z0181/0008)

Foreign material for comparison:
Late Badenian (Early Serravallian) – Gârbova de Sus (= Felsö-Orbó), Romania
MAFI: 2 specimens [MAFI Ech 266 and 315 (reference material of *E. barcinensis* in Vadász, 1915)]

Dimensions: see Tab. 10

Description:
Size and shape: The test is large and has an oval, antero-posteriorly elongated outline. The maximum width lies subcentrally or slightly posterior. Test width ranges from 92.5 to 101.7 % TL in the studied specimens, with a mean of 95.8 %. In profile the test is low arched. Test height ranges from 26 to 35 % TL, with a mean of 30.7 % in the studied material. The maximum height coincides with the apical disc. The ambitus lies low and is relatively thin. The oral surface is very slightly concave.
Apical disc: The apical disc lies slightly anterior of the centre, about 42 to 45 % TL from the anterior margin. The apical disc is monobasal with four circular gonopores. The madreporite is pentagonal and crowded with numerous small madreporic pores. The gonopores lie at the apices of the madreporite. The ocular pores are small, often barely visible.
Ambulacra: Adapically the ambulacra are petaloid, moderately broad, straight and only slightly closed distally. The frontal petal extends about 60 to 70 % of the corresponding test radius, the anterior paired petals about 68 to 77 % and the posterior paired petals about 68 to 73 %. The poriferous zones within the frontal and the posterior paired petals are subequal in length. In the anterior paired petals, in contrast, the anterior poriferous zones (IIb and IVa) extend only about 81 to 91 % of the posterior ones (IIa and IVb). The ambulacral pores in the petals are conjugated anisopores, which are strongly conjugated and slightly oblique. The interporiferous zones are about four times as wide as a single poriferous zone at their broadest point. The poriferous zones are only very slightly depressed and the interporiferous zones are not inflated.

Adorally the ambulacra form broad, slightly depressed phyllodes. They consist of four series of unipores in each ambulacrum. Two adradial series consisting of large, closely spaced unipores and two perradial series consisting of smaller, slightly elongated unipores, which are more widely spaced. The buccal pores are slightly larger than the other phyllodal pores. In the adoral third of the phyllodes there is a double row of shallow pits (sphaeridial pits) running along the central suture.
Interambulacra: Interambulacra and ambulacra are virtually flush with each other adapically. They are covered by a homogenous tuberculation made up of crenulate, perforate tubercles with sunken areoles. The primary tubercles are very closely spaced on the aboral surface, their areoles nearly touching each other. On the oral surface the situation is similar, but the tubercles are larger and towards the peristome they become less closely spaced. No medial "naked zone" could be observed in interambulacrum 5, as reported in most extant species of *Echinolampas*. Adorally the interambulacra form very weakly inflated bourrelets.
Peristome: The peristome lies slightly anterior of the centre, about 45 % TL from the anterior margin. It has an oval, transversely elongated outline and is moderately large (usually 13 to 15 % TW).
Periproct: The periproct is situated inframarginally in interambulacrum 5. It is rather large (16 to 19 % of the test width wide) and has an oval, transversely elongated shape.

Differential diagnosis (see Figs. 43, 45-48):
As outlined above, *E. barcinensis* is characterised by its rounded to subpentagonal, only very slightly rostrate outline, low profile, weakly developed bourrelets, large periproct, weakly inflated interporiferous zones and only very slightly sunken poriferous zones.

For the difference to *E. schultzi* see below under that species.

For the difference to *E. sayni* see below under that species.

For the difference to *E. hemisphaerica* see below under that species (and Fig. 44). Additionally, a throughout discussion on the distinguishing characters between those two species can be found in Lambert (1906a: 90-93).

E. sp. 1, from the Badenian (Langhian-Early Serravallian), differs from *E. barcinensis* by its more elongated distinctly oval outline, higher profile, smaller peristome and periproct and much more sunken poriferous zones.

E. manzonii and *E.* aff. *manzonii*, both from the Early Badenian (Langhian) differ from *E. barcinensis* by their very unequal poriferous zones in the paired petals (among other features, but this is the most easily recognised).

Discussion:

The material described here clearly belongs to the species *E. barcinensis* LAMBERT, 1906, a species established for specimens from the Middle Miocene of Spain and Portugal. A specimen (NHMW 2002z0181/0011) from Portugal in the collection of the Naturhistorisches Museum Wien is virtually indistinguishable from the Austrian specimens. In 1927 LAMBERT (1927a) confirmed his earlier assumption that *E. hemisphaerica* var. *maxima* DE LORIOL, 1896 is conspecific with *E. barcinensis*. Material ascribed to *E. hemisphaerica* var. *maxima* by VADÁSZ (1915: 208-209, fig. 95) probably also belongs to *E. barcinensis*, but could not be ascertained as these specimens could not be located in the MAFI collection.

The species *E. subconoideus* VADÁSZ and *E. transylvanicus* VADÁSZ are tentatively referred into the synonymy of *E. barcinensis*. None of the features outlined by VADÁSZ (1915) allow to confidently distinguish them from the co-occurring *E. barcinensis*.

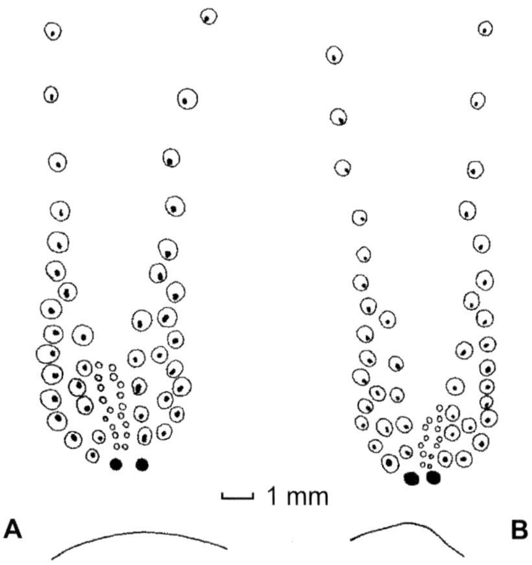

Figure 45: Phyllodes I (A) and IV (B) of *Echinolampas hemisphaerica* (LAMARCK, 1816) (Müllendorf, Bgld; NHMW 1997z0178/1758).

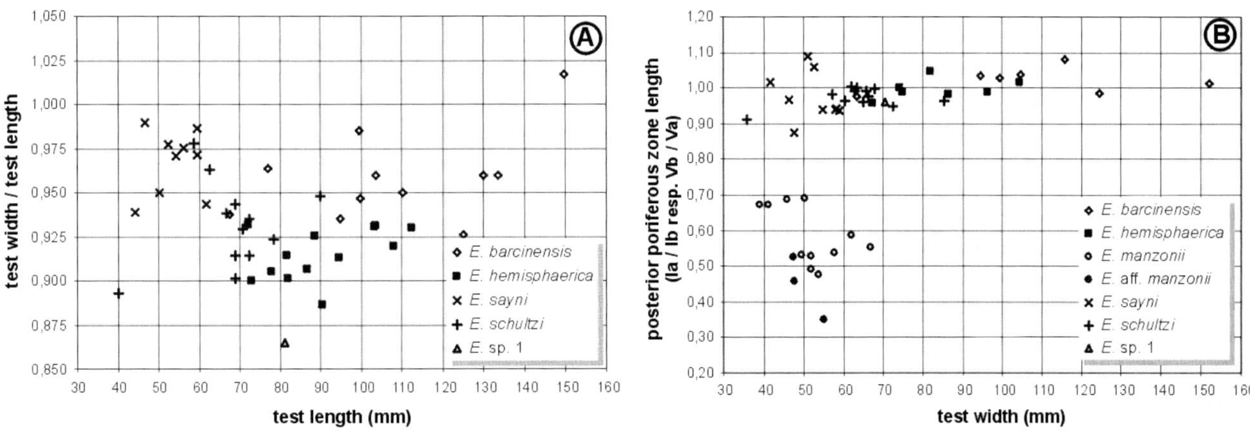

Figure 46: Morphometric comparison between the discussed *Echinolampas* species: Variation of relative test width (A; note the tendency of *E. barcinensis* and *E. sayni* to have wider test than *E. hemisphaerica* and *E. schultzi* respectively) and poriferous zone inequality in the posterior petals (I and V) expressed as ratio between the posterior and anterior poriferous zone length vs. test length (B; note the high variability in *E. sayni* and the clear separation of *E. manzonii* and *E.* aff. *manzonii* from the other *Echinolampas* species).

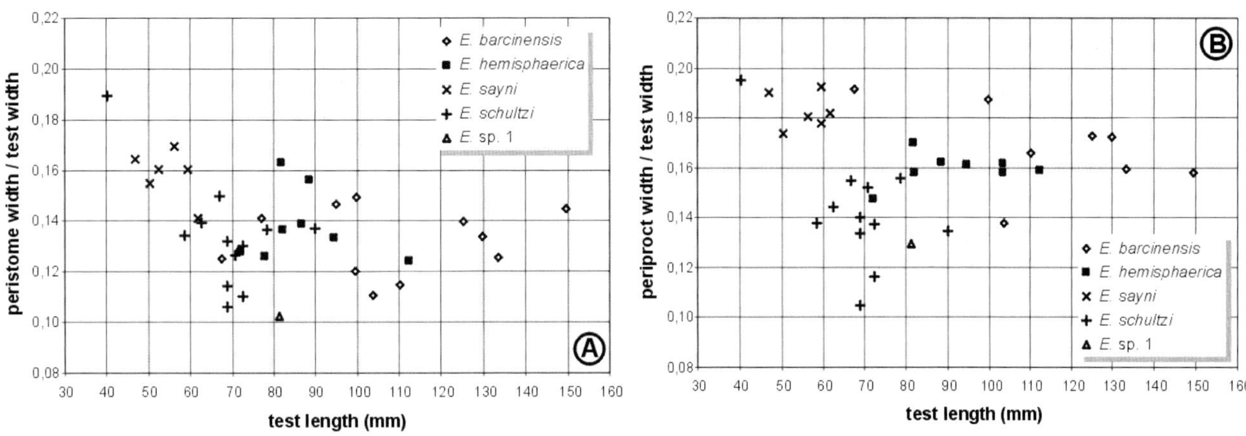

Figure 47: Morphometric comparison between the discussed *Echinolampas* species: Variation of peristome (A) and periproct (B) width in relation to test width. Although, both peristome and periproct show allometric growth and are relatively larger in small specimens than in larger ones (also observed for the extant *E. crassa* by THUM & ALLEN, 1976) a tendency for higher ratios can be observed in *E. barcinensis* and *E. sayni* in comparison to *E. hemisphaerica* and *E. schultzi*.

Species	Inventory Number	Location	TL	TW	TH	PsW	PcW	Tub PPP	FP	FP cr	APP ant	APP post	APP cr	PPP ant	PPP post	PPP cr	Tub1	Loc	Tub2	Loc	Tub3	Loc	Tub4	Loc
E. barcinensis	NHMW 1997z0178/1743	Müllendorf. Bgld	67.4	63.2	24.4	7.9	12.1	8	20	34.2	19.2	23.7	34.8	27.7	27	40.4	216	I	200	1	112	I	124	1
E. barcinensis	NHMW 2003z0031/0001	Müllendorf. Bgld	99.8	94.5	31.4	14.1	17.7	9	31.7	49.9	31	39.5	53.5	42.4	43.8	60.6	172	I	132	1	64	I	84	1
E. barcinensis	NHMW 2002z0181/0011	Schrattenberg. NÖ	103.7	99.5	33.1	11	13.7	9	32.63	55.7	29.8	40.6	56.3	40.7	41.8	59.3	128	I	104	1	-	-	84	1
E. barcinensis	NHMW 2002z0181/0007	Mannersdorf. Bgld	110.3	104.8	31.6	12	17.4	9	40	56.7	35.9	39.9	57.2	43.2	44.8	61.8	156	I	120	1	64	I	64	1
E. barcinensis	NHMW 1978/2020/21	Mannersdorf. Bgld	125.1	115.8	35.4	16.2	20^2	10	40.1	63.4	43.6	47.8	65.9	52.4	56.6	73.2	124	I	100	1	56	I	68	1
E. barcinensis	NHMW 1978/2020/22	Mannersdorf. Bgld	133.4	128	35^2	16.1	20.4	11	-	-	-	-	-	-	-	-	124	I	108	1	48	I	76	1
E. barcinensis	NHMW 2002z0181/0006	Mannersdorf. Bgld	149.5	152^2	45^2	22	24^2	10	48.4	78.7	47.5	54.2	73	56.1	56.8	84	132	I	98	1	48	I	64	1
E. barcinensis	NHMW 2003z0031/0002	Müllendorf. Bgld	77.1	74.3	25.9	10.5	-	-	-	-	-	-	-	-	-	-	-	-	-	-	-	-	-	-
E. barcinensis	NHMW 2003z0031/0003	Müllendorf. Bgld	94.9	88.7	27.7	13	-	-	-	-	-	-	-	-	-	-	-	-	-	-	-	-	-	-
E. barcinensis	NHMW 2002z0181/0014	Hof. Bgld	99.6	98.1	31.3	11.8	-	-	-	-	-	-	-	-	-	-	-	-	-	-	-	-	-	-
E. hemisphaerica	NHMW 1997z0178/2424	Budatéhény. HU	72.1	67.2	37.3	8.6	9.9	9	30.8	45.9	31.2	34.1	45.7	37.4	35.8	48.4	216	I	192	1	104	I	112	1
E. hemisphaerica	NHMW 2002z0181/0010	Müllendorf. Bgld	88.5	81.9	42.5	12.8	13.3	11	33.2	51.3	31.4	33.6	50.3	40.3	42.2	57.2	180	I	164	1	76	I	96	1
E. hemisphaerica	NHMW 1997z0178/1758	Müllendorf. Bgld	94.5	86.3	36.9	11.5	13.9	11	34.9	52	33.1	38.7	51.8	41.5	40.7	57.7	152	I	128	1	56	I	84	1
E. hemisphaerica	NHMW 1997z0178/1744	Müllendorf. Bgld	112.4	104.5	47.6	13	16.6	12	45.8	63.7	41.7	50.4	62.4	55.2	56.1	69.5	180	I	144	1	64	I	92	1
E. hemisphaerica	NHMW 1904.VIII.53	Kalksburg. Wien	103.5	96.4	48.75	-	15.6	10	37.5	62	37.5	41.5	61.5	46.9	46.3	66	136	V	136	1	64	V	88	4
E. hemisphaerica	NHMW 2002z0181/0016	Brunn/Steinfeld. NÖ	78	70.6	36.9	8.9	-	-	-	-	-	-	-	-	-	-	-	-	-	-	-	-	-	-
E. hemisphaerica	NHMW 2002z0181/0017	Neckenmarkt. Bgld	86.6	78.5	34.3	10.9	-	-	-	-	-	-	-	-	-	-	-	-	-	-	-	-	-	-
E. hemisphaerica	NHMW 2002z0181/0020	Kalksburg. Wien	90.5	80.2	37.9	-	-	-	-	-	-	-	-	-	-	-	-	-	-	-	-	-	-	-
E. hemisphaerica	NHMW 1973/1615/180a	Ritzing. Bgld	73.1	65.8	40.5	-	-	-	-	-	-	-	-	-	-	-	-	-	-	-	-	-	-	-
E. hemisphaerica	NHMW 1904.VIII.50	Kalksburg. Wien	108.1	99.4	45.1	-	-	-	-	-	-	-	-	-	-	-	-	-	-	-	-	-	-	-
E. hemisphaerica	NHMW 1904.VIII.53	Kalksburg. Wien	103.4	96.2	48.8	-	15.2	-	-	-	-	-	-	-	-	-	-	-	-	-	-	-	-	-
E. schultzi	NHMW 1914.VII.20/1	Gauderndorf. NÖ	40.2	35.9	18.3	6.8	7	8	13.9	21.3	12.4	18.7	23	20.2	18.4	25.9	392	I	296	1	-	-	208	1
E. schultzi	NHMW 1997z0178/0466a	Gauderndorf. NÖ	58.6	57.3	29.2	7.7	7.9	-	22	31.7	19.9	27.5	34.4	29.1	28.6	38.3	-	-	-	-	-	-	-	-
E. schultzi	NHMW 1851.VI.65	Gauderndorf. NÖ	62.6	60.3	27.8	8.4	8.7	12	23.3	33.7	24	28.7	37.2	32	30.8	41.2	320	I	240	1	-	-	-	-
E. schultzi	NHMW 1997z0178/0466b	Gauderndorf. NÖ	66.8	62.7	28.5	9.4	9.7	12	24.2	34.4	23.7	32	36.6	31.1	30.8	41.4	320	I	240	1	152	I	168	1
E. schultzi	NHMW 1869.III.40	Gauderndorf. NÖ	68.8	62.9	30.6	8.3	8.8	12	23.4	37.5	22.8	28.7	38.5	33.4	33.3	43.7	324	I	260	1	-	-	-	-
E. schultzi	NHMW 2003z0036/0001	Gauderndorf. NÖ	68.9	65	31.7	6.9	6.8	11	-	39.2	24.8	30.6	38.2	35.4	34	42.4	368	I	272	1	132	I	172	1
E. schultzi	NHMW 1869.III.38b	Gauderndorf. NÖ	68.9	62.1	32	7.1	8.3	12	24	36.2	22.5	27.9	39.6	31.3	31.4	43.6	328	I	272	1	140	I	184	1
E. schultzi	NHMW 1997z0178/0511	Gauderndorf. NÖ	70.7	65.7	31.2	8.3	10	13	26	39.3	24	31	40.4	32.3	32	45.1	328	I	264	1	112	I	-	-
E. schultzi	NHMW 1904.VIII.24	Gauderndorf. NÖ	72.4	67.7	30.2	8.8	9.3	14	-	37.4	-	-	38.7	37.7	37.6	46.4	328	I	264	1	120	I	152	1
E. schultzi	NHMW 2003z0037/0001	Maria Dreieichen. NÖ	72.4	66.2	32.3	7.3	7.7	-	27.2	41.2	24.6	33.1	40.5	35.7	34.8	44.9	-	-	232	1	144	I	172	1
E. schultzi	NHMW 2003z0036/0003	Gauderndorf. NÖ	78.5	72.5	31.9	9.9	11.3	16	27.8	43.9	25.8	36.2	44.7	36	34.2	48.4	336	I	272	1	112	I	164	1
E. schultzi	NHMW 2003z0036/0002	Gauderndorf. NÖ	90	85.3	35.8	11.7	11.5	14	-	-	38.8	42.6	50.4	43	41.5	53.9	248	I	204	1	-	-	-	-
E. sayni	NHMW 1914.VII.20/2	Grübern. NÖ	52.4	51.2	18.2	8.2	-	7	15	23.4	14.9	19.9	28.8	21.4	23.3	32	264	V	264	4	136	I	192	1
E. sayni	NHMW 1914.VII.20/5	Grübern. NÖ	46.8	46.3	15.3	7.6	8.8	7	15	21.5	13	19.1	23.3	20.5	19.8	29.7	240	I	216	1	128	I	152	1
E. sayni	NHMW 2003z0035/0001	Grübern. NÖ	50.3	47.8	17.9	7.4	8.3	7	17.2	26.4	16.3	20.9	27.1	23.4	20.5	29.5	250	V	240	4	112	I	152	1
E. sayni	NHMW 1914.VII.20/4	Grübern. NÖ	56.3	54.9	19.2	9.3	9.9	8	20.3	26.2	19.2	26.2	29.6	30.3	28.5	36.7	224	V	232	4	100	I	132	1
E. sayni	NHMW 1850.IX.83	Gauderndorf. NÖ	61.7	58.2	25.7	8.2	10.6	8	21.4	33.2	19.7	26.5	34.7	30.9	29.1	38.6	216	I	184	1	108	I	140	1
E. sayni	NHMW 2003z0035/0002	Grübern. NÖ	44.3	41.6	14	-	-	5	10.4	21	9.7	16.5	22.4	19	19.3	27.9	216	V	208	1	-	-	-	-
E. sayni	NHMW 1914.VII.20/3	Grübern. NÖ	54.3	52.7	22	-	-	6	17.7	27.3	16.6	25.6	28.5	26.7	28.3	35.4	192	I	168	1	-	-	-	-
E. sayni	NHMW 1914.VII.20/6	Grübern. NÖ	59.5	58.7	19.1	9.4	11.3	7	19.2	27.5	19.7	27.8	32.3	29.5	27.7	36.3	192	I	148	1	96	I	104	1
E. sayni	NHMW 1866.XI.57	Gauderndorf. NÖ	59.6	57.9	23.1	-	10.3	6	19.5	28.1	17.4	23	28.5	31.1	29.2	39.4	176	I	144	1	-	-	-	-

Species	Inventory Number	Location	TL	TW	TH	PsW	PcW	Tub PPP	FP	FP cr	APP ant	APP post	APP cr	PPP ant	PPP post	PPP cr	Tub1	Loc	Tub2	Loc	Tub3	Loc	Tub4	Loc
E. manzonii	NHMW 2003z0009/0001	Hart. Eisenstadt. Bgld	52.9	48.7	25.5	-	11.1	-	-	-	-	-	-	-	-	-	-	-	-	-	-	-	-	-
E. manzonii	NHMW 2003z0009/0002	Hart. Eisenstadt. Bgld	67.1	59.4	25.9	-	-	-	-	-	-	-	-	-	-	-	-	-	-	-	-	-	-	-
E. manzonii	NHMW 2003z0009/0003	Hart. Eisenstadt. Bgld	66.2	57.8	21[2]	-	-	3	18	27.4	13.3	20.4	28.8	23.5	12.6	39.1	96	I	104	1	-	-	96	1
E. manzonii	NHMW 2003z0009/0004a	Hart. Eisenstadt. Bgld	51.2	39.3	24.5	7.6	7.7	-	-	-	-	-	-	-	-	-	-	-	-	-	-	-	120	1
E. manzonii	NHMW 2003z0009/0004b	Hart. Eisenstadt. Bgld	-	49.5	22.1	-	11.8	2	-	-	-	-	-	22.8	12.1	36	-	-	96	1	-	-	120	1
E. manzonii	NHMW 2003z0009/0005	Hart. Eisenstadt. Bgld	-	67[2]	34.5	-	-	4	32.2	45.4	20.4	31.5	42.9	32.2	17.8	-	104	I	112	1	-	-	96	1
E. manzonii	NHMW 1859.L.799a	Hart. Eisenstadt. Bgld	42.7	39	20	7.8	7.4	-	12.3	23.4	9.9	15.2	21.2	15.8	10.6	26.3	-	-	-	-	-	-	-	-
E. manzonii	NHMW 1859.L.799b	Hart. Eisenstadt. Bgld	42.6	38	20	5.5	6.6	-	14.3	21.7	-	-	-	-	-	-	-	-	-	-	-	-	-	-
E. manzonii	W1	Hart. Eisenstadt. Bgld	49.9	44.8	20	8.6	9.8	3	15	22.2	10.9	18.2	24.4	-	-	30.5	120	I	120	1	-	-	-	-
E. manzonii	W2	Hart. Eisenstadt. Bgld	47.1	41	23.1	8.6	10.7	2	-	22.7	11	16.4	21.8	19.8	13.3	31.5	-	-	-	-	-	-	-	-
E. manzonii	W3	Hart. Eisenstadt. Bgld	58.7	45.9	20.4	-	10.6	3	20.1	29.5	11	18.7	28.9	21.2	14.6	35.5	96	I	96	1	-	-	112	4
E. manzonii	W4	Hart. Eisenstadt. Bgld	63.5	52	23.4	9[2]	-	3	19.2	32.9	12.6	23.6	33.9	24.8	13.1	37.1	104	V	88	4	-	-	88	4
E. manzonii	W5	Hart. Eisenstadt. Bgld	57.6	50.3	31.6	-	10.2	-	18.1	35.6	-	-	-	23.9	16.5	37.1	-	-	-	-	-	-	-	-
E. manzonii	W6	Hart. Eisenstadt. Bgld	64.8	62.1	-	-	-	3	-	-	14.3	23.1	30.6	25.1	14.7	35.6	72	I	64	1	-	-	-	-
E. manzonii	W18	Hart. Eisenstadt. Bgld	63.2	53.7	23.5	-	-	3	19.4	31.2	10.3	21.6	31.8	25.4	12.1	41.2	96	V	96	4	-	-	-	-
E. manzonii	W19	Eisenstadt. Bgld	-	51.9	-	-	-	2	13.5	25.9	8.7	17.3	26.6	19.5	9.6	30.5	104	V	88	1	-	-	-	-
E. manzonii	NHMW 1997z0178/1717a	Hart. Eisenstadt. Bgld	51.2	42	22.1	-	9.8	-	-	-	-	-	-	-	-	-	-	-	-	-	-	-	-	-
E. aff. *manzonii*	W17	Stotzing. Bgld	55.9	55[2]	20.3	8	-	2	18.5	28.9	8.1	19.3	30.1	22	7.7	32.1	120	I	112	1	96	I	128	1
E. aff. *manzonii*	W7	Stotzing. Bgld	51.4	47.8	22.8	9.2	10	2	18.7	25.6	10.1	18.7	29.1	19.9	9.1	34.4	112	I	112	1	88	I	144	1
E. aff. *manzonii*	NHMW 1997z0178/1717b	Hart. Eisenstadt. Bgld	52.7	47.5	19.7	8.9	11	2	18.4	27.9	10.8	17.8	26.5	18.9	9.9	29.2	120	V	-	-	-	-	-	-
Echinolampas sp. 1	NHMW 1869.III.38a	Baden. NÖ	81.3	70.3	39.5	7.2	9.1	12	34.1	49.3	29.4	35.5	46.3	41.1	39.5	53.6	304	I	288	1	128	I	176	1
C. elegans	NHMW 2002z0181/0004	Canetto. Corsica	96.6	94.6	52.7	-	-	-	-	-	-	-	-	-	-	-	-	-	-	-	56	64	-	52
C. elegans	W14	Stotzing. NÖ	111.0	109.2	61.4	11.9	-	47[1]	5	61[1]	78.5[1]	53.0	60.9	78.0	IV	63.2	58.8	85.4	13.3	I	48	60	40	52
C. elegans	NHMW 2003z0004/0001	Retznei. Stmk	120.0	-	72.0	-	-	-	-	-	-	-	-	-	-	-	-	-	-	-	56	64	-	-
C. elegans	W12	Stotzing. NÖ	121.0	120.5	70.0	-	-	-	5	58.9	81.4	60.6	65.4	88[1]	IV	71.8	64.8	95.0	13.8	V	40	48	-	-
C. elegans	W13	Stotzing. NÖ	124.5	116[1]	64.5	11.9	13.6	50.6	6	61[1]	78.4	61.6	67.6	83.7	IV	66.8	62.6	90.4	12.8	V	48	48	52	60
C. elegans	W16	Retznei. Stmk	130.0	122[2]	73.0	-	-	-	-	-	-	56.9	64.3	94.6	II	-	-	89.6	11[1]	I	40	56	-	-
C. elegans	NHMW 2003z0004/0002	Retznei. Stmk	132.0	-	78.0	-	-	-	-	-	-	-	-	-	-	-	-	-	-	-	-	-	52	84
C. elegans	NHMW 2003z0005/0003	Stotzing. NÖ	139.9	135[1]	83.1	-	-	56.6	6	-	-	66.5	77.5	101.7	IV	80.2	73.1	103.2	13.7	V	48	52	-	-
C. elegans	NHMW 2003z0005/0002	Stotzing. NÖ	145.0	-	86.0	-	-	-	-	-	-	-	-	-	-	-	-	-	-	-	57	63	56	59
C. subpentagonalis	W15	Retznei. Stmk	113.5	101.0	54.0	-	-	-	4	50.2	71.5	42.8	46[1]	70.1	IV	-	-	76.1	10.9	V	48	56	-	-
C. subpentagonalis	NHMW 2002z0181/0005	Ewitsch. Stmk	117.0	-	48.1	-	-	-	-	-	-	-	-	-	-	-	-	-	-	-	48	56	-	-
C. subpentagonalis	W8	Stotzing. NÖ	124.9	115.6	49[2]	12.7	14.5[1]	48.2	5	46.6	71.1	49.0	51.7	76.6	IV	57.2	49.9	77.2	11.7	V	56	56	52	60
C. subpentagonalis	W10	Stotzing. NÖ	130.8	125.6	59.0	-	15.3	46.5[1]	6	66.7	82.6	63.9	70.0	87.0	II	71.3	67.5	87.8	14.4	V	60	68	-	68
C. subpentagonalis	W11	Stotzing. NÖ	142.8	132.8	69.9	-	16.1	57.5	6	67.2	85.0	58.3	68.5	93.4	II	68.0	65.2	99.2	13.8	I	52	60	40	60
C. subpentagonalis	NHMW 2003z0005/0001	Stotzing. NÖ	145.0	134.0	58.0	14.6	-	59.8	5	-	-	59.3	68.5	92.2	IV	74.2	69.3	94.8	14.5	V	45	41	36	48
C. subpentagonalis	W9	Stotzing. NÖ	147.1	144.4	63.3	15.6	20.1	58.8	-	68.6	91.8	63.6	68.9	91.6	IV	78.4	70.1	92.8	15.0	V	36	48	46	56
C. cf. *subpentagonalis*	NHMW 1970/1396/898	Hornstein. Bgld	113.0	109.8	49.7	10.9	13.6	44.4	6	48.1	74.3	47.3	52.3	76.6	IV	55.0	51.0	75.8	13.3	V	72	84	44	64
C. cf. *ugolinii*	NHMW V.b.ß.k.12	Steiermark	119.0	1092	65.9	12.0	-	-	8	67.3	81.6	66.7	74.3	84.52	-	-	-	-	14.4	-	76	88	44	72

Abbreviations for variables: see Fig. 55; Loc = Location of measurement; Footnotes: 1 = poor observability, values may vary +/- 0.5 mm; 2 = approximation

Table 10: Dimensions of the *Echinolampas* and *Conolampas* specimens used in the present study.

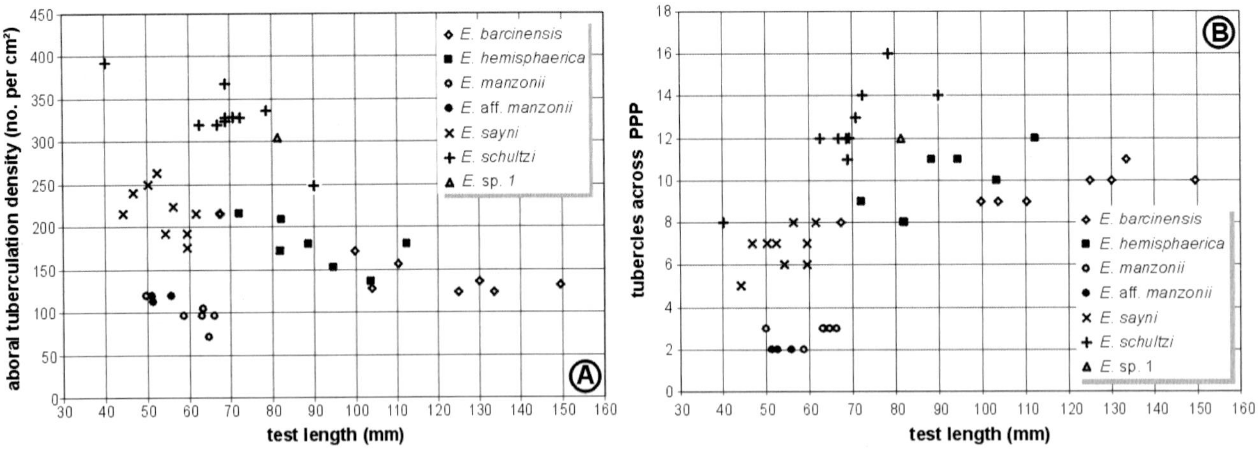

Figure 48: Morphometric comparison between the discussed *Echinolampas* species: Variation of aboral tubercle density (A; counted in the interporiferous zones of the petals I or V) and number of tubercles across the interporiferous zone of the posterior paired petals (B). Again, a distinct allometric component (see fig. 4.4) is evident. However, separation of the individual taxa is relative straightforward (e.g. large difference between *E. sayni* and *E. schultzi*,...).

Occurrence:

Austria: Early to Late Badenian (Langhian to Early Serravallian)
Vienna Basin: Hof am Leithagebirge, NÖ ([NHMW]); Hornstein (=Szarvkö), Bgld (Vadász, 1915; [MAFI]); Mannersdorf, NÖ ([NHMW]); Müllendorf (Mühlendorfer Kreide AG quarry), Bgld ([NHMW]); Schrattenberg, NÖ ([NHMW])
Styrian Basin: Retznei (Weissenegg Fm., Lafarge quarry), Styria ([coll. Meindl])

Paratethys (non-Austrian occurrences): Early to Late Badenian (Langhian to Early Serravallian)
Great Hungarian Basin (Pannonian Basin): ? Budapest, Pest, Hungary (Kókay et al., 1984); ? Budapest-Rákos, Pest, Hungary (Mihály, 1969)
Fore-Carpathian Basin: Niechobrz, south-eastern Poland [Szörényi, 1953 (locality is of Early Badenian age, see discussion in Bitner & Pisera, 2000: 7-8)]
Transylvanian Basin: Gârbova de Sus (= Felsö-Orbó), Romania (Vadász, 1915; [MAFI]); Livezile (= Ùrháza, = Vládháza), Alba, Romania (Vadász, 1915); ? Moldoveneşti (= Várfalva), Cluj, Romania (Vadász, 1915); ? Pietroasa (= Csegez), Romania (Vadász, 1915); region east of Taşad, Romania (Paucă, 1936)

Mediterranean: ? Burdigalian to Serravallian
Western Mediterranean: Ploaghe, Province Sassari, Sardinia (Lovisato, 1910); Viloví dans le Penedés, Catalogne, Spain (Lambert, 1906a, 1927a)
Eastern Mediterranean: Gubba Cyrenaica, Lybia (Taviani, 1939)

Atlantic Ocean: Middle ? Miocene
Iberian Atlantic coast: Alro dos Buxos, Trafaria, Portugal (de Loriol, 1896); Cazal das Rolas, Portugal (de Loriol, 1896); De Forno do Tijolo à Pragal, Portugal (de Loriol, 1896); Marvilla, Portugal (de Loriol, 1896); Palma, north of Alcacer do Sal, SW Portugal (de Loriol, 1896)

Echinolampas hemisphaerica (Lamarck, 1816)

(Figs. 43.2, 44, 45; Pl. 51, Figs. 1-2)

1829 *Clypeaster Linkii* nobis – Goldfuss: 133; pl. 42, figs. 4a-c
1835 *E.[chinolampas] hemisphæricus* Ag. (*Clypeaster hemisph.* Lam.)– Agassiz: 187
1835 *E.[chinolampas] Linkii* Ag. (*Clypeaster Link.* Goldf.) – Agassiz: 187
? 1836 *Clypeaster hemisphæricus.* – Grateloup: 146-147; No. II/7
pp 1847a *[Echinolampas] hemisphæricus* Agass. – Agassiz & Desor: 165
1847a *[Echinolampas] Linkii* Agass. – Agassiz & Desor: 166
1848 *Echinolampas hemisphæricus* (Ag.) – Gras: 52
1852 *E.[chinolampas] hemisphæricus* (Ag.) – Gras: 47
1857 *E.[chinolampas] hemisphæricus*, Ag. – Pictet: 210
1857 *E.[chinolampas] Linkii*, Ag. – Pictet: 210
pp 1858 *Echinolampas hemisphaericus* Agass. – Desor: 307
1858 *[Echinolampas] Linkii* Agass. – Desor: 309
? 1864 *Echinolampas hemisphæricus*, Lamarck – Wright: 480-481
1869 *Echinolampas hemisphæricus* Lamarck (*Clypeaster*) – des Moulins: 315
1869 *Echinolampas hemisphaericus* Lam. var. *Linki* Goldf. – Fuchs: 194
1869a *Echinolampas hemisphaericus* Lamarck sp. Varietas *Linkii* Goldfuss. – Laube: 183
pp 1869a *Echinolampas hemisphaericus* Lamarck sp. Varietas *Rhodensis* Laube. – Laube: 183
1870 *Echinolampas hemisphaericus* var. *Linki* Goldf. – Laube: 314
pp 1870 *Echinolampas hemisphaericus* var. *Rhodensis* Lbe. – Laube: 314
1871 *Echinolampas hemisphaericus* Lamarck. Varietas *Linkii* Goldfuss. – Laube: 65, 66; pl. 18, fig. 3
pp 1871 *Echinolampas hemisphaericus* Lamarck. Varietas *Rhodensis* Laube. – Laube: 65, 66; pl. 18, fig. 2
? 1873 *Echinolampas hemisphaericus*, Lamk – Manzoni: 10-11, 19
1875 *Clypeaster hemisphaericus* Lmck. – Quenstedt: 494-495
1877 *Echinolampas hemisphaericus* Lam. var. *Linki* Goldf. – Karrer: 312
1877 *Echinolampas haemisphaericus* Lam. var. *Linkii* Goldf. – Lóczy: 63
1879 *Echinolampas hemisphæricus*, Lamck. – Hermite: 252

non 1880a *Echinolampas hemisphaericus* Lk. – Manzoni: 186, pl. 1, figs. 1-3 [specimens from the Pliocene of Castelarquato belong to *E.* (*Macrolampas*) *hoffmanni* Desor according to Borghi (1994: 7); specimens from the "Molassa serpentinosa" were referred to *E. italica* by Lambert (1906a: 93)]
1881 *Echinolampas hemisphæricus*, Lamk. – Manzoni: 174
1882a *Echinolampas hemisfericus*, Agas. – Mazzetti: 123
1883 *E. Linkii* – Pomel: 62
1883 *E. hemisphæricus* – Pomel: 62
1887b *Echinolampas hemisphaericus*, Lamarck, var. *Rhodi*, Laube. – Koch: 269
1887b *Echinolampas Laurillardi*, Agissiz. – Koch: 269
1888b *Echinolampas haemisphaericus*, var. *Rodi*, Lbe – Nemes: 22, 32
pp 1891 *Echinolampas hemisphæricus* (Lamarck), 1816. – Gregory: 605-606
1894 *Echinolampas hemisphaericus* var. *Rodi* Laube. – Mártonfi: 153
non 1897 *Echinolampas hemisphaericus* Lmk. sp. – Vinassa de Regny: 152 [= *E.* (*Macrolampas*) *hoffmanni* Desor fide Borghi (1994: 7)]
1899 *Echinolampas hemisphaericus* Lmk. sp., var. *Linkii* Goldf. – Roth von Telegd: 95
1906 *Echinolampas hemisphaericus* Lam. var. *Linkii*, Goldfs. – Vadász: 329, 331-332
pp v 1915 *Echinolampas hemisphaericus* Lam. sp. – Vadász: 206-209; fig. 94
? 1915 *Echinolampas angustipetalus* n. sp. – Vadász: 215; fig. 101; pl. 10 (4), fig. 4
1931 *Echinolampas hemisphaericus* Lam. – Janoschek: 83, 84
1939 *Echinolampas hemisphaericus* Lam. sp.– Kapounek: 72
non 1953 *Echinolampas hemisphaericus* (Lamarck), 1816. – Szörényi: 34, 84-85; pl. 3, figs. 2, 2a-b [misidentified *Conolampas* species]
1965 [*Echinolampas*] *hemisphaerica* Lamarck 1816 (*Clypeaster*) – Roman: 283
1969 *Echinolampas hemisphaericus* Lamarck – Mihály: 256
1969 *Echinolampas lecointrae* Lambert – Mitrović-Petrović: 122; pl. 1, figs. 1, 1a-b
1969 *Echinolampas italicus* Lambert – Mitrović-Petrović: 123; pl. 2, figs. 1, 1a
1969 *Echinolampas wrighti* Gregory – Mitrović-Petrović: 123; pl. 2, figs. 2, 2a
1970 *Echinolampas hemisphærica* (Lamarck) – Montenat & Roman: 116-117
1974 *Echinolampas* (*Echinolampas*) *hemisphaerica* (Lamarck) 1816 – Chavanon: vol.1: 184-186; vol. 2: 147-148; figs. 120-16, 169-171; pl. 12, figs. 7a-b
1974a *Echinolampas hemisphericus* – Rose: 345; fig. 3
? 1974b *Echinolampas hemisphericus* – Rose: 353; fig. 3
? 1980 *Echinolampas hemisphaerica* (Lamarck) – Challis: 149-152; pl. 49, figs. a-c, pl. 50, fig. a [pl. 51, figs. a-c depict a specimen from Martignas, Gironde France, which undoubtedly belongs to *E. hemisphaerica*, for the records from the Messinian of Malta see discussion below]
1984 *Echinolampas hemisphaeicus* Lamk. – Kókay et al.: 288
1985 *Echinolampas hemisphaericus* Lamarck, 1816 – Mihály: 244
1997 *Echinolampas laurillardi* Agassiz – Majcen et al.: 106; pl. 5, fig. 1
pp 1998 *Echinolampas hemisphaerica* (Lamarck, 1816) – Philippe: 86-90; pl. 17, figs. 4, 6-8
? 2001 *Echinolampas hemisphaerica* – Néraudeau et al.: 52; tab. 1
v. 2001 *Echinolampas hemisphaericus* (Lamarck) – Schmid et al.: 23

Type-material:
The provenance and current whereabouts of Lamarck's (1816) material is unknown. Subsequent publications do not agree on a single type region. Two regions, the Rhône Basin (more specifically the region of Tricastin) and the Aquitaine Basin (the area of Dax, Landes and Bordeaux, Gironde) are cited.

Material:
Badenian (Langhian-Early Serravallian) – Brunn am Steinfeld, NÖ, Austria
NHMW: 1 specimen (NHMW 2002z0181/0016)
Badenian (Langhian-Early Serravallian) – Kalksburg, Wien, Austria
NHMW: 3 specimens (NHMW 1904.VIII.50, 1904.VIII.53, 2002z0181/0020)
Badenian (Langhian-Early Serravallian) – Neckenmarkt (Galgenberg), Bgld, Austria
NHMW: 1 specimen (NHMW 2002z0181/0017)
Badenian (Langhian-Early Serravallian) – Ritzing, Bgld, Austria
NHMW: 1 specimen (NHMW 1973/1615/180a)
Late Badenian (Early Serravallian) – Müllendorf (Mühlendorfer Kreide AG quarry), Bgld, Austria
NHMW: 4 specimens (NHMW 1997z0178/1744, 1997z0178/1758, 2002z0181/0009-10)

Foreign material for comparison:
Burdigalian – Martignas, Gironde, France
NHMW: 1 specimen (NHMW C 6041)
Early Badenian (Langhian) – Kovácsszénnája (= Kovacéna), Baranya, Hungary
MAFI: 1 specimen [MAFI Ech 436 (reference material of Vadász, 1915)]
Badenian (Langhian-Early Serravallian) – Budatétény, Budapest, Hungary
NHMW: 1 specimen (NHMW 1997z0178/2424)
Late Badenian (Early Serravallian) – Buituri (= Bujtur), Romania
MAFI: 1 specimen [MAFI Ech 267 (reference material of Vadász, 1915; specimen very poor, determination tentative at best)]

Dimensions: see Tab. 10

Description:
Size and shape: The test is large and has an oval, antero-posteriorly elongated outline. Test width ranges from 88.6 to 93.2 % of TL with a mean of 91.5 % in the studied specimens, but test width may range up to 97 % TL in other areas/time slices [Rhône Basin (Philippe, 1998), Aquitaine Basin (Chavanon, 1974)]. The anterior margin is rounded, the posterior margin distinctly rostrate. The maximum width lies posterior of the apical disc, about 62 to 66 % of TL from the anterior margin. In profile the test is domed, nearly hemispherical, the maximum height coinciding with the posterior part of the apical disc. Test height ranges from 39.0 to 51.7 % TL, with a mean of 45.0 %. The ambitus is rounded and tumid.
Apical disc: The apical disc lies slightly anterior of the centre, about 42 to 43 % TL from the anterior margin. The apical disc is monobasal with four circular gonopores. The madreporite is pentagonal and crowded with numerous small madreporic pores. The gonopores lie at the apices of the madreporite. The ocular plates are small, bearing a small, circular ocular pore each.

Ambulacra: Adapically the ambulacra are petaloid, moderately broad, straight and only slightly closed distally. The frontal petal extends about 60 to 72 % of the corresponding test radius, the anterior paired petals about 68 to 80 (mean 73) % and the posterior paired petals about 70 to 80 %. In contrast to the other petals, the poriferous zones within the anterior paired petals are not of equal length. The anterior poriferous zones (IIb and IVa) are shorter than the posterior ones (IIa and IVb) extending about 80 to 94 % of the length of the petal. Poriferous zones Ib, IIa, IVb and Va are distinctly arched, whereas the other poriferous zones are nearly straight. The ambulacral pores in the petals are conjugated anisopores, which are strongly conjugated and slightly oblique. The interporiferous zones within the petals are distinctly inflated and up to four times as wide as a single poriferous zone. The poriferous zones are slightly depressed.

Adorally the ambulacra form slightly depressed phyllodes. They consist of four series of unipores in each ambulacrum. Two adradial series consisting of large, closely spaced unipores with rather large attachment area and two perradial series consisting of smaller, slightly elongated unipores, which are more widely spaced. The two buccal pores are slightly larger than the other phyllodal pores. In the adoral third of the phyllodes there is a double row of shallow pits running along the central suture. These are interpreted as sphaeridial pits.

Interambulacra: Adapically the interambulacra are slightly inflated between the petals. On the aboral surface they are covered with small, crenulate, perforate tubercles which are deeply sunken and closely spaced. The areoles are inclined towards the margin of the test. On the oral side the interambulacra are slightly inflated and form small, inflated bourrelets in interambulacra 2 and 3 adorally. There is no "naked zone" along the central suture between peristome and periproct in interambulacrum 5, as reported in other species of *Echinolampas*.

Peristome: The peristome lies slightly anterior of the centre, about 42-44 % TL from the anterior margin. It is oval to subpentagonal and transversely elongated. It is moderately large, being usually 13-16 % TW in width.

Periproct: The periproct is situated inframarginally in interambulacrum 5 and lies nearly completely on the oral surface. It is large, about 15 to 16 % TW wide and has an oval, transversely elongated shape.

Differential diagnosis (see Figs. 43, 45-48)**:**

As outlined above, *E. hemisphaerica* is characterised by its subpentagonal, rostrate outline, high, globose profile, distinctly inflated interporiferous zones of the petals, weakly developed bourrelets, and relatively wide poriferous zones.

For the difference to *E. schultzi* see below under that species.

For the difference to *E. sayni* see below under that species.

E. sp. 1, from the Badenian (Langhian-Early Serravallian), differs from *E. hemisphaerica* by its more elongated distinctly oval outline, smaller peristome and periproct and much more sunken poriferous zones.

E. barcinensis, a co-occurring species, differs from *E. hemisphaerica* by its much lower profile, lower test height (Fig. 44), distinctly less rostrate shape, larger periproct, less inflated petals and less sunken poriferous zones. Additionally, the number of tubercles across the interporiferous zone is distinctly lower.

E. manzonii and *E.* aff. *manzonii*, both from the Early Badenian (Langhian) differ from *E. hemisphaerica* by their very unequal poriferous zones in the paired petals (among other features, but this is the most easily recognised).

E. hoffmanni Desor in Agassiz & Desor, 1847 from the Pliocene of Sicily and Northern Italy has many similarities with the present species. According to Borghi (1994: 7) the two species might be closely related. When the measurements provided by Borghi (1994) are compared with those obtained from the *E. hemisphaerica* from the Central Paratethys it becomes evident that the Pliocene *E. hoffmanni* is less elongated (TW mean 97.6 % TL vs. 91.5 % TL in *E. hemisphaerica*) and has a smaller peristome (peristome width mean 10.7 % TL vs. 12.7 % TL) and periproct (periproct width mean 12.2 % TL vs. 14.7 % TL). These examples illustrate that albeit the two are obviously very similar it is possible to distinguish them. Due to the limited information available it is currently impossible to decide whether these differences might represent shifts due to different environmental conditions or if they represent differences at species level.

Discussion:

Roman (1965: 283) included the following species and subspecies into the synonymy of *E. hemisphaerica*: *E. linkii* (Goldfuss, 1826), var. *rhodensis* Laube, 1869, *E. subhemisphaerica* Pomel, 1887, *E. pomeli* Peron & Gauthie in Cotteau et al., 1891, *E. deydieri* Lambert, 1913, var. *cyrenaica* Desio, 1929, var. *bardiensis* Desio, 1929 and tentatively var. *cubensis* Palmer in Sanchez Roig, 1949. Unfortunately he did not provide a formal synonymy or discussed his reasons for synonymising the species with *E. hemisphaerica*. Following Lambert (1906a: 90-93; 1927a: 24) he excludes var. *maxima* de Loriol, 1896 from the synonymy and states that it belongs into the synonymy of *E. barcinensis* Lambert, 1906.

This species was re-described in detail by Philippe (1998: 86-90), who gave an extensive synonymy and revised all *Echinolampas* species reported from the Rhône Basin earlier. He also included *E. sayni* Lambert, 1913a into the synonymy of *E. hemisphaerica*, an opinion with which the present author does not agree with (see below under that species). The Austrian material fits rather well with the French material of *E. hemisphaerica* [both from the Rhône Basin (Philippe, 1998) and the Aquitaine Basin (Chavanon, 1974); see e.g. Fig. 44) and one can confidently regard them as conspecific. The quite variable morphology of this species, the uncertainty about its provenance and lack of a sufficient description during most of the 19th century led to the establishing of many junior synonyms of this species. Moreover, there are a number of similar and/or related other species on the originality of which there is no common agreement (e.g. *E. hoffmanni*, ...). Recently Néraudeau et al. (1999: 356) even made the extant Western African *E. rangii* des Moulins, 1869 a junior synonym of *E. hemisphaerica*. Néraudeau et al. (1999) state that this is based on "biometric and statistical" analysis, but neither data nor analysis are presented.

Records of *E. hemisphaericus* from the Upper Coralline Limestone (Messinian, see Kienel et al., 1995) by Wright (1864), Rose (1974b) and Challis (1980, and records therein) are doubtful. The Maltese specimens described and figured by Challis (1980) have much lower profiles and test heights than usually observed in *E. hemisphaerica*, and are also less anteroposteriorly elongated. They show large similarities with the species *E. barcinensis* discussed above. Records of *E. hemisphaerica* from the *Globigerina* Limestone (Aquitanian to Langhian; see Mazzei, 1985 and references therein) by Cottreau (1913a) remain to be confirmed.

Laube (1869a, 1871) established a new subspecies (variety) of *E. hemisphaerica*, namely var. *rhodensis*, based on Pliocene material from the Island of Rhodes. Additionally he attributed specimens from the Badenian of Sóskut to this subspecies. The latter, however, is not conspecific with the material from Rhodes, but belongs to *E. hemisphaerica*. Laube's subspecies *rhodensis* is here elevated to species rank and is probably the senior synonym of *E. orbignyi* Desor, 1876, a species also based on Pliocene material from Rhodes. The latter was considered a probable junior synonym of *E. hoffmanni* by Borghi (1994) based on comparison of biometric data of specimens from the Pliocene of Northern Italy, Sicily and the measurements of *E. orbigny* from Rhodes given by Airaghi (1930). To fully resolve the relation between those three species, the Miocene species *E. hemisphaerica* and the extant *E. rangii* more refined investigations are necessary.

Occurrence:

Austria: Early ? to Late Badenian (Langhian-Early Serravallian)
Vienna Basin: Baden, NÖ (Goldfuss, 1829; Desor, 1858); Brunn am Gebirge, NÖ (Laube, 1869a, 1871; Manzoni, 1873); Brunn am Steinfeld, NÖ ([NHMW]); Kalksburg, Vienna (Fuchs, 1869; Laube, 1869a, 1871; Manzoni, 1873; Karrer, 1877; Gregory, 1891; [NHMW]); Müllendorf (Mühlendorfer Kreide AG quarry), Bgld (Kapounek, 1939; [NHMW]); Vienna (Agassiz & Desor, 1847a; Pictet, 1857); Vienna Basin (Lóczy, 1877; Roman, 1965)
Eisenstadt-Sopron Basin: Großhöflein, Bgld (Laube, 1869a, 1871; Manzoni, 1873); St. Margarethen (Kummer quarry), Bgld (Schmid et al., 2001)
Oberpullendorf Bay: Neckenmarkt, Bgld (Janoschek, 1931; [NHMW]); Ritzing, Bgld (Quenstedt, 1875; [NHMW]);

Paratethys (non-Austrian occurrences): Early to Late Badenian (Langhian-Early Serravallian)
Eisenstadt-Sopron Basin: Sopron, Györ-Moson-Sopron, Hungary (Lóczy, 1877)
Great Hungarian Basin (Pannonian Basin): Bia, Pest, Hungary (Lóczy, 1877; Vadász, 1915); Budapest-Gyakorló, Hungary (Kókay et al., 1984; Mihály, 1985); Budapest Rákos, Hungary (Vadász, 1906, 1915; Mihály, 1969); Budatétény (= Tétény), Budapest, Hungary (Laube, 1869a, 1871; Manzoni, 1873; Lóczy, 1877; Vadász, 1915; [NHMW]); Kovácsszénnája (= Kovacéna), Baranya, Hungary (Vadász, 1915); Paptelek, Romania (Vadász, 1915); Sóskut, Hungary (Laube, 1869a, 1871; Gregory, 1891)
Fore-Carpathian Basin: not recorded [earlier records (Szörényi, 1953; Roman, 1965) are erroneous]
Transylvanian Basin: Buituri (= Bujtur), Romania (Koch, 1887b; Nemes, 1888b; Mártonfi, 1894; ? Vadász, 1915); Cacovalerii (= Cacova), Romania (Koch, 1887b); Gârbova de Sus (= Felsö-Orbó), Romania (Koch, 1887b); Minişu des Sus (= Felménes), Romania (Lóczy, 1877); Moldoveneşti (= Várfalva), Cluj, Romania (Koch, 1887b); Pietroasa (= Csegez), Romania (Roth von Telegd, 1899; Vadász, 1915)
Styrian Basin: Slovenia (Majcen et al., 1997)
Zala, Sáva and Dráva Basins: Derventa and Ugljevik, Bosnia & Herzegovina (Mitrović-Petrović, 1969)

Mediterranean: Burdigalian to Tortonian, ? Messinian*
Western Mediterranean: Bologna, Italy (Manzoni, 1881); Corsica, France (Roman, 1965); Eastern Algeria (Roman, 1965); Montese, Italy (Mazzetti, 1882a); Isère, France (Gras, 1848, 1852; Gregory, 1891); Santa Ponsa de Ferrerias, Menorca, Spain (Hermite, 1879); Sardinia, Italy (Roman, 1965); Sierra de Las Atalayas, near Alicante, Valenica, Spain (Montenat & Roman. 1970)
Central Mediterranean: Calabria, Southern Italy (Roman, 1965); Cyrenaica, Libya (Roman, 1965; Rose, 1974a); ? Greensand, Maltese Islands (Gregory, 1891); ? Upper Coralline Limestone, Maltese Islands (Wright, 1864; Laube, 1871; Gregory, 1891; Wigglesworth, 1964; Rose, 1974b; Challis, 1984)
Eastern Mediterranean: Anatolia, Turkey (Roman, 1965); Cyprus (Manzoni, 1873; Gregory, 1891); Egypt (Roman, 1965); Turkey (Gregory, 1891)
Rhône Basin: Burdigalian: Secteur es étangs (Fos-sur-Mer, Istres, Saint-Mitre-les-Ramparts), France (Philippe, 1998); Bordure sud des Alpilles (Fontvieille), France (Philippe, 1998); Bassin de Mus-Sommières (environs de Sommières), France (Philippe, 1998); Secteur de Montpellier (Celleneuve, Juvignac), France (Philippe, 1998); Bassin d'Avignon (les Angeles, Barbentane), France (Philippe, 1998); Vallée du Jabron (Châteauneuf-Miravail), France (Philippe, 1998); Bassin de Vairéas-Visan [Baumes-de-Transit; Chamaret, Grignan ?, Montségur-sur-Lauzon, St. Paul-trois-Châteaux (Agassiz & Desor, 1847a; Desor, 1858; Wright, 1864; Laube, 1871; Manzoni, 1873; Gregory, 1891), Saint-Restitut; Solérieux], France (Philippe, 1998); Sillon préalpin [Bassin de Crest: Cerst; Vercors: Bouvante, Oriol-en-Royans, Pont-en-Royans, Saint-Jean-en-Royans (Agassiz & Desor, 1847a; Gras, 1848, 1852; Desor, 1858; Wright, 1864; Laube, 1871; Manzoni, 1873)], France (Philippe, 1998)
Tortonian: Bassin d'Aix (Rognes), France (Philippe, 1998); Bordure Sud-Luberon (Ansouis, Cadenet, Cucuron, Vaugines), France (Philippe, 1998)

Atlantic Ocean: Early Miocene
Aquitane Basin: Bordelaise, France (Roman, 1965); Dax, Landes, France (Grateloup, 1836; Agassiz & Desor, 1847a); Faluns bleus de Narosse, near Dax, France (Grateloup, 1836; Desor, 1858; Wright, 1864; Chavanon, 1974); Landes, France (Manzoni, 1873; Gregory, 1891; Roman, 1965); Leognan, Gironde, France (Chavanon, 1974); Martignas, Gironde, France (Chavanon, 1974; Challis, 1980; [NHMW]); Martigues, Landes, France (Agassiz & Desor, 1847a; Desor, 1858; Wright, 1864; Laube, 1871; Manzoni, 1873; Gregory, 1891)
Caribbean: ? Cuba [Roman, 1965 (Late Oligocene or Early Miocene)]
Iberian Atlantic coast: Portugal (Roman, 1965)

* concerning the occurrences from the Messinian of the Maltese Islands see above under "Discussion"

Echinolampas manzonii Pomel, 1883

(Figs. 49, 50; Pl. 52, Figs. 1-3)

	1855	*Echinolampas Richardi*, Desmarest. – Wright: 124-125 [misidentified *E. manzonii* according to Gregory (1891) and Challis (1980)]
	1864	*Echinolampas Laurillardi*, Agassiz. – Wright: 480 [misidentified *E. manzonii* according to Gregory (1891) and Challis (1980)]
	1880a	*Echinolampas depressa* Gray. – Manzoni: 186-187, pl. 1, figs. 4-15
	1881	*Echinolampas depressa*, Gray. – Manzoni: 174
.	1882	*Echinolampas angulatus*, Mérian. – de Loriol: 13-16; pl. 2, figs. 1-10; pl. 2, figs. 1-2
non	1882	*Echinolampas angulatus*, Mérian. – de Loriol: 17-18; pl. 2, figs. 11, 11a-b
	1882a	*Echinolampas depressus*, Gray. – Mazzetti: 123
*	1883	*E. Manzonii* (*E. depressa* Manz. non Gray) – Pomel: 62
	1891	*Echinolampas manzoni*, n. sp. – Gregory: 606-607
?	1900	*Echinolampas angulatus* Mérian. – Verri & de Angelis d'Ossat: 259
	1907	*Echinolampas angulatus* Mérian. – Nelli: 263-265.
	1908a	*Echinolampas angulatus* Mérian. – Stefanini: 451-453; no. 10.
	1908b	*Echinolampas angulatus* Mér. – Stefanini: 79-83; pl. 13 [1], figs. 10-13
pp	1913a	*Echinolampas scutiformis* Leske, var. *angulatus* Merian. – Cottreau: 23, 64, 111-112; pl. 11, figs. 7, 7a
non	1913a	*Echinolampas scutiformis* Leske, var. *angulatus* Merian. – Cottreau: 64, 111-112; pl. 12, figs. 8-10
v.	1915	*Echinolampas angulatus* Mér. sp. – Vadász: 217
	1941	*Echinolampas angulatus* Mérian. – Meznerics: 90-91; pl. 2, fig. 6; pl. 3, fig. 1

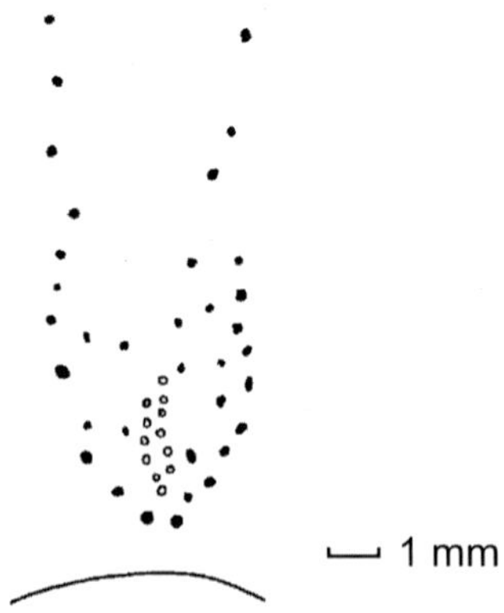

Figure 49: Phyllode I of *Echinolampas manzonii* POMEL, 1883 (Hartl Fm., Eisenstadt, Bgld; WANZENBÖCK coll. W4).

#	1953	*Echinolampas atrophus podoloicus* n. ssp. – SZÖRÉNYI: 34-35, 85-86; pl. 6, figs. 4, 4a-b
	1965	[*Echinolampas*] *manzonii* POMEL 1883 – ROMAN: 289
	1974b	*Echinolampas manzonii* – ROSE: 353; fig. 3
	1975	*Echinolampas manzoni* GREGORY, non POMEL – ROSE: 79, tab. 12
	1980	*Echinolampas manzoni* GREGORY – CHALLIS: 156-158; pl. 55, figs. a-c; pl. 56, fig. b
	2000	*Echinolampas* – PILLER: 87
?	2001	*Echinolampas manzonii* – NÉRAUDEAU et al.: 52; tab. 1
v.	2003	*Echinolampas manzonii* POMEL – KROH et al.: 92

Type-material:
Echinolampas manzonii POMEL, 1883:
Lectotype: the specimen figured by MANZONI (1880a) on plate 1, figures 14-15 (GREGORY, 1891: 607); current whereabouts unknown
Locus typicus: Colline di Bologna, Italy
Stratum typicum: Molassa serpentinosa
Age: Miocene
Remarks: GREGORY (1891: 607) designated MANZONI's (1880a) figure (pl. 1, fig. 14) as type. Additionally, he designated a specimen ("Brit. Mus., 24,621") from the British Museum of Natural History as type. This specimen, however, comes from a "nodule seam" of the *Globigerina* Limestone [probably either from the Burdigalian Qammieh conglomerate bed of ROSE et al., 1992 (=Qolla l-Bajda conglomerate bed of CHALLIS, 1980; = C_1 phosphate conglomerate of PEDLEY et al., 1976) or the Langhian Xwieni conglomerate bed of CHALLIS, 1980 and ROSE et al., 1992 (= C_2 phosphate conglomerate of PEDLEY et al., 1976), or from one of the subsidiary phosphate conglomerate horizons within the Upper *Globigerina* Limestone of an unnamed locality in Malta (written comm. Andrew B. SMITH, 20.05.2003). Thus that specimen is not available for lectotype selection, as it was not part of MANZONI's original material and GREGORY's action is therefore invalid.

Echinolampas atrophus podoloicus SZÖRÉNYI, 1953:
Holotype: the specimen figured by SZÖRÉNYI (1953: pl. 6, figs. 4, 4a-b); collection of the University Lwów, Ukraine [not seen]
Locus typicus: Podjarków, near Lwów, Western Ukraine
Biozone: Lower lagenid zone
Age: Early Badenian (Langhian), Middle Miocene (written comm. Anna WYSOCKA & Andrzej RADWAŃSKI, 02.04.2004)
Remarks: SZÖRÉNYI gave No. 1 as the inventory number of the sole specimen of this species. The same number (1), however, is listed under *Fibularia sandalina* and *Scutella paulensis* too.

Material:
Early Badenian (Langhian) – Eisenstadt (Hartl Fm., Hartl hill), Bgld, Austria
NHMW: 9 specimens (NHMW 1859.L.799; 1997z0178/1717a; 2003z0009/0001-7)
MAFI: 1 specimen [MAFI Ech 265 (reference material of VADÁSZ, 1915)]
WANZENBÖCK coll.: 7 specimens (W1-W6, W18)
Early Badenian (Langhian) – Eisenstadt (Hartl Fm., shooting ground), Bgld, Austria
WANZENBÖCK coll.: 1 specimen (W19)

Dimensions (in mm):

Inv. No.	TL	TW	TH	Remarks
NHMW 1997z0178/1717a	52.7	47.6	20.3	
NHMW 2003z0009/0001	52.9	48.7	25.5	
NHMW 2002z0009/0002	67.1	59.4	~26	deformed
W1	49.5	44.9	20.4	
W2	58.1	46.2	20.9	deformed
W3	47.5	40.2	23.3	
W4	63.4	52.5	24.2	deformed
W5	57.5	50.9	31.6	

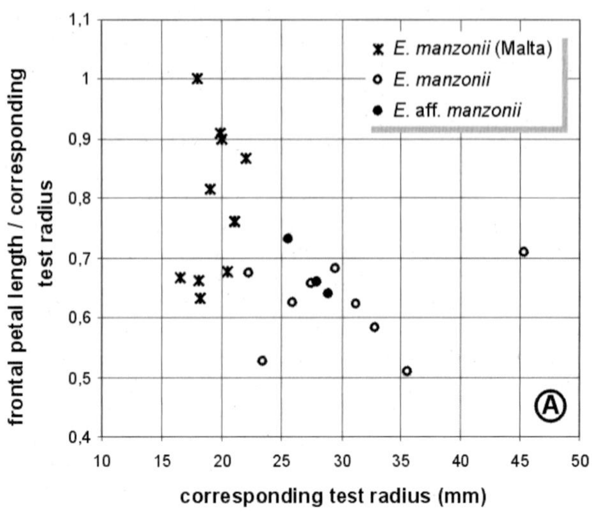

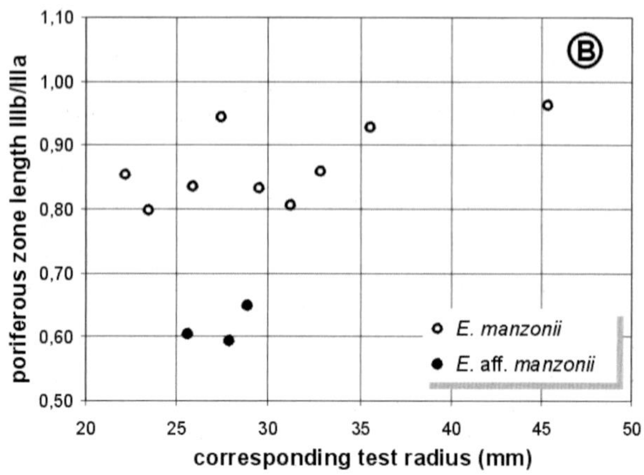

Figure 50: Morphometric comparison between *Echinolampas manzonii* and *E.* aff. *manzonii*: Variation of relative length of the frontal petal (A) and inequality of the poriferous zones within the frontal petal (B). Note the large variability of frontal petal length, especially when specimens from the Maltese Islands are considered (data provided by courtesy of M. GATT, Rabat, Malta). The inequality of the poriferous zones is much more pronounced in *E.* aff. *manzonii* (also in the other petals, not shown here).

Description:
Size and shape: The test is of medium size, antero-posteriorly elongated and distinctly rostrate. The anterior margin is rounded, the posterior margin bluntly pointed. The maximum width lies posterior of the apical disc, about 60 % of TL from the anterior margin. In profile the test is domed, the maximum height coinciding with the position of the apical disc. The ambitus is rounded and tumid. The oral surface is strongly depressed around the peristome. The test width ranges from 79.5 to 96.1 % TL (mean = 87.3 %), the test height from 35.9 to 55 % TL (mean = 43.5 %).
Apical disc: The apical disc lies slightly anterior of the centre, about 40 to 45 % TL from the anterior margin. The apical disc is monobasal with 4 large, circular gonopores. The madreporite is pentagonal and crowded with numerous small madreporic pores. The gonopores lie at the apices of the madreporite. The ocular plates are small, bearing a small, circular ocular pore each.
Ambulacra: The ambulacra are petaloid and relatively narrow. The ambulacral pores within the petals are elongate isopores, outside the petals they are small, slit-shaped unipores. The frontal petal is the shortest and extends about 50-70 % of the corresponding test radius. The anterior paired petals extend about 65-75 % of the corresponding test radius. The posterior paired petals are longest (referring to the absolute values) and extend about 60-70 % of the corresponding test radius. The poriferous zones of all petals are strongly unequal. In the frontal petal this is not very pronounced, poriferous zone IIIb being about 80-96 % of IIIa. In the anterior and posterior paired petals, this is strongly expressed, IIb/IVa are about 47 to 67% of IIa/IVb (anterior paired petals) and Ia/Vb are 47 to 69% of Ib/Va (posterior paired petals). Adorally the ambulacra form distinct phyllodes. These consist of large rounded unipores. The shape of the phyllodes is shown in Fig. 49.
Interambulacra: The interambulacra are slightly inflated between the petals. On the oral surface the interambulacra are slightly inflated and form weak bourrelets between the phyllodes. A weakly developed, granular "naked zone" is present along the central suture adorally in interambulacrum 5.
Peristome: The peristome lies slightly anterior of the centre, about 44 % TL from the anterior margin. It lies in a depression and is oval (transversely elongated) to slightly subpentagonal. It measures 9 x 5 mm in a 52 mm TL specimen.
Periproct: The periproct is situated inframarginally in interambulacrum 5. It is large (11 x 6 mm long in a 52 mm TL specimen), distinctly rostrate and has an oval, transversely elongated shape.

Differential diagnosis (see Figs. 43, 45-48, 50):
This species differs from *E. richardi* Desmarest in Brongniart, 1829 [= *E. laurillardi* Agassiz & Desor (obj. syn.)] by its greater inequality of the poriferous zones of the paired petals (Gregory, 1891: 607), especially in the posterior paired petals, the more elongated shape and stronger developed rostrum.

E. manzonii is very similar to *E. scutiformis* (Leske, 1778) and the species have been synonymised by some authors (e.g. Stefanini, 1908a: 451-453; Cottreau, 1913a: 111-112), whereas others considered these two species as clearly separated (e.g. Gregory, 1891: 607; Challis, 1980: 158). Consequently there is considerable confusion and it is often unclear which references belong into the synonymy of these species. Gregory (1891: 607) stated that the same characters that distinguish *E. manzonii* from *E. laurillardi*, distinguishes it also from *E. scutiformis*. Comparing the present material from Austria to French specimens of *E. scutiformis* from St. Restitute (Drôme) and Martigues in the collection of the NHMW (1849.V.43, 1852.II.1519 and 1858.XL.452) and to the detailed description and illustrations by Philippe (1998: 93-98; text-figs. 2a-f, 3a-f; pl. 18, figs. 1-9) it is apparent that the two species are different. Based on this material, the following differences can be observed: The petals are distinctly longer in *E. scutiformis* (extending between 72-78 % of the corresponding test radius in the frontal petal, 70-78 % in the anterior paired petals and 64-80 % in the posterior paired petals); the inequality of the poriferous zones is less pronounced in *E. scutiformis* (the anterior poriferous zone extends usually 69-74 % of the posterior poriferous zone in the anterior paired petals and the posterior poriferous zone extends about 74-76 % of the anterior poriferous zone in the posterior paired petals); the periproct lies fully on the oral side in *E. scutiformis*, whereas it is distinctly inframarginal in *E. manzonii*. Furthermore, the oral interambulacra are less swollen between the ambulacra in *E. scutiformis* and the test is less rostrate.

E. aff. *manzonii* differs from *E. manzonii* by its even more asymmetric frontal petal, where poriferous zone IIIb is only 60 to 65% of IIIa, whereas in *E. manzonii* IIIb is 80-96 % of IIIa. Additionally, the posterior poriferous zones of the posterior paired petals are usually shorter in *E.* aff. *manzonii* than in *E. manzonii*.

Discussion:
Contrary to the statement of Gregory (1891: 606, footnote) and later usage by Rose (1975) and Challis (1980) *E. manzonii* Pomel, 1883 is a valid name according to the ICZN rules (Article 12.2.1).

This species is characterised by strong inequality of the poriferous zones in the petals, its oval, elongated outline and the distinct rostrum.

According to Philippe (1998: 98) *E. angulata* Mérian in Agassiz & Desor, 1847 is a junior synonym of *E. scutiformis* (Leske, 1778), since the specimens corresponding to the diagnosis of Mérian are subadult specimens of *E. scutiformis*. The specimens from Camerino (Italy) erroneously attributed to *E. angulata* by de Loriol (1882: 13-18), belong to *E. manzonii*, on account of the structure of their petals. However, since de Loriol referred also to the type specimens of *E. angulata* Mérian this led to some confusion. Therefore, it is very difficult to judge whether or not a record of either species belongs into the synonymy of this species or the very similar *E. scutiformis*, unless accompanied by a description and illustrations (the synonymy list is restricted to such records where a decision was possible). Hence, the distribution data given below may not represent the total temporal and spatial distribution.

The Austrian material was compared with material from the C1 and C2 phosphate conglomerate horizons in the *Globigerina* Limestone of Malta (Early Burdigalian and Langhian respectively). The Maltese material is very similar to the Austria specimens, but usually smaller. Due to the fact that the relative length of the petals decreased during growth, the petals of the smaller Maltese specimens usually extend farther to the margin. The Maltese specimens also show an even larger variation in test height (biometric data of Maltese specimens provided by Michael Gatt, Mdina, Malta).

E. manzonii was placed into the genus *Miolampas* by Pomel (1883: 62) together with the extant *E. depressa* Gray, 1851. This genus was subsequently placed into the synonymy of *Echinolampas* by Kier (1962: 112). While the author does not propose to resurrect the genus *Miolampas*, *E. manzonii* should be referred to this genus in case it is re-established in future. *Miolampas* Pomel, 1883 [type-species: *E. depressa* Gray, 1851, by subsequent designation (Lambert, 1918: 44)] and the similar genus *Planilampas* Mortensen, 1948a [type-species: *E. sternopetala* Agassiz & Clark, 1907, by original designation; Mortensen (1948a) included also a new species, *P. keiensis*, in this genus] differ from typical *Echinolampas* by their strong inequality of the poriferous zones (especially in ambulacrum III) and the different nature of the aboral tuberculation (tubercles very closely spaced with areoles nearly touching in *Echinolampas*, *vs.* widely spaced tubercles with an average distance of *c.* one times the primary spine tubercle diameter in species formerly attributed to *Miolampas* or *Planilampas*). Kier (1962) does not give any specific reasons for synonymising these two genera with *Echinolampas* besides the general remark that "... all these sections and subgenera are based on characters too variable to be of generic distinction." (KIER, 1962: 107).

The subspecies *E. atrophus podolicus* established for a single specimen from the Early Badenian of the Western Ukraine by SZÖRÉNYI (1953) is placed into the synonymy of *E. manzonii*. Her description and illustrations provide no features which would be suited to confidently separate her subspecies from *E. manzonii*. In *E. atropus* LAMBERT, 1906 from the Middle Miocene of Menora (Balearic Islands) the poriferous zone Ia, IIb, IVa, and Vb are almost completely reduced, with only a single pore pair developed adapically according to LAMBERT (1906a: 94). Thus the latter is well differentiated from both *E. manzonii* and SZÖRÉNYI's subspecies.

Little is known of the ecology of similar extant species (*E. depressa*, *E. keiensis*, *E. sternopetala*). According to SERAFY (1979) and MOOI (1990b: 696) *E. depressa* lives in relatively coarse carbonate sand composed of fragments of calcareous algae, based on substrate collected together with the specimens and gut content. MOOI (1990b: 696) observed that the gut content of museum specimens supports MORTENSEN's (1948a) suggestion that *Echinolampas* feeds mainly on foraminifers. While *E. depressa* was reported also from shallower habitats (30 to 310 m, SERAFY, 1979) *E. keiensis* and *E. sternopetala* are known up till now only from deeper settings [245 to 400 m (MORTENSEN, 1948a) and 100 to 500 m respectively (SHIGEI, 1986)].

Occurrence:

Austria: Early Badenian (Langhian)

Eisenstadt-Sopron Basin: Eisenstadt (Hartl Fm., Hartl hill), Bgld (VADÁSZ, 1915; KROH et al. 2003; [NHMW]); Eisenstadt (Hartl Fm., Johannesgrotte), Bgld (PILLER, 2000)

Paratethys (non-Austrian occurrences): Early Badenian (Langhian)

Great Hungarian Basin (Pannonian Basin): Kemence, Pest, Hungary (MEZNERICS, 1941); Márkháza, Nógrád, Hungary (MEZNERICS, 1941); Sámsonháza, Hungary (MEZNERICS, 1941)

Fore-Carpathian Basin: Pod'yarkov (= Podjarków), near Lwów, western Ukraine (SZÖRÉNYI, 1953)

Mediterranean: Burdigalian to Langhian; records from the Messinian and Early Pliocene (NÉRAUDEAU et al. 2001, based on ROMAN & SOUDET, 1990) need to be substantiated

Western Mediterranean: Italy: Campobono, region of Camerino (pp DE LORIOL, 1882); ? Città di Castello (VERRI & DE ANGELIS D'OSSAT, 1900); ? Dogana (VERRI & DE ANGELIS D'OSSAT, 1900; NELLI, 1907); La Vignaccia, near Piedebovigliana, region of Camerino (pp DE LORIOL, 1882); ? Monte Cedrone (VERRI & DE ANGELIS D'OSSAT, 1900; NELLI, 1907); ? Monte S. Maria Tiberina (VERRI & DE ANGELIS D'OSSAT, 1900); St-Ilario region of Camerino (pp DE LORIOL, 1882)

Central Mediterranean: Maltese Islands: Qammieh conglomerate bed (= Qolla l-Bajda conglomerate bed, = C_1 phosphate conglomerate) and Xwieni conglomerate bed (= C_2 phosphate conglomerate) (WRIGHT, 1864; GREGORY, 1891; pp STEFANINI, 1908a; pp STEFANINI, 1908b; pp COTTREAU, 1913a; ROMAN, 1965; ROSE, 1974b, 1975; CHALLIS, 1980) records from the Blue Clay (WRIGHT, 1864) and Greensand (WRIGHT, 1855) are dubious according to CHALLIS (1980) and need to be confirmed; Italy: Maserna, Emilia (pp STEFANINI, 1908b); Molassa di Serra di Guidoni, Bologna (MANZONI, 1881; GREGORY, 1891; pp STEFANINI, 1908b); Montese (Molassa serpentinosa), Colline Bologne (MANZONI, 1880a; MAZZETTI, 1882a; pp STEFANINI, 1908b; ROMAN, 1965); Paullo, Emilia (pp STEFANINI, 1908b); Salto, Emilia (pp STEFANINI, 1908b); San Marino (MANZONI, 1880a; NELLI, 1907); S. Maria Vigliana (Molassa serpentinosa) (MANZONI, 1880a; pp STEFANINI, 1908b; NELLI, 1907)

Echinolampas aff. *manzonii* POMEL, 1883

(Fig. 51; Pl. 52, Fig. 4-6)

Material:

Early Badenian (Langhian) – Stotzing (sandpit Mayer), Bgld, Austria

WANZENBÖCK coll.: 2 specimens (W7, W17)

? Early Badenian (Langhian) – Eisenstadt (Hartl Fm., Hartl hill), Bgld, Austria

NHMW: 1 specimen (NHMW 1997z0178/1717b)

Dimensions (in mm):

Inv. No.	TL	TW	TH
NHMW 1997z0178/1717b	52.7	47.5	19.7
W7	51.4	47.8	22.8
W17	55.9	~55	~20

Description:

Size and shape: The test is of medium size, with oval, anteroposteriorly elongated outline. The anterior margin is rounded, the posterior margin bluntly pointed. The maximum width lies subcentrally. In profile the test is arched, the maximum height coinciding with the position of the apical disc. The ambitus is rounded and tumid, albeit less tumid than in *E. manzonii*. The oral surface is depressed around the peristome.

Apical disc: The apical disc lies slightly anterior of the centre, about 45 % TL from the anterior margin. The apical disc is monobasal with 4 large, circular gonopores. The madreporite is pentagonal and crowded with numerous small madreporic pores. The gonopores lie at the apices of the madreporite. The ocular plates are small, bearing a small, circular ocular pore each.

Ambulacra: The ambulacra are petaloid and relatively narrow. The ambulacral pores within the petals are elongate isopores, outside the petals they are small, slit-shaped unipores. The frontal petal is the shortest and the posterior paired petals longest. They extend about 65-70 % of the corresponding test radius. The poriferous zones of all petals are strongly unequal. In the frontal petal IIIb is about 60 to 65 % of IIIa. In the anterior and posterior paired petals, this is even more strongly expressed, IIb/IVa are about 42 to 54% of IIa/IVb (anterior paired petals) and Ia/Vb are 35 to 45% of Ib/Va (posterior paired petals). Adorally the ambulacra form distinct phyllodes consisting of large rounded unipores.

Interambulacra: The interambulacra are slightly inflated between the petals. The bourrelets between the phyllodes are very weakly expressed.

Peristome: The peristome lies slightly anterior of the centre, about 45 % TL from the anterior margin. It lies in a shallow depression and has an oval, transversely elongated shape.

Periproct: The periproct lies inframarginally, very close to the posterior margin. In specimen W17, however, it lies fully on the oral side, but that may in part be due to the distortion of the test by sedimentary compaction. It has an oval, transversely elongated shape and it is moderately rostrate.

Differential diagnosis:

For the difference between this species and *E. manzonii* see above under that species. From the other *Echinolampas* species discussed in this work this species differs by the pronounced inequality of its poriferous zones.

E. scutiformis (LESKE, 1778) differs by its higher, slightly conical profile and distinctly lesser inequality of its poriferous zones.

Discussion:

Only two specimens from a private collection and a single specimen from the Hartl Fm., Eisenstadt tentatively referred to this species were available. Albeit this species is very similar to *E. manzonii* in many aspects, it shows considerable differences (see above, and Figs. 50.A-B). However, because there is a large overlap between this species and *E. manzonii* in many features it is named in open nomenclature and placed into the affinity of *E. manzonii*. More material is necessary to explore the intraspecific variability of this species and *E. manzonii* and to test if *E.* aff. *manzonii* is just an end-member of the morphospace of *E. manzonii*.

Occurrence:

Austria: Early Badenian (Langhian)
Eisenstadt-Sopron Basin: Eisenstadt (Hartl Fm., Hartl hill), Bgld ([NHMW])
Vienna Basin: Stotzing (sandpit Mayer), Bgld ([NHMW])

Echinolampas sayni Lambert, 1913

(Fig. 43; Pl. 53, Figs. 1-6; Pl. 54, Figs. 1a-c)

pp 1847a [*Echinolampas*] *hemisphæricus* Agass. – Agassiz & Desor: 165
pp 1858 *Echinolampas hemisphaericus* Agass. – Desor: 307
#?pp 1869a *Echinolampas angustistellatus* Laube. – Laube: 183
? pp 1870 *Echinolampas angustistellatus* Lbe. – Laube: 314
? pp 1871 *Echinolampas angustistellatus* Laube. – Laube: 67; pl. 18, fig. 4
*. 1913a *Echinolampas sayni* Lambert. – Lambert: 148-149; pl. 11, figs. 3-5
? v 1915 *Echinolampas* cfr. *Sayni* Lamb. – Vadász: 215-216
pp 1965 [*Echinolampas*] *hemisphaerica* Lamarck 1816 (*Clypeaster*) – Roman: 283-284
. 1974 *Echinolampas hemisphaerica sayni* Lamb. – Roman: 332
. 1984 *Echinolampas sayni* Lambert, 1913 – Negretti: 113-114; pl. 8, figs. 2-6
1990 *Echinolampas hemisphaerica* (Lamarck, 1816) – Philippe et al.: 244
pp 1998 *Echinolampas hemisphaerica* (Lamarck, 1816) – Philippe: 86-90; pl. 17, fig. 5

Type-material:

Echinolampas sayni Lambert, 1913:
Syntypes: specimens figured by Lambert (1913: pl. 11, figs. 3-5), nos. L 19 670 and 19 671; Lambert collection, Muséum national, Paris, France (Philippe, 1998: 290)
Locus typicus: Cap Couronne, Martigues, France
Age: Burdigalian

? *Echinolampas angustistellatus* Laube, 1869:
Syntypes: specimens housed at the k. k. Hof-Mineraliencabinet (today the Naturhistorisches Museum Wien) and k. k. geol. Reichsanstalt (today the Geological Survey of Austria) according to Laube (1871); see below
Type area: Eggenburg Basin (localities Eggenburg and Gauderndorf) and Horn Basin (Maria Dreieichen, east of Horn), Lower Austria, Molasse zone
Age: Eggenburgian (Early Burdigalian)
Remarks: The overwhelming part of Laube's types and reference material was not marked as such in the collections. Only in few cases new labels were attached to the specimens in the collections. Hence tracing of this material is difficult in most cases. Only a single specimen (NHMW 1866.XX.16) in the NHMW and GBA collections labelled *E. angustistellatus* could be located. Whether or not the hand-writing on the label is that of Laube could not be ascertained as no sample of Laube's hand-writing (e.g. letters,...) was available for comparison [neither at the archive of the Geological Survey of Austria (Tillfried Cernajsek, pers. comm. April 2003), nor at the archives of the Naturhistorisches Museum Wien (Christa Riedl-Dorn, pers. comm. April 2003)]. The potential syntype (NHMW 1866.XX.16) is a moderately well preserved specimen of *E. schultzi*. A second specimen (NHMW 1850.IX.83) attributed here to *E. sayni*, could be a potential syntype of *E. angustistellatus*. Although the latter is not accompanied by a label with the name *E. angustistellatus* it is the only specimen in the NHMW and GBA collections that has a large similarity to Laube's figure of the aboral surface of *E. angustistellatus* [similarities include: the shape of the petals, the outline, the missing apical disc, the crushing of the test in interambulacra 1 and 2 and ambulacrum II, as well as two round penetrations of the test [cassid predation ?] in poriferous zone Vb and along the central suture of aboral interambulacrum 5 [in Laube's figure they are located in different areas, but their relation to each other is identical; a similar mistake happened in his figure of *S. leithanus*: the two strongly visible arched sutures in interambulacrum 1 of the type were drawn in both interambulacrum 1 and 5 in Laube's figure]; all specimens in Laube, 1871 are mirror images, as can be seen by comparison of the type material of *Spatangus austriacus* (Pl. 75, Fig. 2a) and *Schizaster leithanus* (Pl. 66, Fig. 1a) with Laube's figures (pl. 19, fig. 1a, and pl. 18, fig. 7 respectively)]. Why Laube's figure of *E. angustistellatus* does not show a periproct but a tuberculated area instead, is dubious. Considering the problems associated with identifying the type material of *E. angustistellatus* and the fact that the two potential syntypes belong to different species *E. angustistellatus* is here considered as *nomen dubium*.

Material:

Early Eggenburgian (Early Burdigalian) – Maria Dreieichen, NÖ, Austria
GBA: 1 specimen, tentatively referred to *E. sayni* (GBA 2004/1/27)
Late Eggenburgian (Early Burdigalian) – Eggenburg, NÖ, Austria
GBA: 2 specimens (GBA 2004/1/21, 2004/1/25), and 1 specimen, tentatively referred to *E. sayni* (GBA 2004/1/29)
Late Eggenburgian (Early Burdigalian) – Gauderndorf (Zogelsdorf Fm.), near Eggenburg, NÖ, Austria
NHMW: 2 specimens [NMHW 1850.IX.83 (potential syntype and figured specimen of *Echinolampas angustistellatus* Laube, 1869); 1866.XI.57]
GBA: 1 specimen, tentatively referred to *E. sayni* (GBA 2004/1/16)
Late Eggenburgian (Early Burdigalian) – Grübern (Zogelsdorf Fm.), near Maissau, NÖ, Austria
NHMW: 18 specimens (NMHW 1914.VII.20/2-14, 2003z0035/0001-3)

Dimensions: see Tab. 10

Description:

Size and shape: The test is of medium size, its outline is subcircular to slightly antero-posteriorly elongated. Test width ranges of the studied specimens from 93.9 to 98.9 % of TL with a mean of 96.7 %. The anterior and posterior margins are rounded. The maximum width lies posterior of the apical disc, about 53 to 60 % of TL from the anterior margin. In profile the test is low arched, the maximum height coinciding with the position of the apical disc. Test height ranges from 31.6 to 41.6 % TL, with a mean of 35.7 %. The ambitus is rounded and tumid.

Apical disc: The apical disc lies slightly anterior of the centre, about 38 to 42 % TL from the anterior margin. The apical disc is monobasal with 4 circular gonopores. The madreporite is subpentagonal and crowded with numerous small madreporic pores. The gonopores lie at the apices of the madreporite. The ocular pores are small and circular in outline.

Ambulacra: Adapically the ambulacra are petaloid, moderately broad, straight and slightly closing distally. The frontal petal extends about 64 to 70 (mean 66.7) % of the corresponding test radius, the anterior paired petals about 69 to 89 (mean 80.7) % and the posterior paired petals about 67 to 82 (mean 75.7) %. The poriferous zones within the anterior paired petals are not of equal length. The anterior poriferous zones (IIb and IVa) are shorter than the posterior ones (IIa and IVb) extending only 63 to 78 (mean 70.8) % of the length of the petal. In the frontal petal, the right-handed poriferous zone is slightly longer than the left-handed. The poriferous are slightly arched, resulting in a slightly lanceolate shape of the petals. The ambulacral pores in the petals are conjugated anisopores, which are strongly conjugated and slightly oblique. The interporiferous zones within the petals are inflated and up to four times as wide as a single poriferous zone. The poriferous zones are distinctly depressed.

Adorally the ambulacra form moderately depressed phyllodes consisting of four series of unipores in each ambulacrum. Two adradial series consisting of large closely spaced unipores and two perradial series consisting of smaller, slightly vertically elongated unipores, which are more widely spaced. The two buccal pores are of similar size as the pores from the adradial series, but distinctly elongated vertically. In the adoral third of the phyllodes there is a double row of shallow pits running along the central suture (interpreted as sphaeridial pits).

Interambulacra: Adapically the interambulacra are slightly inflated between the petals. On the aboral surface they are covered with small, crenulate, perforate tubercles which are deeply sunken and closely spaced. On the oral surface the tubercles are distinctly larger. From the margin towards the peristome the tubercles are increasingly widely spaced, with distances of up to the diameter of the areole between the tubercles in the area immediately around the peristome. Adorally the interambulacra are only slightly inflated and form indistinct bourrelets. In some of the specimens, usually in those with lower tubercle densities, a narrow, not very prominent "naked zone" can be found medially in interambulacrum 5 (about halfway between peristome and periproct).

Peristome: The peristome lies slightly anterior of the centre, about 42 to 46 % TL from the anterior margin. It lies in a shallow depression. It is oval, transversely elongated and measures 14 to 17 % test width.

Periproct: The periproct is situated inframarginally in interambulacrum 5 and lies fully on the oral surface. It is oval (transversely elongated) and measures 17 to 19 % test width.

Differential diagnosis (see Figs. 43, 45-48)**:**

As outlined above, *E. sayni*, is characterised by its rounded, oval and only weakly rostrate outline, low test height, very weakly developed bourrelets, large peristome (~15 to 17 % of TW) and periproct (~17 to 19 % of TW), the latter of which lies very close to the edge or at of the posterior margin, moderate tubercle densities in the aboral interporiferous zones (usually between 256-288 tubs./cm²) and moderately high number of tubercles across the interporiferous zone in the posterior petals (usually 7 to 8).

For the difference to *E. schultzi* see below under that species.

E. sp. 1, from the Badenian (Langhian-Early Serravallian), differs from *E. sayni* by its more elongated distinctly oval outline, much higher profile, flatter oral surface, smaller periproct and peristome, and higher tubercle densities.

E. hemisphaerica from the Late Badenian (Early Serravallian) differs from *E. sayni* by its more elongated, subpentagonal outline, more globose profile and larger test height.

E. barcinensis from the Late Badenian (Early Serravallian) differs from *E. sayni* by its subpentagonal and more rostrate outline, more pentagonal peristome, lower test height, usually smaller periproct and peristome and position of the periproct which lies higher on the ambitus.

E. manzonii and *E.* aff. *manzonii*, both from the Early Badenian (Langhian) differ from *E. sayni* by their very unequal poriferous zones in the paired petals (among other features, but this is the most easily recognised).

E. richardi from the Burdigalian of the Aquitaine Basin differs by its slightly more subpentagonal outline, shorter petals, less depressed and narrower poriferous zones, and less concave oral surface.

Discussion:

This species is similar to the contemporaneous/co-occurring *E. schultzi* but is well differentiated by the features outlined under the latter species. There are a few specimens of *E. schultzi*, which are also more or less subcircular in outline, but these specimens, do retain a bluntly pointed (rostrate) posterior end, whereas specimens of this *E. sayni* usually have a rounded posterior end. Additionally, these specimens can easily be distinguished from *E. sayni* by their stronger inflated bourrelets, smaller periproct, which is also farther from the posterior margin and higher tuberculation densities.

The Austrian material is attributed to the species *E. sayni* Lambert, 1913 from the Burdigalian of the southern Rhône Basin. This species was placed into the synonymy of *E. hemisphaerica* by Roman (1958: 257; 1965: 283-284), Philippe et al. (1990: 244) and Philippe (1998: 89). An opinion, with which the present author disagrees: *E. sayni* can be well distinguished from *E. hemisphaerica* by its more rounded, not rostrate outline, lesser test height, more strongly sunken poriferous zones and smaller petals (compare Figs. 43, 46.A). Roman (1974: 335) and Negretti (1984: 113-114) considered it as subspecies of *E. hemisphaerica*.

In this species, contrary to the other species of *Echinolampas* described in the present paper, a naked, granular zone may be developed along the central suture of adoral interambulacrum 5. This feature is not present in all specimens of E. sayni and varies continuously from being absent to being small, but well visible. It is thus not possible to separate two groups on base of this feature.

The description of *E. angustistellatus* given by Laube (1871) reveals many similarities to *E. sayni* and it is likely that part of the material attributed to *E. angustistellatus* by Laube, 1871 belongs to this species. For the status of *E. angustistellatus* see above under "Type-material".

Two specimens from the Transylvanian Basin [Săndulesti (= Szind), Romania] were tentatively placed into this species by Vadász (1915: 215-216). These specimens (MAFI Ech 304) are indeed similar to the French and Austrian material of *E. sayni*. Unfortunately the stratigraphic age of the locality is unclear, although Vadász (1915: 216) considers it to belong to the so-called "Obermediterran", which in most cases translates into Badenian (Langhian-Early Serravallian) today. The present author found no evidence supporting or rejecting this dating.

Occurrence:

Austria: Early to Late Eggenburgian (Early Burdigalian)

Molasse Zone: Eggenburg, NÖ ([GBA]); Gauderndorf (Zogelsdorf Fm.), near Eggenburg, NÖ (pp Laube, 1869a, 1870, 1871; [NHMW]); Grübern (Zogelsdorf Fm.), near Maissau, NÖ ([NHMW]); Maria Dreieichen, NÖ ([GBA])

Paratethys (non-Austrian occurrences):

Transylvanian Basin: ? Săndulesti (= Szind), Romania [Vadász, 1915 (specimens poorly preserved, stratigraphy unclear)]

Mediterranean: Burdigalian

Rhône Basin: Littoral de la Nerthe [Beaumaderie (Negretti, 1984; Philippe, 1998); Cap Couronne (pp Agassiz & Desor,

1847a; pp DESOR, 1858; WRIGHT, 1864; LAUBE, 1871; MANZONI, 1873; GREGORY, 1891; LAMBERT, 1913a; ROMAN, 1974; NEGRETTI, 1984; PHILIPPE et al., 1990; pp PHILIPPE, 1998); Verdon (pp PHILIPPE, 1998)], southern France

Echinolampas schultzi KROH, *nomen novum*

(pro *E. laurillardi acuminata* SCHAFFER, 1912, preoccupied by *E. acuminata* ABICH, 1882)
(= *E. laurillardi* sensu LAUBE, 1871, non AGASSIZ in AGASSIZ & DESOR, 1847)

(Fig. 43.5, 51; Pl. 54, Figs. 2-3; Pl. 55, Figs. 1-3; Pl. 56, Figs. 1-2)

1850 *Clypeaster Linkii* – HÖRNES: 668
1866a *Echinolampas Linkii* – SUESS: 94, 95, 99, 101, 103, 112-114, 139
1868 *Echinolampas Kleinii* – FUCHS: 586, 588, 596
1869a *Echinolampas Laurillardi* AG. – LAUBE: 183
? pp1869a *Echinolampas angustistellatus* LAUBE. – LAUBE: 183
1870 *Echinolampas Laurillardi* AGASSIZ. – LAUBE: 314
1870 *Echinolampas angustistellatus* LBE. – LAUBE: 314
1871 *Echinolampas Laurillardi* AGASSIZ. – LAUBE: 66-67; pl. 18, fig. 1
? pp 1871 *Echinolampas angustistellatus* LAUBE. – LAUBE: 67; pl. 18, fig. 4
pp 1873 *Echinolampas Laurillardi*, AGASS – MANZONI: 10-11, 19
1875a *Echinolampas Linkii* – R. HOERNES: 342
1891 *Echinolampas Laurillardi* AGASS. (non *Linkii*) – SUESS: 409
? 1891 *Echinolampas* sp. aff. *Laurillardi* – SUESS: 413, 415
1898 *Echinolampas Laurillardi* AG. – ABEL: 212, 215, 216
non 1899 *Echinolampas Laurillardii*, AGAS. (non LAUBE). – AIRAGHI: 161-162; pl. 6 (1), figs. 7a-b, 8a-b, 9a-b
1900 *Echinolampas Laurillardi* – FUCHS: 861
? 1900 *Echinolampas* – FUCHS: 873-874
1903 *Echinolampas Laurillardii* – HOERNES: 931
? 1904 *Echinolampas* cfr. *Laurillardi* – ABEL: 20
1904 *Echinolampas Laurillardi* – ABEL: 24
1912a *Echinolampas Laurillardi* AG. – SCHAFFER: 189; text-fig. 2; pl. 60, fig 4, 6
1912a *Echinolampas Laurillardi* AG. var. *acuminata* SCHFF. – SCHAFFER: 190; pl. 60, fig. 5
? 1913a *Echinolampas* – SCHAFFER: 6, 67, 73-74, 82
1913a *Echinolampas Laurillardi* AG.– SCHAFFER: 9, 17, 36, 47, 56; 155-157; pl. 10, fig. 9
1924b *Echinolampas Laurillardi* – SCHAFFER: 484
1927a *Echinolampas Laurillardi*. – SCHAFFER: 56
non 1938 *Echinolampas laurillardi* AGASSIZ 1843 – POLJAK: 195; pl. 10, fig. 1 [not conspecific with *E. schultzi*; probably misidentified *E. hemisphaerica*, although the subcircular outline and the long poriferous zones are unusual]
non 1939 *Echinolampas laurillardi* AGASSIZ – TAVIANI: 38; pl. 2 (4), figs. 14a-b [misidentified *E. barcinensis*]
1943 *Echinolampas laurillardi* AG. – SCHAFFER: 521, 523, 525, 527
non 1950 *Echinolampas laurillardi* AG. DES. 1847 – SZÖRÉNYI: 145; pl. 1, fig. 7 (the specimen on which this record is based is very poorly preserved, it might belong to *E. barcinensis* or *E. hemisphaerica*, but this cannot be decided without re-examination; based on the photograph of the aboral surface an attribution to *E. schultzi* can be excluded)
1951 *Echinolampas laurillardi* AG. – SCHAFFER & GRILL: 709, 711, 712, 714
1955 *Echinolampas* sp. – BERNHAUSER: 182
1955 *Echinolampas laurillardi* – THENIUS: 108-109; pl. 12, fig. 10
1957 *Echinolampas laurillardi* AG. – TOLLMANN: 173
1962 *Echinolampas laurilardi* – PAPP & THENIUS in KÜHN: 118
1962a *Echinolampas laurillardi* AGASSIZ – THENIUS: 18; pl. 5, fig. 10
1962b *Echinolampas laurillardi* – THENIUS: 106-107; pl. 7, fig. 10
pp 1965 [*Echinolampas*] *richardi* DESMAREST in BRONGNIART 1829 (*Clypeaster*) – ROMAN: 297
1974 *Echinolampas laurillardi* AGASSIZ – THENIUS: 46-47; fig. 1/21
1971a *Echinolampas laurillardi* AGASSIZ, 1847 – STEININGER: 595; pl.3, fig. 3-4
1971b *Echinolampas laurillardi* AGASSIZ – STEININGER: 119
1971c *Echinolampas laurillardi* AGASSIZ – STEININGER: 129
1975 *Echinolampas* (*Echinolampas richardi*) DESMAREST in BRONGNIART, 1829. – KALABIS: 175
1979 *Echinolampas laurillardi* – THENIUS: 50; pl. 2, fig. 21
1983 *Echinolampas laurillardi* – THENIUS: 118; pl. 6, fig. 21
non 1997 *Echinolampas laurillardi* AGASSIZ – MAJCEN et al.: 106; pl. 5, fig. 1
v. 1997 *Echinolampas* sp. – NEBELSICK et al.: 272-274; figs. 2-3
v. 1999 *Echinolampas laurillardi* – HARZHAUSER & KROH: 221
v. 1999 *Echinolampas laurillardi* AGASSIZ & DESOR, 1847 – KROH & HARZHAUSER: 158-159; pl. 2, figs. 1-3
v. 2002 *Echinolampas laurillardi* – LUKENEDER & HARZHAUSER: 462

Type-material:
Echinolampas laurillardi acuminata SCHAFFER, 1912:
Holotype: figured by SCHAFFER (1912a: pl. 60, fig. 5); the specimen could not be located at the Krahuletz Museum (Eggenburg, NÖ) where it was housed according to SCHAFFER (1912a: 190) and is considered to be lost
Locus typicus: Eggenburg (Kremserberg), NÖ
Stratum typicum: Zogelsdorf Fm.
Age: Late Eggenburgian (Early Burdigalian)

Echinolampas schultzi KROH, *nomen novum*:
Neotype: specimen NMHW 2003z0040/0001 from the train station in Eggenburg; housed at the Naturhistorisches Museum Wien, Geologische Abteilung
Dervatio nominis: In honour of my colleague Ortwin SCHULTZ (Geologische Abteilung at the Naturhistorisches Museum Wien). Without his continuous support and willingness to discuss all the numerous nomenclatorial problems that arose during the preparation of the present volume of the Catalogus Fossilium Austriae this work would not be the same.
Remarks: As the holotype of SCHAFFER (1912a) is lost, a neotype is selected. The historical locality Eggenburg (Kremserberg) is not longer exposed. It consisted of outcrops of the Zogelsdorf Fm. along the road from the centre to the train station and in immediate the vicinity of the latter (F.F. STEININGER, *in litt.*, Sept. 2004; for a section of Zogelsdorf Fm. outcrops exposed during excavations in the 1980ies see NEBELSICK, 1989). Therefore the specimen collected at the train station Eggenburg (NMHW 2003z0040/0001) can be considered as topotype and is available for neotype-selection.

? *Echinolampas angustistellatus* LAUBE, 1869:
Syntypes: specimens housed at the k. k. Hof-Mineraliencabinet (today the Naturhistorisches Museum Wien) and k. k. geol.

Reichsanstalt (today the Geological Survey of Austria) according to LAUBE (1871); see below
Type area: Eggenburg Basin (localities Eggenburg and Gauderndorf) and Horn Basin (Maria Dreieichen, east of Horn), Lower Austria, Molasse zone
Type area: Eggenburg and Horn Basin, Molasse Zone; Eggenburgian (Early Burdigalian)
Remarks: *E. angustistellatus* is here considered as *nomen dubium* because the type-material could not be unequivocally be identified and potential syntypes belong to two different species (see above under *E. sayni*).

Material:
Early Eggenburgian (Early Burdigalian) – Horn, NÖ, Austria
NHMW: 1 specimen (NMHW 2002z0181/0018)
Early Eggenburgian (Early Burdigalian) – Maria Dreieichen, near Horn, NÖ, Austria
NHMW: 4 specimens (NMHW 1849.XIII.39, 1860.L.206, 2003z0037/0001-2)
Late Eggenburgian (Early Burdigalian) – Eggenburg, NÖ, Austria
NHMW: 6 specimens (NMHW 1904.VIII.24, 1972/1551/121, 1997z0178/0466, 2002z0181/0019)
Late Eggenburgian (Early Burdigalian) – Eggenburg (Schindergraben), NÖ, Austria
NHMW: 1 specimen (NMHW 2003z0038/0002)
Late Eggenburgian (Early Burdigalian) – Eggenburg (Zogelsdorf Fm., train station), NÖ, Austria
NHMW: 1 specimen [NMHW 2003z0040/0001 (**neotype**)]
Late Eggenburgian (Early Burdigalian) – Gauderndorf (Zogelsdorf Fm.), near Eggenburg (often referred to as Gauderndorf-Himmelreich), NÖ, Austria
NHMW: 32 specimens (NMHW 1851.VI.65, 1852.XII.7 [A605], 1860.L.445, 1860.L.446, 1860.L.449, 1860.L.744, 1861.L.295, 1866.XX.14, 1866.XX.15, 1866.XX.16, 1866.XX.17, 1869.III, 1869.III.38, 1869.III.39, 1869.III.40, 1869.III.41, 1934.I.168, 592/1964, 1973/1615/355, 1997z0178/0511, 1998z0048/0006, 2002z0181/0021, 2003z0036/0001, 2003z0036/0002-8)
GBA: 25 specimens [GBA 2004/1/15, 2004/1/17-18, 2004/1/20, 2004/1/22-24, 2004/1/26, 2004/1/28, 2004/1/30-44, 2004/43 (figured specimen of NEBELSICK et al., 1997: figs. 2a-b, 3c)]
IPUW: 1 specimen (IPUW 1992/209)
Late Eggenburgian (Early Burdigalian) – Grübern (Zogelsdorf Fm.), near Maissau, NÖ, Austria
NHMW: 5 specimens (NMHW1914.VII.20/1, 1914.VII.21)
Late Eggenburgian (Early Burdigalian) – Limberg (Zogelsdorf Fm., Hengl quarry), NÖ, Austria
NHMW: 1 specimen (NMHW 1998z0048/0049)
Late Eggenburgian (Early Burdigalian) – Unternalb (Retz Fm., Hungerfeld), NÖ, Austria
NHMW: 1 specimen (NMHW 1999z0049/0002)

Dimensions: see Tab. 10

Description:
Size and shape: The test is large, subpentagonal in outline and is slightly elongated antero-posteriorly. The anterior margin is rounded, the posterior margin bluntly pointed. The maximum width lies posterior of the apical disc, about 62 to 66 % of TL from the anterior margin. In profile the test is domed, the maximum height coinciding with the position of the apical disc. The ambitus is rounded and tumid. The test width is about 89.3 to 97.7 % of TL (mean 93.2 %), the height about 39.8 to 49.8 % (mean 44.2 %).
Apical disc: The apical disc lies slightly anterior of the centre, about 42 to 44 % TL from the anterior margin. The apical disc is monobasal with 4 circular gonopores. The madreporite is pentagonal and crowded with numerous small madreporic pores. The gonopores lie at the apices of the madreporite. The ocular plates are small, bearing a small, circular ocular pore each.

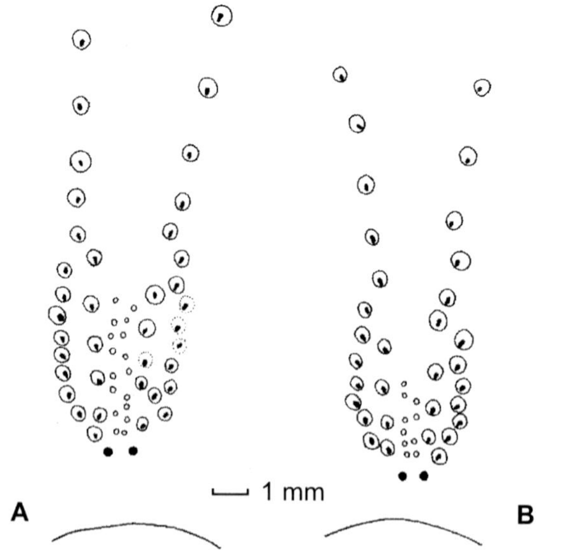

Figure 51: Phyllodes I (A) and III (B) of *Echinolampas schultzi* KROH, nov. nom. (Gauderndorf, NÖ; NHMW A606).

Ambulacra: Adapically the ambulacra are petaloid, moderately broad, straight and moderately closed distally. The frontal petal extends about 60 to 70 (mean 66) % of the corresponding test radius, the anterior paired petals about 70 to 87 (mean 79) % and the posterior paired petals about 71 to 80 (mean 76) %. In contrast to the frontal and posterior paired petals, the poriferous zones within the anterior paired petals are not of equal length. The anterior poriferous zones (IIb and IVa) are shorter than the posterior ones (IIa and IVb) extending only 66 to 80 (mean 75) % of the length of the petal. Poriferous zones Ib, IIa, IVb and Va are distinctly arched, whereas the other poriferous zones are more or the less straight. The ambulacral pores in the petals are conjugated anisopores, which are strongly conjugated and slightly oblique. The interporiferous zones within the petals are inflated and up to five times as wide as a single poriferous zone. The poriferous zones are distinctly depressed. Adorally the ambulacra form moderately depressed phyllodes. They consist of four series of unipores in each ambulacrum. Two adradial series consisting of large, closely spaced unipores with rather large attachment area and two perradial series consisting of smaller, slightly vertically elongated unipores, which are more widely spaced. The two buccal pores are of similar size as the pores from the adradial series, but distinctly elongated vertically. In the adoral third of the phyllodes there is a double row of shallow pits (sphaeridial pits) running along the central suture.
Interambulacra: Adapically the interambulacra are slightly inflated between the petals. They are covered with small, crenulate, perforate tubercles which are deeply sunken and closely spaced. The areoles are inclined towards the margin of the test. The tubercle density in the aboral interambulacra ranges from 200 to 300 tubercles/cm², being highest near the apical system and decreasing gradually towards the margin, where the tubercles are larger. On the oral surface the tubercles are slightly larger and their density ranges from 110 to 210/cm²; the highest densities on the oral side are found near the margin. The tubercles are closely spaced, with the areoles nearly "touching" each other. On the oral side the interambulacra are slightly inflated and form well developed bourrelets adorally. There is no "naked zone" along the central suture between peristome and periproct in interambulacrum 5, as reported in other species of *Echinolampas*.
Peristome: The peristome lies slightly anterior of the centre, about 42-44 % TL from the anterior margin. It lies in a shallow

depression. It is subpentagonal and transversely elongated. It is nearly of the same size as the periproct, being 11 to 14 % of the test width wide. In subadult specimens it can be up to 19 % test width.

Periproct: The periproct lies near the posterior margin in interambulacrum 5, fully on the oral surface and not visible in posterior view. It is usually between 2 and 3 mm from the posterior margin, which is more than in all other species of *Echinolampas* considered here. The periproct is about 12 to 15 % test width wide, except in subadult specimens it can be up to 19 %. The shape of the periproct is oval, transversely elongated and it is slightly rostrate.

Emended diagnosis:

As outlined above, *E. schultzi* is characterised by its subpentagonal, rostrate outline, strongly developed bourrelets, small peristome (~11 to 14 % of TW) and periproct (~12 to 15 % of TW), the latter of which lies fully on the oral surface, high tubercle densities, especially in the aboral ambulacra (usually exceeding 300 tubs./cm²) and large number of tubercles across the interporiferous zone in the posterior petals (8-16 in adult specimens).

Differential diagnosis (see Figs. 43, 45-48):

E. sayni, a co-occurring species, differs from *E. schultzi* by its more rounded, oval and only weakly rostrate outline, lower test height, very weakly developed bourrelets, larger peristome (~15 to 17 % of TW) and periproct (~17 to 19 % of TW), the latter of which lies very close to the edge or at of the posterior margin. Tubercle densities in the aboral ambulacra are distinctly lower (usually between 180-260 tubs./cm²) and there are less tubercles across the interporiferous zone in the posterior petals (usually 5 to 8).

E. sp. 1, from the Badenian (Langhian-Early Serravallian), differs from *E. schultzi* by its more elongated distinctly oval outline, higher profile and flatter oral surface.

E. hemisphaerica from the Late Badenian (Early Serravallian) differs from *E. schultzi* by its more globose profile, less rostrate outline, weakly developed bourrelets, and larger periproct (usually ~ 16% of TW), which lies closer to the edge. Tubercle densities in the aboral interporiferous zones are distinctly lower (usually below 200 tubs./cm²) and there are less tubercles across the interporiferous zone in the posterior petals in specimens of similar size. Additionally the poriferous zone of the paired petals is wider in *E. hemisphaerica* than in *E. schultzi*.

E. barcinensis from the Late Badenian (Early Serravallian) differs from *E. schultzi* by its lower profile, lower test height, larger periproct (~16 to 19 % of TW) which lies very close to the edge or at of the posterior margin. Tubercle densities in the aboral interporiferous zones are distinctly lower (usually below 200 tubs./cm²) and there are less tubercles across the interporiferous zone in the posterior petals in specimens of similar size.

E. manzonii and *E.* aff. *manzonii*, both from the Early Badenian (Langhian) differ from *E. schultzi* by their very unequal poriferous zones in the paired petals (among other features, but this is the most easily recognised).

Discussion:

This species was formerly referred to *Echinolampas laurillardi* by most Austrian authors (e.g. LAUBE, 1869a, 1871; SCHAFFER, 1912a). *E. laurillardi* was proposed by AGASSIZ in AGASSIZ & DESOR (1847a: 165-166), however, for fossil specimens from France recorded under the name *E. richardi* DESMAREST in BRONGNIART, 1829 by DES MOULINS (1837: 188), of which it is a junior objective synonym. Although DES MOULINS (1869: 313) states that there are two different species of *Echinolampas* in the Early Miocene of Léognan, and were confused with each other (he considered *E. laurillardi* and *E. richardi* to be different).

Additional confusion was created by the inversion of the specimen numbers of the type specimen of *E. laurillardi* [AGASSIZ & DESOR (1847a: 165) state that "moulage 34" is the holotype; this however, is a mistake corrected by DESOR (1858), who gave the correct specimen number (moulage 35); see also LAMBERT & JEANNET (1928: 120)].

For the extant specimens from Senegal, originally referred to *E. richardi* by DES MOULINS (1837: 186), DES MOULINS (1869: 316, pl. 20, figs. 1-3) proposed the name *E. rangii* {more information on this species can be found in AGASSIZ [1872-74: 114, 552, pl. 15, figs. 5-11, pl. 15a, figs. 5-6 (under the name *E. hellei*)], MORTENSEN [1948a: 289-292, pl. 5, fig. 7, pl. 11, figs. 16-17] and MOOI [1990a: 79]; a synonymy is given by LAMBERT [1927b: 116-117]; contrary to the statement of COTTEAU [1876: 196] *E. hellei* does not have priority over *E. rangii*, as it is a *nomen nudum* [see LAMBERT, 1927b: 116 (footnote)]}.

ROMAN (1965: 297) synonymised *E. laurillardi* AGASSIZ in AGASSIZ & DESOR, 1847, *E. angustistellatus* LAUBE, 1869 and *E. laurillardi* var. *acuminata* SCHAFFER, 1912a with *E. richardi*, without discussion on the reasons or the problems associated with the type material of *E. laurillardi*.

The specimens from Austria are, however, distinctly different from *E. richardi* (= *E. laurillardi*), when compared to two specimens in the collection of the Naturhistorisches Museum Wien (NHMW 1914.VI.348a-b), from the type locality Léognan in the Aquitaine Basin in Western France. *E. richardi* has shorter petals, which extend only ~ 50 % (frontal petal) respectively 66 % (paired petals) of the corresponding test radius, while these of the Austrian specimens extend 66% (frontal petal) respectively 75-85 % (paired petals). Moreover, there are less tubercles across the interporiferous zones in *E. richardi*, the bourrelets are only weakly inflated, the poriferous zones only slightly sunken and the test is also less rostrate. This has already been observed by LAMBERT (1913a: 148), who stated that the material reported as *E. laurillardi* by LAUBE had nothing to do with *E. laurillardi* AGASSIZ. This, however, was largely ignored by subsequent workers. A synonymy and morphometric data of *E. richardi* can be found in CHAVANON (1974: 192-193; pl. 13, figs. 5a-b).

The next available name is that of SCHAFFER (1912a), who established the subspecies *Echinolampas laurillardi acuminata* SCHAFFER, 1912 for material from the Eggenburgian of Austria (the reasons for excluding *E. angustistellatus* are given below). This subspecies can, however, not be simply elevated to species rank since the name is preoccupied by *Echinolampas acuminata* ABICH, 1882. Therefore it is necessary to establish a *nomen novum* (*E. schultzi*).

E. angustistellatus LAUBE, 1869 might belong in the synonymy of this species, but this is unclear. The type material of *E. angustistellatus* could not be definitely traced in the collections of the Naturhistorisches Museum Wien and the Geological Survey of Austria. Two potential syntypes belong to two different species (*E. sayni* and *E. schultzi*). Hence, *E. angustistellatus* is considered a *nomen dubium* (see above under "Type-material" and under *E. sayni*).

Occurrences in the Badenian of the Vienna Basin and the Oberpullendorf Bay (Ritzing, Bgld) reported by LAUBE (1869a, 1871) and MANZONI (1873) could not be verified. Most probably these records refer to misidentified *E. hemisphaerica*. Likewise, records from the Badenian of other areas of the Central Paratethys (POLJAK, 1938; SZÖRÉNYI, 1950; MAJCEN et al., 1997) are misidentifications.

Palaeoecology:

Predation scars are common; they consist of irregular holes in the test, which are tentatively attributed to predation by cassids. These gastropods drill holes in the calcareous shell of their prey using acid and neurotoxin glands in their proboscis to weaken the shell and paralyse the spines/pedicilaria and finally enlarge the hole using their radula (HUGHES & HUGHES, 1971; LINDER, 1990; KOWALEWSKI & NEBELSICK, 2003 and references therein). One gastropod may drill several holes into the test of their prey in order to reach all parts of the intestines, especially in sand dollars, where the interior of the test is divided by pillars.

This species is common in mobile coarse sandy sediments in high water energy environments (e.g. the Bryozoan-balanid-sands of Retz and Gauderndorf).

Occurrence:

Austria: ? Egerian (Chattian-Aquitanian), Early to Late Eggenburgian (Early Burdigalian)

Molasse Zone: Bayersdorf, NÖ (SUESS, 1866a); Eggenburg, NÖ (pp LAUBE, 1869a, 1871; SUESS, 1891; FUCHS, 1900; HOERNES, 1903; SCHAFFER, 1943; SCHAFFER & GRILL, 1951; THENIUS, 1955; PAPP & THENIUS in KÜHN, 1962; THENIUS, 1962a, b, 1974, 1979, 1983; [NHMW]); Eggenburg (Bahnhof), NÖ (ABEL, 1898; ? SCHAFFER, 1913a; [NHMW]); Eggenburg (Bauerhanslgrube), NÖ (SCHAFFER, 1912a); Eggenburg (Brunnstube), NÖ (SUESS, 1866a; HOERNES, 1875a; ABEL, 1898; SCHAFFER, 1912a; SCHAFFER, 1943; SCHAFFER & GRILL, 1951; STEININGER, 1971a, c); ? Eggenburg (Floriani-bründl), NÖ (SCHAFFER, 1913a); Eggenburg (Kalvarienberg), NÖ (SUESS, 1866a); Eggenburg (Kremserberg), NÖ (SCHAFFER, 1912a, 1913a; KROH & HARZHAUSER, 1999); Eggenburg (Schindergraben), NÖ (SCHAFFER, 1912a; [NHMW]); Gauderndorf (Zogelsdorf Fm.), NÖ (HÖRNES, 1850; SUESS, 1866a; FUCHS, 1868; pp LAUBE, 1869a, 1871; pp MANZONI, 1873; FUCHS, 1900; ABEL, 1904; SCHAFFER, 1912a; TOLLMANN, 1957; NEBELSICK et al., 1997; SCHULTZ, 1998; KROH & HARZHAUSER, 1999; [NHMW]); Gauderndorf (Himmelreich, Zogelsdorf Fm.), NÖ (FUCHS, 1868; [NHMW]); Gauderndorf (Kattauer Straße), NÖ (SCHAFFER, 1913a, 1943; SCHAFFER & GRILL, 1951); Grübern (Zogelsdorf Fm.), NÖ (SUESS, 1891; KROH & HARZHAUSER, 1999; [NHMW]); Klein-Meiseldorf , NÖ (SCHAFFER, 1912a, 1913a); Kühnring, NÖ (ABEL, 1898); Maissau, NÖ (HOERNES, 1875a; SCHAFFER, 1912a); Limberg (Zogelsdorf Fm., Hengl quarry), NÖ ([NHMW]); Maria Dreieichen, NÖ (SUESS, 1866a; pp LAUBE, 1869a, 1871; SCHAFFER, 1912a; SCHAFFER, 1943); Maria Dreieichen-Eichberg, NÖ (STEININGER, 1971a, b); between Mold and Maria Dreieichen, NÖ (SCHAFFER, 1913a); ? Plesching, Upper Austria (SUESS, 1891); ? Plesching-Pfennigberg, Upper Austria (ABEL, 1904); ? Roßberg near Ober-Dürnbach, NÖ (SCHAFFER, 1913a); Unternalb (Hungerfeld, Retz Fm.), NÖ (BERNHAUSER, 1955; HARZHAUSER & KROH, 1999; KROH & HARZHAUSER, 1999; LUKENEDER & HARZHAUSER, 2002; [NHMW]); Zogelsdorf, NÖ (HOERNES, 1875a; SCHAFFER, 1912a; ? SCHAFFER, 1913a; KROH & HARZHAUSER, 1999); "bassin de Vienne, bassin extra-alpin" (= earlier name for the Molasse Zone) (ROMAN, 1965); Außeralpines Wiener Becken (= earlier name for the Molasse Zone) (SCHAFFER, 1924b, 1927a)

Paratethys (non-Austrian occurrences): not reported (records of POLJAK, 1938, SZÖRÉNYI, 1950, and MAJCEN et al., 1997 are erroneous)

Mediterranean: not recorded

Echinolampas sp. 1

(Pl. 56, Figs. 3a-b)

Material:

Badenian (Langhian-Early Serravallian) – Baden, NÖ, Austria
NHMW: 1 specimen (NHMW 2003z0034/0001)
GBA: 1 specimen (no inventory number)

Dimensions: see Tab. 10

Description:

Size and shape: The test is large with distinctly oval, antero-posteriorly elongated outline. The anterior margin is rounded, the posterior margin bluntly pointed. The maximum width lies far posterior of the apical disc, about 66 % of TL from the anterior margin. In profile the test is highly arched, the maximum height coinciding with the position of the apical disc. The ambitus lies low and is rounded and tumid. The oral side is conspicuously flattened and only very slightly sunken around the peristome.

Apical disc: The apical disc lies slightly anterior of the centre, about 45 % TL from the anterior margin. The apical disc is monobasal with 4 large circular gonopores. The madreporite is roughly pentagonal and crowded with numerous small madreporic pores. The gonopores lie at the apices of the madreporite. The ocular pores are minute.

Ambulacra: The ambulacra are petaloid adapically with strongly sunken poriferous zones. The frontal petal extends about 70 % of the corresponding test radius, the paired petals about 76 %. In the frontal and posterior paired petals the length of the poriferous zones is subequal, whereas in the anterior paired petals it is not. The anterior poriferous zones (IIb and IVa) are shorter than the posterior ones (IIa and IVb) extending about 83 % of the length of the petal. The ambulacral pores in the petals are conjugated anisopores, which are strongly conjugated and slightly oblique. The interporiferous zones within the petals are distinctly inflated and up to three times as wide as a single poriferous zone.

Adorally the ambulacra form depressed phyllodes consisting of four series of unipores in each ambulacrum. Two adradial series with larger, more closely spaced unipores and two perradial series consisting of smaller, more widely spaced unipores. In each phyllodes two buccal of about the same size as the pores from the adradial series are present. In the adoral third of the phyllodes there is a double row of shallow sphaeridial pits running along the central suture.

Interambulacra: Adapically the interambulacra are inflated between the petals. The aboral tuberculation consists of rather small, crenulate, perforate tubercles which are closely spaced. On the oral surface the tubercles are only slightly larger and a little bit less closely spaced. No "naked zone" along the central suture between peristome and periproct in interambulacrum 5 is observed. Well developed bourrelets are present in all interambulacra.

Peristome: The peristome lies slightly anterior of the centre, about 43 % TL from the anterior margin. It is rather small (7.2 mm wide in a 81.3 mm TL specimen) and has a subpentagonal outline.

Periproct: The periproct is situated inframarginally in interambulacrum 5. It has an oval, transversely elongated outline and is moderately large, being 9.1 mm wide in a 81.3 mm TL specimen.

Differential diagnosis:

As outlined above, *E.* sp. 1, is characterised by its distinctly oval, antero-posteriorly elongated outline, high profile and flattened oral surface. These features distinguish it from all other species discussed here (for more details see under the respective species).

Discussion:

Unfortunately, only two specimens of this species were available for study. Both are from historical collections (one from the NHMW and one from the GBA) and date back to the 19th century. The labels state Baden as locality in both specimens, the exact outcrops, where they came from, could, however, not be located. The matrix inside the test is greyish to light brown, well cemented, marly sand; similar lithologies were exposed in historical outcrops around the city of Baden. Since no contradicting evidence was found, the statement on the labels is accepted as valid. No similar species has been reported from the Miocene of the Central Paratethys until now and the present material could not be related to any species known to the author. Hence, until more material with better documented collection data is available, the species is left in open nomenclature.

Occurrence:

Austria: Badenian (Langhian-Early Serravallian)
Vienna Basin: Baden, NÖ ([NHMW])

Echinolampas sp. indet.

(Figs. 52, 53)

1877 *Echinolampas* – Fuchs: 662
1877 *Echinolampas* – Karrer: 258
1899 *Echinolampas* sp. – Roth von Telegd: 98
1903b *Echinolampas* – Fuchs: 8, 12.
1935 *Echinolampas* sp. – Grill: 46-52
1955 *Echinolampas* sp. – Tollmann: Tab. 5b
1958b *Echinolampas* – Sieber: 297
1961 *Echinolampas* – Schaffer: 149
1971d *Echinolampas* – Steininger: 135
1985 *Echinolampas* – Tollmann: 501
pp 1987 *Echinolampas* sp. – Mączyńska: 147, 149; pl. 5, fig. 1
1989 *Echinolampas* sp. – Nebelsick: 15
1991a *Echinolampas* sp. – Nebelsick et al.: 88
1991b *Echinolampas* – Nebelsick et al.: 119
1991 *Echinolampas* sp. – Roetzel et al.: 48

Figure 52: *Echinolampas* sp. indet., ? Gauderndorf, NÖ (AO of Schultz, 1998; NHMW 1869.III.38a); A: aboral, B: oral and C: lateral view (specimen coated with ammonium chloride in all figures).

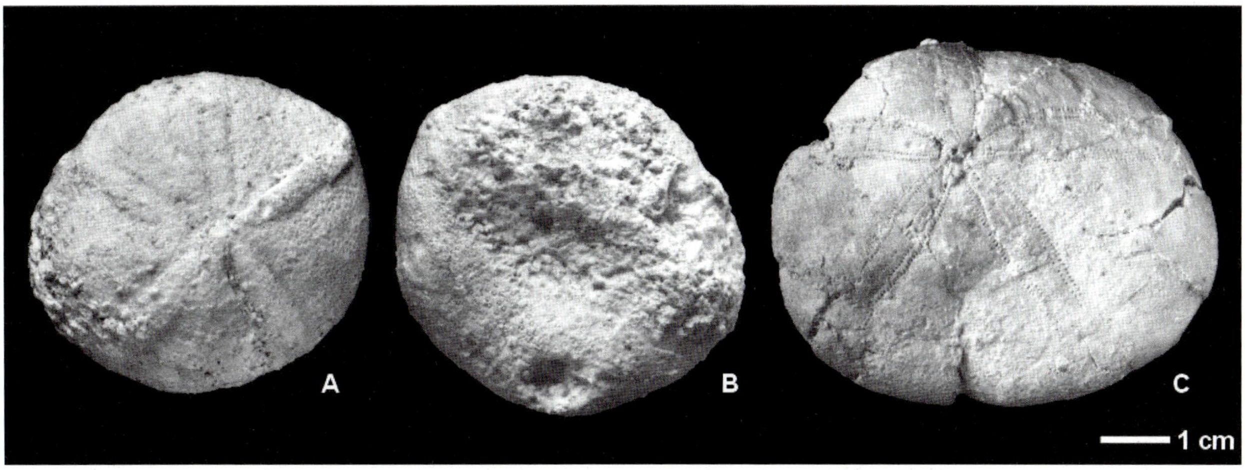

Figure 53: *Echinolampas* sp. indet., Egerian (Chattian – Aquitanian) of Steyregg, OÖ; A: aboral view, and B: oral view of specimen NHMW 2004z0092/0001, C: oral view of specimen NHMW 2004z0092/0002.

1996 *Echinolampas* sp. – MĄCZYŃSKA: 41; pl. 1, figs. 10a-b
v 1998 *Echinolampas laurillardi* AGASSIZ– SCHULTZ: 118; pl. 53, fig. 2
v. 2002b *Echinolampas* sp. – KROH: 12
v. 2003a *Echinolampas* ? sp. – KROH: 165; pl. 4, fig. 15

Material:

Egerian (Chattian-Aquitanian) – Steyregg (sandpit Treul), OÖ (see ROETZEL et al., 1991; KAISER et al., 2001)
NHMW: 5 specimens (NHMW 2004z0092/0001-5)
Late Eggenburgian (Early Burdigalian) – Gauderndorf (Himmelreich), NÖ, Austria
NHMW: 1 specimen (NHMW A606)
Remark: locality not confirmed, attached sediment untypical
Late Eggenburgian (Early Burdigalian) – Limberg (Zogelsdorf Fm., Hengl quarry), NÖ, Austria
NHMW: 10 fragments (NHMW 2003z0002/0009)
Badenian (Langhian-Early Serravallian) – Baden, NÖ, Austria
NHMW: 13 test fragments (NHMW 1846.37.957, 1852.II.1616)
Badenian (Langhian-Early Serravallian) – Oslip, Bgld, Austria
NHMW: numerous test fragments (NHMW 2003z0083/0005)
Late Badenian (Early Serravallian) – Stiefingbach (old mill, Pesendorf 16), Styria Austria
NHMW: 1 specimen (NHMW 2004z0114/0002)
Late Badenian (Early Serravallian) – St. Margarethen, Bgld, Austria
NHMW: 1 test fragment (NHMW 1989/66/5)
Late Badenian (Early Serravallian) – St. Margarethen (Kummer quarry), Bgld, Austria
NHMW: 1 crushed specimen (NHMW 2004z0001/0045)

Foreign material:

Late Badenian (Early Serravallian) – Buituri (= Bujtur), 20 km S of Deva, Romania
NHMW: 1 test fragment (NHMW 1852.II.1603f)

Dimensions (in mm):

Inv. No.	TL	TW	TH
NHMW 2004z0092/0001	45.8	48.1	25.6
NHMW 2004z0092/0002	~65	>54	>24

Discussion:

Being easily recognisable, the genus *Echinolampas* has been often reported in the geological/palaeontological literature of Austria. Records listed under the present heading either lack sufficient information to attach them to any of the species described above or are based on material too fragmentary or poorly preserved for specific determination.

Recently a number of *Echinolampas* specimens from the Egerian (Chattian-Aquitanian) of Steyregg, OÖ were obtained by the NHMW from a private collector (Josef KASTL, Linz). Despite their poor preservation (Figs. 53.A-C) the specimens are nonetheless very interesting as they represent one of the first records and the only available specimens of this genus from the Egerian (material from earlier records could not be traced). They are rather small (none of the specimens is larger than 65 mm in length), have an angular outline with subequal length and width, a concave oral side, marginal periproct (fully on the oral side) and long petals. They are not well enough preserved for specific determination.

The specimen (Figs. 52.A-C) figured as *Echinolampas laurillardi* by SCHULTZ (1998) is problematic. The label accompanying the specimen states that it comes from Himmelreich, near Gauderndorf (Lower Austria) but the preservation and attached sediment are untypical for this locality. Neither is it similar to the sediment of any of the other Austrian localities where echinolampadids were recovered. The specimen could not associated with any of the species documented from the Central Paratethys. The combination of its small peristome with very weakly developed bourrelets, the dense aboral tuberculation, the relatively high test and weakly sunken poriferous zones, as well as the inframarginal periproct (also visible in posterior view) is unparalleled in the other species discussed here. It is thus likely that the specimen does not come from the Central Paratethys at all and is left in open nomenclature here.

Occurrence:

Austria: Egerian (Chattian-Aquitanian) to Late Badenian (Early Serravallian)
Molasse Zone: Eggenburg (Bahnhof), NÖ (FUCHS, 1903b); Eggenburg (Brunnstube), NÖ (NEBELSICK et al., 1991a); Eggenburg, surroundings, NÖ (FUCHS, 1877); Eggenburg region (Zogelsdorf Fm.), NÖ (NEBELSICK, 1989); Gauderndorf, NÖ (STEININGER, 1971d; SCHULTZ, 1998); Limberg (Hengl quarry) NÖ (NEBELSICK et al., 1991b); Steyregg (sandpit Treul), OÖ (ROETZEL et al., 1991; [NHMW]); Plesching (phosphoritic sand), OÖ (GRILL, 1935)
Vienna Basin: Baden, NÖ ([NHMW]); Gainfarn, NÖ (KROH, 2002b); Leitha Mts. (TOLLMANN, 1985); Mödling, NÖ (KARRER, 1877); Müllendorf, Bgld (TOLLMANN, 1955; SCHAFFER, 1961); Niederleis, NÖ (KROH, 2003a); Steinabrunn, near Drasenhofen, NÖ (SIEBER, 1958b);
Eisenstadt-Sopron Basin: Oslip, Bgld ([NHMW]); St. Margarethen, Bgld ([NHMW]); St. Margarethen (Kummer quarry), Bgld ([NHMW])
Styrian Basin: Stiefingbach (old mill, Pesendorf 16), Styria ([NHMW])

Paratethys (non-Austrian occurrences): Badenian (Langhian-Early Serravallian)
Great Hungarian Basin (Pannonian Basin): Hidas, Hungary (ROTH VON TELEGD, 1899)
Transylvanian Basin: Buituri (= Bujtur), Romania ([NHMW])

Unconfirmed records of *Echinolampas* from the Central Paratethys

Echinolampas italicus LAMBERT, 1906

1915 *Echinolampas italicus* LAMB. – VADÁSZ: 212-213; fig. 98

Reported occurrence: Badenian of Livezile (= Ùrháza, =Vládháza), Alba, Romania (VADÁSZ, 1915)

Remarks: The reference material of VADÁSZ (1915) was not seen by the present author and from VADÁSZ's description and illustrations it is not clear how this species might be differentiated from flat specimens of *E. hemisphaerica*.

Echinolampas aff. *italicus* LAMBERT, 1906

1955 *Echinolampas* aff. *italicus* LAMB. – TOLLMANN: Tab. 5b

Reported occurrence: sandpit between Großhöflein and Kleinhöflein, Bgld, Austria (TOLLMANN, 1955)

Remarks: The name *E. italica* was proposed by LAMBERT (1906a) for *Echinolampas hemisphaerica* MANZONI (1880a): 186; pl. 1, figs. 1-3), non LAMARCK (1816). This species has never again been reported from Austria and the material on which this record is based could not be traced. It is unclear to which of the species discussed above this record refers.

Echinolampas dumasi COTTEAU, 1893

v 1915 *Echinolampas Dumasi* COTT. – VADÁSZ: 216; fig. 102; pl. 10 (4), fig. 5

Reported occurrence: Late Badenian (Early Serravallian) of Biatorbágy (= Bia), Pest, Hungary (Vadász, 1915)

Remarks: This record is based on a single specimen in the MAFI collection (Ech 243). It differs from the species discussed above by its strongly inflated oral side, its relatively narrow petals and its periproct which lies inframarginally to marginally. It can, nevertheless, not be ruled out that this specimen is simply an abnormally developed *E. hemisphaerica* and additional specimens are needed to settle this issue.

***Echinolampas dacica* Vadász, 1915**
& *Echinolampas dacica humilis* Vadász, 1915

* 1915 *Echinolampas dacicus* n. sp. – Vadász: 211-212; fig. 97a; pl. 12 (6), fig. 5

* 1915 *Echinolampas dacicus* var. *humilis* Vad. – Vadász: 211-212; fig. 97b; pl. 12 (6), fig. 6

Reported occurrence: Badenia of Livezile (= Ùrháza, =Vládháza), Alba, Romania (Vadász, 1915)

Remarks: The differential diagnostic features to *E. hemisphaerica* outlined by Vadász (1915) do not allow a confident separation of the two species. It seems likely that *E. dacica* may turn out synonymous with *E. hemispherica*. Yet, without having seen the types or any other specimens attributed to this species the author hesitates to synonymise them.

Genus *Conolampas* Agassiz, 1883

Type-species: *Conoclypus sigsbei* Agassiz, 1878, by original designation.

Synonyms of *Conolampas*:
Pygasterides Lovén, 1888,p. 13. Type-species: *P. relictus* Lovén, 1888, p. 13, by monotypy.
Scutolampas Lambert, 1906, p. 33. Type-species: *Conoclypus plagiosomus* Agassiz, 1840b, p. 5 [*nomen nudum*], by original designation.
Hypsoheteroclypus Szörényi, 1953: p. 76. Type-species: *Hypsoclypus doma* Pomel, 1887, p. 163, by original designation.

Potential synonyms of *Conolampas*:
Hypsoclypus Pomel, 1869, p. 25. Type-species: *Galerites semiglobus* Lamarck, 1816, p. 311, by subsequent designation (Stefanini, 1907: 348).
Heteroclypeus Cotteau, 1891, p. 194. Type-species: *Galerites semiglobus* Lamarck, 1816, p. 311, by original designation.

Emended diagnosis: Large campanulate to bell-shaped echinolampadid with straight, long and distally open petals; outline circular to subpentagonal; apical disc monobasal, knob-shaped in some species; straight, narrow poriferous zones; periproct transversely elongated and inframarginal; bourrelets and phyllodes well developed; basicoronal plates of the interambulacra 1 and 4 exsert in adult specimens; aboral tuberculation sparse, rarely exceeding 100 primary tubercles/cm² in adult specimens; distance between aboral primary tubercles more than one times their diameter (Kroh & Mooi, in prep.).
Distribution: ? Oligocene of the Caribbean; Early Miocene to Pliocene of the Mediterranean; Recent: Caribbean area and Indo-Malayan regions (Kroh & Mooi, in prep.).

Discussion: Originally the fossil species of this genus were associated with the genus *Conoclypus* (Agassiz, 1840b: 5; Agassiz & Desor, 1847a: 167-168; Desor, 1858: 318-323,...) or *Conoclypeus* [Manzoni, 1880a: 187, 1881: 174; Mazzetti, 1882a: 123-125; *Conoclypeus* is a *nomen nudum* according to Durham et al. (1966: U633)] because of similarities in general test shape. Later these species were removed from the conoclypids and recognised as having affinities with the echinolampadids by Pomel (1869: 25, 62-63), mainly because of similarities in the poriferous zones and the shape of the periproct (Cotteau, 1891: 194). Munier-Chalmas (fide Stefanini, 1908c: 348), however, showed that there is no Aristotle's lantern in *C. semiglobosus* contrary to the "true" conoclypids. Based on this, and the presence of poriferous zones of unequal length within the petals, Stefanini (1908c) placed most of the Neogene taxa ascribed to the genera *Conoclypus*, *Hypsoclypus* and *Heteroclypeus* into the genus *Echinolampas*.

Subsequently the genera *Hypsoclypus* Pomel, 1869, *Heteroclypeus* Cotteau, 1891, *Scutolampas* Lambert, 1906 and *Hypsoheteroclypus* Szörényi, 1953 were established for species of this group, resulting in many discussions how to separate these genera and which species they should include. Part of these discussions resulted from the action of Lambert (1907a: 54), who invalidly designated *C. plagiosomus* Agassiz, 1840b as type species for *Hypsoclypus* Pomel, 1869. However, according to the ICZN-rules this is not possible since *C. plagiosomus* was not included in the original publication and, moreover, is a *nomen nudum* (see below). Another hotly disputed question was whether *C. plagiosomus* and *C. lucae* are synonyms and which of those two names has priority [Kier (1962: 112) showed that *Conoclypus plagiosomus* is a junior subjective synonym of *C. lucae*. Agassiz's (1840b: 5) *Conoclypus plagiosomus* is a *nomen nudum* since no description, definition or indication is given (ICZN 4th ed., 2000, Article 12.1.). In Agassiz & Desor (1847a: 168) both taxa (*plagiosomus* and *lucae*) are accompanied by a short description and are thus available names (ICZN 4th ed., Articles 11 and 12). Moreover, they count as simultaneously published. Desor (1858: 322) acting as "the first reviser" stated that *C. lucae* and *C. plagiosomus* are conspecific and gave preference to *C. lucae*, which is thus the valid name for the species (ICZN 4th ed., Article 24.2.). In the copy of Agassiz & Desor (1847a) in the NHMW-library, *ucæ* instead of *lucæ* is written on page 168, however, on page 361 the name is correctly spelled. Desor (1858: 322) acting as "the First Reviser" choose the spelling *lucae*, which is therefore the correct one (ICZN 4th ed., Article 32.2.1.). Initially Lambert (1907a: 54) also considered *C. lucae* and *C. plagiosomus* as synonyms, but treated them as different species later. This erroneous opinion of Lambert (1913a: 131-133), Lambert & Thiéry (1921: 376) and Lambert & Jeannet (1928: 124, 178) was based on a specimen labelled erroneously as coming from the "Helvétien" of La Couronne, France. According to Roman & Strougo (1988: 147) the specimen figured by Lambert (1913a: pl. 10, figs. 7-9) is of Eocene age and belongs to *Echinolampas africana* var. *fraasi* de Loriol, 1880.]. The action of Stefanini (1908c: 348), who designated *C. semiglobus* as type species, rendered the discussion of the differences between *Hypsoclypus* and *Heteroclypeus* futile, since the type-species of both taxa is identical due to this decision.

Kroh & Mooi (in prep.) showed that the type-species of the above mentioned genera, as well as a number of closely related species are congeneric with the extant species of the genus *Conolampas*, a deeper water echinolampadid. Main arguments for placing these species in *Conolampas* are the aboral tuberculation (compare Mooi, 1990a), which is very distinct from the tuberculation of *Echinolampas*, the exsert condition of basicoronal plates in the interambulacra 1 and 4 (in adult specimens) and the narrower, straight or divergent poriferous zones in the petals. In addition, the great size and general shape of the test, particularly the very low ambitus, and the nearly flat oral surface are identical with these features in other members of the genus *Conolampas* (Mooi, 1990a). Although not unique to *Conolampas*, the well developed bourrelets and phyllodes and tiny naked zone in the posterior interambulacrum are consistent with this genus.

Pygasterides Lovén, 1888 is a genus established for a species based on juvenile material of *Conolampas sigsbei* Agassiz, 1878 (compare Mortensen, 1948a).

Remarks: In the Miocene of the Mediterranean and Paratethys more than 20 species of this group were reported (AIRAGHI 1900; STEFANINI, 1908c; ROMAN, 1965). This inflation of nominal taxa was caused by the variable gross morphology of this group and the praxis of basing species on minor variations in overall test morphology. However, specimens from single localities and narrow time horizons [e.g. specimens from the Early Badenian of the Vienna Basin or the Early Badenian of Pod'yarkov (= Podjarków), in the Ukraine] show a wide variation in gross test morphology. SZÖRÉNYI (1953: 78-84) described and illustrated eight different "*Hypsoheteroclypus*" species from a single locality on base of a single specimen each; most of these specimens show minor variations in overall test morphology only and it is highly questionable whether all of these species are justified. The specimen SZÖRÉNYI (1953: 84-85) determined as *Echinolampas hemisphaerica* also belongs to *Conolampas*.

Analysis of the Austrian material revealed four different morphotypes in two different geological horizons. One of these morphotypes is based on a single specimen [NHMW V.b.ß.k.12 (6983)] of unknown stratigraphic and doubtful geographic origin [the label states "(angeblich Nattheim) wahrscheinlich Steiermark"]. Nattheim is a locality in Germany famous for its beautiful Jurassic fossils but there are no marine sediments of the Neogene in this area. This specimen is similar to the specimens figured by STEFANINI (1908c: pl. 13, fig. 1a-c) as *Echinolampas ugolinii* and differs from the other specimens by its longer petals, higher tubercle density, broader posterior paired petals and higher number of tubercles across the interporiferous zone. Since no other specimens similar to this one turned up in any Neogene deposit of the Central Paratethys, it is probably not from this area at all. Therefore it is excluded from further consideration. Most specimens come from the Early Badenian (Langhian) time-slice, where this genus was widely distributed throughout the Central Paratethys (Austria, Hungary and Ukraine). These specimens represent two morphotypes, which are very similar and differ mainly in their relative test height. The fourth morphotype is represented by a single specimen in the Austrian material and comes from Late Badenian (Early Serravallian) sediments. It differs from the Early Badenian morphotypes by its hemispherical profile, higher tuberculation density, shorter petals and even more inflated madreporite.

Four species of *Conolampas* are known today, only two of which co-occur (*C. diomedeae* and *C. malayana*). These two species are well separated by distinct morphological features (i.e. the length of the poriferous zones, which extend nearly up to the ambitus in *C. diomedeae*, whereas they extend only about 1/2 of the corresponding test radius in *C. malayana*). MORTENSEN (1948a) used the following morphological features to separate the extant species of *Conolampas*: 1) Respiratory pore morphology (isopores or anisopores); 2) Lengths of the petals (1/2-2/3 of the corresponding test radius or extending nearly to the ambitus; 3) position of the periproct; 4) position of the peristome; and 5) width of the petals.

Ecology and biogeography: Today four extant species of *Conolampas* are known, which occur at depths between 120 to 800 meters in the Caribbean (*C. sigsbei*), the Maldives (*C. murrayana*) and the Philippines (*C. diomedeae* and *C. malayana*) (MOOI, 1990a). As in *Echinolampas*, little is known on the biology of the extant species of *Conolampas*. For the Indo-Pacific species only the collection depths are known [*C. diomedeae*: 265 m (MORTENSEN, 1948a), 195-181 m (DAVID & DE RIDDER, 1989); *C. malayana*: 245 to 400 m (MORTENSEN, 1948a); *C. murrayana*: 229 m (MORTENSEN, 1948a)]. *C. sigsbei*, from the Caribbean Sea and the eastern margin of the Gulf of Mexico, is better studied. It is found between 120 and 800 m as epibenthic browser on substrate composed of fine carbonate debris and small foraminifers. As it moves over the surface it leaves a trail as wide as its test (based on direct observations from submersibles by David PAWSON and Charles MESSING, in MOOI, 1990b: 698).

Predation scars can commonly be observed on specimens of this genus (Pl. 60, Fig. 2d). These scars consist of irregularly circular holes in the test, which are tentatively attributed to boring predation by cassid gastropods (for more information on drilling predation see KOWALEWSKI & NEBELSICK, 2003 and references therein). This genus is generally rather rare, and seems to be associated with slightly deeper habitats. It is usually found in limestone (rudstone) or coarse calcareous sandstone. An exception of this is a single specimen of *C. subpentagonalis* found in the sediments of the Hartl Formation (Eisenstadt, Austria), which was interpreted as very shallow, higher energy deposits (KROH et al., 2003).

Conolampas subpentagonalis (GREGORY, 1891)

(Pl. 58, Figs. 1, 3; Pl. 59, Figs. 1a-b; Pl. 60, Figs. 1-2)

. 1869a *Conoclypus plagiosomus* AG. – LAUBE: 183
1870 *Conoclypus plagiosomus* AG. – LAUBE: 314
1871 *Conoclypus plagiosomus* AGASSIZ (Teste WRIGHT). – LAUBE: 67-68; pl. 19, fig. 3
? 1877 *Conoclypeus* – FUCHS: 667
? 1878 *Conoclypus plagiosomus* AG. – HILBER: 563, 575
* pp 1891 *Heteroclypeus subpentagonalis*, n. sp. – GREGORY: 599-600
#. 1900 *Heteroclypus Nevianii*, n. f. – AIRAGHI: 177-178; pl. 1, figs. 5-6
? 1907 *Echinolampas plagiosomus* AG. sp. – VOGL: 245, 306
v. 1915 *Echinolampas (Heteroclypeus) subpentagonalis* GREG. – VADÁSZ: 201-202; text-fig. 91
v. 1915 *Echinolampas (Heteroclypeus) semiglobus* LAM. – VADÁSZ: 203
1915 *Echinolampas (Heteroclypeus) Nevianii* AIR. – VADÁSZ: 204-205; text-fig. 92
? 1926 *Echinolampas subpentagonalis* GREG. – STRAUSZ: 213, 368
. 1938 *Heteroclypeus cotteaui* LAMBERT 1907 – POLJAK: 194-195; pl. 9, figs. 1, 1a
? 1950 *Hypsoclypus subpentagonalis* GREG. – SZÖRÉNYI: 145
? # 1950 *Hypsoclypus egregyensis* n.sp. – SZÖRÉNYI: 143-145; pl. 1, fig. 6; pl. 2, figs. 1, 1a
? # 1953 *Hypsoheteroclypus plagiosomus corsicanus* n. ssp. – SZÖRÉNYI: 28-29, 78-79; pl. 4, figs. 1, 1a-b
1953 *Hypsoheteroclypus montesiensis* (MAZZETTI), 1881, (STEFANINI), 1907 – SZÖRÉNYI: 30-31, 80-81; pl. 5, figs. 3, 3a-b
? 1953 *Echinolampas hemisphaericus* (LAMARCK), 1816. – SZÖRÉNYI: 34, 84-85; pl. 3, figs. 2, 2a-b
? 1962 *Echinolampas lucae* (DESOR) – KIER: 111-113; text-fig. 93; pl. 30, fig. 5; pl. 31, fig. 1; pl. 32, fig. 1
1963 *Conoclypus plagiosomus* AGASSIZ – MÜLLER, 515; figs. 677B a-b
1965 [*Echinolampas (Hypsoclypus)*] *egregyensis* SZÖRÉNYI 1950 (*Hypsoclypus*) – ROMAN: 306
1965 [*Echinolampas (Hypsoclypus)*] *nevianii* AIRAGHI 1900 (*Heteroclypus*) – ROMAN: 307
1965 [*Echinolampas (Hypsoclypus)*] *subpentagonalis* GREGORY 1891 (*Heteroclypeus*) – ROMAN: 308
1981 *E. subpentagonalis* GREG. – HALMAI: 106
v pp 2002 *Conoclypus* sp. – KAZÁR: 153; fig. 1
2003b *Echinolampas subpentagonalis* (GREGORY, 1891) – KROH: 251 [literature review]

Type-material:

Heteroclypeus subpentagonalis GREGORY, 1891:

Holotype: the specimen figured by LAUBE (1871: pl. 19, fig. 3); Geological Survey of Austria. The holotype could not be located at the Geological Survey in Vienna. It is also not found in

the Catalogue of this institution dating back to the beginning of the second world war and has to be considered lost.
Type-area: Großhöflein, Bgld or Zirknitz, Styria (both Austria)
Age: presumably Early Badenian (Langhian), Middle Miocene

Heteroclypus nevianii AIRAGHI, 1900:
Holotype: specimen figured by AIRAGHI (1900: pl. 1, figs. 5-6); Museo civico di Milano, Italy [not examined]
Locus typicus: Vena presso Monteleone Calabro, Southern Italy
Age: Middle? Miocene

? *Hypsoheteroclypus plagiosomus corsicanus* SZÖRÉNYI, 1953:
Holotype: the specimen figured by SZÖRÉNYI (1953: pl. 4, figs.1, 1a-b); collection of the University Lwów, Ukraine, specimen no. 6 [not seen]
Locus typicus: Pod'yarkov (= Podjarków), near Lwów, Western Ukraine
Biozone: Lower lagenid zone
Age: Early Badenian (Langhian), Middle Miocene (written comm. Anna WYSOCKA & Andrzej RADWAŃSKI, 02.04.2004)
Remarks: As in other species established by SZÖRÉNYI (1953) the specimen number of the holotype is ambiguous. Specimen no. 6 is mentioned under *Scutella paulensis* as well as under this species.

Material:
Early Badenian (Langhian) – Eisenstadt (basal Hartl Fm.), Bgld, Austria
WANZENBÖCK coll.: 1 specimen
Early Badenian (Langhian) – Retznei [Weissenegg Fm., Lafarge quarry (formerly Perlmoser)], Styria, Austria
WANZENBÖCK coll.: 1 specimen (W15)
Early Badenian (Langhian) – Stotzing (sandpit Mayer), Bgld, Austria
NHMW: 1 specimen (NHMW 2003z0008/0001)
WANZENBÖCK coll.: 4 specimens (W8-W11)
Early ? Badenian (Langhian) – Ewitsch, Styria, Austria
NHMW: 1 specimen (NHMW 2002z0181/0005)
Badenian (Langhian-Early Serravallian) – Eisenstadt, Bgld, Austria [possibly from the so-called "Terebratelfundstelle", Großhöflein near Eisenstadt, based on adhering sediment; certainly not from Müllendorf (as the younger label states) or the Hartl Fm., Eisenstadt]
GBA: 1 specimen (no inventory no.)

Foreign material for comparison:
Ottnangian (Late Burdigalian) – Fót, Hungary
MAFI:2 specimens [MAFI Ech 242 (figured specimen of *E. subpentagonalis* in VADÁSZ, 1915: fig. 91)], 268 (reference material of *E. subpentagonalis* in VADÁSZ, 1915)]
Early Badenian (Langhian) – Mátraverebély, Nógrád, Hungary
MAFI:1 specimen [MAFI Ech 433 (reference material of *E. semiglobus* in VADÁSZ, 1915)]

Dimensions: see Tab. 10

Description (based on the Austrian specimens):
Size and shape: The test is large with up to 15 cm test length. The outline is subcircular to very slightly subpentagonal. Both anterior and posterior margin are rounded. The maximum width and height coincide with the position of the apical disc. In some specimens the maximum width may lie slightly posterior of the apical disc. The profile is low conical with slightly inflated slopes. The ambitus is thin and rounded and it lies very low on the test. The oral surface is nearly flat.
Apical disc: The apical disc lies subcentrally to slightly anterior of the centre. The apical disc is monobasal with four large, circular gonopores and five small ocular pores. The madreporite is pentagonal in outline and highly swollen (knob-shaped) and bears numerous small hydropores, as well as small tubercles.
Ambulacra: The ambulacra are semi-petaloid on the aboral surface. They consist of two slightly diverging rows of closely spaced conjugated anisopores in each ambulacrum. The adradial pores in each pair are large, slightly oblique and teardrop-shaped. Adapically the two poriferous zones of each ambulacrum converge towards each other and the apex, and both poriferous and interporiferous zones are distinctly depressed in this area. The poriferous zones of the "petals" are very slightly depressed. In the paired ambulacra, the poriferous zones are of slightly different length: They are longer in the posterior half of the anterior paired ambulacra (IIa and IVb) and the anterior half of the posterior paired ambulacra (Ib and Va). The frontal petal is slightly curved towards the left and extends about 65 to 80 % of the corresponding test radius (measured along the slope of the test, not in planar view). The anterior paired petals are similar in length to the frontal petal, the posterior paired petals are slightly longer (70-85 % TL). Outside the petals only minute unipores with sunken attachment area and adoral positioned neural canal are found. The tuberculation on the aboral interporiferous zones is not very dense; the spacing between the primary tubercles is about one and a half to two times their diameter. Miliary tubercles are loosely spread among them. In the poriferous zones, only small miliary tubercles are present, they are situated on the ridges between adjacent pore pairs. Halfway along the test height there are 5 to 6 primary tubercles across the interporiferous zones.
On the oral side the ambulacra are depressed around the peristome and form well developed phyllodes (Pl. 60, Fig. 2f). The poriferous zones bear small unipores with sunken attachment area and adorally positioned neural canal. The unipores of the phyllodes are slightly larger. The tuberculation of the ambulacra on the oral surface is very similar to those on the interambulacra. The phyllodes consist of three series of unipores in each half-ambulacrum. An "outer" or adradial series consisting of large, closely spaced unipores with rather large attachment area. An "inner" or perradial series consisting of smaller, slightly vertically elongated unipores, which are more widely spaced. Between those two a third series consisting of widely spaced unipores is found. The two buccal pores are of similar size as the pores from the adradial series, but distinctly elongated vertically. In the adoral third of the phyllodes there is a double row of shallow pits (sphaeridial pits) running along the central suture.
Interambulacra: Around the apex the interambulacra are distinctly inflated. The tuberculation on the aboral side is fairly homogenous. The primary tubercles are small and widely spaced. The mean distance between them is about one and a half to three times their diameters. Miliary tubercles are loosely spread among them. Directly at the ambitus the spacing of the primary tubercles becomes very tight suddenly, with the areoles of the tubercles nearly touching each other. On the oral side the spacing of the primaries is more widely again, except near the margin. Along the central suture of interambulacrum 5 there is a zone of especially widely spaced tubercles [similar to the "naked zone" KIER (1962) noted in many species of *Echinolampas*]. The primary tubercles of the oral side are about one and a half to two times larger than those of the aboral surface. Data on the tubercle density of both ambulacra and interambulacra can be found in Tab. 10. On the oral side the interambulacra are highly inflated around the peristome and along their interradial sutures, forming well developed bourrelets.
Peristome: The peristome lies centrally to slightly anterior of the centre. It is oval (transversely elongated) to slightly subpentagonal in shape. It is always smaller than the periproct, being about 10-11 % of TL wide. Inside the test, no auricles are present.
Periproct: The periproct is situated very close to the margin in interambulacrum 5, but lies fully on the oral surface. It is large, being 12-14 % of TL wide. The shape of the peristome is oval, transversely elongated.

Differential diagnosis:
C. semiglobus (LAMARCK) from the Aquitanian of the Aquitaine Basin differs from this species by its broader petals, by its small

periproct (8-10% TW vs. 12-14 % TW in *C. subpentagonalis*), denser aboral tuberculation, higher number of tubercles across the interporiferous zones (10 to 12 vs. 5 to 6 in *C. subpentagonalis* and *C. elegans*), its more strongly developed floscelle and the presence of primary tubercles on the ridges between the respiratory pores (based on a specimen from the type region Dax, France housed at the NHMW under the number 1898.I.14)

C. lucae (Desor in Agassiz & Desor, 1847) [syn. *C. plagiosomus* (Agassiz in Agassiz & Desor, 1847)] from the Burdigalian of the Rhône Basin differs by its moderately arched profile (not at all conical as in *C. subpentagonalis* or *C. elegans*), lack of an inflated madreporite and long petals which abut only shortly before they reach the ambitus (based on the description and illustration in Philippe, 1998).

C. pignatarii (Airaghi, 1900) [syn. *C. hemsphaericus* (Gregory, 1891) and *C. melitensis* (Lambert & Thiéry, 1921)] from the Serravallian ? to Messinian of the Central Mediterranean differs from this species by its high (TH ~ 55 % TL), hemispherical profile, much broader petals and its smaller periproct (8-10% TW vs. 12-14 % TW in *C. subpentagonalis*) (based on the description of Stefanini, 1907).

Discussion:

This species differs from *C. elegans* mainly by its lower relative test height. In a scatter-plot of test length versus test height two groups, which align along two different regression lines can be observed (Fig. 54). Due to the fact that both this species and *C. elegans* are relatively rare and often deformed by compaction multivariate statistics proved to be problematic. A preliminary principal component analysis (in the following: PCA) with a reduced data set using the variables TL, TH, APP_ant, APP_post, APP_Cr, PPP_Cr, PPP_W, Tubs1 and Tubs2 (for an explanation of the variables see Fig. 55) was carried out. Unfortunately only eleven specimens could be used for the analysis, since the other cases had too many missing values. The PCA resulted in 3 factors controlling the variance in the data-set. The first factor had high loads of most variables except these concerning tubercle density and test height. The second factor was controlled by the tubercle density variables and the third factor by test height (see Fig. 56.B). A 3D-scatter plot of the resulting factors shows a separation of the data-set into two groups as assumed by the initial ad-hoc hypothesis based on careful examination of the specimens. The two point clouds lie very close together, partially overlapping, however, and the number of cases is very low, effectively limiting the success of this kind of analysis (Fig. 56.A). PCA's with even more reduced data-sets, facilitated more cases but gave no useful results due to the low number of variables employed (PCA calculated with SPSS 11.0.1).

Therefore another approach was used to test the separation of the two groups. Test length and height data from the examined specimens and from the specimens published in the literature on Central Paratethyan and Northern Italian specimens (Stefanini, 1907; Vadász, 1915; Poljak, 1938 and Szörényi, 1953) were used to generate a scatter-plot and calculate two regression lines (Fig. 54). These were tested using the method proposed by Zorn (1972: 373). The test shows a significant difference at a 99% confidence level (Z = 4.293; Tab. 11). The Kolmogorov-Smirnov-Test was employed to test if the distributions of TL and TH conform to normal distributions. Both variables show a normal distribution according to this test, but this might be an artefact due to the low number of cases (n=45). The independent sample t-test confirms that there is a significant difference between the two groups in the variable TH (Sign. = 0.003), but not in TL (Sign. = 0.185) (t-test calculated with SPSS 11.0.1).

Therefore, the two observed morphotypes are considered as distinct species for the time being, but this should be critically tested with more well preserved specimens. Unfortunately no data on the test height variability in the only common extant *Conolampas* species (*C. sigsbei*) is available in the literature.

Another possible explanation for the observed differences in relative test height would be the "flatness-rule" (German: "Flachheitsregel") of Ernst (1970: 54-60). According to this

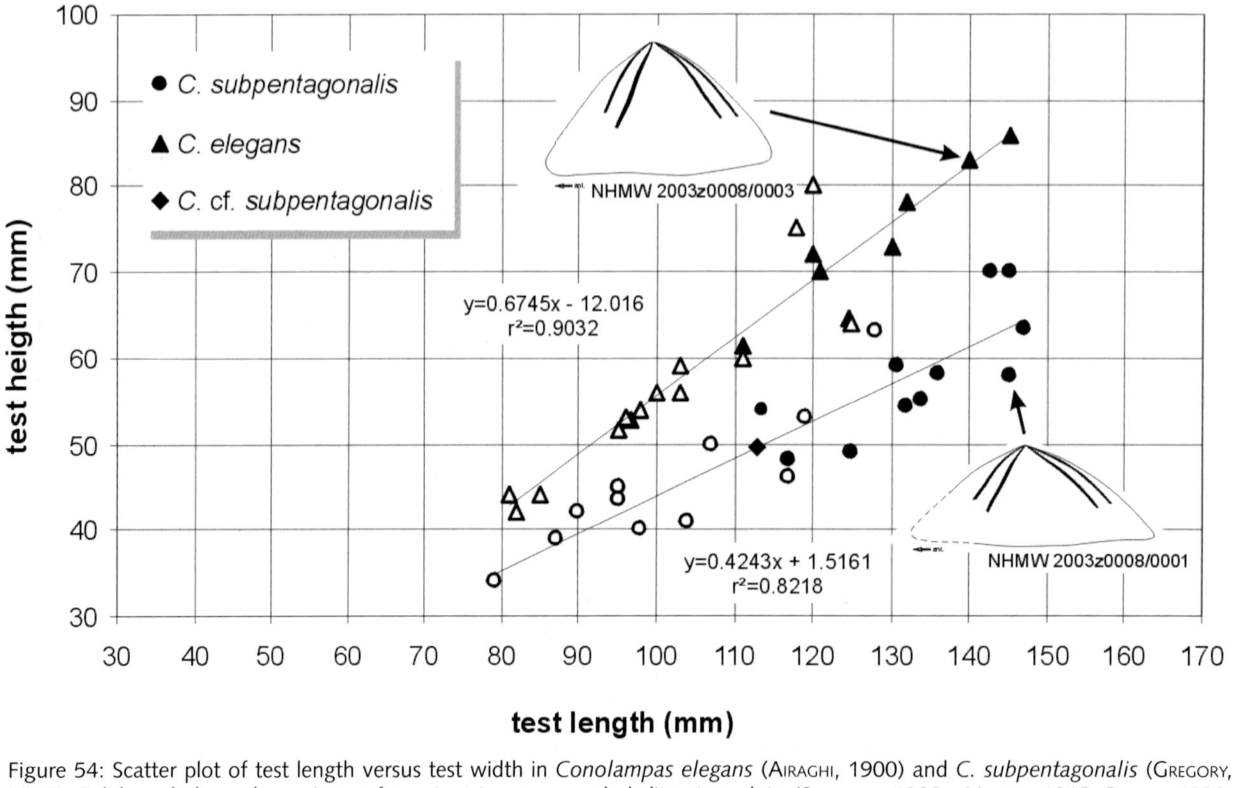

Figure 54: Scatter plot of test length versus test width in *Conolampas elegans* (Airaghi, 1900) and *C. subpentagonalis* (Gregory, 1891). Solid symbols mark specimens from Austria, open symbols literature data (Stefanini, 1908c; Vadász, 1915; Poljak, 1938; Szörényi, 1953).

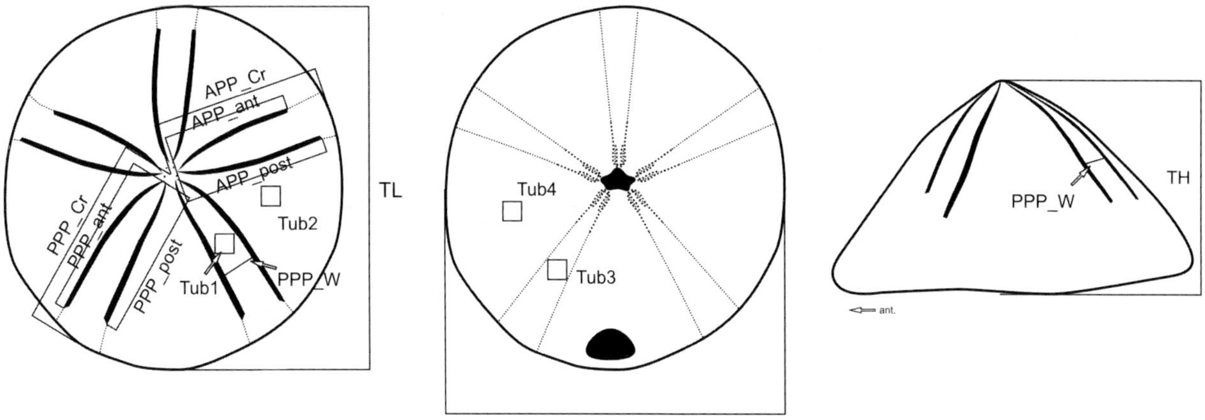

Figure 55: Schematic drawing of *Conolampas elegans* showing the variables used in the PCA of *Conolampas*. TL, TH, APP_ant, APP_post, APP_Cr, PPP_Cr, and PPP_W are linear measurements (note that the variables concerning the poriferous zones and the corresponding test radius were not measured in planar view; instead they were measured parallel to the test surface); Tubs1 to 4 are tubercle counts performed in the area indicated by the respective square or (in case of poor preservation of that area) on the respective location on the other half of the bilaterally symmetrical test.

theory the test height of epibenthic irregular echinoids is correlated with the amount of terrigenous material in the substrate (high morphotypes in pure limestone, low morphotypes in terrigenous sediments). Ernst observed this in six Late Cretaceous genera (*Galerites*, *Conulus*, *Echinocorys*, *Offaster*, *Cardiotaxis* and *Galeola*) in the chalk facies of Northern Europe. However, even Ernst was not sure how to interpret this phenomenon and did not treat it consistently in all these genera. In some he placed the higher and lower group in a single species, in others in different species. Moreover, in some of these genera these groups have only barely overlapping variation ranges and few intermediary forms (e.g. *Echinocorys*, see Ernst, 1970: fig. 5) as in the considered *Conolampas* material. Unfortunately, this "flatness-rule" was never tested with extant irregular echinoids. Although Ernst (1973a) did a lot morphometrics on extant Mediterranean echinoids these were mostly epibenthic regular echinoids (*Arbacia*, *Paracentrotus* and *Sphaerechinus*). In these the test height varies considerably between 35 to 65 % test diameter and seems to be controlled (at least partially) by water energy and turbulence. As the two morphotypes of *Conolampas* occur within the same layer and sediment type a correlation to sediment grain size and chemistry seems unlikely in the present case.

Nichols (1959: 405) discussed the possibility that the existence of high and low types might be related to water depth. This, however, seems to be unlikely in the case of *Conolampas*,

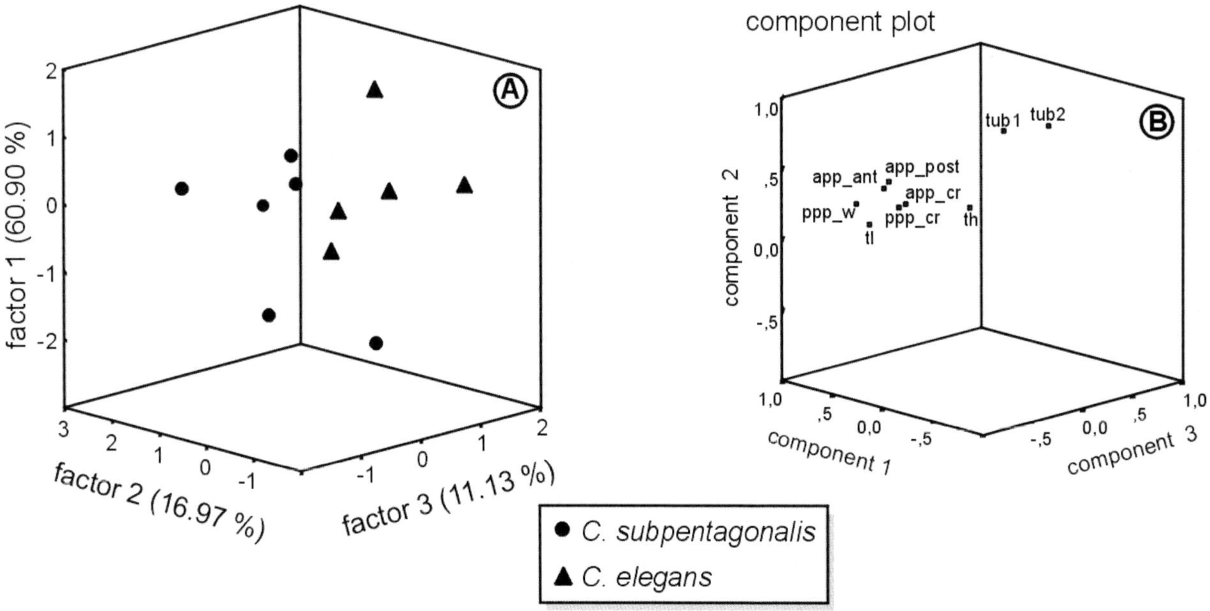

Figure 56: PCA of *Conolampas elegans* and *C. subpentagonalis*: 3-D scatter plot (A) of the three factors calculated by the PCA (percentages of variance explained by the individual factors is found in brackets behind the factor numbers) and component plot (B) showing the factor loading for the used variables (TL, TH, APP_ant, APP_post, APP_Cr, PPP_Cr, PPP_W, Tubs1 and Tubs2). Factors 1 to 3 explain 89 % of the variance in the present sample.

	n	b	r^2	SB
Conolampas elegans	22	0.6745	0.9032	0.04474127
Conolampas subpentagonalis	23	0.4243	0.8218	0.03734762
n=no. of specimens; b=gradient of the regression line; r^2=coefficient of determination; SB=standard deviation of r				

Table 11: Results of the comparison of the regression lines of test length versus test width in *Conolampas elegans* (AIRAGHI, 1900) and *C. subpentagonalis* (GREGORY, 1891) after the method of ZORN (1972) (regressions calculated with MS Excel 97).

since the two morphotypes occur together in the same horizon in the outcrops Stotzing and Retznei.

It is widely agreed, that the Austrian morphotype belongs into a different species than the specimens from the Burdigalian of the Rhône Basin (GREGORY, 1891; ROMAN, 1965; PHILIPPE, 1998) and this could be confirmed by the present study. Unfortunately LAUBE's specimens were lost, among them the holotype of *C. subpentagonalis*, necessitating the proposal of a neotype.

PHILIPPE (1998: 98-99) included *Heteroclypeus subpentagonalis* GREGORY, 1891 into the synonymy of *Conolampas lucae* (DESOR in AGASSIZ & DESOR, 1847) [under the name *Echinolampas (Hypsoclypus) plagiosomus* (AGASSIZ, 1840b)], yet he explicitly excluded the material of LAUBE (1871). This action, however is not possible, since the type specimen of *H. subpentagonalis* is LAUBE's specimen (compare GREGORY, 1891: 599-600).

GREGORY (1891: 599-600) based the name *subpentagonalis* on the pentagonal shape shown by the figure in LAUBE (1871). LAUBE, however, stated explicitly that his specimens had a circular outline and several of the figures given in his work do not correspond exactly to the material on which they are based.

Occurrence:

Austria: Early Badenian (Langhian)

Vienna Basin: Stotzing (sandpit Mayer), Bgld (pp KAZÁR, 2002; [NHMW]); Vienna Basin (? FUCHS, 1877)

Eisenstadt-Sopron Basin: Eisenstadt (basal Hartl Fm., Hartl hill), Bgld ([NHMW]); Großhöflein, Bgld (LAUBE, 1869a, 1871; p. p. GREGORY, 1891; MÜLLER, 1963); Kleinhöflein (= Nagy-Höflány), Bgld (VADÁSZ, 1915)

Styrian Basin: Ewitsch, Styria ([NHMW]); ? Kainberg, Styria (HILBER, 1878); Retznei (Weissenegg Fm., Lafarge quarry), Styria ([NHMW]); Zirknitz, Styria (LAUBE, 1869a, 1871; p. p. GREGORY, 1891)

Paratethys (non-Austrian occurrences): Ottnangian (Late Burdigalian) – Early Badenian (Langhian), ? Late Badenian (Early Serravallian)

Great Hungarian Basin (Pannonian Basin): Fót, Hungary (? VOGL, 1907; VADÁSZ, 1915; HALMAI, 1981; KROH, 2003b; [MAFI]); ? Magyaregregy, Leányköi árok, Mecsek Mts., Hungary (SZÖRÉNYI, 1950; ROMAN, 1965); Mátraverebély, Nógrád, Hungary (VADÁSZ, 1915; [MAFI]); ? Fót (Somlyó Mt.), Hungary (STRAUSZ, 1926; KROH, 2003b)

Fore-Carpathian Basin: Pod'yarkov (= Podjarków), near Lwów, western Ukraine (SZÖRÉNYI, 1953)

Transylvanian Basin: Gârbova de Sus (= Felsö-Orbó), Romania (VADÁSZ, 1915; ROMAN, 1965)

Zala, Sáva and Dráva Basins: surroundings of Beograd, Serbia (ROMAN, 1965); Donja Velešnja near Kostajnica, southern Croatia (POLJAK, 1938); Donja Voča, in the Ivanšćica Mts., Croatia (POLJAK, 1938); Giznik, Samobora Gora, Slovenian-Croatian border region (POLJAK, 1938); Kirin, Petrova Gora, in Gliskom Pokuplju, Croatia (POLJAK, 1938); Mesarski Brijeg, near Daruvara, Psunj Mts., Croatia (POLJAK, 1938)

Mediterranean: Middle Miocene

Central Mediterranean: Vena presso Monteleone Calabro, southern Italy (AIRAGHI, 1900; ROMAN, 1965); ? *Globigerina* Limestone, Maltese Islands (GREGORY, 1891; ROMAN, 1965)

Conolampas elegans (AIRAGHI, 1900)

(Pl. 57, Figs. 1-2; Pl. 58, Figs. 2, 4)

? 1880a *Conoclypeus plagiosomus* AG. – MANZONI: 187; pl. 2, figs. 23

? 1881 *Conoclypeus plagiosomus*, AG. – MANZONI: 174

? 1882a *Conoclypeus plagiosomus*, AGAS. – MAZZETTI: 123-124

?* 1882a *Conoclypeus montesiensis*, MAZZETTI. – MAZZETTI: 124-125; Pl. 2, Fig. 3

*. 1900 *Heteroclypus elegans*, n. f. – AIRAGHI: 176-177; pl. 1, figs. 3-4

. 1905a *Heteroclypus semiglobosus* (LAM.) COTT. – AIRAGHI: 214-216; figs. 1-2

. 1908c *Echinolampas Montesiensis* (MAZZETTI) STEF. – STEFANINI: 370-375; pl. 13, figs. 3, 4a-b, 5a-b, 6

? v 1915 *Echinolampas (Heteroclypeus)* cfr. *elegans* AIR. – VADÁSZ: 203-204

v. 1915 *Echinolampas (Heteroclypeus) hungaricus* n. sp. – VADÁSZ: 205-206; text-fig. 93

1938 *Hypsoclypus plagiosomus* AGASSIZ (*Conoclypeus*) 1840 – POLJAK: 193-194

1953 *Hypsoheteroclypus plagiosomus subpentagonalis* (GREGORY), 1892, SZÖRÉNYI – SZÖRÉNYI: 29, 79-80; pl. 4, figs. 2, 2a-b

1953 *Hypsoheteroclypus plagiosomus lamberti* (CHECCHIA-RISPOLI), 1917. SZÖRÉNYI – SZÖRÉNYI: 30, 80; pl. 5, figs. 1, 1a, 2a-b

? 1953 *Hypsoheteroclypus hungaricus* (VADÁSZ), 1915. – SZÖRÉNYI: 31-32, 81-82; pl. 3, figs. 3, 3a-b

1953 *Hypsoheteroclypus pyramidalis* (ABICH), 1859. – SZÖRÉNYI: 32-33, 83; pl. 2, figs. 4, 4a-b

1953 *Hypsoheteroclypus vicinoconoideus* n.sp. – SZÖRÉNYI: 33, 83-84; pl. 6, figs. 2, 2a-b

? 1955 *Echinolampas hungaricus* VADAS – TOLLMANN: Tab. 5b [probably refers to a specimen erroneously attributed to the locality Müllendorf in the GBA coll.]

1965 [*Echinolampas (Hypsoclypus)*] *elegans* AIRAGHI 1901 (*Heteroclypus*) – ROMAN: 306

1965 [*Echinolampas (Hypsoclypus)*] *vadaszi* ROMAN (*nov. nom.*) – ROMAN: 309

1993 *Conoclypus plagiosomus* AGASSIZ – HIDEN: 17-18; fig.

v pp 2002 *Conoclypus* sp. – KAZÁR: 153; fig. 1

Type-material:

Heteroclypus elegans AIRAGHI, 1900:

Holotype: specimen figured by AIRAGHI (1900: pl. 1, figs. 3-4); Museo civico di Milano, Italy [not examined]

Locus typicus: Portotorres, Sardinia

Age: Middle ? Miocene

Echinolampas (Heteroclypeus) hungaricus VADÁSZ, 1915:
Holotype: MAFI Ech 219; housed at the Museum of the Hungarian Geological Survey
Locus typicus: Mátraverebély, Nógrád, Hungary
Age: Early Badenian (Langhian), Middle Miocene

Hypsoheteroclypus vicinoconoideus SZÖRÉNYI, 1953:
Holotype: the specimen figured by SZÖRÉNYI (1953: pl. 6, figs. 2, 2a-b); collection of the University Lwów, Ukraine, specimen no. 44 [not seen]
Locus typicus: Pod'yarkov (= Podjarków), near Lwów, Western Ukraine
Biozone: Lower lagenid zone
Age: Early Badenian (Langhian), Middle Miocene (written comm. Anna WYSOCKA & Andrzej RADWAŃSKI, 02.04.2004)

Echinolampas (Hypsoclypus) vadaszi ROMAN, 1965:
Type-material: see above under *hungaricus* VADÁSZ, 1915
Remarks: this name was established by ROMAN (1965: 309) because *E. hungaricus* VADÁSZ, 1915 is a junior homonym of *E. hungaricus* (DORNYAY, 1913).

Material:
Early Badenian (Langhian) – Retznei [Weissenegg Fm., Lafarge quarry (formerly Perlmoser)], Styria, Austria
NHMW: 3 specimens (NHMW 2003z0007/0001-3)
WANZENBÖCK coll.: 1 specimen (W16)
Early Badenian (Langhian) – Stotzing (sandpit Mayer), Bgld, Austria
NHMW: 2 specimens (NHMW 2003z0008/0002-3)
WANZENBÖCK coll.: 3 specimens (W12-14)

Foreign material for comparison:
Middle ? Miocene – Cannetto, Corsica, France
NHMW: 1 specimen (NHMW 2002z0181/0004)
Early Badenian (Langhian) – Mátraverebély, Nógrád, Hungary
MAFI:1 specimen [MAFI Ech 219 (**holotype** of *E. hungaricus* VADÁSZ, 1915)]
? Late Badenian (Early Serravallian) – Tótmarokháza, Nógrád, Hungary
MAFI:1 specimens [MAFI Ech 269 (reference material of *E.* cf. *elegans* in VADÁSZ, 1915)]

Description (based on the Austrian specimens):
Size and shape: The test is large with up to 14.5 cm test length. The outline is slightly oval, antero-posteriorly elongated. The anterior margin is rounded, the posterior margin bluntly pointed. The maximum width lies slightly posterior of the apical disc, about halfway along the long axis of the test. The profile is high conical with slightly inflated slopes. The apical angle is wider around the apical disc but becomes more acute at about one fifth of the test height (i.e. there is a kink in the slopes). The ambitus is thin and rounded and it lies very low on the test. The oral surface is nearly flat.
Apical disc: The apical disc lies slightly anterior of the centre, about 44 to 46 % TL from the anterior margin. The apical disc is monobasal with four large, circular gonopores and five small ocular pores. The madreporite is pentagonal in outline and moderately swollen (not as strong as in *C. subpentagonalis*) and bears numerous small hydropores, as well as small tubercles.
Ambulacra: The ambulacra are semi-petaloid on the aboral surface. They consist of two slightly diverging rows of closely spaced, slightly oblique conjugated anisopores in each ambulacrum. The poriferous zones of the "petals" are nearly flush with the interambulacra and interporiferous zones. In the paired ambulacra, the poriferous zones are of slightly different length: they are longer in the posterior half of the anterior paired ambulacra (IIa and IVb) and the anterior half of the posterior paired ambulacra (Ib and Va). The "petals" extend about 65 to 80% of the corresponding test radius. Beyond, the ambulacra bear only minute unipores with sunken attachment area and adorally positioned neural canal. The tuberculation on the aboral interporiferous zones is not very dense. The spacing between the primary tubercles is about one and a half to two times their diameter. There are between five to six tubercles across the interporiferous zone (halfway along TH). Miliary tubercles are loosely spread among them. In the poriferous zones, only small miliary tubercles are present, they are situated on the ridges between adjacent pore pairs.
On the oral side the ambulacra are depressed around the peristome and form well developed phyllodes. They consist of three series of unipores in each half-ambulacrum. The adradial and perradial series are rather closely spaced. The middle series is loosely spaced. In the adoral third of the phyllodes there is a double row of shallow pits running along the central suture (sphaeridial pits). The tuberculation of the ambulacra on the oral surface is very similar to those on the interambulacra.
Interambulacra: Around the apex the interambulacra are distinctly inflated between the petals, but not as strong as in *C. subpentagonalis*. The tuberculation on the aboral side is fairly homogenous. The primary tubercles are small and widely spaced. Their mean distance is about one and a half to three times their diameters. Miliary tubercles are loosely spread among them. Directly at the ambitus the spacing of the primary tubercles becomes very closely suddenly, with the areoles of the tubercles nearly touching each other. On the oral side the spacing of the primaries is more widely again, except near the margin. Along the central suture of interambulacrum 5 there is a zone of especially widely spaced tubercles [similar to the "naked zone" KIER (1962) noted in many species of *Echinolampas*]. The primary tubercles of the oral side are about one and a half to two times larger than those of the aboral surface. Data on the tubercle density can be found in Tab. 10.
On the oral side the interambulacra are highly inflated around the peristome and along their interradial sutures, forming well developed bourrelets.
Peristome: The peristome lies slightly anterior of the centre. It is oval (transversely elongated) to slightly subpentagonal in shape. It is always smaller than the periproct, being about 10-11 % TL wide. Inside the test, no auricles are present.
Periproct: The periproct is situated very close to the margin in interambulacrum 5 and lies fully on the oral surface. It is large, being 12-14 % TL wide. The shape of the peristome is oval, transversely elongated and it is slightly rostrate.

Differential diagnosis:
See above under *C. subpentagonalis*.

Discussion:
STEFANINI (1908: 370) considered *Conoclypus montesiensis* MAZZETTI, 1882a to be a senior synonym of *Heteroclypus elegans* AIRAGHI, 1900. The figured specimen of MAZZETI (1882a), however, is rather poorly preserved and the present author is not able to decide on base of this illustration and description if *C. montesiensis* is really conspecific with *elegans* and thus prefers to keep AIRAGHI's name until the identity of the two species can be established or rejected on base of a re-examination of the type specimen or topotypic material.

Occurrence:
Austria: Early Badenian (Langhian); ? Late Badenian (Early Serravallian)
Vienna Basin: ? Müllendorf (Mühlendorfer Kreide AG quarry), Bgld {TOLLMANN, 1955 [probably refers to a GBA specimen from Eisenstadt (oldest label; probably locality Großhöflein, near Eisenstadt; certainly not from the Hartl Fm. in Eisenstadt city) erroneously labelled as coming from Müllendorf in younger labels]}; Stotzing (sandpit Mayer), Bgld (pp KAZÁR, 2002; [NHMW])
Styrian Basin: Retznei (Weissenegg Fm., Lafarge quarry), Styria (HIDEN, 1993; [NHMW])

Paratethys (non-Austrian occurrences): Early Badenian (Langhian); ? Late Badenian (Early Serravallian)

Great Hungarian Basin (Pannonian Basin): Mátraverebély, Nógrád, Hungary (VADÁSZ, 1915; [MAFI]); Nógrád, Hungary (ROMAN, 1965); ? Tótmarokháza, Nógrád, Hungary (VADÁSZ, 1915; [MAFI])

Fore-Carpathian Basin: Pod'yarkov (= Podjarków), near Lwów, western Ukraine (SZÖRÉNYI, 1953); Poland (ROMAN, 1965)

Zala, Sáva and Dráva Basins: Mesarski Brijeg, near Daruvara, Psunj Mts., Croatia (POLJAK, 1938)

Mediterranean: Middle (?) Miocene

Western Mediterranean: Cannetto, Corsica, France ([NHMW]); Porto Torres, Sardinia, Italy (AIRAGHI, 1900, 1905a; ROMAN, 1965); Ozieri, Sardinia, Italy (AIRAGHI, 1905a);

Central Mediterranean: Molasse di Contese, Italy (STEFANINI, 1908c); Molassa di Serra de' Guidoni, Bologna, Italy (? MANZONI, 1881; STEFANINI, 1908c); ? Montese, Italy (MANZONI, 1880a; MAZZETTI, 1882a); ? San Marino (MANZONI, 1880a); ? S. Maria Vigliana, Italy (MANZONI, 1880a; STEFANINI, 1908c)

Conolampas cf. *subpentagonalis* (GREGORY, 1891)

(Pl. 61, Figs. 1a-e)

? 1908c *Echinolampas plagiosomus* (AGASS.) DE LORIOL 1880. – STEFANINI: 363-366; pl. 12, figs. 1a-c

v. 1998 *Echinolampas subpentagonalis* (GREGORY) – SCHULTZ: 118; pl. 53, fig. 3

Material:

Late Badenian (Early Serravallian) – Hornstein, Bgld, Austria
NHMW: 1 specimen (NHMW 1970/1396/898)

Description:

Size and shape: The test is large with 113 mm test length. The outline is oval (antero-posteriorly elongated) with rounded anterior and posterior end. The maximum width and height coincide with the position of the apical disc. The profile is domed, with inflated slopes, being nearly hemispherical. The ambitus is thin and rounded and it lies very low on the test. The oral surface is flattened.

Apical disc: The apical disc lies subcentrally and is monobasal with four large, subcircular gonopores. The ocular pores are indistinct. The madreporite is highly swollen (knob-shaped) and bears numerous small hydropores.

Ambulacra: The ambulacra are semi-petaloid on the aboral surface. They consist of two slightly diverging rows of closely spaced conjugated anisopores pores in each ambulacrum. Both poriferous and interporiferous zones are distinctly depressed adapically. The poriferous zones of the "petals" are very slightly depressed. The poriferous zones Ia, IIb, IVa and Vb are slightly shorter than their counterparts. The frontal petal is slightly curved towards the left and extends about 65 % of the corresponding test radius (measured along the slope of the test, not in planar view). The anterior paired petals extend c. 68 % of the corresponding test radius, the posterior paired petals c. 72.5 %. The tuberculation on the aboral interporiferous zones is not very dense. The spacing of the primary tubercles is about one to three times their diameter. Miliary tubercles are loosely spread among them. In the poriferous zones, only small miliary tubercles are present, they are situated on the ridges between adjacent pore pairs. Halfway along the test height there are six primary tubercles across the interporiferous zones. On the oral side the ambulacra are depressed around the peristome and form well developed phyllodes, very similar to those of *C. subpentagonalis* (see above).

Interambulacra: Around the apex the interambulacra are distinctly inflated. The tuberculation on the aboral side is fairly homogenous. The primary tubercles are small and widely spaced. Their mean distance is about one to three times their diameters. Miliary tubercles are loosely spread among them. The "naked zone" along the central suture of interambulacrum 5 is not very pronounced, being more a zone of especially widely spaced tubercles. Tubercle densities of the oral surface are similar to those in *C. subpentagonalis*. Those of the aboral surface are much significantly higher than in *C. subpentagonalis* (compare Tab. 10). Adorally the interambulacra form well developed bourrelets.

Peristome: The peristome lies slightly anterior of the centre and is distinctly subpentagonal.

Periproct: The periproct is situated very close to the margin in interambulacrum 5, but lies fully on the oral surface. It is larger than the periproct and distinctly oval (transversely elongated).

Discussion:

A single specimen from the Late Badenian of Hornstein differs from the typical *C. subpentagonalis* of the Early Badenian by its hemispherical profile, higher tuberculation density, shorter petals and even more inflated madreporite. It is similar to the specimen figured as *Echinolampas plagiosomus* by STEFANINI (1908c: pl. 12, figs. 1a-c).

The specimen comes from calcareous marls with abundant larger foraminifers (*Planostegina* and *Amphistegina*) outcropping at the western slopes of the Leith Mts., around the village Hornstein. SIEHL (1957: 70) dated these sediments as belonging to the Late Badenian *Bolivina/Bulimina* zone.

When the present specimen is added to the data-set used for the PCA employed to explore the differences between *C. elegans* and *C. subpentagonalis* (see above under the discussion of the latter species) only two factors are extracted from the data set. Like in the earlier analysis factor 1 is controlled by all variables except the tubercle density, factor 2 is controlled mainly by tubercle density and posterior petal width. The resulting scatter plot shows that the specimen of *C.* cf. *subpentagonalis* lies well outside the morphospace occupied by *C. elegans* and *C. subpentagonalis*, but the distinction between the latter is even more blurred (Fig. 57). As stated above, larger data-sets are necessary to enhance the success of such analyses.

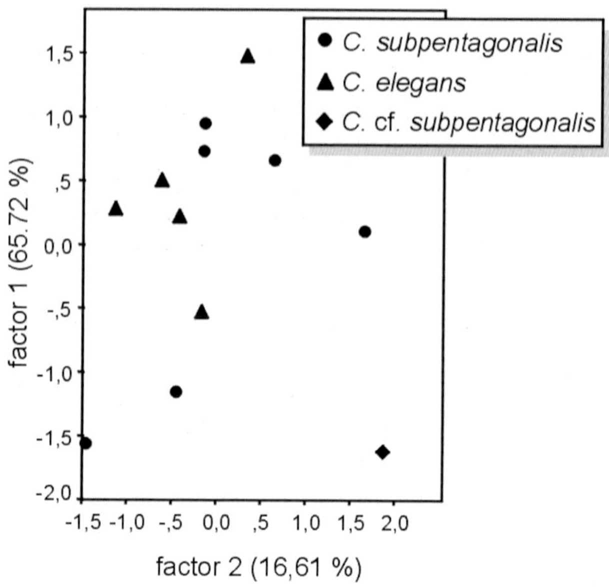

Figure 57: PCA of *Conolampas*: scatter plot of the two factors calculated by the PCA [using the same variables as in the previous analysis (Fig. 55) but with the inclusion of *C.* cf. *subpentagonalis*). Factors 1 and 2 explain 82.33 % of the variance in the present sample.

Occurrence:

Austria: Late Badenian (Early Serravallian)
Vienna Basin: Hornstein, Bgld (Schultz, 1998; [NHMW])

Conolampas ? sp. indet.

pp 1873 *Conoclypus plagiosomus*, Agass – Manzoni: 10-11, 20 [Paratethyan material]
1882 *Conoclypus semiglobus* Des. – Hilber: 237
1887b *Conoclypus plagiosomus*, Agiss. – Koch: 269-270
1891 *Conoclypus plagiosomus* Ag. – Roth von Telegd: 151
1892 *Conoclypus plagiosomus* Ag. – Schafarzik: 196
1905 *Conoclypus plagiosomus*, Ag – Gaál: 344, 362
1905 *Conoclypus*, sp. indet., Ag – Gaál: 344
1926 *Conoclypeus* sp. – Strausz: 213, 368
1953 *Hypsoheteroclypus acuminatus* (Abich), 1882. – Szörényi: 32, 82; pl. 6, figs. 1, 1a-b
1969 *Echinolampas ugolinii* Stef. – Mitrović-Petrović: 124; pl. 3, fig. 2
1981 *Conoclypeus* sp. – Halmai: 106
1981 *Echinolampas plagiosomus* Ag. – Halmai: 106
1988a *Hypsoheteroclypus* sp. – Kókay: 264

Material:

Badenian (Langhian-Serravallian) –Teischl quarry, (? Hartl Fm., Eisenstadt, Burgenland), Austria
NHMW: 1 test fragments (NHMW 2004z0001/008d)

Discussion:

In the literature a number of potential *Conolampas* records exist that lack sufficient information for attribution to either of the species discussed above. They are listed here under *Conolampas* ? sp. for sake of completeness.

Occurrence:

Austria: Badenian (Langhian-Early Serravallian)
Eisenstadt-Sopron Basin: Eisenstadt (Hartl hill), Bgld ([NHMW]); Großhöflein, Bgld (Manzoni, 1873)
Styrian Basin: Zirknitz, Styria (Manzoni, 1873)

Paratethys (non-Austrian occurrences): Karpatian (Late Burdigalian) to Badenian (Langhian-Early Serravallian)
Great Hungarian Basin (Pannonian Basin): borehole at Alcsútdoboz, near Budapest, Hungary (Kókay, 1988a); Fót, Hungary (Halmai, 1981); Fót (Somlyó Mt.), Hungary (Strausz, 1926); Hor. Strháre (= Felsö-Esztergály), Slovak Republic (Roth von Telegd, 1891; Gaál, 1905); Sámsonháza, Hungary (Schafarzik, 1892); Tótmarokháza, Hungary (Gaál, 1905)
Fore-Carpathian Basin: Podjarków, Ukraine (Szörényi, 1953); Stulsko valley, E of Mykolaiv (= Mikołajów), Ukraine (Hilber, 1882)
Transylvanian Basin: Felsö-Orbó, Romania (Koch, 1887b; Gaál, 1905)
Zala, Sáva and Dráva Basins: Vujići, Bosnia & Herzegovina (Mitrović-Petrović, 1969)

Family Pliolampadidae Kier, 1962
Genus *Pliolampas* Pomel, 1888

Type-species: *Echinolampas gauthieri* Cotteau, 1880, by monotypy.
Diagnosis: Medium sized cassiduloid with elongated test; apical disc monobasal with three (or sometimes four) gonopores. Petals well developed, open, with broad, equal poriferous zones. Periproct inframarginal; peristome anteriorly displaced and elongated. Bourrelets well developed. Phyllodes broad with few large pores. No naked zone in adoral interambulacrum 5 (Kier, 1962).
Distribution: Eocene to Pliocene – circum-Mediterranean, Malay Archipelago (Kier, 1962)

Pliolampas vassalli (Wright, 1855)

(Fig. 58; Pl. 62, Figs. 1-2)

? 1854 *Echinolampas Elongatula*, Millet. – Millet: 178; no. 393 [*nomen nudum*]
* 1855 *Pygorhynchus Vassalli*, Wright, n. sp. – Wright: 271-272
1858 [*Echinanthus*] *Vassalli* Syn. *Pygorhynchus Vassalli* Wright – Desor: 296
1864 *Pygorhynchus Vassali*, Wright. – Wright: 479; pl. 22, figs. 6a-c
? # 1865 *Echinolampas elongatula* – Millet: 611, no. 270 [reference not seen; cited from Cottreau, 1912]
? 1866 *Echinolampas elongatula* – Millet: 35, no. 270 [reference not seen; cited from Cottreau, 1912]
non 1877 *Pygorhynchus collombi*, Desor, 1857 – Cotteau in Locard: 284-286; pl. 11, figs. 6-10
non 1877 *Pygorhynchus Vassali*, Wright. – Duncan: 51, 67
? 1883 *Echinolampas elongatula* Millet, 1854. – Cotteau: 29-31; pl. 4, figs. 6-8
pp 1891 *Breynella vasalli* (Wright), 1855. – Gregory: 602
non 1895 *Pliolampas Vassali* (*Plesiolampas*), (Wright) Pomel, 1888. – Cotteau: 38
1909 *Tristomanthus Vassali* Wright (*Pygorhynchus*) – Lambert: 134
? 1912 *Echinolampas elongatula* Millet, 1854. – Cottreau: 256, 256a-b; 15 figs. [includes photographs of the holotype]
1913a *Tristomanthus Vassali* Wr. (*Pygorynchus*). – Cottreau: 23
1913a [*Pliolampas*] *M.*[*illetia*] *Vassali* Wright (Rygorhynchus). – Cottreau: 66
v pp 1915 *Milletia angulosa* Mazz. sp. – Vadász: 218; pl. 9 (3), figs. 12-13
1921 *P.*[*liolampas*] *Vassali* Wright (*Pygorhynchus*) – Lambert & Thiéry: 372
1948a *Breynella Vassali* (Wright) – Mortensen: 248; figs. 234a-c
? 1941 *Pliolampas angulosus* Mazetti. – Meznerics: 91
? 1953 *Pliolampas vassalli* (Wright), 1855. – Szörényi: 21, 69-70; pl. 3, figs. 5, 5a-b
? 1953 *Tristomanthus podjarkovi* n. sp. – Szörényi: 25-261, 74-75; pl. 4, figs. 4, 4a-b
1962 *Pliolampas vassalli* (Wright) – Kier: 195-196; pl. 37, figs. 1-4; text-fig. 164
? 1969 *Echinolampas aequizonatus* Gregory – Mitrović-Petrović: 124; pl. 3, figs. 1, 1a-b
1974b *Pliolampas vassalli* – Rose: 353; fig. 3 [table]
1975 *Pliolampas vassalli* (Wright) – Rose: 79; tab. 12
1980 *Pliolampas vassalli* (Wright) – Challis: 168-170; pl. 62, figs. A-C; pl. 63, figs. A-C

Type-material:

Pygorhynchus vassalli Wright, 1855:
Lectotype: designated by Gregory (1891: 602); British Museum of Natural History (Geology) E 1581 (figured in Challis, 1980: pl. 6, figs. a-c)
Locus typicus: Maltese Islands
Stratum typicum: *Globigerina* Limestone
Age: Aquitanian to Langhian (Early or Middle Miocene)

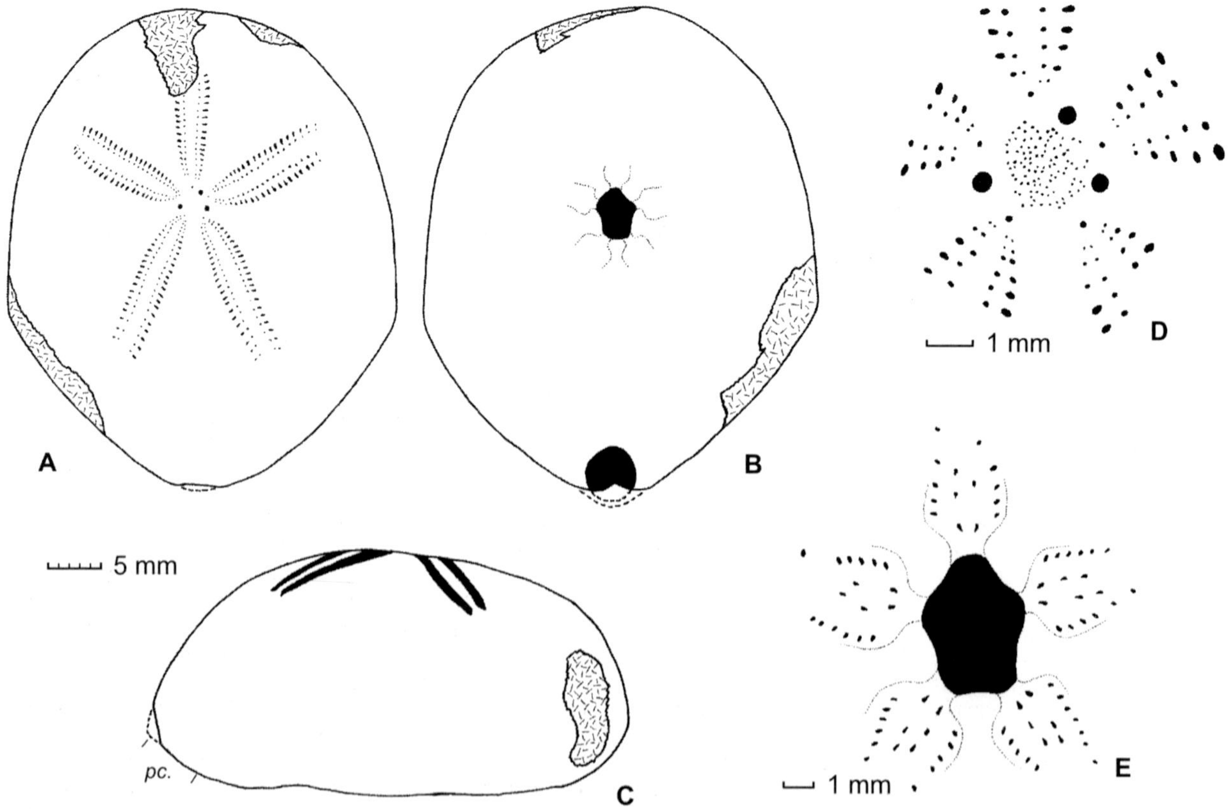

Figure 58: *Pliolampas vassalli* (Wright, 1855): aboral (A), oral (B), and lateral (C) view (damaged areas hatched), detail of the apical disc (D) and the floscelle (E) (sandpit Mayer, Stotzing, Bgld, Wanzenböck coll. W 25); pc. = periproct.

Material:
Early Badenian (Langhian) – Stotzing (sandpit Mayer), Bgld, Austria
Wanzenböck coll.: 1 specimen (W25)

Foreign material for comparison:
Early Badenian (Langhian) – Pod'yarkov (= Podjarków), near Lwów, western Ukraine
NHMW: 1 specimen (NHMW 1859.XLV.557)
Early Badenian (Langhian) – Mátraverebély (Meszestetö), Nógrád, Hungary
MAFI: 2 specimens [MAFI Ech 438 (reference material of *Milletia angulosa* Mazz. in Vadász, 1915)]

Dimensions (in mm):

Inv. No.	TL	TW	TH
Wanzenböck coll. W25	41.1	34.6	20.4
NHMW 1859.XLV.557	35.8	26.6	18.1

Description:
Size and shape: Test of medium size, elongated, with rounded anterior and pointed posterior end. In profile the test is distinctly inflated with rounded anterior and steeply sloping posterior end. The maximum height lies slightly posterior of the centre on a weakly developed keel in interambulacrum 5. The maximum width lies posteriorly, coinciding with the end of the posterior petals in the present specimens. The adoral surface is depressed around the peristome. The width is about 84.2 % of TL, the height about 49.6 %.
Apical disc: The apical disc lies anteriorly, about 41 % from the anterior margin. It is monobasal, with 3 circular gonopores and 5 small ocular pores (Fig. 58.D). There is no genital pore 3 developed. The madreporite has a pentagonal shape and is crowded by numerous small madreporic pores.
Ambulacra: Adapically the ambulacra are petaloid, straight, open distally and relatively small. They consist of conjugate anisopores, with rounded inner and elongated, teardrop-shaped outer pores. Adjacent pores are separated by low ridges bearing a single row of small tubercles. The interporiferous zones are about as wide as a single poriferous zone. Outside the petals only small, slit-like, vertically elongated unipores are present. The anterior petals are slightly flexed anteriorly.
The frontal petal extends about half the corresponding test radius, whereas the paired petals extend about two third of the corresponding test radius. The anterior paired petals form an obtuse angle of about 115°, the posterior paired petals an acute angel of about 50°. Adorally the ambulacra form broad phyllodes consisting of large rounded unipores, arranged in two series in each half ambulacrum, an outer series consisting of 5-6 pores and an inner series consisting of 1-2 pores (Fig. 58.E).
Interambulacra: The interambulacra are loosely covered by perforate, crenulate primary tubercles in sunken areoles. Between them dense secondary/miliary tuberculation is present. The aboral tubercle density is about 150/cm², orally the tubercles are larger, showing a mean density of 110/cm². The size of the tubercles increases from the ambitus towards the peristome. Around the peristome, between the phyllodes, the primary tubercles are very widely spaced. Adorally the ambulacra form moderately well developed bourrelets.
Peristome: The peristome is pentagonal, elongated antero-posteriorly and lies anterior of the centre, about 43 % TL from the anterior margin. It is about 4.6 mm long and 3.2 mm wide in the 41.1 mm test length specimen.
Periproct: The periproct lies inframarginally in interambulacrum 5 and is strongly overhanging. It is distinctly rostrate. Only the posterior/aboral margin of the periproct reaches nearly halfway up the ambitus but is still visible in oral view (due to the

rostrate nature). The periproct has an oval, vertically elongated shape and is 5.5 mm long and 4.2 mm wide in the 41.1 mm TL specimen. Below the periproct there is a slight subanal depression.

Differential diagnosis:
This species differs from *P. gauthieri* (Cotteau, 1880) from the Burdigalian of Saint-Restitut, France, the type species of *Pliolampas*, by its smaller petals, larger test width (TW of *P. gauthieri* = 77 % TL), slightly less anterior position of the apical disc, position of maximum height (coincides with the apical disc in *P. gauthieri*) and profile outline (gently sloping posterior end in *P. gauthieri*) (compare Cotteau, 1880: 227-228, pl. 32, figs. 5-8; and Kier, 1962: 194-195, pl. 36, figs. 1-3, text-fig. 162).

P. pioti Gauthier in Fourtau, 1900 from the Late (?) Burdigalian of Gebel Geneffe, Egypt differs by its longer (but see below in discussion) and broader petals, distinctly oval outline (the characteristic kink between the anterior and posterior margin is not developed) more central position of the peristome and the position of the periproct which lies higher on the ambitus than in *P. vassalli* (compare Gauthier in Fourtau, 1900: 712-714, pl. 3, figs. 7-10; Fourtau, 1920: 62, pl. 4, figs. 5, 5a-b; Kier, 1962: pl. 36, figs. 4-7).

"Echinanthus" marginatus Mazzetti, 1882 and *"E." angulosus* Mazzetti in Mazzetti & Pantanelli, 1883 (for *"Echinanthus"* sp. figured in Mazzetti, 1882a) from the Middle Miocene of Northern Italy are based on material of such poor preservation that a comparison is impossible.

Pliolampas titanensis Nelli, 1907 from the Miocene of San Marino is very similar to *P. vassalli* (compare Nelli, 1907: 2268-269; pl. 9, figs. 10, 10a-b). According to Nelli (1907: 269) it differs from the latter by its longer posterior paired petals and the higher number of pores within the petals.

Studeria corsica (Cotteau, 1877) which co-occurs with *P. vassalli* in the locality Stotzing (Bgld), differs from that species by its ovoid outline, broader petals, marginal periproct, which is visible only in posterior view and central position of the peristome.

Discussion:
The studied specimen is very similar to the lectotype of *Pliolampas vassalli* (Wright, 1855) and differs only slightly from that specimen. In the lectotype the gonopores are larger [could easily be related to sexual dimorphism or other intraspecific variation; in the very similar *P. elongatula* specimens with both, large and small gonopores are known (see Cottreau, 1912)] and the frontal and anterior paired petals extend farther to the ambitus (about three quarters corresponding test radius). This may be due to the fact, that the studied specimens are larger (TL = 41.1 and 35.8 mm respectively) than the lectotype (TL = 29.3 mm) and since all other features are virtually identical, the Austrian specimens is placed into the species *P. vassalli*.

Cotteau (in Locard, 1877: 284-286) placed *P. vassalli* into the synonymy of *Pygorhynchus collombi* Desor, 1857. As Gregory (1891: 602) already pointed out, this cannot be accepted, since *P. collombi* is clearly not conspecific, moreover, not even congeneric with *P. vassalli*, since the former species has a distinct naked zone in interambulacrum 5, a non-rostrate periproct, which is visible only in posterior view, a different outline and different petal morphology. On the other hand, Gregory's (1891: 602) statement that *"Echinanthus" corsicus* Cotteau, 1877 belongs into the synonymy of *P. vassalli* is also dubious (see also Lambert, 1907a: 58). *"E." corsicus* has a marginal periproct, lying higher on the ambitus than in *P. vassalli*, it is only very slightly rostrate and the shape of the test in profile is also different [judging by the description and illustration given by Cotteau (1877: 282-284, pl. 11, figs. 1-5)].

The specimens figured as *P. vasalli* by Airaghi (1905b: 49-50; pl., fig. 7-10) are not conspecific with Wright's type-material according to Lambert (in Cottreau, 1913a: 66, under *Tristomanthus corsicus*) and probably belong to *"Echinanthus" corsicus* Cotteau, 1877.

P. vassalli has also been reported from the Early Badenian of the Ukraine (Szörényi, 1953), however, the description and figures are not sufficient to decide, whether or not the described specimen belongs into the species. Moreover, Szörényi (1953) reports two species of *Pliolampas* and seven of *Studeria* (under the junior synonym *Tristomanthus*), including two new ones, from the same locality, each represented by a single specimen. As Philippe (1998: 105) pointed out, the high number of closely related species reported from a single locality is rather surprising and a revision of this material is urgently needed. A specimen from the Early Badenian of Pod'yarkov (= Podjarków, near Lwów, western Ukraine) in the collection of the Naturhistorisches Museum Wien (NHMW 1859.XLV.557) is clearly conspecific with *P. vassalli*.

Part of the material referred to *Milletia angulosa* Mazz. by Vadász (1915: 218) belong to the species discussed here (two specimens from the Early Badenian of Mátraverebély (Meszestetö), Nógrád, Hungary; MAFI Ech 438). The remaining reference material of Vadász (1915) including the figured specimen [MAFI Ech 246 from Mátraverebély and MAFI Ech 449 from Tótmarokháza (= Márkháza), both in Nógrád, Hungary], as well as the material referred to as *Pliolampas* sp. by Vadász (1915) (MAFI Ech 305 from Mátraverebély) are poorly preserved and could not be referred to any definitive species or genus.

Pliolampas elongatula (Millet, 1865) from the Serravallian ["facies Savignéen" in Cottreau, 1912; dated as Serravallian by Demarq (in Steininger et al., 1985: vol. 1: 26, vol. 2: 55, area no. 309 Loire-Bretagne)] of western France, the type-species of *Milletia* Duncan, 1877 (a genus considered as junior synonym of *Pliolampas* by Kier, 1962: 196-197), is remarkably similar to *P. vassalli*. The type-material of this species was re-illustrated by Cottreau (1912), who gave 1854 as publication date. In Millet (1854), the species is only mentioned without description or illustration and thus represents a *nomen nudum*. According to Cottreau (1912) it was first described in Millet (1865: 611), contrary to the statement of Kier (1962: 196), who maintained that in this publication there is no description in that paper either. *P. elongatula* differs from *P. vassalli* by its slightly longer posterior paired petals (as far as can be judged from the photographs in Cottreau, 1912), the more posterior position of the characteristic kink between anterior and posterior margin, as well as the less pointed posterior end. As these features show high intraspecific variation (and, in fact, the topotype figured by Cottreau, 1912 is nearly indistinguishable from a typical *P. vassalli*), it is likely that the two species are conspecific or at least closely related.

Cotteau (1877: 284) listed *P. vassalli* in the synonymy of the taxon *"Pygorhynchus collombi* Desor, 1857". Besides being the younger name anyway (thus *P. vassalli* would have priority, if they were conspecific), this *"Pygorhynchus collombi"* is very different from *P. vassalli*, and not even congeneric. If Cotteau's figures and descriptions are correct [and correspond to the type material (in Desor, 1855-1858 there is only a very short description)] this species has a transversely elongated periproct (not known in any pliolampadid), a well developed naked zone along the central suture of adoral interambulacrum 5 (also missing in most pliolampadid genera) and petals which are closing distally and consist of numerous, closely spaced pores (a condition very different from that observed in *Pliolampas*, *Studeria* and most other pliolampadid genera).

Occurrence:

<u>Austria</u>: Early Badenian (Langhian)

Vienna Basin: Stotzing (sandpit Mayer), Bgld ([Wanzenböck coll.])

<u>Paratethys (non-Austrian occurrences)</u>: Early Badenian (Langhian)

Great Hungarian Basin (Pannonian Basin): ? Márkháza, Nógrád, Hungary (Meznerics, 1941); Mátraverebély (Meszestetö), Nógrád, Hungary (pp Vadász, 1915; [MAFI])

Fore-Carpathian Basin: Pod'yarkov (= Podjarków), near Lwów, western Ukraine (? Szörényi, 1953; [NHMW])

Zala, Sáva and Dráva Basins: ? Vujići, Bosnia & Herzegovina (Mitrović-Petrović, 1969)

Mediterranean: ? Aquitanian, ? Burdigalian, Langhian [in her revision Challis (1980) was not able to ascertain pre-Langhian occurrences]

Central Mediterranean: *Globigerina* Limestone, Maltese Islands (Wright, 1855; Desor, 1858; pp Gregory, 1891; Lambert, 1909; Cottreau, 1913a, Kier, 1962); "nodular beds" (= phosphate conglomerates) within the *Globigerina* Limestone, Maltese Islands (Wright, 1864); Lower (?) to Middle *Globigerina* Limestone, Maltese Islands (Rose, 1974b, 1975); Xwieni conglomerate bed Islands (= C_2 phosphate conglomerate of Pedley et al., 1976), Upper *Globigerina* Limestone, Maltese (Challis, 1980)

Atlantic Ocean: Serravallian [under the name *P. elongatula* (Millet, 1865)]

Bretagne – Loire Basin: ? Briand, Maine-et-Loire, western France (Cottreau, 1912); ? Chavagnes, Maine-et-Loire, western France (Millet, 1854, 1865, 1866; Cotteau, 1883; Cottreau, 1912); ? Doué-la-Fontaine, Maine-et-Loire, western France (Millet, 1854, 1865, 1866; Cotteau, 1883; Cottreau, 1912); ? Martigné, Maine-et-Loire, western France (Millet, 1854, 1865, 1866; Cotteau, 1883; Cottreau, 1912); ? Saumurois, Maine-et-Loire, western France (Cottreau, 1912)

Genus *Studeria* Duncan, 1889

Type-species: *Catopygus elegans* Laube, 1869; by monotypy (Duncan, 1889: 185).
Diagnosis: Small pliolampadid, with anteriorly eccentric monobasal apical disc with three pores; no pore in left anterior genital plate; petals long, straight and open, adjacent pore pairs widely spaced; peristome antero-posteriorly elongated; periproct vertically elongated and situated marginally; bourrelets strongly developed, phyllodes slightly widened; no "naked zone" in interambulacrum 5 (summarised from Kier, 1962).
Distribution: Oligocene to Miocene – Europe and Australia (Kier, 1962); Recent – Arafura Sea, south of Papua New Guinea (Kier, 1962; Mooi, 1990a)
Remarks: The genera *Pliolampas* Pomel, 1888 and *Studeria* Duncan, 1891 are very similar. *Studeria* differs from the former by its marginal periproct and more strongly developed bourrelets (Kier, 1962: 217). Several authors suggested that the two genera should be regarded as synonymous (e.g. Lambert & Thiéry, 1921: 371-372; Pomel, 1888: 446). Lately Philippe (1998: 102) expressed his doubts if the two genera should really be separated. Here the classification of Kier (1962) is followed, since that question must be investigated at a broader scale and cannot be decided on base of the limited material discussed here. For *Studeria recens* (Agassiz, 1879), the species most closely related (in morphological terms) to the species discussed below, a separate genus (*Hypselolampas*) was established by Clark (1917: 104), but this genus was synonymised with *Studeria* by Kier (1962: 217). Lately Mooi (1990a: 82) stated that "... there is some uncertainty that *S. recens* is a *Studeria*, ...". In a recent paper on the cassiduloid ancestry of the clypeasteroids by Smith (2001) indeed some evidence turned up that *S. recens* is not congeneric with the fossil members of this genus and that Clark's genus might be justified. In Smith's cladistic analysis *Studeria recens* (Ag.) and *S. elegans* (Laube) do not form a monophyletic group. *S. recens* is a sister taxon to a clade including *Oolopygus*, *S. elegans*, *Notolampas*, *Oligopodia* and *Neolampas* (Smith, 2001: fig 2, semistrict consensus tree). Its phylogenetic position, however, is not well resolved, depending on the taxa included in the analysis *Studeria* is either closer to the echinolampadids and cassidulids or the neolampadids (see Suter, 1994b; Smith, 2001: fig. 3).

***Studeria corsica* ? (Cotteau, 1877)**

(Fig. 59; Pl. 62, Figs. 3-4)

?	1840b	[*Pygorhynchus*] *subcylindricus* Ag. – Agassiz: 5 [*nomen nudum*]
? *	1847a	[*Pygorhynchus*] *subcylindricus* Agass. – Agassiz in Agassiz & Desor: 161 [*nomen dubium*]
?	1858	[*Pygorhynchus*] *subcylindricus* Agass. – Desor: 298
*	1877	*Echinanthus corsicus*, Cotteau, 1876 – Cotteau in Locard: 282-284; pl. 11, figs. 1-5
	1895	*Pliolampas Vassali* (*Plesiolampas*), (Wright) Pomel, 1888. – Cotteau: 38
	1907a	*Tristomanthus corsicus* Cotteau (*Echinanthus*), 1877. – Lambert: 58
?	1921	[*Pliolampas*] *S.*[*tuderia*] *subcylindrica* Agassiz (*Pygorhynchus*) – Lambert & Thiéry: 373

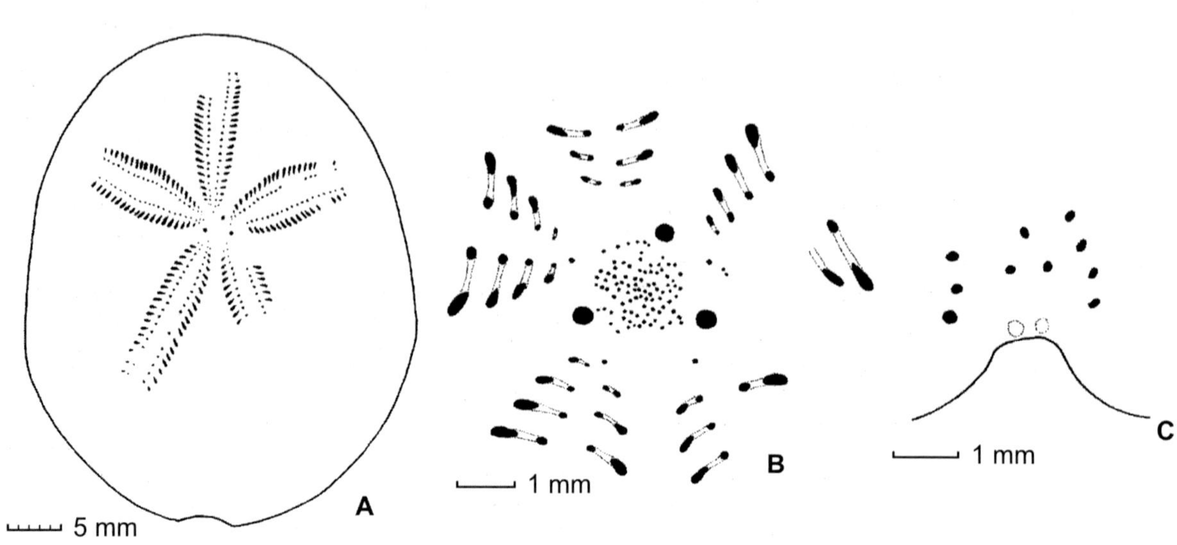

Figure 59: *Studeria corsica* ? (Cotteau, 1877): aboral view (A), detail of the apical disc (B) and phyllode (C) (Lafarge quarry, Retznei, Stmk, NHMW 2002z0173/0001).

? 1928 *Tristomanthus subcylindricus* AGASSIZ (*Pygorhynchus*) – LAMBERT & JEANNET: 154; pl. 1, figs. 1-4
? 1905b *Pliolampas Silvestrii* n. sp. – AIRAGHI: 51-52, pl., figs. 1-4
non 1953 *Tristomanthus subcylindricus* (AGASSIZ), 1846. – SZÖRÉNYI: 23, 71-72; pl. 5, figs. 4, 4a-b

Type-material:
Echinanthus corsicus COTTEAU, 1877:
Syntypes: collection of PERON and COTTEAU; current whereabouts unknown
Locus typicus: Bonifacio, Corsica, France
Age: Late Burdigalian to Serravallian [based on the dating by ORSZAG-SPERBER (in STEININGER et al., 1985: Vol. 1: 27, Vol. 2: 57, area 20b Corsica, Bonifacio Basin)]

Material:
Early Badenian (Langhian) – Retznei [Weissenegg Fm., Lafarge quarry (formerly Perlmoser)], Styria, Austria
NHMW: 1 specimen (NHMW 2002z0173/0001)
Early Badenian (Langhian) – Stotzing (sandpit Mayer), Bgld, Austria
WANZENBÖCK coll.: 1 deformed specimen (W26)

Foreign material for comparison:
Late Burdigalian to Serravallian – Monte Balistro, near Bonifacio, Corsica, France
NHMW: 1 slightly crushed specimen (NHMW 1857.XV.124)

Dimensions (in mm):

Inv. No.	TL	TW	TH
NHMW 2002z0173/0001	41.4	34.6	21.2
Wanzenböck coll. W26	~ 45	~38	24.2

Description:
Size and shape: Test of medium size, antero-posteriorly elongated, with distinctly ovate outline. Posterior margin indented by the groove of the periproct. In profile the test is distinctly inflated with rounded anterior and obliquely truncated posterior margin. The maximum height lies centrally, the maximum width of the test lies posteriorly, approximately at the point, where the posterior petals end. The adoral surface is strongly depressed around the peristome. The width is about 83.5 % of TL, the height about 51.2 %.
Apical disc: The apical disc lies anteriorly, about 35 % TL from the anterior margin. It is monobasal, with 3 relatively small gonopores. Genital pore 3 is not developed (Fig. 59.B). The madreporite has a pentagonal shape and is crowded by numerous small madreporic pores. The ocular pores are small and rounded.
Ambulacra: Adapically the ambulacra are petaloid, straight and open distally. They consist of conjugate anisopores, with rounded inner and elongated, teardrop-shaped outer pores. Adjacent pores are separated by low ridges. Adapically the pores are quite closely spaced, in the most distal part of the petals, they become increasingly less closely spaced. The interporiferous zones are smaller than a single poriferous zone. In the anterior paired petals alone they may be nearly as wide as a single poriferous zone. Outside the petals only small unipores are present.

The frontal petal extends about two third of the corresponding test radius, whereas the paired petals extend about half of the corresponding test radius to the ambitus. The anterior paired petals form an obtuse angle of about 135°, the posterior paired petals an acute angel of about 50°. Adorally the ambulacra form broad phyllodes consisting of large rounded unipores, arranged in two series in each half ambulacrum, an outer series consisting of four to five pores and an inner series consisting of one to two pores (Fig. 59.C).
Interambulacra: The interambulacra are covered by perforate, crenulate primary tubercles in sunken areoles. Between them a dense secondary/miliary tuberculation is developed. The aboral tubercle density is about 180/cm², orally the tubercles are larger, showing a mean density of 140/cm². Adorally the ambulacra form weakly developed bourrelets.
Peristome: The peristome is sub-pentagonal, and lies in a moderately deep depression in the centre of the oral surface. It is about 3.7 mm wide in the specimen from Retznei (NHMW 2002z0173/0001).
Periproct: The periproct lies marginally in interambulacrum 5 and is visible only in posterior view. It is situated halfway up the ambitus at the end of a shallow groove connecting it with the oral surface. The periproct has an oval, vertically elongated shape and is 5.4 mm long and 4.1 mm wide in specimen NHMW 2002z0173/0001.

Differential diagnosis:
S. aremorica (BAZIN, 1884) from the Miocene of the Bretagne differs by its longer petals, which nearly extend up to the ambitus and its less anteriorly positioned apical disc (compare BAZIN, 1884: 40-41, pl. 1, figs. 26-30).

S. caralitana (LAMBERT, 1907) from the Early or Middle Miocene of Sardinia, differs from this species by its strongly anteroposteriorly elongated outline (TW ~ 70 % of TL), high, cylindrical profile (TH ~ 60 % TL), narrow petals, anteriorly eccentric peristome, posteriorly eccentric maximum height, pronounced keel in aboral interambulacrum 5 and position of the periproct, which lies lower on the ambitus (compare LAMBERT, 1907a: 37-38, pl. 3, figs. 8-11).

S. elegans (LAUBE, 1869), the type-species of the genus *Studeria* (not *S. recens*, see KIER, 1962: 217), from the Miocene of Southern Australia, differs from this species by its less elongated outline (width is about 90 % TL), its longer petals, which extend nearly up to the ambitus, its pointed posterior end, and its less anterior positioned apical system (about 40 % TL from the anterior margin) (compare LAUBE, 1869b: 190-191; figs. 8, 8a).

S. meslei (GAUTHIER in COTTEAU, 1877) from the Burdigalian of the Rhône Basin differs from this species by less elongated shape (width is about 85 to 88 % TL), its longer petals, which extend nearly up to the ambitus, its pointed posterior end, its less anterior positioned apical system (about 40 % TL from the anterior margin), its slightly wedge shaped test profile, with lower ambitus and the position of its peristome, which is inframarginal and visible in oral view [compare the description and illustrations of COTTREAU (1913a: 112, pl. 12, figs. 11, 11a-c), and PHILIPPE (1998: 102-104; pl. 19, figs. 1-4)].

S. pantanellii (STEFANINI, 1908) from the Middle Miocene of northern Italy, which is similar to *S. spratti* (see below) differs by its small, regularly oval outline, subcentral maximum width and its slightly overhanging periproct [compare STEFANINI, 1908b: 76-77, pl. 13 (1), figs. 6a-b].

S. recens (AGASSIZ, 1879) an extant species from the Arafura Sea between Papua New-Guinea and Australia, differs by its even more anterior positioned apical system (about 30 % TL from the anterior margin), its highly inflated test (test height about 70 % TL), its distinct keel in aboral interambulacrum 5 and its wider interporiferous zone of the frontal petal (compare AGASSIZ, 1881: 123-124, pl. 20, figs. 17-21, MOOI, 1990a: 82, fig. 10d).

S. silvestrii (AIRAGHI, 1905) from Miocene of S. Maria Tiberina (Umbria, western Italy) differs from the present specimen by its less anteriorly situated maximum width, less ovoid outline, anteriorly eccentric peristome and its periproct which lies lower on the ambitus.

S. spratti (WRIGHT, 1964) from the Burdigalian to Langhian Middle and Upper *Globigerina* Limestone of Malta differs by its pointed posterior end, sub-centrally to anteriorly positioned maximum width, anteriorly eccentric peristome, slightly overhung periproct and regular spacing of the pore pairs within the petals [compare WRIGHT, 1864: 490, pl. 21, figs. 6a-d (only figured); COTTREAU, 1913a: 113, pl. 11, figs. 8, 8a, 9-10; CHALLIS, 1980: 173-176, pl. 66, figs. a-c, pl. 67, figs. a-c].

S. subcarinata (GOLDFUSS, 1826) from the Late Oligocene of Doberg, Germany, differs from this species by its less elongated outline (width is about 88 to 90 % TL), conical profile, its less anterior positioned apical system (about 44 to 50 % TL from the anterior margin), its well developed bourrelets and its more anteriorly positioned peristome (based on the description and figures in KIER, 1962: 218; pl. 41, figs. 6-9 and conspecific material at the NHMW).

SZÖRÉNYI (1953) described seven species of *Studeria* (under the name *Tristomanthus*, a synonym of *Studeria* according to KIER, 1962) from the Miocene of Pod'yarkov (= Podjarków) in the Ukraine, two of them new. PHILIPPE (1998: 104) suggested, that most of her specimens might belong to *S. meslei*, the descriptions and figures, however, are insufficient and it is necessary to examine the material to judge this. Her two new species *"T." podjarkovi* and *"T." podolicus* are clearly different from the material discussed here, the former differs by its narrower petals, its inframarginal periproct and its different profile with its bluntly pointed posterior end; the latter differs by its less elongated outline, inframarginal periproct and its longer petals, which nearly reach the ambitus. Some of her specimens might belong to *P. vassalli*, the occurrence at Pod'yarkov of which is documented by a specimen in the collection of the NHMW (see above under that species).

The *Pliolampas* species *P. elongatula* (MILLET, 1865), *P. gauthieri* (COTTEAU, 1880), *P. pioti* GAUTHIER in FOURTAU, 1900, *P. subcarinata* COTTEAU, 1895, *P. titanensis* NELLI, 1907 and *P. vassalli* (WRIGHT, 1855) can easily be distinguished from the present species by their pointed posterior end and inframarginal periproct, which is well visible in oral view.

"Echinanthus" marginatus MAZZETTI, 1882a and *"E." angulosus* MAZZETTI in MAZZETTI & PANTANELLI, 1883, as stated above, are based on very poor material hampering comparison, even if the material described and illustrated by STEFANINI [1908b: 78-79, pl. 13 (1), figs. 7-9] is taken into account. STEFANINI (1908b) referred the two species to *Milletia*, a junior synonym of *Pliolampas* according to KIER (1962).

"Echinanthus" camerinensis DE LORIOL, 1882 from Camerino, Italy is a very small sub-globular, slightly elongated form with narrow, open petals and a slightly overhung, marginal periproct. Probable the specimen illustrated by DE LORIOL is a juvenile. The present species differs by its more elongate test, lower test height and broader petals.

Discussion:

From the known previously described species the Austrian material is most similar to *S. corsica* (COTTEAU, 1877) and the extant *S. recens* (AGASSIZ, 1879). *S. corsica* differs by its slightly narrower posterior end wider anterior end, its wider angle between the posterior paired petals (about 75°; according to the figure in COTTEAU, but see below) and its more anterior position of the peristome (about 43 % TL from anterior margin) (compare COTTEAU, 1877: 282-284, pl. 11, figs. 1-5). Yet if compared to a specimen from a locality next to or identical with the type locality (NHMW 1857.XV.124, from Monte Balistro, near Bonifacio, Corsica; Pl. 62, Fig. 4a-d), most of these differences (except the more anterior position of the peristome) are not present. It is well known that COTTEAU's artist made figures which are nice but often not very accurate [e.g. he almost invariably figured double pores in the phyllodes or the ambulacral plates beyond the petals, regardless of the actual situation in the figured material (KIER, 1962: 197, 1966: U492); likewise the plating patterns of fossil spatangoids and clypeasteroids are often wrongly depicted]. Based on the comparison with the mentioned specimen and the description of COTTEAU (1877), the present specimen is tentatively attributed to *S. corsica*. There is also a distinct similarity to *S. ficheuri* (POMEL, 1887) from the Burdigalian of Algeria and Spain [providing the illustration and description in LAMBERT (1906a: 95-100, pl. 5, fig. 11) corresponds to the type-material].

Studeria subcylindrica (AGASSIZ in AGASSIZ & DESOR, 1847), likewise, is very similar to the present specimens, judging by the plaster cast figured by LAMBERT & JEANNET (1928: pl. 1, figs. 1-4). The whereabouts of the holotype of this species is, however, unknown and neither type locality nor type stratum are known. A plaster cast of the holotype with the number P31 is housed at the Musée d'Histoire Naturelle de Neuchâtel. DESOR (1858: 298) expressed his doubts about the specimen's provenance and according to LAMBERT & JEANNET (1928: 154) the specimen is not from the "Calcaire grossier, des environs de Paris" as stated by AGASSIZ & DESOR (1847a: 161), but from an unknown Miocene locality. The original diagnosis of AGASSIZ (in AGASSIZ & DESOR, 1847a, p. 161: "Espèce renflée, subcylindrique, à ambulacres étroits, à anus proéminent, subrostré [Species reinflated, subcylindric, with narrow ambulacra, prominent anus, subrostrate]" indicates the presence of a rostrum in *S. subcylindrica*, a feature not present in the Austrian specimens. Since its holotype is lost *S. subcylindrica* and the information on the characters of this species in the literature are insufficient, it would be necessary to select a neotype (ICZN 4th ed., 2000, Article 75.1.). Unfortunately that is not possible as the type locality and area is unknown. Thus *S. subcylindrica* is considered as *nomen dubium*.

GREGORY (1891: 602) placed *Echinanthus corsicus* COTTEAU, 1877 in the synonymy of *Breynella vassalli* (WRIGHT, 1855) [*Breynella* is a junior synonym of *Pliolampas* according to KIER (1962: 195)], an opinion which was subsequently accepted by COTTEAU (1895: 38). Yet, as LAMBERT (1907a: 58) pointed out, the two are clearly different species and can be separated by their different outline (oval to egg-shaped with rounded anterior and posterior ends in *S. corsica* – outline with distinct kinks, distinctly pointed, rostrate posterior end in *P. vassalli*), different posterior paired petals (increasingly widely spaced pores in *S. corsica* – regularly spaced pores in *P. vassalli*), and position of the periproct (marginal, vertical, halfway up the ambitus, not visible in oral view in *S. corsica* – inframarginal, overhanging, well visible in oral view in *P. vassalli*).

Occurrence:

Austria: Early Badenian (Langhian)
- Vienna Basin: Stotzing (sandpit Mayer), Bgld ([coll. WANZENBÖCK])
- Styrian Basin: Retznei (Weissenegg Fm., Lafarge quarry), Styria ([NHMW])

Mediterranean: Late Burdigalian to Early Serravallian
- Western Mediterranean: Bonifacio, Corsica, France (COTTEAU, 1877); Capo Sant'Elia, Sardinia, Italy (COTTEAU, 1895; LAMBERT, 1907a); Monte Balistro, near Bonifacio, Corsica, France ([NHMW])

Series Atelostomata ZITTEL, 1879
Order Spatangoida CLAUS, 1876
Suborder Hemiasterina FISCHER, 1966
Family Hemiasteridae CLARK, 1917
Genus *Hemiaster* AGASSIZ in AGASSIZ & DESOR, 1847

Type-species: *Spatangus bufo* BRONGNIART, 1822; by subsequent designation (SAVIN, 1903).
Diagnosis: see FISCHER (1966)
Distribution: Aptian to Recent – cosmopolitan (FISCHER, 1966)

"Hemiaster" kalksburgensis LAUBE, 1869

(Fig. 60)

1869a *Hemiaster Kalkburgensis* LAUBE sp. ined. – FUCHS: 194
* 1869a *Hemiaster Kalksburgensis* LAUBE. – LAUBE: 183
1870 *Hemiaster Kalksburgensis* LBE. – LAUBE: 314
1871 *Hemiaster Kalksburgensis* LAUBE. – LAUBE: 69; pl. 18, fig. 5

Fig. 5.

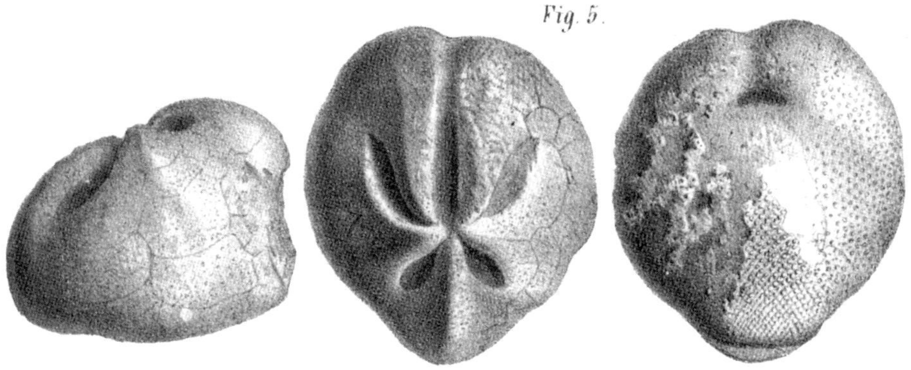

Figure 60: *Hemiaster kalksburgensis* Laube, 1969: reproduction from Laube (1871: pl. 18, fig. 5, specimen from Kalksburg, Vienna); current whereabouts of the specimen unknown; state unclear, probably a misidentified schizasterid (see text).

	1877	*Hemiaster Kalksburgensis* Laube – Karrer: 312
	1883	[*Opissaster*] *Kalksburgensis* – Pomel: 38
?	1906	*Hemiaster kalksburgensis* Lbe. – Vadász: 333
	1913a	*S.*[*chizaster*] *kalksburgensis* Laube (*Hemiaster*). – Cottreau: 68
	1925	*S.*[*chizaster*] *Kalksburgensis* Laube (*Hemiaster*) – Lambert & Thiéry: 526

Type-material:

Holotype: a single specimen originally housed in the collection of Felix Karrer (Laube, 1871)
Locus-typicus: Kalksburg, Vienna, Austria
Age: Badenian (Langhian-Early Serravallian), Middle Miocene
Remarks: The collection of Felix Karrer was inherited by the Naturhistorisches Museum Wien (Geologische Abteilung) in 1903. Although, the type of *H. kalksburgensis* was registered under the inventory number 1904.VIII.45, the specimen could not be located in the collection.

Discussion:

Based on Laube's (1871) description and illustrations it is difficult to understand why he considered the specimen he had at hand different from his *Schizaster karreri* (he did not compare the two species). In the absence of the type material and lacking sufficient information on the diagnostic characters of this species it has to be considered as *nomen nudum*. It is very likely that it is conspecific with the co-occurring *Schizaster karreri*, but this is an assumption that remains to be verified. At least, the attribution to the genus *Schizaster* seems to be more or less secured and has already been proposed by Lambert & Thiéry (1925).

Occurrence:

Austria: Badenian (Langhian-Early Serravallian)
Vienna Basin: Kalksburg, Vienna (Fuchs, 1869; Laube, 1869a, 1870, 1871; Karrer, 1877)

Paratethys (non-Austrian occurrences): ? Late Badenian (Early Serravallian)
Great Hungarian Basin (Pannonian Basin): ? Budapest-Rákos, Pest, Hungary (Vadász, 1906)

Genus *Ditremaster* Munier-Chalmas, 1885

Type-species: *Hemiaster nux* Desor, 1853; by subsequent designation (Cotteau, 1887: 422)
Diagnosis: Sub-globular with faint frontal sinus. Apical disc ethmolytic with 2 gonopores. Paired ambulacra petaloid. Posterior pair very short, about 0.3 times the length of the anterior ones (modified from Fischer, 1966).
Distribution: Eocene to Pliocene – cosmopolitan (Fischer, 1966)

Ditremaster scillae (Wright, 1855)

(Fig. 61; Pl. 63, Figs. 1a-c)

*	1855	*Hemiaster Scillæ*, Wright, n.sp. – Wright: 191-193; pl. 7, figs. 1a-f
	1858	[*Hemiaster*] *Scillae* Wright – Desor: 375
	1864	*Hemiaster Scillæ*, Wright. – Wright: 483-484
# v.	1869a	*Hemiaster rotundus* Laube. – Laube: 183
v.	1870	*Hemiaster rotundus* Lbe. – Laube: 314
v.	1871	*Hemiaster rotundus* Laube. – Laube: 68-69; pl. 18, fig. 6
v.	1883	[*Trachyaster*] *rotundus* – Pomel: 38
	1883	[*Opissaster*] *Scillæ* – Pomel: 38
.	1891	*Hemiaster scillæ*, Wright, 1855. – Gregory: 611
?	1905a	*Hemiaster ovatus* (Sism.) Air. – Airaghi: 216 [fide Lambert (1907a)]
non	1907a	*Opissaster Scillæ* Wright (*Hemiaster*) – Lambert: pl. 3, fig. 12-13 [according to Lambert (1909: 98) this is a *Brissopsis consobrinus*]
	1908a	*Opissaster Scillae* (Wright). – Stefanini: 470-471; pl. 17, fig. 7
?	1909	*Opissaster Scillæ* Wright (*Hemiaster*) – Lambert: 80-81
	1909	*Opissaster Scillæ* Wright (*Hemiaster*) – Lambert: 135
.	1911	*Hemiaster scillæ*, Wright, 1855. – Gregory: 673
.	1913a	*Opissaster Scillæ* Wright (*Hemiaster*) – Cottreau: 70
? v	1915	*Schizaster sardiniensis* Cott. – Vadász: 223-224; text-fig. 113; pl. 10 (4), fig. 9
	1924	*Trachyaster* (*Opissaster*) *Scillæ* Wright (*Hemiaster*) – Lambert & Thiéry: 509
v.	1924	*Trachyaster* (*Opissaster*) *rotundus* Laube (*Hemiaster*) – Lambert & Thiéry: 509
v.	1925	*Rotundaster rotundus* Laube (*Hemiaster*) – Lambert & Thiéry: 527
	1974a	*Hemiaster scillae* – Rose: 345, fig. 3 [table]
.	1974b	*Hemiaster scillae* – Rose: 351, fig. 3 [table]
.	1975	*Hemiaster scillae* Wright – Rose: 79, tab. 12
v.	1975	*Hemiaster rotundus* Laube, 1871 – Stojaspal: A192
.	1979	*Ditremaster scillae* (Wright) – Challis: 255
.	1979a	*Opissaster scilla* (Wright) 1855 – Menesini: 60

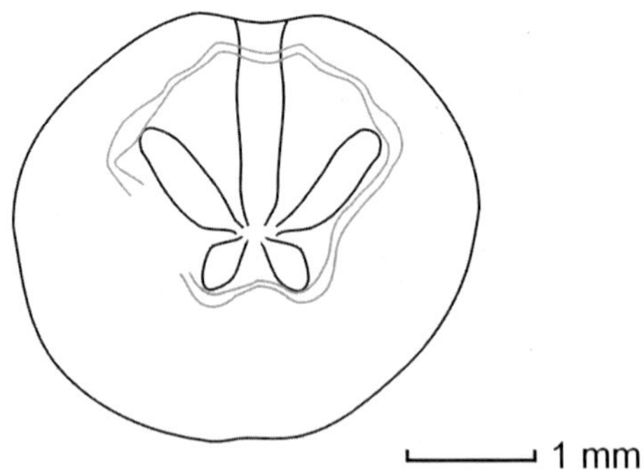

Figure 61: *Ditremaster scillae* (Wright, 1855): outline of the peripetalous fasciole (grey) (Badenian, Sievering, Vienna, GBA 1871/3/1). Note that the test outline is deformed due to compaction, originally the specimen was subcircular.

1979b *Opissaster scilla* (Wright) 1855 – Menesini: 802, 804
1984 *Ditremaster scillae* – Boggild & Rose: 64; fig. 3
1990 *Opissaster scillae* (Wright) – Ragaini: 1-13; figs. 1-10
1994 *Ditremaster scillae* – Néraudeau: 329, tab. 4

Type-material:
Hemiaster scillae Wright, 1855:
Syntypes: the material on which Wright's description is based (formerly in the collection of the Earl of Ducie) is housed at the Bristol Museum, UK under the numbers Cb 4624 – 4630 (Challis, 1980)
Locus typicus: Malta
Stratum typicum: *Globigerina* Limestone (Bed 4 of Wright, 1855)
Age: Aquitanian to Langhian, Miocene

Hemiaster rotundus Laube, 1869:
Holotype (Pl. 63, Figs. 1a-c): specimen figured by Laube (1871: pl. 18, fig. 6); Geological Survey Austria, Type collection, specimen no. 1871/3/1
Locus typicus: Sievering, Vienna, Austria
Age: Badenian (Langhian-Early Serravallian), Middle Miocene

Material:
Badenian (Langhian-Early Serravallian) – Sievering, Vienna, Austria
GBA: 1 specimen in the type collection (holotype of *Hemiaster rotundus* Laube, 1869)
? Late Badenian (Early Serravallian) – Gârbova de Sus (= Felsö-Orbó), Romania
MAFI: 1 specimen [MAFI Ech 306 (figured specimen of *Schizaster sardiniensis* in Vadász, 1915)]

Dimensions (in mm):

Inv. No.	TL	TW	TH
GBA 1871/3/1	> 33	36.2	25.3
(anterior part deformed)			

Description:
Size and shape: Test of small size with subcircular outline. The anterior margin is rounded with a faint frontal notch. The posterior margin is rounded to very bluntly pointed. The maximum width lies halfway along the test. In profile the test is high and slightly wedge-shaped. The maximum height lies just posterior of the apical disc in interambulacrum 5.
Apical disc: The apical disc is not preserved in the specimen studied (in Maltese specimens it is ethmolytic with 2 gonopores). The apical disc lies posterior of the centre, approximately two third of TL from the anterior margin.
Ambulacra: Adapically the ambulacra form moderately deep petals. Ambulacrum III is shallower than the paired petals and forms a very shallow frontal sinus. It closes slightly distally, where the poriferous zones end. The pores (oblique partitioned isopores with an axially positioned neural canal) are arranged in two straight rows. The interporiferous zone is very broad and bears few tubercles (no primaries).

The paired petals are relatively short and straight. The posterior paired petals are very short, nearly half as long as the posterior ones. The anterior paired petals form an angle of nearly 90°, while the posterior petals diverge less strongly, forming an angle of about 80°. The poriferous zones of the paired petals consist of very narrow, slit-like elongate isopores. The interporiferous zones are about half as wide as a single poriferous zone and bear only secondary tubercles.

Outside the peripetalous fasciole, the ambulacral pores are minute unipores. The adoral part of the ambulacra is not well preserved, so nothing can be said about the phyllodes. On the oral surface ambulacra I and V form long narrow peri-plastronal areas lacking primary tubercles.

The tuberculation of the ambulacra is rather sparse and conspicuous naked zones running from the tips of the petals down the ambitus are present.
Interambulacra: The interambulacra are strongly inflated adapically, forming high keels along the sunken petals, except in interambulacrum 5, where only a low rounded keel is present. The aboral tuberculation of the interambulacra is heterogeneous. It is very dense with small tubercles within the peripetalous fasciole, except along the sunken part of ambulacrum III, where the tubercles are large. At the ambitus the tubercles are rather large and widely spaced. On the oral side the tuberculation becomes even more sparse. Only on the plastron a regular fan-shaped pattern is observed. The nature of the tubercles is difficult to assess due to abrasion, but most of them seem to be perforate, crenulate tubercles. The plastron is moderately large and mesamphisternous. The labrum is partly missing, but seems to have been about twice its width in length. There are two distinct ridges along the midline of each half-interambulacrum running from the apex to the margin of the oral side, they are most prominent in the interambulacra 1 and 4.
Peristome: The peristome is very poorly preserved in the present specimen. It is situated approximately 20 to 30 % TL from the anterior margin.
Periproct: The periproct is situated marginally, high on the vertical posterior face of the test. Below the peristome there is a large flattened, vertical area with few tubercles in its centre. The shape of the periproct seems to have been subcircular.
Fascioles: Only a peripetalous fasciole, with a broad fasciole band is present. It is widest at the tips of the petals, narrowing in the interambulacra. The peripetalous fasciole is indented in the interambulacra 1 and 4 and crosses the remaining interambulacra more or less straight (Fig. 61).

Differential diagnosis:
This species differs from *Aliaster cotteauii* (Wright, 1855) by its small size, faint frontal sinus, sparse adoral tuberculation, slit-like respiratory pores and most importantly by the shape of the peripetalous fasciole.

Schizasterids present in the Neogene of Austria differ by the presence of a latero-anal fasciole, longer flexed petals, and a heart-shaped outline.

There are no other species with which the present taxon could be confused in the Central Paratethys and few in the Miocene of the Mediterranean. Among these *Hemiaster* ? *vadosus* Gregory, 1891 from the Maltese Islands, which has a similar size, outline and peripetalous fasciole can be distinguished by its ethmophract apical disc with 4 gonopores and more wedge-shaped profile. Nearly the same is valid for

Trachyaster globosus Pomel, 1869 from the Pliocene of Algeria only that the latter has an ethmolytic apical disc. *Opissaster cotteri* de Loriol, 1896 (referred by Néraudeau, 1994 to *Ditremaster*) from the Miocene (?) of Portugal is much larger, has longer, slightly outwards flexed petals, a deeply sunken ambulacrum III with overhanging walls aborally and a very different peripetalous fasciole. *Opissaster polygonalis* Pomel, 1883 and *O. declivis* Pomel, 1887 differ by their lower profile, elongated outline, and presence of a latero-anal fasciole. *Opissaster ? bleicheri* Peron & Gauthier in Cotteau et al., 1891 and *Aliaster jourdyi* (Péron & Gauthier in Cotteau et al., 1891) both from the Miocene of Algeria differ also rather well from *D. scillae*. The former has a strongly wedge shaped profile and less diverging petals, the latter a different peripetalous fasciole.

Discussion:

Laube's (1871: pl. 18, fig. 6) figure of the type of *Hemiaster rotundus* is relatively poor. It evokes the impression that the specimen is more wide than long but that is an artefact of the deformation of the anterior part of the corona. Furthermore, he illustrates a well preserved peristome, while the type is badly damaged in that area and the peristome poorly visible due to the deformation of the anterior part of the corona.

Based on a re-examination of the holotype of *H. rotundus* Laube, 1869 (the only specimen of this nominal species) and comparison with specimens of *Ditremaster scillae* (Wright, 1855) from the Aquitanian to Langhian *Globigerina* Limestone of the Maltese Islands *H. rotundus* is considered as junior synonym of *D. scillae*. In fact, Laube (1871: 69) pointed out the similarity of these two species already. The features used by him to distinguish between those two species (more rounded posterior end, less anterior peristome and more elongated outline in *D. scillae*) were partly artefacts of the deformation of the Austrian material and partly misinterpretations of the then available figures and descriptions of *D. scillae*. If one compares actual specimens, there is no doubt of their conspecifity. Moreover, the holotype of *H. rotundus* fits very well in the morphological range illustrated for *D. scillae* in a study on the morphometrics of this species by Ragaini (1990; based on material from the Maltese Islands).

Pomel (1883: 38) placed this species in the genus *Opissaster*. An opinion rejected by Gregory (1891: 611), who stated that his specimen of *H. scillae* had 3(?) or 4 genital pores and an ethmophract apical disc, although poorly preserved. Therefore he argued that *H. scillae* should be "allowed to remain in *Hemiaster*". The type species of *Opissaster*, *O. polygonalis* Pomel, 1883, however, has only two gonopores and Pomel (1883: 37) considered that as main feature of the genus. Lambert (1906a: 102) made *Ditremaster* Munier-Chalmas, 1885 a junior subjective synonym of *Opissaster* Pomel, 1883. Lambert (1906a: 102), however, rejected Pomel's inclusion of the species *H. cotteauii* Wright, 1855 and *H. scillae* Wright, 1855 into the genus *Opissaster*.

Despite Gregory's and Lambert's rejection, *H. scillae* was considered as *Opissaster* by subsequent authors (e.g. Stefanini, 1908a, Ragaini, 1990....). In 1907 Lambert himself (Lambert, 1907a: 80), contrary to his former statement placed *H. scillae* in the genus *Opissaster*.

Rose (1975: 78, Tab. 11) suggested that *Hemiaster vadosus* Gregory, 1891 might be a junior synonym of the present species. This, however, is improbable since *H. vadosus* has an ethmophract apical disc with four gonopores (see Gregory, 1891:pl. 2, fig. 6d). Néraudeau (1994: 329, tab. 4), following Lambert & Thiéry (1924:509) placed the latter species in the genus *Ditremaster*, which is also a poor choice considering the structure of the apical disc.

Challis (1980: 194-199) was the first to place *H. scillae* into the genus *Ditremaster* on base of the ethmolytic apical disc with two gonopores and its shallow frontal sinus. Since this work is an unpublished PhD thesis, this was no valid action in sense of the ICZN, but subsequent authors followed it nevertheless (e.g. Néraudeau, 1994: 329, tab. 4).

Néraudeau (1994) considered the two genera *Ditremaster* and *Opissaster* to be distinct, albeit closely related. According to him *Opissaster* is more "schizasteriform" in having a more elongate outline, lower profile and less diverging, slightly flexed anterior petals. Certain species of *Opissaster* acquired, furthermore, a latero-anal fasciole at the end of the Palaeogene.

Néraudeau's (1994: 332) statement that "The *Opissaster* reported in the Miocene of Austria-Hungary by Laube (1871) were in fact *Schizaster* (Lambert & Thiéry, 1909-1925) and only one *H.* (*Trachyaster*) is known in the Miocene of Czechoslovakia (Seneš, 1955)." cannot be accepted. First, Laube (1871) did never report any *Opissaster* species (he reported two species of the genus *Hemiaster*). Second, while *H. kalksburgensis* is most probably a *Schizaster*, *H. rotundus* is certainly a hemiasterid, albeit today placed into the synonymy of *Ditremaster scillae*.

The specimen referred to *Schizaster sardiniensis* by Vadász (1915) was re-examined at the MAFI (Ech 306). It has an ethmolytic apical disc with two gonopores, a very shallow frontal notch, and a small petalodium with very short posterior paired petals. Although the specimen is rather low in profile it is tentatively referred to *D. scillae* here. In Maltese specimens test height varies between 61 and 89 % TL with a mean of 79 % (Ragaini, 1990: tab. 2). In Vadász's specimen the height is around 50 % TL, probably resulting from deformation.

Occurrence:

Austria: Badenian (Langhian-Early Serravallian)
Vienna Basin: Sievering, Vienna (Laube, 1869a, 1871; Stojaspal, 1975; [GBA])

Paratethys (non-Austrian occurrences): Late Badenian (Early Serravallian)
Transylvanian Basin: ? Gârbova de Sus (= Felsö-Orbó), Romania (Vadász, 1915; [MAFI])

Mediterranean: Aquitanian – Langhian
Western Mediterranean: ? Bingia Fargeri (Fangarion), Sardinia, Italy (Lambert, 1907a); ? Valtorta, Sardinia, Italy (Airaghi, 1905a)
Central Mediterranean: Maltese Islands: Lower Coralline Limestone (Rose, 1975); *Globigerina* Limestone [Mgarr ix-Xini bed (=*Scutella* bed), Lower *Globigerina* Lst. to Upper *Globigerina* Lst., and phosphate conglomerates therein] (Wright, 1855, 1864; Desor, 1858; Gregory, 1891, 1911; Stefanini, 1908a; Lambert, 1909; Cottreau, 1913a; Rose, 1974b, 1975; Challis, 1979, 1980; Menesini, 1979a, b; Boggild & Rose, 1984; Ragaini, 1990; Néraudeau, 1994; [NHMW]); Cyrenaica, Libya: Ain Sciahat (Gregory, 1911); Cyrene and Derna, (Rose, 1974a, b)

Family Pericosmidae Lambert, 1905
Genus *Pericosmus* Agassiz, 1847

Type-species: *Hemiaster* (*Pericosmus*) *latus* Desor in Agassiz & Desor, 1847; by subsequent designation (de Loriol, 1875: 114).
Diagnosis: Apex sub-central to somewhat anterior; petals moderately broad, straight, subequal; anterior sinus moderately deep; peripetalous fasciole passing above the periproct; marginal fasciole generally complete, passing below periproct; apical disc ethmolytic with 3 gonopores. (modified from Fischer, 1966)
Distribution: Eocene to Recent – circumtropical (Fischer, 1966)
Ecology and biogeography: Although ten, possibly twelve, extant species of *Pericosmus* have been described, little is known on the ecology of this genus. Today the genus has an Indo-malayan distribution, ranging south as far as southern

Queensland (Australia) and occur in depth between 18 to 486 metres, most commonly around 200 m (MORTENSEN, 1951; MCNAMARA, 1984). It is essentially a tropical genus, although its range extends into the northern part southern warm temperate zone east of Australia. Its distribution, however, seems to be limited by winter surface temperatures between 15° and 20° C. *P. porphyrocardius* MCNAMARA, 1984 is the only extant species based on large numbers of specimens. It was trawled from depths between 309 and 420 metres, from muddy sediments and found in association with *Lovenia gregalis* ALCOCK, 1893, and species of *Stereocidaris* and *Araeosoma* (MCNAMARA, 1984). MCNAMARA & PHILIP (1984) assumed similar habitats for the other extant species and stated that they are ill-adapted to burrowing in these fine sediments. In contrast, the Oligocene and Miocene *Pericosmus* species of Australia lived in coarse-grained bioclastic calcareous sands, being adapted to relatively shallow-water, high hydrodynamic environments and more effective burrowing (MCNAMARA & PHILIP, 1984: 355).

Pericosmus latus (DESOR in AGASSIZ & DESOR, 1847)

(Fig. 62; Pl. 63, Figs. 2-3)

1840b *Micraster latus* AG. – AGASSIZ: 2 [*nomen nudum*]
non 1843 *Micraster latus* AG. – SISMONDA: 29-30; pl. 1, fig.13 [a *Micraster* from the Cretaceous of Nice, according to LAMBERT (1909:83-84)]
* 1847b *Hemiaster* (*Pericosmus*) *latus* DESOR – AGASSIZ & DESOR: 19; pl.16, fig. 1, 1a
. 1855 *Pericosmus latus*, DESOR. – WRIGHT: 193-195
1857 *P.*[*ericosmus*] *latus*, DESOR – PICTET: 198
pp 1858 *Pericosmus latus* AGASS. – DESOR: 396
. 1864 *Pericosmus latus*, AGASSIZ – WRIGHT: 490
1873 *Pericosmus latus*, AGASS – MANZONI: 10-11, 20
1877 *Pericosmus latus*, AGASSIZ, 1847 – COTTEAU *in* LOCARD: 310-312
non 1875a *Schizaster Grateloupii* SISM. – HOERNES: 385, 389
non 1875b *Schizaster Grateloupii* SISM. – HOERNES: 211
1882a *Pericosmus latus*, AGASS, – MAZZETTI: 115
pp 1891 *Pericosmus latus* (AGASSIZ), 1840 – GREGORY: 613-615
1895 *Pericosmus latus*, AGASSIZ, 1847. – COTTEAU: 40
\# 1900 *Pericosmus Lyonsi* GAUTHIER, 1898 – GAUTHIER *in* FOURTAU: 716-717; pl. 4, figs. 10-11
1902 *Pericosmus latus* AGASSIZ, 1847 – FOURTAU: 108-109
. 1908a *Pericosmus latus* (AGASS.) – STEFANINI: 481-482
1909 *Pericosmus latus* AGASSIZ, 1847 – LAMBERT: 83-85; pl. 6, fig. 1-2
. 1913a *Pericosmus latus* AGASSIZ. – COTTREAU: 71, 120-122; fig. 36; pl. 15, fig. 5, 5a
1913a [*Pericosmus latus* AGASSIZ] variété *minor* – COTTREAU: 71, 120-122; pl. 15, fig. 6, 6a
1915a *Pericosmus latus* AGASSIZ, 1847. – LAMBERT: 171
v. 1915 *Pericosmus latus* AG. sp. – VADÁSZ: 227; fig. 117
1920 *Pericosmus latus*, AGASSIZ 1847 – FOURTAU: 73-74
1924 *P.*[*ericosmus*] *latus* AGASSIZ (*Micraster*) – LAMBERT & THIÉRY: 512
1939 *Pericosmus latus* AGASSIZ – TAVIANI: 44-45; pl. 3 (5), fig. 5
? 1955 *Periscomus* cf. *latus* AG. – SENEŠ: 6
? 1956 *Pericosmus latus* AGASSIZ – SOCIN: 14
? 1967 *Pericosmus* cf. *latus* AG. – CICHA et al.: 93
. 1975 *Pericosmus latus* DESOR – ROSE: 79, tab. 12
. 1979 *Pericosmus latus* DESOR 1840 – CHALLIS: 256
. 1979a *Pericosmus latus* (AGASSIZ) 1840 – MENESINI: 60
. 1979b *Pericosmus latus* (AGASSIZ) 1840 – MENESINI: 802
. 1980 *Pericosmus* (*Pericosmus*) *latus* (AGASSIZ). – CHALLIS: 204-209; pl. 79, figs. A-C; pl. 80, figs. A-C; pl. 81, figs. A-C
. 1984 *Pericosmus latus* – BOGGILD & ROSE: tab. 61, 64; figs. 2-3
1992 *Pericosmus lyonsi* GAUTHIER – BLONDEL & PHILIPPE: 438; pl. 1, figs. 3a-b
. 1998 *Pericosmus latus* (AGASSIZ, 1840) – PHILIPPE: 169-171; pl. 19, figs. 5a-c
v. 2002 *Pericosmus* sp. – KAZÁR: 153; fig. 1

Type-material:
Syntypes: collection MICHELIN (AGASSIZ & DESOR, 1847b: 19); a plaster cast ("moulage" M23) of this specimen is housed in the collection of the Museé d'Histoire Naturelle de Neuchâtel (LAMBERT & JEANNET, 1928: 144, M.23); CHALLIS (1980: pl. 80, figs. A-C) figured two specimens from the collection of the Université de Paris-Sud, Orsay, which are probably syntypes of this species
Locus typicus: Bonifacio, Corsica, France
Age: Burdigalian (PHILIPPE, 1998: 170)

Material:
Early Badenian (Langhian) – Stotzing (sandpit Mayer), Bgld, Austria
NHMW: 6 specimens (NHMW 2004z0093/0001-5, 2004z0093/0023)
WANZENBÖCK coll.: numerous specimens (no inventory nos.)

Foreign material for comparison:
Early Badenian (Langhian) – Totmárokházá, Nógrád, Hungary
MAFI: 1 specimen [MAFI Ech 308 (reference material of VADÁSZ, 1915)]
Late Badenian (Early Serravallian) – Gârbova de Sus (= Felsö-Orbó), Romania
MAFI: 2 specimens [MAFI Ech 309 and 310 (reference material of VADÁSZ, 1915)]

Dimensions (in mm):

Inv. No.	TL	TW	TH
NHMW 2004z0093/0001	72.5	74.1	37.3
NHMW 2004z0093/0002	60.6	>58	>24
NHMW 2004z0093/0004	79.5	75.6	~46
NHMW 2004z0093/0005	>74	75.3	>32

Description:
Size and shape: Test of medium to large size with rounded, slightly heart-shaped outline. Test length and width subequal. Anterior margin rounded with moderately deep frontal sinus. Posterior margin bluntly pointed. Maximum width halfway along test length. In profile test high, low arched to domed. Posterior end obliquely truncated. Maximum height coinciding with inflated interambulacra anterior and posterior of apical disc. Test sloping steeply from the apex to the anterior margin and gently to the posterior margin. Ambitus tumid to rounded. Oral surface flat to slightly convex, sloping gently from a central elevation on the plastron to the margin.
Apical disc: Apical disc ethmolytic with three gonopores (no gonopore in genital plate 2; Fig. 62.B) and situated centrally.
Ambulacra: Ambulacrum III straight, open and increasingly depressed towards the anterior margin. The pores are minute, oblique partitioned anisopores, which are arranged in straight rows. Interporiferous zone wide, sparsely tuberculated with small perforated tubercles.

Paired ambulacra petaloid, deeply sunken and closed distally. The anterior pair forms an obtuse angle of about 110° to 120°; the posterior pair an acute angle of approximately 70°. The posterior paired petals flex slightly laterally at their distal ends in some specimens. Anterior paired petals about 14 % to 20 % larger than posterior paired petals. Anterior paired petals extend about two third and posterior pair about half of corresponding test radius. The pores within the paired petals are elongated isopores. Outside the petals only small, slit shaped unipores, situated near the adoral suture of each ambulacral plate are present.

On the oral surface ambulacra I and IV each form broad, nearly naked peri-plastronal areas, with very few irregularly spread secondary tubercles. Each ambulacral plate bears one small unipore near the adoral suture.

Adorally the ambulacra form slightly sunken phyllodes bearing large constricted unipores with extensive periporal area.

Interambulacra: On the apical surface the interambulacra are inflated, forming keels between the petals. Tuberculation is moderately dense. Primary tubercles are small, crenulate, perforate with undercut mamelon, concave platform and inclined areoles. Very small secondary tubercles are spread among them. The largest tubercles are situated inside the peripetalous fasciole, bordering the petals, especially around the apical disc and along ambulacrum III.

On the oral surface the interambulacra are flattened and covered with loosely arranged, large primary tubercles. Secondary tuberculation rather dense. Towards the margin the primary tubercles decrease in size and there are few or no secondary tubercles present near the margin. The primary tubercles are perforate, crenulate, have a raised platform and large areoles which are inclined towards the peristome. The plastron is mesamphisternous and up to one third as wide as the test (Fig. 62.A). The junction between the sternal plates forms an antero-posteriorly elongated keel. The tuberculation is dense and forms a radiating pattern from a central elevation. Very few secondary tubercles are present. Two distinct elevations on each side of a very sparsely tuberculated subanal depression are visible at the posterior end of the plastron.

Peristome: The peristome is situated near the anterior margin, about 20 % of TL from the anterior margin. It is rather small, kidney-shaped, transversely elongated, with a short, bluntly pointed labrum. The anterior margin of the peristome is depressed.

Periproct: The periproct which is situated high on the ambitus, is overhanging and has an oval, transversely elongated shape. A broad, sparsely tuberculated subanal depression is present.

Fascioles: The peripetalous fasciole band is rather narrow. There are distinct indentations of the peripetalous fasciole in interambulacra 1, 4 and 5. In interambulacra 2 and 3 the fasciole is indented several times, forming a stair-shaped pattern. Here it splits in several bands which cross ambulacrum III at several different positions (usually between the apical disc and about halfway between apex and margin). A narrow marginal fasciole band is present. It is situated at or slightly above the ambitus except in ambulacrum III, where it is indistinct and below the periproct where it lies on the oral surface and is somewhat broader.

Differential diagnosis:

Pericosmus orbignyi Cotteau, 1877 from the Miocene of Corsica can be distinguished from this species by its slightly lower test profile, its stronger inflated interambulacra between the petals, its flatter oral surface, its larger petals and its peripetalous fasciole, which is not or only slightly indented.

P. peroni Cotteau, 1877 another species from Corsica can be held apart from the species considered here by its oval, transversely elongated outline (width > length), its high, angular test profile, the stronger diverging anterior petals and its transversely truncated posterior margin.

P. agassizi (Sismonda, 1841) from the Late Oligocene of the Colli Torinesi and Sardinia differs from *P. latus* by its more angular outline, more strongly diverging anterior petals, its different profile (with vertically truncated posterior end) and different outline of the peripetalous fasciole (compare Lambert, 1909: 85-86; pl. 7, figs. 6-7).

P. lovisatoi Lambert, 1909 and *P. petasatus* Lambert, 1909, both from the Late Oligocene of Sardinia, differ from *P. latus* by their longer and more strongly flexed petals (compare Lambert, 1909: 87-88; pl. 6, figs. 3-4; pl. 9, fig. 1)

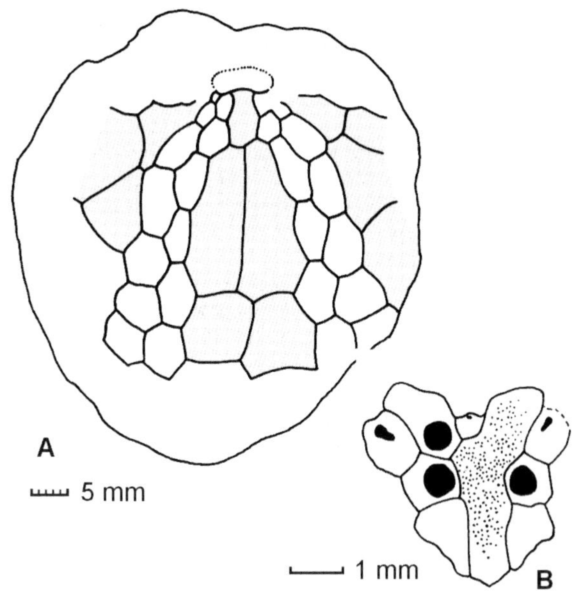

Figure 62: *Pericosmus latus* (Desor in Agassiz & Desor, 1847b): oral plating (A) and apical disc (B) (Early Badenian, Stotzing, Bgld, NHMW 2004z0093/0002).

Discussion:

The Austrian specimens clearly belong to the well known and widely distributed species *Pericosmus latus* (Desor in Agassiz & Desor, 1847). The species *Pericosmus lyonsi* Gauthier was placed into the synonymy of *P. latus* (Desor, 1847) by Cottreau (1913a), Fourtau (1920), and Philippe (1998) because they considered the type species as an aberrant morphotype of this species, a fact which was already noted by Gauthier himself according to Cottreau (1913a). Philippe (1998; 169-171) also placed *P. pasqualii* Gauthier in Fourtau, 1900 into the synonymy of *P. latus*, which would extend the range of this species down to the Eocene.

The correct authorship of the species *P. latus* has been subject to some confusion in the past, resulting in quite surprising combinations as e.g. *Pericosmus latus* Desor, 1840 (Challis, 1979). The name *Micraster latus* Agassiz, 1840b is a *nomen nudum* according to the ICZN rules (ICZN 4th ed., 2000, Article 12.1.). The next author who published a description for this name is Sismonda (1843: 29-30; pl. 1, fig.13), which is thus the valid author for *M. latus* (ICZN 4th ed., Articles 11. and 12.). Sismonda's description, however, is based on Cretaceous specimens from Nice (Lambert, 1909) which are micrasterids not related to the species discussed here. In 1847 Desor (in Agassiz & Desor, 1847b: 19), validly established the taxon *"Hemiaster (Pericosmus) latus"* using Agassiz's (1840b) material as type. The correct authorship for the present taxon is thus *Pericosmus latus* (Desor in Agassiz & Desor, 1847) and this species should not be confused with the valid species *Micraster latus* Sismonda, 1843.

Until now *P. latus* was not recorded form the Neogene of Austria, although its presence in the Central Paratethys was documented by Vadász (1915) already. Laube (1869a) established the species *Pericosmus affinis* on base of a single specimen from Sievering, Vienna (now lost) and it seems possible that this specimen was a misidentified *P. latus*. A detailed comparison of Laube's (1871) description and illustration of *P. latus* specimens from the Central Paratethys and the Mediterranean, however, reveals many discrepancies and it seems more likely that *P. affinis* belongs into the synonymy of *Aliaster cotteauii* (see below under that species for details).

Philippe (1998: 169) placed Hoernes' (1875a, b) record of *Schizaster grateloupi* Sismonda in the synonymy of *P. latus*.

Based on a re-examination of the specimen on which Hoernes' record is based (GBA 1875/01/77) it is placed into *Linthia* ? *hlinnensis* (Seneš, 1955) (see below under that species).

Occurrence:

Austria: Early Badenian (Langhian)
Vienna Basin: Stotzing (sandpit Mayer), Bgld (Kazár, 2002; [NHMW])

Paratethys (non-Austrian occurrences): ? Karpatian (Late Burdigalian), Early to Late Badenian (Langhian-Early Serravallian)
Great Hungarian Basin (Pannonian Basin): Tótmarokháza, Nógrád (Vadász, 1915; [MAFI])
Transcarpathian Basin: ? Hlinné, eastern Slovakia (Seneš, 1955; Cicha et al., 1967)
Transylvanian Basin: Gârbova de Sus (= Felsö-Orbó), Romania (Vadász, 1915; [MAFI])

Mediterranean: Aquitanian to Messinian
Western Mediterranean: Bonifacio, Corsica, France (Agassiz & Desor, 1847b; Pictet, 1857; Desor, 1858; Manzoni, 1873; Cotteau in Locard, 1877; Gregory, 1891); Cap Sant'Elia, Cagliari, Sardinia, Italy (Lambert, 1909); Emilia, Northern Italy (Cottreau, 1913a); Isili, Sardinia, Italy (Cotteau, 1895); Montese, Italy (Mazzetti, 1882a; Gregory, 1891); Monte Titano, Italy (Manzoni, 1873; Gregory, 1891); Saint-Florent, Corsica, France (Cotteau in Locard, 1877; Gregory, 1891; Cottreau, 1913a); Ste Colombe, near Vence, Alpes-Maritimes, France (Lambert, 1906b, 1915a; Cottreau, 1913a)
Rhône Basin: Bassin d'Apt, Boillons, Goult, France (Philippe, 1998)
Central Mediterranean: Bologna, Italy (Gregory, 1891); Jebel Korbous, Tunisia (Blondel & Philippe, 1992); Maltese Islands: *Globigerina* Limestone (Lower *Globigerina* Lst. to Upper *Globigerina* Lst., and phospate conglomerates therein) (Wright, 1855, 1864; Desor, 1858; Manzoni, 1873; Gregory, 1891; Stefanini, 1908a; Cottreau, 1913a; Rose, 1975; Challis, 1979, 1980; Menesini, 1979a, b; Boggild & Rose, 1984), records from the Blue Clay are dubious according to Challis (1980) and need to be confirmed
Eastern Mediterranean: Gebel Aouebet, Egypt (Gauthier in Fourtau, 1900; Fourtau, 1920); Gebel Chebrewet, Eastern Desert, Egypt (Fourtau, 1902, 1920);Gebel Damasq, Egypt (Gauthier in Fourtau, 1900; Fourtau, 1920); Gebel Geneffé, Eastern Desert, Egypt (Gauthier in Fourtau, 1900; Fourtau, 1902, 1920); Umm er Rzem, Cyrenaica, Libya (Taviani, 1939)

Atlantic Ocean: ? Miocene
Canary Islands: ? Madeira [Gregory, 1891 (needs to be confirmed)]

Indian Ocean: ? Early Miocene
Eastern African coast: ? Ehil, Somalia [Socin, 1956 (specific identification needs to be confirmed)]

Family Schizasteridae Lambert, 1905
Genus *Schizaster* Agassiz, 1836

Type-species: *Schizaster studeri* Agassiz, 1840, by subsequent designation (ICZN 1954b, opinion 209, p. 385, 8th May 1954).
Diagnosis: Test cordiform with deep anterior sulcus and slightly pointed posterior end; apical disc ethmolytic, with 2, 3 or 4 gonopores (see remarks); ambulacrum III and petaloid paired ambulacra moderately to deeply sunken aborally; anterior petals longer and more flexed than posterior petals; periproct small and marginal, on near-vertical or obliquely truncate posterior face; peristome opening facing anterior; kidney-shaped; labral plate short and wide; not extending beyond the first ambulacral plate; broad between labral and sternal plates; plastron mesamphisternous with large and symmetric sternals and biserially offset episternals; aboral tuberculation fine, uniform and dense; oral tuberculation coarser, but also dense and uniform; peripetalous and latero-anal fascioles well-developed; latero-anal fasciole branches off about one-third up from the ends of the anterior petals (modified from Mortensen, 1951; Fischer, 1966 and Smith *"The Echinoid Directory"*, 07.05. 2004).
Distribution: Eocene to Recent – cosmopolitan (Fischer, 1966)
Remarks: Smith (*"The Echinoid Directory"*; 07.05.2004) proposed to restrict the use of the genus *Schizaster* to include only those forms with four genital pores as in the type-species *S. studeri*. For the forms with two genital pores he suggested the use of the genus *Ova* Gray, 1825 [type species: *Spatangus canaliferus* Lamarck 1816; by subsequent designation (ICZN op. 209, 1954), non *Ova* Pomel 1887, p. 701; (= *Nina* Gray, 1855, p. 60)]. Within this genus he recognises two subgenera: one (*Ova*) for forms with pore-pairs of aboral ambulacrum III offset in double column in each half ambulacrum, and the other one (*Aplospatangus* Lambert, 1906) for forms with the pores in single series. Although this classification seems to be quite reasonable, there is a problem: in the type species of *Aplospatangus* Lambert, 1906 *Schizaster eurynotus* Sismonda, 1841 (*S. eurynotus* Agassiz, 1840 is a *nomen nudum*; see below), as well as in a number of other species, the number of gonopores varies between two (the posterior ones; Lambert, 1907a: 68) and four (Challis, 1980: 221). Yet in these cases where four gonopores are present the anterior ones are very small reaching only about 25 % the diameter of the posterior ones (in *S. eurynotus*: Challis, 1980: 221; and pers. comm. M. Gatt, Malta, 22.12.2003).

McNamara (1995) documented similar variation in gonopore number in a population of the extant *Schizaster* (*S.*) *compactus* (Koehler, 1914) from Western Australia. He found that 55.6 % had four gonopores (n=27), 29.6 % had three and 14.8% had two, with no correlation between gonopore number and test size. Similar to the situation in the fossil *S. eurynotus* the anterior gonopores, where present, are very small in *S.* (*S.*) *compactus* (McNamara, 1995: fig. 3). They range between 10 and 33 % of the posterior ones and in species with three gonopores the odd gonopore may be either the left or the right one. At least two further Australian schizasterids show the same phenomenon (McNamara & Philip, 1980a, b; McNamara, 1995). Based on these observations McNamara (1995) proposed an emended diagnosis for the subgenus *Schizaster* (*S.*) and provided differential diagnostic characters to distinguish it from the two other subgenera [*S.* (*Ova*) and *S.* (*Paraster*)] he recognised.

Although the findings of McNamara (1995) do not invalidate the proposed classification of Smith, they illustrate the potential problems. Here the classification of McNamara (1995) is favoured over that of Smith as it seems to better reflect the relationship between the taxa on one hand, and provides more stability of the nomenclature on the other (Smith's concept would necessitate to change the generic attribution of a large number of taxa without providing a better understanding of the relationship within this group).

Subgenus *Schizaster* Agassiz, 1836

Diagnosis: Ambulacrum III moderately to deeply sunken aborally, with numerous pore pairs (more than 15 in each poriferous zone), which are arranged in single, regular series and are aligned transversely, or only slightly obliquely (from McNamara, 1995).
Distribution: Eocene to Recent – cosmopolitan (Fischer, 1966)
Ecology and biogeography: Modern representatives of the subgenus *Schizaster* (*S.*) usually occupy soft bottoms in depths between 4 and 900 metres. Four living species are currently

referred to this subgenus: *S.* (*S.*) *compactus* KOEHLER (Bengal Sea and Western Australia), *S.* (*S.*) *edwardsi* COTTEAU (Gulf of New Guinea), *S.* (*S.*) *lacunosus* (LINNÉ) (Indo-malayan region and western Indian Ocean), and *S.* (*S.*) *orbignyanus* AGASSIZ (Caribbean) (see MORTENSEN, 1951; MCNAMARA & PHILIP, 1980b). They are deposit-feeders and capable of burrowing up to c. 25 cm deep into the sediment. Little is known on their biology otherwise.

Schizaster (*S.*) *eurynotus* SISMONDA, 1841

(Fig. 63; Pl. 63, Figs. 4-5; Pl. 64, Figs. 1a-c; Pl. 66, Figs. 1a-c)

non 1840b *Schizaster eurynotus* AG. – AGASSIZ: 2, nos. P. 86 and X. 93 (*nomen nudum*)
* 1841 *Schizaster eurynotus* AG. – SISMONDA: 22
pp 1841 *Schizaster canaliferus* AG.– SISMONDA: 20-22 [fossil Maltese material, fide CHALLIS (1980: 223-224)]
? 1843 *Schizaster eurynotus* AG. – SISMONDA: 31-32; pl. 2, figs. 2-3
1847b [*Schizaster*] *eurynotus* AGASS. – AGASSIZ & DESOR: 21; no. P86
1855 *Schizaster eurynotus*, AGASSIZ. – WRIGHT: 262-264
1857 *Sch.*[*izaster*] *eurynotus*, AG. – PICTET: 199
pp 1858 [*Schizaster*] *Scillae* AGASS. et DESOR – DESOR: 389-390
pp 1864 *Schizaster Scillæ*, DESMOULINS. – WRIGHT: 484 [fide CHALLIS (1980: 220)]
\# v. 1869a *Schizaster leithanus* LAUBE. – LAUBE: 183
1869a *Schiz. Scillae* DESMOULINS. – LAUBE: 184
v. 1870 *Schizaster leithanus* LBE. – LAUBE: 314
v. 1871 *Schizaster Leithanus* LAUBE. – LAUBE: 69-70; pl. 18, fig. 7
1871 *Schizaster Scillae* DESMOULINS sp. – LAUBE: 71
1874 *Schizaster eurynotus* – QUENSTEDT: pl. 89, fig. 6
1875 *Schizaster eurynotus* – QUENSTEDT: 672
1877 *Schizaster scillae* DESM. – KARRER: 170
1883 *Schizaster Scillai*, AG. – GUTZWILLER: 46 [fide COTTREAU (1913a: 68)
? 1884 *Schizaster Leithanus* LAUB. – COPPI: 191, no. 277
pp 1891 *Schizaster scillae* (DESMOUL.), 1837. – GREGORY: 617-618.
1906a *Schizaster* (*Aplospatangus*) *eurynotus* AGASSIZ (in SISMONDA), 1841. – LAMBERT: 117-118, fig. 3(4)
1907a *Schizaster eurynotus* AGASSIZ, 1840. – LAMBERT: 67-69; pl. 5, figs. 6-7
? # 1907a *Schizaster Capederi* LAMBERT. – LAMBERT: 74; pl. 10, figs. 3-4
1907 *Schizaster Scillae* DESM. – SCHAFFER: 28
1908a *Schizaster eurynotus* AGASS. – STEFANINI: pl. 17, figs. 13, 13b
. 1913a *S.* [*chizaster*] *eurynotus* AG. (in SISMONDA) – COTTREAU: 68, 114-115; fig. 29; pl. 14, fig.1, 1a, 2-4, 5, 5a, 6
v. 1915 *Schizaster eurynotus* AG. – VADÁSZ: 220; fig. 107; pl. 11 (5), figs. 3, 7
? 1915 *Schizaster calceolus* LAMBERT. – VADÁSZ: 220; text-fig. 108.
? 1915 *Schizaster* cfr. *lajtanus* LBE. – VADÁSZ: 224 (two poorly preserved specimens)
1920 *Schizaster eurynotus*, AGASSIZ 1840 – FOURTAU: 78-79
1928 *Schizaster Scillae* DESM. – BOBIES: 48
1928 *Schizaster eurynotus*, AG. – JEANNET: 44-47; pl. 4, figs, 25-26; pl. 5, figs. 6-8
1936 *Schizaster eurynotus* AGSSIZ. – PAUCĂ: 143, 196-197; pl. 1, figs. 5-6

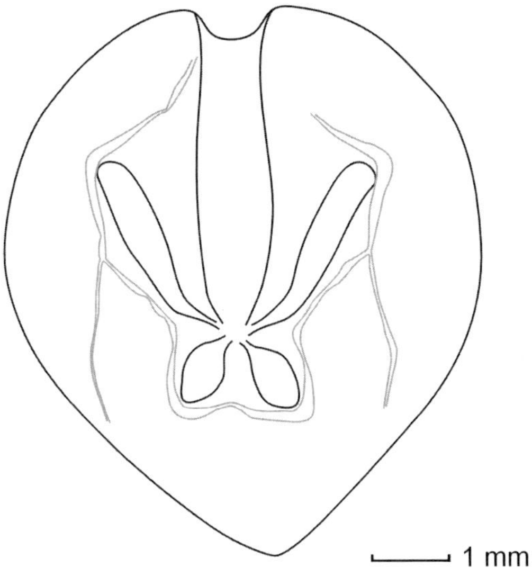

Figure 63: *Schizaster* (*S.*) *eurynotus* SISMONDA, 1841: outline of the peripetalous and latero-anal fascioles (grey) (Early Badenian, Stotzing, Bgld, WANZENBÖCK coll. W31).

1939 *Schizaster eurynotus* AGASSIZ – TAVIANI: 46-47; pl. 3 (5), figs. 8a-b
1942 *Schizaster Scillae* DESM. – SCHAFFER: 130
? 1955 *Schizaster eurynotus* AG. – SENEŠ: 6
? 1956 *Schizaster eurynotus* AGASSIZ – SOCIN: 11
1963 *Schizaster eurynotus* AGASSIZ – COMASCHI CARIA: 30; pl. 3, figs. 8
pp 1964 *Schizaster eurynotus* AGASSIZ – WIGGLESWORTH: 73-78; pl. 12
? 1967 *Schizaster eurynotus* AG. – CICHA et al.: 93
1967 *Schizaster eurynotus* AGASSIZ – MENESINI: 147-149; pl. 46 (1), figs. 1, 1a
1974a *Schizaster eurynotus* – ROSE: 345; fig. 3
pp 1974b *Schizaster eurynotus* – ROSE: 345; fig. 3
1975 *Schizaster* (*Ova*) *eurynotus* – BROMLEY & ASGAARD: 279
1975 *Schizaster eurynotus* AGASSIZ – MENESINI: 96-101; fig. 1b
pp 1975 *Schizaster eurynotus* AGASSIZ – ROSE: 79; tab. 12
v. 1975 *Schizaster leithanus* LAUBE 1871 – STOJASPAL: A193
. 1979 *Schizaster eurynotus* AGASSIZ 1840 – CHALLIS: 256
1979a *Schizaster eurynotus* AGASSIZ 1840 – MENESINI: 61
? 1979a *Schizaster cavernosus* POMEL 1887 – MENESINI: 61 [fide CHALLIS (1980: 221)]
1979b *Schizaster eurynotus* AGASSIZ 1840 – MENESINI: 803
? 1979b *Schizaster cavernosus* POMEL 1887 – MENESINI: 803 [fide CHALLIS (1980: 221)]
. 1980 *Schizaster eurynotus* AGASSIZ – CHALLIS: 220-224; pl. 83, fig. A-C; pl.86, fig. C; pl. 88, fig. A-C; pl. 89, fig. A-C
. 1984 *Schizaster eurynotus* – BOGGILD & ROSE: 61, 64, 65; figs. 2-3
1994 *Schizaster eurynotus* AGASSIZ – RAGAINI: 288, 289; tab. 1-2
. 1998 *Schizaster eurynotus* L. AGASSIZ, 1841 – ROSE & WATSON: 813; fig. 5a-d
? 1998 *Schizaster* (*Shizaster*) *eurynotus* AGASSIZ, 1840 – PHILIPPE: 172-175; pl. 20, figs. 1-3

Type-material:

Schizaster eurynotus SISMONDA, 1841:

Type-material: whereabouts unknown, not indicated in original publication

Locus typicus: Colli Torinesi, Italy
Age: Middle (?) Miocene
Remarks: the specimen figured by SISMONDA (1843: pl.2, figs. 2-3) is peculiar, while the outline, profile, and shape of the paired petals fit very well with *S. eurynotus*, the figure shows multiple rows of pores in the poriferous zones of aboral ambulacrum III and a peripetalous fasciole, which is not indented in interambulacra 2 and 3. SISMONDA's figure might thus be a composite image based on specimens of *S. eurynotus* and the extant *S. canaliferus*.

Schizaster leithanus LAUBE, 1869:
Holotype: specimen figured by LAUBE (1871: pl. 18, fig. 7); Geological Survey of Austria, Type collection, specimen no. 1871/3/2
Locus typicus: Mörbisch (Merwisch) near Rust, Bgld, Austria
Age: Badenian (Langhian-Early Serravallian), Middle Miocene
Remarks: LAUBE's figure is not very accurate (compare Pl. 66, Fig. 1a-c)

Material:
Early Badenian (Langhian) – Stotzing (sandpit Mayer), Bgld, Austria
NHMW: 12 specimens (NHMW 2004z0093/0006-17)
WANZENBÖCK coll.: numerous specimens
Badenian (Langhian-Early Serravallian) – Mörbisch (formerly Merwisch) near Rust, Bgld, Austria
GBA: 1 specimen [GBA 1871/3/2 (**holotype** of *Schizaster leithanus* LAUBE, 1869)]
Late Badenian (Early Serravallian) – Müllendorf (Mühlendorfer Kreide AG quarry), Bgld, Austria
NHMW: 1 specimen (NHMW 2004z0094/0001)

Foreign material for comparison:
Late Badenian (Early Serravallian) – Gârbova de Sus (= Felsö-Orbó), Romania
MAFI: 3 specimens [MAFI Ech 241 (reference material of VADÁSZ, 1915)]

Dimensions (in mm):

Inv. No.	TL	TW	TH
NHMW 2004z0093/0008	~60	56.2	>40
NHMW 2004z0093/0009	72.3	65.6	~46
NHMW 2004z0093/0010	~70	~67	-
NHMW 2004z0093/0013	67.7	61.4	~42

Description:
Size and shape: Test of medium to large size ranging from 60 to 80 mm TL in the studied material. The outline is heart-shaped and elongated antero-posteriorly. The anterior margin is rounded with a distinct frontal sinus. In some specimens a slightly asymmetry is observed. In these specimens one side – usually the right one – projects further anterior than the other. The posterior margin is sharply pointed. The maximum width lies slightly anterior of the centre. In profile the test is strongly wedge-shaped with its maximum height roughly halfway along highly arched keel in interambulacrum 5. A distinct rostrum overhangs the periproct.
Apical disc: The apical disc lies posterior of the centre, about 60 to 65 % of TL from the anterior margin. It is lacking in all examined specimen, but was reported to be ethmolytic with two to four gonopores (the anterior of which are only about 25 % the diameter of the posterior ones) in Mediterranean specimens (LAMBERT, 1907a; CHALLIS, 1980).
Ambulacra: The ambulacra are all petaloid and rather deeply sunken. Ambulacrum III is deeply sunken with overhanging walls. It closes slightly distally, where the pores become smaller and less closely spaced. The pores are arranged in two straight rows. They are slightly oblique partitioned isopores with an axially positioned neural canal. There may be up to 40 pore pairs in each row. The interporiferous zone is very broad and bears many secondary, but few primary tubercles.

The anterior paired petals are long, relatively narrow and strongly flexed anteriorly. Their tips tend to flex slightly laterally. The posterior paired petals are much shorter, extending about 38 to 45 % of the lengths of the anterior petals. Anterior paired petals form an acute angle of about 60-70° with each other, the posterior ones an angle between 60 to 65°. The posterior paired petals are broadest at their midpoint, whereas the anterior petals are broadest near the distal flexure. The poriferous zones of the paired petals consist of conjugate elongate isopores and tend to close distally. The interporiferous zones are slightly smaller than a single poriferous zone and bear only secondary tubercles.

Outside the peripetalous fasciole, the ambulacral pores are minute, vertically elongated constricted unipores. Adorally the ambulacra form distinct phyllodes consisting of large partitioned anisopores with a large, transversely elongated interporal partition.

On the oral surface ambulacra I and V form long narrow peri-plastronal areas lacking primary tubercles. Each bears four partitioned isopores on each side of the posterior end of the plastron.
Interambulacra: The interambulacra are strongly inflated adapically, forming high keels along the sunken petals. The aboral tuberculation is fairly homogenous, consisting of small perforate crenulate primary tubercles with distinct elevated bosses and inclined areoles. The tubercles become larger towards the ambitus and within the peripetalous fasciole. On the oral side primary tubercles are larger and more strongly crenulate. They are not so densely packed as on the aboral surface and more secondary tubercles are found between them. The largest tubercles are situated on the anterior half of the oral side, sideways of the peristome. The plastron is large, mesamphisternous and nearly half as wide as the test (~45 % of TW). It shows a confluent, fan-shaped tuberculation radiating from the posterior margin. The primary tubercles increase in size from the posterior end of the plastron towards the anterior end. The labrum is moderately wide and T-shaped.
Peristome: The peristome is situated in anterior half of the oral side, about 15 to 20 % of TL from the anterior margin. It is kidney-shaped with a protuberant rounded labrum bearing a distinct rim. It is 10.0 mm wide and 3 to 4 mm long in a 61.5 mm test length specimen.
Periproct: The periproct is situated marginally, high on the vertical posterior face of the test. It is overhung by a distinct rostrum and has an oval, vertically elongated shape. Below the peristome there is a shallow subanal depression. The periproct is about as large as the peristome. It is 10.5 x 6.7 mm in a 71 mm specimen.
Fascioles: The peripetalous fasciole is broad, especially at the anterior end of the test. It is widest at the tips of the petals and in ambulacrum III; narrowing in the interambulacra. The peripetalous fasciole is slightly indented in interambulacrum 5, distinctly indented in interambulacra 2 and 3 (Fig. 63). The degree of indentation in interambulacra 1 and 4 varies from deeply to slightly indented. The latero-anal fasciole branches off from the peripetalous fasciole about halfway along petal II and III running laterally towards the periproct and forming a deep, narrow "V" beneath it.

Differential diagnosis:
This species differs from the similar *Schizaster scillae* (DES MOULINS, 1837) by its antero-posteriorly elongation, its shorter posterior paired petals, its more strongly wedge-shaped profile and its different peripetalous fasciole outline (compare LAMBERT, 1907a and CHALLIS, 1980).

Schizaster parkinsoni (DEFRANCE) and *Schizaster desori* WRIGHT, 1855, two common Early to mid-Miocene representatives of the genus are rather easily distinguished from *S. eurynotus* by their subequal length and width, shallower frontal ambulacrum III and less strongly wedge shaped profile. Furthermore, the shape of the peripetalous and latero-anal fascioles helps to distinguish these forms (compare CHALLIS, 1980; ROSE & WATSON, 1998).

Schizaster karreri LAUBE, 1869 is usually much smaller and can readily distinguished from *S. eurynotus*: *S. karreri* has a less rostrate, near-vertically truncated posterior end, a distinct subanal heel, a different peripetalous fasciole outline (not indented at all in interambulacra 2 and 3), longer posterior paired petals and a less posteriorly displaced apical disc.

For the difference to *Schizaster laubei* HOERNES, 1875a see below under that species.

Discussion:

The specimens considered here are clearly conspecific with the Early to mid-Miocene specimens from the Mediterranean area named *Schizaster eurynotus* AGASSIZ, 1840b by SISMONDA (1841) or ROSE & WATSON (1998). There are, however, some taxonomic problems associated with this name and species:

1) The name *Schizaster eurynotus* AGASSIZ, 1840b is a *nomen nudum* according to the ICZN rules (ICZN 4th ed., 2000, Article 12.1.).
2) The next author who published a description for this name is SISMONDA (1841), which is thus the valid author for *S. eurynotus* (ICZN 4th ed., Articles 11. and 12.).
3) SISMONDA (1841: 22) based his description on species from the Early to mid-Miocene of the Colli Torinesi in Italy, which is thus the type area for this species.
4) AGASSIZ's specimens (which are plaster casts) are not conspecific with the material of the type area.
5) Hence, the name *S. eurynotus* has been used in two different senses: a) in the sense of AGASSIZ (1840b) [but this is a *nomen nudum*] from the Burdigalian of France (e.g. PHILIPPE, 1998) and b) in the sense of SISMONDA (1841) (e.g. ROSE & WATSON, 1998).

LAMBERT, obviously being aware that the name *S. eurynotus* AGASSIZ, 1840b is a *nomen nudum*, often referred to the taxon as "*Schizaster eurynotus* AGASSIZ (in SISMONDA), 1841" (e.g. LAMBERT, 1906a: 115; LAMBERT, 1915a: 159).

PHILIPPE (1998) reported *S. eurynotus* from the Burdigalian of the Rhone Basin, but his specimens differ from typical *S. eurynotus* by their oval outline with less pointed posterior end, lack of a pronounced rostrum, lack of the lateral flexure of the tips of the anterior paired petals and its different peripetalous fasciole, which is straight in interambulacra 2 and 3 and only slightly indentation in interambulacra 1 and 4 (PHILIPPE, 1998: pl. 20, figs 2a-b). These specimens in fact resemble large specimens of *S. karreri* to some extent.

Schizaster capederi LAMBERT, 1907 from Sardinia is closely similar to *S. eurynotus*, as stated also by LAMBERT (1907a) himself. The species is based on a laterally deformed specimen missing the posterior end. According to LAMBERT (1907a: 74) it differs from *S. eurynotus* by its "sillon antérieur plus allongé" and the presence of four gonopores. Concerning the latter feature the reader is referred to the introductory remarks under the genus *Schizaster* (see above). The former feature may also result from the lateral compression to which LAMBERT's specimen was obviously subject and it seems very likely that *S. capederi* and *S. eurynotus* are conspecific.

LAUBE (1869a) established the species *Schizaster leithanus* for a single specimen from the Badenian (Langhian-Early Serravallian) Leitha Limestone of Mörbisch (formerly Merwisch; near Rust, Bgld, Austria). The type could be located in the type collection of the Geological Survey of Austria (specimen no. 1871/3/2). LAUBE's (1871) illustration is misleading, as it depicts a nearly circular specimen. This, however, is due to the poor preservation of the specimen, which lacks the posterior end and nearly the complete oral surface. When compared to well preserved specimens of *S. eurynotus* from Stotzing (Bgld, Austria) and the Maltese Islands no features could be found which would allow to separate the two species. Hence, *S. leithanus* is considered as junior synonym of *S. eurynotus* SISMOMDA, 1841. The characteristic difference to *S. eurynotus* (under the name *S. scillae*, following DESOR, 1858) mentioned by LAUBE (1871: 71) "durch den Mangel eines Fortsatzes nach hinten" (by the lack of a posterior projection) is obviously no original feature but simply due to poor preservation (see Pl. 66, Fig. 1a-c). Whether the two specimens of "*Schizaster* cfr. *lajtanus* LBE." mentioned by VADÁSZ (1915: 224) really belong to *S. eurynotus* is doubtful. Yet the mention of a strongly eccentric apex fit well with this species.

LAUBE (1869a, 1871) followed DESOR (1858) in placing *S. eurynotus* into the synonymy of *S. scillae*. Although his material of *S. scillae* could not be located his description indicates that he had specimens of *S. eurynotus*. This is also supported by the fact that all known specimens of *Schizaster* (museum material as well as specimens collected by private collectors recently) from the Leitha Limestone exposed in Baden belong to *S. eurynotus*. Additionally LAUBE's comparison materials [partly preserved at the NHMW] from the Miocene of Malta which are labelled *S. scillae*, are actually misidentified *S. eurynotus*. All other references of *S. scillae* from the Miocene of Austria are just re-iterations of LAUBE's record and not based on new material. *S. scillae* (DES MOULINS, 1837), a species based on the figure in SCILLA (1670, 1752) was re-described by LAMBERT (1907a) from the Tortonian of Sardinia and by CHALLIS (1980) from the Messinian of Malta. Both describe a large species which differs from *S. eurynotus* by its longer posterior paired petals.

Schizaster eurynotus has also been reported from the Middle Miocene of Venezuela by JEANNET (1928). Despite the lack of records from other places in the Caribbean and in the area between Venezuela and the Iberian peninsula (the western-most occurrence of this species, apart from JEANNET's record) JEANNET's determination seems to be correct. At least, nothing in his description or illustrations gives reason to doubt it and this could be a fossil example of a transatlantic taxon (for other examples see PODDUBIUK & ROSE, 1984). Today there are also a number of purportedly transatlantic echinoid taxa [e.g. *Brissus unicolor* (LESKE), *Echinometra lucunter* (LAMARCK), *Tripneustes ventricosus* (LAMARCK), ...] (see e.g. FELL, 1967; PAWSON, 1978). In most cases it is unknown if these species maintain gene-flow across the Atlantic and molecular genetics has yet to tackle this question. A notable exception are the recent studies by the working-group around H.A. LESSIOS: in a study on the East Pacific and Atlantic species of *Echinometra* MCCARTNEY et al. (2000) discussed also the amphi-atlantic *E. lucunter*. According to their results gene-flow between the Caribbean stock of *E. lucunter* and the rest of the Atlantic Ocean is restricted, with all investigated specimens from outside the Caribbean sharing a single substitution in COI that distinguished them from the Caribbean individuals. In *Eucidaris tribuloides* (LAMARCK), in contrast, very little genetic differentiation between Western and Eastern Atlantic populations could be documented by LESSIOS et al. (1999). Only the populations on the Central Atlantic islands Ascension and St. Helena (formerly considered either as subspecies of *E. tribuloides* or as separate species) form a resolved subclade within the Atlantic clade. In *Diadema antillarum*, the situation is yet again different. According to the results of LESSIOS et al. (2001b) the differences in the mitochondrial DNA of the East and West Atlantic lineages is as different as between several other accepted morphospecies and they estimate that the two were separated between 1.02 and 1.86 my ago. The populations from the Central Atlantic islands were found to nest as monophyletic entity within the Western Atlantic lineage of *D. antillarum* and LESSIOS et al. (2001b) support PAWSON's (1978) decision to treat them as subspecies of *D. antillarum*.

Occurrence:

Austria: Early to Late Badenian (Langhian-Early Serravallian)
Vienna Basin: Baden (Leitha Limestone), NÖ (LAUBE, 1869a, 1871; GREGORY, 1891 (erroneously reported from the Baden Tegel)]; Müllendorf (Mühlendorfer Kreide AG quarry), Bgld ([NHMW]); Rauchstallbrunngraben, near Baden, NÖ (KARRER, 1877; SCHAFFER, 1907, 1942; BOBIES, 1928); Stotzing (sandpit Mayer), Bgld (KAZÁR, 2002; [NHMW])
Eisenstadt-Sopron Basin: Mörbisch, Bgld (LAUBE, 1869a, 1871; STOJASPAL, 1975 [GBA])

Paratethys (non-Austrian occurrences): ? Karpatian (Late Burdigalian), Late Badenian (Early Serravallian)
Swiss Molasse Basin: St. Gallen, Switzerland (GUTZWILLER, 1883; COTTREAU, 1913a)
Great Hungarian Basin (Pannonian Basin): ? Hont, Börzsöny, Hungary (VADÁSZ, 1915)
Transcarpathian Basin: ? Hlinné, eastern Slovakia (SENEŠ, 1955; CICHA et al., 1967)
Transylvanian Basin: Gârbova de Sus (= Felsö-Orbó), Romania (VADÁSZ, 1915; [MAFI]); region east of Taşad, Beiuş Basin, Romania (PAUCĂ, 1936)

Mediterranean: Aquitanian to Messinian
Western Mediterranean: Algeria (COTTREAU, 1913a); Corsica, France (AGASSIZ & DESOR, 1847b; LAMBERT, 1906a): Bonifacio (GREGORY, 1891; COTTREAU, 1913a), Santa Manza (WRIGHT, 1855; DESOR, 1858; QUENSTEDT, 1874, 1875; GREGORY, 1891); France, mainland: Peripignan (AGASSIZ & DESOR, 1847b; WRIGHT, 1855; DESOR, 1858; LAUBE, 1871), Vence, Alpes-Maritimes, (LAMBERT, 1906a; COTTREAU, 1913a); Italy, mainland: ? Monte Pelegrino (DESOR, 1858; LAUBE, 1871), Nice (PICTET, 1857; LAMBERT, 1906a), Palermo, Sicily (DESOR, 1858; LAUBE, 1871; GREGORY, 1891), Piemont (COTTREAU, 1913a), Posano, near Volterra, Prov. di Pisa (MENESINI, 1967), Reggio Emilia (COTTREAU, 1913a), Rosignano, Toscana (STEFANINI, 1908a); Salento, Apulia (RAGAINI, 1994); Sardinia, Italy (PICTET, 1857; GREGORY, 1891; LAMBERT, 1906a; COTTREAU, 1913a): Cagliari (AGASSIZ & DESOR, 1847b; WRIGHT, 1855), Cap Sant'Elia, Coroneddu (near Bosa), Cadréas (Bonorva), ? Fontanaccia, ? Montevecchio, Soddi and Zuri (LAMBERT, 1907a); Funtanazza (COMASCHI CARIA, 1963)
Central Mediterranean: Italy, mainland: ? Asti (DESOR, 1858; LAUBE, 1871; GREGORY, 1891); Maltese Islands (DESOR, 1858; LAUBE, 1871; COTTREAU, 1913a): *Globigerina* Lst. indifferenciated (WRIGHT, 1855; MENESINI, 1975; BOGGILD & ROSE, 1984); Lower *Globigerina* Lst. (MENESINI, 1979a, b); Middle *Globigerina* Lst. (ROSE, 1974b, 1975; CHALLIS, 1980; ROSE & WATSON, 1998); Xwieni Conglomerate Bed [basal Upper *Globigerina* Lst. (= C_2 phosphate conglomerate of PEDLEY et al., 1976) (CHALLIS, 1979, 1980; ROSE & WATSON, 1998); Upper *Globigerina* Lst. (ROSE, 1974b, 1975); Blue Clay (ROSE, 1974b, 1975; CHALLIS, 1980); Greensand (ROSE, 1974b, 1975; ? MENESINI, 1979a, b; CHALLIS, 1980); records from the Upper Coralline Lst. (ROSE, 1974b, 1975) are dubious and are probably misidentified *S. scillae*
Eastern Mediterranean: Armenia (COTTREAU, 1913a); Wadi Ramlieh, near Marsa Matrouh, Egypt (FOURTAU, 1920); Turkey (COTTREAU, 1913a); Greece: Attika (= Attique) (COTTREAU, 1913a), Morea (= Morée) (DESOR, 1858; LAUBE, 1871; GREGORY, 1891); Libya: Cyrene, Cyrenaica (ROSE, 1974a); Umm er Rzem, Cyrenaica (TAVIANI, 1939)

Atlantic Ocean: Middle Miocene
Caribbean: Ojo de Agua, Capadare, San José de la Costa and Cerro de Chichiriviche, Venezuela (JEANNET, 1928)

Indian Ocean: ? Early Miocene
Eastern African coast: ? Ehil, Somalia [SOCIN, 1956 (specific identification needs to be confirmed)]

Schizaster (*S.*) *karreri* LAUBE, 1869

(Figs. 64-65; Pl. 65, Figs. 1-7)

	1869a	*Schiz. Parkinsoni* DEFR. – LAUBE: 183
*	1869a	*Schiz. Karreri* LAUBE. – LAUBE: 184
	1869	*Schizaster Karreri* LAUBE sp. ined. – FUCHS: 194
	1869	*Schizaster Parkinsoni* DEFR. – FUCHS: 194
	1870	*Schizaster Karreri* LBE. – LAUBE: 314
	1870	*Schizaster Parkinsoni* DEFR. – LAUBE: 314
	1871	*Schizaster Parkinsoni* DEFRANCE sp. – LAUBE: 70
	1871	*Schizaster Karreri* LAUBE. – LAUBE: 70-71; pl. 16, fig. 6
	1877	*Schizaster Karreri* LAUBE – KARRER: 312
	1877	*Schizaster Parkinsoni* DEFR. – KARRER: 312
	1877	*Schizaster karreri* LAUBE. – LÓCZY: 63
?	1884	*Schizaster Karreri* ? LAUB. – COPPI: 191, no. 279
	1887b	*Schizaster* cfr. *Karreri*, LAUBE. – KOCH: 270
pp	1891	*Schizaster parkinsoni* (DEFRANCE), 1827. – GREGORY: 616-617 [Austrian material only]
	1891	*Schizaster Karreri* LAUBE. – ROTH VON TELEGD: 151
	1894	*Schizaster* cfr. *Karreri*. LAUBE. – LÖRENTHEY: 61
#	1895	*Schizaster Lovisatoi*, COTTEAU, 1895. – COTTEAU: 45-46; pl. 5, figs. 9-10
#	1895	*Schizaster sardiniensis*, COTTEAU, 1895. – COTTEAU: 46-47; pl. 5, figs. 11-12
?	1896	*Schizaster Karreri* LAUBE. – BOTTO-MICCA. 359
#	1902a	*Schizaster bouziguensis* – DE LORIOL: 26-28; pl. 4, figs. 1a-c [not seen; fide PHILIPPE (1998)]
	1905	*Schizaster Karreri*, LBE. – GAÁL: 344, 362
? #	1906a	*Schizaster* (*Aplospatangus*) *calceolus* – LAMBERT: 118
#	1906a	*Schizaster morgadesi* LAMBERT – LAMBERT: 119; pl. 8, figs. 3-4
#	1906a	*Schizaster gymnesiæ* LAMBERT – LAMBERT: 119-120; pl. 10, figs. 1-2
#	1906b	*Schizaster venetiensis* – LAMBERT: 45-46; pl 7, figs. 3-5 [reference not seen; fide PHILIPPE (1998)]
v. #	1906	*Schizaster Karreri* LBE. var. *hungaricus* nov. var. – VADÁSZ: 333-335; pl. 10, figs. 3a-b
v. #	1906	*Schizaster Lovisatoi* COTTEAU var. *rákosensis* nov. var. – VADÁSZ: 335-336; pl. 10, figs. 4a-b
	1907a	*Schizaster calceolus* LAMBERT, 1907. – LAMBERT: 69-70; pl. 5, fig. 8
	1907a	*Schizaster sardiniensis* COTTEAU, 1895. – LAMBERT: 71-72; pl. 4, figs. 8-10
	1907	*Schizaster Parkinsoni* DEFR. – SCHAFFER: 35
? pp	1909	*Schizaster parkinsoni* DEFRANCE (*Spatangus*), 1827. – LAMBERT: 66-67 ; pl. 5, figs. 3-4 [fide LAMBERT, 1909]
#	1909	*Schizaster decipiens* LAMBERT. – LAMBERT: 74-75; pl. 6, figs. 6-7
#	1909	*Schizaster Ilottoi* LAMBERT. – LAMBERT: 75-76 [new name for the specimens figured as *S. parkinsoni* in 1907; fide LAMBERT, 1928 & PHILIPPE, 1998]
	1913a	*Schizaster Lovisatoi* COTTEAU. – COTTREAU: 118-120; pl. 13, figs. 2-7
?	1915	*Schizaster calceolus* LAUB. – MÁJER: 29, 82
v.	1915	*Schizaster rákosensis* LAMB. – VADÁSZ: 218-219; text-figs. 103-105; pl. 10 (4) , fig. 7
v.	1915	*Schizaster hungaricus* VAD. – VADÁSZ: 219; text-fig. 106; pl. 10 (4) , fig. 8
non	1915	*Schizaster calceolus* LAMBERT. – VADÁSZ: 220; text-fig. 108. [probably a misidentified *S. eurynotus*]
.	1915	*Schizaster Ilottoi* LAMBERT. – VADÁSZ: 221; text-fig. 109; pl. 10 (4), fig. 6
.	1915	*Schizaster barcinensis* LAMB. var. ? – VADÁSZ: 221-222; text-fig. 110; pl. 11 (5), fig. 4
.	1915	*Schizaster ventiensis* LAMB. – VADÁSZ: 222; text-fig. 111; pl. 9 (3), fig. 19

. 1915 *Schizaster Karreri* Laube. – Vadász: 223; pl. 12 (6), fig. 8
v. 1915 *Schizaster* cfr. *Bellardi* Ag. – Vadász: 224; text-fig. 114
1915a *Schizaster Jacquemeti* Lambert. – Lambert: 163-165; pl. 13, figs. 2-5
1915a *Schizaster ventiensis* Lambert, 1906. – Lambert: 165
? 1919 *Schizaster Lovisatoi* Cott. – Stefanini: 147-148; pl. 14 (8), fig. 5a-b [fide. Philippe (1998: 175)]
1923 *Schizaster Bellardii* Agassiz & Des. 1847 – di Giorgio: 121-122; pl. 2 (1), fig. 2
1923 *Schizaster ventiensis* Lambert 1906 – di Giorgio: 122; pl. 2 (1), fig. 3a-b
1927a *Schizaster Lovisatoi* Cotteau 1895 – Lambert: 29
1931 *Schizaster Karreri* Laube – Janoschek: 84
1936 *Schizaster calceolus* Lambert – Paucă: 143, 195-196; pl. 2, figs. 3-4
? 1939 *Schizaster Lovisatoi* Cotteau – Stchepinsky: 10; pl. 2, fig. 4-6
1939 *Schizaster Lovisatoi* Cotteau – Taviani: 45-46; pl. 3 (5), figs. 6-7
1942 *Schizaster Parkinsoni* Defr. – Schaffer: 94
1953 *Schizaster karreri* Laube, 1871. – Szörényi: 41, 93; pl. 2, fig. 5
1953 *Schizaster sardiniensis* Cotteau, 1895. – Szörényi: 42, 94; pl. 2, figs. 6, 6a
1953 *Schizaster ilottoi* Lambert, 1908. – Szörényi: 42, 94; pl. 2, figs. 7, 7a
1960 *Schizaster ventiensis* Lambert 1906 – Kojumdgieva & Strachimirov: 232; pl. 9, figs. 4a-c
1962 *Schizaster parkinsoni* Defr. – Bachmayer: 378
? 1963 *Schizaster lovisatoi* Cotteau – Comaschi Caria: 29-30; pl. 3, figs. 5-6
? 1966 *Schizaster* cfr. *karreri* Laube – Kókay: 83
? 1967 *Schizaster sardiniensis* Cott. – Cicha et al.: 93
1967 *Schizaster lovisatoi* Cotteau – Menesini: 149-151; pl. 46 (1), figs. 2, 2a, 4, 3a, 4, 4a, 5, 5a, 6, 6a; pl. 47 (2), figs. 1, 1a, 2, 2a, 3. 3a, 4, 4a, 5, 5a
1968 *Schizaster "parkinsoni"* – Küpper: pl. 4, fig. 33
pp 1970 *Schizaster* cf. *karreri* Laube – Bałuk: pl. 1, fig. 8
1970 *Schizaster "parkinsoni"* – Thenius: p. 211; pl. 3, fig. 33
? 1972 *Schizaster lovisatoi* Cotteau, 1895. – Marcopoulou-Diacantoni: 138-140; pl. 15 (1), fig. 3 [fide Philippe (1998: 175)]
? 1972 *Schizaster calceolus* Lambert. – Marcopoulou-Diacantoni: 156 [fide Philippe (1998: 175)]
1974 *Schizaster "parkinsoni"* Defrance – Thenius: 70-71; fig. 13/25
? 1977 *Schizaster ventiensis* Lambert, <u>1836</u> – Mączyńska: 197; pl. 8 figs. 1a-g [fide Philippe (1998: 175)]
1979 *Schizaster lovisatoi* Cotteau, 1895 – d'Alessandro et al.: 93; pl. 17, fig. 10
1979 *Schizaster "parkinsoni"* – Thenius: 52; pl. 3, fig. 25
? 1981 *Schizaster scillae* Desor – Mitrović-Petrović: 181; figs. 12a-b
1983 *Schizaster "parkinsoni"* – Thenius: 119; pl. 7, fig. 25
1984 *Schizaster rákosensis* Vadász – Kókay et al.: 290
1984 *Schizaster hungaricus* Vadász – Kókay et al.: 290
1984 *Schizaster lovisatoi* Cotteau, 1895 – Negretti: 115-116, pl. 9, figs. 1-6
1985 *Schizaster rákosiensis* Vadász, 1906– Mihály: 244; pl. 5, figs. 1-2
1985 *Schizaster hungaricus* Vadász, 1906– Mihály: 244
? 1987 *Schizaster ventiensis* Lambert, 1906 – Mączyńska: 149 [fide Philippe (1998: 175)]
1987 *Schizaster karreri* Laube, 1871 – Mączyńska: 149
1989 *Schizaster lovisatoi* Cotteau – Philippe: 31; tab. 4
1990 *Schizaster lovisatoi* Cotteau, 1895 – Philippe et al.: 245
1992 *Schizaster* cf. *lovisatoi* Cotteau – Blondel & Philippe: 438, 441; pl. 2, fig. 4
1998 *Schizaster (Schizaster) lovisatoi* Cotteau, 1895 – Philippe: 175-178; pl. 20, figs. 4-7
1998 *Schizaster karreri* Laube – Schultz: 120; pl. 54, figs. 2
1999 *Schizaster karreri* – Piller & Harzhauser: 223; fig. 24

Type-material:

<u>*Schizaster karreri* Laube, 1869</u>:
Holotype: Laube (1871: 69) chose an exceptionally well preserved specimen form "Bid [sic] im Pester Comitat" [= Biatorbágy (formerly Bia), Pest, Hungary] as holotype. That specimen was figured by Vadász [1915: pl. 12 (6), fig. 8], who stated that it was housed in the collection of the "Wiener Hofmuseum" (old name for the NHMW). Unfortunately, the holotype could not be located at the NHMW. The figured specimen of Laube (1871: pl. 16, fig. 6; here Fig. 65.A-C) is a different specimen (NHMW 1865.XXX.73), which is labelled as coming from "Sóskút im Stuhlweissenburger Comitat". The two localities approximately five kilometres apart (despite belonging to different counties in the 19th and early 20th century).
Locus typicus: Biatorbágy (= Bia), Pest, Hungary
Age: Badenian (Langhian-Early Serravallian), Middle Miocene

<u>*Schizaster lovisatoi* Cotteau, 1895</u>:
Holotype: the specimen figured by Cotteau (1895: pl. 5, figs. 9-10); private collection of Lovisato (not seen, current whereabouts unknown)
Locus typicus: Castelsardo, Sassari, Sardinia, Italy
Stratum typicum: "couches arénacée inférieure aux couches langhiennes à *Vaginella depressa*"
Age: Langhian, Middle Miocene

<u>*Schizaster sardiniensis* Cotteau, 1895</u>:
Syntypes: private collection of Lovisato (not seen, current whereabouts unknown)
Locus typicus: Cuccuruddu, below Chemerule, Thiesi, Sardinia, Italy
Stratum typicum: "Calcaire argileux"
Age: Middle Miocene

<u>*Schizaster bouziguensis* de Loriol, 1902</u>:
Holotype: figured by de Loriol (1902a: pl. 4, figs. 1a-c); a cast of the type is housed at the Centre des Sciences de la Terre, Lyon under the number FSL 27 470 (Philippe, 1998: 323)
Locus typicus: Bouzigues, France
Age: Middle Miocene
Remark: Re-examined by Philippe (1998) and placed into the synonymy of *S. lovisatoi* Cotteau, 1895, which in turn is here placed into the synonymy of *S. karreri*.

<u>? *Schizaster calceolus* Lambert, 1906</u>:
Holotype: whereabouts unknown
Locus typicus: ? Bonorva, Sardinia, Italy
Age: "Langhien" of Lambert, Early or Middle Miocene

<u>*Schizaster morgadesi* Lambert, 1906</u>:
Holotype: specimen figured by Lambert (1906a: pl. 8, figs. 3-4), private collection of J. Almera (not seen)
Locus typicus: Calafell, Province Barcelona, Spain
Age: "Helvétien", Early or Middle Miocene

<u>*Schizaster gymnesiae* Lambert, 1906</u>:
Holotype: specimen figured by Lambert (1906a: pl. 10, figs. 1-2), private collection of J. Almera (not seen)

Locus typicus: Ciudaadèla, Menorca, Balearic Islands, Spain
Age: "Helvétien", Early or Middle Miocene

Schizaster lovisatoi rakosensis VADÁSZ, 1906:
Holotype: specimen MAFI Ech 127
Locus typicus: Budapest-Rákos, Pest, Hungary
Age: Late Badenian (Early Serravallian), Middle Miocene
Remarks: The holotype falls well within the variation observed in *S. karreri*. No features could be found, which would allow a confident separation of this form.

Schizaster karreri hungaricus VADÁSZ, 1906:
Holotype: specimen MAFI Ech 128
Locus typicus: Budapest-Rákos, Pest, Hungary
Age: Late Badenian (Early Serravallian), Middle Miocene
Remarks: As in the former nominal (sub)species, none of the characters shown by the type allow to distinguish it from *S. karreri*.

Schizaster decipiens LAMBERT, 1909:
Holotype: the specimen figured by LAMBERT (1909: pl. 6, figs. 6-7); current whereabouts unknown
Locus typicus: Cuccuruddu, Thiesi, Sardinia, Italy
Stratum typicum: Calcaire marneux
Age: "Helvétien", Early or Middle Miocene

Schizaster ilottoi LAMBERT, 1909:
Holotype: the specimen figured by LAMBERT (1907a: pl. 5, figs. 3-4) as *S. parkinsoni*; current whereabouts unknown
Locus typicus: Cadreas, grande tranchée de Bonorva, Sardinia, Italy
Age: "Langhien" of LAMBERT, Early or Middle Miocene

Schizaster jacquemeti LAMBERT, 1915:
Syntypes: casts of the two syntypes of this species are housed at the Centre des Sciences de la Terre, Lyon under the number FSL 27 467 and in the collection LAMBERT (Muséum national, Paris) (PHILIPPE, 1998: 323-324)
Locus typicus: la Couronne, Martigues, France
Age: "Langhien" of LAMBERT, = Burdigalian, Early Miocene (PHILIPPE, 1998)

Material:
Early Badenian (Langhian) – Baden (Badener Tegel), NÖ, Austria
NHMW: 1 specimen (NHMW 1933.X.108)
Early Badenian (Langhian) – Retznei [Weissenegg Fm., Lafarge (formerly Perlmoser) quarry Rosenberg], Styria, Austria
NHMW: 2 specimens (NHMW 2004z0001/0033-34)
Early Badenian (Langhian) – Wagna (brickyard Wagna), Styria, Austria
NHMW: 5 specimens (NHMW 2004z0001/0035-39)
Badenian (Langhian-Early Serravallian) – Haschendorf, near Neckenmarkt, Bgld, Austria
NHMW: 1 specimen (NHMW 1848.III.62)
Badenian (Langhian-Early Serravallian) – Kalksburg, Vienna, Austria
NHMW: 9 specimens (NHMW 1866.XLVII.37, 1868.VIII.59, 1997z0178/1670, ../1671, 2004z0001/0032a-b, ../0061, ../0064-65)
IPUW: 1 specimen (no inventory no.)

Foreign material for comparison:
Badenian (Langhian-Early Serravallian) – Soskút, Hungary
NHMW: 1 specimen [NHMW 1865.XXX.73 (figured specimen of *Schizaster karreri* in LAUBE, 1871)]
Remarks: According to VADÁSZ (1915: 105, footnote) there are only strata of Sarmatian age outcropping at Soskút and material from the marine Badenian comes from the nearby Leitha Limestone of Bia.
Badenian (Langhian-Early Serravallian) – Piliny (Köhegy), Hungary
MAFI: 1 specimen [MAFI Ech 153 (figured specimen of *Schizaster* cf. *bellardi* in VADÁSZ, 1915)]
Late Badenian (Early Serravallian) – Budapest-Rákos, Pest, Hungary
MAFI: 2 specimens [MAFI Ech 127 (**holotype** of *Schizaster lovisatoi rákosensis* VADÁSZ, 1906), ECH 128 (**holotype** of *Schizaster karreri hungaricus* VADÁSZ, 1906)]

Dimensions (in mm):

Inv. No.	TL	TW	TH
NHMW 1865.XXX.73	46.1	45.3	26.7
NHMW 1866.XLVII.37	63.9	58.6	37.8
NHMW 1933.X.108a	43.3	39.6	-
NHMW 2004z0001/0032a	50.8	52.4	28.2
NHMW 2004z0001/0032b	41.4	39.8	-
NHMW 2004z0001/0033	50.9	48.7	35.3
NHMW 2004z0001/0034	34.8	35.5	30.9
NHMW 2004z0001/0035	43.3	41.3	>24.0
NHMW 2004z0001/0036	27.2	25.5	19.1
NHMW 2004z0001/0037	30.5	29.9	25.2
NHMW 2004z0001/0038	~21	~20	17.7
NHMW 2004z0001/0039	49.2	46.4	>27.5
NHMW 2004z0001/0061	48.2	44.1	28.4

Description:
Size and shape: Test ranging from *c.* 21 to 63 mm test length in the studied material. Test outline ranges from polygonal-subcircular in smaller to elongated antero-posteriorly in larger specimens. The posterior end may be bluntly pointed, giving the test a cordiform appearance. There is a well-developed, moderately deep frontal notch. In some specimens a slightly left-right asymmetry is observed (primary, not resulting from post-sedimentary compaction). In these specimens one side – usually the right one – projects further anterior than the other. The maximum width lies subcentrally. In profile the test is moderately to strongly wedge-shaped with its maximum height roughly halfway along a highly arched keel in interambulacrum 5. Wedge-shapedness and test height vary considerable, even if post-sedimentary deformation effects are taken into account. In profile the posterior end is vertically truncated with the posterior part of the plastron forming a slightly projecting subanal heel. A small rostrum overhangs the periproct.
Apical disc: The apical disc lies slightly posterior of the centre, its exact position varies between c. 50 to 60 % of TL from the anterior margin. It is ethmolytic with 2 gonopores (Fig. 64.D-E). There are no gonopores in the anterior genital plates.
Ambulacra: The ambulacra are petaloid and rather deeply sunken. Ambulacrum III is very deeply sunken with overhanging walls. It closes slightly distally, where the pores change from partitioned isopores to minute unipores. The pores are arranged in straight single rows in each half ambulacrum. Ambitally ambulacrum III is less deeply sunken forming a moderately deep frontal notch.

The anterior paired petals are long, relatively narrow and strongly flexed anteriorly. In large specimens (TL > 50 mm) the tips may very slightly flex laterally. The posterior paired petals are much shorter, usually extending less than 50 % of the length of the anterior petals. In larger specimens, however, (e.g. NHMW 1866.XLVII.37) they may extend up to 60 % of the anterior petals. Similar allometric growth with an increase of posterior petal length during ontogeny was observed in extant *S. (S.) compactus* by MCNAMARA (1995).

Anterior paired petals form an acute angle of about 65 to 75°, the posterior ones an angle between 55 to 65°. The posterior paired petals are broadest at their midpoint, whereas the anterior petals are broadest in their distal third. The pores of the paired petals are conjugated elongate isopores. The interporiferous zones are slightly smaller than a single poriferous zone and bear only secondary tubercles.

Outside the peripetalous fasciole, the ambulacral pores are minute (? constricted) unipores. Adorally the ambulacra form

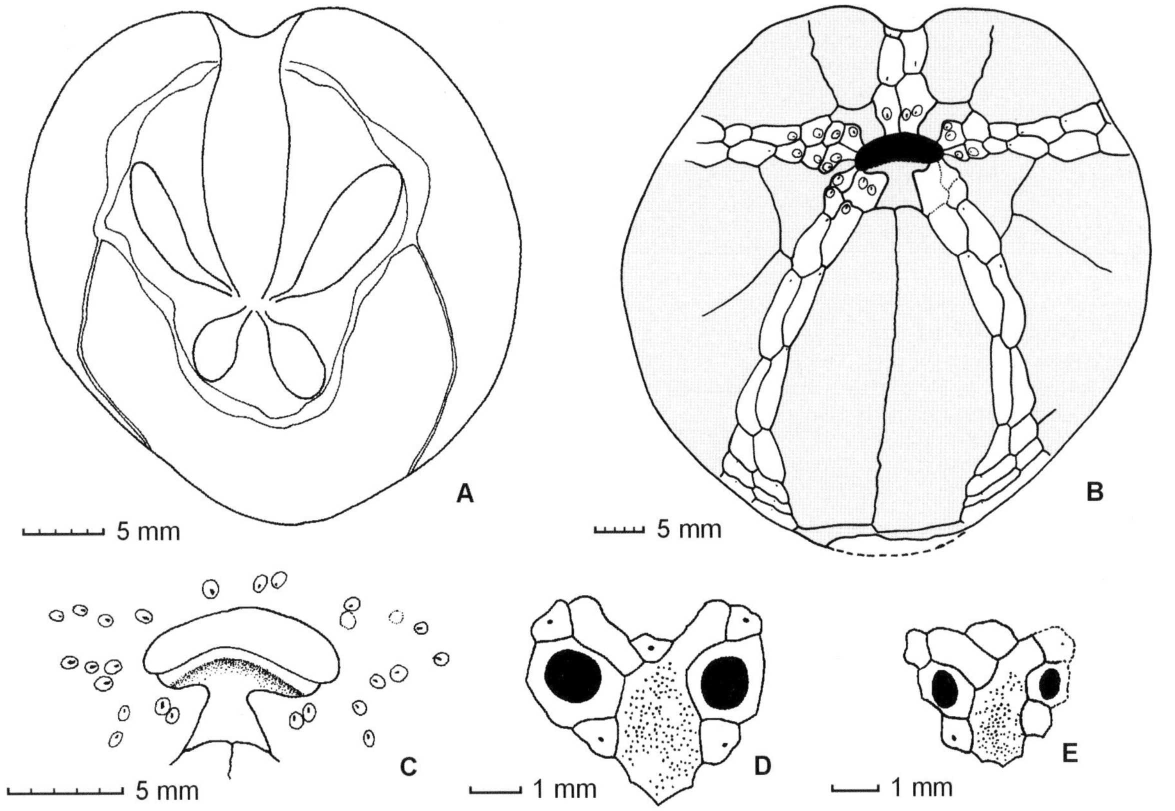

Figure 64: *Schizaster (S.) karreri* LAUBE, 1869: outline of the peripetalous and latero-anal fascioles (A: Wagna, Styria, NHMW 2004z0001/0037), oral plating (B: Kalksburg, Vienna, NHMW 2004z0001/0032a), labrum shape and phyllodes (C: Wagna, Styria, NHMW 2004z0001/0035), apical disc (D: Kalksburg, Vienna, NHMW 1866.XLVII.37; E: Sóskut, Pest, Hungary, NHMW 1865.XXX.73). Interambulacra shaded; poorly visible sutures stippled.

distinct phyllodes consisting of large unipores with extended periporal partition.

On the oral surface ambulacra I and V form long narrow peri-plastronal areas lacking primary tubercles. Each bears four partitioned isopores on each side of the posterior end of the plastron.

Interambulacra: The interambulacra are highly inflated adapically, forming high keels between the sunken petals. The aboral tuberculation is fairly homogenous, consisting of small perforate crenulate primary tubercles with distinct elevated bosses and inclined areoles. On the oral side primary tubercles are larger, more strongly crenulate and less densely arranged. The largest tubercles are situated in the antero-lateral areas of the oral surface. The plastron is large, mesamphisternous and nearly half as wide as the test (~45 % of TW). It shows a confluent fan-shaped tuberculation radiating from the posterior margin. The primary tubercles increase in size from the posterior end of the plastron towards the anterior end. The labrum is moderately wide and roughly T-shaped with a triangular posterior part. The contact between labrum and sternal plates is broad.

Peristome: The peristome lies in the anterior half of the oral side, about 15 to 20 % of TL from the anterior margin. It is kidney-shaped with a protuberant rounded labrum bearing a distinct rim. It measures 8.2 *vs.* 3.7 mm in a 51 mm TL specimen.

Periproct: The periproct is situated marginally, high on the near-vertical posterior face of the test. It is overhung by a slight rostrum. Its outline varies from horizontally elongated in specimens with low test height to vertically elongated shape in tall specimens. Below the peristome there is a distinct subanal depression, followed by a distinct subanal heel. The periproct is slightly smaller than the peristome. Its largest diameter is *c.* 4 mm in a 41 mm TL specimen and *c.* 6 mm in a 51 mm TL specimen.

Fascioles: A peripetalous and a latero-anal fasciole are developed. The peripetalous fasciole is moderately broad and fairly well visible. The latero-anal fasciole is very narrow and often indistinct. It branches off the peripetalous fasciole behind the anterior paired petals, about two third their length from the apical disc. The peripetalous fasciole is not indented in interambulacrum 5, slightly indented in interambulacra 1 and 4, and only faintly indented to nearly straight (Fig. 64.A). Both fascioles belong to the orthofasciole *sensu* NÉRAUDEAU et al (1998b).

Differential diagnosis:

This species can be easily distinguished from *S. eurynotus* SISMONDA, 1841 by its less cordiform outline, subcentral maximum width, longer posterior paired petals, less posteriorly displaced apical disc, presence of a distinct subanal heel, lack of a prominent, hood-like rostrum and weak indentation of its peripetalous fasciole in interambulacra 1 and 4.

Schizaster parkinsoni (DEFRANCE) from the Early to Mid-Miocene of the Maltese Islands and other Mediterranean regions differs by the presence of 4 genital pores (the anterior pore have a diameter of up to 50 % of the posterior ones), more strongly diverging anterior paired petals, longer posterior paired petals and a peripetalous fasciole, which is strongly indented in interambulacra 1 and 4 (compare CHALLIS, 1980 and ROSE & WATSON, 1998).

Schizaster desori WRIGHT, 1855 another species from the Early to Mid-Miocene of the Maltese islands has a shallower frontal notch, a narrower frontal ambulacrum and an overhanging, obliquely truncated posterior end with well developed rostrum. Additionally the peripetalous fasciole is deeply embayed in interambulacra 1 and 4, the divergence of the anterior paired petals is stronger and the anterior ambitus is less tumid in side view (compare Wright, 1855 and Challis, 1980).

Schizaster scillae (des Moulins, 1837) differs by its much longer posterior paired petals and very different overall appearance (compare Lambert, 1907a and Challis, 1980).

For the difference to *Schizaster laubei* Hoernes, 1875a see below under that species.

Discussion:

The material discussed here clearly belongs to *Schizaster karreri* Laube, 1869 Although the holotype could not be located in the collection of the NHMW, a photograph of that specimen is published in Vadász (1915) and abundant material from the locality Kalksburg and the Styrian Basin was available.

In the Mediterranean area this species is known under the name *Schizaster lovisatoi* Cotteau, 1895. Indeed, none of the differential diagnostic characters which distinguish the two species according to Cotteau (1895: 46; more elongated, hexagonal outline, higher test, more strongly wedge-shaped profile, ...) allow a confident separation of the two forms, especially if the intraspecific variation is taken into account. Especially a larger (TL = 63.9 mm), elongated specimen (NHMW 1866.XLVII.37) is extremely similar to the figured specimen of Cotteau (1895) (see Pl. 65, Figs. 7a-c). Many of the features Cotteau (1895) used to distinguish *S. lovisatoi* from *S. karreri* are related to ontogenetic changes [similar to those documented by McNamara (1995) for *S.* (*S.*) *compactus*] and the comparison of specimens of different size. Allometric growth also accounts for the differences between *S. lovisatoi* and *S. sardiniensis* Cotteau, 1895, which is virtually identical with smaller specimens (TL between 25 and 35 mm) of *S. karreri*.

Likewise, Vadász (1915) under-estimated the intraspecific variation in *Schizaster karreri*. Based on a re-examination of his specimens (at least those, which are still available at the MAFI) seven out of the sixteen *Schizaster* species recognised by him are placed in the synonymy of *S. karreri*. Four others are based on very poorly preserved material, and are very difficult to determine at species level, if possible at all. Three are referred to *S. eurynotus*, one is tentatively referred to *Ditremaster scillae* and the remaining one (*S. latipetalus*) is probably a separate species (this is the only one of Vadász's *Schizaster* species that does not come from Badenian strata).

In 1927 Lambert (1927a: 28-29) placed several species (*S. morgadesi*, *S. gymnesiae*, *S. ventiensis*, *S. decipiens* and *S. ilottoi*) he had established earlier into the synonymy of *S. lovisatoi*. Philippe (1998), in a revision of the Miocene echinoids of the Rhône Basin, shared this opinion and, furthermore added *S. calceolus* and *S. jacquemeti* as junior synonyms. Among the Austrian material attributed to *S. karreri* here there are several specimens which come very close to the morphotype described as *S. ilottoi* by Lambert. Apart from the outer shape, which is more elongated in these specimens and a slightly deeper frontal notch, there are few differences to typical *S. lovisatoi*. These differences are attributed to allometric growth as outlined above. Such elongated specimens (among them NHMW 1866.XLVII.37) were misidentified as *S. parkinsoni* by Laube (1869a 1871), which clearly is a different species (for the diagnostic characters distinguishing these two forms see above under "Differential diagnosis"). Later records of *S. parkinsoni* from Austria are based on Laube's paper and not on new material.

The material reported under the name *Schizaster ventiensis* Lambert, 1906 from the uppermost part of the Korytnica Clays by Mączyńska (1977, 1987) might possibly belong to this species (see also Philippe, 1998: 175), but is too poorly preserved to decide.

Occurrence:

Austria: Early to Late Badenian (Langhian-Early Serravallian)

Vienna Basin: Baden Tegel, Baden, NÖ (Piller & Harzhauser, 1999; [NHMW]); Hagenbrunn (Veitsberg), NÖ (Bachmayer, 1962); Kalksburg, Wien (Fuchs, 1869; Laube, 1869a 1871; Karrer, 1877; Schaffer, 1907, 1942; Thenius, 1970, 1974, 1979, 1983; Schultz, 1998; [NHMW]); Vienna Basin (Lóczy, 1877; Küpper, 1968)

Oberpullendorf Bay: Haschendorf (= Hasfalva), Bgld (Laube, 1869a 1871); Neckenmarkt (Schanzgraben and Schneckenberg), Bgld (Janoschek, 1931)

Styrian Basin: Retznei (Weissenegg Fm., Lafarge quarry Rosenberg), Styria ([NHMW]); Wagna (brickyard Wagna), Styria ([NHMW])

Paratethys (non-Austrian occurrences): Early to Late Badenian (Langhian-Early Serravallian)

Great Hungarian Basin (Pannonian Basin): Biatorbágy (= Bia, = Bid), Pest, Hungary (Laube, 1869a 1871; Lóczy, 1877; Vadász, 1915); Budapest, Pest, Hungary (Kókay et al., 1984); Budapest-Rákos, Pest, Hungary (Vadász, 1906, 1915; Mihály, 1985; [MAFI]); ? Herend-Márkó, Veszprém, Hungary (Kókay, 1966); Hont, Börzsöny, Hungary (? Májer, 1915; Vadász, 1915); Hor. Strháre (= Felsö-Esztergály), Slovak Republic (Roth von Telegd, 1891; Gaál, 1905;); Piliny, Nógrád, Hungary (Vadász, 1915; [MAFI]); Sóskut, Pest, Hungary (Laube, 1869a 1871; Lóczy, 1877; [NHMW]); Budapest-Budatétény (= Tétény), Hungary (Lóczy, 1877)

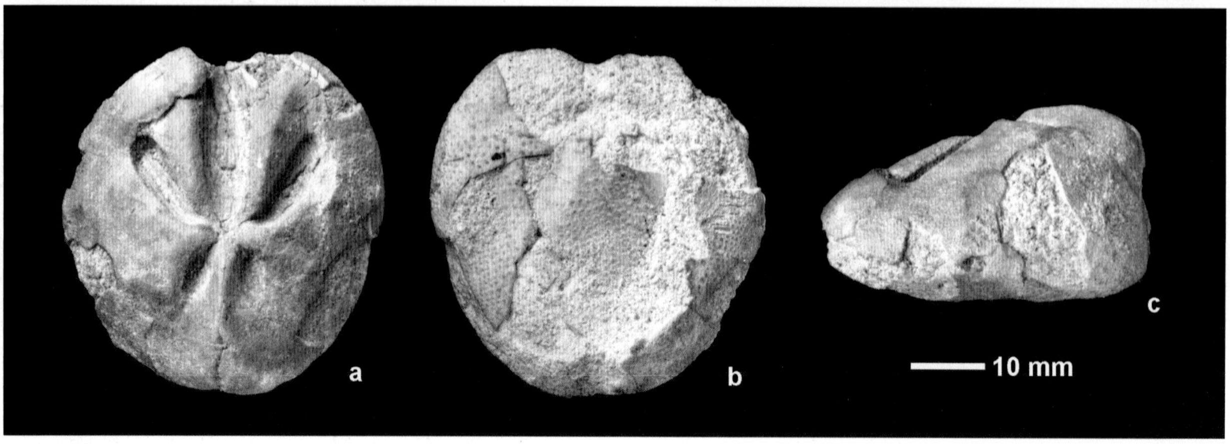

Figure 65: *Schizaster* (*S.*) *karreri* Laube, 1869: figured specimen of Laube (1871: pl. 16, fig. 6) from Sóskut, Pest, Hungary (NHMW 1865.XXX.73).

Fore-Carpathian Basin: Korytnica, Poland (MĄCZYŃSKA, 1977, 1987); Krzywcze Gorne (= Krzywczyce), western Ukraine (SZÖRÉNYI, 1953); Lwów, western Ukraine (SZÖRÉNYI, 1953); Pod'yarkov (= Podjarków), near Lwów, western Ukraine (SZÖRÉNYI, 1953)
Nowy Sącz Basin: Niskowa, western solpes of the Nowy Sącz depression, Poland (BAŁUK, 1970)
Transcarpathian Basin: ? Hlinné, eastern Slovakia (CICHA et al., 1967)
Transylvanian Basin: Buituri (= Bujtur), Romania (KOCH, 1887b; VADÁSZ, 1915); Gârbova de Sus (= Felsö-Orbó), Romania (KOCH, 1887b; LÖRENTHEY, 1894; GAÁL, 1905; VADÁSZ, 1915; [MAFI]); Minişu des Sus (= Felménes), Arad, Romania (LÓCZY, 1877; VADÁSZ, 1915); region east of Taşad, Beiuş Basin, Romania (PAUCĂ, 1936)
Zala, Sáva and Dráva Basins: unnamed locality in former Yugoslavia (MITROVIĆ-PETROVIĆ, 1981)
Lom Basin: Opansko bardo, Northern Bulgaria (KOJUMDGIEVA & STRACHIMIROV, 1960)

Mediterranean: Burdigalian to Tortonian
Western Mediterranean: Balearic Islands, Spain: Ciudaadèla, Menorca (LAMBERT, 1906a); Corsica, France: Bonifacio (COTTREAU, 1913a); France, mainland: Ste-Colombe, near Vence, Alpes-Maritimes (LAMBERT, 1915a); Italy, mainland: ? Carcare, Piemonte (BOTTO-MICCA, 1896), ? Cornino, Veneto (STEFANINI, 1919), Lauro, near Monte Gargano (D'ALESSANDRO et al., 1979), ? Montagnana, Modena (COPPI, 1884), ? Meduno, Veneto (STEFANINI, 1919), Posano, near Volterra, Prov. Pisa (MENESINI, 1967), ? Sottomonte, Veneto (STEFANINI, 1919), ? Vito d'Asio, Veneto (STEFANINI, 1919); Sardinia, Italy (LAMBERT, 1906a; COTTREAU, 1913a): Bonorva, Sassari (LAMBERT, 1907a, 1909), Biugia Targesi, Fangario (LAMBERT, 1907a), Cadreas, Sassari (LAMBERT, 1909), ? Cala Campu Sali (COMASCHI CARIA, 1963), Cap Sant'Elia (LAMBERT, 1907a), Castelsardo, Sassari (COTTEAU, 1895), Cuccuruddu, below Chemerule, Thiesi (COTTEAU, 1895; LAMBERT, 1909), Monte Alvu (LAMBERT, 1907a), Portotorres (DI GIORGIO, 1923); Spain, mainland: Calafell, Prov. Barcelona (LAMBERT, 1906a, 1927a), Monjos , Prov. Barcelona (COTTREAU, 1913a; LAMBERT, 1927a), Panadas, Prov. Barcelona (COTTREAU, 1913a)
Central Mediterranean: Jebel Korbous, Tunisia (BLONDEL & PHILIPPE, 1992)
Eastern Mediterranean: Cyrenaica, Libya: Bardia (TAVIANI, 1939), Bir Acroma (TAVIANI, 1939), Uadi el Hanaui (TAVIANI, 1939), Umm er Rzem (TAVIANI, 1939); ? Island of Crete, Greece: Pitsidhia, Iráklion and Achladhia, Sitia (MARCOPOULOU-DIACANTONI, 1972); Turkey: ? Kozağaç, E of Celâlli, near Sivas (STCHEPINSKY, 1939)
Rhône Basin: Burdigalian: Littoral de Nerthe (Cap Couronne, Carro near Carry, Martigues) (COTTREAU, 1913a; LAMBERT, 1915a; NEGRETTI, 1984; PHILIPPE et al., 1990; PHILIPPE, 1998), Secteur es étangs (Fos-sur-Mer, Istres) (PHILIPPE, 1998), Bas-Languedoc (Béziers, Bouzigues) (PHILIPPE, 1998), Bassin d'Apt-Reillanne-Forcalquier (Pierrerue, Saint-Michael-l'Observatoire, Villemus, Reillanne) (LAMBERT, 1915a; PHILIPPE, 1998), Bassin de Valréas-Visan (Bollène, Lacoste, St.-Paul-Trois-Châtteaux, St-Restitut) (LAMBERT, 1915a; PHILIPPE, 1998), Burdigalian or Langhian: Secteur es étangs (Fos-sur-Mer, Istres) (PHILIPPE, 1998), Bas-Languedoc occidental (Bouzigues, Loupian, Sète) (PHILIPPE, 1998), Bassin de Mus-Sommières (Aigues-Vives) (PHILIPPE, 1998), Vallée du Jabron (Châteauneuf-Miravail) (PHILIPPE, 1998), Serravallian: Bas-Languedoc occidental (Bouzigues) (PHILIPPE, 1998), Bassin de Faucon-Mollans-Malaucène (Mérindol-les-Oliviers) (PHILIPPE, 1998), Tortonian: Bordure Sud-Luberon (Cucuron) (PHILIPPE, 1998), Bassin de Valréas-Visan (Cairanne) (PHILIPPE, 1998)

Schizaster (*S.*) *laubei* HOERNES, 1875

(Figs. 66-68; Pl. 66, Figs. 2a-b; Pl. 67, Figs. 1-2)

v.	1869a	*Schiz.* spec. – LAUBE: 184
v.	1871	*Schizaster* spec. – LAUBE: 71
* v.	1875a	*Schizaster Laubei* nov. sp. – HOERNES: 385, 387-389; pl. 12, fig. 3; pl. 15, fig. 1a-d
v.	1875b	*Schizaster Laubei* nov. sp. – HOERNES: 211
v.	1877	*Schizaster Laubei* HOERN. jun. – FUCHS: 663
non	1882	*Schizaster Laubei* nov. spec. – BITTNER: 96; pl. 11, fig. 1, 1a-b [= *S. postalensis*, see BITTNER (1891: 140)]
non	1883	*Schizaster Laubei*, BITTNER – MAZZETTI & PANTANELLI: 80 [= *S. desori*, see STEFANINI (1909: 22-24)]
?	1884	*Schizaster* sp. (*Sch. Laubei* R. HOERN.?) – BITTNER: 457
?	1884	*Schizaster* sp. – BITTNER: 487
?	1896	*Schizaster Laubei* HÖRNES. – BOTTO-MICCA. 356-357
	1913	*Schizaster* cf. *Laubei* HÖRN. – DREGER: 70
	1913	*Schizaster Laubei* R. HOERN. – HILBER: 231
	1925	*Schizaster Laubei* – SCHWINNER: 197
	1927	*Schisaster Laubei* HÖRN. R. – HORUSITZKY: 25, 165
	1930	*Schizaster laubei* R. HOERN. – VENDL: 50
	1936	*Schizaster laubei* R. HÖRN. – MEZNERICS: 119
	1943	*Schizaster laubei* R. HÖRN. – VEIT: 6
	1951	*Schizaster laubei* R. HOERN. – SCHAFFER & GRILL: 707
	1953a	*Schizaster laubei* HÖ.R. – SIEBER: 207
?	1955	*Schizaster laubei* R.HÖRN. – SENEŠ: 6
?	1955	*Schizaster ventiensis* LAMB. – SENEŠ: 6, 26
?	1955	*Schizaster sardiniensis* COTT. – SENEŠ: 6
	1956	*Schizaster laubei* R. HÖRN. – SIEBER: 317
	1960	*Schizaster latipetalus* VAD. – GIVULESCU & DUŞA: 936-937; figs. 4, 6 [misidentification; the legends of fig. 3 and 6 were confused]
v.	1962	*Schizaster laubei* – PAPP & THENIUS in KÜHN: 330
?	1967	*Schizaster laubei* R. HOERN. – CICHA et al.: 93
?	1970	*Schizaster* cf. *laubei* HOERN. – BAGASARYAN: 17, 23, 128; pl. 7, figs. 3-4
?	1971	*Schizaster* cf. *laubei* R. HÖRNES – SENEŠ: 201
?	1974	*Schizaster* cf. *laubei* R. HOERNES – KALABIS: 314
v.	1975	*Schizaster laubei* R. HOERNES 1875 – STOJASPAL: A193
	1978	*Schizaster laubei* HOERN. – CICHA: 170
?	1980	*Schizaster* aff. *Laube* HÖRNES? – WANK & STOJASPAL: 448
?	1981	*Schizaster* cf. *laubei* R. HÖRNES – WANK: 382; pl. 3, fig. 5
	1991	*Schizaster laubei* HOERNES – RUPP et al.: 40
v.	1998	*Schizaster laubei* HOERNES – SCHULTZ: 120; pl. 54, fig. 1
	2003	*Schizaster laubei* – WEIDLINGER: 199, unnumbered fig.
non	2003	*Schizaster laubei* – WEIDLINGER: 199, unnumbered fig. [misidentified *Brissopsis ottnangensis*]
	2003b	*Schizaster laubei* HOERNES, 1875 – KROH: 251

Type-material:
Schizaster laubei HOERNES 1875:
Syntypes: 1 specimen figured by HOERNES (1875a: pl. 12, fig. 3; pl. 13, figs. 1a-d) in the collection of the Naturhistorisches Museum Wien (Geologische Abteilung; NHMW 1854.II.149), and 3 specimens in the collection of the Geological Survey of Austria (GBA 1875/01/76a-c)
Lectotype (Figs. 66.A, 67; Pl. 67, Figs. 1a-c, 2a-b): the best-preserved specimen (NHMW 1854.II.149a-c) figured by HOERNES is proposed as lectotype

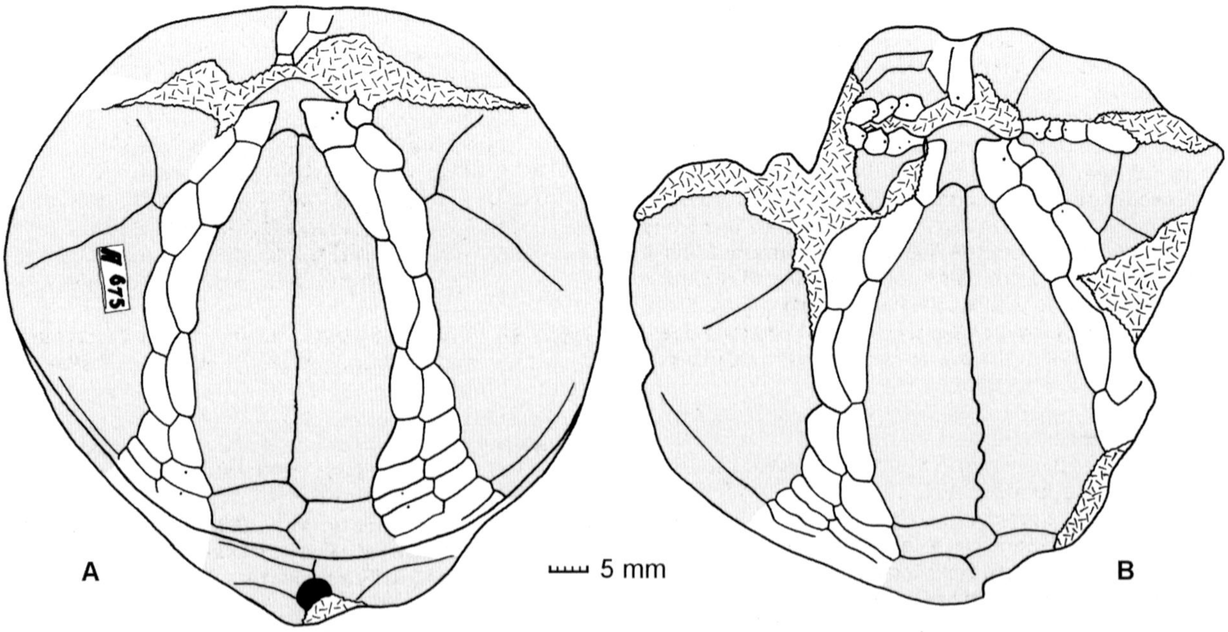

Figure 66: *Schizaster* (*S.*) *laubei* Hoernes, 1875a: oral plating (A: lectotype, Ottnang, OÖ, NHMW 1854.II.149b; B: Wetzelsdorf in der Weststeiermark, Stmk, NHMW 2004z0096/0002). Interambulacra shaded; poorly visible sutures stippled; damaged areas hatched.

Locus typicus: Ottnang, OÖ, Austria
Stratum typicum: Ottnang Schlier
Age: Ottnangian (Late Burdigalian), Early Miocene

Material:
Ottnangian (Late Burdigalian) – Ottnang, OÖ, Austria
NHMW: 2 specimens [NHMW 1854.II.149a-c (**holotype**; 1 mould of the aboral side, 1 partial mould of the oral side and 1 internal cast, all of a single specimen), NHMW A677]
GBA: 3 specimens [GBA 1875/01/76a (mould and internal cast) , 1875/01/76 (internal cast), 1875/01/76c (mould and internal cast)]
Early ? Badenian (Langhian) – Bad Vöslau, NÖ, Austria
NHMW: 2 internal casts (NHMW 2004z0103/0004)
Early Badenian (Langhian) – Gainfarn, NÖ, Austria
NHMW: 1 specimen (NHMW 2004z0154/0001)
Early Badenian (Langhian) – Weitendorf, Styria, Austria
NHMW: 2 specimens (NHMW 2004z0097/0003-4)
Early Badenian (Langhian) – Wetzelsdorf in der Weststeiermark, Styria, Austria
NHMW: 3 specimens (NHMW 2004z0096/0001-3)
? Badenian (Langhian-Early Serravallian) – borehole Matzen 325 (1909-1917 m), Austria
NHMW: 1 specimen (NHMW 2004z0101/0001)

Foreign material for comparison:
Badenian (Langhian-Early Serravallian) – Mogyoród, Hungary
NHMW: 3 specimens (NHMW 1997z0178/2091)

Dimensions (in mm):

Inv. no.	TL	TW	TH
NHMW 1854.II.149b	77.4	77.9	>>26

All other specimens are crushed and reliable dimensions can only be given for the lectotype.

Description:
Size and shape: Test of large size ranging from 50 to 90 mm test length in the studied material. The outline is broadly heart shaped to slightly antero-posteriorly elongated. The anterior margin is rounded with moderately deep frontal sinus. The posterior margin is bluntly pointed. The maximum width lies anterior of the centre, about at the level of the end of the anterior paired petals. In profile the test is strongly wedge-shaped with its maximum height along the keel in interambulacrum 5. A distinct rostrum overhangs the periproct.
Apical disc: The apical disc is preserved only in the type specimen. It is ethmolytic with 4 gonopores. The anterior gonopores are slightly smaller than the posterior ones. The madreporite extends well beyond the posterior ocular plates. The apical disc lies posterior of the centre, about 60 to 65 % of TL from the anterior margin.
Ambulacra: The ambulacra are all petaloid and deeply sunken. Ambulacrum III has overhanging walls and closes slightly distally. The pores are numerous and arranged in two straight rows, with up to 35 pore pairs in the largest specimens. The anterior paired petals are moderately long, relatively broad and strongly flexed anteriorly. Contrary to *S. eurynotus* their tips are not flexed laterally. The posterior paired petals are shorter and extend about 60 % of the length of the anterior petals. The anterior paired petals form an acute angle of about 70-80° with each other, the posterior ones an angle between 50-60°. Both anterior and posterior paired petals are broadest at about two third their length from the apical disc. The poriferous zones of the paired petals consist of conjugate elongate isopores. The interporiferous zones are slightly smaller than a single poriferous zone and bear only secondary tubercles. On the oral surface ambulacra I and V form long narrow peri-plastronal areas lacking primary tubercles.
Interambulacra: The interambulacra are strongly inflated adapically, forming high keels along the sunken petals. The aboral tuberculation is fairly homogenous, consisting of small perforate crenulate primary tubercles with distinct elevated bosses and inclined areoles. The tubercles become larger towards the ambitus. On the oral side primary tubercles are larger. The plastron is large and belongs to the mesamphisternous type. It is about 40 % of TW wide and shows a confluent fan shaped tuberculation radiating from the posterior margin. The labrum is moderately wide and rather short, with a triangular proximal part (see Figs. 66.A-B).
Peristome: The peristome is situated in the anterior half of the oral side close to the anterior margin of the test. It is kidney-shaped with a protuberant rounded labrum bearing a distinct rim and faces anteriorly.

Periproct: The periproct is situated marginally, high on the vertical posterior face of the test. It is overhung by a distinct rostrum and has an oval, vertically elongated shape. Below the peristome there is a shallow subanal depression.
Fascioles: The peripetalous fasciole is very broad, especially in the anterior part of the test. It is widest at the tips of the petals (up to 3.5 mm) and is narrower in the interambulacra. The peripetalous fasciole forms a slight outwards angle in interambulacrum 5, is deeply indented in interambulacra 2 and 3 and forms a more or less straight line in interambulacra 1 and 4 (Figs. 66.A-B). The latero-anal fasciole branches off from the peripetalous fasciole at about the first third along petal II and III running laterally towards the periproct and forming a deep, narrow V beneath it. It is much narrower than the peripetalous fasciole.

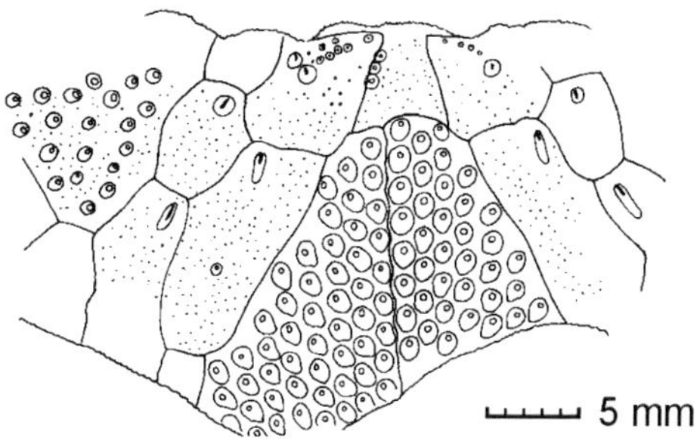

Figure 67: *Schizaster (S.) laubei* HOERNES, 1875a: adoral tuberculation (lectotype, Ottnang, OÖ, NHMW 1854.II.149a).

Differential diagnosis:
This species differs from *Schizaster scillae* (DES MOULINS, 1837) from the Messinian of Malta by its longer and differently shaped posterior petals, the presence of relatively large anterior gonopores (those of *S. scillae* are minute according to CHALLIS, 1980) and a more deeply embayed (in interambulacra 1 and 4) peripetalous fasciole. Otherwise, the two species are, however, rather similar and might be closely related. Unfortunately few good descriptions and illustrations of *S. scillae* are published, hampering comparison (compare LAMBERT, 1907a and CHALLIS, 1980).

Schizaster parkinsoni (DEFRANCE) a common Early to mid Miocene species in the Mediterranean region differs by its more strongly diverging anterior petals, which nearly form a right angle and its less wedge-shaped profile (compare CHALLIS, 1980; ROSE & WATSON, 1998).

Schizaster desori WRIGHT, 1855, another common Early to mid Miocene species, differs by its laterally flexed anterior paired petals (respectively their tips), minute anterior gonopores, shorter posterior paired petals, more elongated outline and shallower frontal sinus (compare CHALLIS, 1980).

Schizaster eurynotus SISMONDA, 1841 differs by its less diverging anterior petals with laterally flexed tips, deeper frontal sinus, more elongated outline, shorter posterior petals, indentation of the peripetalous fasciole in interambulacra 2 and 3, and the minute anterior gonopores (if present at all).

Schizaster karreri LAUBE, 1869 differs by the presence of a distinct subanal heel, lack of anterior gonopores, less rostrate posterior end, its more tumid anterior margin (in lateral view), shorter posterior paired petals, subcentral maximum width, slightly more central apical disc, and the broader aboral ambulacrum III.

Discussion:
This species is common in the "Schlier"-facies of the Lower to Middle Miocene in the Central Paratethys. Although the outline of the peripetalous fasciole and the shape of the anterior paired petals is very similar to *Schizaster karreri* LAUBE, 1869 it is unlikely that the two are conspecific, even if changes due to allometric growth (see above under *S. karreri*) are taken into account. The features outlined above (under "Differential diagnosis") and in particular the presence of large anterior gonopores (and the complete lack of anterior gonopores in *S. karreri*) differentiate the two species. Although, anterior gonopores may be developed in species which usually lack them (e.g. *S. eurynotus* or *S. compactus*), they are minute when present.

Schizaster latipetalus VADÁSZ, 1915 from the Lower Miocene "Schlier"-facies of Northern Hungary might seem related at the first sight. Re-examination of the holotype (which is an internal cast), however, showed that the two are not conspecific. *S. latipetalus* has much longer and broader petals, nearly reaching the ambitus, only two gonopores, and a very different overall shape (the specimen is only slightly distorted).

From the Mediterranean no similar *Schizaster* species has been reported until now, although many species are very poorly illustrated and comparison is difficult [e.g. the *Schizaster* and *Hemiaster* species figured in MAZZETTI & PANTANELLI (1883), ...]. There is only a single record of *S. laubei* from the Miocene of the Colli Torinesi by BOTTO-MICCA (1896). This record, however, lacks sufficient information to evaluate its correctness. For the differences to *S. scillae* see above.

Occurrence:

Austria: ? Eggenburgian (Early Burdigalian), Ottnangian (Late Burdigalian), Early Badenian (Langhian)

Molasse Zone: borehole Eisenhub 2, OÖ (SCHAFFER & GRILL, 1951); Ort, near Bad Füssing, OÖ (WEIDLINGER, 2003); Ottnang, OÖ (LAUBE, 1869a, 1871; HOERNES, 1875a, b; FUCHS, 1877; VEIT, 1843; SIEBER, 1956; PAPP & THENIUS in KÜHN, 1962; STOJASPAL, 1975; RUPP et al., 1991; SCHULTZ, 1998)

Vienna Basin: Gainfarn, NÖ ([NHMW]); Steinberggebiet, Zistersdorf, NÖ (SIEBER, 1953a)

Lavant Basin: ? Mühldorf, Lavanttal, Ktn (WANK & STOJASPAL, 1980; WANK, 1981)

Styrian Basin: Jahring, probably Slovenia (MEZNERICS, 1936); Kollitschberg, near Weißenegg, Styria (HILBER, 1943); Perbersdorf, Styria (SCHWINNER, 1925); Repnik, probably Slovenia (MEZNERICS, 1936); region Wildon-Leibnitz, Styria (DREGER, 1913)

Paratethys (non-Austrian occurrences): Eggenburgian (Early Burdigalian) to Badenian (Langhian-Early Serravallian)

Molasse Zone: ? borehole ČČ-3, Vel'ká Čausa, near Prievidza, western Slovak Republic (SENEŠ, 1971) ; Židlochovice, Moravia, Czech Republic (CICHA, 1978)

Great Hungarian Basin (Pannonian Basin): Györ, Györ-Monson-Sopron, Hungary (VENDL, 1930); Mogyoród, near Fót, Pest, Hungary (HORUSITZKY, 1927; KROH, 2003b)

Transcarpathian Basin: ? Hlinné, eastern Slovakia (SENEŠ, 1955; CICHA et al., 1967; KROH, 2003b)

Transylvanian Basin: Coştei (= Kostéj), Timiş, Romania (GIVULESCU & DUŞA, 1960)

Zala, Sáva and Dráva Basins: ? Trbovlje (formerly Trifail), Slovenia (BITTNER, 1884); ? La'sko (formerly Tüffer; "Schlier" facies), Slovenia (BITTNER, 1884)

Eastern Paratethys: ? Гориса, Хорщи, Western Georgia (BAGASARYAN, 1970)

Mediterranean: Middle Miocene

Western Mediterranean: ? Colli Torinesi, Italy (BOTTO-MICCA, 1896)

Schizaster sp.

1855 *Schizaster* – ROLLE: 354
1871 *Schizaster* – STUR: 562, 563
1877 *Schizaster* sp. – HILBER: 259, 260, 262
1877 *Schizaster* sp. – KARRER: 305

	1881	*Schizaster* – FRANZENAU: 32, 84
	1897	*Schizaster* spec. ind. – SCHAFFER: 535, 547
	1905	*Schizaster* sp. – GAÁL: 343, 344, 362
	1906	*Schizaster* sp. – SCHAFFER: 63
	1908	*Schizaster* – SCHAFFER: 30
	1915	*Schizaster* spec. – TOULA: 672
	1927	*Schisaster* sp. – HORUSITZKY: 25, 165
	1931	*Schizaster* spec. indet. – JANOSCHEK: 83
	1939	*Schizaster* sp. – KALABIS: 90; fig. 2.
	1940	*Schizaster* sp. – JASKÓ: tab. 11
	1943	*Schizaster* sp. – VEIT: 10
	1965	*Schizaster* – KOLLMANN: 516
	1967	*Schizaster* sp. – CICHA et al.: 87, 88
	1969	*Schizaster* sp. – MIHÁLY: 257
	1970	*Schizaster* – THENIUS: p. 215
	1977	*Schizaster* – THENIUS: 130; fig. 51
	1979	*Schizaster* – THENIUS: 109; fig. 57
	2001	*Schizaster* – GONCHAROVA et al.: 518
	2002	*Schizaster* – GONCHAROVA et al.: 197, 200
	2002	*Schizaster* sp. – KAZÁR: 153; fig. 1
v	2002b	*Schizaster* sp. – KROH: 11
	2004	*Schizaster* – BOJAR et al.: 98
	2004	*Schizaster* – RADWAŃSKI & WYSOCKA: 384

Material:

Early Badenian (Langhian) – Aflenz, Styria, Austria
NHMW: 8 specimens (NHMW 1978/2023/1-8)
Early Badenian (Langhian) – Bad Vöslau, NÖ, Austria
NHMW: 2 internal casts (NHMW 1933.X.108)
Badenian (Langhian-Early Serravallian) – Haschendorf, near Neckenmarkt, Bgld, Austria
NHMW: 1 specimen (NHMW 1848.III.63)

Discussion:

The genus *Schizaster* occurs abundantly in pelitic siliciclastics and carbonate mud throughout the Neogene of Austria. Many records are based on poorly preserved and/or fragmented material that does not allow specific determination. In most cases, these remains will probably belong to one of the three *Schizaster* species discussed above.

Occurrence:

Austria: Ottnangian (Late Burdigalian), Badenian (Langhian-Early Serravallian)
Vienna Basin: borehole Lichtenwarth 1, NÖ (VEIT, 1943); Gainfarn, NÖ (KROH, 2002b); Ottakring, Vienna (SCHAFFER, 1906); Perchtoldsdorf (formerly Berchtoldsdorf), NÖ (KARRER, 1877); Rauchstallbrunngraben, near Baden, NÖ (KALABIS, 1939); Stotzing (sandpit Mayer), Bgld (KAZÁR, 2002); Vienna (THENIUS, 1970); Vienna Basin (THENIUS, 1977, 1979)
Oberpullendorf Bay: Neckenmarkt, Bgld (JANOSCHEK, 1931)
Styrian Basin: Grubthal (Weinleiten), near Gamlitz, Styria (HILBER, 1877); Retznei (Weissenegg Fm., Lafarge quarry), Styria (BOJAR et al., 2004); Spielfeld, Styria (ROLLE, 1855; STUR, 1871); Wagna (brickyard), Styria (KOLLMANN, 1965)

Central Paratethys (non-Austrian occurrences): Oligocene, Ottnangian (Late Burdigalian) to Late Badenian (Early Serravallian)
Vienna Basin: Devínska Nová Ves, Slovak Republic (SCHAFFER, 1897, 1908; TOULA, 1915); Dubovce, *c.* 7 km ESE Holíč, Slovak Republic (CICHA et al., 1967); Radošovce, *c.* 9 km ESE Holíč, Slovak Republic (CICHA et al., 1967)
Great Hungarian Basin (Pannonian Basin): Budapest-Rákos, Pest, Hungary (FRANZENAU, 1881; MIHÁLY, 1969); Hor. Strháre (= Felsö-Esztergály), Slovak Republic (GAÁL, 1905); Mogyoród, near Fót, Pest, Hungary (HORUSITZKY, 1927); Sajóvárkony, near Ózd, Borsod-Abaúj-Zemplén, Hungary (JASKÓ, 1940)
Fore-Carpathian Basin: Świniary, Poland (RADWAŃSKI & WYSOCKA, 2004)
Zala, Sáva and Dráva Basins: Dráva valley, near Melling, E of Maribor (Marburg), Slovenia (STUR, 1871)

Eastern Paratethys: Mid-Tarkhanian (Late Burdigalian) to Chokrakian (Langhian)
Mid-Tarkhanian: Dzirul massif (Gorisa and Khorshi areas); western and eastern Ciscaucasia; Krasnovodsk plateau, Fore-Kopetdag depression (GONCHAROVA et al., 2001)
Chokrakian: Georgia and Transcaspian region (GONCHAROVA et al., 2002)

Genus *Aliaster* VALDINUCCI, 1974

Type-species: *Opissaster lovisatoi* COTTEAU, 1895; by original designation; regarded as junior synonym of *Hemiaster cotteauii* WRIGHT, 1855 in the present paper.
Emended diagnosis: Test subglobular with subcircular outline; high cushion- to wedge-shaped profile, obliquely or vertically truncated posterior end, ethmolytic apical disc with (2 or) 4 genital pores (see remarks below), deeply sunken frontal am-

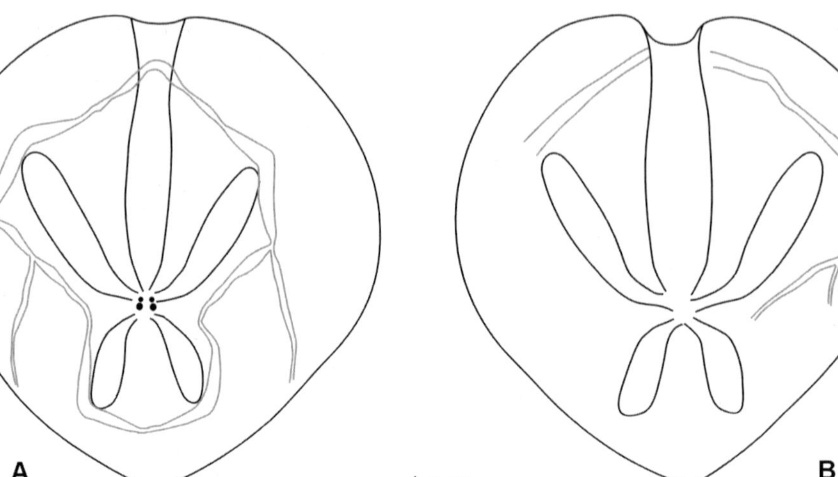

Figure 68: *Schizaster* (*S.*) *laubei* HOERNES, 1875a: fasciole outline (A: lectotype, Ottnang, OÖ, NHMW 1854.II.149a; B: Ottnang, OÖ, GBA 1875/1/76)].

bulacrum, very shallow frontal sinus; sunken paired petals, the posterior ones of which are distinctly shorter than the anterior ones; a peripetalous fasciole only, which is deeply indented in interambulacra 1 and 4 and forms a right angle in interambulacra 2 and 3 (modified from VALDINUCCI, 1974a).
Distribution: Early Miocene to Pliocene – Mediterranean region (VALDINUCCI, 1974a)
Remarks: The number of genital pores is not known for many of the species included in this genus by VALDINUCCI (1974a). Some reports state there "seem to be not more than two" (e.g. COTTEAU et al., 1891) despite noting the poor preservation of the apical disc. In fact, in those species where the apical disc is known [*A. cotteauii* (WRIGHT, 1855), *A. lovisatoi* (COTTEAU, 1895), *A. aichnoi* (CHECCHIA-RISPOLI, 1927)] it is ethmolytic with 4 gonopores. Whether or not there are species with 2 gonopores in this genus remains to be shown.

The genus *Trachyaster* POMEL, 1869 (type-species: *Trachyaster globosus* POMEL, 1869, p. 14, by monotypy) differs from *Aliaster* by its more "hemiasteriform" shape with vertically truncated posterior end, more diverging paired petals and its peripetalous fasciole which is hardly indented in the interambulacra. The genus *Opissaster* POMEL, 1869 (type-species: *Opissaster polygonalis* POMEL, 1883, p. 37, by original designation) differs by its more elongated, oval outline, its petaloid area which is displaced anteriorly, an apical disc with 2 gonopores, at least in the type species, occasional presence of a latero-anal fasciole, and its peripetalous fasciole, which is only very slightly indented between the frontal and the anterior paired petals.

Traditionally both *Trachyaster* and *Opissaster* were referred to the family Hemiasteridae. *Aliaster* was likewise placed into this family by VALDINUCCI (1974a). SMITH & JEFFERY (2000), however, transferred the genus *Trachyaster* from the Hemiasteridae to the family Schizasteridae. Here the genus *Aliaster* is referred to the Schizasteridae too, on base of the ethmolytic apical disc and overall "schizasteriform" test shape (it can indeed be difficult to distinguish *A. cotteauii* from e.g. *Schizaster eurynotus*, when deformed or otherwise not optimally preserved).

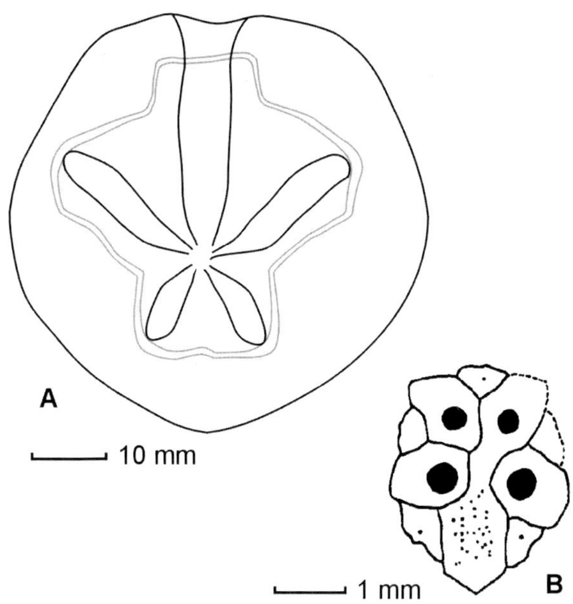

Figure 69: *Aliaster cotteauii* (WRIGHT, 1855): A: outline of the peripetalous fasciole (grey); B: apical disc (Early Badenian, Stotzing, Bgld, WANZENBÖCK coll. W20)

Aliaster cotteauii (WRIGHT, 1855)

(Fig. 69; Pl. 64, Figs. 2-3)

* 1855 *Hemiaster Cotteauii*, WRIGHT. – WRIGHT: 190-191; pl. 7, figs. 2a-d
1858 [*Hemiaster*] *Cotteauii*, WRIGHT – DESOR: 375
. 1864 *Hemiaster Cotteaui*, WRIGHT. – WRIGHT: 483
? # 1869a *Pericosmus affinis* LAUBE. – LAUBE: 183
? 1870 *Pericosmus affinis* LBE. – LAUBE: 314
? 1871 *Pericosmus affinis* LAUBE. – LAUBE: 68; pl. 17, fig. 2.
? 1882a *Hemiaster Cotteaui*, WRIGHT. – MAZZETTI: 117
1883 [*Opissaster*] *Cotteaui* – POMEL: 38
non 1883 *P.*[*ericosmus*] *affinis*, LAUB. – MAZZETTI & PANTANELLI: 67 [= *Trachyaster airaghii* acording to STEFANINI (1909: 75)]
1891 *Hemiaster cotteaui*, WRIGHT, 1855. – GREGORY: 610-611
. # 1895 *Opissaster Lovisatoi*. COTTEAU, 1895. – COTTEAU: 48; pl. 5, figs. 6-8
? # 1895 *Opissaster Mariæ*. LOVISATO, 1895 – COTTEAU: 49; pl. 5, figs. 4-5
. pp 1906a *Opissaster almerai* LAMBERT, 1906. – LAMBERT: 100; pl. 9, figs. 4-6 (specimens from the Miocene of Sardinia)
non 1906a *Opissaster almerai* LAMBERT, 1906. – LAMBERT: 100; pl. 7, fig. 5 (specimens from the Miocene of Spain)
1909 *Opissaster Lovisatoi* COTTEAU, 1895. – LAMBERT: 77-78
1909 *Opissaster Almerai* LAMBERT, 1906. – LAMBERT: 78-80
1909 *Opissaster Cotteaui* WRIGHT (*Hemiaster*), 1855. – LAMBERT: 80
. 1909 *Trachyaster Lovisatoi* (COTT.) – STEFANINI: 73-74; pl. 2 (4), figs. 9a-b
1913a *T.* [*rachyaster*] *Lovisatoi* COTT. (*Opissaster*). – COTTREAU: 67
1913a *O.* [*pissaster*] *Cotteaui* WRIGHT (*Hemiaster*). – COTTREAU: 70
? 1913a *Pericosmus affinis* LAUBE – COTTREAU: 70
1915a *Trachyaster Lovisatoi* COTTEAU (*Opissaster*), 1895. – LAMBERT: 157-158
1915 *Trachyaster Lovisatoi* COTT. sp. – VADÁSZ: 225; fig. 115
v. 1915 *Trachyaster Cotteaui* WRIGHT sp. – VADÁSZ: 226; fig. 116
1924 *O.*[*pissaster*] *Cotteaui* WRIGHT (*Hemiaster*) – LAMBERT & THIÉRY: 510
1924 *T.*[*rachyaster*] *Lovisatoi* COTTEAU (*Opissaster*) – LAMBERT & THIÉRY: 510
. 1927a *Trachyaster Lovisatoi* COTT. – CHECCHIA-RISPOLI: figs. 3, 3a-c
? 1955 *Trachyaster cotteaui* WRIGHT. ex. aff. – SENEŠ: 6, 26
? 1955 *Trachyaster* (*Schizaster*)? sp. – SENEŠ: 6
? 1967 *Trachyaster cotteaui* WRIGHT. ex. aff. – CICHA et al.: 93
. 1974a *Aliaster* (*Trachyaster*) *lovisatoi* COTTEAU – VALDINUCCI: 463-468; pl. 1, figs. 1a-c, 2a-c, 3a-c
. 1974a *Aliaster* (*Opissaster*) *cotteaui* WRIGHT 1855 – VALDINUCCI: 468-471
1974b *Hemiaster cotteaui* – ROSE: 353, fig. 3
1975 *Hemiaster cotteaui* WRIGHT – ROSE: 79; tab. 12
pp 1980 *Hemiaster (Trachyaster) cotteaui* (WRIGHT) – CHALLIS: 190-194; pl. 71, fig. c; pl. 73, figs. a-c; pl. 74, fig. A
1984 *Hemiaster cotteaui* – BOGGILD & ROSE: 61, 64; figs. 2, 3
1994 *Opissaster cotteaui* – NÉRAUDEAU: 329; tab. 4
1994 *H. (Trachyaster) lovisatoi* – NÉRAUDEAU: 329; tab. 4
v. 2002 *Hemiaster* sp. – KAZÁR: 153; fig. 1
v. 2003 *Trachyaster ? cotteaui* (WRIGHT, 1855) – KROH & NEBELSICK: 166-167; fig 5c

Type-material:
Hemiaster cotteauii WRIGHT, 1855:
Holotype: BMNH E1584 (figured by WRIGHT, 1855: pl. 7, fig. 2a-d); British Museum of Natural History, London
Locus typicus: Malta
Stratum typicum: *Globigerina* Limestone
Age: Aquitanian-Langhian, Early to Middle Miocene
Remarks: According to ROSE (1974b, 1975) and CHALLIS (1980) the occurrence is restricted to the Lower *Globigerina* Limestone.

Opissaster lovisatoi COTTEAU, 1895:
Syntypes: private collection of LOVISATO and collection of COTTEAU (not seen)
Type-area: Sardinia, Italy [two localities are mentioned by COTTEAU (1895): Mont San Michele, near Cagliari and Cadreas, Bonorva]

? *Opissaster mariae* COTTEAU, 1895:
Holotype: the specimen figured by COTTEAU (1895: pl. 5, figs. 4-5); private collection of LOVISATO (not seen)
Locus typicus: Cagliari, Sardinia
Stratum typicum: "Calcaire compact (tramezzario)"
Age: Early or Middle Miocene

? *Pericosmus affinis* LAUBE, 1869:
Holotype: a single specimen [figured by LAUBE (1871: pl. 17, fig. 2)] from the collection of the "k. k. geol. Reichsanstalt" (Geological Survey of Austria); specimen lost
Locus typicus: Sievering, Vienna
Stratum typicum: Leitha limestone
Age: Badenian (Langhian-Early Serravallian)
Remarks: The holotype of this species could not be located in the collections of the Geological Survey of Austria. According to Franz STOJASPAL (former head of the palaeontological collection, pers. comm. 2003) part of the collection was destroyed during the Second World War, among it also part of the type-collection. The holotype of *P. affinis* is thus considered lost.

Material:
Early Badenian (Langhian) – Stotzing (sandpit Mayer), Bgld, Austria
NHMW: 1 specimen (NHMW 2004z0093/0018)
WANZENBÖCK coll.: 2 specimens (W20, W21)
Late Badenian (Early Serravallian) – Müllendorf (Mühlen-dorfer Kreide AG quarry), Bgld, Austria
NEITZ coll.: 1 specimen (no inv. no.)

Foreign material for comparison:
Late Badenian (Early Serravallian) – Gârbova de Sus (= Felsö-Orbó), Romania
MAFI: 1 specimen [MAFI Ech 245 (figured as *T. cotteaui* by VADÁSZ, 1915)]

Dimensions (in mm):

Inv. No.	TL	TW	TH
W20	56.1	62.2	40.4
W21	67.5	66.5	-

Description:
Size and shape: Test large, subcircular to slightly heart shaped. Anterior margin rounded with very shallow frontal sinus. Posterior margin rounded to slightly bluntly pointed. The maximum width lies slightly posterior of the centre, coinciding with the apical disc. In profile the test is high, arched to slightly wedge shaped with tumid ambitus and obliquely truncated posterior end. The maximum height lies slightly posterior of the centre, around the apical disc.
Apical disc: The apical disc is situated slightly posterior of the centre. It is ethmolytic with four large genital pores. The ocular pores are small and have a circular outline. The anterior gonopores are slightly smaller than the posterior ones (Fig. 69.B).
Ambulacra: Ambulacrum III deeply sunken with undercut walls adapically. On the aboral surface it is straight and closes slightly outside the peripetalous fasciole. Towards the ambitus it becomes distinctly shallower, forming only a very shallow frontal sinus. The pores are oblique partitioned anisopores with laterally positioned neural canal (compare SMITH, 1980c) and are arranged in two straight rows. The interporiferous zone is broad with dense secondary tuberculation. Outside the peripetalous fasciole the pores are minute constricted unipores (compare SMITH, 1980c). Adorally ambulacrum III forms a slightly depressed phyllode consisting of three unipores with extensive periporal area.

The paired ambulacra are petaloid and deeply sunken. The anterior pair forms an angle of around 105°, the posterior pair an angle of about 60 to 70°. The distal ends of the anterior paired petals flex slightly laterally. The posterior petals are about half as long as the anterior pair. The pores are conjugated anisopores and separated from each other by low ridges. The interporiferous zone is narrow and smaller than the single poriferous zone. The pores are smaller and more rounded at the distal ends of the petals.

Adorally, ambulacra II and IV form distinct, large phyllodes, which lack primary and bear few secondary tubercles. Each bears up to nine constricted unipores with extensive periporal areas. The pores are slit-like, broadest at their adoral end, where a shallow but distinct neural canal is found. Ambulacra I and V form sparsely tuberculated peri-plastronal areas on the oral surface. Adorally they form small, slightly depressed phyllodes similar to those of ambulacra II and IV.
Interambulacra: On the aboral surface, the interambulacra are highly inflated around the apical disc and along the petals. They all form high, arched keels, except in interambulacrum 5. The tuberculation is dense and homogeneous. Small perforate, crenulate primary tubercles with inclined areoles forming a confluent pattern are present. Secondary tubercles are sparse. The primary tubercle slowly become larger towards the petals and the margin of the test. The largest tubercles are found at the border of ambulacrum III.

The primary tubercles are larger and more widely spaced on the oral surface, with simple secondary tubercles spread among them. The primary tubercles have a rounded perforate mamelon with undercut neck, distinct crenulate platforms and flat to slightly concave, inclined areoles. The largest tubercles are found on the anterior half of the margin of the oral side and near the peristome.

The plastron is mesamphisternous, broad, with a rounded anterior end. Its tuberculation is denser than on the remaining oral surface. Only primary tubercles, forming a fan-shaped pattern, radiating from the posterior end are present; they increase in size from the posterior to the anterior end.
Peristome: A large peristome is situated near the anterior margin of the oral side (about 20 % of TL from anterior margin). It is kidney-shaped and has a large protuberant rounded labrum with a distinct rim.
Periproct: The periproct is large, marginally situated and overhung by a small rostrum. It is oval in shape and vertically elongated.
Fascioles: The peripetalous fasciole band is very broad, especially at the distal ends of the petals and in ambulacrum III. The fasciole is deeply indented between the anterior and posterior paired petals; somewhat lesser between the anterior paired petals and the frontal petal and only slightly between the posterior petals (Fig. 69.A). No lateral or subanal fascioles are present.

Differential diagnosis:
Aliaster ? almerai LAMBERT, 1906 from the Miocene of Vence in France can be distinguished from the species considered here by its spherical form (length, width and height are all sub-

equal); proportionally higher interambulacral plates on the ambitus, a shallower ambulacrum III, which is relatively wide at the ambitus; a very shallow frontal sinus; a more acute angle between the anterior paired petals; shorter posterior paired petals and the presence of only two gonopores in the apical disc (compare LAMBERT, 1906a, 1906b, 1909). Sardinian material attributed to *Aliaster almerai* (LAMBERT, 1909) (under the genus name *Opissaster*) belongs to *A. lovisatoi* (COTTEAU, 1895) according to LAMBERT (1915a: 158-159; under the name *Trachyaster*), a species considered synonymous to *A. cotteauii* (WRIGHT, 1855) here. LAMBERT caused also considerable confusion by figuring Sardinian material of *A. cotteauii* under the name "*Opissaster almerai*" in his monograph on the echinoids of the Province Barcelona, Spain (LAMBERT, 1906a). The specimens figured on pl. 9, fig. 4-6 in that work (LAMBERT, 1906a), originate from Sardinia and belong to *A. cotteauii*, whereas the specimen figured on pl. 7, fig. 5 in LAMBERT (1906a) originates from Spain and is conspecific with *A.* ? *almerai* (see LAMBERT, 1915a: 157-159).

Aliaster insignis (POMEL, 1887), and its junior synonym *A. jourdyi* (PÉRON & GAUTHIER in COTTEAU et al., 1891) (according to LACHKHEM & ROMAN, 1995: 11) from the Messinian of Algeria and Morocco are very similar to *A. cotteauii*, and indeed LAMBERT (1909: 80) synonymised *A. jourdyi* with *A. cotteauii*. LACHKHEM & ROMAN (1995), however, list that species as *A. insignis*. The information available to the present author is insufficient to decide if both *A. jourdyi* and *A. insignis*, one of them or non are synonymous to *A. cotteauii*. Fact is that they differ by their slightly smaller petaloid area, lower test height and narrower frontal petal.

Another species attributed to the genus *Aliaster* by VALDINUCCI (1974a) is *Opissaster cotteri* DE LORIOL, 1896, which is known only from a single poorly preserved specimen. Based on DE LORIOL's description this species seems to differ by its longer, laterally flexed posterior petals and more strongly indented peripetalous fasciole from *A. cotteauii*.

Aliaster aichinoi (CHECCHIA-RISPOLI, 1927) from the Middle Miocene of Cyrenaica, Libya differs by its strongly flexed anterior petals, more elongated test, and less diverging paired petals.

Another species somewhat similar to *A. cotteauii* considered here is *Gregoryaster coranguinum* (GREGORY, 1891), which cooccurs with *A. cotteauii* in Malta. *G. coranguinum* differs by its highly arched profile, long paired petals, which are subequal in length; two distinct protuberances on the posterior end of the plastron; an ethmophract apical disc which is situated centrally or slightly anterior of the centre and the not indented anterior part of the peripetalous fasciole (compare GREGORY, 1891 and CHALLIS, 1980).

Discussion:

The Austrian material is clearly conspecific with material from the Lower to Middle Miocene *Globigerina* Limestone of Malta described as *Hemiaster cotteauii* WRIGHT, 1885. The generic placement of this species has been subject to some discussion. First it was transferred to the genus *Opissaster* by LAMBERT (1909: 80), later to the genus *Trachyaster* by CHALLIS (1980: 190). Recently it was transferred back to *Opissaster* by NÉRAUDEAU (1994: 329, tab. 4). KROH & NEBELSICK (2003) placed it tentatively into the genus *Trachyaster*. This is supported by the presence of relatively short petals, only a peripetalous fasciole, the shallow frontal sinus, the posteriorly eccentric, ethmolytic apical disc and the short posterior paired petals, which are about half as long as the anterior paired petals. Other features, however, contradict this generic placement (e.g. the peripetalous fasciole is hardly indented between the paired petals in the type species of *Trachyaster*, *T. globosus* POMEL, 1869, whereas it is deeply embayed in *H. cotteauii*). The genus *Opissaster* can be ruled out because it is characterised among others by an ovoid outline and a deep frontal sinus (FISCHER, 1966). Moreover, most authors (e.g. LAMBERT, 1915a; LAMBERT & THIÉRY, 1924) included only forms with 2 to 3 gonopores. These features cannot be found in the species considered here. Since neither the placement in *Trachyaster* nor in *Opissaster* seemed to be feasible and other genera were not available, VALDINUCCI (1974a) established the genus *Aliaster* for the species which seem to fall "in between" *Opissaster* and *Trachyaster*. Although, SMITH ("*The Echinoid Directory*", state 25.05.2003) regards *Aliaster* as junior synonym of *Opissaster*, the present author does not share this opinion based on the very different shape and other features discussed above (under the remarks to the genus).

COTTREAU (1913a) placed *A. jourdyi* (PÉRON & GAUTHIER) and *A. mariae* (LOVISATO) in the synonymy of *A. cotteauii* (referred to as *Opissaster cotteauii* in COTTREAU, 1913a).

CHALLIS (1980: 193) opinion that *Hemiaster vadosus* GREGORY, 1891 from the *Globigerina* Limestone of Malta represents juveniles of *A. cotteauii* cannot be accepted here. *H. vadosus* has an ethmophract apical disc and a distinctly different test shape (GREGORY, 1891 and pers. obs.), whereas *A. cotteauii* has an ethmolytic apical disc.

Pericosmus affinis LAUBE, 1869 a species based on a single specimen from Sievering (now lost) is possibly a junior synonym of *Aliaster cotteauii*. Although LAUBE (1869a, 1871) placed the species in the genus *Pericosmus* and compared it with *P. latus* (a species here shown to occur in Austria too) his description and illustration reveal many discrepancies. In particular the extremely shallow frontal notch (well developed in *Pericosmus*), the adapically deeply sunken, narrow ambulacrum III (shallow adapically and moderately deep and rather broad ambitally in *Pericosmus*), the shape of the peripetalous fasciole, the posteriorly eccentric maximum height and the short posterior paired petals do not fit with the genus *Pericosmus*. These discrepancies were also pointed out by STEFANINI (1909: 20), who remarked that *P. affinis* could probably be a *Trachyaster*. The only feature in LAUBE's description which does not fit with *Aliaster cotteauii* is the presence of a thin marginal fasciole, the course of which is, unfortunately, not mentioned in the description and not visible in the illustration. Without a re-examination of the type it is currently impossible to settle this question. Since the type is lost *P. affinis* LAUBE, 1869 is tentatively placed in the synonymy of *A. cotteauii* based on the description of LAUBE (1871).

In most papers this species is referred to as *cotteaui*, in contrast to the original spelling *cotteauii* by WRIGHT (1855). According to the ICZN (4th ed., Articles 33.4. and 33.5.) the original spelling is to be used and *cotteaui* has to be regarded as "incorrect subsequent spelling" (not as emendation) regardless whether the change from –ii to –i was intentional or not.

Occurrence:

Austria: Early to Late Badenian (Langhian-Early Serravallian)

Vienna Basin: Müllendorf (Mühlendorfer Kreide AG quarry), Bgld ([NEITZ coll.]); ? Sievering, Vienna (LAUBE, 1869a, 1871); Stotzing (sandpit Mayer), Bgld (KAZÁR, 2002; [WANZENBÖCK coll.])

Paratethys (non-Austrian occurrences): ? Karpatian (Late Burdigalian) – Late Badenian (Early Serravallian)

Transcarpathian Basin: ? Hlinné, eastern Slovakia (SENEŠ, 1955; CICHA et al., 1967)

Transylvanian Basin: Gârbova de Sus (= Felsö-Orbó), Romania (VADÁSZ, 1915; [MAFI])

Mediterranean: Chattian to Serravallian

Western Mediterranean: Algeria: Prov. of Oran, (COTTREAU, 1913a); France: Vence, Alpes-Maritimes (LAMBERT, 1915a); Italy, mainland: Emilia (COTTREAU, 1913a), ? Montese (MAZZETTI, 1882a), ? Salto, Emilia (STEFANINI, 1909), ? Rocca Malatina, Emilia (STEFANINI, 1909), Serra die Guidoni, Emilia (STEFANINI, 1909); Morocco (NÉRAUDEAU, 1994); Sardinia, Italy: (COTTREAU, 1913a; VALDINUCCI, 1974a; NÉRAUDEAU, 1994): Bosa, Planargia (LAMBERT, 1909), Cadreas, Bonorva (COTTEAU, 1895; LAMBERT, 1909), Cagliari (? COTTEAU, 1895; LAMBERT, 1909), Cap Sant'Elia (LAMBERT, 1909), Modulo, Planargia (LAMBERT, 1909), Mont San Michele, near Cagliari (COTTEAU, 1895), Monte Alvu, near Bosa (LAMBERT, 1909), Planargia (LAMBERT, 1906a), Portotorres (CHECCHIA-RISPOLI, 1927a), Sant Michele (LAMBERT, 1909), Tresnuraghes, Cagliari (LAMBERT, 1909);

Central Mediterranean: Maltese Islands (DESOR, 1858; COTTREAU, 1913a; NÉRAUDEAU, 1994): Lower Coralline Lst. (WRIGHT, 1864), *Globigerina* Lst. undifferentiated (WRIGHT, 1855, 1864; GREGORY, 1891; BOGGILD & ROSE, 1984); Lower *Globigerina* Lst. (ROSE, 1974b, 1975; CHALLIS, 1980)

Eastern Mediterranean: Attique, Greece (COTTREAU, 1913a); Gebel Gharra, Eastern Desert, Egypt (KROH & NEBELSICK, 2003)

Genus *Linthia* DESOR, 1853

Type-species: *Linthia insignis* DESOR, 1853; by original designation.
Diagnosis: Test usually small, heart-shaped, anterior ambulacrum not petaloid, depressed with groove at the margin; apical disc ethmolytic with four gonopores; posterior paired petals only slightly shorter than anterior ones; peripetalous fasciole sharply indented between the petals; latero-anal fasciole present or absent (modified from KIER, 1984; and MCNAMARA, 1985a).
Distribution: Late Cretaceous to Pliocene – cosmopolitan (FISCHER, 1966)
Remarks: FISCHER (1966: U576) distinguished two subgenera within the genus *Linthia*: *Linthia (L.)* and *Linthia (Lutetiaster)* (Type-species: *Spatangus subglobosus* LAMARCK, 1816; by original designation). This distinction was based mainly on periproct shape (vertically or transversely elongated) and position of the apical disc (central to anteriorly or central to posteriorly). Few workers followed this classification, as it proved to be very difficult to apply.

Linthia ? *hlinnensis* (SENEŠ, 1955)

(Fig. 70; Pl. 68, Figs. 1a-b)

v. 1875a *Schizaster Grateloupii* SISM. – HOERNES: 385, 389
v. 1875b *Schizaster Grateloupii* SISM. – HOERNES: 211
v. 1877 *Schizaster Grateloupi* SISM. – FUCHS: 663
* 1955 *Trachyaster hlinnensis* nov. sp. – SENEŠ: 6, 26-27; pl. 1, fig. 3
1956 *Schizaster grateloupi* SISM. – SIEBER: 317
1967 *Trachyaster hlinnensis* SENEŠ – CICHA et al.: 93
1978 *T.[rachyaster] hlinnensis* SENEŠ, 1955 – KIER & LAWSON: 100
1991 *Schizaster grateloupi* SISMONDA – RUPP et al.: 40
1994 *H. (Trachyaster) hlinnensis* – NÉRAUDEAU: 329, tab. 4
2003b *Linthia ? hlinnensis* (SENEŠ, 1955) – KROH: 251

Type-material:
Trachyaster hlinnensis SENEŠ, 1955:
Holotype: the specimen figured by SENEŠ (1955: pl. 1, fig. 3); housed in the collection of coal exploration at Turč Teplice (SENEŠ, 1955: 26) (not seen)
Locus typicus: Hlinné, Eastern Slovakia
Age: "Helvetian" [Karpatian (Late Burdigalian), Early Miocene, pers. com. I. CICHA, 11.04.2003]

Material:
Ottnangian (Late Burdigalian) – Ottnang, OÖ, Austria
GBA: 1 natural internal cast (steinkern) (GBA 1875/01/77, reference specimen of HOERNES, 1975a, b)

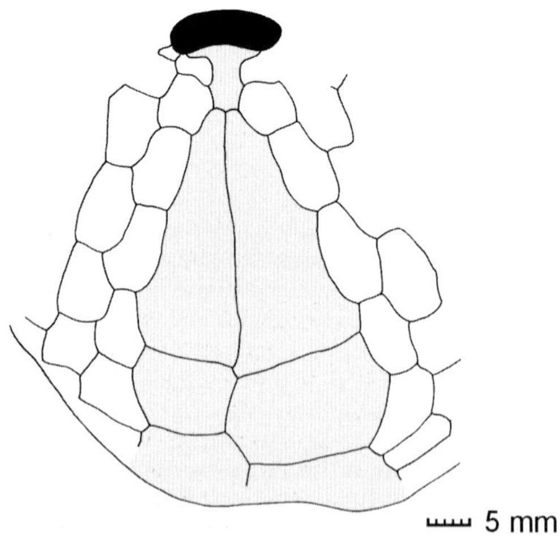

Figure 70: *Linthia ? hlinnensis* (SENEŠ, 1955): plastron structure (Ottnang, OÖ, GBA 1875/01/77).

Dimensions (in mm):

Inv. No.	TL	TW	TH
GBA 1875/01/77	75.7	68.9	-

Description:
Size and shape: Test of large size with elongated, broadly heart-shaped outline. The anterior margin is rounded with a moderately deep frontal sinus. The posterior margin is transversely truncated. The maximum width lies slightly anterior of the centre. In profile the test might have been low arched or slightly wedge shaped, sedimentary compaction, however, obliterated any detailed information.
Apical disc: The structure of the apical disc is unclear. Four gonopores can be observed nevertheless. The apical disc lies slightly posterior of the centre, about 55 % of TL from the anterior margin.
Ambulacra: Adapically the ambulacra form broad, sunken petals. Ambulacrum III is slightly shallower than the paired petals and forms a very moderately deep frontal sinus. The pores of ambulacrum III are arranged in two straight rows, adapically they are closely spaced.
The paired petals are moderately long, straight and extend about 65 to 70 % of the corresponding test radius. The posterior paired petals are shorter than the anterior ones, being about 80 % their length. The anterior paired petals form an obtuse angle of nearly 100°, while the posterior petals form an acute angle of about 65°. The poriferous zones of the paired petals consist closely spaced elongate isopores. The interporiferous zones are very narrow.
Interambulacra: All interambulacra except interambulacrum 5 are distinctly inflated adapically, forming keels along the sunken petals. The plastron is moderately large and mesamphisternous (Fig. 70). The labrum is about twice as long as wide, extends to the second ambulacral plate and projects only slightly over the peristome. Adorally it bears a distinct rim.
Peristome: The peristome lies close to the anterior margin at the oral side of the test, about 20 % TL from it. It is facing downwards and is transversely elongated, kidney-shaped, measuring 13.4 x 5 mm in the present specimen.
Periproct: The periproct is poorly preserved in the specimen examined. It lies high on the vertically truncated posterior end of the test.
Fascioles: not preserved

Differential diagnosis:
Apart from outer shape and gross morphology this species differs from the co-occurring *Schizaster laubei* HOERNES, 1875a by

its transversely truncated margin, straight and more diverging anterior paired petals, longer posterior paired petals and its longer and differently shaped labrum. Albeit the two taxa are easily distinguished in whole specimens, partial internal casts may be more difficult to distinguish.

Discussion:
The only specimen available is a natural internal mould (steinkern) from the Ottnangian (Late Burdigalian) of Ottnang, OÖ. This specimen was originally referred to *Schizaster grateloupi* SISMONDA, 1841 by HOERNES (1875a: 385, 389). It differs, however, from the latter species by its more elongated outline and different angles between the paired petals. LAMBERT & THIÉRY (1924: 504) placed the latter species in the "section" (subgenus) *Hemiaster* (*Gregoryaster*). Subsequent authors followed HOERNES' determination uncritically. In 1955 SENEŠ (1955: 26) established the species *Trachyaster hlinnensis* for specimens from a similar sedimentary environment of Karpatian (Late Burdigalian) age. The specimen from Ottnang considered here is clearly conspecific with *T. hlinennsis* on base of overall test morphology and shape of the petals. The generic attribution of this species is, however, much more difficult since only internal moulds are known. The genus *Trachyaster*, in which this species was originally placed, is characterised (among other features) by an ethmolytic apical disc, short posterior petals and a subcircular, wedge-shaped test. The morphology of the petalodium of *hlinnensis* alone excludes an inclusion in the genus *Trachyaster*. Without information on the fascioles and tuberculation it is very difficult to confidently refer this species to a particular genus. Within the Spatangoida the families Hemiasteridae, Pericosmidae and Schizasteridae would be possible (on base of plastron architecture). The Pericosmidae, however, can be excluded, since *Lambertona* and *Victoriaster* have a very different morphology and *Pericosmus* has a different structure of the adapical ambulacrum III (there are fewer, higher ambulacral plates). Despite the overall similarity with some species of *Brissopsis*, brissids can be excluded because the have an ultramphisternous plastron (with near-symmetric sternal and episternal plates), whereas the species considered here has a mesamphisternous plastron (with biserially offset episternals).

The species considered here is tentatively referred to the subgenus *Linthia* based on the similarity in petal morphology, plastron structure and gross morphology. It has to be pointed out, that this has to be tested with better preserved material because fasciole morphology and apical disc structure are crucial in hemiasterids and schizasterids.

Occurrence:

Austria: Ottnangian (Late Burdigalian)
Molasse Zone: Ottnang, OÖ (HOERNES, 1875a, b; FUCHS, 1877; SIEBER, 1956; RUPP et al., 1991; [NHMW])

Paratethys (non-Austrian occurrences): Karpatian (Late Burdigalian)
Transcarpathian Basin: Hlinné, eastern Slovakia (SENEŠ, 1955; CICHA et al., 1967; KROH, 2003b)

Linthia ? *summesbergeri* nov. sp.

(Fig. 71; Pl. 68, Figs. 2-7)

Type-material:
Holotype (Pl. 68, Fig. 2): specimen NHMW 2003z0026/1187a (cast of the aboral side on a slab together with 3 other specimens, two of which are designated as paratypes)
Paratypes: NHMW 2003z0026/1178, 1187b (cast of the aboral side), 1187c (cast of the oral side), 1190 (cast of the aboral side), 1212 (cast of the oral side; 1213 is the corresponding mould), 1226 (cast of the oral side)
Locus typicus: Weikerlsee, in the eastern part of Linz, OÖ, Austria
Stratum typicum: grey shales of the Ebelsberg Formation
Age: Egerian (Chattian to Aquitanian), Late Oligocene to Early Miocene
Remarks: HOCHULI (1978: 31) dated the Ebelsberg Fm. at Ebelsberg as belonging into the late Pollen biozone Ng. Z I., which can be correlated with the calcareous Nannoplankton biozone NP25, respectively Foraminifera biozone P22 (Late Oligocene, Chattian; HOCHULI, 1978: 4, 31; STEININGER et al., 1989). The upper part of the Ebelsberg Fm. at section Ebelsberg (unpublished data, pers. comm. F. RÖGL, 26.11.2003) and Asten (RÖGL et al., 1979: 1047), however, yielded Early Miocene nannofossils. Similarly, at section Hinzenbach (near Eferding, Upper Austria) the upper parts of the Ebelsberg Fm. yielded a pollen flora in-

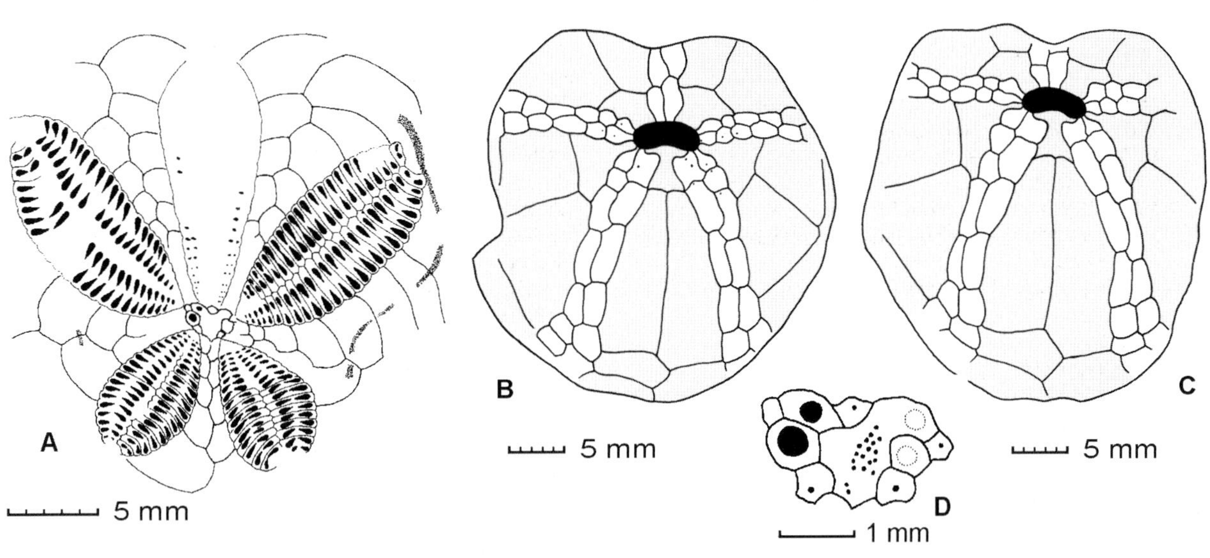

Figure 71: *Linthia* ? *summesbergeri* nov. sp.: petal structure and fasciole outline (dotted) (A: holotype, NHMW 2003z0026/1187a), oral plating (B: NHMW 2003z0026/1212, C: NHMW 2003z0026/1226), and apical disc (D: holotype, NHMW 2003z0026/1187a). All specimens from the Egerian Ebelsberg Fm. at Weikerlsee, Linz E, OÖ; all drawn from natural casts of the surface (internal view) and are presented mirrored here for easier comparison (corresponding to external view).

dicative of Pollen biozone Ng. Z. II (HOCHULI, 1978: 32), which correlates with the Late Egerian to Early Eggenburgian (HOCHULI, 1978; STEININGER et al., 1989). The geographic situation, the flora and additional data to the sections Ebelsberg, Weikerlsee, Asten and Pucking can be found in KOVAR (1982: 9-11).

Dervatio nominis: Named after my colleague Herbert SUMMESBERGER.

Diagnosis:

Small, cordate schizasterid with broad petals, the anterior paired petals nearly reaching the ambitus, ethmolytic apical disc, mesamphisternous plastron, broad, arrowhead-shaped labrum, and moderately deep frontal sinus.

Material:

Egerian (Chattian – Aquitanian) – Weikerlsee (Ebelsberg Fm.), Linz, OÖ, Austria

NHMW: 95 moulds and internal casts (NHMW 2003z0026/1169-1188, 1190-1218, 1220-1224, 1226-1228, 1231-1233)

Egerian (Chattian – Aquitanian) – Ebelsberg Fm., Pucking or Ebelsberg or Weikerlsee or Asten, OÖ, Austria

NHMW: 1 specimen (NHMW 2003z0026/1394)

Dimensions (in mm):

Inv. No.	TL	TW	TH*
NHMW 2003z0026/1178	38.0	35.6	-
NHMW 2003z0026/1187a	20.8	20.1	-
NHMW 2003z0026/1190	15.5	15.0	-
NHMW 2003z0026/1196	>27	~32	-
NHMW 2003z0026/1212	33.7	31.8	-
NHMW 2003z0026/1226	31.4	30.5	-

*as all specimens are preserved either as moulds or casts and are strongly flattened no height measurements were possible

Description:

Size and shape: Test of small to medium size, usually between 20 and 40 mm in length. Outline cordate, antero-posteriorly elongated. A shallow but distinct frontal notch is developed. Test profile unknown, all specimens flattened due to sediment compaction.

Apical disc: Apical disc subcentrally, ethmolytic, with 4 (?; preservation poor, unclear if there are genital pores in genital plates G1 and G2) circular gonopores, the anterior ones of which are smaller. The madreporite separates both genitals 1 and 4, as well as oculars I and V (Fig. 71.D). (Based on specimen NHMW 2003z0026/1187a in which some gonopores and madreporic pores are preserved as casts).

Ambulacra: Paired ambulacra petaloid with closely spaced elongate isopores. Petals very broad, slightly sunken, with laterally flexed tips (Fig. 71.A). Anterior paired petals nearly reach the ambitus and diverge at an obtuse angle between 100 and 120°. Posterior paired petals about 60 % of anterior ones in length, diverging at about 60 to 70°. Ambulacrum III nonpetaloid, strongly widening towards the anterior margin, but becoming smaller again just before reaching the ambitus. Pores in ambulacrum III are closely spaced partitioned isopores.

Adorally phyllodes are developed, nature and number of pores, however, unclear due to unfavourable preservation. Ambulacra I and V form moderately wide naked peri-plastronal areas adorally.

Interambulacra: Aborally the interambulacra 2 and 3 form well developed keels, whereas only weak keels are developed in interambulacra 1, 4 and 5. Plastron mesamphisternous with T-shaped labrum, unequal sternal plates and biserially offset episternals (Figs. 71.B-C). Contact between labrum and sternum broad, giving the posterior part of the labrum an arrowhead-shaped (triangular) appearance. Labrum extending to the second ambulacral plate, although sometimes barely so (e.g. Fig. 71.B).

Aboral tuberculation largely unknown. Orally the interambulacra bear moderately large tubercles, which are largest at the plastron, where they are arranged in a fan-shaped pattern.

Peristome: The peristome is transversely elongated, kidney-shaped with a slightly projecting labrum. It measures 6.1 *vs.* 2.4 mm in a 33.7 mm TL specimen and lies *c.* 27-30 % from the anterior margin.

Periproct: Due to its position on the ambitus the periproct is poorly preserved in all available specimens.

Fascioles: The holotype shows traces of a narrow fasciole band in interambulacrum 1 and ambulacrum 2 (Fig. 71.A). Thus the presence of a peripetalous fasciole seems to be highly probable. Other fascioles not ascertained.

Differential diagnosis:

Linthia burdigalensis (TOURNOUER, 1869), from the Calcaire a Astérides near Bordeaux, France, differs by its shorter and differently shaped petals.

L. calabra (SEGUENZA, 1880) from the Oligocene of Calabria, Southern Italy differs strongly by its narrow and short petals.

L. montecchiana OPPENHEIM, 1902 from the Lower Oligocene of Northern Italy differs by its much smaller, globular shape, lack of a frontal notch and short petals.

L. peolae BOTTO-MICCA, 1896 from Monte Castello, Piemonte, Italy, a species similar in appearance to *L. insignis* DESOR, 1853, differs from the present species by its elongate outline and narrow petals.

L.? lorioli AIRAGHI, 1899 from the Oligocene of Italy differs strongly by its large size, deep frontal notch and narrow, elongated petals (this species actually might belong to *Pericosmus*, based on the description and photographs provided by AIRAGHI, 1899).

L. cevensis BOTTO-MICCA, 1896 from the Miocene of Ceva, Piemonte, Italy differs by its shorter and narrower petals (see BOTTO-MICCA, 1896; and AIRAGHI, 1901).

This form can easily be differentiated from *Brissopsis ottnangensis*, which occurs in a similar facies of the Ottnangian by its mesamphisternous plastron with biserially offset episternals and asymmetric sternals. Incomplete specimens of these two taxa can be extremely similar when preserved as external or internal casts.

Discussion:

Lacking any information on the fascioles generic and family placement of this taxon is difficult. The mesamphisternous plastron indicates that it is a member of the suborder Hemiasterina. While palaeostomatids have an ethmophract apical disc and pericosmids and aeropsids can be ruled out based on overall test morphology, hemiasterids and schizasterids are possible candidates. Fortunately, known hemiasterids with ethmolytic apical disc are sufficiently different. Within the Schizasteridae the material seems to be most closely related with *Linthia* based on the subcentral, ethmolytic apical disc, the long petals, mesamphisternous plastron and strong divergence of anterior petals. There is also a marked similarity to the marsupiate genera *Abatus* and *Tripylus*, which differ from *Linthia* mainly by the loss of the latero-anal fasciole band in adults and the presence of just three (rarely two) gonopores (no gonopores in genital plate G2). Especially the broad petals are reminiscent of these genera. Both are, however, restricted to the Antarctic and sub-Antarctic region today. Fossil representatives are known from the Paleocene of Madagascar (LAMBERT, 1933) and the Eocene of Seymour Island, Antarctic Peninsula (MCKINNEY et al., 1988). Since there is only incomplete information on the fascioles and since it is unclear if all 4 gonopores are developed, the present material is tentatively referred to *Linthia*.

This species is clearly distinct from all known Oligocene and Miocene species of *Linthia*. LAMBERT & THIÉRY (1925: 518-521)

list 8 species from this interval and none have been described since that time (KIER & LAWSON, 1978; Zoological Record). Differential diagnoses to these taxa are given above.

Occurrence:

Austria: Egerian (Chattian-Aquitanian)
Molasse Zone: Weikerlsee (Ebelsberg Fm.), Linz, OÖ ([NHMW])

Genus *Prenaster* DESOR, 1853

Type-species: *Prenaster alpinus* DESOR, 1853; by original designation

Diagnosis: Test oval (antero-posteriorly elongated) without frontal notch; posterior end truncated; ethmolytic apical disc, situated strongly eccentric anteriorly; ambulacrum III non-petaloid, narrow, flush with test aborally, with minute pores; paired ambulacra petaloid, weakly sunken; anterior petals strongly divergent, up to 180°; plastron mesamphisternous, with short but wide labrum and large, symmetrical sternal plates; peristome strongly eccentric anteriorly; periproct high on posterior face; peripetalous and latero-anal fascioles present; fascioles meet behind the anterior petals and pass subambitally around the anterior margin; peripetalous fasciole may be indented or straight behind anterior petals and passes ambulacra about 3 to 4 plates below the end of the petals (modified from FISCHER 1966 and SMITH "*The Echinoid Directory*", 02.11.2003).

Distribution: Eocene to Recent – Europe, Northern Africa, Madagascar, Australia (SMITH "*The Echinoid Directory*", 02.11.2003)

Remarks: FISCHER (1966: U576) regarded *Savinaster* LAMBERT, 1911 and *Protenaster* POMEL, 1883 as subgenera of *Prenaster*. Based on a careful revision of extant and fossil species of *Protenaster*, MCNAMARA (1985b:312-313) rejected the placement of this taxon as subgenus of *Prenaster*. He showed that *Protenaster* is a morphologically very conservative genus and has a fossil record from the Late Eocene onwards. According to him *Protenaster* can be distinguished from *Prenaster* by its possession of deeper petals; a deeper ambulacrum III, in which well developed pore pairs occur; an anterior notch; and a peripetalous fasciole which is indented between the anterior and posterior petals, and which does not extend below the ambitus anteriorly (MCNAMARA, 1985b: 313). SMITH ("*The Echinoid Directory*", 02.11.2003) placed *Parabrissus* BITTNER, 1880, *Saviniaster* LAMBERT, 1911, and *Waurnia* MCNAMARA & PHILIP, 1984 into the synonymy of *Prenaster*.

Ecology and biogeography: *Prenaster enodatus* (CHESHER, 1968), the only extant species of *Prenaster*, was dredged south of the Grand Bahama Island from a mud bottom in 366 metres depth (CHESHER, 1968).

Prenaster fuchsi (LAUBE, 1871) is morphologically very similar to the australasian species *Protenaster australis* (GRAY, 1851). That genus differs mainly by its sunken petals, presence of a frontal notch, different position of its periproct and supra-ambital passing peripetalous fasciole. *P. australis* burrows up to 10 cm deep in near-shore, calcareous sand bottoms or coarse shell grit in shallow depth between 0 and 20 metres water depth. Many specimens are recorded as having been collected from reef platforms or from beaches off which limestone reefs occur. They live buried up to 10 cm deep in fine calcareous sand (74.5 % of the sediment in the grain size fraction between 0.125 and 0.25 mm, 17.5 % between 0.25 and 0.5 mm) (MCNAMARA & BRYCE, 1983; MCNAMARA, 1985b; MISKELLY, 2002).

Prenaster fuchsi (LAUBE, 1871)

(Figs. 72-73; Pl. 69, Figs. 1-2)

v. 1869a *Brissomorpha Fuchsi* LAUBE. – LAUBE: 184 [*nomen nudum*]
v. 1870 *Brissomorpha Fuchsi* LBE. – LAUBE: 314
v.* 1871 *Brissomorpha Fuchsi* LAUBE. – LAUBE: 73; pl. 19, fig. 1
v. 1875 *Brissomorpha Fuchsi* LAUBE – QUENSTEDT: 678
1877 *Brissomorpha* – FUCHS: 662
1900 *Brissomorpha Fuchsi* LAUBE – HESSE: 259-260; fig. 6
v. 1912a *Brissomorpha Fuchsi* LAUBE. – SCHAFFER: 192; text-fig. 4; pl. 58, figs. 20-21
1924 *Brissomorpha Fuchsi* LAUBE – LAMBERT & THIÉRY: 442
v. 1966 *Brissomorpha fuchsi* – FISCHER: U616; fig. 504-2a-b

Type-material:

Lectotype (Pl. 69, Figs. 1a-d): NHMW 1866.XX.18 (LAUBE, 1871: 73; pl. 19, fig. 1)
Paralectotype (Pl. 69, Figs. 2a-c): NHMW 1866.XX.19
Locus typicus: Gauderndorf, NÖ
Stratum typicum: Gauderndorf Fm.
Age: Late Eggenburgian (Early Burdigalian), Early Miocene
Remarks: Both specimens mentioned in the paper of LAUBE (1871: 73) could be traced in the collection Naturhistorisches Museum Wien (previously k. k. Hof-Mineraliencabinet). The better preserved one is chosen as lectotype.

Material:

Eggenburgian (Early Burdigalian) – Gauderndorf (Zogelsdorf Fm.), NÖ, Austria
NHMW: 2 specimens (NHMW 1866.XX.18 – 19)

Dimensions (in mm):

Inv. No.	TL	TW	TH
NHMW 1866.XX.18	84.3	75.9	48.3
NHMW 1866.XX.19	71.1	60.0	42.1

Description:

Size and Shape: Test large, ranging from 71.1 to 84.3 mm in the studied material. The outline of the test is oval, antero-posteriorly elongated. The anterior margin is rounded, without frontal notch, the posterior margin bluntly pointed. In profile the test is rhomboidal and rather high. The maximum width lies about halfway along the length. The test width is about 85-90 % of TL, the height 57-59 % of TL.

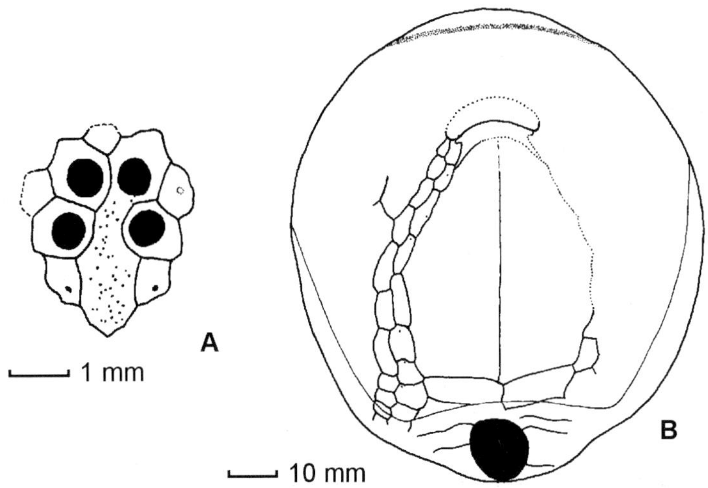

Figure 72: *Prenaster fuchsi* (LAUBE, 1871): apical disc (A) and plastron structure (B) (Eggenburgian, Gauderndorf, NÖ, lectotype, NHMW 1866.XX.18).

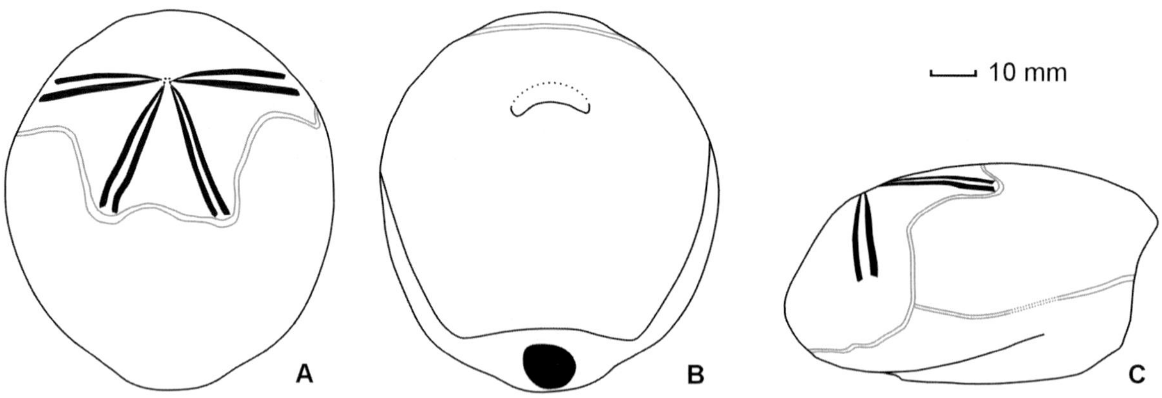

Figure 73: *Prenaster fuchsi* (LAUBE, 1871): fasciole outlines (A: aboral, B: oral, and C: left lateral view; Eggenburgian, Gauderndorf, NÖ, lectotype, NHMW 1866.XX.18).

Apical disc: The apical disc is situated strongly anteriorly, about 20 % TL of the anterior margin. It is ethmolytic with four large, circular gonopores (Fig. 72.A).
Ambulacra: Frontal ambulacrum nonpetaloid, bearing numerous small, oblique, partitioned isopores. A frontal sinus is not developed. The paired ambulacra are petaloid. The pores within the petals are conjugated iso- to anisopores (outer pore oval to teardrop-shaped in distal pore pairs). The uppermost 15 to 20 pore pairs of the anterior poriferous zones of the anterior paired petals are very small. Except in the phyllodes, where large unipores with extended periporal area can be observed, only minute pores (? unipores) are present outside the petals. The poriferous zones are slightly depressed within the petals. The interporiferous zones are narrow and covered by small tubercles similar to those of the interambulacra. The anterior paired petals diverge at about 180° and extend nearly to the ambitus, the posterior paired petals are shorter (about 80 % of the APP), extending only about halfway towards the ambitus and form an acute angle of about 50°. On the oral side ambulacra I and V from narrow naked peri-plastronal areas.
Interambulacra: The interambulacra are covered with small perforate, crenulate primary tubercles to which relatively short, spatulate spines are attached. On the aboral surface the tuberculation is relatively homogenous, whereas it is differentiated on the oral surface. There larger tubercles are found along the anterior-lateral margin and on the plastron where they form a radiating pattern. Interambulacrum 5 forms a rounded, raised keel on the aboral side, which overhangs the posterior end of the test forming a prominent beak. The plastron is mesamphisternous (Fig. 72.B) and very broad. The labrum is also broad, but short and projects slightly over the peristome.
Peristome: The peristome is kidney shaped and moderately large, measuring 13.4 *vs.* ~7 mm in the smaller specimen. It lies strongly anteriorly, about 21 % TL from the anterior margin.
Periproct: The periproct is large, measuring 18.7 *vs.* 11.7 mm in the larger specimen and has an oval, dorso-ventrally elongated outline. It is situated below the beak formed by the aboral interambulacrum 5. The distinctly obliqueness of the periproct observed in both specimens available, does not seem to be related to post-sedimentary compression.
Fascioles: Both, peripetalous and latero-anal fascioles are present. The fasciole band of both is moderately wide and corresponds to the orthofasciole type (*sensu* NÉRAUDEAU et al., 1998b). The peripetalous fasciole is indented between anterior and posterior paired petals and runs down towards the ambitus along the anterior paired petals. It crosses ambulacrum III, interambulacra 2 and 3 just below the ambitus (see Figs. 73.A-C). The latero-anal fasciole is not well visible in both specimens since it is partly covered by spines still attached to the test. It branches of from the peripetalous fasciole just behind the posterior petals and forms a V below the periproct.
Spines: The spine microstructure of this species was described by HESSE (1900: 260, fig. 6) based on spines detached from the type specimens. According to him the spines show a hollow central lumen and 23 to 30 radial septa.

Differential diagnosis:
Prenaster nelsoni (MCCOY, 1882) from the Late Oligocene Waurn Ponds Limestone of Australia (MCNAMARA & PHILIP, 1984) differs by its more widely spaced pore pairs, possession of large double pores beyond the peripetalous fasciole and lack of a hood overhanging the periproct. Otherwise the two species are remarkably similar.

Prenaster enodatus (CHESHER, 1968), an extant species from the Caribbean Sea differs by its less well developed and smaller petals, more reduced anterior poriferous zone of the anterior paired petals, extremely thin peripetalous fasciole, which is not indented between the anterior and posterior paired petals (CHESHER, 1968).

Prenaster excentricus (WRIGHT, 1855) from the Messinian Upper Coralline Limestone of the Maltese Islands differs by its lower profile, slightly more sunken petals, shorter and more closed posterior petals, lack of a hood overhanging the periproct and its horizontally elongated periproct (compare WRIGHT, 1864; CHALLIS, 1980).

Pseudobrissus corsicus (COTTEAU, 1877) from the Miocene of Corsica and the Burdigalian of the Rhône Basin differs by its strongly sunken petals and presence of a slight frontal notch. It shares the presence of peripetalous and latero-anal fascioles, their course and a hood-like projection above the periproct (compare COTTEAU, 1877; PHILIPPE, 1998).

Prenaster jeanneti PIJPERS, 1933 from the Late Eocene of Bonaire differs by its even more anteriorly displaced apical disc, backwards flexed anterior paired petals, more posterior peristome, and lack of a rostrum or hood (compare KIER, 1984).

Prenaster parvus PALMER in SÁNCHEZ ROIG, 1949 from the Late Eocene of Cuba differs by its stronger divergence of the anterior paired petals (flexed backwards), more deeply sunken, well developed petals, lack of a rostrum or hood, and the presence of well developed pore pairs in aboral ambulacrum III (compare KIER, 1984).

Brissus ? *cabrerai* (SÁNCHEZ ROIG, 1953) from the Late Eocene of Cuba differs by its more elongate outline, less diverging posterior paired petals, more backwards flexed anterior paired petals and lack of a rostrum or hood. KIER (1984: 82-83) placed this species in the genus *Brissus*, based on a "short track of subanal fasciole present forming lobe below periproct" on the otherwise very weathered test. As all other features are reminiscent of *Prenaster*, the present author is of the opinion that this "short track" might as well have been part of the subanal portion of a latero-anal fasciole.

Discussion:

Re-investigation of the type material of *Brissomorpha fuchsi* LAUBE, 1871 (the only known specimens of this species) with the help of light microscopy after careful cleaning revealed the presence of a latero-anal fasciole branching off the peripetalous fasciole just behind the end of the anterior paired petals. Combined with an ethmolytic apical disc (Fig. 72.A) and a mesamphisternous plastron (Fig. 72.B) this feature is typical for the schizasterid genus *Prenaster*. Thus a relation to the Asterostomatidae, as suggested by FISCHER (1966: U616) or the Palaeopneustidae, as suggested by MORTENSEN (1950) can be ruled out.

Brissomorpha LAUBE, 1869 [Type-species: *Brissomorpha fuchsi* LAUBE, 1871, p. 73; by subsequent designation (LAUBE, 1871: 72)] differs from *Prenaster* mainly by its large size, hood-like projection over periproct and the large, oblique periproct. These characters are best considered of specific value. Otherwise many of the species currently placed within *Prenaster* too would have to be placed into several monotypic genera [e.g. *P. nelsoni* (MCCOY, 1882) in *Wraunia*]. Records of the genus *Brissomorpha* from Algeria and Indonesia listed by FISCHER (1966) are based on misidentifications (see below).

BÖHM (1882: 364-365; pl. 2, figs. 1a-d) described another species of this genus *Brissomorpha mojsvari* from the northern coast of Madura (north of Java). The specimens, however, does not show the features characteristic of *Brissomorpha* (respectively *Prenaster*), but instead seems to belong to the genus *Eupatagus*. [JEANNET & MARTIN (1937: 277) referred this species to the genus *Maretia*. BÖHM (1882), however, explicitly mentioned the presence of a peripetalous fasciole (contrary to JEANNET & MARTIN, who assume that a peripetalous fasciole is lacking based on the distribution of the aboral primary tubercles in BÖHM's figure). Thus this species is tentatively referred to *Eupatagus* here.]

POMEL (1887: 27) mentioned a *Brissomorpha welschi* from the Miocene of Algeria, according to LAMBERT & THIÉRY (1924: 442), however, this is a *nomen nudum*.

Based on illustrations and descriptions given in the literature LAUBE (1871: 72) suggested that *Brissus depressus* LEYMERIE & COTTEAU, 1856 from the Eocene of Bise (Aude) (COTTEAU, 1863: 140, pl. 7, fig. 10) and *Brissus tuberculatus* WRIGHT,1864 from the Messinian of Malta (WRIGHT, 1864: 486; pl. 22, fig. 1) should probably also be referred to *Brissomorpha*. Concerning *B. depressus*, this statement is difficult to evaluate, since the specimens described by COTTEAU (1863) are poorly preserved. *B. tuberculatus*, however, was revised by STEFANINI (1908b: 118-119), who referred it to the spatangid genus *Trachypatagus* [for good illustrations of this genus see VALDINUCCI (1974b)]. The holotype of this species is located in the British Museum (Natural History) under the number E.1866 (CHALLIS, 1980: 275).

Occurrence:

Austria: Late Eggenburgian (Early Burdigalian)

Molasse Zone: Gauderndorf, near Eggenburg, NÖ (LAUBE, 1869a, 1871; QUENSTEDT, 1875; FUCHS, 1877; HESSE, 1900; SCHAFFER, 1912a)

Genus *Pseudobrissus* LAMBERT, 1905

Type-species: *Brissus corsicus* COTTEAU in LOCARD, 1877; by original designation (LAMBERT in DONCIEUX, 1905)

Emended diagnosis: Test ovate without frontal sinus; apical disc anteriorly displaced, ethmolytic with four gonopores; ambulacrum III flush with the remaining surface adapically, narrow and with very small, oblique partitioned isopores; paired ambulacra petaloid aborally; petals strongly sunken, long and parallel-sided; anterior paired petals strongly divergent (around 180°), extending almost to the ambitus; posterior paired petals long and only weakly divergent; plastron structure unknown; peristome strongly anteriorly displaced; periproct high on truncate posterior face, overhung by a (slight?) rostrum; peripetalous and latero-anal fascioles present; peripetalous fasciole broadly indented in interambulacra 1 and 4, and running at the ambitus anteriorly (based on COTTEAU, 1877; PHILIPPE, 1998; SMITH *"The Echinoid Directory"*, 03.11.2003; and the new specimen described below).

Distribution: Early to Middle Miocene – Mediterranean and Paratethys

Remarks: This genus seems to be closely related to *Protenaster* which can be distinguished from *Pseudobrissus* by the presence of a frontal sinus (albeit often shallow) and well developed pores in aboral ambulacrum III (see MCNAMARA, 1985b for details on *Protenaster*).

Based on the present specimen from the Middle Miocene of the Vienna Basin (Austria, Central Paratethys) the presence of a latero-anal fasciole could be confirmed. Additionally the previously unknown structure of the apical disc is shown to be ethmolytic. These two features support the placement in the family Schizasteridae suggested by FISCHER (1966) and later authors.

Pseudobrissus corsicus (COTTEAU in LOCARD, 1877)

(Fig. 74-75; Pl. 70, Figs. a-d)

* 1877 *Brissus corsicus*, COTTEAU, 1877 – COTTEAU in LOCARD: 325-327; pl. 16, figs. 1-2
1887 *Brissus corsicus*, COTTEAU – PARONA: 306
1895 *Brissus corsicus*, COTTEAU, 1877 – COTTEAU: 50-51
1899 *Brissus corsicus*, COTT. – AIRAGHI: 164-165
1901 *Brissus corsicus*, COTT. – AIRAGHI: 197
1905 *Pseudobrissus corsicus* (COTTEAU) – LAMBERT in DONCIEUX: 155
1906d *Pseudobrissus corsicus* – LAMBERT: 63-64
1909 *Pseudobrissus corsicus* COTTEAU (*Brissus*), 1877. – LAMBERT: 81
1913a *Pseudobrissus corsicus* COTT. – COTTREAU: 72
1915a *Pseudobrissus corsicus* COTTEAU (*Brissus*), 1877. – LAMBERT: 185-186
1951 *Pseudobrissus corsicus* (COTTEAU) – MORTENSEN: 352-353; fig. 167
1966 *Pseudobrissus corsicus* (COTTEAU) – FISCHER: U578; fig. 462.3
1974 *Pseudobrissus corsicus* (COTT.) – ROMAN: 332
1989 *Pseudobrissus corsicus* (COTTEAU) – PHILIPPE: 31, tab. 1
1990 *Pseudobrissus corsicus* (COTTEAU, 1877) – PHILIPPE et al.: 246
1998 *Pseudobrissus corsicus* (COTTEAU, 1877) – PHILIPPE: 182-184; pl. 19, figs. 7a-b, 8a-b

Type-material:

Brissus corsicus COTTEAU, 1877:
Holotype: the specimen figured by COTTEAU (1877: pl. 16, figs. 1-2); collection PÉRON
Locus typicus: Bonifacio, Corsica, France

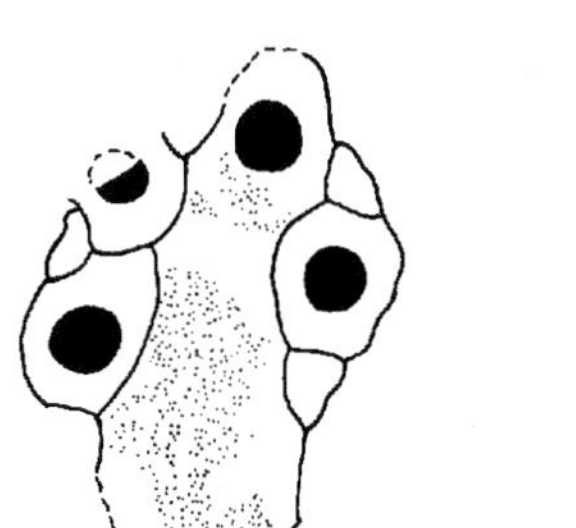

Figure 74: *Pseudobrissus corsicus* (COTTEAU in LOCARD, 1877): apical disc (Badenian, Wöllersdorf, NÖ, KNÖPPEL coll.).

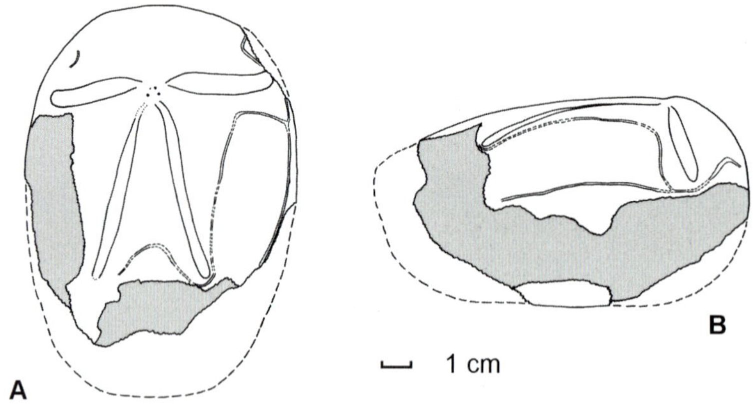

Figure 75: *Pseudobrissus corsicus* (Cotteau in Locard, 1877): fasciole outline in aboral (A) and right lateral view (B) [Badenian, Wöllersdorf, NÖ (Knöppel coll.)].

Age: Burdigalian, Early Miocene (according to Philippe, 1998)
Remarks: The marine sedimentation in the Bonifacio Basin started in the Burdigalian and continues into the Early Serravallian. A radiometric age of the basal volcanites gave an age of 17.8 Ma (Orszag-Sperber in Steininger et al., 1985: Vol. 1: 27, Vol. 2: 57, area 20b Corsica, Bonifacio Basin).

Material:
Badenian (Langhian-Early Serravallian) – Wöllersdorf, NÖ, Austria
R. Knöppel coll.: 1 specimen (a cast of this specimen is housed in the collection of the NHMW under the number 2005z0008/0001)

Dimensions (in mm)**:**
As the specimen is crushed and lacks the posterior end, as well as most of the oral surface, it is difficult to provide accurate measurements. The preserved fragment is 11.5 cm long, 9.5 cm wide, and 5.5 cm high. Before fragmentation and deformation the test is estimated to have been around 13 cm long.

Description:
Size and Shape: Test large, with oval, antero-posteriorly elongated outline. No frontal sinus. In profile the test is rectangular with rounded or vertically truncated anterior end.
Apical disc: The apical disc is situated strongly anteriorly, about 3 cm from the anterior margin. It is ethmolytic with four large circular gonopores, the anterior ones of which are slightly smaller (Fig. 74).
Ambulacra: Frontal ambulacrum nonpetaloid, narrow, bearing numerous small, strongly oblique to vertically elongated, partitioned isopores. Aborally ambulacrum III is flush with the test, below the ambitus it forms a shallow groove leading to the peristome. Aborally the paired ambulacra are petaloid, parallel-sided and rather narrow. The petals are strongly sunken, with large elongate iso- to anisopores (outer pores oval to teardrop-shaped in each pair) and narrow interporiferous zones. The anterior paired petals more or less form a straight line, diverging at about 180°. They extend nearly to the ambitus and their tips are slightly flexed anteriorly. The posterior paired petals are even longer (*c.* 120 % of the length of the anterior paired petals) and form an acute angle of about 35°.
Interambulacra: The interambulacra are covered with small perforate primary tubercles. On the aboral surface the tuberculation is relatively homogenous, only in the interambulacral columns 2b and 3a larger, more widely spaced tubercles are found. The tubercles are non- or very weakly crenulate. Oral surface not preserved.
Peristome: Not preserved in the present specimen; strongly anteriorly displaced in French specimens (Philippe, 1998).
Periproct: Not preserved in the present specimen; high on the truncated posterior face in French specimens (Philippe, 1998).
Fascioles: Both peripetalous and latero-anal fascioles are present. The fasciole band varies in width, but is usually relatively broad and of the orthofasciole type (*sensu* Néraudeau et al., 1998b). The peripetalous fasciole is strongly indented in interambulacrum 5 and runs at the ambitus anteriorly in interambulacra 2 and 3, and ambulacrum III. In interambulacra 1 and 4 it is also strongly indented, half way up the anterior petals, however, it changes its course and forms a straight line running towards the midpoint of the posterior petals. It follows the margin of the distal half of the posterior paired petals closely (Fig. 74.A). The latero-anal fasciole branches of the peripetalous fasciole just behind the tips of the anterior paired petals and runs towards the posterior end just above the ambitus (Fig. 74.B). Its subanal part is not preserved in the present specimen.

Discussion:
Despite the numerous entries in the synonymy list this species is extremely rare. Until now only seven specimens of this species are known [one from Bonifacio, Corsica, France (Cotteau, 1877), one from San Michele, Sardinia (Parona, 1887; Cotteau, 1895; Lambert, 1909), four from the southern Rhône Basin (Lambert, 1915a, Philippe, 1998) and one from Italy (Airaghi, 1899, 1901)]. Just three of these specimens have been figured (by Cotteau, 1877 and Philippe, 1998). As in most large spatangoids all described specimens of *P. corsicus* are more or less crushed, thus obscuring some important features. The present specimen from the Middle Miocene of Austria is the first in which the structure of the apical disc could be observed and in which the presence of a latero-anal fasciole could be ascertained. These features confirm the placement in the family Schizasteridae suggested by earlier authors.
The present specimen comes from a coarse maerl sediment (tightly packed debris of coralline red algae branches with a greenish, fine grained marl matrix). As it is morphologically very similar to *Protenaster australis* it might have inhabited similar environments (see above under *Prenaster*).

Occurrence:

Austria: Badenian (Langhian-Early Serravallian)
Vienna Basin: Wöllersdorf, NÖ ([Knöppel coll.])

Mediterranean: ? Late Oligocene, Burdigalian to Langhian
Western Mediterranean: Corsica, France: Bonifacio (Cotteau, 1877); Italy, mainland: Dego (Airaghi, 1899, 1901); Rhône Basin, France: Carro, à Martigues, Bouches-du-Rhône (Lambert, 1915a; Philippe et al., 1990; Philippe, 1989, 1998); cap Couronne, à Martigues (Philippe, 1998); Sardinia, Italy: San Michele, near Cagliari (Parona, 1887; Cotteau, 1895; Lambert, 1909)

Suborder Micrasterina Fischer, 1966
Family Brissidae Gray, 1855
Genus *Brissus* Gray, 1825

Type-species: *Spatangus brissus* var. *unicolor* Leske, 1778; by subsequent designation (ICZN, opinion 209, p. 385, 8th May 1954)
Diagnosis: Test ovoid, lacking frontal sinus; apical system anteriorly displaced, ethmolytic, with four gonopores; ambulacrum III flush with the test aborally, with small, simple isopores; paired ambulacra petaloid and distinctly sunken; anterior pair transversely orientated, diverging at about 180°; petals straight

sided, with narrow interporiferous zones; periproct on truncated posterior face, oral rim formed by interambulacral plates 5a5 and 5b5; peristome transversely elongate and strongly displaced anteriorly; plastron ultramphisternous, with near-symmetrical sternal and episternal plates; labrum wide, but short, in broad contact with sternal plates; no enlarged primary tubercles; well developed peripetalous and subanal fasciole; subanal fasciole bilobed, usually enclosing four penicillate tube feet on each side; peripetalous fasciole indented in all interambulacra, although to a variable degree in interambulacra 2 and 3 (modified from Fischer, 1966; Mortensen, 1951; and Smith *"The Echinoid Directory"*, 03.11.2003).

Distribution: Eocene to Recent – cosmopolitan (Fischer, 1966)

Ecology and biogeography: Extant species of the genus *Brissus* live in muddy sand, coarse sand, shell grit, sand-filled pockets of reef rock, under rock slabs or within sea grass (Kier & Grant, 1975; Riedl, 1983; Hendler et al., 1995; Miskelly, 2002). They may be buried up to 20 cm deep (Miskelly, 2002). Most species live at shallow depth between 0 and 250 metres, but are also found much deeper occasionally (Mortensen, 1951; Hendler et al., 1995). They occur widely throughout the tropical and warm temperate climate zones.

Brissus abeli (Reidl, 1941)

(Figs. 76-77; Pl. 73, Figs. 1-4)

	1937	*Plagiobrissus abeli* nov. spec. – Reidl: 36-39 (*nomen nudum*; unpublished thesis)
v *	1941	*Plagiobrissus abeli* nov. spec. – Reidl: 24-28; figs. 1-2
?	1953	*Brissus jacquementi* Lambert, 1915. – Szörényi: 41, 92-93; pl. 4, figs. 5, 5a-b
#	1961	*Brissus* (*Allobrissus*) *miocaenicus* n. sp. – Schaffer: 149-155; fig. 1a-d; pl. 1, fig. 1-6; pl. 2, fig. 1-9
	1978	*Brissus* (*A.*) *miocaenicus* Schaffer, 1961 – Kier & Lawson: 111
	1978	*P.*[*lagiobrissus*] *abeli* Reidl, 1941 – Kier & Lawson: 118

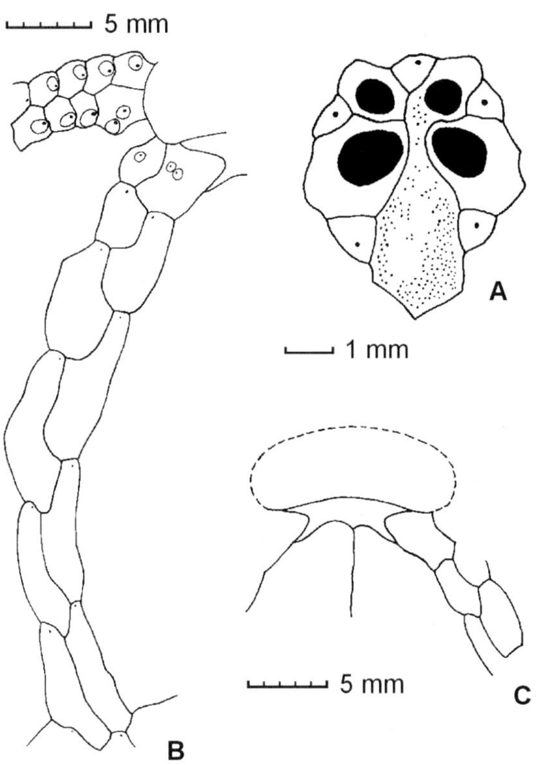

Figure 76: *Brissus abeli* (Reidl, 1941): apical disc (A: NHMW 1997z0178/1742b), plating of adoral ambulacra I and II (B: NHMW 2002z0180/0004), and labrum shape (C: NHMW 1997z0178/1742a); all from the Late Badenian of Müllendorf, Bgld.

Type-material:

Plagiobrissus abeli Reidl, 1941:

Holotype: figured in Reidl (1941: fig 1-2, right-hand specimen in both figs.), presumably lost; reported to be in the collection of the Institute of Palaeontology, University of Vienna (Reidl, 1941: 25) but cannot be traced at this institution.

Paratype: No. IPUW Reidl-50, figured in Reidl (1941: fig 1-2, left-hand specimen in both figs.); housed at the Institute of Palaeontology, University of Vienna; a cast of this specimen is housed at the NHMW (no. 2002z0180/0006).

Locus typicus: Müllendorf (Mühlendorfer Kreide AG quarry), Bgld, Austria

Stratum typicum: Leitha limestone

Age: Late Badenian (Early Serravallian), Middle Miocene

Brissus (Allobrissus) miocaenicus Schaffer, 1961:

Holotype: No. IPUW, 1561, figured in Schaffer (1961: pl. 1, figs. 1-2), presumably lost; reported to be in the collection of the Institute of Palaeontology, University of Vienna (Schaffer, 1961: 149) but cannot be traced at this institution.

Locus typicus: Müllendorf (Mühlendorfer Kreide AG quarry), Bgld, Austria

Stratum typicum: Leitha limestone

Age: Late Badenian (Early Serravallian), Middle Miocene

Material:

Late Badenian (Langhian – Serravallian) – Mannersdorf, Bgld, Austria

NHMW: 2 specimens (NHMW 1978/2020/23, 2002z0181/0002)

Late Badenian (Langhian – Serravallian) – Müllendorf (Mühlendorfer Kreide AG quarry), Bgld, Austria

NHMW: 13 specimens (NHMW 1987/13/4, 1988/19a-b, 1997z0178/1742a-d, 2002z0180/0001-5, 2002z0181/0001)

IPUW: 1 specimen [IPUW Reidl-50 (**paratype** of *Plagiobrissus abeli* Reidl, 1941)]

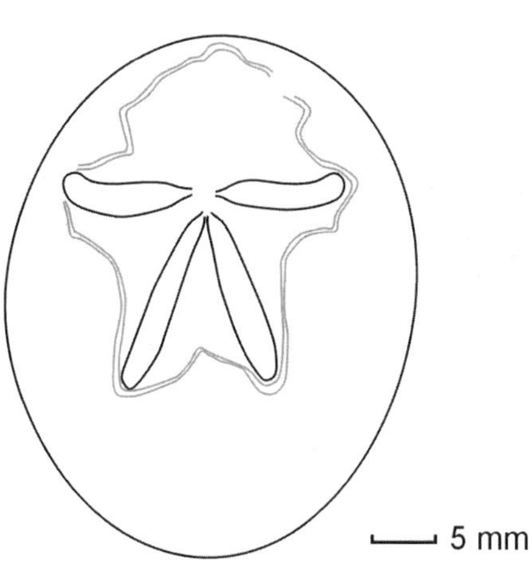

Figure 77: *Brissus abeli* (Reidl, 1941): peripetalous fasciole outline (grey) (Late Badenian of Müllendorf, NHMW 1997z0178/1742d).

Dimensions (in mm):

Inv. No.	TL	TW	TH
NHMW 1997z0178/1742a	55.2	46.0	26.3
NHMW 1997z0178/1742b	55.2	46.2	21.5
NHMW 1997z0178/1742c	52.4	43.3	20.2
NHMW 1997z0178/1742d	37.2	29.9	15.3
NHMW 2002z0180/0002	~49.5	40.8	-
NHMW 2002z0180/0003	~ 46	-	24.8
NHMW 2002z0180/0005	> 62	56.5	> 26.5
NHMW 2002z0181/0001	51.4	~ 43	-
IPUW Reidl-50 (paratype)	56.0	44.5	22.3

All specimens are more or less slightly deformed.

Description:

Size and shape: Test of medium to large size, ranging from 37 to more than 62 mm TL. The outline of the test is oval, antero-posteriorly elongated. A frontal sinus is not present. The maximum width lies sub-centrally but may be slightly displaced anteriorly or posteriorly. In profile the test is slightly wedge-shaped with tumid anterior and obliquely truncated posterior end. The maximum height lies on the raised keel in interambulacrum 5 posterior of the tips of the posterior paired petals. The test width ranges from 79.5 to 83.7 % TL and the test height from ~ 42 to 48 % TL.

Apical disc: The apical disc lies anterior of the centre, about 30 to 38 % TL from the anterior margin. It is ethmolytic with four large, radially elongated gonopores, the anterior pair of which is distinctly smaller than the posterior pair (Fig. 76.A). The ocular pores are small and circular.

Ambulacra: The frontal ambulacrum is nonpetaloid and nearly flush with the test aborally. It bears minute, vertically oriented partitioned isopores. The paired ambulacra are petaloid, consisting of closely spaced elongated anisopores. The petals are strongly depressed, narrow and extend approximately two third of the corresponding test radius. The anterior paired petals form an obtuse angle between 160° and 170°, the posterior petals an acute angle of 40 to 50°. In juvenile or subadult specimens the anterior paired petals diverge at a larger angle of about 170 to 180° (compare Pl. 73, Fig. 4a). The distal parts of the anterior paired petals are slightly flexed anteriorly, those of the posterior petals are flexed laterally. The interporiferous zones are very small, about 0.25 times the width of a single poriferous zone and lack primary tubercles.

Adorally the ambulacra form moderately large phyllodes with large unipores with extensive periporal area. Ambulacra I and V form narrow, naked peri-plastronal areas on the oral side (Fig. 76.B). Within the subanal fasciole 5 enlarged constricted unipores are found in each ambulacrum.

Interambulacra: The interambulacra are densely covered by perforate, crenulate primary tubercles with inclined areoles. The aboral tuberculation is rather homogenous, only adapically within the peripetalous fasciole there are slightly enlarged tubercles. On the oral side, the tubercles are distinctly larger. On the plastron they are arranged in a fan-shaped pattern radiating from a posterior elevation.

Adapically the interambulacra are inflated between the petals. Interambulacrum 5 forms a broad rounded keel posterior of the tips of the posterior paired petals. The distinctly inflated plastron is ultramphisternous with very short, crescentic labrum (Fig. 76.C) and large sternum.

Peristome: The peristome lies close to the anterior margin and its anterior margin lies about 13 to 15 % TL from it. The peristome is kidney-shaped, transversely elongated with a slightly projecting labrum and measures 12.2 *vs.* 4.0 mm in a 55.2 mm TL specimen.

Periproct: The periproct lies on the overhanging posterior face, about halfway up the ambitus. It is oval, vertically elongated to slightly teardrop shaped and measures 9.1 *vs.* 6.3 mm a 55.2 TL mm specimen. The area immediately beneath the periproct is sparsely tuberculated, bearing mainly secondary and miliary tubercles.

Fascioles: Both a peripetalous and subanal fasciole are present. The fasciole band is well distinguished from the remaining tuberculation and corresponds to the orthofasciole type of Néraudeau et al. (1998b). The peripetalous fasciole is deeply embayed in the interambulacra 1 and 4 and forms a zigzagging pattern in the interambulacra 2 and 3 (see Fig. 77). The subanal fasciole is bilobed to slightly heart-shaped.

Differential diagnosis:

This species differs from the extant transatlantic species *Brissus unicolor* (Leske, 1778), the fossil record of which dates back at least to the Early Pliocene (Mortensen, 1951: 512; Donovan & Veale, 1996: 635), by the smaller angle between its anterior paired petals (180 to 190° in *B. unicolor*), its oblique posterior end and stronger wedge-shape.

B. aegyptiacus Gauthier in Fourtau, 1900 is based on a single, badly damaged specimen from the Early Miocene of Gebel Genefe in Egypt. It differs from *B. abeli* by its more diverging anterior petals but otherwise comparison is difficult due to the poor preservation of the Egyptian material (compare Gauthier in Fourtau, 1900: 718-720. pl. 3, figs. 11-12).

B. bastiae Oppenheim, 1902 from the Oligocene to Early Miocene of Northern Italy and the Rhône Basin differs by its more elongated outline (TW ~ 70 to 75 % TL), the larger angle between its anterior paired petals (> 180°) and its more anteriorly situated apical disc (~ 20 to 25 % TL from anterior margin) (compare Oppenheim, 1902a: 261-262, fig. 21; Lambert, 1915a: 179-180, pl. 14, figs. 14-16; Philippe, 1998: 184-185, pl. 21, fig. 5).

B. cordieri Agassiz in Agassiz & Desor, 1847 from the Early (?) Miocene of France was never properly described nor figured. The only figure available is that of Lambert (1915a: 180-181; pl. 14, figs. 12-13), which shows a specimen lacking its aboral side and most of the oral side too. Until better preserved specimens are described and figured a comparison is impossible and this species has to be considered as *nomen dubium*.

B. cottreaui Lambert, 1915 from the Pliocene of the Rhône Basin differs by its larger angle between the anterior paired petals (~ 180°), more anterior situated apical disc (~ 28 % TL from the anterior margin), more elongated test (TW ~ 75% TL) and bluntly pointed posterior end. Furthermore, the whole petalodium is smaller and positioned more anteriorly than in *B. abeli* (compare Lambert, 1915a: 182-185; pl. 15, figs. 1-3).

B. gregoryi Stefanini, 1909 from the Burdigalian (?) and Messinian of Malta differs by its larger angle between the anterior paired petals (~ 180°) and characteristic low profile (~ 25 % TL) (compare Gregory, 1891: 620-621, pl. 2, fig. 2; Challis, 1980: 239-243, pl. 96, figs. a-c, pl. 98, figs. a-c, pl. 99, fig. a).

B. jacquemeti Lambert, 1915 is a junior synonym of *B. bastiae* Oppenheim, 1902a (Philippe, 1998: 184; see above).

B. oblongus Wright, 1855 from the Upper Coralline Limestone (Messinian) of Malta differs by its larger angle between the anterior paired petals (~ 180°), more anterior situated apical disc (~ 25 % TL from the anterior margin), less anterior positioned peristome (~ 20 % TL from anterior margin) and more elongated test (TW ~ 75% TL) (compare Wright, 1855: 184-185, pl. 5, figs. 2a-c; Challis, 1980: 243-246, pl. 97, figs. a-c, pl. 99, figs. b-c).

Discussion:

Reidl (1941: 27) originally placed his species in the genus *Plagiobrissus*, arguing that it could not be placed in the genus *Brissus* because of its continuous fasciole and enlarged tubercles between the petals. However, neither the type material nor topotypic specimens show tubercles as in the genus *Plagiobrissus*, and the continuous fasciole is no reason for excluding this species from *Brissus*, since it is present in nearly all species including the type species. *B. miocaenicus* Schaffer, 1961 is clearly a junior synonym of *Brissus abeli* (Reidl, 1941). The type material of both species comes from the same. Obviously

SCHAFFER (1961) did not know REIDL's publication since it is not discussed or cited in his paper.

SZÖRÉNYI (1953: 92-93) reported a specimen of *B. jacquemeti* LAMBERT, 1915 from the Early Badenian (Langhian) of the Ukraine. This species (now considered to be a junior synonym of *B. bastiae* OPPENHEIM, 1902, see PHILIPPE, 1998: 184-185), however, differs from the Ukrainian specimen by its more strongly diverging anterior petals and more anteriorly positioned apical disc. Based on SZÖRÉNYI's description and poor illustration her specimen is tentatively placed into *B. abeli*. Her specimen differs from Austrian specimens of *B. abeli* by its more elongated test and slightly more diverging anterior paired petals. This last feature, however, could be an effect of allometric growth, since the Ukrainian specimen is rather small (TL = 30 mm) and could therefore represent a juvenile or subadult specimen.

Occurrence:

Austria: Late Badenian (Early Serravallian)
Vienna Basin: Mannersdorf, NÖ ([NMHW]); Müllendorf (Mühlendorfer Kreide AG quarry), Bgld (REIDL, 1937, 1941; SCHAFFER, 1961; [NMHW])

Paratethys (non-Austrian occurrences): Early Badenian (Langhian)
Fore-Carpathian Basin: ? Zbarazh (= Zbaraz), Podolia, western Ukraine (SZÖRÉNYI, 1953)

Genus *Brissopsis* AGASSIZ in AGASSIZ & DESOR, 1847

Type-species: *Brissus lyrifera* FORBES, 1841; by subsequent designation (DESOR, 1858: 378).
Diagnosis: Test ovate, somewhat depressed, with shallow frontal notch; apical disc ethmolytic with four gonopores; ambulacrum III slightly depressed with differentiated pores; paired ambulacra petaloid; petals sunken, distally closed; anterior paired petals more strongly diverging than posterior ones, which are parallel in some species; both anterior and posterior paired petals may be confluent and may have rudimentary or completely reduced pores in the adapical plates; periproct high on the truncated posterior end; peristome transversely elongated, and anteriorly displaced; plastron ultramphisternous, with symmetric sternal and episternal plates; contact between 1st and 2nd pair of episternal plates constricted; labrum broadly T-shaped, not extending beyond first ambulacral plate in most species; peripetalous fasciole indented between anterior and posterior petals; subanal fasciole bilobed (modified from MORTENSEN, 1851; and SMITH *"The Echinoid Directory"*, 13.05. 2004).
Distribution: Eocene to Recent – cosmopolitan, except polar regions
Ecology and biogeography: Extant species of *Brissopsis* occur on mud and silt bottoms, where they lived buried between 40 and 100 mm deep (NICHOLS, 1959a; CHESHER, 1968; KIER & GRANT, 1975). Most species occur in relatively deep environments (down to 3200 m; AGASSIZ, 1904; MORTENSEN, 1951), but some are known from shallow depth [e.g. *B. elongata* MORTENSEN, which occurs between 13 and 72 m off Belize (KIER, 1975), or *B. lyrifera* (FORBES) and *B. luzonica* (GRAY) which occur between 5-1400 and 10-1000 metres respectively (MORTENSEN, 1951)]. Due to its habitat (mainly deep-water, fine siliciclastic sediments) the distribution of *Brissopsis* is not as strongly affected by temperature as in other genera. It occurs widely throughout the tropical, warm temperate and cold temperate climate zones.

Brissopsis ottnangensis HOERNES, 1875

(Figs. 78-81; Pl. 71, Figs. 1-6; Pl. 72, Figs. 1-7)

v. * 1875a *Brissopsis ottnangensis* nov. sp. – HOERNES: 385-386, 389-391; pl. 12, fig. 4; pl. 15, figs. 2-7
v. 1875b *Brissopsis ottnangensis* nov. sp. – HOERNES: 211
? 1879 *Brissopsis ottnangensis* R. HÖRN. – MANZONI: 154
? 1882 *Brissopsis Ottnangensis*, R. HOERNESS. – DE LORIOL: 24-25; pl. 3, figs. 4-7 [according to LAMBERT (1915a: 175-175) the specimens figured in fig. 7 (from Drôme, France) belongs to *B. urrensis* FONTANNES]
1884 *Brissopsis ottnangensis* – HOERNES: 305
1884 *Brissopsis ottnangensis* R. HÖRN. – TOULA: 246
1886 *Brissopsis Ottnangensis* – KITTL: 22
1891 *Brissopsis ottnangensis* R. HOERN. – SUESS: 416
1892a *Brissopsis, Ottnangensis* R. HOERN. – PROCHÁZKA: 734, 736, 743
1897 *Brissopsis ottnangensis* R. HOERN. – SCHAFFER: 535, 547
1900 *Brissopsis Ottnangensis* – COMMENDA: 159
1900 *Brissopsis Ottnangensis* Rud. HOERN. – HOLLER: 71
1902 *Brissopsis Ottnangensis* – ABEL: 1175
1903 *Brissopsis attnangensis* – HOERNES: 936
1903 *Brissopsis Ottnangensis* – HOERNES: 947
v. 1904 *Brissopsis* spec.– ABEL: 97, 99
1908 *Brissopsis ottnangensis* R. HÖRN. – SCHAFFER: 30
1913 *Brissopsis* cf. *ottnangensis* HÖRN. – DREGER: 70
1914 *Brissopsis ottnangensis* R. HÖRN. – TOULA: 203, 209
1915 *Brissopsis* cf. *ottnangensis* R. HÖRNES – TOULA: 640, 642, 666; pl. 39, figs. 19-20
? 1915 Stachel aus der *Brissomorpha*-Gruppe – TOULA: 672
1915 *Brissopsis consobrinus* LAMBERT. – VADÁSZ: 230-231; pl. 9 (3), fig. 15
v. 1915 *Brissopsis* (*Brissoma*) *ottnangensis* R. HOERN. – VADÁSZ: 229-230; pl. 9 (3), fig. 16
1927 *Brissopsis Ottnangensis* HÖRN. R. – HORUSITZKY: 25, 165
1930 *Brissopsis ottnangensis* R. HOERN. – VENDL: 43
1930 *Brissopsis* (*Brissoma*) *ottnangensis* R. HOERN. – VENDL: 50
1935 *Brissopsis ottnangensis* – SIEBER: 96
1936 *Brissopsis* (*Brissoma*) *ottnangensis* R. HÖRN. – MEZNERICS: 119
1940 *Brissopsis ottnangensis* HÖRN. – JASKÓ: 305, pl. 9, figs. 19-20
1940 *Brissopsis* (*Brissoma*) cfr. *ottnangensis* HOERNES. – JASKÓ: tab. 11
1943 *Bryssopsis ottnangensis* R. HOERN. – JANOSCHEK in SCHAFFER: 433-434, 448
1943 *Brissopsis ottnangensis* R. HÖRN. – VEIT: 6
1951 *Bryssopsis ottnangensis* R. HOERN. – JANOSCHEK in SCHAFFER: 548, 567
1951 *Brissopsis ottnangensis* R. HOERN. – SCHAFFER & GRILL: 707
1953a *Brissopsis ottnangensis* HÖ.R. – SIEBER: 207
? 1955 *Brissopsis* (*Brissoma*) *ottnangensis* R.HÖRN. – SENEŠ: 6
? 1955 *Brissopsis* cf. *consobrinus* LAMB. – SENEŠ: 6, 26 ; pl. 1, fig. 1-2
1956 *Brissopsis ottnangensis* R. HÖRN. – SIEBER: 312
1960 *Brissopsis* (*Brissoma*) *ottnangensis* R. HOERN. – GIVULESCU & DUŞA: 937-938; figs. 3, 5 [the legends of fig. 3 and 6 were confused]
1962 *Brissopsis ottnangensis* – PAPP & THENIUS in KÜHN: 330
1967 *Brissopsis* (*Brissoma*) *ottnangensis* R. HOERN. – CICHA et al.: 87
1967 *Brissopsis* (*Brissoma*) *ottnangensis* R. HOERN. – CICHA et al.: 92
? 1967 *Brissopsis* cf. *consobrinus* LAMB. – CICHA et al.: 93
pp 1970 *Brissopsis* cf. *crescenticus* WRIGHT – BAŁUK: 113; pl. 1, fig. 7

. 1975 *Brissopsis ottnangensis* R. Hoernes 1875 – Stojaspal: A192
1976 *Brissopsis ottnangensis* R. Hoern. – Rögl & Müller: 222
1978 *Brissopsis ottnangensis* Hoern. – Cicha: 170
1980 *Brissopsis* cf. *ottnangensis* Hörnes? – Wank & Stojaspal: 448
1981 *Brissopsis* cf. *ottnangensis* R. Hörnes – Wank: 382; pl. 3, fig. 6
1985 Brissopsis ottnangensis – Hámor: tab. 25, tab. 28
1985 *Brissopsis ottnangensis* Hoernes – Tollmann: 452
1991 *Brissopsis ottnangensis* Hoernes – Rupp et al.: 40
v 1998 *Brissopsis* cf. *ottnangensis* R. Hoernes – Tomašových: 361, 385
v. 2002b *Brissopsis ottnangensis* Hoernes – Kroh: 11
2003 *Schizaster laubei* – Weidlinger: 199, unnumbered fig. [misidentified *Brissopsis ottnangensis*]
v. 2003b *Brissopsis ottnangensis* Hoernes, 1875 – Kroh: 251; pl. 1, fig. 4

Type-material:
Lectotype: 1 natural internal cast (GBA 1875/1/78a)
Paralectotypes: 380 moulds and internal casts housed in the type collection of the Geological Survey of Austria under the number GBA 1875/1/78 and 5 specimens housed at the Naturhistorisches Museum Wien (NHMW 1854.XXIII.41)
Locus typicus: Ottnang, OÖ, Austria
Stratum typicum: Ottnang Schlier
Age: Ottnangian (Late Burdigalian), Early Miocene
Remarks: A lectotype has been chosen, as the syntypes encompassed also a few specimens of schizasterids.

Material:
Eggenburgian (Early Burdigalian) – Pfarrkirchen (Haller Schlier), OÖ, Austria
NHMW: 2 specimens on a slab (NHMW 1977/1891a+b)
Ottnangian (Late Burdigalian) – Antiesen river, near Antiesenhofen, near Schärding/Inn, OÖ, Austria
NHMW: 22 specimens (NHMW 1971/1485a-j, 1985/67/22a+b, 1989/83/39, 2003z0090/0002)
Ottnangian (Late Burdigalian) – Antiesenhofen, near Schärding/Inn, OÖ, Austria
NHMW: 1 specimen (NHMW 2003z0090/0005)
Ottnangian (Late Burdigalian) – Großwiesenhardt, near St. Marienkirchen, OÖ, Austria
NHMW: 3 specimens (NHMW 1988/40/19, 2003z0090/0004)
Ottnangian (Late Burdigalian) – Ottnang, OÖ, Austria
NHMW: 8 moulds and internal casts [NHMW 1854.XXIII.41 (**paralectotypes**), 1928.No.23, 1997z0178/0580, 2004z0102/0001-2]
GBA: 381 moulds and internal casts (GBA 1875/1/78 **type-material** of *B. ottnangensis* Hoernes, 1875a)
Ottnangian (Late Burdigalian) – Pielach river, south of Völlerndorf, NÖ, Austria
GBA: 1 specimen (no inventory number)
Ottnangian (Late Burdigalian) – St. Pölten (train station), NÖ, Austria
NHMW: 1 specimen (NHMW 2004z0100/0001)
Early Badenian (Langhian) – Bad Vöslau, NÖ, Austria
NHMW: 5 specimens (NHMW 2004z0103/0001-3)
Early Badenian (Langhian) – Gainfarn, NÖ, Austria
NHMW: 1 specimen (NHMW 2004z0154/0004)
Early Badenian (Langhian) – Retznei [Weissenegg Fm., Lafarge (formerly Perlmoser) quarry Rosenberg], Styria, Austria
NHMW: 1 specimen (NHMW 2004z0098/0002)
Early Badenian (Langhian) – Retznei [Weissenegg Fm., Lafarge (formerly Perlmoser) main quarry], Styria, Austria
NHMW: 1 specimen (NHMW 2004z0098/0001)
Early Badenian (Langhian) – Weitendorf (below basalt layer), Styria, Austria
NHMW: 2 specimens (2004z0097/0001-2)
Early Badenian (Langhian) – Wetzelsdorf in der Weststeiermark, Styria, Austria
NHMW: 7 specimens (2004z0096/0004-10)
Badenian (Langhian-Early Serravallian) – Kalksburg, Vienna, Austria
NHMW: 1 specimen (NHMW 1981/55/12)
Late Badenian (Early Serravallian) – Walbersdorf, Bgld, Austria
GBA: 45 fragmentary moulds and internal casts (no inventory numbers)

Foreign material for comparison:
Ottnangian (Late Burdigalian) – Mitterdorf near Griesbach, Bavaria, Germany
NHMW: 30 fragmentary moulds and internal casts (no inventory numbers)
Karpatian (Late Burdigalian) – Strachotín, *c.* 11 km N Mikulov, southern Moravia, Czech Republic
Univ. Brno: numerous specimens (no inv. nos.; see Pospíchalová in Kroh, 2003b)
Badenian (Langhian-Early Serravallian) – Mogyoród, Pest, Hungary
NHMW: 2 specimens (NHMW 1997z0178/2092)
Late Badenian (Early Serravallian) – Devínska Nová Ves (brickyard near train station), Slovak Republic
NHMW: 10 specimens (NHMW 2004z0095/0001-10)
Slovak National Museum: 11 specimens [SNM z24174 to z24182; reference material of *B.* cf. *ottnangensis* in Tomašových, 1998]

Dimensions (in mm):

Inv. No.	TL	TW	TH
GBA 1875/1/78a (lectotype)	31.9	30.3	-
GBA 1875/1/78b	31.7	30.5	-
GBA 1875/1/78c	15.9	15.0	5.1
[specimen figured in Hoernes (1875a: pl. 12, figs. 4a-b)]			
GBA 1875/1/78d	17.1	15.9	-
[specimen figured in Hoernes (1875a: pl. 15, figs. 2a-b)]			
GBA 1875/1/78g	42.2	41.6	> 15
GBA 1875/1/78j	16.6	15.4	> 5
GBA 1875/1/78k	19.6	17.9	-

Description:
Size and shape: Test of small to medium size, commonly between 15 and 45 mm in length. Outline oval, slightly anteroposteriorly elongated. Anterior margin rounded with shallow frontal sinus, posterior margin transversely truncated. The profile is low and slightly wedge shaped with obliquely truncated posterior end and distinct subanal heel (Pl. 72, Fig. 1a-c). The ambitus is round and tumid. The test width is about 92-96% of TL.
Apical disc: The apical disc is ethmolytic with four circular gonopores. The anterior gonopore pair is more closely spaced and the pore diameter is smaller. Ocular pores are barely visible. The madreporite is very long and separates both genitals 1 and 4, as well as oculars I and V (Figs. 78.A-B). It is crowded by many small madreporic pores. The position of the apical disc changes during the growth of this species. In small specimens (around 15 mm TL) the distance to the anterior margin varies from 46 to 55 % TL, in large specimens (30 to 40 mm TL) it is distinctly larger, usually between 58 and 60 %.
Ambulacra: On the aboral side ambulacrum III is slightly widened halfway between apex and ambitus (Fig. 78.C). Adapically it is slightly sunken, increasingly so towards the ambitus. The pores are small, oblique partitioned isopores and are arranged in two straight rows. The paired ambulacra are petaloid and bear closely spaced elongate isopores (pores teardrop-shaped, subequal in size, the perradial pore being slightly larger in some cases, tapering against each other; pores surrounded by a narrow rim of stereom, which is highest at the

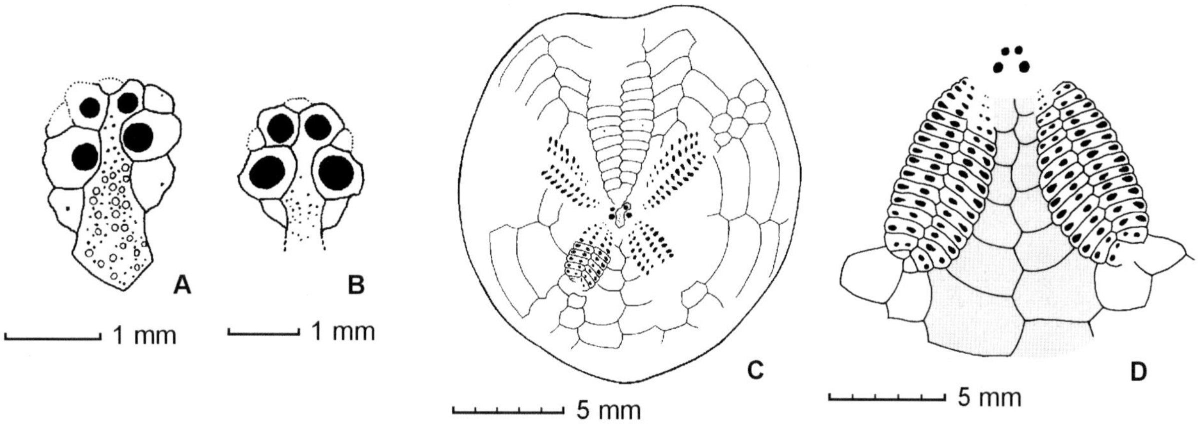

Figure 78: *Brissopsis ottnangensis* Hoernes, 1875: apical disc (A: Badenian, Kalksburg, Vienna, Austria, NHMW 1981/55/12; B: Badenian, Devínska Nová Ves, Slovak Republic, SNM z24182), aboral view (C: Ottnangian, Ottnang, OÖ, Austria, paralectotype GBA 1875/1/78c), and structure of the posterior petals (D: Badenian, Bad Vöslau, NÖ. Austria, NHMW 2004z0103/0001).

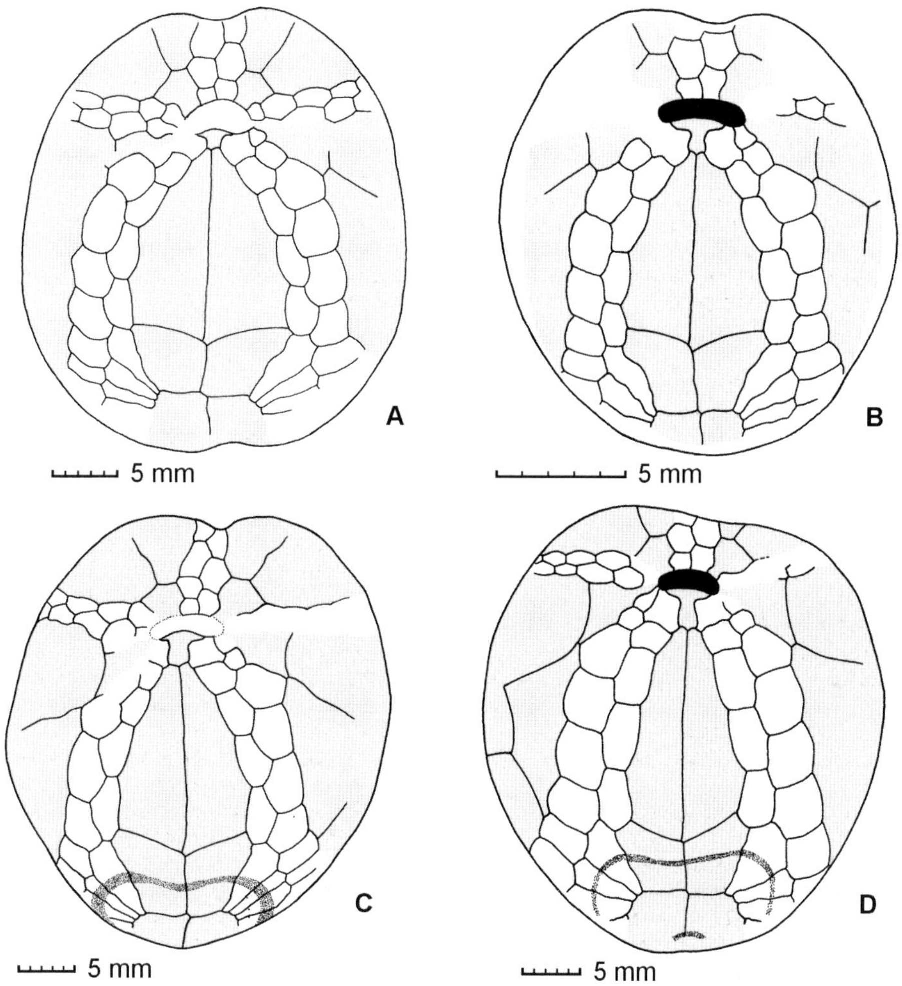

Figure 79: *Brissopsis ottnangensis* Hoernes, 1875: oral plating (A: Ottnangian, Ottnang, OÖ, Austria, paralectotype GBA 1875/1/78a; B: Ottnangian, Ottnang, OÖ, Austria, paralectotype GBA 1875/1/78c; C: Badenian, Retznei, Styria, Austria, NHMW 2004z0098/0002; D: Badenian, Devínska Nová Ves, Slovak Republic, SNM z24174). Interambulacra shaded, poorly visible sutures omitted or stippled.

sides of the pores facing each other; interporal area is crossed by a narrow ridge of stereom, which links the two pores). In the adapical third of the anterior poriferous zones of the anterior paired petals the pores are distinctly reduced in size. In the posterior poriferous zones and in both poriferous of the posterior paired petals this is not the case (Fig. 78.D). The posterior paired petals are usually between 70 and 85 % of the length of the anterior ones. In juveniles and sub-adult specimens (TL ~ 15 mm) they can be even shorter (~ 65 %). Anterior paired petals extending about 60 to 70 % of the corresponding test radius, while the posterior paired petals extend about 50 %. The anterior paired petals diverge at an angle of about 85 to 115°, the posterior ones at about 50 to 75°. Neither, anterior nor posterior petals are confluent (Figs. 78.C-D).

Ambulacra II, III and IV are sparsely tuberculated ambitally and bear vertically elongated, slit-shaped unipores. On the oral side ambulacra I and V form broad, naked peri-plastronal areas bearing few secondary and miliary tubercles. The pores in these areas are antero-posteriorly elongated, slit-shaped unipores. Five ambulacral plates extend into the subanal fasciole, bearing four enlarged pores on each side (Fig. 80.C-D). Adorally all ambulacra form small phyllodes consisting of five to six slit-shaped unipores with an extensive attachment area in each ambulacrum (Fig. 80.A).

Interambulacra: Aborally the interambulacra are strongly inflated between the petals. Aboral tuberculation is very dense and homogenous. The tubercles become larger towards the ambitus and on the oral side. On the plastron the tubercles are arranged in a fan-shaped pattern, radiating from an elevation in the posterior part of the plastron. The plastron belongs to the holamphisternous type (Figs. 79.A-D) and forms a strong keel along its central suture. The labrum is short, T-shaped, slightly wider than long and projects slightly into the peristome. Usually it does not extend beyond the first ambulacral plate of the adjacent ambulacra (Fig. 80.A). In a single specimen, however, it extends to the second ambulacral plate (Fig. 80.B). According to Turner & Norlund (1988) this atypical condition may occur in *Brissopsis* species which are usually characterised by a labrum that does not extend beyond the first ambulacral plates.

Plastron width is relatively variable. In the material from the type-locality it increases during ontogeny due to allometric growth. In the smallest specimens (TL ~ 15 mm) it ranges from 33 to 37 % of TW and increases to 41 % in the largest speci-

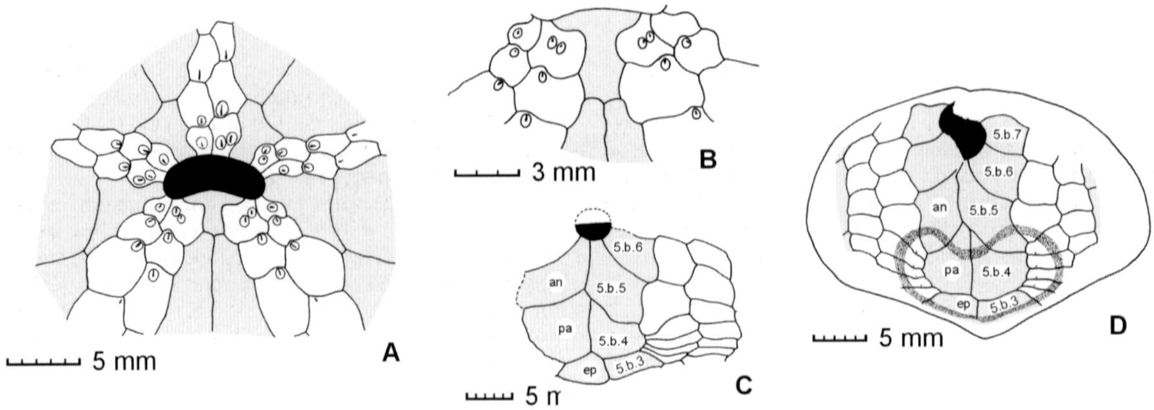

Figure 80: *Brissopsis ottnangensis* Hoernes, 1875: labrum shape, typical condition (A: Badenian, Devínska Nová Ves, Slovak Republic, SNM z24181), rare condition with labrum extending to 2nd ambulacral plate on each side (B: Badenian, Devínska Nová Ves, Slovak Republic, SNM z24180), anal plating (C: Ottnangian, Ottnang, OÖ, Austria, NHMW 1997z0178/0520; D: Badenian, Retznei, Styria, Austria, NHMW 2004z0098/0002). Interambulacra shaded, poorly visible sutures omitted or stippled, subanal fasciole dotted. Numbering of anal and preanal plates follows the Lovénian system; Abbreviations: ep = episternal plate, pa = preanal plate, an = anal plate.

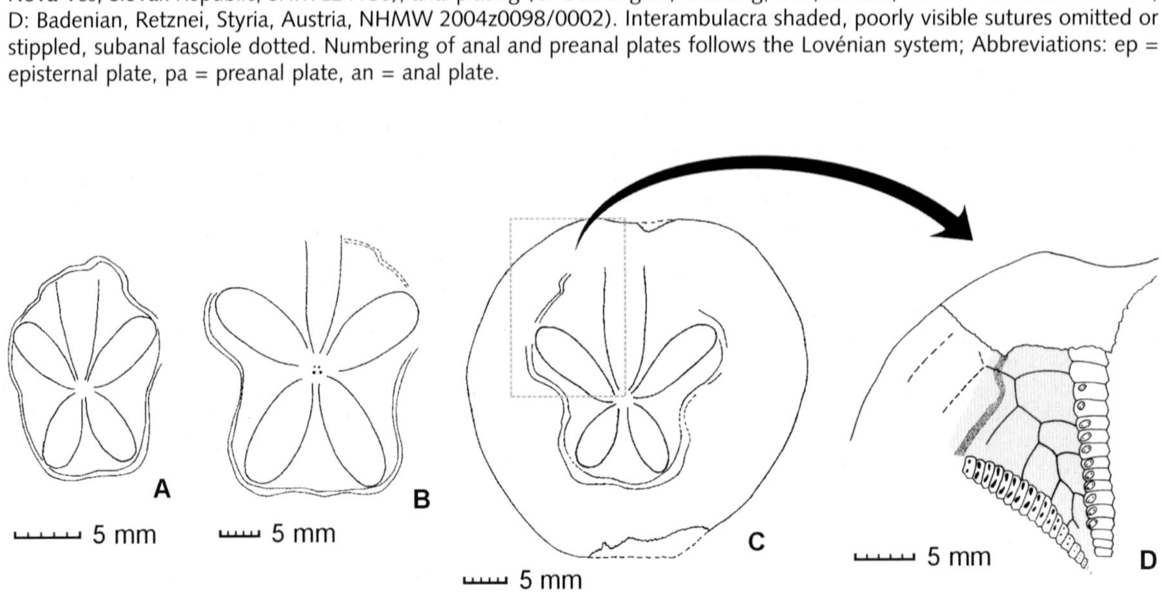

Figure 81: *Brissopsis ottnangensis* Hoernes, 1875: peripetalous fasciole shape (A: Badenian, Devínska Nová Ves, Slovak Republic, SNM z24179; B: Badenian, Devínska Nová Ves, Slovak Republic, SNM z24182; C-D: A: Badenian, Kalksburg, Vienna, Austria, NHMW 1981/55/12).

mens. In specimens from the Badenian of Devínska Nová Ves a markedly narrower plastron (~ 30 % of TW in 30 mm specimens) is observed.

The shape and number of the preanal and anal plates is similar to the condition observed in *B. atlantica* by Turner & Norlund (1988: fig. 3 A), both in material from the type locality Ottnang (Fig. 80.C) and in younger (Badenian) specimens (Fig. 80.D). There is only one pair of preanal plates (plates 5.a.4, 5.b.4), and the first anal plates (5.a.5, 5.b.5) are distinctly elongated vertically, tapering and form only a narrow portion of the periproctal margin.

Peristome: The peristome is situated anteriorly on the oral side of the test, about 15 to 25 % of TL from the anterior margin. It is kidney-shaped and measures 2.5 *vs.* 7 mm in a 40 mm TL specimen. The labrum is only slightly projecting.

Periproct: The periproct lies high on the posterior face of the test. It is nearly circular in shape but poorly preserved in most available specimens. In some specimens, especially in smaller ones it is partially visible in aboral view due to the obliquely truncated posterior end.

Fascioles: Both peripetalous and subanal fascioles are present. They belong to the orthofasciole type (*sensu* Néraudeau & al. 1998b). The peripetalous fasciole is indented between the anterior and posterior paired petals (Figs. 81.A-C) and has a distinct kink in interambulacra 2 and 3 (Fig. 81.D). The subanal fasciole is kidney shaped (Fig. 80.D).

Differential diagnosis:

Brissopsis crescenticus Wright, 1855 from the Early Miocene of Malta differs from this species by its confluent petals, with distinctly laterally flexed petal tips, much less strongly diverging posterior petals, and more elongated outline [based on topotypic material collected from the Lower *Globigerina* Limestone in Malta; see also Challis, 1980; the material from the Burdigalian of the Rhône Basin attributed to *B. crescenticus* by Philippe (1998) is not typical of that species, if at all conspecific].

The Pliocene to extant *B. atlantica* Mortensen, 1907, which is the descendant of *B. crescenticus* according to Mortensen (1951), likewise differs by its confluent posterior petals and laterally flexed anterior petals (among other features; compare Mortensen, 1913; Borghi, 1997; Lacour & Néraudeau, 2000).

The extant *B. elongata* Mortensen, 1907 from the Caribbean and the South Atlantic differs by its confluent posterior petals, different arrangement of the anal and preanal plates and straight labrum margin (compare Turner & Norlund, 1988).

The Late Miocene to extant species *Brissopsis lyrifera* (Forbes, 1841) is very similar to *B. ottnangensis* and the latter species certainly belongs into the *Brissopsis lyrifera* – group as identified and characterised by Lacour & Néraudeau (2000). Main differences to *B. lyrifera* are the shape of the individual petals (more lanceolate in *B. lyrifera*, wider and more lobate in *B. ottnangensis*), fasciole shape and overall test shape [extant *B. lyrifera* are more elongated (TW of ~ 85-91% TL) than *B. ottnangensis*] (compare data and illustrations found in Mortensen, 1951; Borghi, 1997; Lacour & Néraudeau, 2000).

The Cuban species *B. aguayoi* Sánchez Roig, 1952 from the Oligocene to Miocene and *B. jimenoi* Cotteau, 1875 from the Miocene differ by their strongly confluent posterior and laterally flexed anterior petals (compare Kier, 1984). Likewise the three known fossil Australian *Brissopsis* species *B. tatei* Hall, 1907, *B. australis* McNamara et. al. 1986, and *B. praeluzonica* Fell, 1964 are also easily distinguished by the same petal configuration as the Cuban species (see McNamara et. al. 1986).

Discussion:

Brissopsis duciei Wright, 1855 from the Burdigalian to Messinian of Malta (see Challis, 1980) is very similar to *B. ottnangensis* in many aspects. It differs from the present species by its more diverging anterior petals (up to 120°, even in large specimens), more circular outline (TW ranging from 96 to 103 % TL), and less strongly indented peripetalous fasciole in interambulacra 1 and 4. Unfortunately no specimens have been available for direct comparison. It is also unclear how the Burdigalian specimens of *B. duciei* look like, as all illustrations depict specimens from the Messinian Upper Coralline Limestone. In any case the two species seem to be closely related and further investigations involving material from the Burdigalian of Malta are necessary to see if the two forms might be end-members of a species with broader intraspecific variation.

Brissopsis specimens from the Badenian of Austria differ from *B. ottnangensis* from the type-area by its slightly more diverging anterior (~ 90-115°) and slightly less diverging posterior (50-60°) paired petals. Yet these features show considerable variation in the "population" from the type-locality too and the overlap is large. Moreover, the different preservation hampers comparison and distorts metric features (Ottnangian specimens come from pelitic sediments, are deformed and usually lack the original test; Badenian specimens, in contrast come from carbonates and retain the original shell). The key features used for the differentiation of extant *Brissopsis* species (petal shape and structure, labrum shape, number and shape of the preanal and anal plates, shape of the fascioles and number of ambulacral plates extending into the subanal fasciole) are identical in the material from both time slices. Thus they are treated as conspecific here.

Vadász (1915) identified this species with *B. consobrinus* Lambert, 1909 from the "Helvétien" (? Middle Miocene) of Sardinia, which is indeed extremely similar to the present taxon. Without knowledge on the plating pattern, width of the plastron and other features not mentioned by Lambert an attribution to that species can only be tentative.

Several of the *Brissopsis* species and subspecies of Botto-Mica (1896) are similar to *B. ottnangensis*. All of them are, however, based on internal moulds and plating patterns, plastron structure and shape of the fascioles are unknown. Because of this lack of information it is impossible to judge their relation to *B. ottnangensis* without re-examination of Botto-Mica's material.

Occurrence:

Austria: Eggenburgian (Early Burdigalian) to Late Badenian (Early Serravallian)

- Molasse Zone: Antiesenhofen, near Schärding/Inn, OÖ ([NHMW]); borehole Eisenhub 2, OÖ (Schaffer & Grill, 1951); Großwiesenhardt, near St. Marienkirchen, OÖ ([NHMW]); Haller Schlier, Pfarrkirchen, OÖ ([NHMW]); Mechters, near St. Pölten, NÖ (Abel, 1904); Ort, near Bad Füssing, OÖ (Weidlinger, 2003); Ottnang, OÖ (Hoernes, 1875a, b; Kittl, 1886; Commenda, 1900; Toula, 1914; Veit, 1943; Schaffer & Grill, 1951; Sieber, 1956; Papp & Thenius in Kühn, 1962; Stojaspal, 1975; Rupp et al., 1991; [GBA]; [NHMW]); Pichl, WNW of Wels, OÖ (Suess, 1891); Pielach river, south of Völlerndorf, NÖ ([NHMW]); Platt, near Zellerndorf, NÖ (Sieber, 1935; Kroh, 2003b); St. Pölten, NÖ (Abel, 1904; [NHMW]); Schlierfazies der Innviertler Serie, OÖ (Tollmann, 1985)
- Vienna Basin: Bad Vöslau, NÖ ([NHMW]); borehole at Aderklaa (Badenian, "Walbersdorfer Fazies"), NÖ (Veit, 1943); borehole at Oberlaa (Badenian, "Walbersdorfer Fazies"), Vienna (Veit, 1943); Gainfarn, NÖ (Kroh, 2002b; [NHMW]); Kalksburg, Vienna ([NHMW]); Steinberggebiet, near Zistersdorf, NÖ (Sieber, 1953a)
- Eisenstadt-Sopron Basin: Forchtenau, Bgld (Vendl, 1930); Mattersburg, Bgld (Vendl, 1930); Walbersdorf, near Mattersburg, Bgld (Hoernes, 1884; Toula, 1884; Kittl, 1886; Procházka, 1892a; Schaffer, 1897; Hoernes, 1903; Vendl, 1930; Janoschek in Schaffer, 1943, 1951; Rögl & Müller, 1976; [GBA])

Lavant Basin: Mühldorf im Lavanttal, Ktn (Wank & Stojaspal, 1980; Wank, 1981)
Styrian Basin: Retznei (Weissenegg Fm., Lafarge quarry Rosenberg), Styria ([NHMW]); Weitendorf, Styria ([NHMW]); Wetzelsdorf, near Preding, Styria (Holler, 1900; [NHMW]); area Wildon-Leibnitz, Styria (Dreger, 1913)

Paratethys (non-Austrian occurrences): ? Oligocene, Ottnangian (Late Burdigalian) to Late Badenian (Early Serravallian)
Bavarian Molasse Basin: Mitterdorf, near Griesbach, Bavaria ([NHMW])
Molasse Zone: Ostrava (= Ostrau), Czech Republic (Kittl, 1886; Schaffer, 1897); Židlochovice, Moravia, Czech Republic (Cicha, 1978)
Vienna Basin: Devínska Nová Ves (brickyard), Slovak Republic (Schaffer, 1897; Abel, 1902; Hoernes, 1903; Schaffer, 1908; Toula, 1915; Janoschek in Schaffer, 1943; Tomašových, 1998; [NHMW]); Dubovce, c. 7 km ESE Holíč, Slovak Republic (Cicha et al., 1967); Radošovce, c. 9 km ESE Holíč, Slovak Republic (Cicha et al., 1967); Strachotín, c. 11 km N Mikulov, Czech Republic (Pospíchalová in Kroh, 2003b)
Eisenstadt-Sopron Basin: Sopron, Györ-Monson-Sopron, Hungary (Vendl, 1930)
Styrian Basin: Ferental ("Schlier" facies), probably Slovenia (Meznerics, 1936); Jahring ("Schlier" facies), probably Slovenia (Meznerics, 1936)
Great Hungarian Basin (Pannonian Basin): Etes (= Ettes), Nógrád, Hungary (Vadász, 1915); Fót Fm. and Garáb Schlier Fm., Nógrád, Hungary (Hámor, 1985); Györ, Györ-Monson-Sopron, Hungary (Vendl, 1930); Karacs-Lapujtö (= Lapujtö), Nógrád, Hungary (Vadász, 1915); Köalja-hegy, near Sajóvárkony, Borsod-Abaúj-Zemplén, Hungary (Jaskó, 1940); ? Letkés, Pest (Vadász, 1915); Mogyoród, near Fót, NW Budapest, Pest, Hungary (Horusitzky, 1927; Kroh, 2003b [NHMW]); Sajóvárkony, Borsod-Abaúj-Zemplén, Hungary (Jaskó, 1940); Vaszar, Baranya, Hungary (Vadász, 1915); Zagyvapálfalva (= Pálfalva), Nógrád, Hungary (Vadász, 1915)
Nowy Sącz Basin: Niskowa, western solpes of the Nowy Sącz depression, Poland (Bałuk, 1970)
Transcarpathian Basin: Hlinné, eastern Slovak Republic (Seneš, 1955; Cicha et al., 1967)
Transylvanian Basin: Chechiş, Romania (Givulescu & Duşa, 1960); Coştei (= Kostéj), Timiş, Romania (Givulescu & Duşa, 1960); Gârbova de Sus (= Felsö-Orbó), Romania (Vadász, 1915); Ilteu, Romania (Givulescu & Duşa, 1960); Lăpugiu des Sus (= Felsölapugy, = Lapugy), Hunedoara, Romania (Vadász, 1915); Tărăţel (= Cerecel), Romania (Vadász, 1915; Givulescu & Duşa, 1960); Tusa (= Tusza), Romania (Givulescu & Duşa, 1960);
Zala, Sáva and Dráva Basins: La'sko (formerly Tüffer; "Schlier" facies), Slovenia (Hoernes, 1903)

Mediterranean: ? Miocene
Western Mediterranean: ? Colline di Bologna, Italy (Manzoni, 1879); ? Ponti, near Camerino, Italy (de Loriol, 1882)

Brissopsis sp.

	1855	*Brissopsis* – Rolle: 354
	1871	*Brissopsis* – Stur: 562, 563
?	1880	Stacheln von *Brissopsis* – Rzehak: 302
	1884	*Brissopsis* sp. – Bittner: 457
	1893	*Brissopsis* sp. – Toula: 288
	1894	*Brissopsis* – Dreger: 74
	1913b	*Brissopsis* – Schaffer: 58
	1927a	*Brissopsis* – Schaffer: 82
	1936	*Brisopsis* (?) sp. – Paucă: 143
	1938	*Spatangus* cfr. *euglyphus* Laube 1867 – Poljak: 199-200; pl. 8, fig. 2 [misidentified *Brissopsis* sp.]
	1939	*Brissopsis* sp. – Kapounek: 78
	1939	*Brissopsis* sp. – Langer: 353
	1943	*Brissopsis* – Schaffer: 518
	1951	*Bryssopsis*– Janoschek in Schaffer: 575
	1952	*Brissopsis* sp. – Flügel et al.: 178
	1965	*Brissopsis* – Kollmann: 516
	1977	*Brissopsis* sp. – Ebner & Gräf: 161
	1981	*Brissopsis* sp. – Halmai: 105
	1987	*Brissopsis* sp. – Krainer: 150
	1990	*Brissopsis* sp. – Friebe: 245, 254
v.	2001	*Brissopsis* sp. – Schmid et al.: 13

Material:
Ottnangian (Late Burdigalian) – Antiesen river, near Antiesenhofen, near Schärding/Inn, OÖ, Austria
NHMW: 10 specimens (NHMW 1971/1485k-l, 1985/67/24-25, ../23a+b, 1989/83/36-38, ../40, 2003z0090/0001)
Ottnangian (Late Burdigalian) – Antiesenhofen, near Schärding/Inn, OÖ, Austria
NHMW: 11 specimens (NHMW 2003z0090/0006)
Ottnangian (Late Burdigalian) – Großwiesenhardt, near St. Marienkirchen, OÖ, Austria
NHMW: 9 fragments (NHMW 1986/101/24a+b, ../25a+b, 2003z0090/0003)
Ottnangian (Late Burdigalian) – St. Marienkirchen, OÖ, Austria
NHMW: 1 specimen (NHMW 1988/40/20)
Early Badenian (Langhian) – Gainfarn, NÖ, Austria
NHMW: 3 specimens (NHMW 2004z0154/0002, ../0006)

Discussion:
The records listed here either lack sufficient information to allow specific determination or are based on fragmentary material that is poorly preserved.

Occurrence:

Austria: Ottnangian (Late Burdigalian), Early to Late Badenian (Langhian-Early Serravallian)
Molasse Zone: Antiesenhofen, near Schärding/Inn, OÖ ([NHMW]); Gaindorf, NÖ (Schaffer, 1913b); Großwiesenhardt, near St. Marienkirchen, OÖ ([NHMW]); Ottnang, OÖ (Schaffer, 1943); Ravelsbach, NÖ (Schaffer, 1913b)
Vienna Basin: deep well Eichhorn 1, E of Zistersdorf, NÖ (Janoschek in Schaffer, 1951); Gainfarn, Nö ([NHMW]); Jungenberg, near Bisamberg, Vienna (Langer, 1939)
Eisenstadt-Sopron Basin: St. Margarethen (Kummer quarry), Bgld (Kapounek, 1939; Schmid et al., 2001); Walbersdorf, Bgld (Schaffer, 1927a)
Styrian Basin: Retznei (Lafarge quarry), Styria (Friebe, 1990); Spielfeld, Styria (Rolle, 1855; Stur, 1871); Teufelsberg, near St. Veit an der Drann, Styria (Dreger, 1894); Wagna (brickyard), Styria (Kollmann, 1965); Weitendorf, Stmk (Flügel et al., 1952; Ebner & Gräf, 1977; Krainer, 1987)

Paratethys (non-Austrian occurrences): Ottnangian (Late Burdigalian) to Late Badenian (Early Serravallian)
Molasse Zone:? Židlochovice (= Groß-Seelowitz), Moravia, Czech Republic (Rzehak, 1880); Kralice nad Oslavou (= Kralitz), Moravia, Czech Republic (Toula, 1893)
Vienna Basin: Devínska Nová Ves, Slovak Republic (Schaffer, 1927a)
Great Hungarian Basin (Pannonian Basin): Fót, Hungary (Halmai, 1981)
Transylvanian Basin: region east of Taşad, Beiuş Basin, Romania (Paucă, 1936)

Zala, Sáva and Dráva Basins: Dráva valley, near Melling, E of Maribor (Marburg), Slovenia (Stur, 1871); La'sko (formerly Tüffer) ("Schlier" facies), Slovenia (Bittner, 1884); Dolje, Popov Dol, Croatia (Poljak, 1938)

Genus *Meoma* Gray, 1851

Type-species: *Meoma grandis* Gray, 1855; by original designation (Pomel, 1869: 131).

Diagnosis: Test ovate, with slight anterior sulcus; high profile; apical disc ethmolytic, with four gonopores; genital plate 2 projects beyond posterior oculars; ambulacrum III narrow and weakly sunken aborally, with small, simple isopores; paired ambulacra petaloid, narrow and prominently sunken; petals relatively straight with narrow interporal zone, extending to near ambitus; anterior petals strongly divergent; periproct large, on short, steeply undercut posterior face; peristome transversely elongated, kidney-shaped with projecting labrum; labral plate short and wide, not extending beyond first ambulacral plate; plastron fully tuberculated, ovate in outline, formed of sternal and episternal plates; episternal plates not strongly tapered to rear; spines and tubercles within peripetalous fasciole often slightly larger, but tubercles not sunken; peripetalous fasciole indented in interambulacra 1 and 4; subanal fasciole bilobed, well-developed only in juvenile, in adult only the adoral portion remains (modified from Kier, 1984; and Smith *"The Echinoid Directory"*, 13.05.2004).

Distribution: Middle Eocene to Miocene – circumtropical, and temperate regions of Australia and New Zealand

Remarks: *Meoma* has similarities to *Macropneustes*, but can be distinguished by its deeply sunken, narrow petals and the shape of the peripetalous fasciole (indented in interambulacra 1 and 4)

According to Chesher (1969, 1970: 755) *Brissus cruciatus* Agassiz (in Agassiz & Desor, 1847), the type-species of *Schizobrissus* Pomel, 1869, is based on a single, poorly preserved specimen, that cannot be determined confidently even at family level. He thus concluded that *Schizobrissus* cannot be regarded a recognisable taxon. [According to Philippe (1998: 197) no other material was ever recorded from the type-locality, the Island of Capri (Thyrrenian Sea)].

Fischer's (1966) subdivision of *Meoma* into three subgenera was based mainly on the extent of the subanal fasciole. Yet Kier & Grant (1965: fig. 13) and Chesher (1969) showed that this feature is ontogenetically controlled and shows strong intraspecific variation in some species. In recent papers (e.g. Chesher, 1970; Kier, 1984; Žítt, 1985; McNamara et al., 1986) this subdivision was not used.

Ecology and biogeography: Modern species of *Meoma* inhabit clean, coarse sand or shell grit between 2 and 200 metres depth (Kier, 1984; Hendler et al., 1995). They are often associated with turtle grass beds and/or patch reefs and are non-selective or selective deposit feeders. *M. ventricosa* lives buried, with up to 4 cm of sediment above their apex. Most animals, however, are not completely buried, or just barely so. Specimens found within turtle grass beds live epibenthic and cover themselves with shell fragments and leaves of sea grass (Kier & Grant, 1965: 48). *M. ventricosa* is preyed upon the sea-star *Oreaster reticulatus* (Linné), sea turtles, stingrays, bony fish and cassid gastropods (Kier & Grant, 1965; Chesher, 1969; Gladfelter, 1978).

Within one year specimens of *M. ventricosa* can reach 88 mm TL and grow to 134 mm in four years. The largest known specimens are up to 200 mm in TL and come from open sand areas or deeper water (Chesher, 1969).

All extant species occur in the tropical zone (Caribbean and central East Pacific), but one species occurs in the Oligocene of southern Australia (McNamara et al., 1986) and the Oligocene to Early Miocene of New Zealand (Henderson, 1975). Additionally there are numerous records of *Meoma* from the Miocene of the Mediterranean.

Meoma locardi (Cotteau in Locard, 1877)

(Fig. 82; Pl. 74, Figs. 1-2)

*	1877	*Linthia locardi*, Tournouer, 1874 – Cotteau in Locard: 288-289; pl. 12, figs. 1-2
	1909	*Macropneustes saheliensis* (Pomel). – Stefanini: 58-60; pl. 1 (5), figs. 1-2
v.	1915	*Schizobrissus cruciatus* Ag. – Vadász: 228-229; text-fig. 118
	1995	*Meoma (Schizobrissus) locardi* (Cotteau, 1877) – Lachkhem & Roman: 15-16; pl. 5, fig. 1
	1998	*Meoma (Schizobrissus) locardi* (Tournouer in Cotteau, 1877) – Philippe: 194-195; pl. 21, figs. 10-14

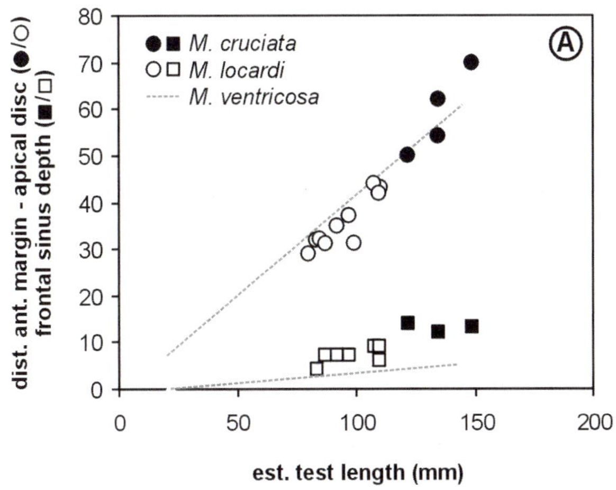

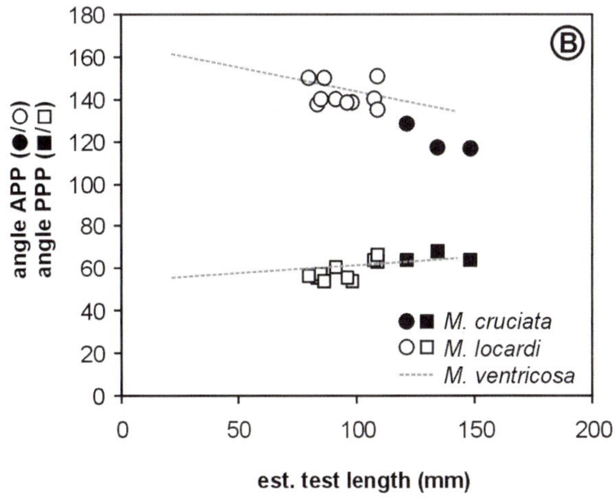

Figure 82: Ontogenetic changes in *Meoma*: distance between anterior margin and apical disc (A, circles), depth of the frontal sinus (A, rectangles), divergence of the anterior (B, circles) and posterior paired petals (B, rectangles). The grey, hatched line represents the degree of ontogenetic change present in extant *Meoma ventricosa* (based on Kier & Grant, 1965 and additional material in the NHMW collection). Data on the fossil specimens based on NHMW material and culled from the literature (Cotteau, 1877; Lambert, 1909; Stefanini, 1909; Vadász, 1915; Lachkhem & Roman, 1995; Philippe, 1998), attribution to *M. locardi* and *M. cruciata* represents the opinion of the respective authors. Note that the distinction between those two species seems to be based mainly on test length. Changes in the four features reported to be relevant for the distinction of the two species could also be interpreted as growth-related changes within a single species similar to those observed in extant *M. ventricosa* (see text).

Material:

Badenian (Langhian-Early Serravallian) – "Gallenegg", near Zagorje ob Savi (formerly Sagor, Krain), Slovenia

NHMW: 1 specimen, only aboral surface preserved (NHMW C.1967)

Badenian (Langhian-Early Serravallian) – Trbovlje (formerly Trifail), Slovenia

GBA: 1 partial internal mould (no inv. no.)

Late Badenian (Early Serravallian) – Mátraszölös (formerly Mátraszöllös), Nógrád, Hungary

MAFI: 1 specimen, with poorly preserved oral surface [MAFI Ech 157 (figured specimen of Vadász, 1915)]

Dimensions (in mm):

Inv. No.	TL	TW	TH
GBA specimen	> 90	94	~54
MAFI Ech 157	110	100	~50
NHMW C.1967	122	106	40

Description:

Size and shape: The test is large, antero-posteriorly elongated with a roughly oval to slightly heart-shaped outline. The anterior margin shows a distinct frontal notch (depth 9 mm in the smallest and 14 mm in the largest specimen). In profile the test is low conical to highly domed, with a steep anterior slope. Posterior end obliquely truncated. The maximum height lies around the apical disc.

Apical disc: The apical disc is ethmolytic with 4 large, circular gonopores. It lies anterior of the centre, about 45 % TL from the anterior margin.

Ambulacra: The paired ambulacra are petaloid adapically. The petals are deeply sunken, subequal in length and bear many closely spaced elongate isopores. The interporiferous zone is narrow, barely reaching the width of a single poriferous zone in the widest, central part of the petals. The paired petals are straight (PPP), respectively slightly flexed anteriorly (APP), and extend about 55 to 60 % of the corresponding test radius. The anterior paired petals form an obtuse angle of around 128° in the largest, 135° in the middle and 140° in the smaller specimen (measured in planar aboral view). The posterior paired petals form an acute angle of about 50 to 65°. Ambulacrum III is nonpetaloid, and is increasingly depressed from the apex to the margin, where it forms a deep frontal sinus. The pores are poorly preserved and very small.

Interambulacra: The interambulacra are inflated between the petals, forming rounded keels between them. Other details are obliterated.

Peristome: Not preserved.

Periproct: Not preserved.

Fascioles: Not preserved.

Discussion:

Despite the poor preservation of the oral side of all three specimens the aboral side provides enough features to identify them with *Meoma locardi*.

Only three specimens of this species were available from the whole Central Paratethys. Although not coming from Austria they are discussed here to give some remarks on the systematics of this species.

Traditionally two species of *Meoma* have been recorded from the Miocene of the Mediterranean area: *M. cruciata* (Agassiz in Agassiz & Desor, 1847) and *M. locardi* (Cotteau in Locard, 1877) [sometimes also attributed to Tournouër, because Cotteau (1877) used a name proposed to him in a letter from that man; yet according to the ICZN Cotteau, 1877 is the correct author]. The problems around the name *cruciata* were outlined by Chesher (1969, 1970; see above) and obviously this name has to be considered as *nomen dubium* [holotype unrecognisable at family level (according to Chesher, 1970); type-locality imprecisely known, no topotypic material available]. Interestingly the holotype was never figured and its whereabouts are unknown [Cotteau (1877: 290) was not able to find it, nor anyone else, as far as I am aware of]. A plaster cast ("moulage" T.75) of the holotype is housed in the collection of the Museé d'Histoire Naturelle de Neuchâtel (Lambert & Jeannet, 1928: 198, T.75).

Most authors (e.g. Cotteau, 1877; Lachkhem & Roman, 1995; Philippe, 1998; ...) distinguished the two species on base of the depth of the frontal notch (deeper in *cruciata*), the position of the apical disc (more anterior in *locardi*) and the shape and divergence of the anterior paired petals (anterior flexed and strongly divergent in *locardi*, straight and less strongly divergent in *cruciata*). Both forms have the same temporal and spatial distribution and Lachkhem & Roman (1995:16) stated that "It is astonishing that two species of the same kind coexist in the same layer [in Melilla, Morocco], but it is also the case in Corsica, in the Rhône Basin and undoubtedly also in Sardinia. This could be attributed to sexual dimorphism, which is difficult to prove or they could be sympatric species." [translated from the French text]. Usually specimens larger than 115 mm TL were classified as *cruciata*, while smaller ones were named *locardi*. Opinions on specimens close to this "critical point", vary [e.g. the record of Vadász (1915), who calls his specimen *cruciata*, was placed into the synonymy list of *locardi* by Lachkem & Roman (1995:15-16) and then again in cf. *cruciata* by Philippe (1998: 197)].

In this context the description of the morphological changes occurring during the ontogeny of *Meoma ventricosa* by Kier & Grant (1965: 38-47, figs. 8-14; plate 9) is very elucidating. According to them the following changes occur during the growth from 22 to 160 mm TL: a) the test becomes lower, more angular, and the depth of the frontal notch increases, b) the petals become depressed, c) the posterior truncation tilts, so that the periproct becomes visible in oral view, d) the posterior petals become much longer (slightly exceeding the anterior petals in the largest specimens, f) the petals become narrower, g) the petals become curved, h) the divergence of the anterior paired petals decreases (from 162 to 135°) and that of the posterior ones increases (from 56 to 65°), i) the peristome becomes smaller relative to overall size and the labrum extends anteriorly, j) the spines become more equal in size and proportionally shorter, k) the subanal fasciole becomes discontinuous, l) the apical disc moves posteriorly towards the centre, m) genital pores appear at stage between 40 and 60 mm TL.

From the data provided by Kier & Grant (1965) it is evident that the features used as diagnostic on species level in the fossil *Meoma* species of the Mediterranean are entirely size related (Fig. 82). This can indeed be demonstrated by plotting, e.g. the anterior paired petal divergence against test length (Fig. 82.B, circles). It is thus doubtful to consider *M. locardi* and *M. cruciata* as two different species, and it is likely that they rather correspond to different growth stages of a single species. Yet the present author hesitates to synonymise them without having examined the type material of either species.

M. saheliensis (Pomel, 1887) was placed into the synonymy of *M. locardi* by Lachkem & Roman (1995:15-16) and Philippe (1998: 194). *M. lata* (Wright, 1855) from the Messinian of Malta is also highly similar to the present species (see Wright, 1855; Challis, 1980).

Occurrence:

Paratethys (non-Austrian occurrences): Badenian (Langhian-Early Serravallian)

Great Hungarian Basin (Pannonian Basin): Mátraszölös, Nógrád, Hungary (Vadász, 1915; [MAFI])

Zala, Sáva and Dráva Basins: "Gallenegg", near Zagorje ob Savi (formerly Sagor), Slovenia ([NHMW]); Trbovlje (formerly Trifail), Slovenia ([GBA])

Mediterranean: Burdigalian to Messinian, ? Pliocene

Western Mediterranean: Algeria (Cotteau et al., 1891); Corsica, France: Santa Manza (Cotteau in Locard, 1877);

Morocco: Idorhotsene, Melilla (LACHKHEM & ROMAN, 1995); Rhône Basin, France: les Baux, Barbentane, Gordes, Goult and Villeneuve-lès-Avignon (PHILIPPE, 1998); Reggio Emilia, Italy: Serra die Guidoni, Montese, Salto and Semelano (STEFANINI, 1909); Sardinia, Italy: Santa-Lucia, near Pozzomaggiore (COTTEAU, 1895), Cagliari and Is Mirrionis (LAMBERT, 1909)

Family Spatangidae GRAY, 1825
Genus *Spatangus* GRAY, 1825

Type-species: *Spatangus purpureus* MÜLLER, 1776; by subsequent designation (ICZN, opinion 209, p. 385, 8th May 1954; see also ICZN, opinion 608, 21 March 1961).
Diagnosis: Test chordate, with shallow or moderately deep frontal notch; domed aboral and flattened oral surface; apical disc ethmolytic, with four gonopores; ambulacrum III narrow with small, simple isopores; paired ambulacra petaloid and flush with test; anterior paired petals flexed anteriorly, rudimentary pore-pairs in adapical half of anterior column; other pore-pairs large and conjugate; Posterior petals expanding adapically and converging distally; periproct on short steeply undercut posterior face; peristome transversely elongated, kidney-shaped, with projecting labrum; labrum anvil-shaped, in broad contact with sternal plates; sternal plates near-symmetrical, relatively broad and fully tuberculate; plastron formed of sternals and episternals; episternals not contracting towards posterior; aboral tuberculation heterogeneous with scattered coarse tubercles (not sunken in camellae); well-developed subanal fasciole, bilobed or shield-shaped; No other fascioles developed (modified from SMITH *"The Echinoid Directory"*, 13.05.2004).
Distribution: ? Eocene, Oligocene to Recent – cosmopolitan (FISCHER, 1966)
Remarks: There are more than 80 nominal species in the subgenus *Spatangus* (*S.*) (compare LAMBERT & THIÉRY, 1909-1925 and KIER & LAWSON, 1978) alone, most of which were reported from Miocene deposits of Europe. Classification of *Spatangus* species is notoriously difficult and hence the synonymy of many species is rather complicated. Due to the fact that most of the ancient species were based upon criteria as overall test shape, arrangement of primary tubercles and the like, there are many synonymous species. MORTENSEN (1951) considered the following features to be useful in specific determination within the genus (for extant species): 1) shape of the subanal plastron; 2) presence of primary tubercles outside the petals; 3) test profile (including test height); 4) shape of the posterior end; 5) the number of pore pairs within the subanal fasciole; and 6) presence of a keeled sternum. A recent investigation by NÉRAUDEAU et al. (1998a) showed that especially the shape of the subanal fasciole (1), the abundance of primary tubercles within the adapical interambulacra (2), depth of the frontal sinus (3) and the plastron architecture (4) are important to distinguish the two *Spatangus* species [*S. purpureus* (MÜLLER) and *S. subinermis* POMEL] present in the Pliocene to Holocene of the Northern Mediterranean Sea. During the investigations done on the Neogene echinoids of Austria the plastron architecture and distribution of primary tubercles proved to be most useful to classify the different species.

Spatangus austriacus LAUBE, 1869

(Figs. 83, 84, 85.A, C, 86.A-B; Pl. 75, Figs. 1-3, Pl. 76, Figs. 1-4)

* v pp 1869a *Spatangus austriacus* LAUBE. – LAUBE: 184
v pp 1870 *Spatangus austriacus* LBE. – LAUBE: 314
v pp 1871 *Spatangus austriacus* LAUBE. – LAUBE: 73; pl. 19, fig. 2, 2a
non 1879 *Spatangus austriacus*, LAUBE. – MANZONI: 160; pl. 2, figs. 10-15; pl. 3, figs. 19-22; pl. 4, figs. 40-41 [fide MAZZETTI (1882b), AIRAGHI (1908: 256), STEFANINI (1908b: 112), LAMBERT (1924 in LAMBERT & THIÉRY, 1909-1925: 462)]
non 1880a *Spatangus austriacus* LAUBE. – MANZONI: 185, 188
non 1881 *Spatangus austriacus*, LAUBE. – MANZONI: 174
non 1882b *Spatangus Austriacus*, LAUBE. – MAZZETTI: 129
non 1882 *S.[patangus] austriacus* LAUB. – PANTANELLI: 45 [book review]
non 1883 *S.[patangus] austriacus* LAUB. – MAZZETTI & PANTANELLI: 66
? 1884 *Spatangus austriacus* LAUB. – COPPI: 191, no. 281
? 1887 *Spatangus austriacus* LAUBE – NEVIANI: 194, 208, no. 208
? non 1887b *Spatangus austriacus*, LAUBE. – KOCH: 270
? 1887 *Spatangus austriacus*, LAUBE. – PARONA: 307
? 1897 *Spatangus austriacus* LAUBE. – ANGELIS D'OSSAT: 291, 292-293
pp 1897 *Spatangus austriacus*, LAUBE – COTTEAU: 194
non 1897 *Spatangus austriacus* LAUBE. – VINASSA DE REGNY: 155
non 1901 *Spatangus austriacus* LBE. – AIRAGHI: 215-216; pl. 27 (9), fig. 8 [= *S. taramellii* AIRAGHI, 1908; fide AIRAGHI (1908: 255)]
non 1908b *Spatangus austriacus* LAUBE. – STEFANINI: 109-112
v. 1912a *Spatangus Austriacus* LAUBE. – SCHAFFER: 191-192; text-fig. 3; pl. 59, fig. 1-3; pl. 60, fig. 7
? non 1930 *Prospatangus austriacus* LAUBE. – VENDL: 50
non 1936 *Spatangus austriacus* LAUBE – PAUCĂ: 143, 197; pl. 2, figs. 1-2
non 1938 *Spatangus austriacus* LAUBE 1871 – POLJAK: 197-198; pl. 10, fig. 2
? non 1953 *Prospatangus austriacus* (LAUBE), 1871. – SZÖRÉNYI: 37-38, 88-89; pl. 8, figs. 3, 3a-b
? 1963 *Spatangus austricacus* LAUBE – COMASCHI CARIA: 27-29
? 1967 *Spatangus austriacus* LAUBE – MENESINI: 149-151; pl. 48 (3), figs. 1, 1a
v. 1971a *Spatangus austriacus* LAUBE, 1871 – STEININGER: 595; pl. 2, fig. 3; pl. 3, fig. 5
1971b *Spatangus austriacus* LAUBE – STEININGER: 119
1971c *Spatangus austriacus* LAUBE – STEININGER: 129
1974 *Spatangus austriacus* LAUBE – THENIUS: 47
? non 1979 *Spatangus austriacus* LAUBE, 1871 – MĄCZYŃSKA: 31-32, pl. 3, figs. 1-2; pl. 4, fig. 1; pl. 5, figs. 1-2
? non 1988 *Spatangus austriacus* LAUBE, 1871 – MĄCZYŃSKA: 62; pl. 4, figs. 4a-f
non 1996 *Spatangus austriacus* – MACHALSKI: 26
non 1997 *Spatangus austriacus* LAUBE – MAJCEN et al.: 106; pl. 5, fig. 2
v. 1998 *Spatangus austriacus* LAUBE – SCHULTZ: 120; pl. 54 fig. 3
v. 1999 *Spatangus austriacus* – HARZHAUSER & KROH: 221

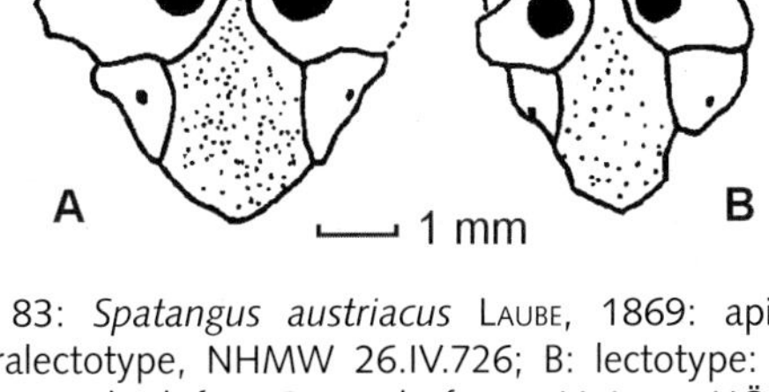

Figure 83: *Spatangus austriacus* LAUBE, 1869: apical disc (A: paralectotype, NHMW 26.IV.726; B: lectotype: NHMW 1866.I.1286; both from Bayersdorf, near Maissau, NÖ). Poorly visible sutures and genital pores omitted or stippled.

v.	1999	*Spatangus austriacus* Laube, 1871 – Kroh & Harzhauser: 160-161; pl. 1, figs. 1-2; pl. 3, fig. 3
v non	2001	*Spatangus austriacus* (Laube) – Schmid et al.: 13; pl. 1, fig. 5
v.	2002	*Spatangus austriacus* – Lukeneder & Harzhauser: 462
non	2004	*Spatangus austriacus* Laube, 1871 – Radwański & Wysocka: 385, 386; pl. 6, figs. 1a-b

Type-material:

Spatangus austriacus Laube, 1869:

Lectotype: NHMW 1866.I.1286, figured by Laube (1871: 73; pl. 19, fig. 2); Naturhistorisches Museum Wien, Geologische Abteilung

Paralectotype: NHMW 26.IV.726, figured by Laube [1871: 73; pl. 19, fig. 2a (erroneously labelled 1a at the plate)]; Naturhistorisches Museum Wien, Geologische Abteilung

Locus typicus: Bayersdorf, near Maissau, NÖ, Austria

Stratum typicum: Zogelsdorf Fm.

Age: Late Eggenburgian (Early Burdigalian), Early Miocene

Remarks: Laube (1869a: 184; 1871: 73) mentioned also an additional specimen from the Badenian of Großhöflein (= Nagyhöflány), Bgld, Austria (then belonging to Hungary), which was located at the Museum of Pest (probably the Hungarian National Museum). Unfortunately, neither Vadász (1915: 237) nor the present author were able to locate that specimen. It is, however, very likely that this specimen is not conspecific with the others, since *S. austriacus* is restricted to the Eggenburgian (see below).

Material:

Late Eggenburgian (Early Burdigalian) – Bayersdorf (Zogelsdorf Fm.), near Maissau, NÖ, Austria

NHMW: 2 specimens [NHMW 1866.I.1286 (**lectotype**) and 26.IV.726 (paralectotype)]

Late Eggenburgian (Early Burdigalian) – Eggendorf, NÖ, Austria

NHMW: 1 specimen (NHMW 1998z0048/0095)

Late Eggenburgian (Early Burdigalian) – Eggenburg (Kalvarienberg, Zogelsdorf Fm.), NÖ, Austria

IPUW: 1 specimen (no inventory no.)

Late Eggenburgian (Early Burdigalian) – Grübern (Zogelsdorf Fm.), near Maissau, NÖ, Austria

NHMW: 17 specimens (NHMW A2254, 1914.VII.19a-j; 2002z0159/0002-7)

IPUW: 3 specimens (no inventory no.)

Late Eggenburgian (Early Burdigalian) – Unternalb (Retz Fm., Hungerfeld), near Retz, NÖ, Austria

NHMW: 4 specimens (NHMW 1999z0049/0001, NHMW 1999z0049/0004)

Dimensions (in mm):

Inv. No.	TL	TW	TH
NHMW 1866.I.1286 (LT)	80.4	77.2	34.4
NHMW 26.IV.726 (PLT)	>75	>50	34.5
NHMW 1998z0048/0095	48.7	51.6	17.4
NHMW 1999z0049/0001	124.1	~115	53.3
NHMW 1914.VII.19a	89.6	87.6	>31
NHMW 1914.VII.19b	78.5	76.5	~ 27
NHMW 1914.VII.19c	66.0	64.2	> 18
NHMW 1914.VII.19d	62.8	60.2	22.4
NHMW 1914.VII.19e	55.3	53.1	18.3
NHMW 1914.VII.19i	54.2	>47	>18.5
NHMW 1914.VII.19j	44.5	42.0	14.4
NHMW 2002z0159/0005	57.2	~55.5	30.2
NHMW 2002z0159/0006	70.6	64.4	21.9

Description:

Size and shape: Test large to very large, heart-shaped and slightly elongated antero-posteriorly. The anterior margin is rounded with a distinct frontal notch, which increases in depth during growth. The posterior margin is bluntly pointed to transversely truncated. Maximum width coincides with the position of the apical disc. In profile the test is arched to wedge-shaped with a steep anterior slope and a gentle posterior slope. The anterior end is rounded, the posterior low and obliquely truncated. Slightly posterior of the apical system there is a distinct kink in lateral view. The maximum height lies posterior of the apical disc in interambulacrum 5, about one-third of the distance apical disc-posterior margin from it. The oral side is flattened, with a sunken peristome and inflated plastron. The test width ranges from 92.2 to 97.7 % TL (Mean=95.6 %), the height from 31.0 to 52.8 % TL (Mean=38.2 %).

Apical disc: The apical disc lies slightly anterior of the centre, about 42 to 45 % TL from the anterior margin. It is ethmolytic with four large gonopores, the anterior ones of which are slightly smaller (Figs. 83.A-B).

Ambulacra: Ambulacrum III nonpetaloid, slightly depressed, increasingly so towards the anterior margin. The pores are strongly oblique, minute partitioned isopores, lying in small depression in the centre of each plate. The interporiferous zone is wide and bears only secondary and miliary tubercles. The paired ambulacra are petaloid, depressed and moderately closed distally. The anterior paired petals form an obtuse angle of about 130 to 145°, the posterior paired petals an acute angle of about 60° (ranging from 57 to 64°). The anterior paired petals are flexed anteriorly distally, the posterior paired petals slightly laterally at their distal tips. The petals are broad and show allometric growth [i.e. width 10.9 to 11.2 % TL in subadult specimens (TL 50-65 mm), 9.9 to 10.4 %TL (TL ~ 80 mm) in adult specimens and 9.3 % in the largest specimen available (TL = 124.1)]. Their shape changes from lobate in subadult to lanceolate in adult specimens. They extend two third to three quarter of the corresponding test radius. Within the petals the pores are elongated isopores. The poriferous zones are slightly depressed, whereas the interporiferous zones, which are about 2 (subadult specimens) to 2.5 (large specimens) times as wide as a single poriferous zone, are slightly inflated. The pores of the uppermost 5 to 8 plates in the anterior plate row of ambulacra II and IV are minute partitioned isopores. Outside the petals only minute microunipores are present, except in the phyllodes where large teardrop- to slit-shaped unipores with extensive periporal area are found.

Adorally well developed phyllodes are present, bearing 7 pores in ambulacrum III, 11 to 12 in each of the anterior paired ambulacra and 7 to 8 in the posterior paired ambulacra. Ambulacra I and V form moderately broad peri-plastronal areas lacking primary tubercles. In many subadult specimens the ambulacral plates are distinctly inflated on the oral side (e.g. Pl. 75, Fig. 3; Pl. 76, Fig. 3b).

Interambulacra: Adapically the interambulacra are slightly inflated between the petals. They bear few crenulate, perforate primary tubercles with inclined areoles, which are arranged in chevron-shaped groups (Fig. 85.A, C) at the adapical border of each plate in the interambulacra 1, 4 and 5. In the interambulacra 2 and 3 the primary tubercles are situated along the borders of ambulacrum III and along the adapical plate borders in the posterior plate rows (2a and 3b). Towards the margin of the test the primary tubercles are less numerous and smaller. The remaining surface of the interambulacra is covered by small, evenly distributed, crenulate perforate secondary tubercles with inclined areoles.

Adorally the interambulacra are densely covered with crenulate, perforate primary tubercles. ly few secondary and miliary tubercles are scattered among them. The tubercles increase in size from the margin towards the peristome. On the sternum and on the first pair of episternal plates the tubercles are arranged in a fan-shaped pattern radiating from a posterior elevation on the sternum respectively central elevation on the

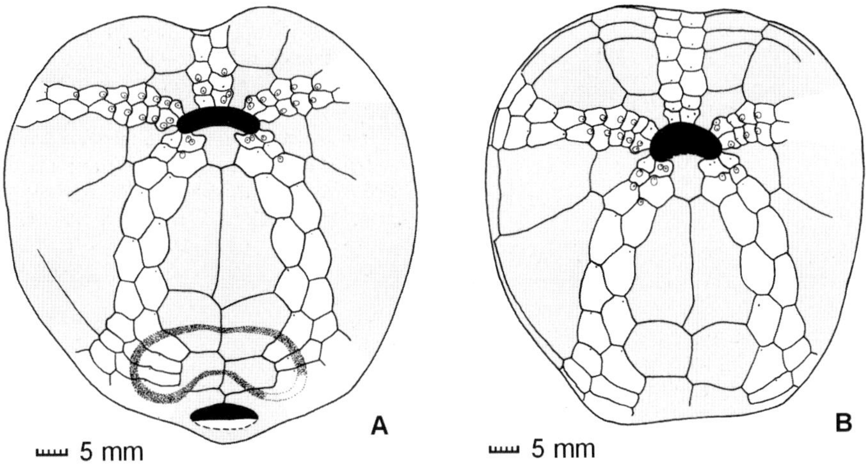

Figure 84: *Spatangus austriacus* LAUBE, 1869: oral plating (A: Grübern, NÖ, NHMW 1914.VII.19c; B: Grübern, NÖ, NHMW 2002z0159/0002). Interambulacra shaded.

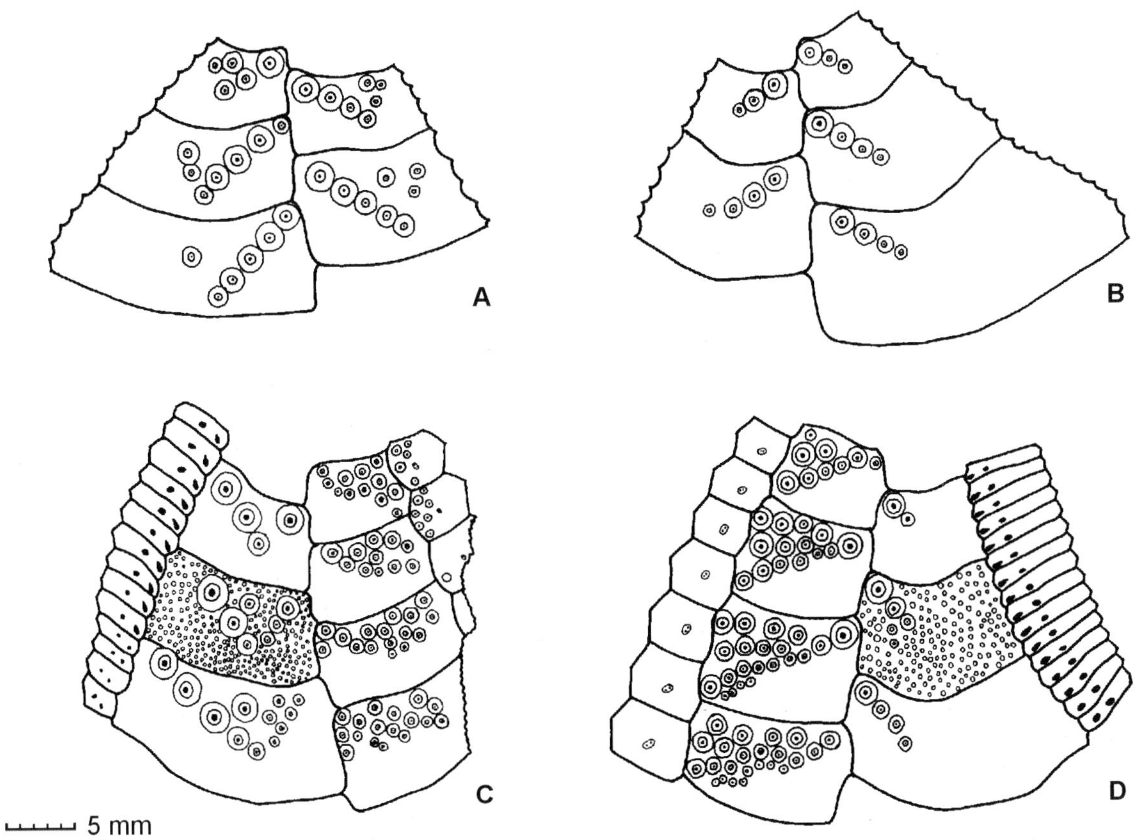

Figure 85: *Spatangus austriacus* LAUBE, 1869 (A, C) and *Spatangus* sp. 2 (B, D): comparison between the tuberculation on the adapical plates in the posterior (A, B) and anterior paired interambulacra (C, D). Secondary tuberculation has been omitted on all but one plate in each figure; specimens are of similar size, their test length range from 78 mm to *c.* 90 mm. *S. austriacus* has a much more strongly developed primary tuberculation. In interambulacral columns 1a, 1b, 2a, 3b, 4a, and 4b a marked en-chevron arrangement is developed in *S. austriacus*, whereas in *S.* sp. 2 there is only a single oblique row of primary tubercles. (A: interambulacrum 1, NHMW A2254, Grübern, NÖ; B: interambulacrum 4, NHMW 1984/138/139, Kummer quarry, St. Margarethen, Bgld; C: interambulacrum 2, NHMW 2002z0159/0004, Grübern, NÖ; C; D: interambulacrum 3, NHMW 1976/1844/45, Kummer quarry, St. Margarethen, Bgld).

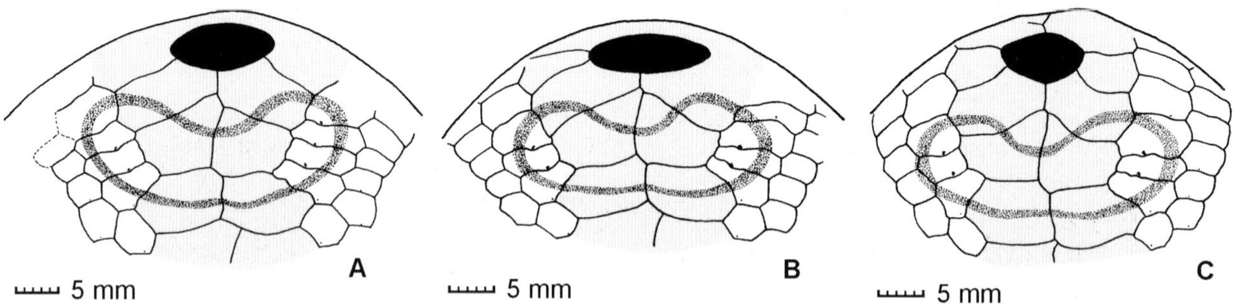

Figure 86: *Spatangus austriacus* Laube, 1869 (A, B) and *Spatangus* cf. *austriacus* Laube, 1869 (C): shape of the subanal fasciole and plating of interambulacrum 5 and the adjacent ambulacra below the periproct (A: Grübern, NÖ, NHMW 1914.VII.19c; B: Grübern, NÖ, NHMW 1914.VII.19d; C: Maria Dreieichen, NÖ, NHMW 1860.L.205). Interambulacra shaded, subanal fasciole densely stippled.

episternal plates. The plastron is holamphisternous with a broad sternum (sternum width ranging from 85 to 95 % sternum length) and slightly elongate labrum (labrum width ~ 82 to 88 % labrum length) (see Fig. 84). The labrum extends to the second ambulacral plate.

Peristome: The peristome is situated about 25 % of TL from the anterior margin. It is kidney-shaped with a slightly projecting labrum.

Periproct: The periproct is situated inframarginally and overhung by a small rostrum. It is rather large and oval (transversely elongated). There is a distinct subanal depression, which is sparsely tuberculated.

Fascioles: Only a subanal fasciole is present. The fasciole band is moderately broad, slightly depressed and corresponds to the orthofasciole type of Néraudeau et al. (1998b). The fasciole is broad and bilobed to kidney-shaped. Within the fasciole up to three enlarged constricted unipores (or partitioned isopores?) are found in each ambulacrum (Figs. 86.A-B).

Differential diagnosis:

This species differs from *S.* sp. 2 from the Badenian of Austria by its different plastron architecture (compare Fig. 84 and Fig. 89), especially in the shape of the labrum which is antero-posteriorly elongated in *S. austriacus* extending to the second ambulacral plates, whereas it is transversely elongated and not extending beyond the first ambulacral plates in *S.* sp. 2. The sternum of *S.* sp. 2 is more "serrated" and more elongated (sternum width 82 to 88 % sternum length, vs. 85 to 95 % in *S. austriacus*). Moreover, the sternum of *S. austriacus* seems to get relatively broader during ontogeny, whereas in *S.* sp. 2 it shows the reverse trend. Additionally there are strong differences in aboral tuberculation (Fig. 85).

S. delphinus Defrance, 1827 from the Burdigalian of the Rhône Valley differs (but see remarks section) from *S. austriacus* by its smaller, more lanceolate petals, higher test (based on the values given by Philippe, 1998: 206, the test height ranges from 32 to 68 % TL, with a mean of 46 %) and the position of the maximum height, which coincides with the apical disc (based on Philippe, 1998: 204-209).

Discussion:

This species is closely related to *Spatangus delphinus* Defrance, 1827 from contemporaneous deposits of the Rhône Basin. Due to the fact that no topotypic specimens of *S. delphinus* were available for examination during this study and the plating pattern of this species is not known, it is difficult to decide if these two species are conspecific or not. In case the plating patterns (in particular plastron architecture and labrum shape) turn out to be identical the two can likely be regarded synonymous.

S. austriacus is rather rare in the study area, only few localities are known. Most specimens are rather poorly preserved and often deformed. The specimens from the immediate vicinity of Eggenburg are slightly, but consistently different from the type material of Laube (1869a) and are discussed below under *Spatangus* sp. 1.

In his description Laube (1871: 73) mentioned three specimens of this species, two of them from the Eggenburgian of Bayersdorf, NÖ (the figured specimens) and one from the Badenian of Großhöflein (= Nagyhöflány), Bgld, Austria (then belonging to Hungary). The Badenian specimen could not be located, and it is unclear if it is conspecific with the Eggenburgian specimens. All Badenian *Spatangus* specimen from Austria are clearly different (but see remark under *S. hungaricus*).

Reports of *S. austriacus* from the Badenian of Poland Mączyńska (1979: 31-32; 1988: 62), the Ukraine (Szörényi, 1953: 37-38, 88-89) and Romania (Koch, 1887b: 270) are doubtful. The descriptions are largely inconclusive. The illustrations suggest that they might not be conspecific with Laube's type material (see under *S.* sp. 2). Most Badenian literature records of *S. austriacus* have therefore been tentatively placed in the synonymy of *S.* sp. 2.

This species has also been reported from many localities in Northern Italy [e.g. Manzoni (1879: 160, 1881:174), Mazzetti (1882b: 129), Stefanini (1908b: 109-112)], where a wide range of morphotypes were placed into *S. austriacus*. Unfortunately illustrations and descriptions rarely include oral plating patterns. Thus nearly all specimens need to be re-examined to decide whether or not they are conspecific with *S. austriacus*. In these specimens, where the plastron structure is known (e.g. specimen in Manzoni, 1879: pl. 3, fig. 21), it is often markedly different from *S. austriacus*. However, since a re-examination of all these specimens is out of the scope of this study, this problem must be solved later. Moreover, all of the figured specimens attributed to *S. austriacus* originally were placed into other (new) species by subsequent authors [Mazzetti (1882b: 21, pl. 2, fig. 2) established the new species *S. subconicus* for the specimen figured on pl. 3, figs. 19-20, the same specimens together with the one on pl. 4, figs. 40-41, were used as types for *Prospatangus fabianii* Lambert, 1924 (in Lambert & Thiéry, 1909-1925: 462), the latter figure (pl. 4, fig. 41) again served as type for *S. destefanii* Stefanini, 1908 (Stefanini, 1908b: 112, pl. 16, fig. 3). For the specimens on the figures 10-15 on plate 2 and 21-22 on plate 3 Lambert (in Lambert & Thiéry, 1909-1925: 462) established the species *Prospatangus airaghii* Lambert, 1924. Nelli (1903: 194) and Airaghi (1908: 256), in contrast, placed all of Manzoni's specimens and most subsequent reports into the synonymy of *S. manzonii* Simonelli, 1884. Similarly the specimens reported under *S. austriacus* from the Piemont area by Airaghi (1901: 215-216; pl. 27 (9), fig. 8) were later placed into a new species (*S. taramellii*) by the same author (Airaghi, 1908: 255)].

Although the available material is rather limited, it spans a broad size range including subadult specimens as well as very large ones. This allowed to observe some of the morphological changes which occur during growth due to the allometric growth of this species: the depth of the frontal sinus increases; the heart-shaped outline becomes more pronounced; the petals close and become more lanceolate in shape; the infla-tion of the oral ambulacral plates lessens; the sternum gets broader; and the wedge shape in profile becomes more evident.

Occurrence:

Austria: Early ? to Late Eggenburgian (Early Burdigalian); Badenian records omitted because identifications are doubtful

Molasse Zone: Achberg, near Maria Dreieichen ("Scutellensande"), NÖ (Steininger, 1971a, b); Bayersdorf, near Maissau, NÖ (Laube, 1869a, 1871; Schaffer, 1912a; [NHMW]); Eggenburg, NÖ (Thenius, 1974); Eggenburg (Brunnstube), NÖ (Steininger, 1971a, c); Eggenburg (Kremserberg), NÖ (Schaffer, 1912a); Eggendorf, NÖ ([NHMW]); Fels am Wagram, NÖ (Steininger, 1971a); Grübern, NÖ (Schaffer, 1912a; Schultz, 1998; [NHMW]); Klein-Meiseldorf, NÖ (Schaffer, 1912a); Unternalb, near Retz, NÖ (Harzhauser & Kroh, 1999; Kroh & Harzhauser, 1999; Lukeneder & Harzhauser, 2002; [NHMW])

Mediterranean: occurrence uncertain; most Italian records have been placed into different species by subsequent authors

Western Mediterranean: Italy, mainland: ? Gaiano, Modena (Coppi, 1884), ? Montagnana, Modena (Coppi, 1884), ? Montardone, Modena (Coppi, 1884), ? Montepenna, Modena (Coppi, 1884), ? Mt. Fuscaldo, Calabria (Neviani, 1887), ? Quatro di Giovanzano, Sambuci (Angelis d'Ossat, 1897),? Sasso, Modena (Coppi, 1884), ? Tagliata, Modena (Coppi, 1884), ? Posano, Volterra, Prov. Pisa (Menesini, 1967); Sardinia, Italy: ? Calada Bianca (Comaschi Caria, 1963), ? Funtanazza (=Fontanazzo) (Parona, 1887; Comaschi Caria, 1963)

Spatangus cf. *austriacus* Laube, 1869

(Fig. 86.C, 87; Pl. 77, Figs. 1-3)

v. 1869a *Spatangus euglyphus* Laube. – Laube: 183
v. 1870 *Eupatagus euglyphus* Lbe. – Laube: 314
v. 1871 *Spatangus euglyphus* Laube. – Laube: 73
non 1938 *Spatangus* cfr. *euglyphus* Laube 1867 – Poljak: 199-200; pl. 8, fig. 2 [misidentified *Brissopsis* sp.]

Material:

Early Eggenburgian (Early Burdigalian) – Maria Dreieichen (Loibersdorf Fm.), NÖ, Austria
NHMW: 1 specimen [NHMW 1860.L.205 (reference material of Laube, 1869a, 1871)]

Late Eggenburgian (Early Burdigalian) – Bayersdorf, near Maissau, NÖ, Austria
NHMW: 1 specimen (NHMW 1934.I.72 [material of Laube, 1871])

Late Eggenburgian (Early Burdigalian) – Grübern, near Maissau, NÖ, Austria
NHMW: 1 specimen (NHMW 1914.VII.19k)

Dimensions (in mm):

Inv. No.	TL	TW	TH
NHMW 1860.L.205	57.8	55.1	27.7
NHMW 1914.VII.19k	49.3	45.5	>15.3
NHMW 1934.I.72	56.7	54.8	25.2

Description:

Size and shape: Test of medium size, ranging from 49.3 to 57.8 mm in the studied specimens. The outline is oval, antero-posteriorly elongated to slightly cordiform with a shallow frontal notch. The maximum width lies subcentrally. In profile the test is low arched to slightly wedge-shaped, with an obliquely truncated end. The maximum height lies posterior of the apical disc on the raised keel in interambulacrum 5.

Apical disc: The apical disc lies slightly anterior of the centre, about 40 to 44 % TL from the anterior margin. It is ethmolytic with four large gonopores, the anterior ones of which are slightly smaller (Figs. 87.A-B).

Ambulacra: Ambulacrum III nonpetaloid, slightly depressed, increasingly so towards the anterior margin. The pores are strongly oblique, minute partitioned isopores. The interporiferous zone is wide and bears only secondary and miliary tubercles. The paired ambulacra are petaloid, slightly depressed adapically and moderately closed distally. The anterior paired petals form an obtuse angle of about 138 to 145°, the posterior paired petals an acute angle of about 55 to 59° with each other. The petals are lanceolate in shape and extend about

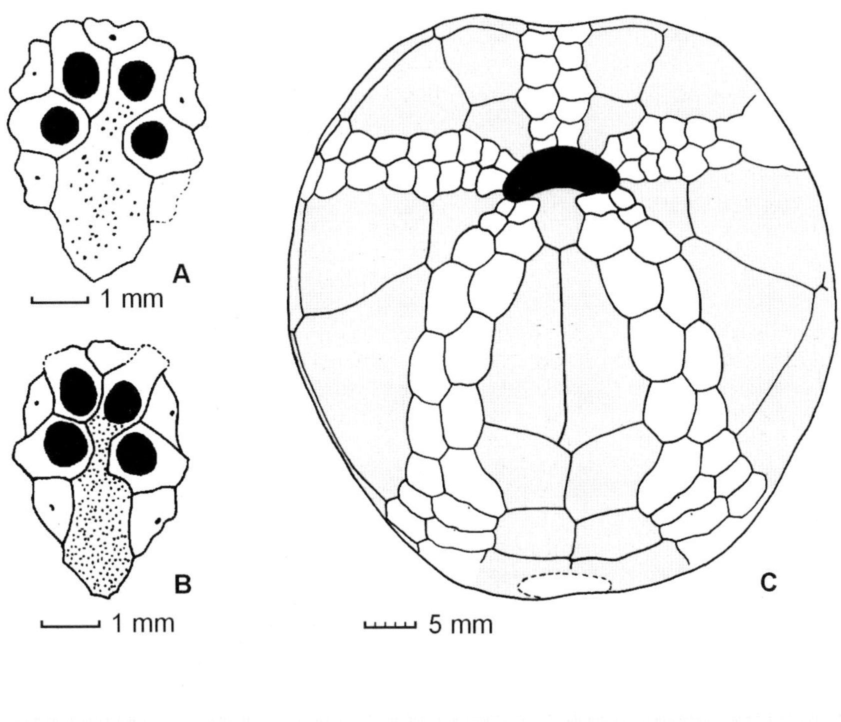

Figure 87: *Spatangus* cf. *austriacus* Laube, 1869: apical disc (A, B) and oral plating (C) (A, C: Grübern, NÖ, NHMW 1934.I.72; B: Maria Dreieichen, NÖ, NHMW 1860.L.205). Interambulacra shaded.

three quarter of the corresponding test radius. Within the petals the pores are closely spaced, elongated isopores. The pores of the uppermost 7 to 11 plates in the anterior plate row of ambulacra II and IV are minute partitioned isopores. Outside the petals only minute microunipores are present, except in the phyllodes where large unipores with extensive periporal area are found. Ambulacra I and V form moderately broad periplastronal areas lacking primary tubercles.
Interambulacra: Adapically the interambulacra are slightly inflated between the petals. They bear few crenulate, perforate primary tubercles with inclined areoles. In the interambulacra 1, 4 and 5 these enlarged tubercles are mainly found along the interradial suture of the adapical plates. In specimen NHMW 1860.L.205 they are restricted to the area between the petals.

Adorally the interambulacra are densely covered with crenulate, perforate primary tubercles. Only few secondary and miliary tubercles are scattered among them. The tubercles increase in size from the margin towards the peristome. On the sternum and on the first pair of episternal plates the tubercles are arranged in a fan-shaped pattern radiating from posterior elevations on the plates. The plastron is distinctly inflated, it is holamphisternous with a broad sternum (sternum width ranging from 83 to 86 % sternum length) and slightly elongated labrum (labrum width 60 to 70 % labrum length) (see Fig. 87.C). The labrum extends to the second ambulacral plate.
Peristome: The peristome is situated about 27 % of TL from the anterior margin. It is kidney-shaped with a slightly projecting labrum.
Periproct: The periproct is situated inframarginally and overhung by a small rostrum. It is rather large and oval (transversely elongated). There is a distinct subanal depression, which is sparsely tuberculated.
Fascioles: Only a subanal fasciole is present. The fasciole band is moderately broad and corresponds to the orthofasciole type of Néraudeau et al. (1998b). The fasciole is broad and has a bilobed outline with a strong indentation below the periproct. Within the fasciole two enlarged constricted unipores are found in each ambulacrum (Fig. 86.C).

Differential diagnosis:
The specimens considered here differ from typical *S. austriacus* by their smaller size, oval outline, lanceolate petals, more elongated sternum and labrum, and the presence of only two enlarged pores within the subanal fasciole.

S. sp. 2 from the Badenian of Austria differs by its different plastron architecture, especially in the shape of the labrum.

Discussion:
The present specimens are generally very similar to *S. austriacus* and show similar proportions when compared with large specimens of that species. Comparison with *S. austriacus* specimens of similar size reveals the following differences (compare Laube, 1868: p. 35-36; pl. 6, figs. 5, 5a-c): moderately closed, broad, lobate petals and a broadly heart-shaped outline with transversely truncated posterior end characterise the subadult specimens of *S. austriacus*. Moreover, *S. austriacus* has three pores within the subanal fasciole, while this species only has two. Based on these differences the specimens are kept separated from *S. austriacus*.

One of the specimens (NHMW 1860.L.205) was also examined by Laube (1871: 73) and referred by him to *Spatangus euglyphus* Laube, 1867, a species based on material from the Aquitanian of Northern Italy. If compared with the descriptions of the Italian material (Laube, 1868: 35-36, pl. 6, figs. 5, 5a-c), it becomes clear that these specimens are not conspecific. *S. euglyphus* is much higher, has a subcircular outline, a narrower plastron, differently shaped labrum and its periproct lies much lower on the test. *S. euglyphus* hence does not occur in Austria, contrary to the statements of Laube (1868, 1869a, 1870, 1871) and Cahuzac & Roman (1994: 362).

Occurrence:

Austria: Early to Late Eggenburgian (Early Burdigalian)
Molasse Zone: Bayersdorf, near Maissau, NÖ ([NHMW]); Grübern, near Maissau, NÖ ([NHMW]); Maria Dreieichen, NÖ (Laube, 1869a, 1871; [NHMW])

Spatangus sp. 1

(Pl. 77, Figs. 4-5)

Material:
Late Eggenburgian (Early Burdigalian) – Eggenburg, NÖ, Austria
NHMW: 2 specimens (NHMW 1970/1396/977, NHMW 2002z0168/0001)
Late Eggenburgian (Early Burdigalian) – Eggenburg (Schindergraben), NÖ, Austria
NHMW: 4 specimens [NHMW 592/1964, NHMW 2002z0169/0001 (2 specimens), NHMW 2002z0169/0002]
Late Eggenburgian (Early Burdigalian) – Eggenburg (Kremserberg), NÖ, Austria
NHMW: 1 specimen (NHMW 2002z0170/0001)

Dimensions (in mm):

Inv. No.	TL	TW	TH
NHMW 1970/1396/977	55.4	55.6	26.1
NHMW 592/1964	62.9	63.5	~ 26
NHMW 2002z0168/0001	73.3	74.9	25.5
NHMW 2002z0170/0001	81.6	78.8	25.3

Discussion:
The poor preservation of the studied specimens precludes a confident specific determination. The present specimens are very similar to *Spatangus austriacus* Laube, 1871, but show some minor differences: the angle between the anterior paired petals is slightly larger (148 to 158°) than in *S. austriacus* (130 to 145°) and the angle between the posterior paired petals smaller (less than 55° *vs.* usually 60°); the test is subcircular or transversely elongated (TL≤TW) instead of antero-posterior elongated (TL>TW); the frontal notch is shallower than in *S. austriacus* specimens of comparable size; and the apical disc is situated more anteriorly (30 to 37 % TL from the anterior margin *vs.* 42 to 45 %). Until better material becomes available, however, the relationship to typical *S. austriacus* morphotypes remains unclear and the specimens are left in open nomenclature.

Occurrence:

Austria: Late Eggenburgian (Early Burdigalian)
Molasse Zone: Eggenburg, NÖ ([NHMW]); Eggenburg (Kremserberg), NÖ ([NHMW]); Eggenburg (Schindergraben), NÖ ([NHMW])

Spatangus sp. 2

(Figs. 85.B, D, 88-90; Pl. 78, Figs. 1-3; Pl. 79, Figs. a-b)

? pp 1869a *Spatangus austriacus* Laube. – Laube: 184
? pp 1870 *Spatangus austriacus* Lbe. – Laube: 314
? pp 1871 *Spatangus austriacus* Laube. – Laube: 73; pl. 19, fig. 2, 2a
? 1887b *Spatangus austriacus*, Laube. – Koch: 270-271
? 1891 *Spatangus* cf. *austriacus* Laube. – Roth von Telegd: 151
1892 *Spatangus austriacus* Lbe. – Schafarzik: 196
? 1893 *Spatangus austriacus* Laube– Toula: 288
? 1905 *Spatangus* cfr *austriacus*, Lbe. – Gaál: 344, 362
? 1930 *Prospatangus austriacus* Laube. – Vendl: 50

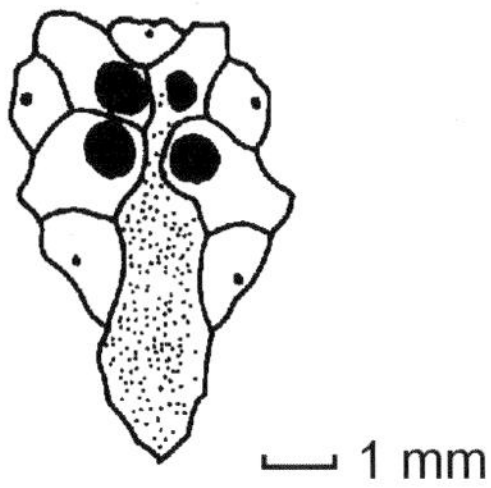

Figure 88: *Spatangus* sp. 2: apical disc (Kummer quarry, St. Margarethen, Bgld, NHMW 1976/1844/54).

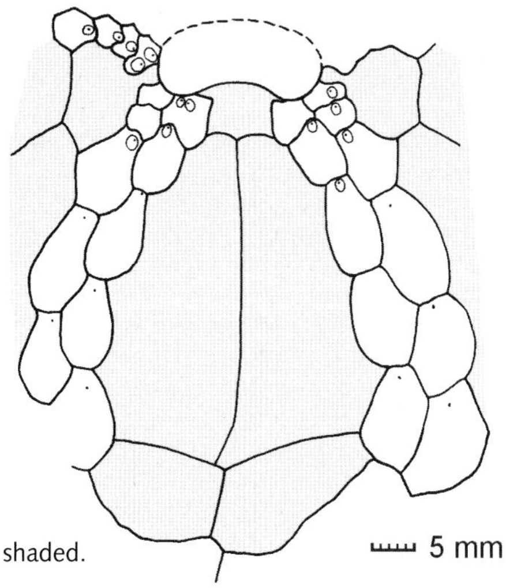

Figure 89: *Spatangus* sp. 2: oral plating (Kummer quarry, St. Margarethen, Bgld, NHMW 2005z0004/0001). Interambulacra shaded.

1936 *Spatangus austriacus* Laube – Paucă: 143, 197; pl. 2, figs. 1-2
1938 *Spatangus austriacus* Laube 1871 – Poljak: 197-198; pl. 10, fig. 2
? 1953 *Prospatangus austriacus* (Laube), 1871. – Szörényi: 37-38, 88-89; pl. 8, figs. 3, 3a-b
? 1969 *Spatangus corsicus* Des. – Mitrović-Petrović: 140; pl. 30, figs. 3, 3a-b
? 1979 *Spatangus austriacus* Laube, 1871 – Mączyńska: 31-32, pl. 3, figs. 1 a-d, 2a-c; pl. 4, figs. 1a-e; pl. 5, figs. 1a, 2a-b
? 1979 *Spatangus hungaricus* (Vadász, 1915) – Mączyńska: 32; pl. 6, figs. 1a-e, 2; pl. 7, fig. 1a-f
? 1988 *Spatangus austriacus* Laube, 1871 – Mączyńska: 62; pl. 4, figs. 4a-f
1996 *Spatangus austriacus* – Machalski: 26
v. 2001 *Spatangus austriacus* (Laube) – Schmid et al.: 13; pl. 1, fig. 5
2004 *Spatangus austriacus* Laube, 1871 – Radwański & Wysocka: 385, 386; pl. 6, figs. 1a-b

Material:
Late Badenian (Early Serravallian) – St. Margarethen (Kummer quarry), Bgld, Austria
NHMW: 11 fragmented coronas (NHMW 1976/1844/45, ../52, ../54-56, 1986/138/130, 1992/151/3, 1992/151/4a+b, 2000z0133/0011, 2005z0004/0001)
Wanzenböck coll.: 1 specimens (W28)

Foreign material for comparison:
Early Badenian (Langhian) – Pod'yarkov (= Podjarków), near Lwów, western Ukraine
NHMW: 1 specimen (NHMW 1859.XLV.556)
Badenian (Langhian-Early Serravallian) – Loska Dolina (formerly Groß-Dollina, Unter-Krain), Slovenia
UGP: 1 specimen (UGP 1877.IV.235)

Dimensions (in mm):

Inv. No.	TL	TL	TH
NHMW 1986/138/130	97	92	-
NHMW 1992/151/3	~148	140-150	-
Wanzenböck coll. W27	~95	~ 93.5	-
Wanzenböck coll. W28	90.3	~ 88	-

Description:
Size and shape: Test of large to very large size, ranging from 90 to 150 mm test length in the studied material. The outline of the test is oval to slightly cordiform, with a shallow frontal sinus and bluntly pointed to transversely truncated posterior end. The maximum width lies subcentrally. Since most specimens are badly crushed, the shape of the test in profile can only be reconstructed based on the curvature of the fragments. It seems, however, that the test was moderately arched with the maximum height around the apical disc. Estimated values for test width are *c.* 95 % TL and *c.* 40 % TL for test height.
Apical disc: The apical disc lies slightly anterior of the centre, about 45 % TL from the anterior margin. It is ethmolytic with four large gonopores and 5 minute ocular pores (Fig. 88). The madreporite is very elongated posteriorly and extend far beyond the posterior ocular plates.
Ambulacra: Ambulacrum III is nonpetaloid, slightly depressed, increasingly so towards the anterior margin, where it forms a shallow frontal sinus. The pores are strongly oblique, minute partitioned isopores in the centre of the plates. They are closely spaced adapically but become increasingly widely spaced towards the ambitus. No primary tubercles are present within ambulacrum III. The paired ambulacra are petaloid, depressed adapically and closed distally. The anterior paired petals form an obtuse angle of about 141°, the posterior paired petals an acute angle of about 60° with each other. The petals are lanceolate in shape and extend about two thirds of the corresponding test radius. Within the petals the pores are elongated isopores. The interporiferous zones are up to four times as wide as a single poriferous zone and slightly inflated. The pores of the uppermost 11 to 14 plates in the anterior plate row of ambulacra II and IV are minute partitioned isopores. Outside the petals only minute pores (partitioned isopores or constricted unipores) are present.

Adorally the ambulacra form broad naked phyllodes with large unipores with an extensive periporal area. In ambulacrum III 8 such pores are present, 9 to 10 in each of the anterior paired ambulacra, and 8 in the posterior paired ambulacra. Ambulacra I and V form moderately broad peri-plastronal areas which are covered by loosely spread secondary and dense miliary tubercles. Primary tubercles are not present.
Interambulacra: Adapically the interambulacra are slightly inflated between the petals. They bear a few enlarged, crenulate, perforate primary tubercles, which are arranged in short oblique rows on each plate (Figs. 85.B, D). In the interambulacra 1, 4 and 5 these enlarged tubercles are mainly found along the interradial suture of the adapical plates and are restricted to the area between the petals (Fig. 85.B). In the interambulacra 2 and 3 they are mostly found on the anterior plate rows (2b and 3a) along the sutures to ambulacrum III (Fig. 85.D). In the posterior plate rows (2a and 3b) there are very few (up to 3) enlarged primary tubercles, which are arranged in an oblique row in the interradial/adapical edge of each plate (Fig. 85.D). The remaining surface of the interambulacra is densely covered by small, evenly distributed, crenulate perforate secondary tu-

bercles. Small miliary tubercles are densely scattered among them.

Adorally the interambulacra are densely covered with crenulate, perforate tubercles which increase in size from the margin towards the peristome. On the sternum and on the first pair of episternal plates the tubercles are arranged in a fan-shaped pattern radiating from posterior elevations on the plates. The plastron is holamphisternous with a broad sternum (sternum width ranging from 82 to 88 % sternum length) and broad labrum (labrum width 102 to 135 % labrum length) (see Fig. 89). The labrum does not extend beyond the first ambulacral plate.

Peristome: The peristome is situated about 28 to 32 % TL from the anterior margin. It is kidney-shaped with a projecting labrum.

Periproct: The periproct is situated inframarginally and overhung by a small rostrum. It is rather large and seems to be oval (transversely elongated).

Fascioles: Only a subanal fasciole is present. The fasciole band is broad and corresponds to the orthofasciole type of Néraudeau et al. (1998b). The fasciole is broad and has a bilobed outline. Within the fasciole four enlarged constricted unipores are found in each ambulacrum.

Differential diagnosis:

This species differs from all other *Spatangus* species of the Neogene of Austria by its plastron architecture, especially in the shape of the labrum which is transversely elongated and does not extend beyond the first ambulacral plate. In specimens where the plastron is not preserved or plate outlines are weakly visible, *S.* sp. 2 can be distinguished form *S. austriacus* by the arrangement and number of the primary tubercles in the adapical interambulacra 1 and 4. In *S.* sp. 2 there are few (usually 3 to 4) primary tubercles arranged in oblique rows, whereas in *S. austriacus* they are more numerous, larger and arranged in chevron-shaped groups on each plate (compare Figs. 85.A, B). The same situation with more numerous coarse tubercles and an en-chevron arrangement is present in the posterior plate rows of interambulacra 2 and 3 (rows 2a and 3 b) of *S. austriacus*. (compare Figs. 85.C, D). Additionally *S.* sp. 2 has four enlarged pores enclosed within the subanal fasciole in each ambulacrum and a less inflated keel in interambulacrum 5.

Discussion:

Badenian specimens of *Spatangus* were earlier attributed to *Spatangus austriacus* Laube, 1869 by most authors (see above) based on the very similar gross morphology of the aboral test. Recently collected material, however, allows to study the architecture of the plastron for the first time. The labrum of the Badenian species is distinctly different from the type material of *S. austriacus* form the Eggenburgian. Additionally the distribution and number of enlarged aboral tubercles is considerably different in the Badenian specimens.

Examination of Museum collections revealed no Badenian specimens of *S. austriacus* (but see under *S. hungaricus*) and hence Badenian literature records of *S. austriacus* are tentatively placed in the synonymy of *S.* sp. 2 here. The descriptions and illustrations of these records, unfortunately, do not reveal the exact morphology of the plastron in those specimens. A specimen from Pod'yarkov (formerly Podjarków), western Ukraine housed at the Naturhistorisches Museum Wien (NHMW 1859.XLV.556) clearly shows the labrum shape typical of *S.* sp. 2 (Fig. 90). This specimen is very similar to those reported by Szörényi (1953) under the name *S. austriacus* from the same locality. Additionally, the specimen from the Early Badenian of Świniary (Poland) illustrated by Radwański & Wysocka (2004: pl. 6, figs. 1a-b) clearly belongs to *Spatangus* sp. 2. Although the labrum is not preserved in that specimen, the aboral tuberculation pattern is a clear evidence.

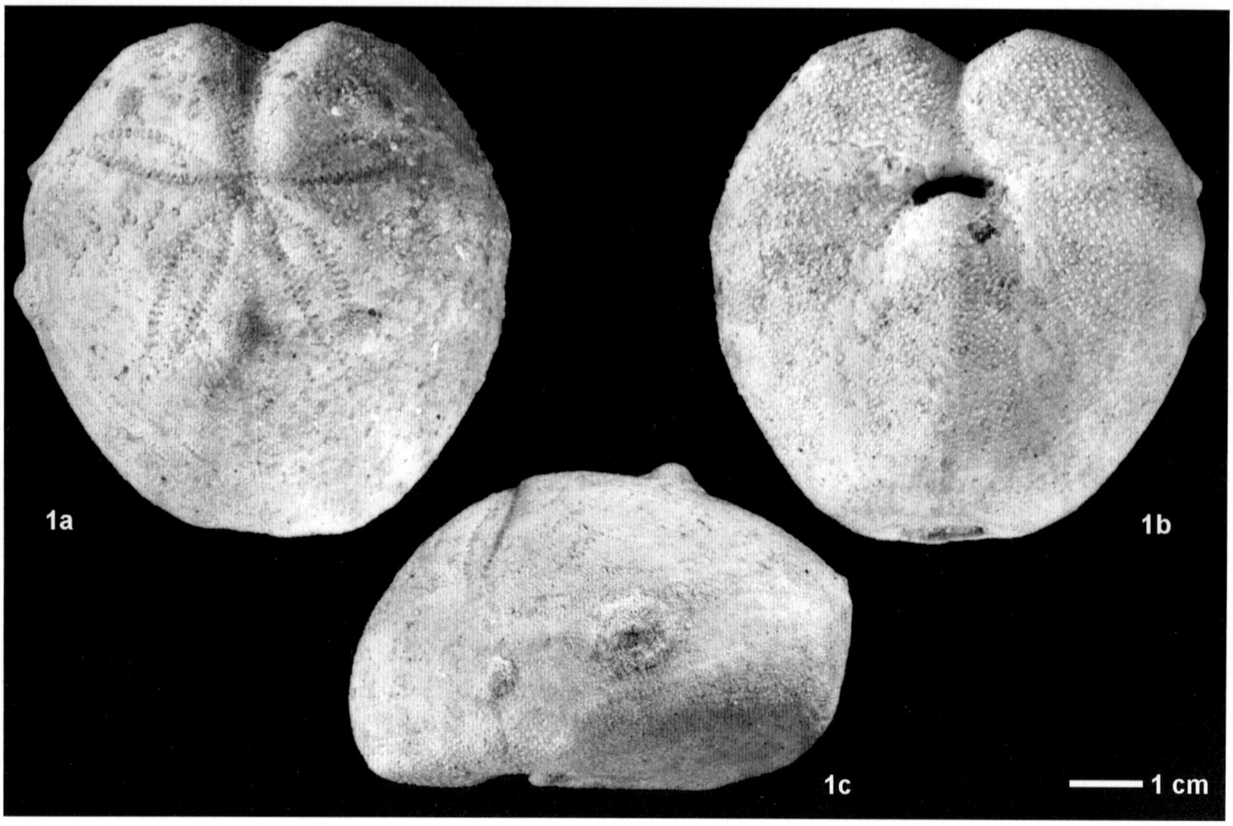

Figure 90: *Spatangus* sp. 2: undeformed specimen from the Early Badenian of Podjarków, near Lwów, western Ukraine (NHMW 1859.XLV.556)

This species could not be identified with any of the numerous established species because lack of information on plating patterns of these species. Hence it is left in open nomenclature at present.

The material figured as *Spatangus austriacus* by Mączyńska (1979) is difficult to evaluate because of its poor preservation and printing quality. The scarcity of large aboral primary tubercles, except adapically in interambulacra 2 and 3 along the border to ambulacrum III is similar to *S.* sp. 2. Additional evidence comes from the plastron shape (Mączyńska, 1979: pl. 5, fig. 1a) which is more reminiscent of that species than *S. austriacus*, although the shape the labrum is unknown. Without being able to re-examine the material the present author suggests to remove it from the synonymy of *S. austriacus* and places it tentatively into that of *S.* sp. 2.

Occurrence:

Austria: Early (?) to Late Badenian (Langhian-Early Serravallian)

Eisenstadt-Sopron Basin: ? Großhöflein, near Eisenstadt, Bgld (Laube, 1869a, 1871); St. Margarethen (Kummer quarry), Bgld (Schmid et al., 2001; [NHMW])

Paratethys (non-Austrian occurrences): Badenian (Langhian-Early Serravallian)

Molasse Basin: ? Kralice nad Oslavou (= Kralitz), Moravia, Czech Republic (Toula, 1893)

Great Hungarian Basin (Pannonian Basin): ? Györ, Györ-Monson-Sopron, Hungary (Vendl, 1930); ? Hor. Strháre (= Felsö-Esztergály), Slovak Republic (Roth von Telegd, 1891; Gaál, 1905); ? Sámsonháza, Nógrád, Hungary (Schafarzik, 1892); ? Tótmarokháza, Nógrád, Hungary (Gaál, 1905)

Fore-Carpathian Basin: ? Huta Lubycka, southeastern Poland (Mączyńska, 1979); ? Podgortsg (= Podhorce), Ukraine (Szörényi, 1953); Pod'yarkov (= Podjarków), near Lwów, western Ukraine (Szörényi, 1953; [NHMW]); Świniary, southern Poland (Mączyńska, 1988; Machalski, 1996; Radwański & Wysocka, 2004); ? Wozniaki, western Ukraine (probably the locality was confused with Woźniki, near Wadowice, Poland; no locality with a similar name could be located in the Ukraine) (Szörényi, 1953)

Transylvanian Basin: ? Gârbova de Sus (= Felsö-Orbó), Romania, Hungary (Koch, 1887b; Gaál, 1905); region east of Taşad, Beiuş Basin, Romania (Paucă, 1936)

Zala, Sáva and Dráva Basins: "Groß-Dollina" (formerly belonging to Unter-Krain, Austria), Slovenia ([UGP]); Zaprešić Breg, Croatia (Poljak, 1938); ? Derventa, Bosnia & Herzegovina (Mitrović-Petrović, 1969)

Spatangus sp. indet.

? 1853 *Spatangus*. – Hörnes: 190 (probably a misidentified schizasterid or brissid)
1867 *Spatangus* sp. – Reuss: 30, 109-110
1877 *Spatangus* – Fuchs: 662
1900 *Spatangus* sp. – Fuchs: 907
1904 *Spatangus* spec. – Abel: 22
1906 *Spatangus* ind. sp. – Vadász: 332-333
1969 *Prospatangus* sp. – Mihály: 257
1979 *Spatangus* sp. A – Mączyńska: 33, pl. 10, figs. 1a-e
1979 *Spatangus* sp. B – Mączyńska: 33, pl. 10, figs. 2a-c
1984 *Prospatangus* sp. – Kókay et al.: 288
1987 *Spatangus* sp. – Mączyńska: 149
1989 *Spatangus* sp. – Nebelsick: 15
1991a *Spatangus* sp. – Nebelsick et al.: 88
? 1996 *Spatangus* ? sp. – Mączyńska: 41; pl. 1, fig. 12

Material:

Late Eggenburgian (Early Burdigalian) – Pulkau (Retz Fm.), NÖ, Austria
NHMW: 1 test fragment (NHMW 1985/81)
Late Eggenburgian (Early Burdigalian) – Unternalb (Retz Fm.), NÖ, Austria
NHMW: 1 test fragment (NHMW 1861.XXXV.169)
Late Badenian (Early Serravallian) – Winden (Nirgl quarry, N of the village), Bgld, Austria
NHMW: 5 test fragments (NHMW 1859.L.831b)
Late Badenian (Early Serravallian) – St. Margarethen (Kummer quarry), Bgld, Austria
NHMW: 1 natural cast (NHMW 1992/151/4b), 2 test fragments (NHMW 2000z0133/0010, 2004z0001/0044)

Discussion:

The record of *Spatangus* from the "Schlier" facies of Ottnang is most probably erroneous and likely a misidentified schizasterid of brissid, the two dominant groups in that locality.

The poor preservation of the specimens referred by Mączyńska (1979) to *Spatangus* sp. A and sp. B respectively does not allow to place them into the affinity of any of the described specimens with certainty.

Occurrence:

Austria: Late Eggenburgian (Early Burdigalian), ? Ottnangian (Late Burdigalian), Late Badenian (Early Serravallian)

Molasse Zone: Eggenburg region, NÖ (Fuchs, 1877; Nebelsick, 1989); Eggenburg (Brunnstube), NÖ (Nebelsick et al., 1991a); Eggenburg (Schindergraben), NÖ (Fuchs, 1900; Abel, 1904); ? Ottnang, OÖ (Hörnes, 1853); Pulkau (Retz Fm.), NÖ ([NHMW])

Eisenstadt-Sopron Basin: St. Margarethen (Kummer quarry), Bgld ([NHMW]); Winden (Nirgl quarry), Bgld ([NHMW])

Paratethys (non-Austrian occurrences): Badenian (Langhian-Early Serravallian)

Great Hungarian Basin (Pannonian Basin): Budapest, Pest, Hungary (Kókay et al., 1984); Budapest-Rákos, Pest, Hungary (Vadász, 1906; Mihály, 1969)

Fore-Carpathian Basin: Huta Lubycka, southeastern Poland (Mączyńska, 1979); Korytnica Basin, Poland (Mączyńska, 1987); Niechobrz, near Rzeszów, southern Poland) (Mączyńska, 1996); Wieliczka, Poland (Reuss, 1867)

Genus *Mariania* Airaghi, 1901

Type-species: *Macropneustes marmorae* Desor in Agassiz & Desor, 1847; by original designation (Airaghi, 1901: 211).

Diagnosis: Test cordate with shallow, broad frontal notch; in profile arched to highly domed with steep sides and flat base; posterior end low and slightly undercut; apical disc ethmolytic with four gonopores; paired ambulacra petaloid; petals long, open, with wide interporiferous zone; peristome close to anterior margin, facing forward; labrum projecting over peristome; sternal pates fully tuberculate; periproct on truncated posterior end; no fasciole; aboral tuberculation heterogeneous with numerous scattered primary tubercles (modified from Fischer, 1966; and Smith "*The Echinoid Directory*", 13.05.2004).

Distribution: Oligocene to Miocene – Southern Europe and Central Paratethys

Remarks: *Mariania* was placed into the affinity of the Palaeopneustidae by Mortensen (1950: 283-284) and later into the family Brissidae by Fischer (1966: U591-592). Recently it was transferred to the family Spatangidae by Philippe (1998: 213-216), who regarded it as a subgenus of *Spatangus*.

Mariania deydieri (COTTEAU, 1897)

(Fig. 91; Pl. 80, Figs. 1a-e)

non 1896 *Spatangus deydieri*, COTTEAU – DOUXAMI: 256-257; pl. 1, figs. 12
* 1897 *Spatangus Deydieri*, COTTEAU, 1894. – COTTEAU: 185-187; pl. 10, figs. 2-4; pl. 11, figs 1-6 [figs. 3-6 on pl. 11 are erroneously designated *Spatangus rissoi* on the plate caption; fide PHILIPPE (1998: 216); according to LAMBERT (1915a: 205) the specimens reported as var. *major* from the Tortonian of Cucuron are *S. delphinus*]
non 1907 *Spatangus Deydieri* COTTEAU (*Spatangus*), 1894 – SAVIN: 11-12 [fide PHILIPPE (1998: 216)]
1913a *S.[patangus] Deydieri* COTT. – COTTREAU: 76
1915a *Mariania Deydieri* COTTEAU (*Spatangus*), 1896. – LAMBERT: 205-206
? 1915a *Mariania Deydieri* COTTEAU (*Spatangus*) – LAMBERT: pl. 15, fig. 10 [in the text (p. 204) this figure is referred to *M. marmorae*; PHILIPPE (1998: 216) list it under *M. deydieri*]
1973 *Mariania deydieri* (COTTEAU, 1896) – PHILIPPE: 195-196; text-fig. 31c
1989 *Mariania deydieri* (COTTEAU) – PHILIPPE: 32, tab. 1
1998 *Spatangus* (*Mariania*) *deydieri* COTTEAU, 1896 – PHILIPPE: 216-219; pl. 25, figs. 4-6

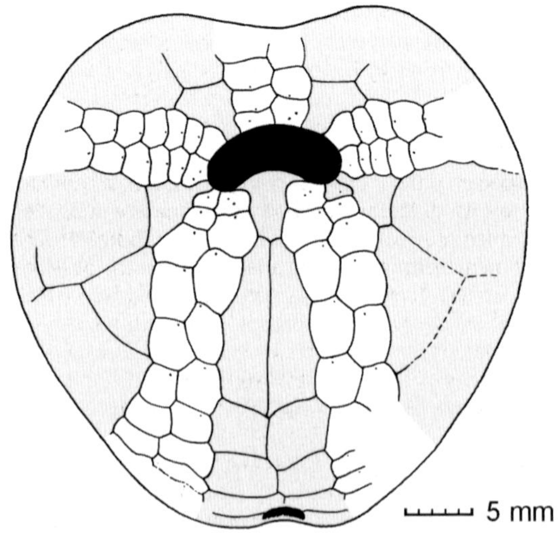

Figure 91: *Mariania deydieri* (COTTEAU, 1897): oral plating (Grübern, near Maissau, NÖ, NHMW 1998z0048/0102). Drawn from the inner surface of the test which was exposed during preparation (mirrored for easier comparison with the other drawing which were made from the outer surface); interambulacra shaded.

Type-material:
Spatangus deydieri COTTEAU, 1897:
Syntypes: 2 syntypes described and figured in COTTEAU (1897: 21-23; pl. 10, figs. 1-6; figures 3-6 erroneously indicated as *Spatangus rissoi* in the legend of the plate); no. MR 3 000 211(casts); Muséum Requien, Avignon, France (PHILIPPE, 1998: 332)
Locus typicus: Reillanne, Alpes-de-Haute-Provence, France
Age: Burdigalian, Early Miocene

Material:
Late Eggenburgian (Early Burdigalian) – Grübern (Zogelsdorf Fm.), near Maissau, NÖ, Austria
NHMW: 1 specimen (NHMW 1998z0048/0102)

Dimensions (in mm):

Inv. No.	TL	TW	TH
NHMW 1998z0048/0102	35.3	36.2	> 9.6

Description:
Size and shape: Test small, broadly heart-shaped with transversely truncated posterior margin and broad but shallow frontal sinus. The maximum width lies at the level of interambulacral columns 1b and 4a. In profile the test is low with a low, obliquely truncated posterior end. The maximum height lies around the apical disc. Oral side flat with slightly inflated plastron. Test width 102.5 % of TL, test height at least 28 % TL, probably up to 34 % (considering the post-mortem deformation).
Apical disc: Not preserved in the studied specimen. It lies slightly anterior of the centre, about 45-47 % from the anterior margin.
Ambulacra: Adapically the paired ambulacra are petaloid. The petals are relatively broad, only slightly closing distally and subequal in length, extending about three third of the corresponding test radius. The anterior paired petals form an obtuse angle of about 125°, the posterior paired petals an acute angle of about 60°. The pores within the petals are elongate isopores. Outside the petals only minute unipores are found (except in the phyllodes). The pores of the uppermost seven to eight plates in the anterior plate rows of ambulacra II and IV are minute partitioned isopores. The interporiferous zones are up to one and a half times as wide as a single poriferous zone and bear only secondary and miliary tubercles. Four to five ambulacral plates border each interambulacral plate in the distal part of the petals.

Ambulacrum III is nonpetaloid and slightly depressed, forming a broad shallow frontal notch. The pores are oblique, minute partitioned iso(?)pores. Only secondary and miliary tubercles are found in ambulacrum III.

On the oral side ambulacra I and V form broad peri-plastronal areas, which bear only miliary and/or small secondary tubercles. The phyllodes are well developed and bear relatively many large unipores with extensive periporal area. In ambulacrum III seven such pores are present, fourteen in each of the anterior paired ambulacra and eight in the posterior paired ambulacra.
Interambulacra: Aborally the interambulacra are inflated between the petals. They are densely covered by closely spaced miliary and secondary tubercles. Primary tubercles are found in all interambulacra, near the apex as well as near the margin. The largest primary tubercles are located along the interradial sutures in interambulacra 1, 4 and 5. In the interambulacra 2 and 3 the largest tubercles are situated along the adradial sutures with the anterior paired petals. The tubercles are crenulate and perforate.

Orally interambulacrum 5 forms a narrow, holamphisternous plastron (Fig. 91). The labrum is elongated and projects into the peristome. The sternum is about 0.5 times wide as long (12.4 mm long, 6.6 mm wide in the studied specimen). The episternal plates are not markedly indented by the interambulacra. The tuberculation on the oral side consists of closely spaced primary tubercles, increasing in size from the margin towards the peristome. On the plastron the primary tubercles form a fan-shaped pattern radiating from a posterior elevation on the sternum and a central elevation on each episternal plate. The labrum, in contrast, bears only few tubercles located around its centre.
Peristome: The peristome is situated anteriorly, about 25 % TL from the anterior margin. It is transversely elongated, kidney-shaped and measures 8.3 mm vs. 4.0 mm in the studied specimen.
Periproct: The periproct lies on the overhanging posterior end of the test. It seems to have been oval, transversely elongated and about 4 mm wide, but is poorly preserved because the test is crushed.
Fascioles: No fascioles discernible.

Differential diagnosis:
This species differs from *M. marmorae* (Desor in Agassiz & Desor, 1847) mainly by its lower test. In that species test height is typically > 50% TL, whereas in *M. deydieri* it is usually between 30 to 36 % TL (based on the available description of the two species).

This species differs from *Spatangus austriacus* and related species by its different distribution of primary tubercles aborally (which are scattered over the whole aboral surface, instead of being restricted to the adapical part), the absence of a subanal fasciole, its long petals and narrow plastron.

Discussion:
Only a single specimen from marly calcarenites of Grübern was available of this taxon. It is highly similar to the material described as *Spatangus (Mariania) deydieri* by Philippe (1998). Although the test of the sole specimen is well preserved and its underside has been carefully exposed by gently air-abrasive techniques, no traces of a subanal fasciole could be found. Instead, judging by the pattern of the tuberculation in comparison with specimens of other spatangids of similar size, no fasciole seems to have been present.

There are some uncertainties about the publication date of this species. While Philippe (1998: 216) gives the year 1896, the front-page of the volume in which Cotteau's paper was published states 1897 as publication date. Unfortunately the situation is complicated by the paper of Douxami (1896), who cites the then unpublished (?) paper of Cotteau and provides a description and illustrations of a *Spatangus deydieri* Cotteau, 1896 (based on the manuscript of Cotteau's paper which was sent to him by Depéret). Unfortunately, Douxami's paper seems to have been published before Cotteau's and, even more regrettable, Douxami's specimen is not conspecific with Cotteau's but represents probably *S. delphinus* according to Philippe (1998: 218). The present author was not able to verify the publication dates of the two papers in question. In case Douxami's paper was really the first to be published either a new name for Cotteau's species is needed or the species of Douxami has to be suppressed by action of the ICZN.

Occurrence:

Austria: Late Eggenburgian (Early Burdigalian)
Molasse Zone: Grübern (Zogelsdorf Fm.), near Maissau, NÖ ([NHMW])

Mediterranean: Burdigalian
Rhône Basin: Apt (Philippe, 1998); Couronne (Philippe, 1973); Forcalquier (Philippe, 1998); Lincel (Philippe, 1998); Mane (Philippe, 1998); Niozelles (Philippe, 1998); Pierrerue (Philippe, 1998); Reillanne (Cotteau, 1897; Lambert, 1915a; Philippe, 1973, 1998); Saint-Martin-de-Castillon (Philippe, 1998); Saint-Michel-l'Observatoire (Philippe, 1998); Viens (Philippe, 1998); Villemus (Philippe, 1998)

Family Loveniidae Lambert, 1905
Genus *Lovenia* Desor in Agassiz & Desor, 1847

Type-species: *Lovenia hystrix* Desor in Agassiz & Desor, 1847; by monotypy (Desor in Agassiz & Desor, 1847b: 10-11); today considered as junior synonym of *Lovenia elongata* (Gray, 1845) (see Dollfuss & Roman, 1981: 110).
Diagnosis: Test elongate, cordate with distinct frontal notch; low profile; apical disc ethmolytic with four gonopores; ambulacrum III nonpetaloid with minute pores; other ambulacra petaloid, widening towards the apex, in most species ambulacral columns Ib and IIa, respectively IVb and Va form a continuous arc; pores are elongate isopores, except inside the internal fasciole, where they are minute or altogether absent; peristome anteriorly displaced and transversely elongated; periproct on low posterior face, sunken in some species; plastron ultramphisternous; labral plate long, with narrow contact to sternal plates; sternal plates tuberculated only on the posterior third; aboral tuberculation heterogeneous with large tubercles sunken in camellae (except in interambulacrum 5); oral tubercles prominent laterally and arranged in distinct rows, areoles deeply sunken with and parapet typically helically spiralled; well developed internal and subanal fascioles; subanal fasciole bilobed (modified from Mortensen, 1951, and Smith *"The Echinoid Directory"*, 28.07.2003).
Distribution: Oligocene to Recent (Henderson, 1975) – cosmopolitan, except polar and sub-polar regions

Lovenia mortenseni Čtyroký, 1965
(Fig. 92)

. 1957 *Hemipatagus* sp. – Čtyroký: 237, tab. 3, 4
* 1965 *Lovenia mortenseni* n. sp. – Čtyroký: 108-118; text-figs. 1-6; pl. 1, figs. 1-4; pl. 2; pl. 3; pl. 4, figs. 1-4; pl. 5, figs. 1-3; pl. 6, figs. 1-4
1967 *Lovenia mortenseni* Čtyroký. – Cicha et al.: 87
. 1975 *Lovenia mortenseni* (Čtyroký, 1957) – Kalabis: 176
. 1978 *L.[ovenia] mortenseni* Čtyroký, 1965 – Kier & Lawson, 121
v. 2003b *Lovenia mortenseni* Čtyroký, 1965 – Kroh: 252; pl. 1, fig. 3

Type-material:
Lovenia mortenseni Čtyroký, 1965:
Holotype: ÚÚG-PČ 15D2 (figured by Čtyroký, 1965: pl. 1, figs. 1-2); Geological Survey, Praha, Czech Republic
Paratypes: ÚÚG-PČ 22D0, 79D4 and 34D0; Geological Survey, Praha, Czech Republic
Locus typicus: Dubovce near Hodonín, Slovak Republic
Stratum typicum: Lower member of the Carpathian Fm.
Age: Karpatian (Late Burdigalian), late Early Miocene

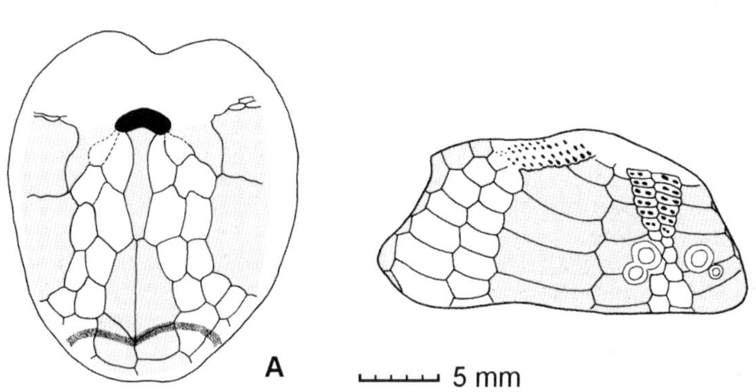

Figure 92: *Lovenia mortenseni* Čtyroký, 1965: oral plating (A) and lateral view (B) (A: paratype, ÚÚG PČ-22D0; B: paratype, ÚÚG PČ-79D4 both from Dubovce, *c.* 7 km ESE Holíč, Slovak Republic). Interambulacra shaded.

Material:
Karpatian (Late Burdigalian) – Oreske, *c.* 14 km ESE Holíč, Slovak Republic
NHMW: 1 specimen (NHMW 2002z0165/0001)
Karpatian (Late Burdigalian) – Strachotín, *c.* 11 km N Mikulov, southern Moravia, Czech Republic
Univ. Brno: numerous specimens (no inv. nos.; see POSPÍCHALOVÁ in KROH, 2003b)

Dimensions (in mm):

Inv. No.	TL	TW	TH	remarks
ÚÚG-PČ 15D2	22.0	21.5	-	holotype
ÚÚG-PČ 22D0	22.0	17.0	-	paratype
ÚÚG-PČ 79D4	23.5	-	11.1	paratype
NHMW	>15.4	>15.7	-	

Due to the fact that all known specimens are mostly deformed internal moulds or natural casts, the measurements may be inaccurate.

Description:
Size and shape: Test small, up to 25 mm in test length with heart-shaped outline. Frontal notch wide and relatively deep. In profile the test is moderately arched, with an obliquely truncated posterior end (Fig. 92.B). The maximum height lies more or less centrally, posterior of the apical disc.
Apical disc: The apical disc is ethmolytic with 4 circular gonopores. It lies anterior of the centre, between 35 and 45 % TL from the anterior margin
Ambulacra: Aborally the paired ambulacra are petaloid and slightly depressed adapically. The petals are subequal in size and extend about two third of the corresponding test radius. In larger specimens the posterior petals are slightly longer. Inside the internal fasciole no ambulacral pores are discernible. Outside the internal fasciole the ambulacral pore pairs are conjugate, elongate isopores. The anterior paired petals form an obtuse angle of about 125°, the posterior paired petals an acute angle of about 50 to 60°. The posterior paired petals are sharply flexed backwards and contain eight to twelve pores in each poriferous zone. The tips of the anterior paired petals are flexed laterally. In the posterior poriferous zone of the anterior paired petals there are usually between seven and nine pore pairs, the anterior plate row bears between five and six. The interporiferous zones are small, being only about two third as wide as a single poriferous zone.

The frontal ambulacrum is nonpetaloid and flush with the interambulacra adapically. Towards the ambitus it becomes increasingly depressed. In ambulacrum III only very small, oblique, partitioned isopores are present.
Interambulacra: The interambulacra 1, 2, 3, and 4 are characterised by the presence of large camellate primary tubercles with well developed internal ampullae. In the interambulacra 1 and 4 their number ranges from 1 to 4, most specimens have 2, whereas in the interambulacra 2 and 3 their number ranges from 1 to 5 and most specimens have 3. These tubercles are crowded at the distal part of the anterior paired petals, but their exact position and number is quite variable. During ontogeny the number of primary tubercles increases. The remaining aboral surface is rather densely covered by secondary and miliary tubercles.

On the oral side interambulacra 1 to 4 are densely covered with camellate primary tubercles, which increase in size from the margin of the test towards the centre. Interambulacrum 5 lacks such primary tubercles and totally lacks tubercles on the labrum and anterior half of the sternum. The plastron is ultramphisternous with a narrow contact between labrum and sternum (Fig. 92.A). The labrum is long and slender, slightly projecting over the peristome. Posteriorly interambulacrum 5 forms a well pronounced subanal heel.
Peristome: The peristome lies anteriorly, about 25 % TL from the anterior margin. It is kidney-shaped and transversely elongated.
Periproct: The periproct is partly obscured in all available specimens. It lies high on the oblique posterior end of the test in a depression.
Fascioles: Both an internal and subanal fasciole are present. The fasciole bands are moderately broad and well distinguished from the remaining tuberculation (orthofasciole *sensu* NÉRAUDEAU et al., 1998b). The internal fasciole crosses the suture of interambulacrum 5 well behind the apical disc. The subanal fasciole is bilobed and consists of a rather broad fasciole band.

Differential diagnosis:
A superficially similar species is *Lovenia anteroalta* (GREGORY, 1891) from the Burdigalian to Langhian of the Maltese Islands and the Burdigalian of the Rhône Basin [considered to be a junior synonym of *L. duncani* (GREGORY, 1891) by CHALLIS (1980), but rejected by PHILIPPE (1998: 223-225)]. According to ČTYROKÝ (1965: 115) *L. anteroalta* differs from this species by its larger test, greater number of pore pairs in the anterior paired petals and the shape of the internal fasciole. Comparison with *L. anteroalta* specimens from Gozo housed at the Naturhistorisches Museum Wien (Geologische Abteilung) revealed that the distinct subanal heel which characterises *L. mortenseni* is not present in *L. anteroalta*.

Most other species of *Lovenia* are well distinguished from *L. mortenseni*, an extensive discussion on this can be found in ČTYROKÝ (1965: 115-117).

Discussion:
Although not yet recovered from Austrian deposits, it is highly likely that this species is also present in contemporaneous deposits in Austria.

The platting patterns illustrated by ČTYROKÝ (1965: figs. 1, 5, and 6) are erroneous. Corrected drawings are provided here (Figs. 92.A-B).

Occurrence:
Paratethys (non-Austrian occurrences): Karpatian (Late Burdigalian)
Vienna Basin: Dubovce, *c.* 7 km ESE Holíč, Slovak Republic (ČTYROKÝ, 1965; KALABIS, 1975; CICHA et al., 1967; KROH, 2003b); Oreske, *c.* 14 km ESE Holíč, Slovak Republic (ČTYROKÝ, 1965, KALABIS, 1975; KROH, 2003b); Radošovce, *c.* 9 km ESE Holíč, Slovak Republic (CICHA et al., 1967); Strachotín, c. 11 km N Mikulov, Czech Republic (POSPÍCHALOVÁ in KROH, 2003b)

Genus *Echinocardium* GRAY, 1825

Type-species: *Echinus cordatus* PENNANT, 1777; by subsequent designation (ICZN, opinion 209, p. 385, 8th May 1954)
Diagnosis: Test ovate to cordate, with or without frontal notch; apical disc ethmolytic with four gonopores; ambulacrum III nonpetaloid, sunken or flush with the test; paired ambulacra petaloid; petals widening towards the apex; rudimentary pores within the internal fasciole, elongate isopores outside; pore zones Ib and IIa, respectively IVb and Va may form continuous lateral arcs; peristome anteriorly displaced, kidney-shaped; periproct on vertically truncated posterior end; labrum short but wide; sternal plates large and fully tuberculate; tuberculation homogenous, except in interambulacral columns 2b and 3a where larger tubercles are developed at the border to ambulacrum III; tubercles never sunken; well developed internal and subanal fascioles; anal branches reaching upwards from the subanal fasciole may be developed (modified from MORTENSEN, 1951 and SMITH "*The Echinoid Directory*", 28.07. 2003).
Distribution: Early Miocene to Recent – cosmopolitan, except polar and sub-polar regions
Ecology and biogeography: Extant species inhabit a wide range

of environments ranging from the intertidal to the mid-shelf and from muddy sand to shell gravel.

Echinocardium depressum (Agassiz in Agassiz & Desor, 1847)

(Fig. 93; Pl. 80, Figs. 2a-d)

* 1847b [*Amphidetus*] *depressus* Agass. – Agassiz in Agassiz & Desor: 12
1852 [*Amphidetus*] *depressus*, Agass. – d'Orbigny: 139; No. 2623
1858 [*Echinocardium*] *depressum* Syn. *Amphidetus depressus* Agass. – Desor: 407
1885 *Echinocardium tuberculatum* Gauthier, 1885. – Gauthier *in* Cotteau: 61-63; pl. 8, figs. 11-14
1885 *Echinocardium depressum* Agassiz, 1840. – Cotteau: 64
1892 *Echinocardium depressum*, Ag. – Depéret in Fontannes: 42
? 1892 *Amphidetus depressus* – Gourret: 130
1913a *E.*[*chinocardium*] *depressum* Ag. (*Amphidetus*) – Cottreau: 78
1915a *Amphidetus depressus* Agassiz, 1847. – Lambert: 208-209; pl. 16, figs. 4-7
1915a *Amphidetus tuberculatus* Gauthier (*Echinocardium*, in Cotteau), 1885. – Lambert: 209; pl. 15, fig. 11
1915a *Amphidetus granifer* Lambert. – Lambert: 210; pl. 16, figs. 1-3
1920 *Echinocardium depressum*, Agassiz 1847. – Fourtau: 91-92; pl. 7, fig. 3
1926 *Amphidetus depressus* Agassiz – Lambert: 252-253
1926 *Amphidetus tuberculatus* (Gauthier) (*Echinocardium*) – Lambert: 253
1973 *Echinocardium tuberculatum* Gauthier, 1885 – Philippe: 199-200
1984 *Echinocardium depressum* (Agassiz), 1847 – Negretti: 124-125; pl. 10, fig. 7
1984 *Echinocardium tuberculatum* Gauthier in Cotteau, 1885 – Negretti: 125-126; pl. 10, figs. 8-15
1989 *Echinocardium depressum* (Agassiz) – Philippe: 32; tab. 1
1990 *Echinocardium depressum* (Agassiz, 1847) – Philippe et al.: 247
1998 *Echinocardium depressum* (Agassiz, 1847) – Philippe: 225-229; pl. 26, figs. 2-6
? 2001 *Echinocardium depressum* (L. Agassiz, 1847) – Radwański & Wysocka: 307

Type-material:
Amphidetus depressus Agassiz in Agassiz & Desor, 1847:
Locus typicus: Cap Couronne, near Martigues, Bouches-du-Rhône, France
Age: Burdigalian, Early Miocene
Remarks: the whereabouts of the type-material, originally in the collection of Michelin, is unknown

Echinocardium tuberculatum Gauthier in Cotteau, 1885:
Holotype: the specimen figured by Cotteau (1885: pl. 8, figs. 11-14); a cast is housed in the collection Lambert, Muséum national, Paris (Philippe, 1998: 335)
Locus typicus: Cap Couronne, near Martigues, Bouches-du-Rhône, France
Age: Burdigalian, Early Miocene

Amphidetus granifer Lambert, 1915:
Holotype: the specimen figured by Lambert (1915a: pl. 16, figs. 1-3); a cast is housed in the collection Lambert, Muséum national, Paris (Philippe, 1998: 335)

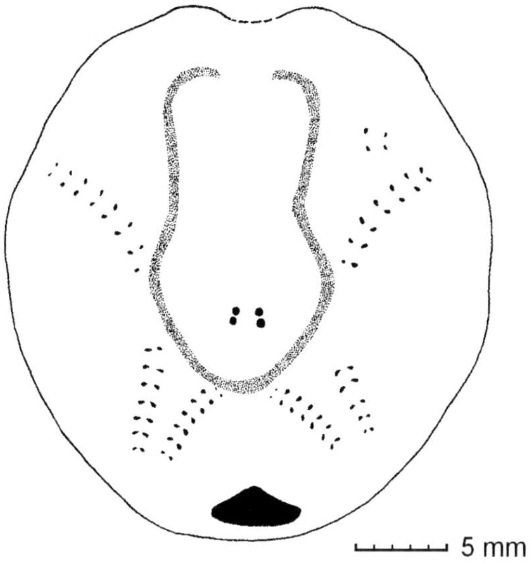

Figure 93: *Echinocardium depressum* (Agassiz in Agassiz & Desor, 1847b): shape of the internal fasciole and the petals (Florianibründl, near Eggenburg, NÖ, Austria, NHMW 2002z0171/0001).

Locus typicus: Cap Couronne, near Martigues, Bouches-du-Rhône, France
Age: Burdigalian, Early Miocene

Material:
Late Eggenburgian (Early Burdigalian) – Eggenburg (Florianibründl), NÖ, Austria
NHMW: 1 specimen (NHMW 2002z0171/0001)

Dimensions (in mm):

Inv. No.	TL	TW	TH
NHMW 2002z0171/0001	28.0	27.3	14.2

Description:
Size and shape: Test of medium size, with subcircular to slightly oval outline and shallow frontal sinus. In profile the test is slightly wedge shaped. Anterior end tumid, posterior end obliquely truncated. The maximum width lies about halfway along the length, the maximum height posteriorly along the rounded keel in interambulacrum 5.
Apical disc: The apical disc lies posterior of the centre, about 55 % TL from the anterior margin. It is ethmolytic and has four circular gonopores.
Ambulacra: The ambulacra are slightly depressed on aboral side between the apical disc and the ambitus. The paired ambulacra are greatly widened adapically and moderately depressed. The frontal ambulacrum, in contrast, is narrow and more depressed. Inside the internal fasciole no pores are found in the paired ambulacra. Outside of it the ambulacral pores are elongated isopores. Outside the petals only minute unipores are present (except in the phyllodes). Ambulacrum III is depressed between the peristome and the anterior margin.
Interambulacra: The interambulacra are slightly inflated between the petals. They are covered by small perforate, crenulate tubercles. The plastron is ultramphisternous (?), and the posterior end of the plastron forms a pointed projection.
Peristome: The peristome lies anterior on the oral side, about 30 % TL from the anterior margin. It is kidney shaped and relatively large, 6.0 mm wide in the studied specimen. The labrum is broad and projects slightly into the peristome.
Periproct: The periproct lies high on the vertical posterior end of the test in interambulacrum 5. It is oval (transversely elongated) to slightly teardrop-shaped. Its adoral margin is round-

ed, the adapical margin pointed. It is distinctly smaller than the peristome, measuring 4.7 mm in width.
Fascioles: The internal fasciole is well developed and belongs to the orthofasciole type of NÉRAUDEAU et al. (1998b). The shape of the internal fasciole can be seen in Fig. 93, note the distinct widening as the fasciole passes laterally of the apical disc. The subanal fasciole lies on the lowermost part of the vertical posterior end and has a subcircular to oval outline.

Differential diagnosis:
The Middle Miocene species *E. leopolitanum* RADWAŃSKI & WYSOCKA, 2001 from the Ukraine differs from this species by its higher, box-shaped profile, less wide petals and different shape of the internal fasciole.

E. peroni COTTEAU, 1877 from the Burdigalian of the Rhône Basin and the Middle Miocene of Corsica differs by its much narrower petals, well developed subanal keel and smaller internal fasciole (for a description of *E. peroni* see PHILIPPE, 1998).

E. intermedium LÓCZY, 1877 from the Middle Miocene of Hungary, which is very similar *to E. peroni* and probably synonymous to it, has much smaller petals, a higher profile, an antero-posteriorly elongated outline and a pronounced subanal heel or rostrum (compare LÓCZY, 1877, and VADÁSZ, 1915).

E. biaense MIHÁLY, 1985 from the Middle Miocene of Hungary is synonymous to *E. peroni* according to RADWAŃSKI & WYSOCKA (2001: 307).

E. deikei DESOR, 1858 from the Burdigalian of Switzerland differs by its more elongate shape and broader ambulacrum III (see DESOR, 1858; DE LORIOL, 1875; PHILIPPE, 1998). The types and other material from the Swiss Molasse are, however, poorly preserved and deformed and comparison is difficult. Material from the Middle Miocene of Poland reported by MĄCZYŃSKA (1979) was reassigned to *E. peroni* by PHILIPPE (1998: 231).

Discussion:
The sole specimen fits best with the contemporary *E. depressum* from the Burdigalian of the Rhône Basin. PHILIPPE (1998: 225-229) synonymised *E. tuberculatum* GAUTHIER in COTTEAU, 1885 and *E. granifer* (LAMBERT, 1915a) with *E. depressum*.

This is the first record of this species from the Neogene of Austria. Apart from the present record *Echinocardium* is known only from three other localities in Austria (see below under *Echinocardium* sp.), two of which are new. The third (TOLLMANN, 1955: tab. 1) could not be confirmed and probably refers to disarticulated spatangoid material. Five other species of the genus *Echinocardium* have been recorded from the Paratethys [all but one of them from Badenian (Langhian to Early Serravallian) strata]. They are briefly discussed in the section dealing with Paratethyal species not recorded from Austria until now (see below). Considering the morphological plasticity of and the difficulties in classifying the extant *Echinocardium cordatum/fenauxi*-complex (see DAVID & LAURIN, 1996; DAVID et al., 1999; CHENULI & FÉRAL, 2003 and references therein) casts some doubts on these four fossil species. They need to be reexamined on base on well preserved material to establish whether they are separate species or not (most are based on few poorly preserved specimens).

Occurrence:

Austria: Late Eggenburgian (Early Burdigalian)
Molasse Zone: Eggenburg (Florianibründl), NÖ ([NHMW])

Paratethys (non-Austrian occurrences): Obere Meeresmolasse (Middle Burdigalian); ? Early Badenian (Langhian)
Swiss Molasse: La Chaux-de-Fonds (LAMBERT, 1926)
Fore-Carpathian Basin: ? Huta Lubycka, Poland (RADWAŃSKI & WYSOCKA, 2001)

Mediterranean: Burdigalian
Western Mediterranean: Ste. Colombe, near Vence, Alpes Maritimes, France (LAMBERT, 1915a); Vence, Alpes Maritimes, France (COTTREAU, 1913a)
Rhône Basin: Apt (PHILIPPE, 1998); Auribeau (PHILIPPE, 1998); les Beaumettes (PHILIPPE, 1998); Bonnieux (PHILIPPE, 1998); Buoux (PHILIPPE, 1998); Cap Couronne près Martigues, Bouches-du-Rhône (AGASSIZ & DESOR, 1847b; D'ORBIGNY, 1852; DESOR, 1858; COTTEAU, 1885; DEPÉRET in FONTANNES, 1892; COTTREAU, 1913a; LAMBERT, 1915a; PHILIPPE, 1972; NEGRETTI, 1984; PHILIPPE et al., 1990); Carry (LAMBERT, 1915a); Grignan (PHILIPPE, 1998); Istres (PHILIPPE, 1998); Lacoste (PHILIPPE, 1998); Martigues, Nerthe (PHILIPPE, 1998); Pierrerue (PHILIPPE, 1998); Plan d'Arren, Bouches-du-Rhône (LAMBERT, 1915a); E of Port de Tamaris to Anse du Verdon, Nerthe (NEGRETTI, 1984; PHILIPPE et al., 1990); Saint-Mitre-les-Remparts (PHILIPPE, 1998); St-Martin-de-Castillon (PHILIPPE, 1998); St Michel-l'Observatoire (PHILIPPE, 1998); Viens (PHILIPPE, 1998)
Eastern Mediterranean: Gebel Tanka, Sinai, Egypt (FOURTAU, 1920)

Echinocardium sp.

(Pl. 80, Figs. 3a-c)

? 1955 *Echinocardium* sp. – TOLLMANN: tab. 1
? 1981 *Echinocardium* sp. – HALMAI: 106

Material:
Early Eggenburgian (Early Burdigalian) – Gösing (Fels Fm.), NÖ, Austria
KM: 7 specimens and 2 fragments (no inv. nos.)
Badenian (Langhian-Early Serravallian) – Oslip (sandpit E of the village), Bgld, Austria
FÜRNKRANZ coll.: 1 fragmentary specimen (no inv. no.)

Dimensions (in mm):

Inv. No.	TL	TW	TH
KM spec. 1	17.6	17.2	10.1
KM spec. 2	19.1	17.4	12.1
KM spec. 3	21.4	18.1	14.3
KM spec. 4	24.1	22.9	14.3
KM spec. 5	24.1	24.0	12.6
KM spec. 6	25.5	24.5	14.4
KM spec. 7	43.5	43.3	21.8

Discussion:
Albeit clearly belonging to the genus *Echinocardium* the material is too fragmentary (Oslip specimen), respectively too poorly preserved (covered by syntaxial cement, sand and still-attached spines in case of the Gösing specimens) to allow specific determination.

The record of TOLLMANN (1955) is based on disarticulated spines and plates. It cannot be judged with certainty whether that material was correctly attributed to *Echinocardium* or refers to indeterminable spatangoid debris. None of the mentioned localities of TOLLMANN yielded material confidently attributable to *Echinocardium* despite extensive sampling (both by scientists and private collectors) since TOLLMANN'S publication. Halmai's (1981) record is based on an unpublished Ph.D. thesis and could not be verified

Occurrence:

Austria: Early Eggenburgian (Early Burdigalian), Badenian (Langhian-Early Serravallian)
Molasse Zone: Gösing (Fels Fm.), NÖ ([KM])
Eisenstadt-Sopron Basin: ? Großhöflein, near Eisenstadt, Bgld (TOLLMANN, 1955); ? Hartl, Eisenstadt, Bgld (TOLLMANN, 1955); ? Müllendorf, Bgld (TOLLMANN, 1955); Oslip, Bgld ([FÜRNKRANZ coll.])

Paratethys (non-Austrian occurrences): ? Karpatian (Late Burdigalian)

Great Hungarian Basin (Pannonian Basin): Kisalag, Hungary (HALMAI, 1981)

Genus *Hemipatagus* DESOR, 1858

Type-species: *Spatangus hoffmanni* GOLDFUSS, 1829; by original designation (DESOR, 1858: 416).
[*nomen correctum* pro *Spatangus hofmanni* GOLDFUSS, 1829 – GRATELOUP (1836: 175) (presumably intentionally) changed the name to *S. hoffmanni*. According to the ICZN rules this is an unjustified emendation. However, as the name *S. hoffmanni* is in prevailing usage since 1836 and is attributed to the original author and date of the original spelling (e.g. DES MOULINS, 1837: 244; AGASSIZ & DESOR, 1847b: 7; ....), it has to be deemed to be a justified emendation according to the ICZN (4th ed., 2000, Article 33.2.3.1.).]
Emended diagnosis: Test cordiform, with distinct frontal notch and transversely truncated posterior end; in profile test high arched to low and slightly wedge-shaped; posterior face low, obliquely to near-vertically truncated; apical disc ethmolytic with four gonopores, madreporite extending beyond posterior oculars; ambulacrum III nonpetaloid, flush with test adapically, increasingly sunken towards the ambitus, with oblique partitioned isopores; paired ambulacra petaloid; petals widening adapically, with large elongate isopores; rudimentary pores in the adapical poriferous zones of the anterior paired petals; adorally ambulacra II and IV strongly constricted near the margin on the oral side in adult specimens; plastron ultramphisternous; labrum elongated, slightly projecting over peristome, with narrow contact to the sternal plates, not extending beyond the 2nd ambulacral plates; tubercles confined to posterior half or third on the sternal plates; V-shaped contact between sternal and episternal plates; peristome anteriorly eccentric, kidney-shaped and facing downwards; periproct high in posterior face, usually subcircular or transversely elongated; aboral tuberculation heterogeneous with large, noncrenulate primary tubercles in interambulacra 1 to 4; the tubercles have strongly sunken areoles or are recessed in camellae in some species; no such tubercles in interambulacrum 5 and the posterior part of interambulacral columns 1a and 4b; conspicuous field of coarse tubercles adjacent to adapical ambulacrum III in interambulacral columns 2b and 3a; oral tubercles prominent laterally and arranged in distinct rows, in some species areoles deeply sunken with parapet typically helically spiralled; well developed, bilobed subanal fasciole; up to four ambulacral plates extending into the subanal fasciole (from KROH submitted).
Distribution: Middle Eocene to Miocene – Europe and North America
Remarks: *Lovenia* differs by the presence of an internal fasciole, higher number of ambulacral plates extending into the subanal fasciole (up to eleven, although in some species there are only three or four as in *Hemipatagus*), and less lanceolate petals; in most species of *Lovenia* ambulacral columns Ib and IIa, respectively IVb and Va form a continuous arc (not observed in any known *Hemipatagus* species); in some species of *Hemipatagus* the aboral and/or oral tubercles are not camellate, while this is always the case in *Lovenia*. *Maretia* differs by its extremely shallow frontal notch, the distribution of the aboral primary tubercles (primary tubercle reach the tips of the posterior paired petals), by the lack of strongly sunken areoles or camellate aboral and oral primary tubercles, by the crenulate nature of the aboral primary tubercles, by the lack of a conspicuous field of coarse tubercles in adapical interambulacral columns 2b and 3a, by its oval to shield-shaped subanal fasciole, by the longer primordial plates of the paired interambulacra (plates 1a1, 2a1, 3a1, and 4a1) (extending to the 3rd or 4th ambulacral plate of the adjacent ambulacra in *Maretia*, while the do not extend beyond the 2nd ambulacral plate in *Hemipatagus* or *Lovenia*), by the labrum which extends to the 3rd or 4th ambulacral plate (not extending beyond the 2nd ambulacral plate in *Hemipatagus* and *Lovenia*) and the less constricted adoral ambulacra 2 and 3.

Hemipatagus ocellatus (DEFRANCE, 1827)

(Fig. 94; Pl. 81, Figs. 1-4; Pl. 82, Figs. 1-2)

#. 1839 *Spatangus Nicoleti* AG. – AGASSIZ: 33; pl. 4, figs. 7-8
1840b *Spatangus ocellatus* DEFR. – AGASSIZ: 2
1846 *S.[patangus] ocellatus*, DEFR. – PICTET: 137
1847b *[Spatangus] ocellatus* DEFR. – AGASSIZ & DESOR: 7
1848 *Spatangus ocellatus* (DEFRANCE) – GRAS: 54
1852 *[Spatangus] ocellatus*, DEFR. – D'ORBIGNY: 139, No. 2631
non 1855 *Spatangus Hoffmanni*, GOLDFUSS. – WRIGHT: 176-178 [= *Lovenia duncani*, see GREGORY (1891: 624), CHALLIS (1980: 297)]
1857 *S.[patangus] ocellatus*, DEFR. – PICTET: 200
1858 *[Spatangus] ocellatus* DEFR. – DESOR: 422
non 1864 *Spatangus ocellatus*, DEFRANCE. – WRIGHT: 487-488; pl. 21, figs. 1a-b [= *Lovenia duncani*, see DUNCAN (1889: 265), GREGORY (1891: 624), CHALLIS (1980: 297)]
? 1873 *Spatangus ocellatus* ? DEFR. – MANZONI: 10-11, 22
1876 *Spatangus ocellatus*, DEFRANCE. – DE LORIOL: 132-134; pl. 23, figs. 2, 2a, 3
? 1883 *S.[patangus] ocellatus*, DEFR. – MAZZETTI & PANTANELLI: 65
1891 *M.[aretia] ocellata* (*Spatangus ocellatus* DEFRANCE) – COTTEAU et al.: 83 (in the discussion of *Maretia soubellensis*)
? 1892 *Spatangus ocellatus*, DEFRANCE. – GOURRET: 130, 142
1909 *Hemipatagus ocellatus* DEFRANCE (*Spatangus*), 1827 – LAMBERT: 108-109
v. 1912a *Spatangus (Maretia) perornatus* SCHFF. – SCHAFFER: 190-191; pl. 59, fig. 4-6
? 1914 *Hemipatagus ocellatus* DEFR. sp. – VADÁSZ: 210-211; fig. 121

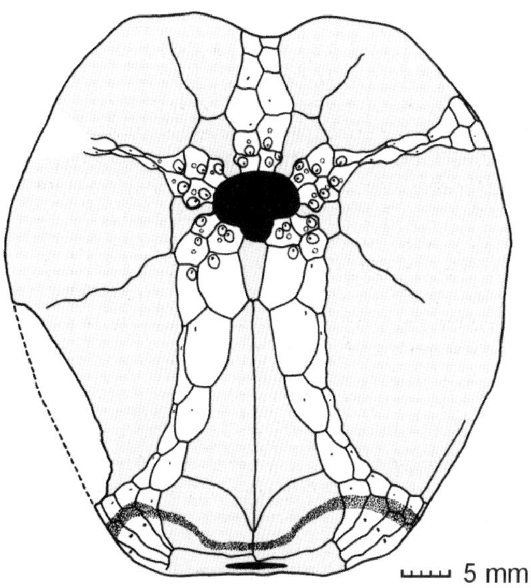

Figure 94: *Hemipatagus ocellatus* (DEFRANCE, 1827): oral plating (Grübern, NÖ, NHMW 1914.VII.22b). Interambulacra shaded.

. 1915a *Hemipatagus Ocellatus* DEFRANCE (*Spatangus*), 1827 – LAMBERT: 189-192; pl. 14, figs. 17-18
? 1915 *Hemipatagus ocellatus* DEFR. sp. – VADÁSZ: 237-238; fig. 123
1928 *Hemipatagus ocellatus* DEFRANCE (*Spatangus*) – LAMBERT & JEANNET: 176 [II, 94], S.42
1951 *Hemipatagus ocellatus* – MORTENSEN: 24; fig. 4d
v. 1971a *Maretia perornatus* SCHAFFER, 1912 – STEININGER: 596; pl. 1, fig. 5-6
1971b *Maretia perornatus* SCHAFFER – STEININGER: 119
1971c *Maretia perornatus* SCHAFFER – STEININGER: 129
1975 *Spatangus ocellatus* (DEFRANCE) – KOTCHETOFF et al.: 80-86; pl. 8, 3 figs.
. 1984 *Maretia ocellata* (DEFRANCE), 1827 – NEGRETTI: 123; pl. 10, fig. 6
. 1989 *Maretia ocellata* (DEFRANCE) – PHILIPPE: 32, tab. 1
. 1990 *Maretia ocellata* (DEFRANCE, 1827) – PHILIPPE et al.: 247
. 1998 *Maretia ocellata* (DEFRANCE, 1827) – PHILIPPE: 220-223; pl. 22, figs. 6, 7, 8a-b, 9a-b, 10a-c
v. 1999 *Maretia perornata* – HARZHAUSER & KROH: 221
v. 1999 *Maretia? perornata* (SCHAFFER, 1912) – KROH & HARZHAUSER: 162-163; pl. 3, fig. 1
v. 2002 *Maretia? perornata* – LUKENEDER & HARZHAUSER: 462

Type-material:

Spatangus ocellatus DEFRANCE, 1827:
Holotype: unknown, collection unknown
Neotype: from the Burdigalian of Taulignan, figured by LAMBERT (1915a: pl. 14, fig. 17); a cast housed in the collection LAMBERT, Muséum national, Paris
Locus typicus: Saint-Paul-Trois-Châteaux, France
Age: Burdigalian, Early Miocene (PHILIPPE, 1998: 220)

Spatangus nicoleti AGASSIZ, 1839:
Holotype: specimen figured by AGASSIZ (1839: pl. 4, figs. 7-8), current whereabouts unknown
Locus typicus: La Chaux-de-Fonds, Canton Neuchâtel, Switzerland
Age: Burdigalian, Early Miocene (PHILIPPE, 1998: 220)

Spatangus (Maretia) perornatus SCHAFFER, 1912a:
Syntypes: 2 syntypes (KM F/0127, F/0128), figured in SCHAFFER (1912a: pl. 59, figs. 4-6); housed at the Krahuletz-Museum, Eggenburg, Austria
Locus typicus: Eggenburg (Kremserberg), NÖ
Stratum typicum: Zogelsdorf Fm.
Age: Late Eggenburgian (Early Burdigalian), Early Miocene

Material:

Early Eggenburgian (Early Burdigalian) – Gösing (Fels Fm.), NÖ, Austria
- NHMW: 2 specimens (NHMW 1998z0048/0063, 1998z0048/0067) and 17 fragments (NHMW 2002z0160/0001)
- KM: 4 specimens (without inventory no.)
- IPUW: 2 specimens (no inventory no.)

Late Eggenburgian (Early Burdigalian) – Eggenburg (Kremserberg, Zogelsdorf Fm.), NÖ, Austria
- KM: 2 specimens [KM F/0127, F/0128 (syntypes of *S. perornatus* SCHAFFER, 1912a)]

Late Eggenburgian (Early Burdigalian) – Grübern (Zogelsdorf Fm.), NÖ, Austria
- NHMW: 4 specimens (NHMW 1914.VII.22a-c, NHMW 2002z0159/0001)

Late Eggenburgian (Early Burdigalian) – Unternalb (Retz Fm.), NÖ, Austria
- NHMW: 1 specimen (NHMW 1999z0049/0003)

Dimensions (in mm):

Inv. No.	TL	TW	TH
NHMW 1997z0049/0003	97.6	89.0	~23
NHMW 1914.VII.22a	63.4	~62 *	15.2
NHMW 1914.VII.22b	57.8	53.8	16.4
NHMW 1914.VII.22c	>55	55.8	-
NHMW 2002z0159/0001	-	54.6	16.3
NHMW 1998z0048/0067	46.2	43.2	~12.5
KM F/0127	72.3	66.9	21.4
KM F/0128	66.4	56.7	>16
KM spec 1 (Gösing)	24.6	23.2	8.2
KM spec 2 (Gösing)	33.0	29.2	10.1
KM spec 3 (Gösing)	33.6	30.8	10.1
KM spec 4 (Gösing)	~50	~47	17.8

* calculated value (based on the bilateral symmetry of the animal)

Description:

Size and shape: Test of medium to large size, ranging from 24 to 100 mm test length in the studied material. The outline of the corona is cordate, slightly elongated antero-posteriorly, with a transversely truncated posterior end. A distinct, moderately deep frontal sinus is present. The maximum width lies slightly anterior of the apical disc. In profile the test is low and slightly wedge-shaped with a vertically truncated to slightly overhanging posterior end. The maximum height lies about halfway between the apical disc and the posterior end.

Apical disc: The apical disc lies slightly anterior of the centre, about 43 to 45 % TL from the anterior test margin. It is ethmolytic with 4 circular gonopores, the anterior pair of which is slightly smaller and more closely spaced. The madreporite is rather large and elongated posteriorly, extending beyond the posterior ocular plates.

Ambulacra: Paired ambulacra petaloid, the anterior paired petals are straight, closed and extend about two-thirds of the corresponding test radius. The posterior paired ambulacra are flexed sharply posteriorly and their distal tips are flexed slightly laterally. Within the paired petals rather large, conjugate, elongate isopores lying in deep, steep-walled depressions are present. The adapical pores, however, are very small, rudimentary conjugated isopores. In the anterior poriferous zones of the anterior paired petals, the ambulacral pores reach their "normal" size from the 11th pore pair on, in the posterior poriferous zone and those of the posterior paired petals only the uppermost three to five pore pairs are reduced in size. Interporiferous zone up to twice as wide as a single poriferous zone at its widest point.

Adapically ambulacrum III is flush with the interambulacra, towards the ambitus it is increasingly depressed and forms a distinct, moderately deep frontal notch. It is nonpetaloid and the pores are distinctly different from those of the paired petals. They are minute, strongly oblique partitioned iso- to anisopores.

Outside the petals only minute microunipores are found in the ambulacra, except in the phyllodes and within the subanal fasciole, where 3 enlarged hourglass-shaped unipores are present in each ambulacrum. The phyllodes are not depressed and consist of six large unipores with wide periporal area (except in ambulacrum III, where only five pores are present).

Only secondary and miliary tubercles are present in the ambulacra. On the oral side ambulacra I and V form moderately broad, naked peri-plastronal areas. Ambulacra II and IV are extremely narrow between the phyllodes and the margin on the oral side of the test (Fig. 94).

Interambulacra: Interambulacra not or only slightly inflated between the petals. Interambulacra 1, 2, 3 and 4 bear very large, perforate, noncrenulate primary tubercles, which are recessed in camellae on the apical surface of the test. The number and position of these enlarged primary tubercles shows considerable variation. Usually they are crowded in the parts bordering the anterior paired petals. Their number ranges from 4 to 60 in interambulacra 1 and 4 and 5 to 35 in interambulacra 2 and 3. The surface between and around these enlarged primary tubercles is densely crowded with miliary tubercles. Secondary tubercles with distinct areoles are scattered among them, but are generally rather rare on the apical surface of the test. Only in interambulacrum 5 along the interradial suture and along the adradial sutures to ambulacrum III in interambulacra 2 and 3 a larger numbers of secondary tubercles is present.

In the interambulacral columns 2b and 3a a conspicuous field of coarse tubercles is developed adjacent to the adradial suture to the adapical part of ambulacrum III (see Pl. 81, Fig. 2b)

On the oral surface interambulacra 1, 2, 3 and 4 are covered densely with camellate primary tubercles arranged in distinct rows. They have deeply sunken, asymmetric areoles and helically spiralled parapets. These tubercles are largest next to the adradial sutures and decrease in size from the centre towards the margin. The plastron is ultramphisternous with a narrow contact between labrum and sternum (Fig. 94). The labrum is triangular and projects slightly into the peristome. Up to two third of the sternal plates are naked, only in the posterior part secondary tubercles are found, which are arranged in a fan-shaped pattern.

Peristome: The peristome is kidney shaped, transversely elongated and lies about 25-28 % TL from the anterior margin. It is moderately large, being about 15 to 16 % TL wide in adult specimens.

Periproct: The periproct lies inframarginally, high on the vertically truncated posterior end of the test. It is oval, transversely elongated and is with approximately 10 % TL distinctly smaller than the peristome. There is a depressed subanal area, which is free of primary and secondary tubercles.

Fascioles: The only fasciole present is a wide bilobed subanal fasciole. The fasciole band is moderately wide and corresponds to the othofasciole type of NÉRAUDEAU et al. (1998b). The existence of an internal fasciole in adult specimens can be excluded. Even in specimens with extremely well preserved aboral tuberculation no traces of fasciole bands were located around the apical disc.

Differential diagnosis:

The closely related *Hemipatagus girundicus* LAMBERT, 1915a from the Burdigalian to Langhian of Bordeaux differs from this species by a broader keel in interambulacrum 5, which is more flattened and more densely tuberculate than in *H. ocellatus* (compare the descriptions in LAMBERT, 1915a: 191; LAMBERT & THIÉRY, 1909-25: 457, pl. 13, fig. 5; LAMBERT, 1927b: 119-121). If this sole feature is enough to keep these species separated is still a matter of debate. CHAVANON (1974: 271-272) and PHILIPPE (1998: 222-223) did not synonymise them, although the latter expressed his doubts and stated that more material is needed to solve this question.

Hemipatagus hoffmanni (GOLDFUSS, 1829) differs by its higher test (test height ranging from 42 to 50 % TL), the bluntly pointed, obliquely truncated posterior end, the laterally flexed anterior petals, the lower number of camellate primary tubercles, which are, moreover, never as large and closely spaced as in *H. ocellatus* [based on the description in GOLDFUSS, 1826-44: 152-153, pl. 47, figs. 3a-c, the type-material in the collection of the Goldfuss Museum (Univ. Bonn) and specimens from the type locality in the collection of the NHMW and the NKM Berlin].

Spatangus ornatus (CUV.) (as described in GOLDFUSS, 1829: 152; pl. 47, figs. 2a-c), which was the name-inspiring species for SCHAFFER'S *S. perornata*, is an *Eupatagus*. Its plastron is fully covered with tubercles and a peripetalous fasciole is present according to AGASSIZ & DESOR (1847b: 9). Moreover, albeit generally similar it has no camellate primary tubercles and cannot easily be confused with the species discussed here.

Discussion:

The generic assignment of this species was subject to considerable debate: originally referred to the genus *Spatangus* (e.g. DEFRANCE, 1827; DE LORIOL, 1876) it was later placed into the genus *Maretia* (COTTEAU et al., 1891), only to be transferred to the genus *Hemipatagus* by LAMBERT (1909) soon afterwards. During this time the question arose, whether *Maretia* and *Hemipatagus* should be considered synonymous (AGASSIZ, 1872: 568, 1873; COTTEAU, 1885-89: 24; FOURTAU, 1920: 83) or not (e.g. DUNCAN, 1889: 222, 252; LAMBERT, 1909: 107, 1915a: 188, 1927b: 87), or even if *Hemipatagus* is synonymous with *Lovenia* (DUNCAN, 1877: 56). In his *"Monograph of the Echinoidea"* MORTENSEN (1951: 23-27) lengthily discussed this matter and finally considered *Hemipatagus* DESOR,1858 a junior synonym of *Maretia* GRAY, 1855. In the *"Treatise on Invertebrate Paleontology"* FISCHER (1966: U609) adopted MORTENSEN'S view and thus it was soon widely accepted in the scientific community, although some authors (e.g. PHILIPPE, 1998: 222) expressed their doubts.

Based on material of the type-species of the genera in question the present author (KROH, in prep) was able to show that *Hemipatagus* is clearly different from *Maretia* and in fact belongs to the Loveniidae, not the Spatangidae as the latter genus. The features with which the two genera can be separated are outlined above under the remarks to the genus.

The studied specimens clearly belong into the genus *Hemipatagus*, and are conspecific with *H. ocellatus* (DEFRANCE, 1827) recently re-described and illustrated by PHILIPPE (1998: 220-223; pl. 22, figs. 6, 7, 8a-b, 9a-b, 10a-c; under the name *Maretia ocellata*).

Based on the examination of the type material housed at the Krahuletz-Museum (Eggenburg, NÖ, Austria) the species *Spatangus (Maretia) perornatus* SCHAFFER, 1912a has to be placed into the synonymy of *H. ocellatus*. According to AGASSIZ & DESOR (1847b: 7) and PHILIPPE (1998: 222) *Spatangus nicoleti* AGASSIZ, 1839 is also a junior synonym of *H. ocellatus*. Although, surprisingly similar and originally determined at first as *Spatangus hoffmanni* (WRIGHT, 1855: 176) and later as *S. ocellatus* (WRIGHT, 1864: 487), the Maltese species *Lovenia duncani* (GREGORY, 1891), is clearly a loveniid, based on the presence of an internal fasciole as shown by CHALLIS (1980).

The material of VADÁSZ (1914, 1915) attributed to *Hemipatagus ocellatus* is rather poorly preserved. The figured specimen could not be located at the MAFI. The only specimens that are preserved there are three weathered fragments with camellate aboral tubercles (MAFI Ech 237) that are not identifiable at genus level.

Occurrence:

Austria: Early to Late Eggenburgian (Early Burdigalian)

Molasse Zone: Achberg (Scutellensande, Loibersdorf Fm.), near Maria Dreieichen, NÖ (STEININGER, 1971a, b); Eggenburg (Brunnstube), NÖ (STEININGER, 1971a, c); Eggenburg (Kremserberg, Zogelsdorf Fm.), NÖ (SCHAFFER, 1912a; [KM]); Fels am Wagram (Fels Fm.), NÖ (STEININGER, 1971a); Gösing (Fels Fm.), NÖ ([KM, NHMW]); Grübern (Zogelsdorf Fm.), NÖ (SCHAFFER, 1912a; [NHMW]); Unternalb (Retz Fm.), NÖ (HARZHAUSER & KROH, 1999; KROH & HARZHAUSER, 1999; LUKENEDER & HARZHAUSER, 2002 [NHMW])

Paratethys (non-Austrian occurrences): Burdigalian, ? Badenian (Langhian-Early Serravallian)

Swiss Molasse: La Chaux-de-Fonds, Jura (AGASSIZ, 1839; DESOR, 1858; MANZONI, 1873; DE LORIOL, 1876; LAMBERT, 1915; LAMBERT & JEANNET, 1928); Les Verrières, Neuchâtel (DE LORIOL, 1876; LAMBERT, 1915a); Neuchâtel (AGASSIZ & DESOR, 1847b; D'ORBIGNY, 1852; PICTET, 1857)

Great Hungarian Basin (Pannonian Basin): ? Mátraverebély, Nógrád, Hungary (VADÁSZ, 1914, 1915)

Transylvanian Basin: ? Gârbova de Sus (= Felsö-Orbó), Romania (VADÁSZ, 1914, 1915)

Mediterranean: Burdigalian, ? Langhian

Western Mediterranean: Italy, mainland (AGASSIZ, 1840b; PICTET, 1846): ? Monteorsello (MAZZETTI & PANTANELLI, 1883), ? Monte Titano (MANZONI, 1873), ? Rocca S. Maria (MAZZETTI & PANTANELLI, 1883), ? Sasso Bolognese (MANZONI, 1873); Rhône Basin, France (PICTET, 1846, 1857): Beaucaire (PHILIPPE, 1998), Bonnieux (PHILIPPE, 1998), Cap Couronne, near Martigues, Bouches-du-Rhône (NEGRETTI, 1984; PHILIPPE et al., 1990), Carry, Bouches-du-Rhône (GOURRET, 1892; LAMBERT, 1915a), Châteauneuf-Miravail (PHILIPPE, 1998), Clansayes (LAMBERT, 1915a; PHILIPPE, 1998), Couremes, near Vence, Alpes-Maritimes (LAMBERT, 1915a), Crest (PHILIPPE, 1998), Grignan (PHILIPPE, 1998), Istres (PHILIPPE, 1998), Lacoste (PHILIPPE, 1998), Martigues, Bouches-du-Rhône (PHILIPPE, 1998), Ménerbes (PHILIPPE, 1998), Montségur-sur-Lauzon (PHILIPPE, 1998), E of Port des Tamaris, Bouches-du-Rhône (NEGRETTI, 1984), Reillanne, Basse-Alpes (LAMBERT, 1915a; PHILIPPE, 1998), Saint-Mitre-les-Remparts (PHILIPPE, 1998), Saint-Paul-Trois-Châteaux, Drôme (AGASSIZ & DESOR, 1847b; GRAS, 1848; D'ORBIGNY, 1852; DESOR, 1858; MANZONI, 1873; LAMBERT, 1915a; PHILIPPE, 1998), St-Jean-en-Royans, Isere (GRAS, 1848), Taulignan (LAMBERT, 1915a; PHILIPPE, 1998); Sardinia, Italy: Monte de Giave (LAMBERT, 1909)

Atlantic Ocean: ? Middle Miocene

Portuguese Atlantic coast: Foz da Fonte, Portugal (KOTCHETOFF et al., 1975)

Spatangoida indet.

(Pl. 82, Fig. 3)

	1877	Kleine Spatangiden – FUCHS: 673
	1942	Spatangidenstacheln. – WINKLER: 106
	1943	Spatangiden – VEIT: 13
	1943	Seeigelstacheln – GRILL: pl. 1; pl. 5
	1952	Stacheln kleiner Echiniden (Spatangiden?) – PAPP: 17
pp	1952	„Seeigelstachel, Echinidenkiefer, Pedicellarien" – KÜHN: 124
	1953	Spatangidenstachel – GRILL: 87, 88, 97, 98
pp	1955	Echiniden: Platten, Zähne, Kieferteile, Echiniden-Stachel indet. – TOLLMANN: tab. 1
	1955	Spatangidarum gen. et spec. indet. – TOLLMANN: tab. 1
?	1971e	*Echinoidea* – STEININGER: 167
	1987	fragments of Spatangoida – MĄCZYŃSKA: 147; pl. 5, fig. 2; pl. 6, figs. 1-2; pl. 7, figs. 1, 3-4
pp	1992	Seeigelstachel – SAUER et al.: fig. 85N
pp v	1999	Undetermined echinoid ossicles – KROH & HARZHAUSER: 164; pl. 9, fig. 1-7
v.	2002a	Hemiasterina indet. – KROH: 309; fig. 2; pl. 2, figs. 1-8
v.	2002a	Spatangoida indet. – KROH: 308-309
v.	2002b	Spatangoida indet. – KROH: 12
v.	2003a	Spatangoida indet. – KROH: 165; pl. 4, fig. 16
	2003b	Spatangoida indet. – KROH: 252
v.	2003	spatangoids [...] indeterminable spines and plate fragments – KROH et al.: 92

Material:

Egerian (Chattian-Aquitanian) – Ebelsberg (Ebelsberg Fm.), OÖ, Austria
NHMW: 1 slab with an accumulation of whole spatangoid spines (NHMW 2003z0026/1382)

Egerian (Chattian-Aquitanian) – Pucking (Ebelsberg Fm.), OÖ, Austria
NHMW: 3 slabs with accumulated spatangoid spines (NHMW 2003z0026/0002, 2003z0026/0010, 2003z0026/0328)

Egerian (Chattian-Aquitanian) – Pucking or Ebelsberg or Weikerlsee or Asten (Ebelsberg Fm.), OÖ, Austria
NHMW: 19 spine aggregates (NHMW 2003z0026/1375-1381, 2003z0026/1383-1393)

Egerian (Chattian-Aquitanian) – Weikerlsee (Ebelsberg Fm.), Linz, OÖ, Austria
NHMW: 2 slabs with accumulated spatangoid spines (NHMW 2003z0026/1229-1230)

Late Eggenburgian (Early Burdigalian) – Eggenburg, NÖ, Austria
NHMW: 1 spine fragment (NHMW 2004z0110/0013)

Late Eggenburgian (Early Burdigalian) – Unternalb (Retz Fm.), SE Retz, NÖ, Austria
NHMW: numerous spine fragments (NHMW 1999z0051/0026)

Ottnangian (Late Burdigalian) – Höbmannsbach, Taufkirchen Bay, ESE Schärding, OÖ, Austria
NHMW: 8 spine fragments (NHMW 1978/1966/12i)

Karpatian (Late Burdigalian) – Teiritzberg, near Stetten, Korneuburg Basin, NÖ, Austria
NHMW: 1 fragmentary specimen and spine fragments (NHMW 1999z0095/0001-2, 1999z0096/0001; 1999z0096/0001a-d)

Karpatian (Late Burdigalian) – Karnabrunn (sample 108/T/B), Korneuburg Basin, NÖ, Austria
NHMW: 12 spine fragments (NHMW 1999z0095/0003)

Early Badenian (Langhian) – Gainfarn, NÖ, Austria
NHMW: 5 test fragments (NHMW 2004z0076/0030, ../0034, ../0040-41, 2004z0154/0003)

Early Badenian (Langhian) – Grund, NÖ, Austria
NHMW: numerous spine fragments (no inv. no.)

Early Badenian (Langhian) – Niederleis, NÖ, Austria
NHMW: 4 test fragments (NHMW 2002z0087/0063-65)

Early Badenian (Langhian) – Bad Vöslau (?), NÖ, Austria
NHMW: 14 spine fragments (NHMW 1997z0178/1329, 1329a-c)

Early Badenian (Langhian) – Stotzing (sandpit Mayer), Bgld, Austria
NHMW: 15 test fragments, 1 spine fragments (NHMW 2004z0050/0010, ../0015)

Badenian (Langhian-Early Serravallian) – Bad Vöslau (brickyard Breyer), NÖ, Austria
NHMW: 35 spines (NHMW 1981/56/4)

Badenian (Langhian-Early Serravallian) – old sandpit between Großhöflein and Kleinhöflein, Bgld, Austria
NHMW: 32 spine fragments (NHMW 2003z0081/0012, ../0029-30), 50 test fragments (NHMW 2003z0081/0011, ../0041)

Badenian (Langhian-Early Serravallian) – Kalksburg, Vienna, Austria
NHMW: numerous fragments (NHMW 1952/14)

Badenian (Langhian-Early Serravallian) – borehole Matzen 325 (1456.5-1464.7 m), NÖ, Austria
NHMW: 1 crushed test (NHMW 2004z0101/0002)

Badenian (Langhian-Early Serravallian) – Perchtoldsdorf (Brunnergasse, highway-ramp), NÖ, Austria
NHMW: 3 test fragments (NHMW 1981/56/9c)

Badenian (Langhian-Early Serravallian) – Steinebrunn (formerly Steinabrunn), NÖ, Austria
NHMW: 5 spine fragments (NHMW 1865.I.612e)

Late Badenian (Early Serravallian) – Winden (Nirgl quarry, N of the village), Bgld, Austria
NHMW: 17 spine fragments (NHMW 1859.L.832a)

Late Badenian (Early Serravallian) – Winden [old quarry, N of the Ludlloch (cave), N of the village], Bgld, Austria
NHMW: 2 ambulacral fragment (NHMW 2003z0082/0033, ../0040), 1 interambulacral fragment (NHMW 2003z0082/0041)

Foreign material for comparison:
Early Badenian (Langhian) – Buituri (= Bujtur), 20 km S of Deva, Romania
NHMW: 23 test fragments and 2 spine fragments (NHMW 1852.II.1603d)
Early Badenian (Langhian) – Lăpugiu des Sus (= Lapugy), Romania
NHMW: 1 test fragment (NHMW 2004z0001/0027d)

Discussion:
The material discussed under this heading is identifiable to order rank only due to the nature of the studied specimens. The unified discussion under the heading "Spatangoida indet." does not imply that they are conspecific or even congeneric.

Disarticulated and fragmented coronal plates and, even more common, spines of spatangoid echinoids are very abundant in the Austrian Miocene. They occur in bulk samples from the Egerian to the Badenian, and occasionally as reworked material in the polyhaline Sarmatian deposits and have sporadically been mentioned in the geological literature (see above). The highly fragmentary nature of the coronal material and the lack of diagnostic characters in the spines preclude a precise determination.

In the Egerian (Chattian – Aquitanian) Ebelsberg Fm. (area of Linz, OÖ) interesting aggregates of echinoid spine fragments occur (Pl. 82, Fig. 3). These aggregates (NHMW 2003z0026/1375-1381) are dense accumulations of spatangoid spine fragments, fish remains, mollusc debris and larger foraminifers. They are here interpreted as feeding remains (? regurgitated material). Possible producers include asteroids, crustaceans and fishes.

Steininger (1971e: 167) reports echinoid spines that he tentatively attributed to *Parmulechinus hoebarthi* (then called *Scutella hoebarthi*). This species was never found at that locality (nor any clypeasteroids) despite intense sampling. Spatangoid debris is known, however, to occur at that locality and it is much more likely that these spines are spatangoid spines.

Occurrence:

Austria: Egerian (Chattian-Aquitanian) to Late Badenian (Early Serravallian)
Molasse Zone: Ebelsberg Fm., Ebelsberg, OÖ ([NHMW]); Ebelsberg Fm., Pucking, OÖ ([NHMW]); Ebelsberg Fm., Weikerlsee, Linz, OÖ ([NHMW]); ? Fels am Wagram (Fels Fm.), NÖ (Steininger, 1971e); Grund, NÖ ([NHMW]); Höbmannsbach, Taufkirchen Bay, ESE Schärding, OÖ ([NHMW]); Unternalb, near Retz, NÖ (Kroh & Harzhauser, 1999; [NHMW])
Waschberg Zone: Auspitzer Mergel (Boudky Fm., NN2, Eggenburgian according to Rögl & Nagymarosy, 2004), Ernstbrunn, NÖ (Grill, 1953)
Korneuburg Basin: borehole Korneuburg 1 (Grill, 1953); Karnabrunn, NÖ ([NHMW]); Teiritzberg, near Stetten, NÖ (Grill, 1953; Kroh, 2002a; [NHMW])
Niederkreuzstetten Bay: Niederkreuzstetten, NÖ (Grill, 1953)
Vienna Basin: Bad Vöslau (brickyard Breyer), NÖ ([NHMW]); Baden Tegel, NÖ (Fuchs, 1877); Brunn am Steinfeld, NÖ (Kühn, 1952); Gainfarn, NÖ (Kroh, 2002b; [NHMW]); Müllendorf, Bgld (Tollmann, 1955); Niederleis, NÖ (Kroh, 2003a; [NHMW]); Perchtoldsdorf, NÖ (Winkler, 1942; [NHMW]); Rauchstallbrunngraben, near Baden, NÖ (Kühn, 1952); Steinebrunn (formerly Steinabrunn), NÖ ([NHMW]); Winden (Nirgl quarry), Bgld ([NHMW]); Vienna Basin (Veit, 1943)
Eisenstadt-Sopron Basin: Eisenstadt, Bgld (Tollmann, 1955); sandpit between Groß- and Kleinhöflein, near Eisenstadt, Bgld ([NHMW]); Großhöflein, near Eisenstadt, Bgld (Tollmann, 1955); Eisenstadt (Hartl Fm., Hartl hill), Bgld (Tollmann, 1955; Kroh et al., 2003); Kleinhöflein, near Eisenstadt, Bgld (Tollmann, 1955); St. Margarethen (Roman quarry), Bgld (Sauer et al., 1992)
Danube Basin: Winden, Bgld ([NHMW])
Lavant Basin: Ettendorf, Lavanttal, Ktn (Papp, 1952)

Paratethys (non-Austrian occurrences): Eggenburgian (Early Burdigalian) – Late Badenian (Early Serravallian)
Fore-Carpathian Basin: Korytnica Clays, Korytnica, Poland (Mączyńska, 1987)
Transylvanian Basin: Buituri (= Bujtur), Romania ([NHMW]); Lapugy, Romania ([NHMW])

Additional spatangoid species reported from the Central Paratethys

Schizaster latipetalus Vadász, 1915

v * 1915 *Schizaster latipetalus* n. sp. – Vadász: 222-223; text-fig. 112; pl. 11 (5), fig. 6

Reported occurrence: Ottnangian (Late Burdigalian) of Etes (= Ettes), Nógrád, Hungary (Vadász, 1915).

Remarks: This species is known only from sub-surface material (2 specimens) recovered from a borehole in Etes. The figured specimen of Vadász is preserved at the MAFI (Ech 229) and was re-examined by the author. Although the specimen is a natural internal cast it is very well preserved and is clearly not conspecific with the *Schizaster* species discussed above (contrary to the numerous other *Schizaster* species recorded by Vadász). *S. latipetalus* is characterised by its long and broad petals, which nearly reach the ambitus, and the presence of just two gonopores.

Brissopsis crescenticus Wright, 1855

1915 *Brissopsis crescenticus* Wright. – Vadász: 230; pl. 9 (3), fig. 14; pl. 10 (4), fig. 10
? 1930 *Brissopsis crescenticus* Wrhigt. – Vendl: 50

Reported occurrence: Late Badenian (Early Serravallian) of Gârbova de Sus (= Felsö-Orbó), Romania (Vadász, 1915).

Remarks: Unfortunately the two specimens on which Vadász's record is based are not preserved at the MAFI, where there were located according to Vadász (1915). Judging from the illustrations and description provided by Vadász (1915) they fit rather well with material of *B. crescenticus* form the type area Malta. Other records of *B. crescenticus* from the Central Paratethys lack sufficient information to evaluate their correctness.

Plagiobrissus hungaricus Vadász, 1915

* 1915 *Plagiobrissus hungaricus* n. sp. – Vadász: 231-232; pl. 10 (4), figs. 11-13

Reported occurrence: Late Badenian (Early Serravallian) of Gârbova de Sus (= Felsö-Orbó), Romania (Vadász, 1915).

Remarks: As stated by Philippe (1998: 201) this species is very similar to *P. imbricatus* (Wright, 1855) from the Burdigalian of the Rhône Basin and the Messinian Upper Coralline Limestone of the Maltese islands. He also summarised the features which might be useful to distinguish the two species, but pointed out that the type-material of *P. hungaricus* needs to be re-examined to judge if the two are synonymous.

Echinocardium deikei Desor, 1858

* 1858 [*Echinocardium*] *Deikei* Desor nov. sp. – Desor: 408

1875 Echinocardium Deikei, Desor. – de Loriol: 122; pl. 22, figs. 6, 6a-b
1883 *Echinocardium Deickei*, Des. – Gutzwiller: 46
non 1979 *Echinocardium deikei* Desor, 1857 – Mączyńska: 33; pl. 9, figs. 2, 4 [= *E. peroni* fide Philippe, 1998]
non 1979 *Echinocardium deikei* Desor, 1857 – Mączyńska: 33; pl. 9, figs. 3 [= deformed schizasterid]
? 1984 *Echinocardium* cf. *deikei* Des. – Kókay et al.: 290
? 1985 *Echinocardium* cf. *deikei* Desor, 1858 – Mihály: 245; pl. 5, fig. 6

Reported occurrence: Swiss Molasse: Obere Meeresmolasse (Middle Burdigalian), region of St. Gallen (Desor, 1858; de Loriol, 1875; Gutzwiller, 1883); ? Panonnian Basin: Late Badenian (Early Serravallian), Budapest-Gyakorló, Pest, Hungary (Kókay et al., 1984; Mihály, 1985)

Remarks: *E. deikei* is a poorly known species based on poorly preserved material. Philippe (1998: 229) compared it to *E. depressum* a contemporary species in the Rhône Basin. He also pointed out that the records of *E. deikei* from the Badenian of Poland by Mączyńska (1979) are misidentified *E. peroni*, respectively deformed schizasterids. According to Radwański & Wysocka (2001: 307) the record of Mihály (1985) is based on material that cannot be identified to species level.

Echinocardium leopolitanum Radwański & Wysocka, 2001

? 1953 *Amphidetus* sp. nov. indet. – Szörényi: 40, 92; pl. 2, figs. 8, 8a [fide Radwański & Wysocka (2001: 306-307)]
* 2001 *Echinocardium leopolitanum* sp.nov. – Radwański & Wysocka: 304-307; pl. 3, figs. 1-4; pl. 4, figs. 1-6; pl. 5, figs. 1-3; pl. 6, figs.1-3.

Reported occurrence: Early Badenian (Langhian): ? Suchodoly (=Suchodol), western Ukraine (Szörényi, 1953); Gleboviti near Bibrka, *c.* 30 km SE Lwów, western Ukraine (Radwański & Wysocka, 2001)

Remarks: This species has recently been described in detail and the reader is referred to the publication of Radwański & Wysocka (2001).

Echinocardium peroni Cotteau, 1877

1906 *Echinocardium* nov. sp. – Vadász: 333
1915 *Echinocardium* nov. sp. ind. – Vadász: 234; text-fig. 119-120
pp 1979 *Echinocardium deikei* Desor, 1857 – Mączyńska: 33; pl. 9, figs. 2, 4 [fide Philippe (1998: 231)]
non 1979 *Echinocardium deikei* Desor, 1857 – Mączyńska: 33; pl. 9, figs. 3
1984 *Echinocardium* n.sp. – Kókay et al.: 290 [later described as *E. biaense* by Mihály (1985)]
* 1985 *Echinocardium biaense* n. sp. – Mihály: 245-246, 262; pl. 5, figs. 3-5 [fide Radwański & Wysocka (2001: 306-307)]
1988 *Echinocardium biaense* Mihály, 1985 – Mączyńska: 62; pl. 5, figs. 1a-e [fide Radwański & Wysocka (2001: 306-307)]
2004 *Echinocardium peroni* Cotteau, 1877 – Radwański & Wysocka: 385, 386; figs. 4

Reported occurrence: Early Badenian (Langhian): Huta Lubycka and Trzęsiny, Poland (Mączyńska, 1979); Monastyrek (= Monastyrz), near Rawa Ruska, Ukraine (Mączyńska, 1979); Świniary, southern Poland (Mączyńska, 1988; Radwański & Wysocka, 2004); Late Badenian (Early Serravallian): Biatorbágy (= Bia), Pest, Hungary (Vadász, 1906, 1915; Mihály, 1985); Budapest-Rákos, Pest, Hungary (Vadász, 1915; Kókay et al., 1984; Mihály, 1985)

Remarks: Philippe (1998: 231) and Radwański & Wysocka (2001: 306-307; 2004: 386) referred the specimens illustrated as *E. deikei* by Mączyńska (1979) to *E. peroni*. One of Mączyńska's specimens (fig. 3), however, is a deformed schizasterid (compare Radwański & Wysocka, 2001: 307). Radwański & Wysocka (2001: 306-307) placed *E. biaense* Mihály, 1985 in the synonymy of *E. peroni* together with one of the specimens illustrated as *E.* nov. sp. by Vadász (1915).

Hemipatagus ? hungaricus (Vadász, 1915)

* v 1915 *Maretia hungarica* n. sp. – Vadász: 237; pl. 9 (3), figs. 17-18
? 1915 *Hemipatagus* nov. sp. – Vadász: 238; pl. 10 (4), fig. 14
. 1952 *Maretia hungarica* Vadász. – Szörényi: 302, figs. a-b

Type-material:
Holotype: specimen MAFI Ech-225, figured by Vadász [1915: pl. 9 (3), figs. 17-18 (drawing)] and Szörényi [1952: figs. a-b (photograph)]; housed at the Museum of the Hungarian Geological Survey, Budapest
Locus typicus: Mátraverebély, Nógrád, Hungary
Age: Late Badenian (Early Serravallian), Middle Miocene
Remark: specimen damaged since the photograph of Szörényi (1952), petal V now missing

Remarks: Lambert & Thiéry (1924: 458) tentatively referred this species to *Atelospatangus* Koch, 1885 because no anterior poriferous zones of petals II and IV are visible in Vadász's drawing. The correctness of this action was questioned by Mortensen (1951: 79), arguing that the drawing might be incorrect. Mortensen was indeed correct, as shown by Szörényi (1952: 302), only the adapical pores of the anterior poriferous zones of petals II and IV are rudimentary. The holotype (the only specimen known) shows many features characteristic of the genus *Lovenia*. An internal fasciole could, however, not be observed. This species is tentatively referred to the genus *Hemipatagus* on base of general shape, petal structure, distribution of the camellate primary tubercles and the lack of an internal fasciole [see above under *H. ocellatus*, and Kroh (in prep) for the reasons of removing *Hemipatagus* from the synonymy of *Maretia*]. It is possible that the specimen illustrated as *Hemipatagus* nov. sp. from the Late Badenian of Felsö-Orbó (Romania) by Vadász (1915: 238, pl. 10, fig. 14) also belongs to this species.

Doubtful records of spatangoid species from Austria and the Central Paratethys

Schizaster desori Wright, 1955

1869a *Schiz. Desori* Wright. – Laube: 184
1871 *Schizaster Desori* Wright. – Laube: 71
1877 *Schizaster Desori* Wright. – Karrer: 181
? 1915 *Schizaster* cfr. *Desori* Wright. – Vadász: 224 [two deformed specimens]

Reported occurrence: Early Badenian (Langhian) Baden Tegel of Baden, NÖ (Laube, 1869a, 1871; Karrer, 1877) and Late Badenian (Early Serravallian) of Gârbova de Sus (= Felsö-Orbó), Romania (Vadász, 1915).

Remarks: This species, which occurs in the Lower and Middle *Globigerina* Limestone of Malta (Aquitanian to Burdigalian), was reported by Laube (1869a, 1871) and Karrer (1877) from the Badenian (Langhian-Early Serravallian) of the Vienna Basin. These records are based on a single specimen from the collection "k.k. polytech. Institut" (the Technical University of Vienna today). According to SCHULTZ (NHMW, pers. comm., 06.05.2004) the palaeontological collection was transferred to the GBA (all the palaeozoological part) and the NHMW (palaeobotanic specimens) in the 1950's. LAUBE's specimen could not be located in either collection. As LAUBE's (1871) description is insufficient and not accompanied by an illustration it is currently impossible to evaluate this record. All other *Schizaster* specimens from the Baden Tegel examined during the present study were either too poor to determine specifically or belong to *S. karreri*.

***Schizaster* cf. *scillae* DES MOULINS, 1837**

1915 *Schizaster* cfr. *Scillae* DESMOUL. sp. – VADÁSZ: 225 [two poorly preserved internal moulds]

Reported occurrence: ? Karpatian (Late Burdigalian) of Zagyvapálfalva (= Pálfalva), Nógrád, Hungary (VADÁSZ, 1915)

Remarks: Specimens not located in the Hungarian collections. VADÁSZ's statement that the specimens are two poorly preserved internal moulds and the fact that none of the other records of *S. scillae* from the Central Paratethys could be confirmed rises some doubts about the correctness of his identification.

***Schizaster* cf. *curtus* POMEL, 1887**

v 1915 *Schizaster* cfr. *curtus* POMEL. – VADÁSZ: 224-225 [a single fragment; indeterminable]

Reported occurrence: Late Badenian (Early Serravallian) of Buituri (= Bujtur), Romania (VADÁSZ, 1915)

Remarks: The specimen (MAFI Ech 307) on which this record is strongly fragmented and indeterminable. The record is thus rejected.

***Schizaster* cf. *parkinsoni* (DEFRANCE)**

1915 *Schizaster* cfr. *Parkinsonii* DEFR. sp.– VADÁSZ: 224 [three poorly preserved internal moulds]

Reported occurrence: Karpatian (Late Burdigalian) of Piliny, Nógrád, Hungary (VADÁSZ, 1915)

Remarks: Like the material of VADÁSZ's *S.* cf. *scillae*, the specimens of *S.* cf. *parkinsoi* mentioned by VADÁSZ could not be located. All other records of this species from the Central Paratethys have turned out to be misidentifications. The Central Paratethyan occurrence of this species thus seems very doubtful

***Hemiaster* sp.**

1882 *Hemiaster* sp. – HILBER: 236
1905 *Hemiaster* sp. – GAÁL: 343, 362

Reported occurrence: Badenian (Langhian-Early Serravallian) Mykolaiv (= Mikołajów), western Ukraine (HILBER, 1882) and Hor. Strháre (= Felsö-Esztergály), Slovak Republic (GAÁL, 1905).

Remarks: It is unclear to which of the taxa discussed above these two records belong to. They might refer to misidentified schizasterids (*Schizaster* or *Aliaster*). From the locality Hor. Strháre *Schizaster karreri* was documented, poorly preserved specimens of which might have been misidentified as *Hemiaster*. At the other locality only a species of *Echinocardium*, a fibulariid and some temnopleurids were recorded by SZÖRÉNYI (1953), who referred to this locality as Mikołajów-Radziejów.

***Pericosmus airgahii* LAMBERT, 1909**

? 1955 *Periscoma airaghii* LAMB. – SENEŠ: 6
? 1967 *Periscoma airaghii* LAMB. – CICHA et al.: 93

Reported occurrence: Karpatian of Hlinné, eastern Slovak Republic

Remarks: Recorded without description or illustration on base of a single specimen (from a locality where all specimens are preserved as deformed internal moulds). Until the specimen is adequately described and/or illustrated the record has to be considered doubtful.

***Macropneustes* ? *compressus* NEMES, 1888a**

* 1888a *Macropneustes* (?) *compressus*, nov. sp. – NEMES: 164, 218; pl. 6, fig. 5

Reported occurrence: "Czereczeler Schlier", Tărăţel (= Czereczel), western Csetrás Mountains, Hunedoara, Romania

Remarks: *Macropneustes* ? *compressus*, NEMES, 1888a from the "Schlier" facies of Tărăţel, Romania is based on a single, poorly preserved specimen. The figure of NEMES (1888a: pl. 6, fig. 5) reveals that the specimen is not suitable for specific determination and that there are similarities to a deformed *Brissopsis ottnangensis*, a species occurring at that locality too (VADÁSZ, 1915).

***Spatangus* cf. *corsicus* DESOR in AGASSIZ & DESOR, 1847**

1915 *Prospatangus* cfr. *corsicus* COTT. – VADÁSZ: 236

Reported occurrence: Late Badenian (Early Serravallian) of Gârbova de Sus (= Felsö-Orbó), Romania (VADÁSZ, 1915)

Remarks: Record based on a poorly preserved specimens, identification doubtful. The presence of this species in the Middle Miocene of the Central Paratethys needs to be confirmed. According to PHILIPPE (1998: 204) *S. corsicus* is a junior synonym of *S. delphinus*.

***Spatangus delphinus* DEFRANCE, 1827**

1915 *Prospatangus delphinus* AG. sp. – VADÁSZ: 236; fig. 122

Reported occurrence: Late Badenian (Langhian-Early Serravallian) of Matraszölös, Nógrád, Hungary (VADÁSZ, 1915)

Remarks: Record based on a single poorly preserved specimen that could not be located in the MAFI collection. Description and illustration do not allow to confidently refer the specimen to any of the species discussed above and the record has to be considered doubtful.

***Spatangus* cf. *delphinus* DEFRANCE, 1827**

1988 *Spatangus* cf. *delphinus* DEFRANCE, 1827 – MĄCZYŃSKA: 61-62; pl. 5, fig. 3

Reported occurrence: Badenian (Langhian-Early Serravallian) of Rybnica, southern Poland (MĄCZYŃSKA, 1988)

Remarks: MĄCZYŃSKA (1988) figured only the oral side of her single specimen and the printing quality is to poor to see the plate outlines. Neither the labrum shape nor the number and arrangement of the aboral primary tubercles are described in the text. It is thus impossible to revise this record without re-examination of the material.

Spatangus desmaresti **Münster in Goldfuss, 1829**

? 1852 *Spatangus Desmarestii* Münst. – Eichwald: pl. 3, figs. 2a-b, 3 a-b
? 1853 *Spatangus Desmarestii* Münst. – Eichwald: 46
? 1968 *Spatangus demarestii* Münst. – Stancu & Andreescu: 466, tab.

Reported occurrence: Early Badenian of Zhukov (= Żukowice, = Zukowce), near Wyżgródek, western Ukraine (Eichwald, 1852, 1853); Badenian (Langhian-Early Serravallian) of Rugi-Delineşti, Romania (Stancu & Andreescu, 1968)

Remarks: *S. desmaresti* is a species from the Oligocene of Doberg, Northern Germany and it seems likely that the records from the Badenian (Middle Miocene) of Romania and the Ukraine are based on misidentifications.

Spatangus fabianii **(Lambert, 1924)**

? 1953 *Prospatangus fabianii* Lambert, 1924 – Szörényi: 39, 90, pl. 8, figs. 5, 5a
? 1979 *Spatangus fabianii* (Lambert, 1924) – Mączyńska: 32; pl. 8, figs. 1-2; pl. 9, fig. 1
? 1988 *Spatangus fabianii* (Lambert, 1924) – Mączyńska: 62; pl. 5, figs. 2a-e

Reported occurrence: Early Badenian (Langhian) of Potelych (= Potylicz) and Kurzany, western Ukraine (Szörényi, 1953), Badenian (Langhian-Early Serravallian) of Świniary, southern Poland (Mączyńska, 1988) and Huta Lubycka, south-eastern Poland (Mączyńska, 1979)

Remarks: The specimens figured by Szörényi (1953) and Mączyńska (1979, 1988) have a peculiar profile with the maximum height around the anteriorly displaced apical disc and very low posterior end unlike any of the species discussed above. Partial spine cover, adhering sediment and poor printing quality obliterate nearly all details, including plate boundaries.

Spatangus fothiensis **(Strausz, 1926)**

* 1926 *Prospatangus fothiensis* nov. sp. – Strausz: 214-215, 368-369; figs. 2a-c
1981 *Prospatangus fótensis* Strausz – Halmai: 106
2003b *Prospatangus fothiensis* Strausz, 1926 – Kroh: 252

Reported occurrence: Karpatian (Late Burdigalian) of Somlyó Mt., near Fót, Pest, Hungary (Strausz, 1926; Halmai, 1981)

Remarks: This is the only record of a species of *Spatangus* from the Karpatian of the Central Paratethys. Unfortunately, the type-material could not be traced in the Hungarian collections (Kroh, 2003b). On Strausz's illustrations details on plating patterns or tuberculation details cannot be seen, even the shape of the petals is not entirely clear. The general form alone does not allow to differentiate the species from others and must thus be considered as *nomen dubium* until re-described.

Spatangus hungaricus **(Vadász, 1915)**

* 1915 *Prospatangus hungaricus* n. sp. – Vadász: 235; text-fig. 121; pl. 11 (5), figs. 1-2
? 1953 *Prospatangus hungaricus* Vadász, 1915. – Szörényi: 38-39, 89-90, pl. 8, figs. 4, 4a-b
non 1979 *Spatangus hungaricus* (Vadász, 1915) – Mączyńska: 32; pl. 6, figs. 1a-e, 2; pl. 7, fig. 1a-f

Reported occurrence: Early Badenian of Pod'yarkov (= Podjarków), near Lwów, western Ukraine (Szörényi, 1953), Late Badenian (Early Serravallian) of Gârbova de Sus (= Felsö-Orbó), Romania (Vadász, 1915)

Remarks: Unfortunately the type-material of this species could not be examined during the present study. The specimen figured by Vadász (1915: pl. 11, figs. 1-2) has many similarities to *S. austriacus*, including a labrum which extends to the second ambulacral plates. As far as can be seen in Vadász's figure the tubercle arrangement is also similar to *S. austriacus*, while the petals seem to be shorter. The specimens figured by Mączyńska (1979: 32) under this name do not fit very well with Vadász's description and illustrations and are probable misidentified. One of the figures (Mączyńska, 1979: pl. 7, fig. 1b) shows a short, broad labrum, which does not extend beyond the first ambulacral plate, very similar to *Spatangus* sp. 2 (see above under that species).

Spatangus hungaricus buekkensis **(Kutassy, 1928)**

* 1928 *Prospatangus hungaricus* Vadász n. var. *bükkensis*. – Kutassy: 5-6; pl. 1, figs. 1a-b

Reported occurrence: Early Miocene of Királd, Hungray (Kutassy, 1928)

Remarks: Based on a single specimen in which most of the oral surface is obscured. The test is slightly transversely elongated and the petals are relatively long, extending about three quarter of the corresponding test radius. The specimen does not resemble any of the species of *Spatangus*.

Spatangus peroni **Cotteau in Locard, 1877**

1915 *Prospatangus peroni* Cott. sp. – Vadász: 235

Reported occurrence: Badenian (Langhian-Early Serravallian) of Tótmarokháza, Nógrád, Hungary (Vadász, 1915)

Remarks: Record based on a single poorly preserved specimen. The specimen could not be located among Vadász's material at the MAFI and the record lacks a sufficient description or illustration. Thus the record is considered doubtful and need to be confirmed.

Spatangus **sp. 3**

1997 *Spatangus austriacus* Laube – Majcen et al.: 106; pl. 5, fig. 2

Reported occurrence: provenance unknown

Remarks: The specimen figured by Majcen et al. (pl. 5, fig. 2) is very large, has a very deep frontal sinus and much more enlarged primary tubercles than any of the species discussed above. Particularly the en-chevron arrangement of the tubercles in interambulacrum 5 is not observed in the other species (they show thin arrangement only in the paired interambulacra, if developed at all). Additionally, the petalodium is very small. This, however, could also be an effect of allometric growth. Unfortunately the locality and exact stratigraphic provenance of the species are unknown and it is probably not from the Central Paratethys at all.

Echinocardium intermedium **Lóczy, 1877**

* 1877 *Echinocardium intermedium* n. sp. – Lóczy: 64-67; pl. 5, figs. 1a-f, 2a-h

	1915	*Echinocardium intermedium* LÓCZY. – VADÁSZ: 233; pl. 3 (9), figs. 20-24
?	1953	*Amphidetus intermedus* (LÓCZY), 1877. – SZÖRÉNYI: 40, 91-92; pl. 8, fig. 6
?	1969	*Echinocardium intermedium* LÓCZY – MIHÁLY: 257; pl. 2, figs. 4-5
?	1984	*Echinocardium intermedium* LÓCZY – KÓKAY et al.: 288
?	1985	*Echinocardium intermedium* LÓCZY, 1877 – MIHÁLY: 244-245; pl. 5, figs. 7-8

Reported occurrence: Early Badenian (Langhian): ? Mykolaiv (= Mikołajów) and Radziechów (= Radekhov, = Radziejów), western Ukraine (SZÖRÉNYI, 1953); Late Badenian (Early Serravallian): Biatorbágy (= Bia), Pest, Hungary (LÓCZY, 1877; VADÁSZ, 1915); ? Budapest-Gyakorló, Pest, Hungary (KÓKAY et al., 1984; MIHÁLY, 1985); ? Budapest-Rákos, Pest, Hungary (MIHÁLY, 1969); Minişu des Sus (= Felménes), Arad, Romania (LÓCZY, 1877; VADÁSZ, 1915)

Remarks: According to RADWAŃSKI & WYSOCKA (2001: 307) the specimen figured by VADÁSZ (1915), which belongs to the syntypes of *E. intermedium* LÓCZY, 1877 does not belong to *Echinocardium* at all based on the apparently closed petals, swollen interambulacral plates and the peculiar outline (but could also be a result of a poor drawing). Unfortunately the specimen could not be located at the MAFI where it was housed according to VADÁSZ (1915). According to RADWAŃSKI & WYSOCKA (2001: 307) the record of MIHÁLY (1985) is based on material that cannot be identified to species level.

Systematic index

The index is arranged in alphabetical order of the species. Species written in **bold** font are valid taxa discussed in the present paper, page and table numbers in **bold** font are the locations where the main information concerning the respective taxon is found. (Incorrect spellings of species names used in the literature have been included in the index on purpose. They are usually marked by underlining the erroneous characters.)
Species mentioned in differential diagnoses or in the discussion have been included in the index too. Authorships, dates and open nomenclature symbols (aff., cf.) have been omitted in sake of easier searching.

Plates 1–82

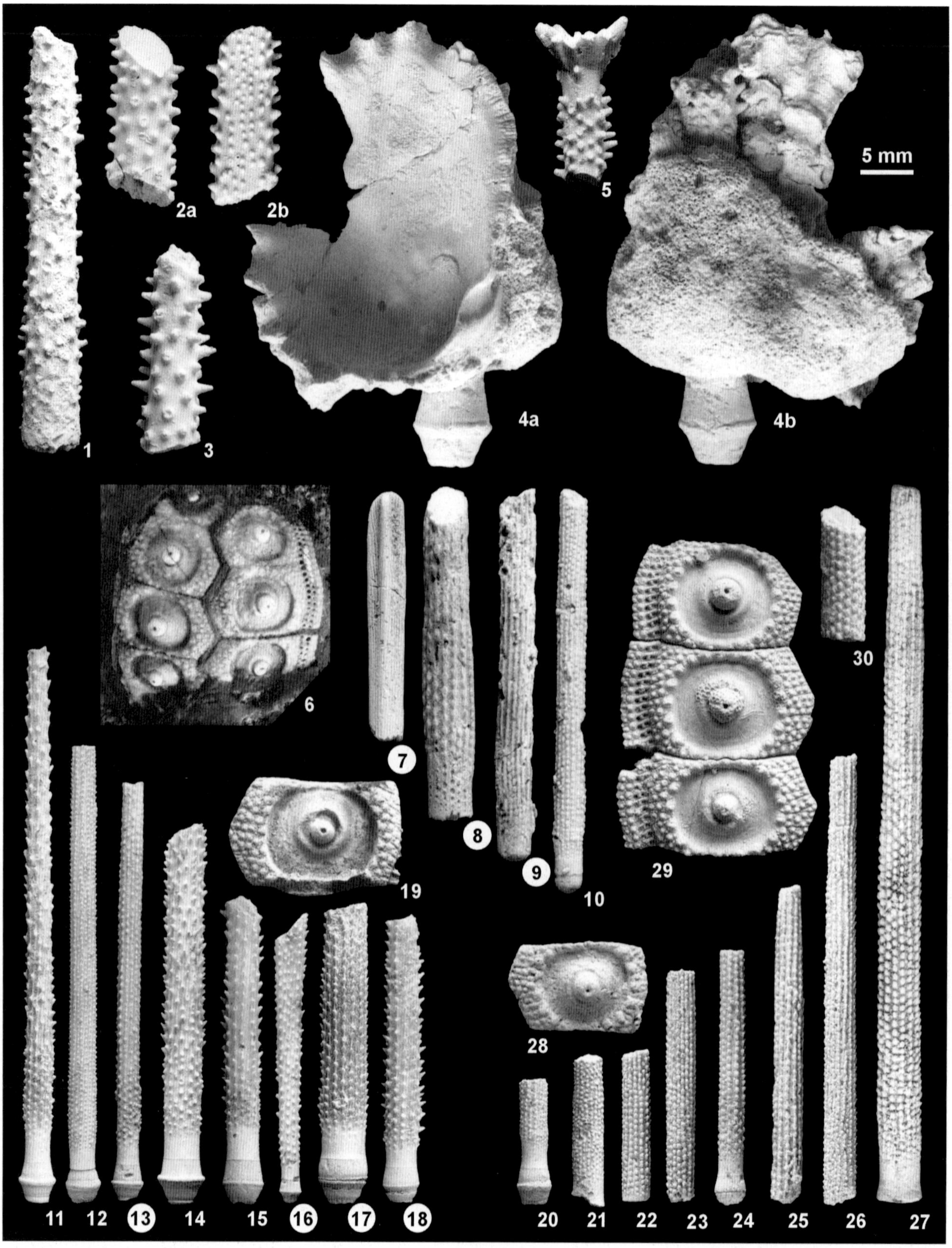

Prionocidaris sismondai **(MAYER, 1864):** 1: Rauchstallbrunngraben, NÖ (NHMW 1997z0178/0904); 2-5: Großhöflein, NÖ [NHMW 2004z0004/0001a-d (2: d, 3: c, 4: a, 5: b)]; **Cidarinae sp. indet. (Ottnangian type):** 6: Allerding, OÖ (coll. DANNINGER); 7-9: Höch, SW Passau, Germany [NHMW 1990/1487/229a-c (7: a, 8: c, 9: b)]; 10: Mitterndorf, OÖ (NHMW 1986/102/37); ***Stylocidaris ? polyacantha*** **(REUSS, 1860):** 11: Bad Vöslau, NÖ (NHMW 2004z0001/0002a); 12, 14-15: Baden, NÖ [NHMW 1997z0178/0899a-c (12: a, 14: b, 15: c)]; 13, 16-18: Hrušovany, CZ [13, 17-18: NHMW 1981/56/3a-c (13: a, 17: b, 18: c); 16: NHMW 1997z0178/0740a); 19: Bad Vöslau, NÖ (NHMW 2004z0001/0002b); ***Stylocidaris ? schwabenaui*** **(LAUBE, 1869):** 20-26: Winden, Bgld [20-21: NHMW 2004z0004/0002a-b (20: a, 21: b), 22-26: NHMW 1997z0178/1936a-e (22: c, 23: d, 24: a, 25: b, 26: e)]; 27: Walbersdorf, Bgld (NHMW 2004z0001/0001); 28: Großhöflein, NÖ (NHMW 2004z0004/0002); 29-30: Winden, Bgld (29: NHMW 1997z0178/1683a-c, 30: NHMW 2004z0004/0002c)

All specimens (except fig. 6) coated with ammonium chloride

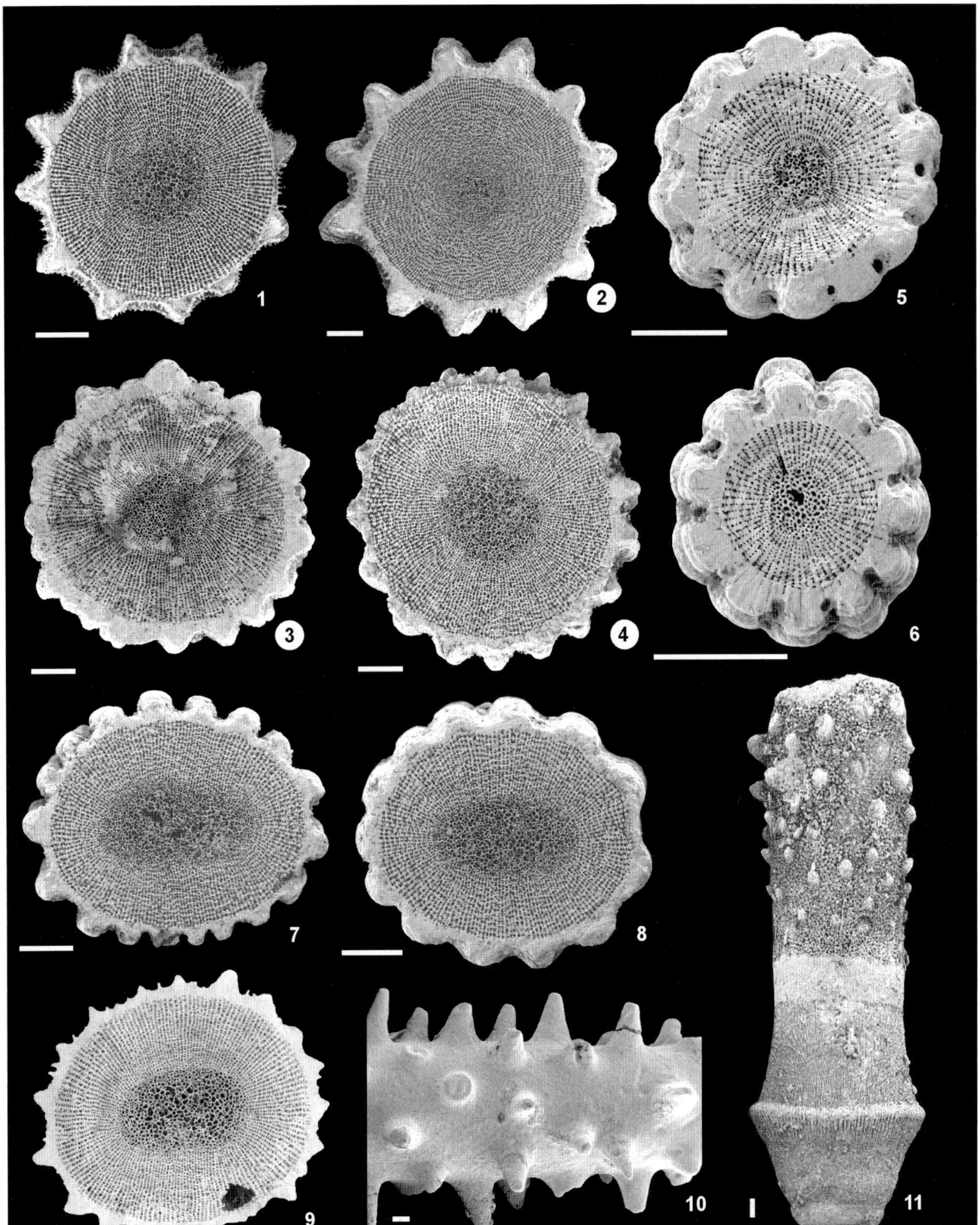

***Stylocidaris* ? *polyacantha* (Reuss, 1860):** 1: primary spine cross section (SEM), Baden, NÖ (NHMW 2004z0001/0063e); 2: primary spine cross section (SEM), Hrušovany (= Grusbach), CZ (NHMW 1981/56/3); ***Stylocidaris* ? *schwabenaui* (Laube, 1869):** 3-4: primary spine cross section (SEM), Devínska Nová Ves, SK (3: NHMW 1868.VIII.27f, 4: NHMW 1868.VIII.27b); ***Eucidaris zeamays* (Sismonda, 1842):** 5: primary spine cross section (SEM), Gainfarn, NÖ (NHMW 2003z0076/0012); 6: primary spine cross section (SEM), Forchtenau, Bgld (NHMW 1997z0178/1328a); **Cidarinae sp. indet. (Ottnangian type):** 7-8: primary spine cross section (SEM), Rainbach, OÖ (7: NHMW 2004z0002/0002d, 8: NHMW 2004z0002/0002c); ***Plegiocidaris* ? *peroni* (Cotteau, 1877):** 9: primary spine cross section (SEM), Unternalb, near Retz, NÖ (NHMW 1999z0051/0015, AO of Kroh & Harzhauser, 1999: pl. 7, fig. 2); ***Prionocidaris sismondai* (Mayer, 1864):** 10: primary spine (detail of shaft, SEM), Großhöflein, NÖ (NHMW 2004z0004/0001c); 11: primary spine (base, SEM), Großhöflein, NÖ (NHMW 2003z0081/0051)

Scale bars equal 0.5 mm

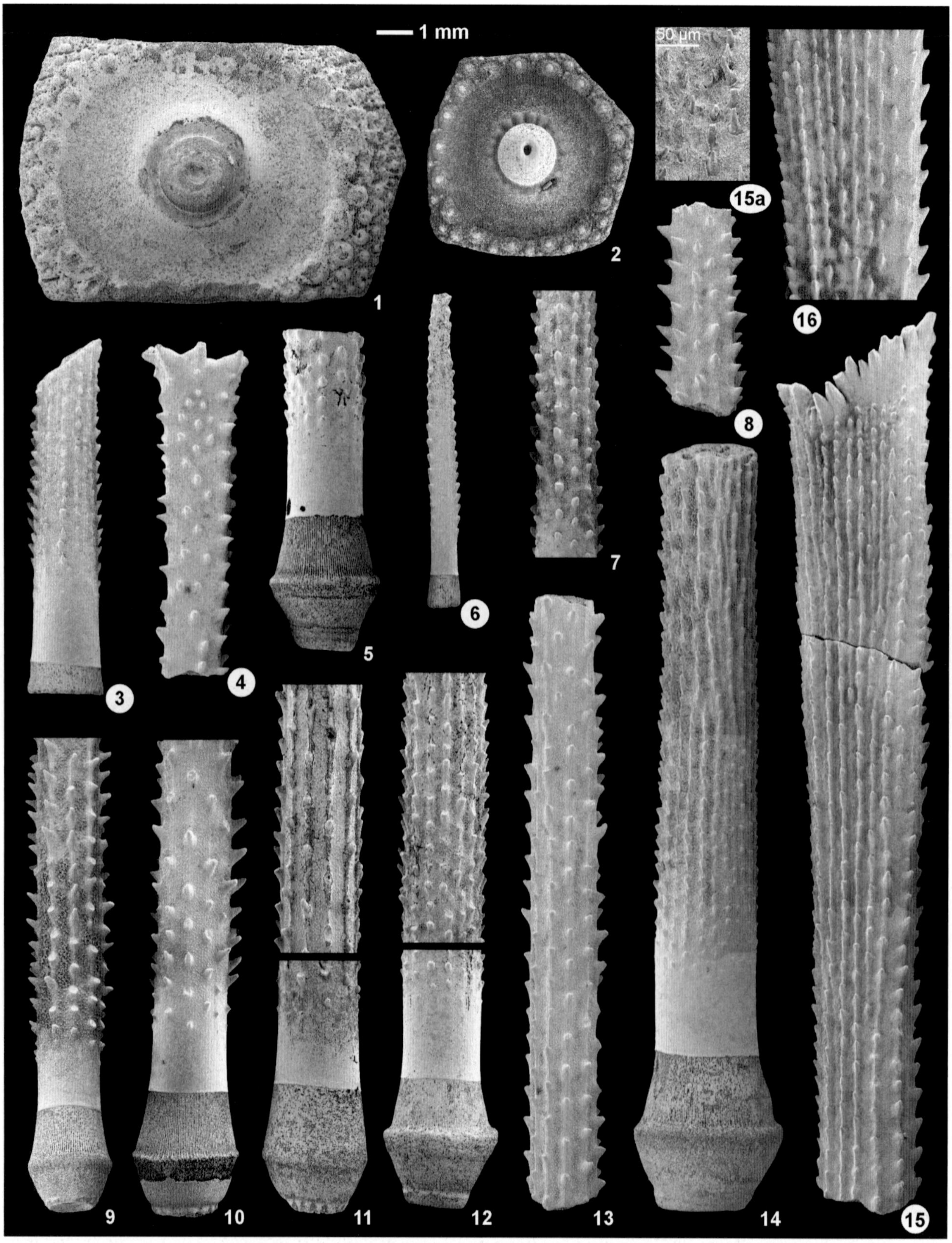

Stylocidaris* ? *polyacantha **(Reuss, 1860):** 1: ambital interambulacral plate, Baden, NÖ (GBA coll.); 2: adapical interambulacral plate, note crenulation, Bad Vöslau, NÖ (NHMW 2004z0001/0012b); 3-4, 6, 8: primary and secondary (6) spines; Kinitz, CZ [NHMW 2004z0001/0017a-d (3: b, 4: a, 6: d, 8: c)]; 5: primary spine (base), Baden, NÖ (NHMW 2004z0001/0063a); 7, 9: primary spine (prox.), Hrušovany, CZ (NHMW 1981/56/3); 10: primary spine (prox.), note crenulation, Bad Vöslau, NÖ (NHMW 2004z0001/0012a); 11, 12: base + shaft of primary spines, note crenulation in fig. 11, Baden, NÖ (11: NHMW 1866.I.131, 12: NHMW 1846.37.952a); 13, 14: primary spines, distal (13) and proximal (14) part, Baden, NÖ (GBA coll.); 15-16: distal part of primary spines, 15a is a detail of the cortical hair of the specimen figured in as 15, Hrušovany, CZ (NHMW 1981/56/3)

All photographs are SEM images; "1 mm" scale bar valid for all figs. except fig. 15a

***Stylocidaris* ? *schwabenaui* (LAUBE, 1869):** 1: adoral test fragment, Winden, Bgld (NHMW 2003z0082/35); 2a-b: ambital test fragment (see also pl. 1, fig. 29), Winden, Bgld (NHMW 1997z0178/01638a); 3: primary spine base, Winden, Bgld (NHMW 2003z0082/32); 4: primary spine fragment, Perchtoldsdorf, NÖ (NHMW 1981/56/9a); 5 primary spine fragment, Winden, Bgld (NHMW 2003z0082/31); 6-7; primary spine (shaft), Devínska Nová Ves, SK (NHMW 1868.VIII.27g-h); 8-11: primary spines, Winden, NÖ (8: NHMW 2003z0082/0038; 9: ../0036; 10: ../0037; 11: ../0039)

***Plegiocidaris* ? *peroni* (COTTEAU, 1877):** 12: primary spine base; Sonndorf, NÖ (NHMW 2003z0076/0001a); 13-14: primary spine fragments, Grübern, NÖ (NHMW 2003z0075/12a+b); 15: primary spine, Unternalb, near Retz, NÖ (NHMW 1999z0051/0015, AO of KROH & HARZHAUSER, 1999: pl. 7, fig. 4); 16-18: primary spines; Sonndorf, NÖ (NHMW 2003z0076/0001a-c)

"1 mm" sale bar valid for figs. 1-14 (SEM images), except fig. 2b; "5 mm" scale bar valid for figs. 15-18 (coated with ammonium chloride)

Cidarinae sp. indet. (Ottnangian type): 1-3: Mitterndorf, OÖ (primary spines; 1: NHMW 1989/58/24b, 2: NHMW 1986/102/37d, 3: NHMW 1986/102/37e); 4-7: Plesching, OÖ (primary spine ornamentation; NHMW 2004z0002/0001c-f); 8-15: Höbmannsbach, OÖ [primary spine ornamentation (8-9), bases (10-11), and secondary spines (12-15); 8: NHMW 1978/1966/12j, 9: ../12k, 10: ../12s, 11: ../12l, 12: ../12y, 13: ../12w, 14: ../12u, 15: ../12x]; 16-17: Rainbach, OÖ (primary spine ornamentation; NHMW 2004z0002/0003a-b); 18-20, 22: Höbmannsbach, OÖ [18: large oral primary spine, NHMW 1978/1966/12f; 19-20: primary spine tips, 19: NHMW 2004z0002/0005, 20: 1978/1966/12m; 22: small (adapical ?) primary spine, NHMW 1978/1966/12n]; 21: Plesching, OÖ [small (adapical ?) primary spine, NHMW 2004z0002/0001g); 23-26: Höbmannsbach, OÖ (oral primary spines, NHMW 1978/1966/12o-r)

All photographs are SEM images

Eucidaris zeamays **(SISMONDA, 1842)** (primary spines)**:** 1-7: Niederleis, NÖ [NHMW 2002z0087/0040-45, 47 (1: ../0043, 2: ../0044, 3: ../0040, 4: : ../0042, 5: ../0041, 6: ../0047, 7: ../0045), AO of KROH, 2003: pl. 1, figs.1-6, 8]; 2a: ornamentation of spine with thorny surface, Niederleis, NÖ (NHMW 2002z0087/0044); 8-11: Badener Tegel, ? Bad Vöslau, NÖ [NHMW 1997z0178/1328a-d (8: a, 9: b, 10: c, 11: d)]; 11a: ornamentation of spine with relatively smooth surface (note preserved cortical hair between the granulae), Baden Tegel, ? Bad Vöslau, NÖ (NHMW 1997z0178/1328d); 12-14: Forchtenau, Bgld (NHMW 2004z0001/0015a); 15-17, 17a: Lăpugiu des Sus, Romania [NHMW 1859.XLV.617B.a; specimen in figure 17 is a growing spine (17a shows spine surface showing lamellae without covering cortex layer)]; 18-21: Stotzing, Bgld (18: NHMW 2004z0050/0030, 19: ../0029, 20: ../0027, 21: ../0028)

"1 mm" scale bar valid for of all photographs (except 2a, 11a and 17); all photographs are SEM images

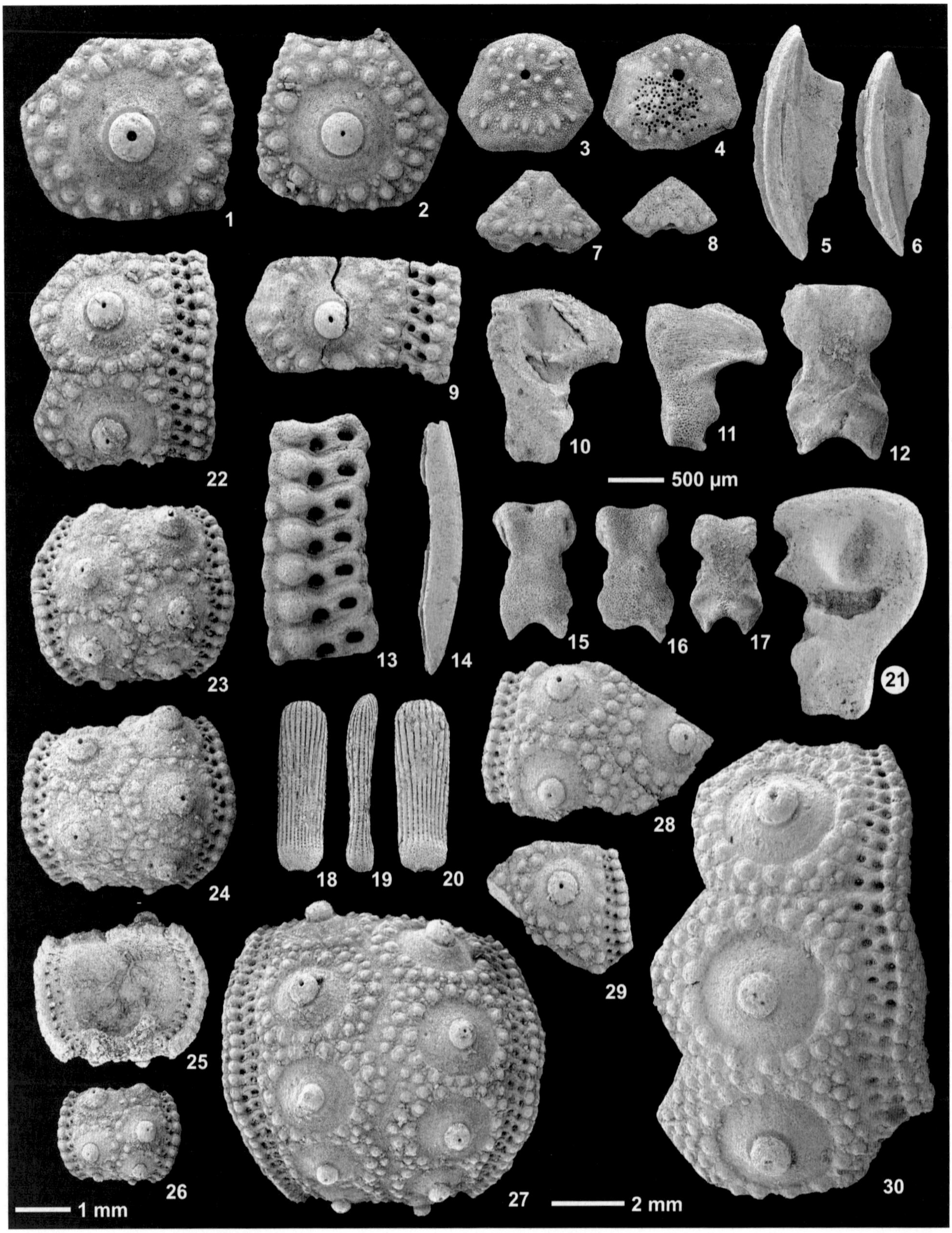

Eucidaris zeamays **(SISMONDA, 1842):** 1-20: Rauchstallbrunngraben, near Baden, NÖ [1-2: adapical interambulacral plates (1: NHMW 2004z0121/0002, 2: ../0001), 3: genital plate (NHMW 2004z0121/0005), 4: madreporite (genital plate 2, NHMW 2004z0121/0006), 5-6: demipyramids (NHMW 2004z0121/0018-19), 7-8: ocular plates (NHMW 2004z0121/0007-08), 9: ambital interambulacral plate with attached ambulacral plates (NHMW 2004z0121/0010), 10-11: epiphyses (NHMW 2004z0121/0020-21), 12, 15-17: rotulae (12: NHMW 2004z0121/0024, 15: ../0023, 16: ../0022, 17: ../0025), 13: ambulacral plates (NHMW 2004z0121/0009), 14: tooth (NHMW 2004z0121/0026), 18-20: scrobicular spines (18: NHMW 2004z0121/0028, 19: ../0029, 20: ../0027)]; 21: epiphysis, Coştei, Romania (NHMW 2004z0084/0002); 22-26: test segments, Großhöflein, Bgld (NHMW 2003z0081/0046-50); 27-30: test fragments, Stotzing, Bgld (27: NHMW 2004z0050/0024, 28: ../0021, 29: ../0020, 30: ../0025)

"1 mm" scale bar valid for figs. 1-9, 22-26, "500 µm" scale bar for figs. 10-21, "2 mm" scale bar for figs. 27-31; all are SEM images

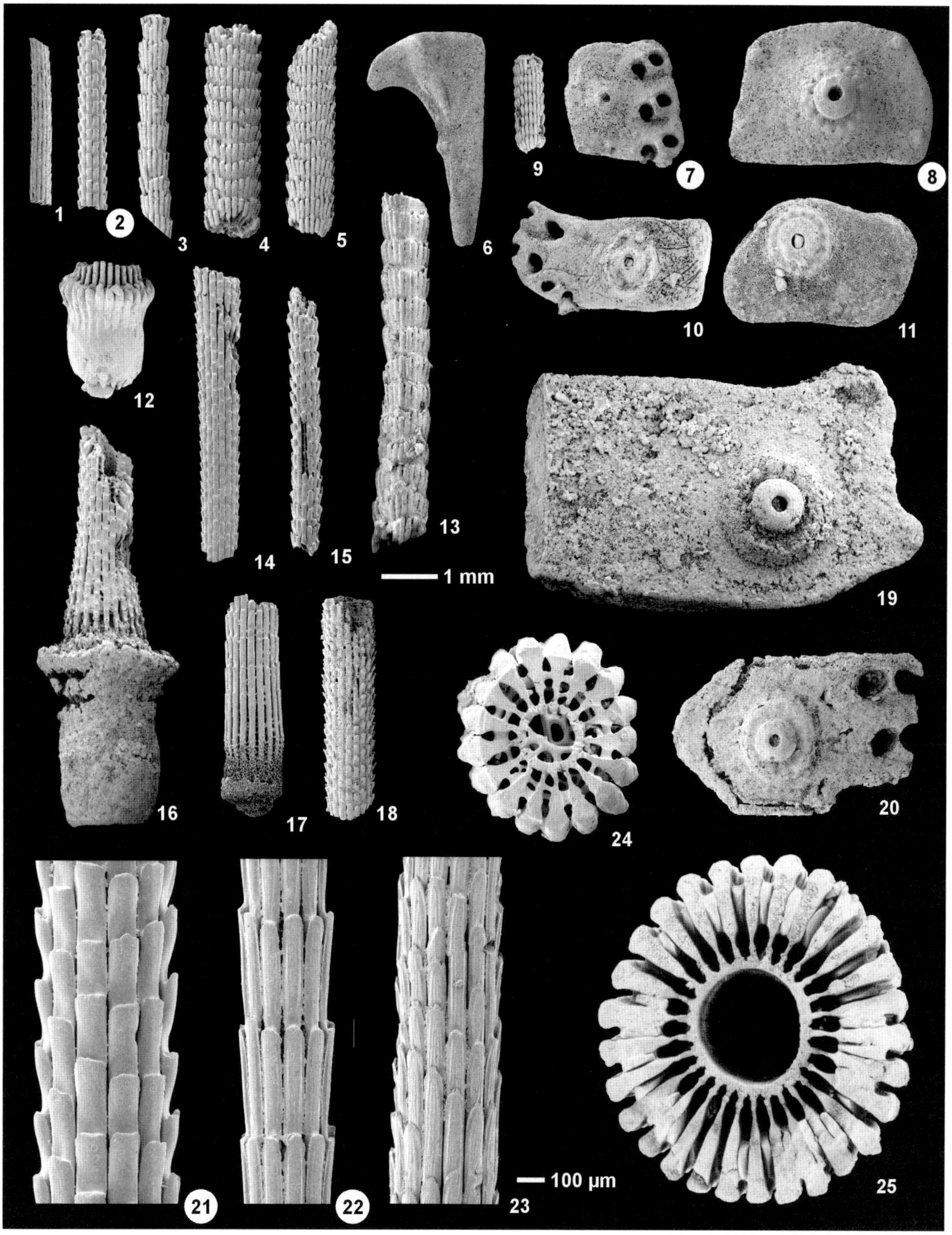

Diadematidae indet.: 1, 23: Höbmannsbach, near Schäring, OÖ (Ottnangian, NHMW 1978/1966/12z); 2, 21-22: Lăpugiu des Sus, Romania (Early Badenian, NHMW 1859.XLV.617B.b); 3-6: Großhöflein, near Eisenstadt, Bgld (Badenian, 3: NHMW 2004z0081/0025, 4: ../0024, 5: ../0023, 6: ../0038); 7-8: Nový Rybik, Czech Republic (Badenian; 7: ambulacral plate, 8: interambulacral plate; NHMW 1858.XLVIII.78b); 9-12, 25: Unternalb, near Retz, NÖ (Late Eggenburgian; 9: spine fragment, NHMW 1999z0051/0022; 10: ambulacral plate, NHMW ../0021; 11: interambulacral plate, ../0019; 12: spine base, ../0022; 25: spine cross section, ../0022); 13: Eggenburg, NÖ (Eggenburgian; NHMW 2004z00110/0007); 14-20: Gainfarn, NÖ [Badenian; 14-18: spine fragments (14: NHMW 2004z0076/0025, 15: ../0026, 16: ../0023, 17: ../0024, 18: NHMW ../0029); 19: interambulacral plate (NHMW 2004z0076/0021); 20: ambulacral plate (NHMW 2004z0076/0022)]; 24: Bad Vöslau, NÖ (Badenian; spine cross section, NHMW 1997z0178/1327)

"1 mm" scale bar valid for figs. 1-20, "100 µm" scale bars for figs. 21-25

Plate 9

Retzneiosoma jaseneki **Kroh, nov. gen. nov. sp.:** 1-12: Lafarge quarry, Retznei, Styria [1: NHMW 2003z0071/0002; 2: holotype, coated with ammonium chloride, NHMW 2003z0071/0001; 3-12: SEM images of tubercles (3-4), spines (5-9), and pedicellariae (10-12) from specimen NHMW 2003z0071/0002 (1)]

"1 cm" scale bar valid for figs. 1-2, "500 µm" scale bars for figs. 3-4, 7, "100 µm" scale bars for figs. 5-6, 8-12

Retzneiosoma jaseneki **Kroh, nov. gen. nov. sp.:** 1-8: Lafarge quarry, Retznei, Styria [1-6: holotype, coated with ammonium chloride, NHMW 2003z0071/0001 (1: lantern, 2-3: aboral surface, half ambulacrum and adjacent interambulacral column in different lighting, 4: interambulacrum apex to ambitus, 5: adapical ambulacrum, 6: inside of oral surface); 7-8: NHMW 2003z0071/0002]

all coated with ammonium chloride, scale bars equal 5 mm

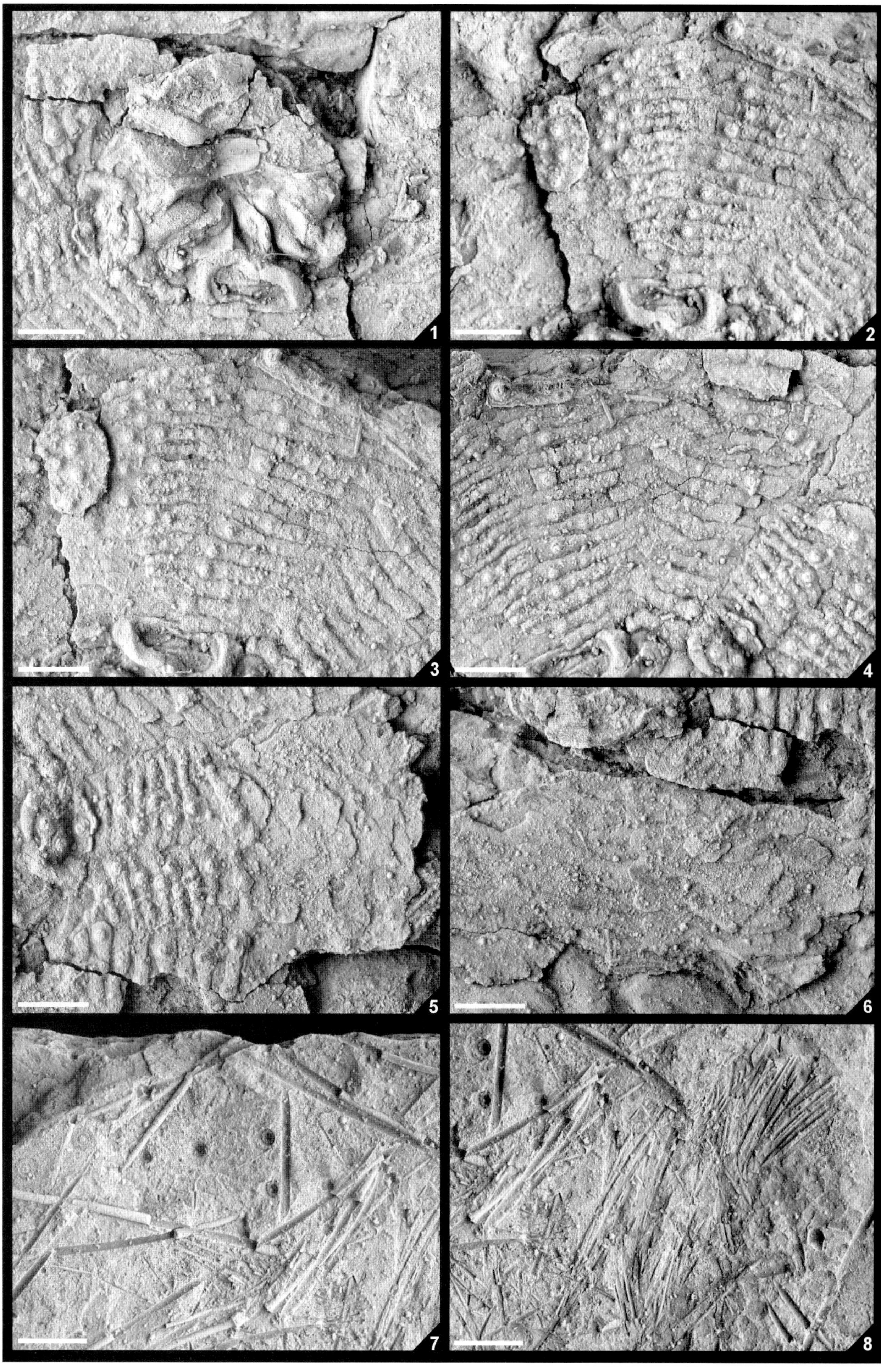
1
2
3
4
5
6
7
8

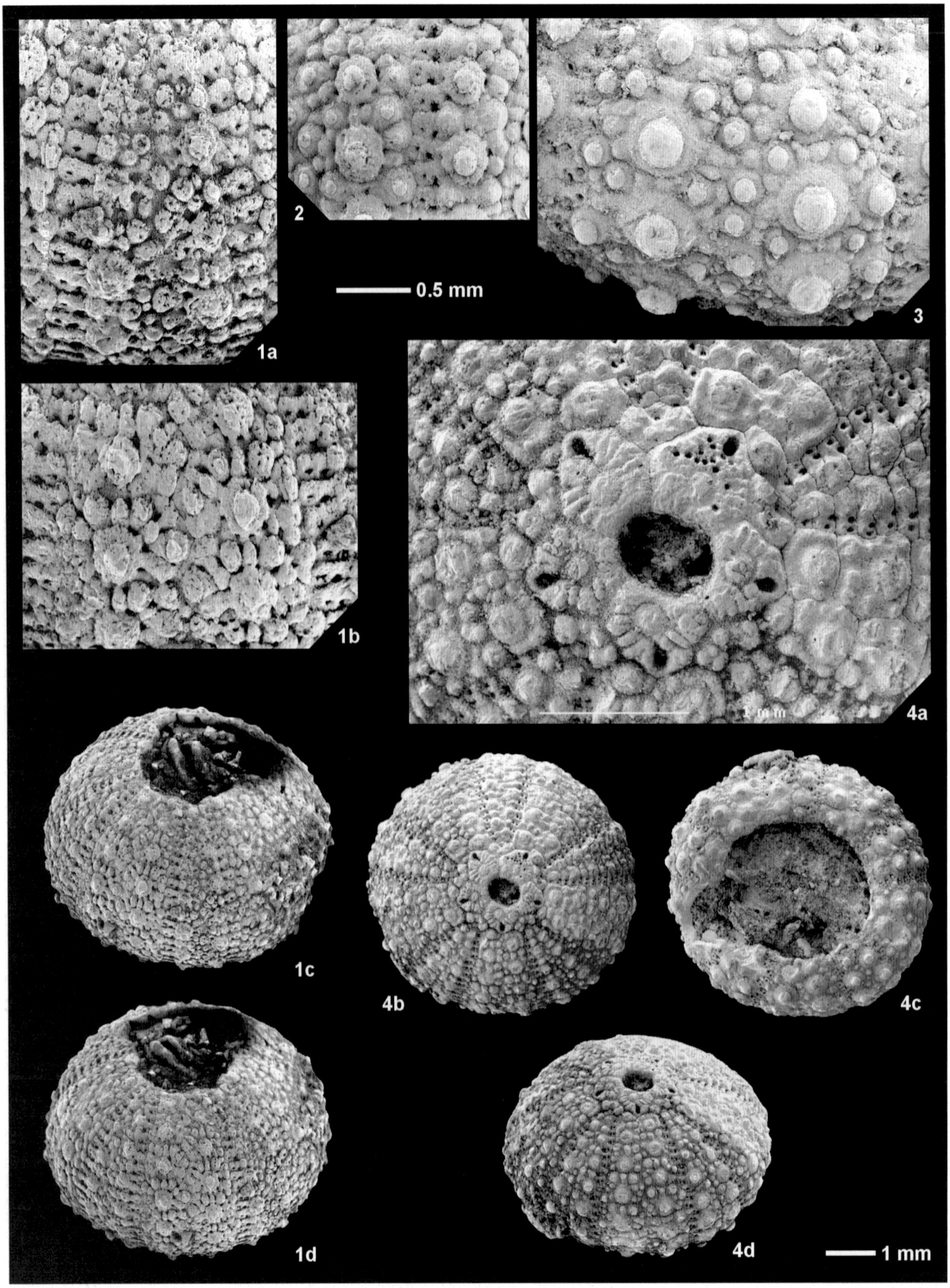

Arbacina catenata **(DESOR, 1846):** 1a-d: Roggendorf, NÖ (NHMW 2004z0001/0018); 2: Unternalb, NÖ (NHMW 1999z0051/0018)
Arbacina* cf. *macrophyma **LAMBERT, 1910:** 3, 4a-d: Müllendorf, Bgld (3: NHMW 2004z0006/0008, 4: NHMW 2004z0006/0001)

Arbacina* cf. *macrophyma **LAMBERT, 1910:** 1-6: Müllendorf (Kreide AG quarry), Bgld (1: NHMW 2004z0006/0011, 2: ../0010, 3-4: ../0009, 5: ../0008, 6: ../0007)
Schizechinus **sp. juv.:** 7-8: Müllendorf (Kreide AG quarry), Bgld (NHMW 2004z0006/0004)

1
0.5 mm
2
0.5 mm
3
0.5 mm
4
6
1 mm
5
0.5 mm
7
0.5 mm
8
1 mm

Plate 13

***Genocidaris* sp.**: 1-10: Rauchstallbrunngraben, near Baden, NÖ (NHMW 2004z0121/0031-0040); 11-12: Großhöflein, Bgld (11: NHMW 2004z0081/0013; 12: NHMW 2004z0081/0014); figure 3b shows a noncrenulate primary tubercle, figure 5b a crenulate one; indentations in the bases of the primary tubercles marked by white arrows, crenulation by a black arrow.

"1 mm" scale bar valid for all figs. except 3b and 5b, "100 µm" scale bars for figs. 3b and 5b

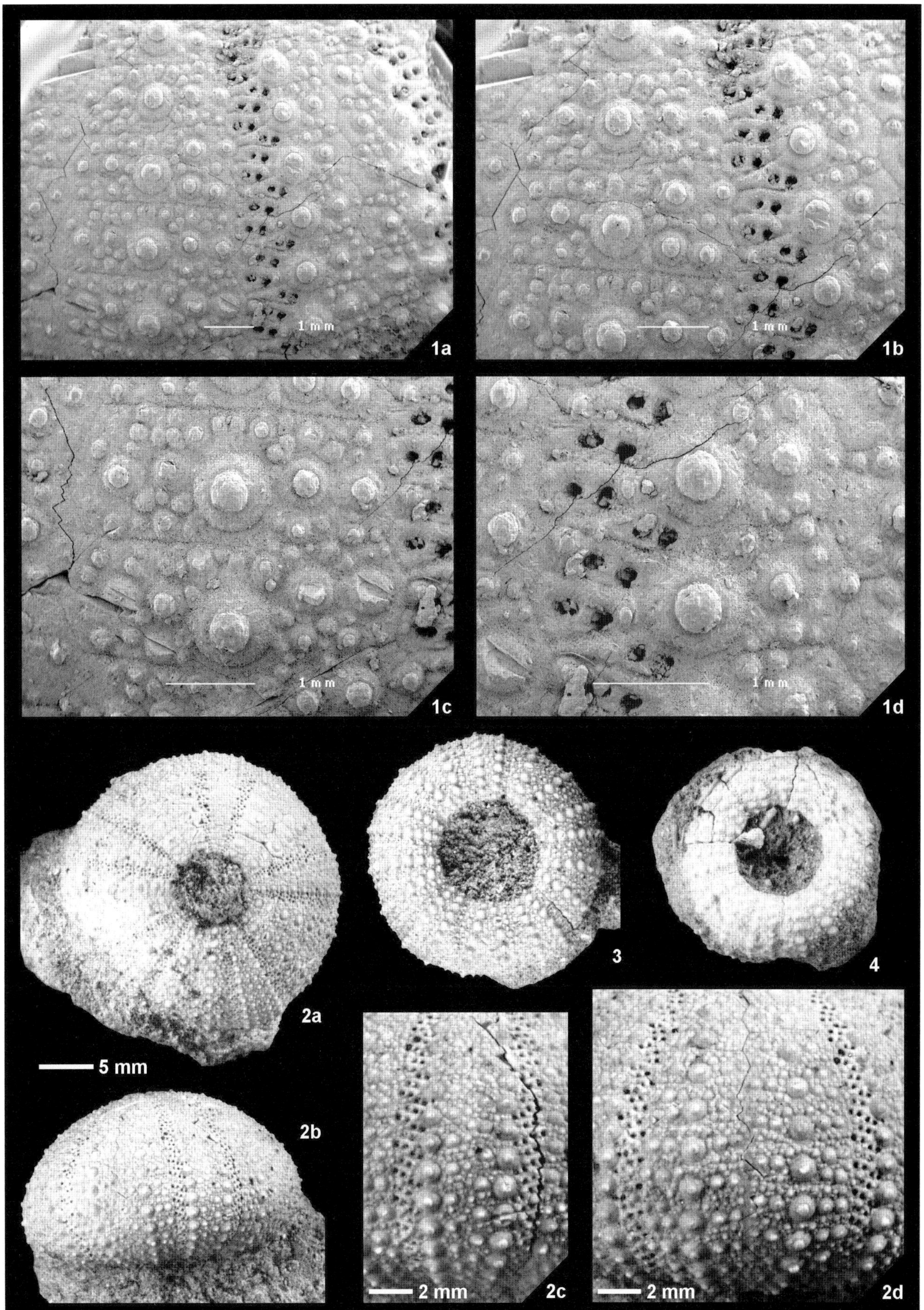

Psammechinus dubius dubius **(AGASSIZ, 1840):** 1-4: Reinprechtspölla, NÖ [1a-d: NHMW 1914.VII.55a (a-b: tuberculation of ambital plates, c: close-up of ambital interambulacra, d: close-up of ambital ambulacra; all SEM images); 2a-d: NHMW 1914.VII.55b (a: aboral, b: lateral, c: detail of an ambulacrum, d: detail of an interambulacrum); 3: NHMW 1914.VII.55c; 4: NHMW 1997z0178/1832d]

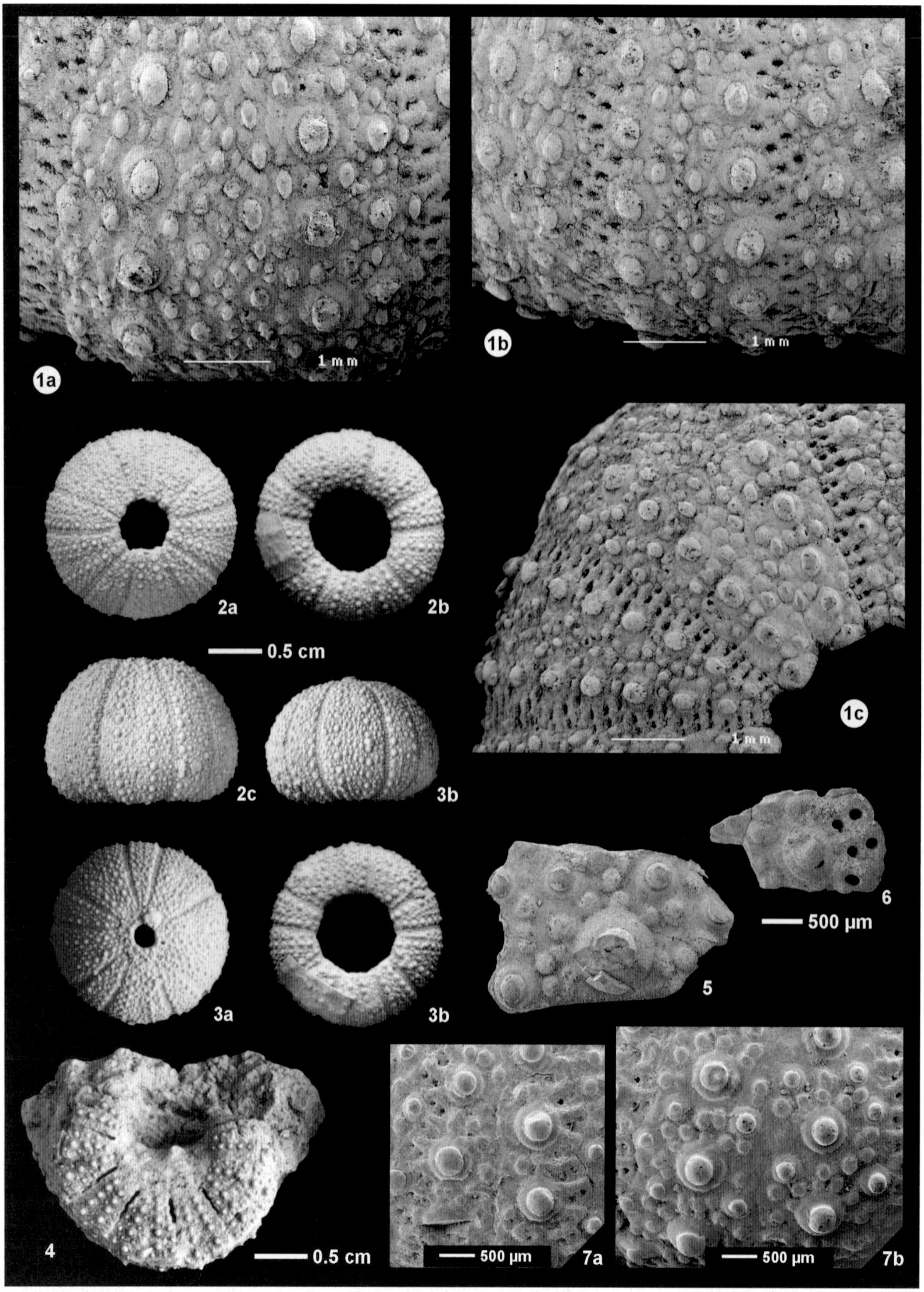

Psammechinus **cf. *dubius gauthieri*** **Cotteau, 1885:** 1a-c: Úvaly (= Garschenthal), Czech Republic (SEM images; NHMW 1859.XLV.633a); 2a-c, 3a-c: Steinebrunn, NÖ (coated with ammonium chloride; 2: NHMW 9023b; 3: NHMW 9023a)
Psammechinus **sp.:** 4-7: Allerding, OÖ [4: Danninger coll.; 5: NHMW 2004z0005/0002a; 6: NHMW 2004z0005/0002b; 7a-b: (cast of a specimen from the Danninger coll.; NHMW 2004z0005/0001); all except fig. 4 are SEM images]

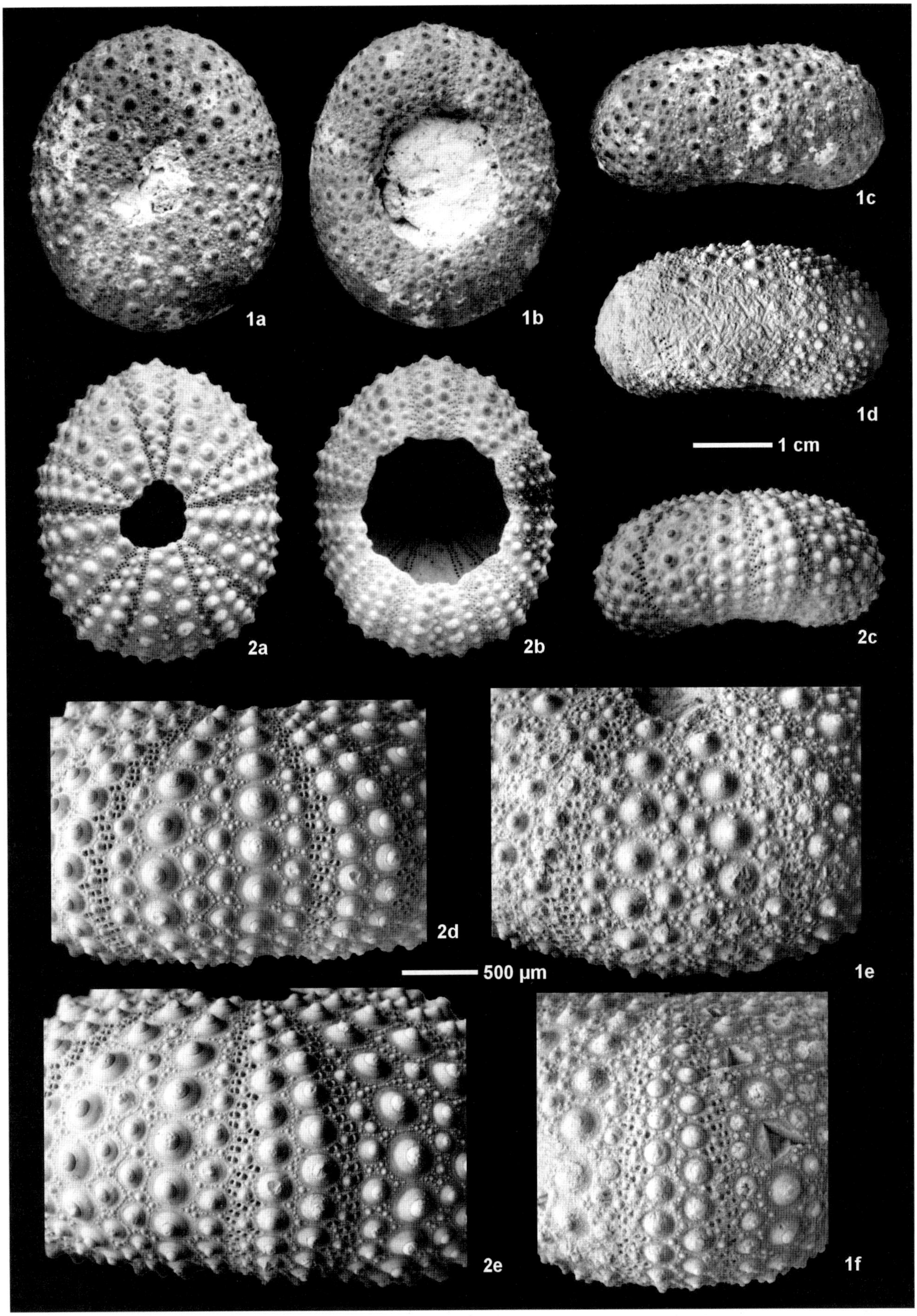

***Echinometra mathaei* (de Blainville, 1825):** 1a-f: Hundsheim, near Deutsch-Altenburg, NÖ [Badenian; NHMW 1939 No.2; note the gnawing traces (*Gnathichnus pentax* Bromley, 1975) in fig. 1d]; 2a-e: Luzon (?), Philippines (extant; NHMW coll.)

"1 cm" scale bar valid for figs. 1a-d, 2a-c; "500 µm" scale bars for figs. 1e-f and 2d-e, the latter are coated with ammonium chloride

Schizechinus hungaricus **(Laube, 1869):** 1a-b: Wiesfleck, Bgld (NHMW 2004z0001/0031); 2a-b: Ritzing, Bgld (NHMW 1848.II.61); 3a-b, 4a-c: Steinebrunn, NÖ (3: NHMW A1050; 4: lectotype of *Echinus dux* Laube, 1869, NHMW 1859.XLV.632); 5-6: Winden, Bgld (NHMW 2003z0082/0029-30)

"1 cm" scale bar valid for figs. 1-4; figs. 1-4 coated with ammonium chloride; figs. 5-6 are SEM images

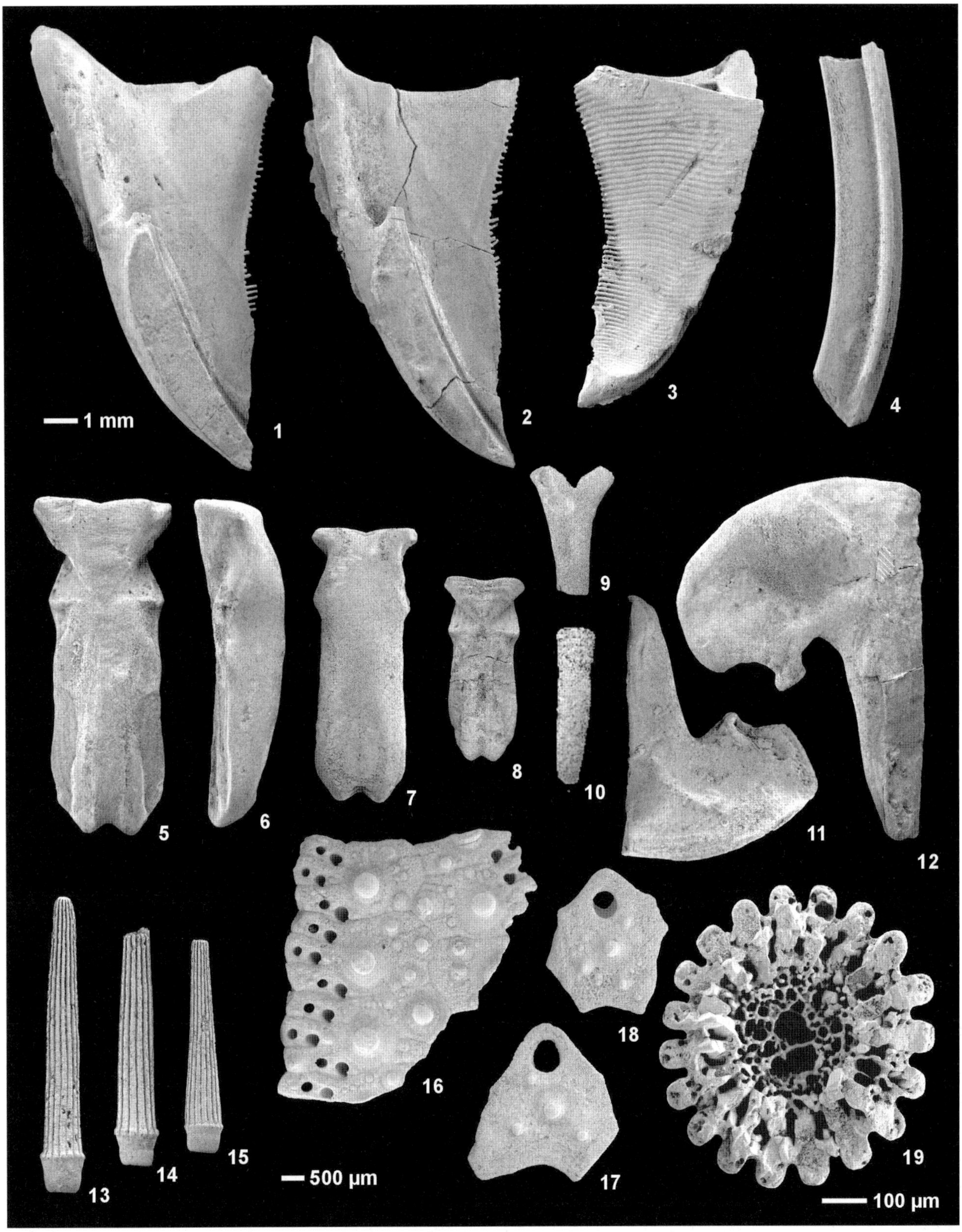

Schizechinus hungaricus **(LAUBE, 1869):** 1-19: Winden, Bgld [1-3: demipyramids (NHMW 2003z0082/0010-12); 4: tooth (NHMW 2003z0082/0013); 5-8: rotulae (NHMW 2003z0082/0014-17); 9-10: compass, distal (9) and proximal (10) part (NHMW 2003z0082/0018-19); 11-12: epiphyses (NHMW 2003z0082/0020-21); 13-15: primary spines (NHMW 2003z0082/0022-24); 16: adapical ambulacral test fragment (NHMW 2003z0082/0025); 17-18: genital plates (NHMW 2003z0082/0026-27); 19: spine cross section (NHMW 2003z0082/0028)

"1 mm" scale bar valid for figs. 1-3; "500 µm" scale bar for figs. 4-18; "100 µm" scale bar for fig. 19; all figs. are SEM images

Tripneustes planus **AGASSIZ in AGASSIZ & DESOR, 1846:** 1-3: Limberg (Hengl quarry), NÖ (NHMW 2003z0002/0001-3); 4: Roggendorf, NÖ (NHMW 2003z0001/0001)
Tripneustes **cf. *ventricosus*** **(LAMARCK, 1816):** 5-13: Gainfarn, NÖ (5: NHMW 2003z0003/0016; 6: .../0018; 7: .../0002; 8: .../0008; 9: .../0009; 10: /0005; 11: .../0006; 12a-b: .../0013; 13: .../0001)

"1 mm" scale bar valid for figs. 1-4, 7-11, 13; "500 µm" for fig. 12b; "200 µm" for figs. 5-6; "100 µm" for fig. 12a; all figs. are SEM images

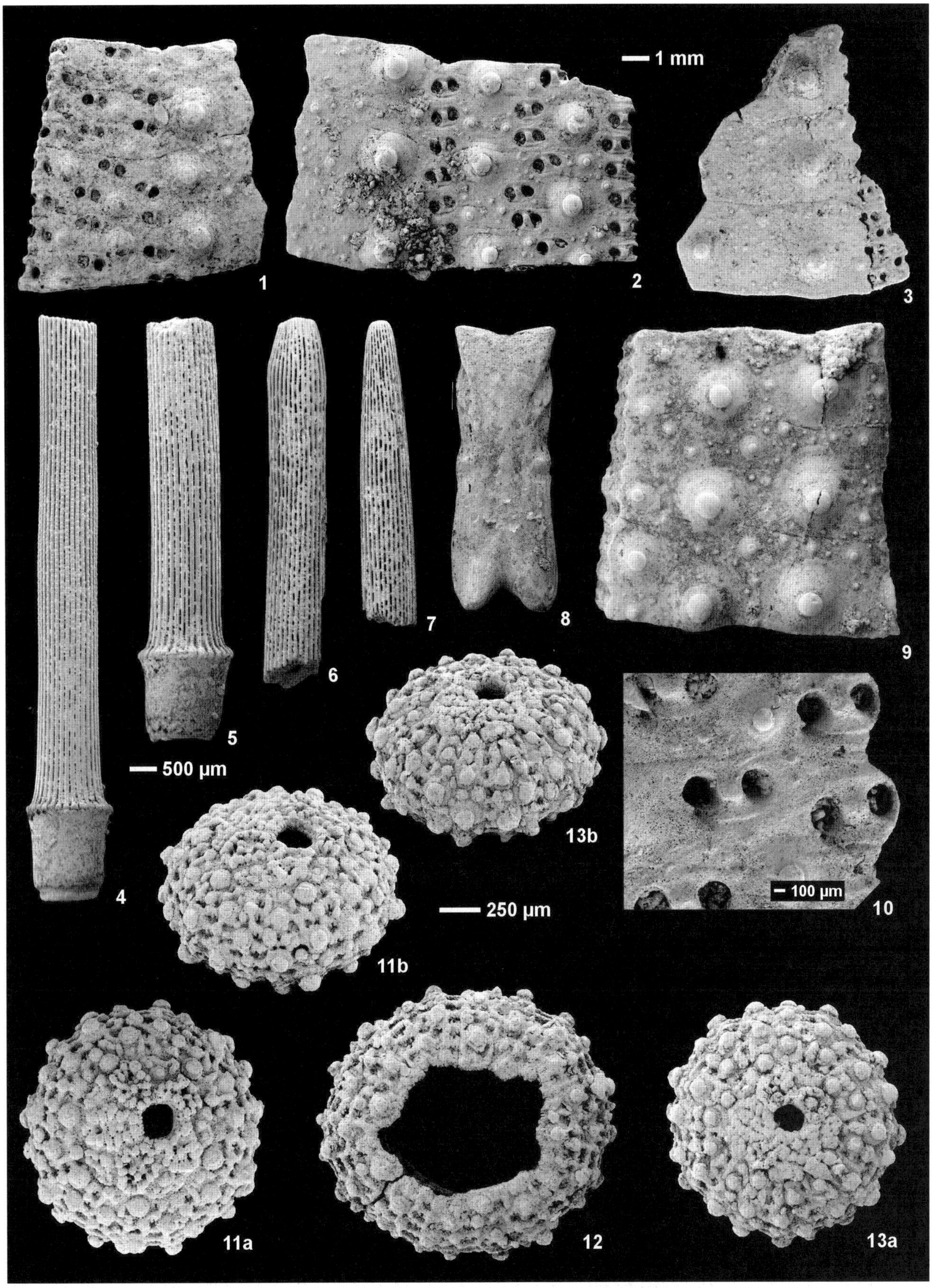

***Tripneustes* cf. *ventricosus* (Lamarck, 1816):** 1-10: Gainfarn, NÖ (1: NHMW 2003z0003/0007; 2: .../0004; 3: NHMW 2004z0076/0069; 4: NHMW 2003z0003/0013; 5: .../0012; 6: .../0011; 7: .../0014; 8: .../0010; 9: .../0003; 10: .../0009)
Echinacea indet. (juveniles): 11-13: Großhöflein, Bgld (NHMW 2003z0081/0016-18)

"1 mm" scale bar valid for figs. 1-3, 8-9; "500 µm" scale bar for figs. 4-7; "250 µm" scale bar for figs. 11-13; all figs. are SEM images

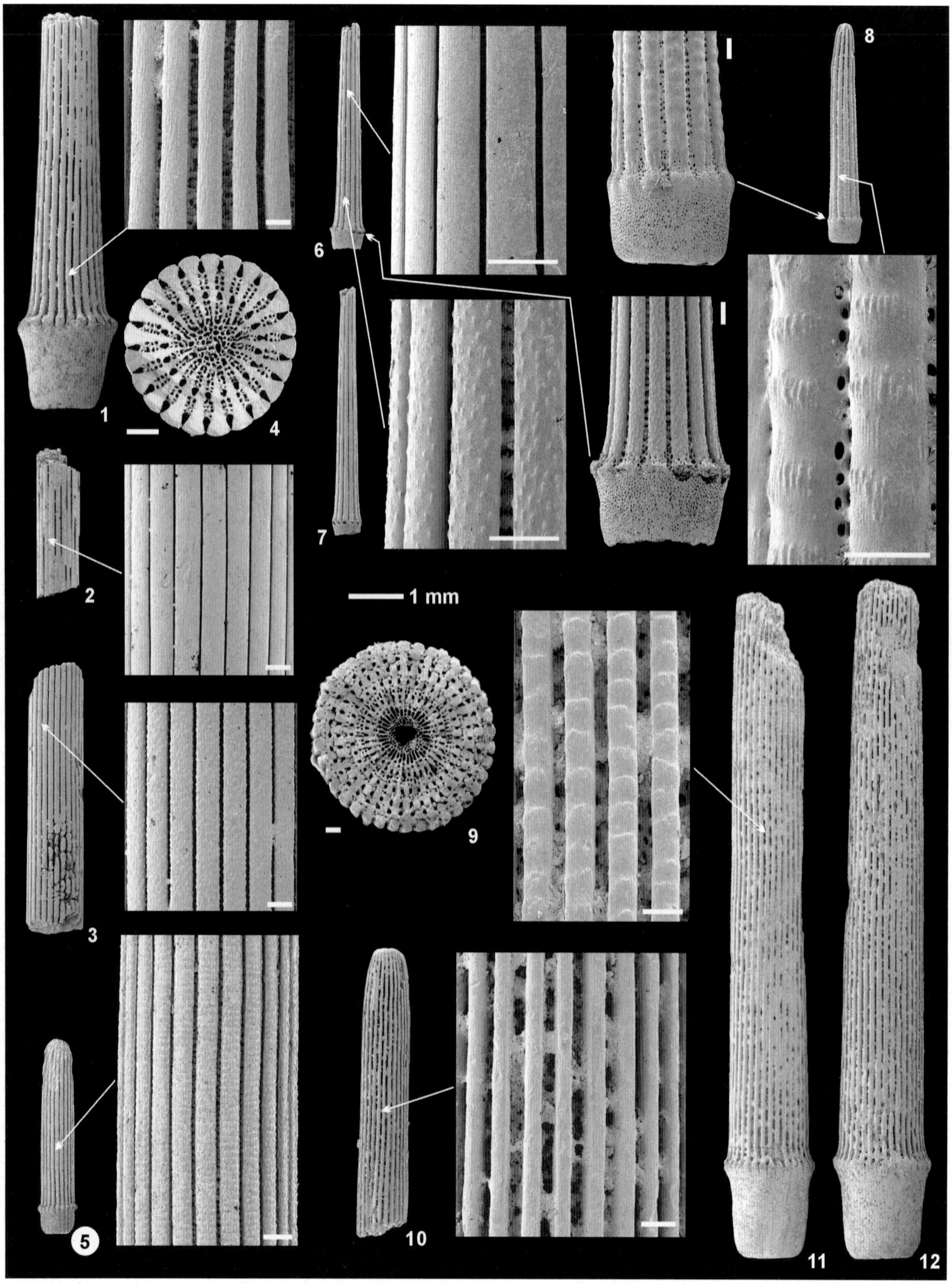

Echinacea indet.: 1: Steinebrunn, NÖ (NHMW 1865.I.612f); 2-4: Steinebrunn, NÖ [probably *Psammechinus* (ornament similar to spines of extant *P. miliaris*); 2: mid-shaft fragment, 3: proximal fragment, 4: cross section (NHMW 1865.I.612g-i)]; 5: Lăpugiu des Sus, Romania [probably *Echinometra* (ornament similar to spines of extant *E. mathaei*), NHMW 1859.XLV.617B.c]; 6-7: Steinebrunn, NÖ [probably *Psammechinus* (ornament similar to spines of extant *P. microtuberculatus*), NHMW 1865.I.612j-k)]; 8: Steinebrunn, NÖ [probably *Schizechinus*, NHMW 1865.I.612l)

Tripneustes **sp.:** 9: Bad Vöslau (NHMW 1997z0178/1325b); 10: Steinebrunn, NÖ (NHMW 1865.I.612m); 11-12: Forchtenau, Bgld (NHMW 2004z0001/0015b)

"1 mm" scale bar valid for the images of the whole specimens; scale bars for the surface details and cross sections equal 100 µm

***Clypeaster campanulatus* forma *campanulatus* (Schlotheim, 1820):** 1a-c: Müllendorf, Bgld (NHMW 2004z0112/0001)
***Clypeaster campanulatus* forma *portentosus* Michelin, 1861:** 2a-b: Eisenstadt, Bgld (syntype of *C. portentosus* Michelin, 1861; NHMW 1846.37.960); 3a-b: Rauchstallbrunngraben, near Baden, NÖ (syntype of *C. alticostatus* Michelin, 1861; NHMW 2003z0055/0001)

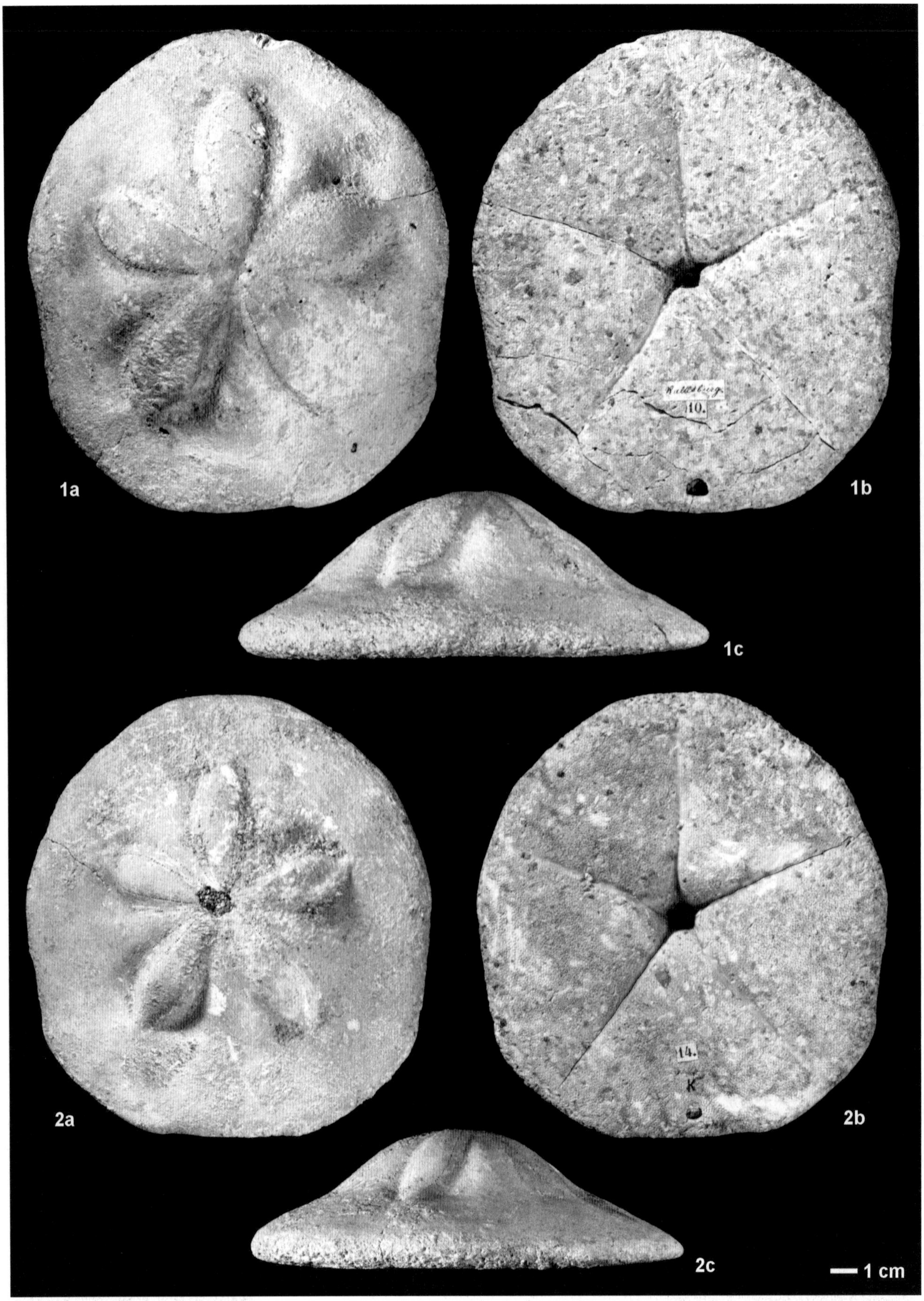

***Clypeaster campanulatus* forma *partschii* Michelin, 1861:** 1a-c, 2a-c: Kalksburg, W (syntypes of Michelin, 1861; 1: NHMW 1857.38.30; 2: NHMW 1858.III.19)

a: aboral, b: oral, c: left lateral view

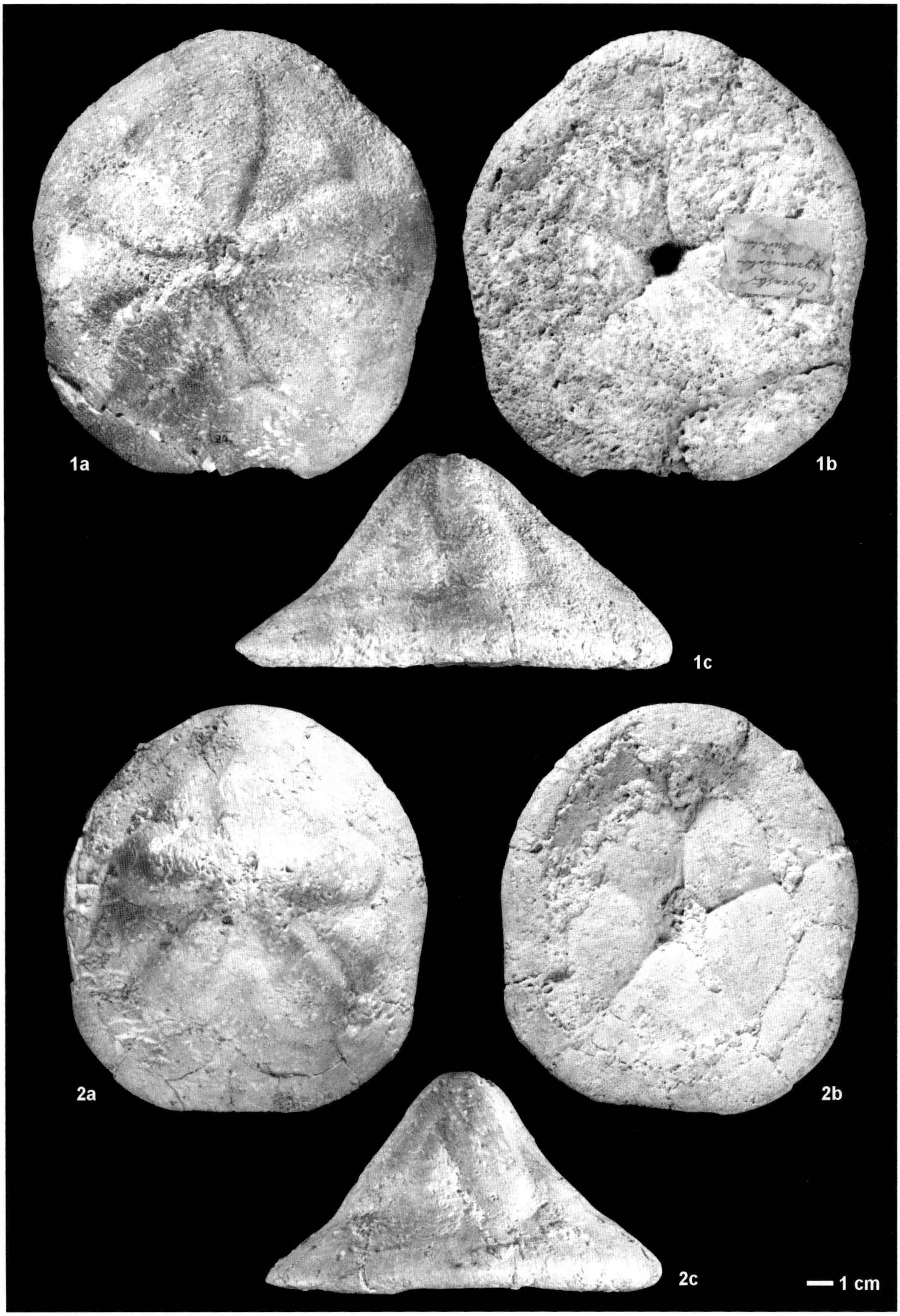

***Clypeaster campanulatus* forma *pyramidalis* Michelin, 1861:** 1a-c: Mitterberg, near Baden, NÖ (holotype of Michelin, 1861; NHMW 1843.LIII.1); 2a-c: Rauchstallbrunngraben, near Baden, NÖ (NHMW 1904.VIII.83)

a: aboral, b: oral, c: right lateral view

***Clypeaster campanulatus* forma aff. *gibbosus* DE SERRES, 1829:** 1a-c: Kalksburg, W (NHMW 1904.VIII.88)
***Clypeaster campanulatus* forma aff. *reidii* WRIGHT, 1855:** 2a-c: Müllendorf, Bgld (NHMW 1997z0178/1754)

a: aboral, b: oral, c: lateral view (1c: left lateral, 2c: right lateral)

Clypeaster campanulatus **forma** ***partschii*** **MICHELIN, 1861:** 1a-b: corroded test revealing plate sutures; Gainfarn, NÖ (NHMW 2003z0003/0020)

Clypeaster campanulatus **(SCHLOTHEIM, 1820):** 2: internal view of aboral surface; Mannersdorf, NÖ (NHMW A3158); 3: horizontal cross section revealing support structures; Kalksburg, W (NHMW A2807); 4: lantern; Mikulov, Czech Republic (NHMW A2812)

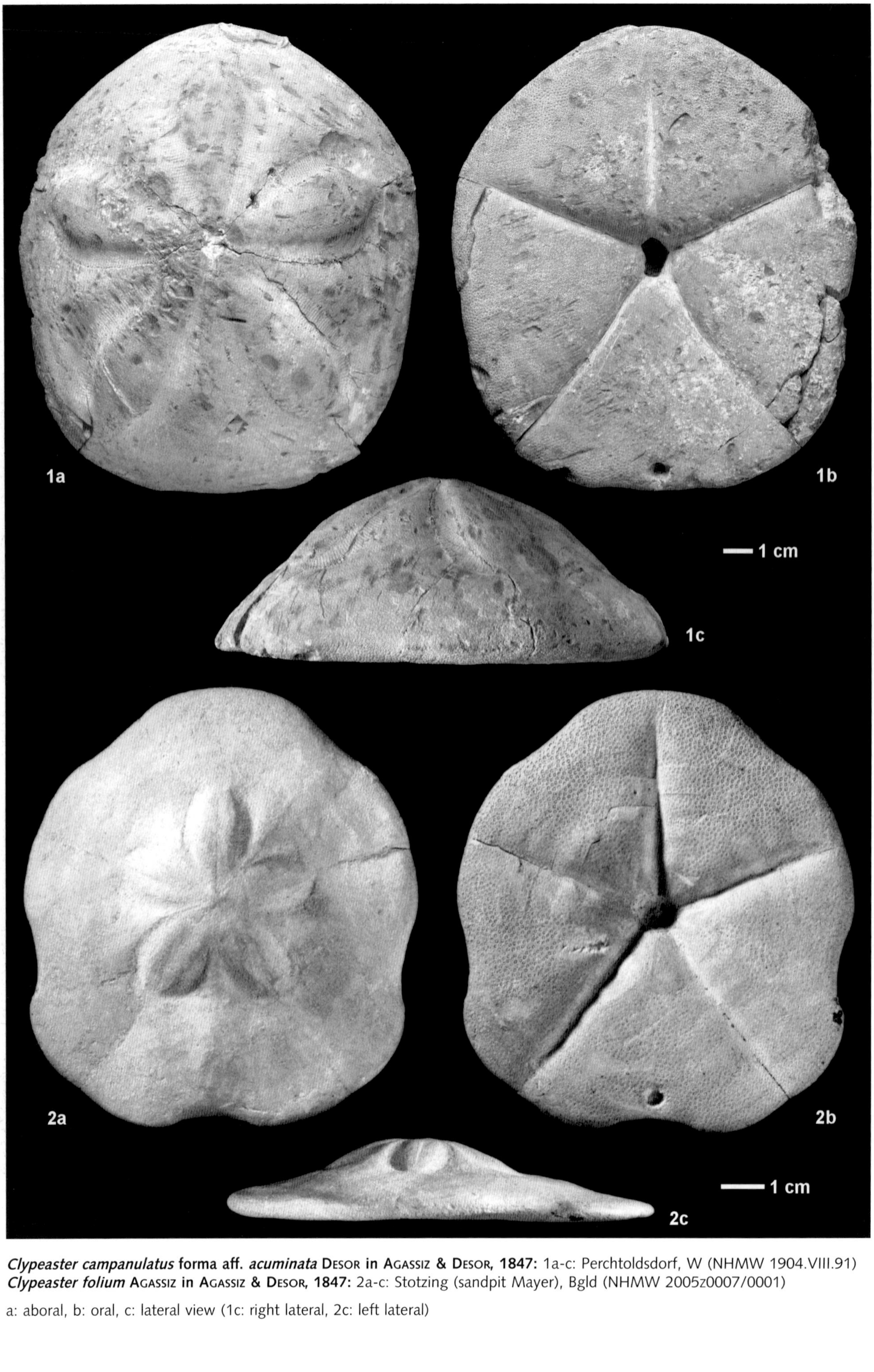

Clypeaster campanulatus **forma aff.** ***acuminata*** **Desor in Agassiz & Desor, 1847:** 1a-c: Perchtoldsdorf, W (NHMW 1904.VIII.91)
Clypeaster folium **Agassiz in Agassiz & Desor, 1847:** 2a-c: Stotzing (sandpit Mayer), Bgld (NHMW 2005z0007/0001)

a: aboral, b: oral, c: lateral view (1c: right lateral, 2c: left lateral)

Clypeaster intermedius **DES MOULINS, 1837:** 1a-c: Roggendorf (Schloßthal), NÖ (holotype of *C. subpartschi* SCHAFFER, 1912; KM 32); 2a-c: Grübern, NÖ (HM 644)

Clypeaster latirostris **Michelin, 1861:** 1a-b: Gauderndorf (Himmelreich), NÖ (NHMW 1869.III.37); 2a-c: Grübern, NÖ (NHMW 1997z0178/0526); 3: ? Gauderndorf, NÖ (paralectotype, NHMW 1843.II.67); 4: Limberg (Hengl quarry), NÖ (NHMW 2004z0115/0003)

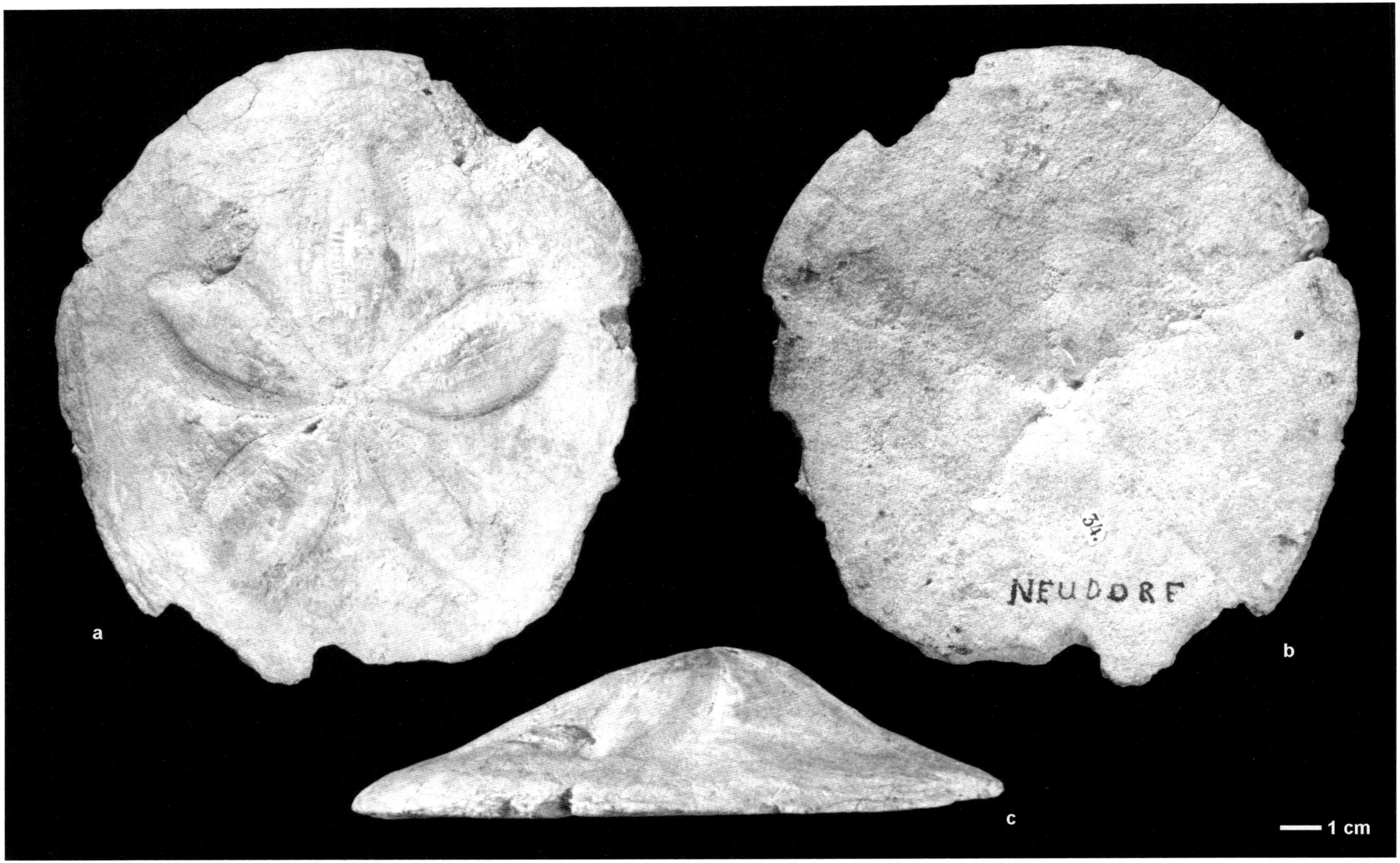

Clypeaster neudorfensis **LAMBERT, 1928:** Devínska Nová Ves, Slovak Republic (AO of *C. melitensis* MICHELIN, 1861; holotype of *C. neudorfensis* LAMBERT, 1928; NHMW 1851.VI.119)

1a
1b
2a
3.
2b
1 cm
3a
3b
3c

Clypeaster calabrus Seguenza, **1880:** 1a-c, 2: Müllendorf, Bgld (1a-c: GBA 2004/1/4; 2: NHMW 1997z0178/1753a); 3: Kalksburg, W (NHMW 2003z0049/0003)
Clypeaster barcinensis ? Lambert, **1906:** 4: Devínska Nová Ves, Slovak Republic (NHMW 2003z0048/0001)

⇦ *Clypeaster calabrus* Seguenza, **1880:** 1a-b: Kalksburg, W (NHMW 2003z0049/0003); 2a-b, 3a-c: Müllendorf, Bgld (2: NHMW 1997z0178/1753a; 3: NHMW 1997z0178/1753b)

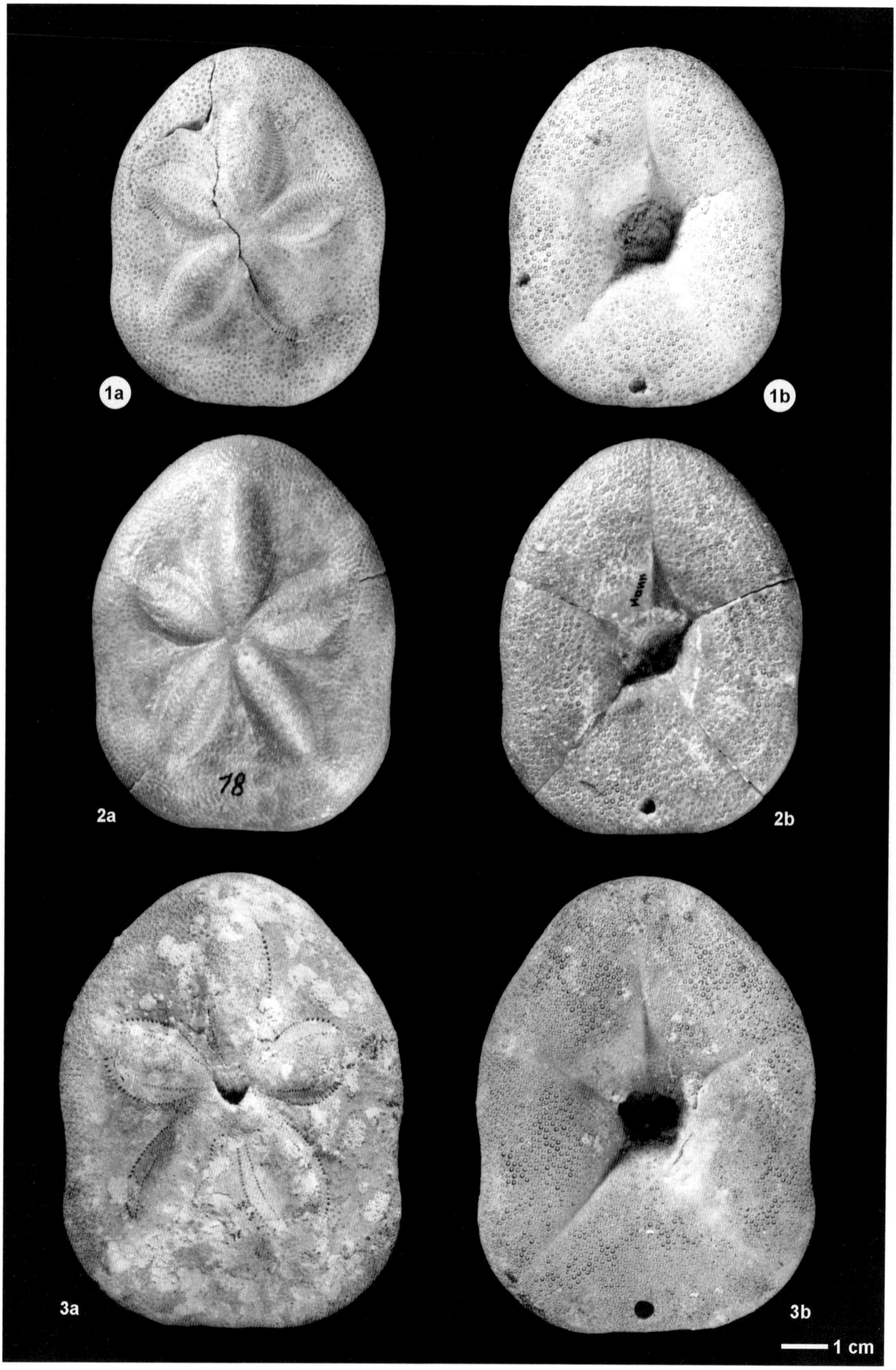
1a
1b
78
2a
2b
3a
3b
1 cm

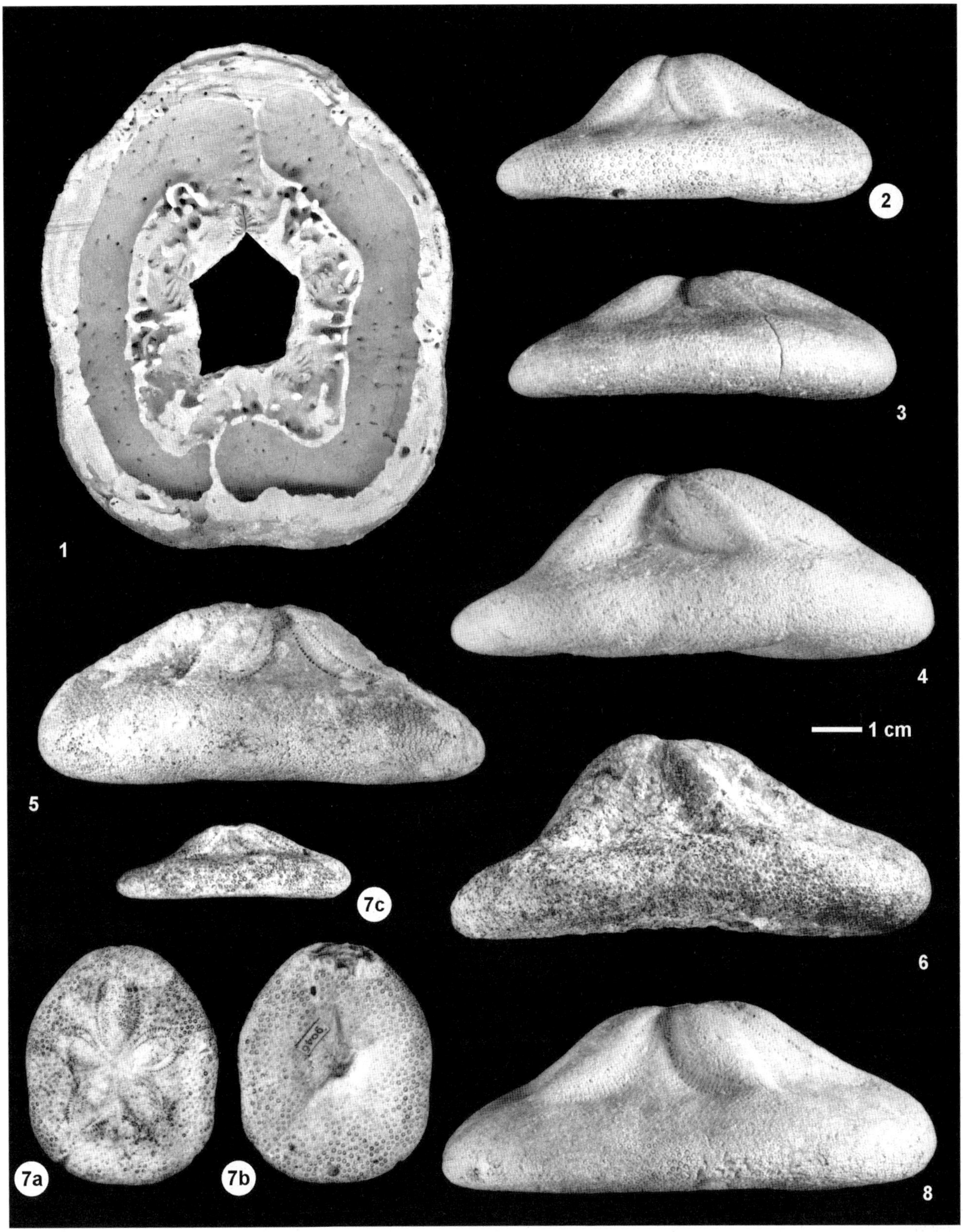

Clypeaster scillae **DES MOULINS, 1837:** 1: aboral surface cut to reveal internal support structures, Großhöflein, near Eisenstadt, Bgld (NHMW 1869.I.664); 2, 7a-c: Kemence, Hungary (2: NHMW 1858.XXXIX.17; 7: NHMW 1865.XXV.17); 3: Mannersdorf, NÖ (NHMW 1978/2020/13); 4, 8: Müllendorf, Bgld (4: NHMW 2004z0112/0002; 8: NHMW A2328); 5: Wiesfleck, Bgld (NHMW 2004z0113/0001); 6: Kalksburg, W (NHMW 1857.38.28)

Clypeaster scillae **DES MOULINS, 1837:** 1a-b: Kemence, Hungary (NHMW 1858.XXXIX.17); 2a-b: Mannersdorf, NÖ (NHMW 1978/2020/18); 3: Wiesfleck, Bgld (NHMW 2004z0113/0001)

Clypeaster scillae DES MOULINS, **1837:** 1a-b, 2a-b: Müllendorf, Bgld (1: NHMW 2004z0112/0002; 2: NHMW A2328)

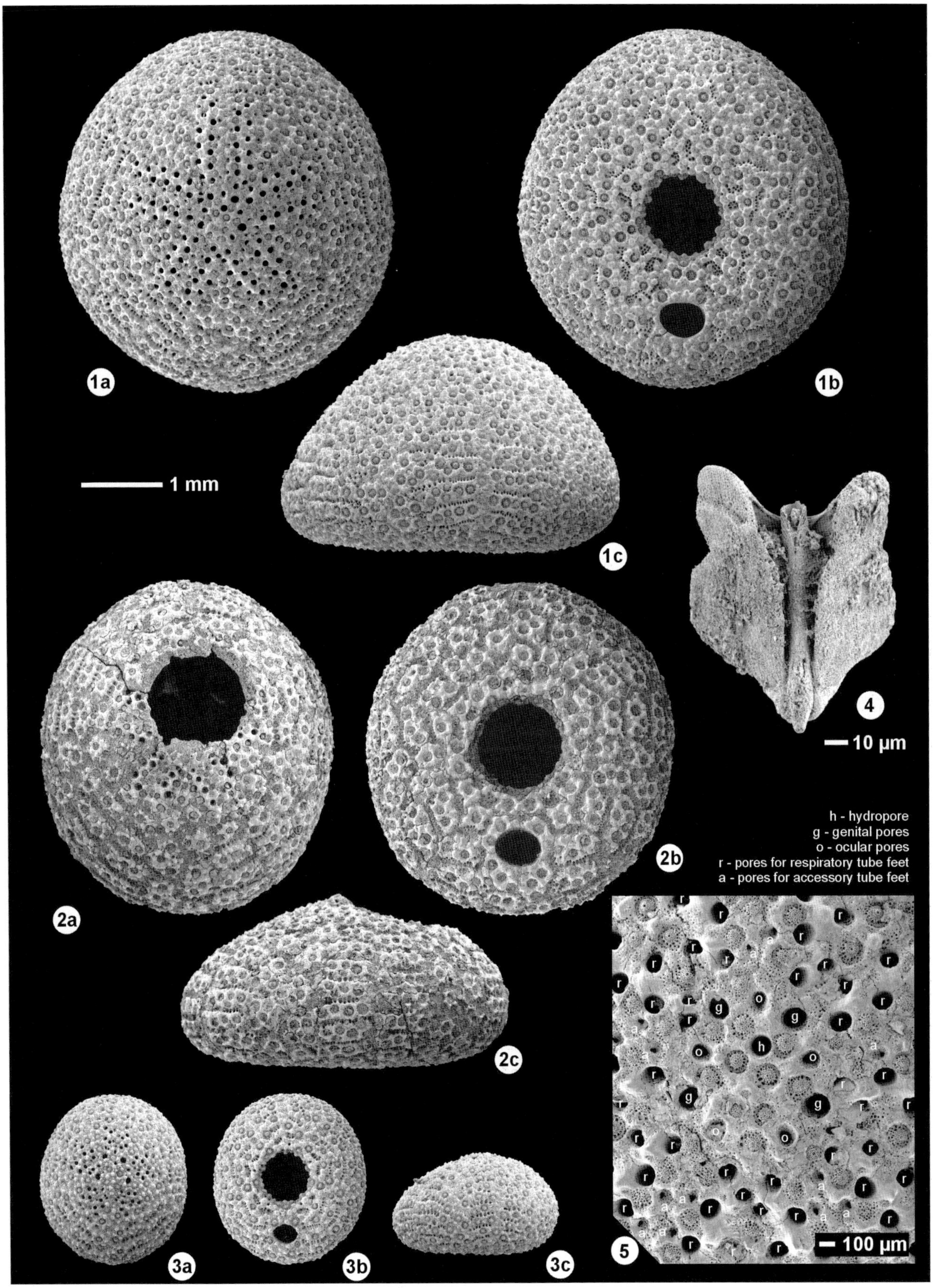

Echinocyamus transylvanicus **Laube, 1869:** 1-3: whole coronas, Lăpugiu des Sus, Romania [1: lectotype (NHMW 1860.XL.529a), 2: paralectotype, note predatory boring (NHMW 1860.XL.529b), 3: paralectotype (NHMW 1859.XLV.615); a: aboral views, b: oral views, c: right lateral views]; 4: pyramid with tooth, Lăpugiu des Sus, Romania (NHMW 1860.XL.529c); 5: apical disc of lectotype, Lăpugiu des Sus, Romania (NHMW 1860.XL.529a)

"1 mm" scale bar valid for all figures except figs. 4-5; all photographs are SEM images

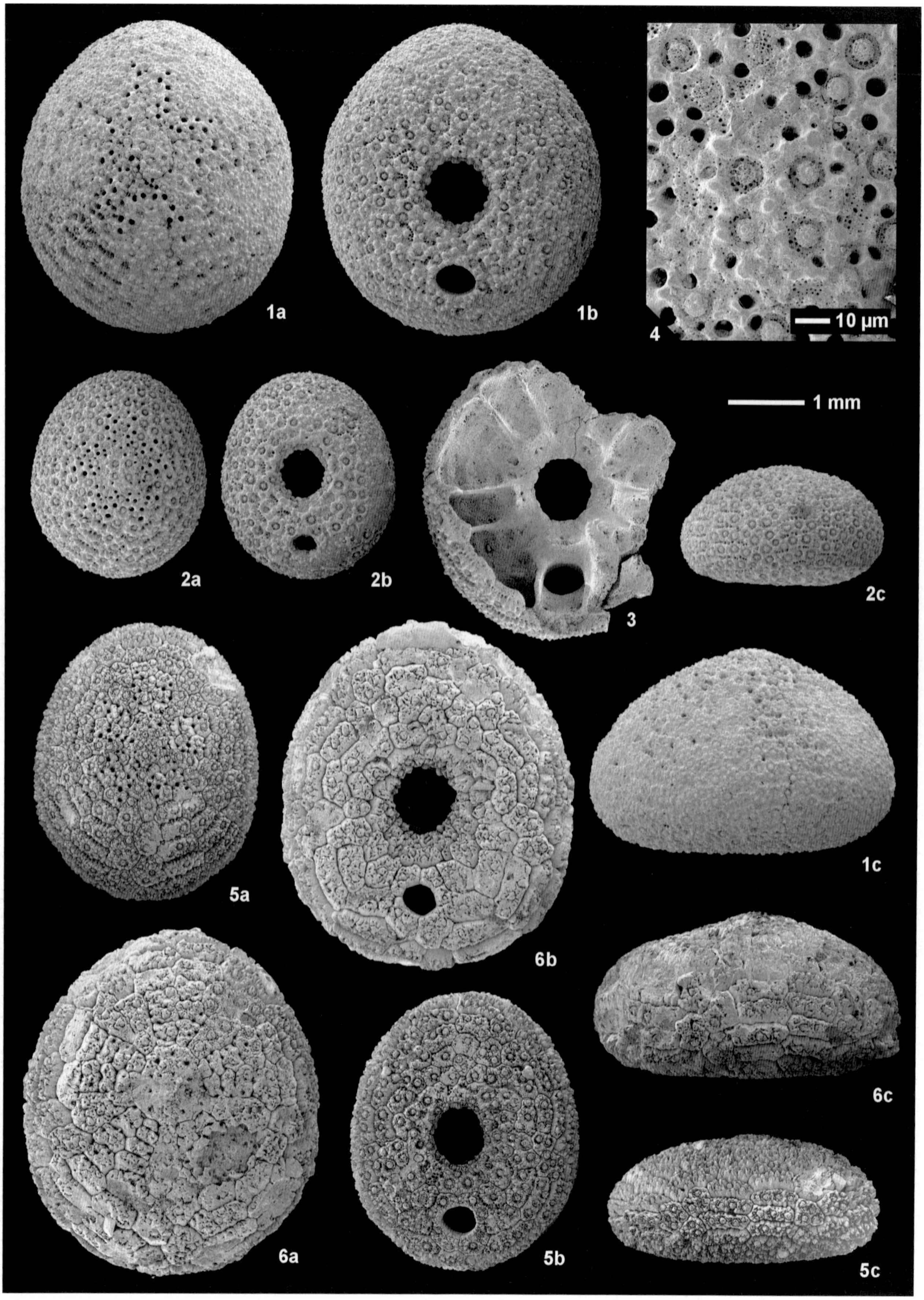

Echinocyamus transylvanicus **Laube, 1869:** 1-3: Baden Tegel, Baden, NÖ (1: NHMW 1852.II.1612b, 2: NHMW 1852.II.1612c, 3: NHMW 1852.II.1612a); 4: apical disc of juvenile specimen figured in 2a-c, note that gonopores are still closed
Echinocyamus pseudopusillus **Cotteau, 1895:** 5-6: Hartl Fm., Eisenstadt, Bgld (5: NHMW 1859.L.800b, 6: NHMW 1859.L.800a)

"1 mm" scale bar valid for all figures except 4; a: aboral views, b: oral views, c: right lateral views; all photographs are SEM images

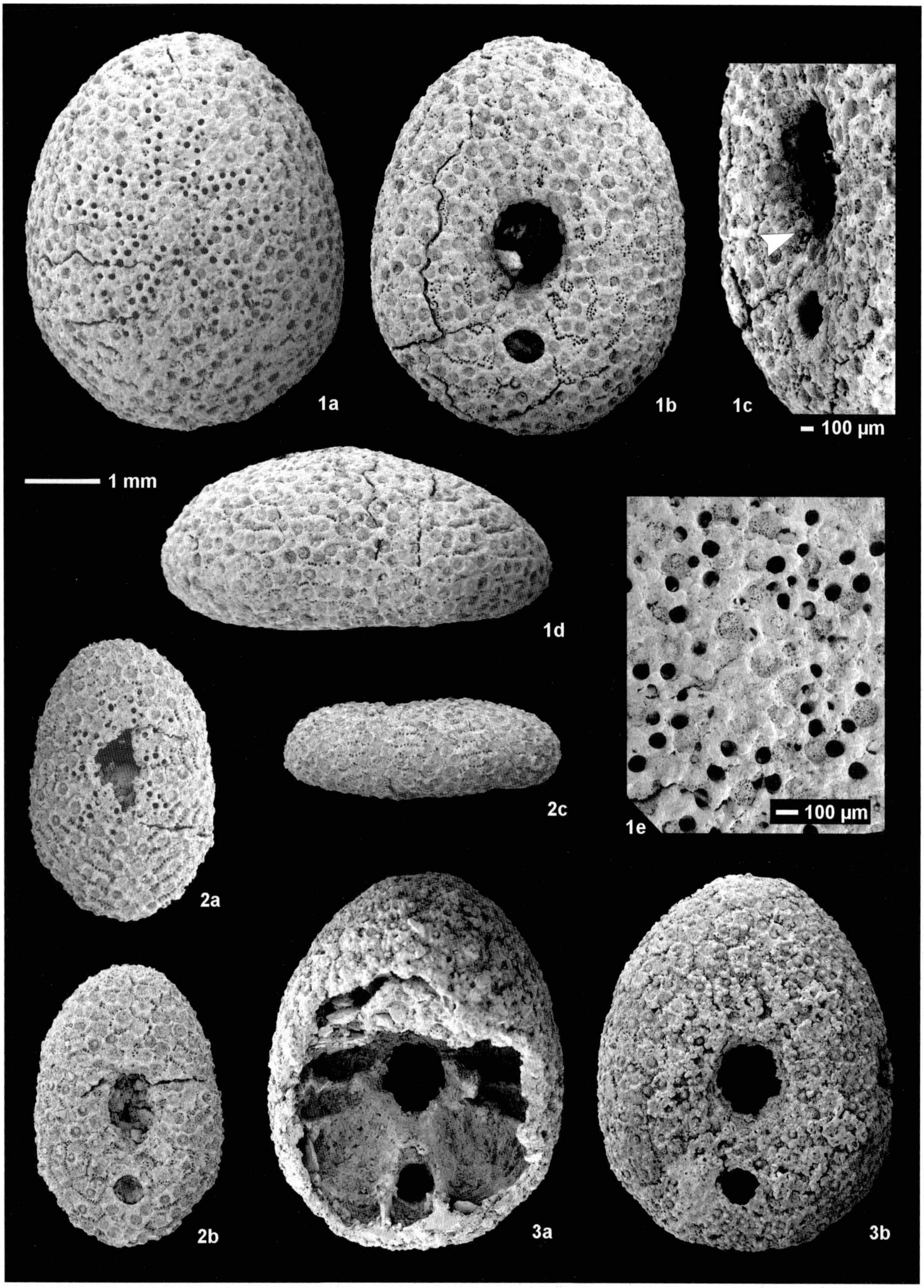

Echinocyamus pseudopusillus **COTTEAU, 1895:** 1a-e: Gainfarn, NÖ [NHMW 2004z0076/0045; a: aboral view, b: oral view, c: oblique oral view, note depressed posterior margin of peristome (arrow), d: left lateral view, e: apical disc]; 3a-b: Gainfarn, NÖ [NHMW 2004z0076/0046, a: aboral (note paired internal partitions), and b: oral view]
Echinocyamus calariensis **? LAMBERT, 1907:** 2a-c: Gainfarn, NÖ (NHMW 2004z0076/0047; a: aboral, b: oral , and c: left lateral views)

"1 mm"-scale bar valid for all figures except 1c and 1e; All photographs are SEM images

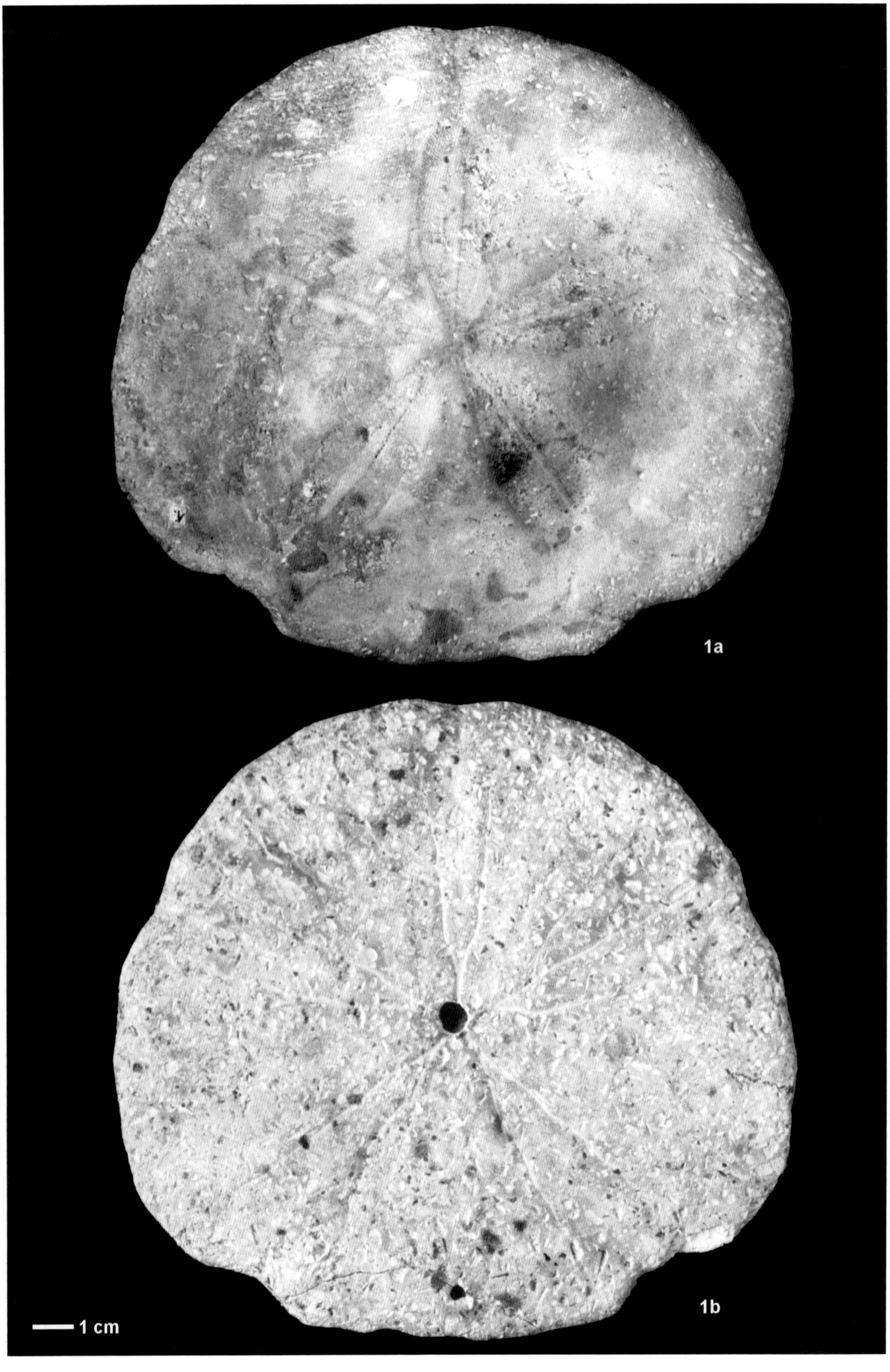

Parascutella gibbercula **(DE SERRES, 1829)**: 1a-b: Kalksburg, Vienna (lectotype of *Scutella vindobonensis* LAUBE, 1869; NHMW 1904.VIII.59; slightly reduced)

***Parascutella gibbercula* (de Serres, 1829)**: 1a-c: Devínska Nová Ves, SK (holotype of *Scutella vindobonensis secunda* Schaffer, 1962; NHMW 1964/589); 2: Kalksburg, Vienna (lectotype of *Scutella vindobonensis* Laube, 1869; NHMW 1904.VIII.59)

1a
1b
1c
2
3
1 cm

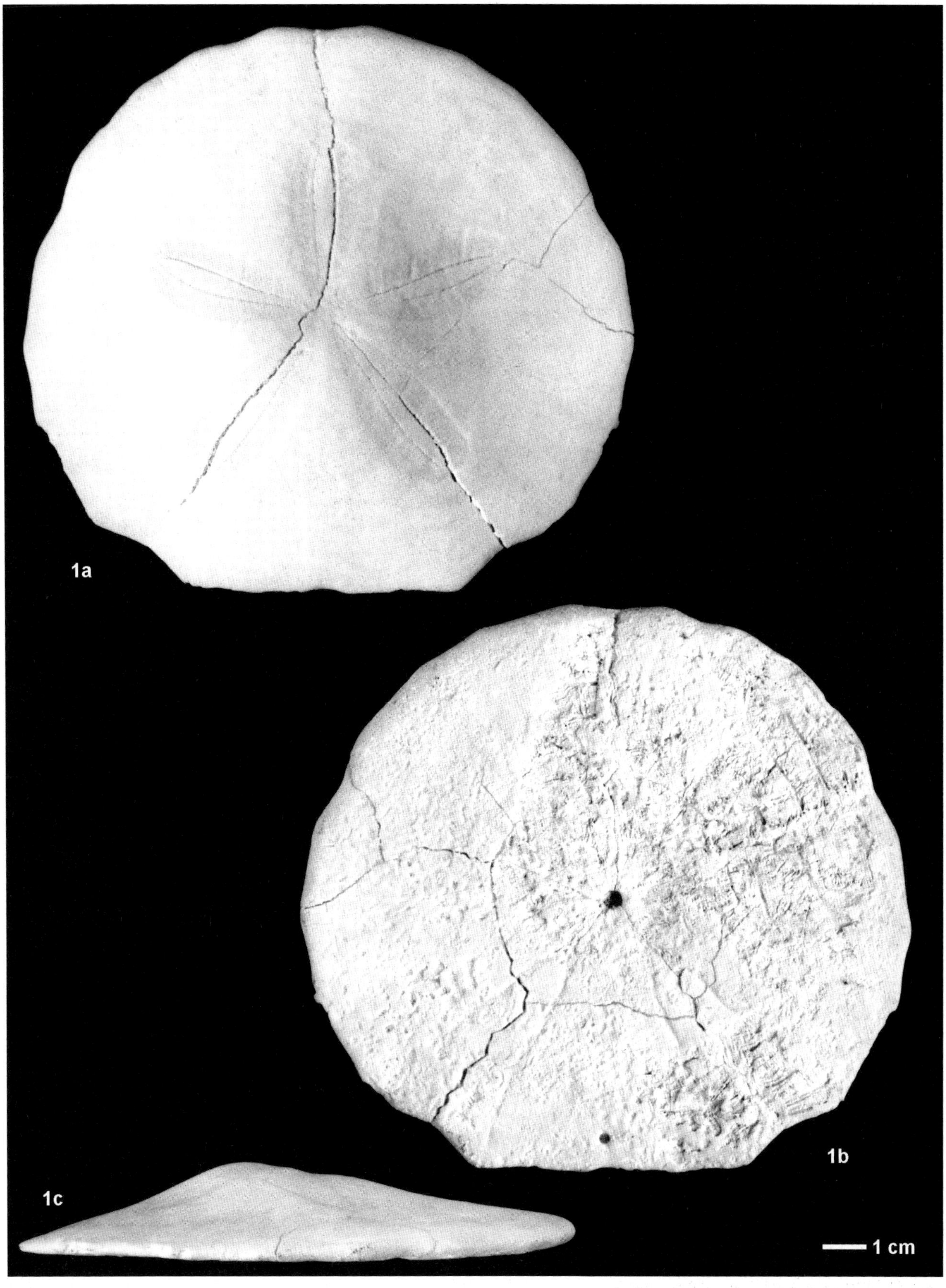

Parascutella gibbercula **(DE SERRES, 1829)**: 1a-c: Müllendorf, Bgld (NHMW 1997z0178/1757; slightly reduced)

↶ ***Parascutella gibbercula*** **(DE SERRES, 1829)**: 1a-c: Neckenmarkt, Bgld (holotype of *Scutella multiconcava* SCHAFFER, 1962; NHMW 390/1961; slightly reduced); 2: Neckenmarkt, Bgld (AO of *Scutella multiconcava* SCHAFFER, 1962; NHMW 1965/590; slightly reduced); 3: oral surface ground to reveal the plating pattern and the microcanal system, Kalksburg, Vienna (NHMW 2004z0111/0001; slightly reduced)

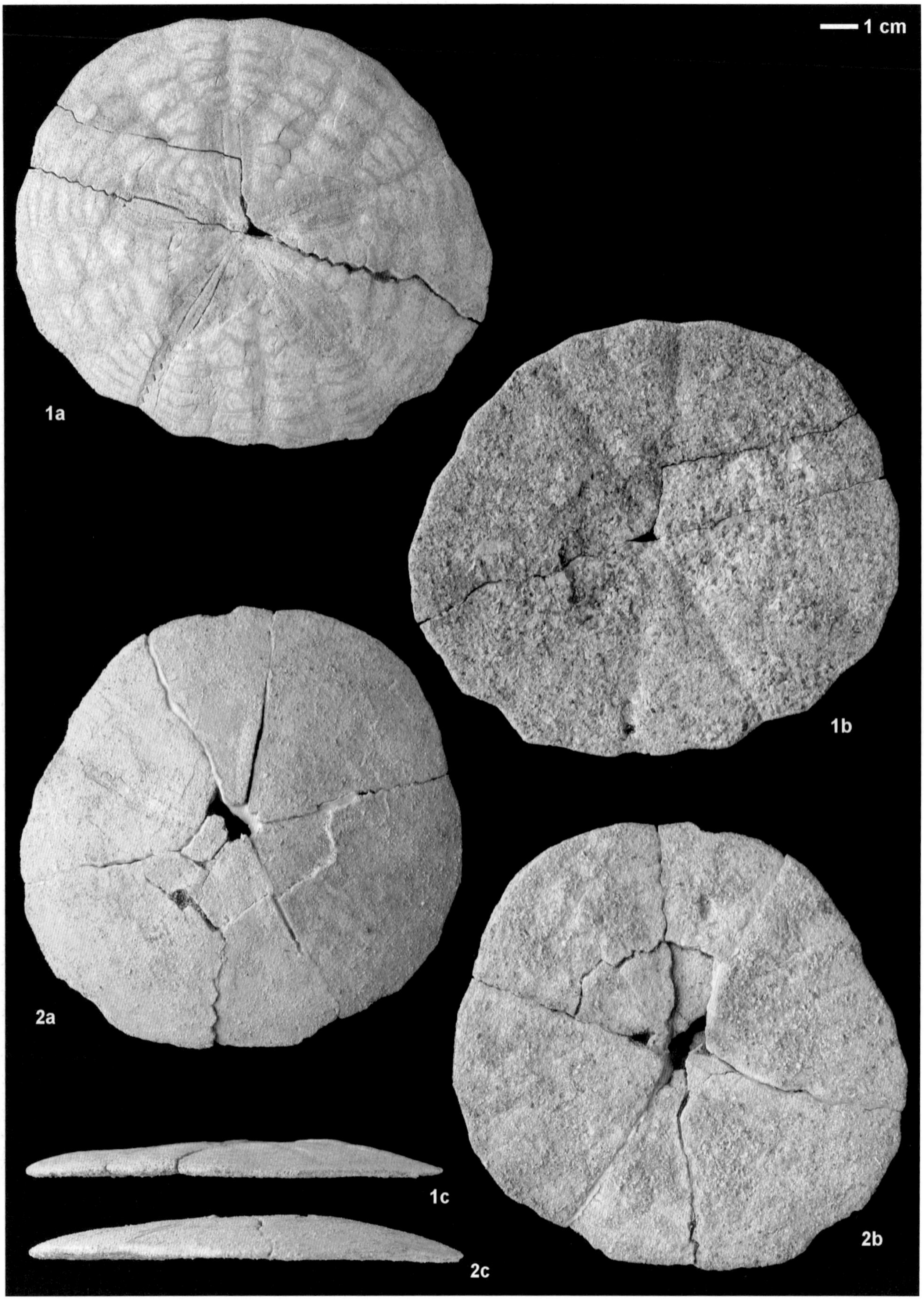

Parascutella paulensis **(Agassiz, 1841)**: 1a-c, 2a-c: Niederkreuzstetten, NÖ (1: holotype of *Scutella media* Schaffer, 1962, NHMW 1861.L.137; 2: NHMW 2004z0077/0001)

a: aboral, b: oral, c: left lateral view

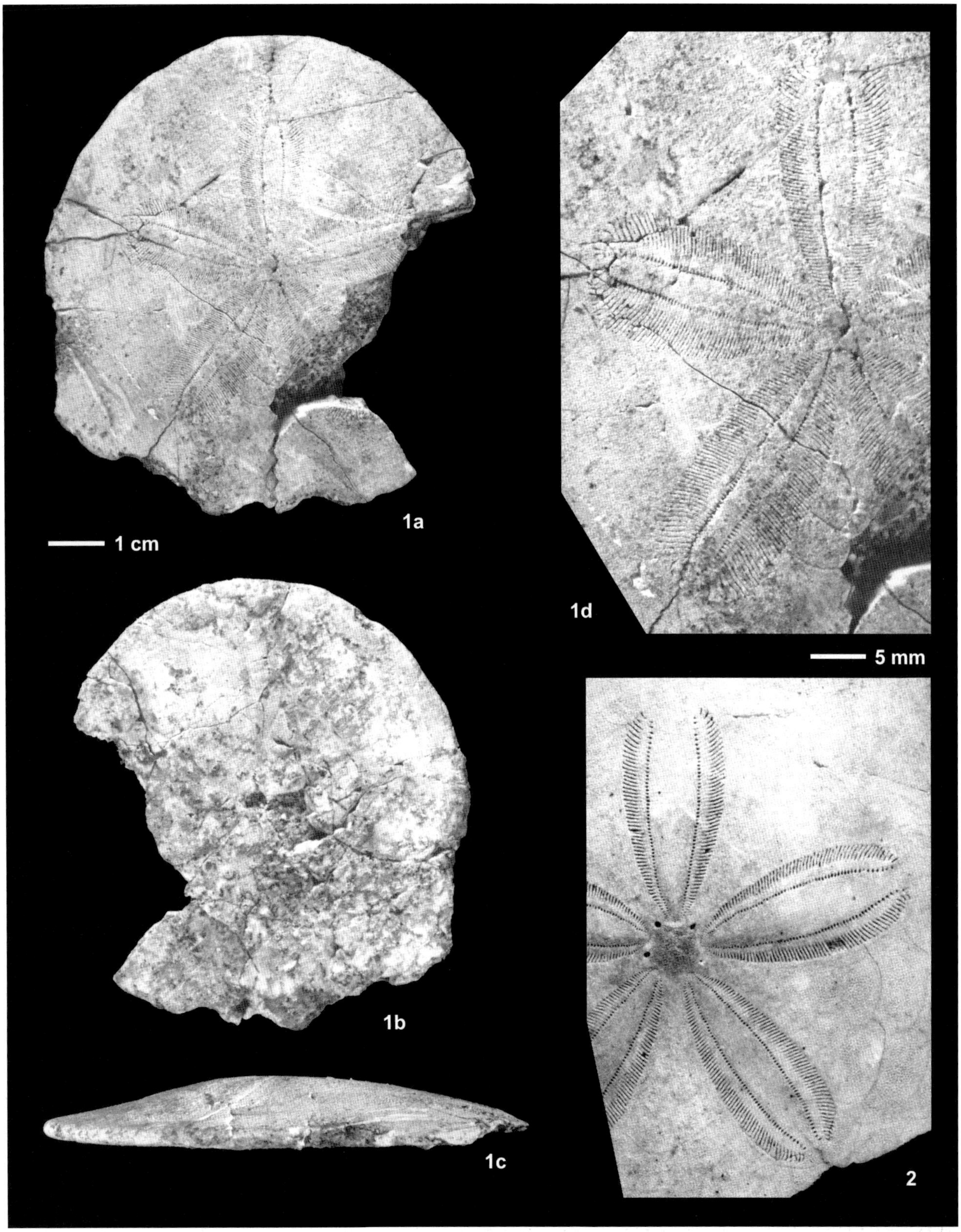

***Parascutella paulensis* (AGASSIZ, 1841)**: 1a-d: Hengl quarry, Limberg, NÖ (NHMW 1998z0048/0050)
***Parmulechinus hoebarthi* (KÜHN, 1936)**: 2: Eichberg, near Maria Dreieichen, NÖ [AO of KÜHN, 1936; NHMW 1935(No. 69)/1/97]

"1 cm"-scale bar valid for figures 1a-c, "5 mm"-scale bar valid for figures 1d and 2

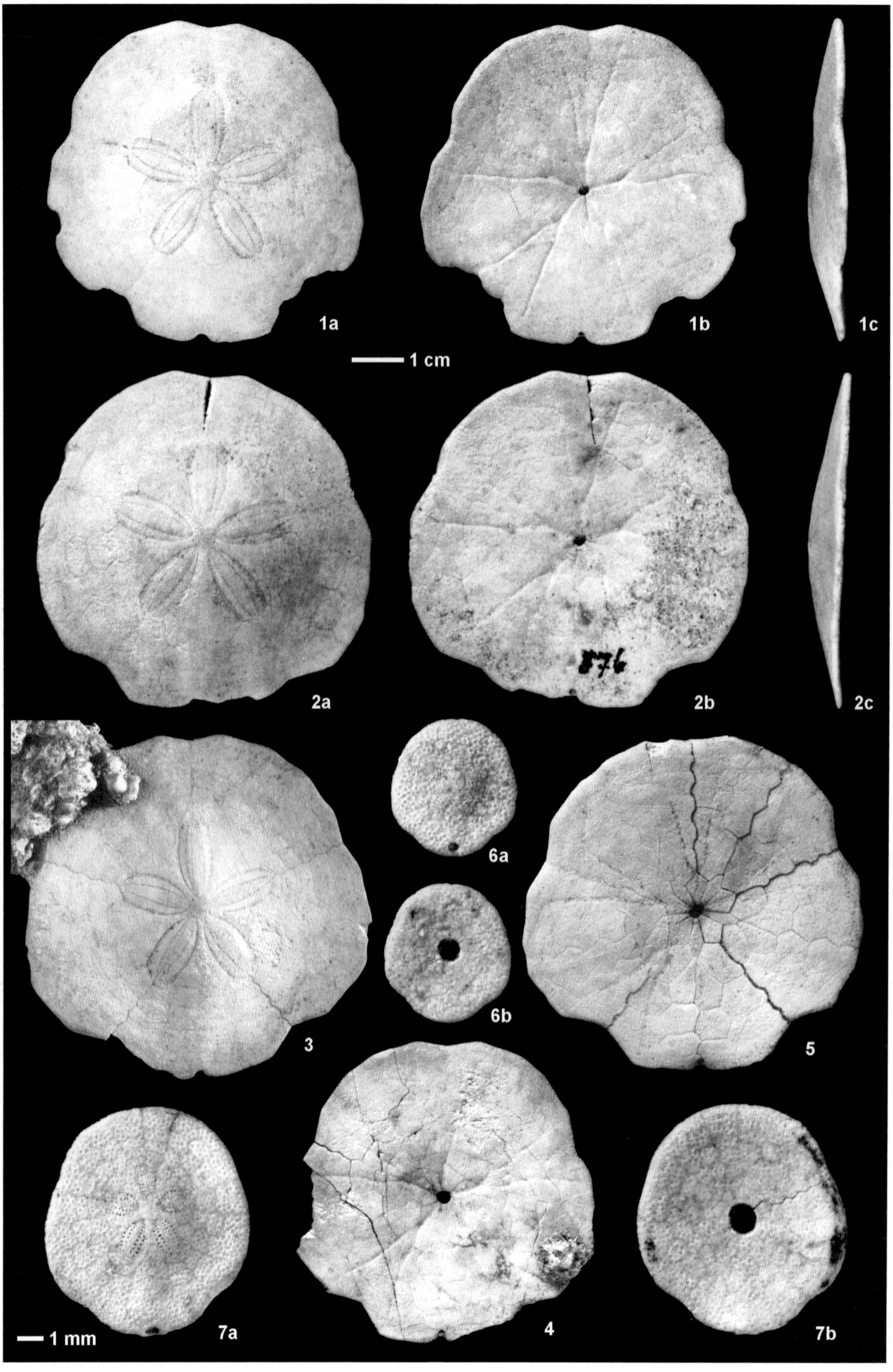
1a
1b
1c
1 cm
2a
2b
2c
3
6a
6b
5
7a
4
7b
1 mm

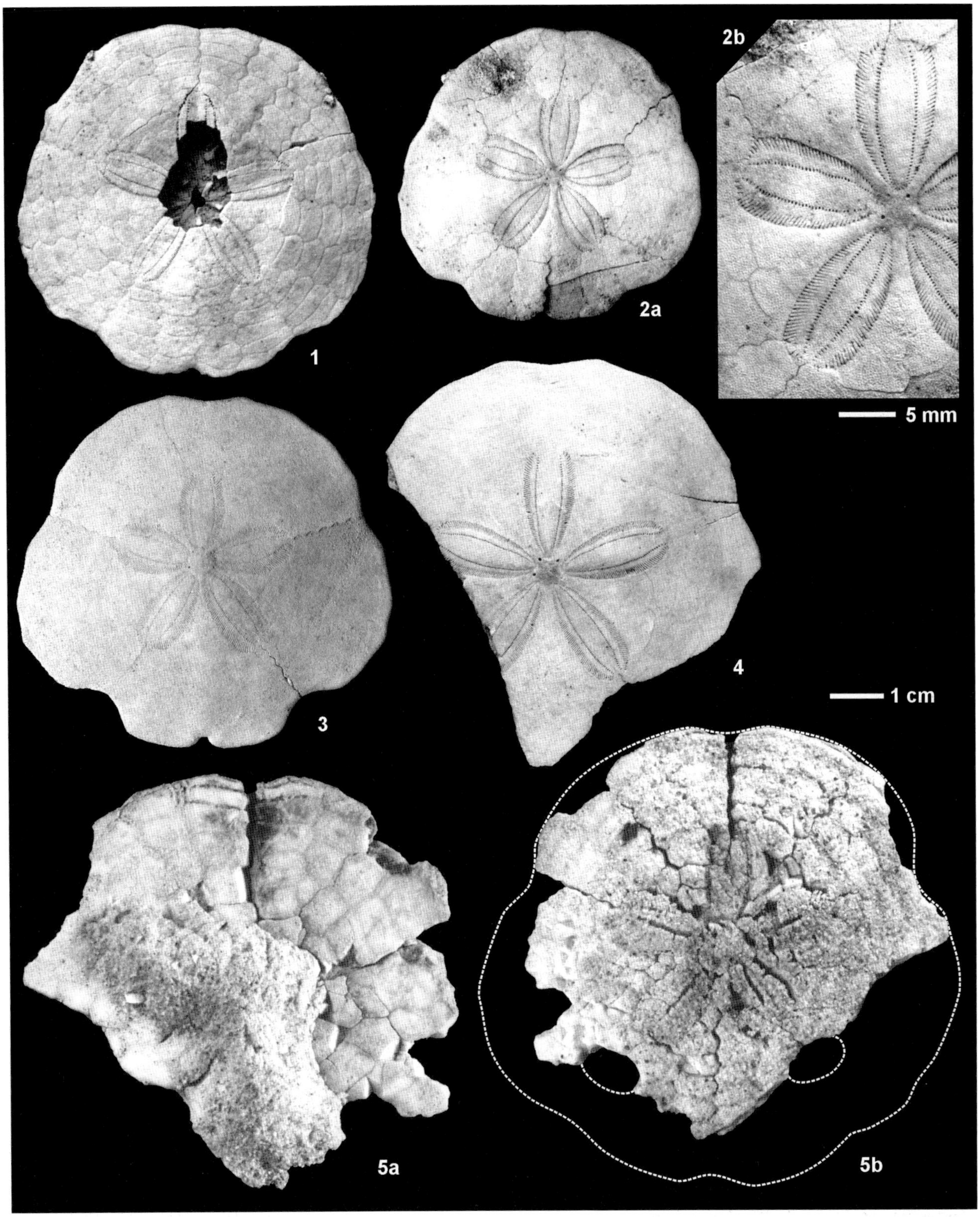

Parmulechinus hoebarthi **(KÜHN, 1936):** 1-4: adult specimens, Eichberg, near Maria Dreieichen, NÖ [1: NHMW 1968/781/10; 2: NHMW 2002z0118/0013; 3: holotype of *Scutella hoebarthi* KÜHN, 1936, HM 577/1; 4: AO of KÜHN, 1936, NHMW 1935(No. 69)/1/97]

Amphiope bioculata **? (DES MOULINS, 1837):** 5a-b: Niederkreuzstetten, NÖ (NHMW 1861.L.136; a: oral, and b: aboral view; reconstructed outline stippled)

"1 cm"-scale bar valid for all figures except figure 2b

Parmulechinus hoebarthi **(KÜHN, 1936):** 1-5: adult specimens, Eichberg, near Maria Dreieichen, NÖ (1: NHMW 2002z0118/008; 2: NHMW 2002z0118/0006; 3: NHMW 1972/1546/9; 4: NHMW 1968/781/7; 5: NHMW 2002z0118/0025); 6-7: juvenile specimens, Eichberg, near Maria Dreieichen, NÖ [6: IPUW 1992/195b; 7: AO of KÜHN, 1936, NHMW 1935(No. 69)/1/95]

"1 cm"-scale bar valid for figures 1 to 5, "1 mm"-scale bar valid for figures 6 to 7; a: aboral, b: oral, c: left lateral view

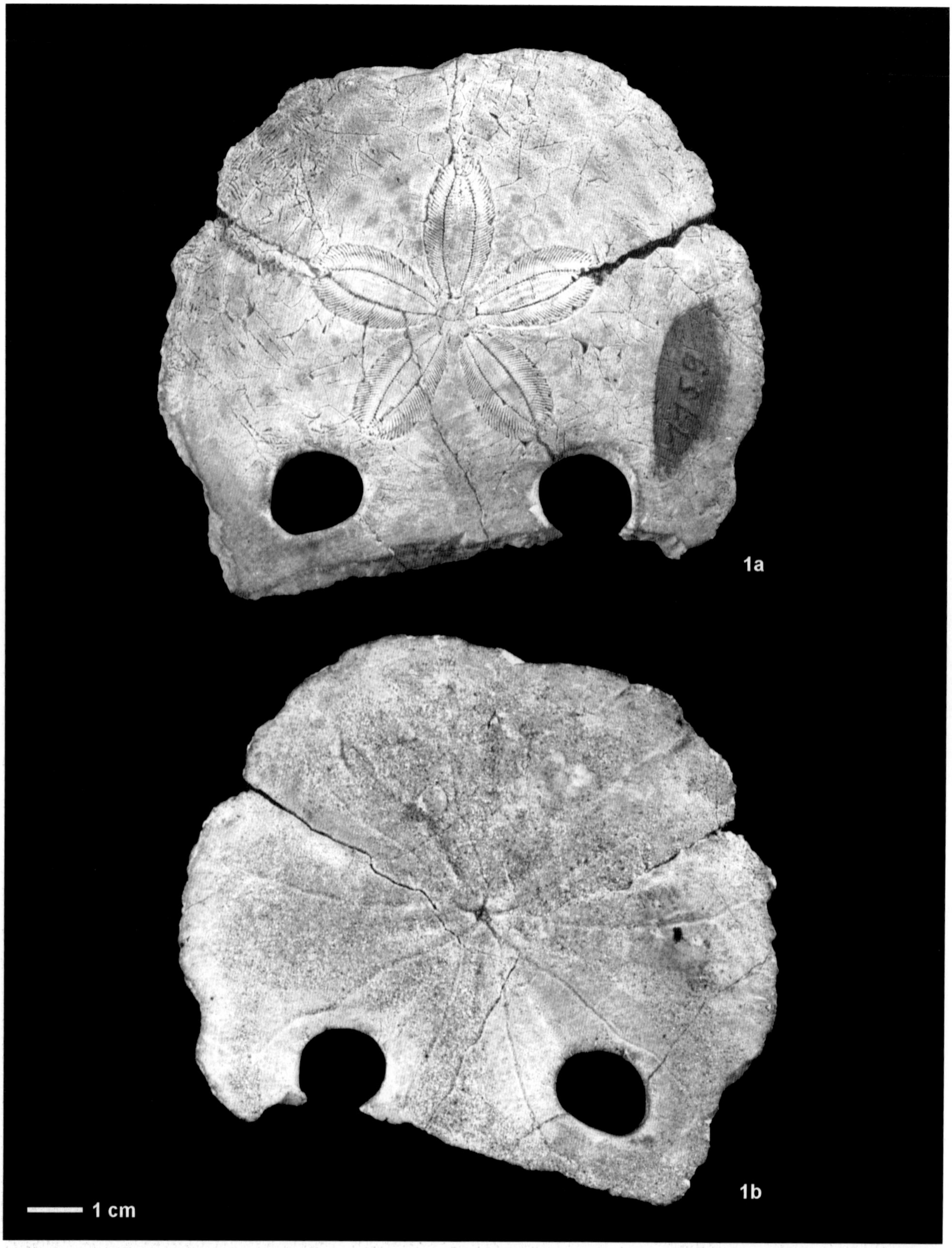

***Amphiope bioculata* (DES MOULINS, 1837):** 1a-b: Seggauberg (= Sekkauer Berg), near Leibnitz, Styria (holotype of *A. styriaca* HOERNES, 1883, UGP 1880.XX.5; a: aboral, b: oral view)

Amphiope bioculata **(DES MOULINS, 1837):** 1a-b, 2a-c: Niederkreuzstetten, NÖ (1: holotype of *A. laubei* LAMBERT, 1912, NHMW 1849.XXIII.39; 2: NHMW 1849.XXIII.40; 1b and 2c: left lateral views); 3: Seggauberg (= Sekkauer Berg), near Leibnitz, Styria (holotype of *A. styriaca* HOERNES, 1883, UGP 1880.XX.5; left lateral view)
Echinoneus **aff.** ***abnormalis*** **DE LORIOL, 1883:** 4a-c, 5: Müllendorf, Bgld (4: NHMW 2002z0172/0001, 5: TEMEL coll.; 4c: right lateral view)

"1 cm" scale bar valid for figs. 1-3; "5 mm" scale bar valid for figs. 4-5

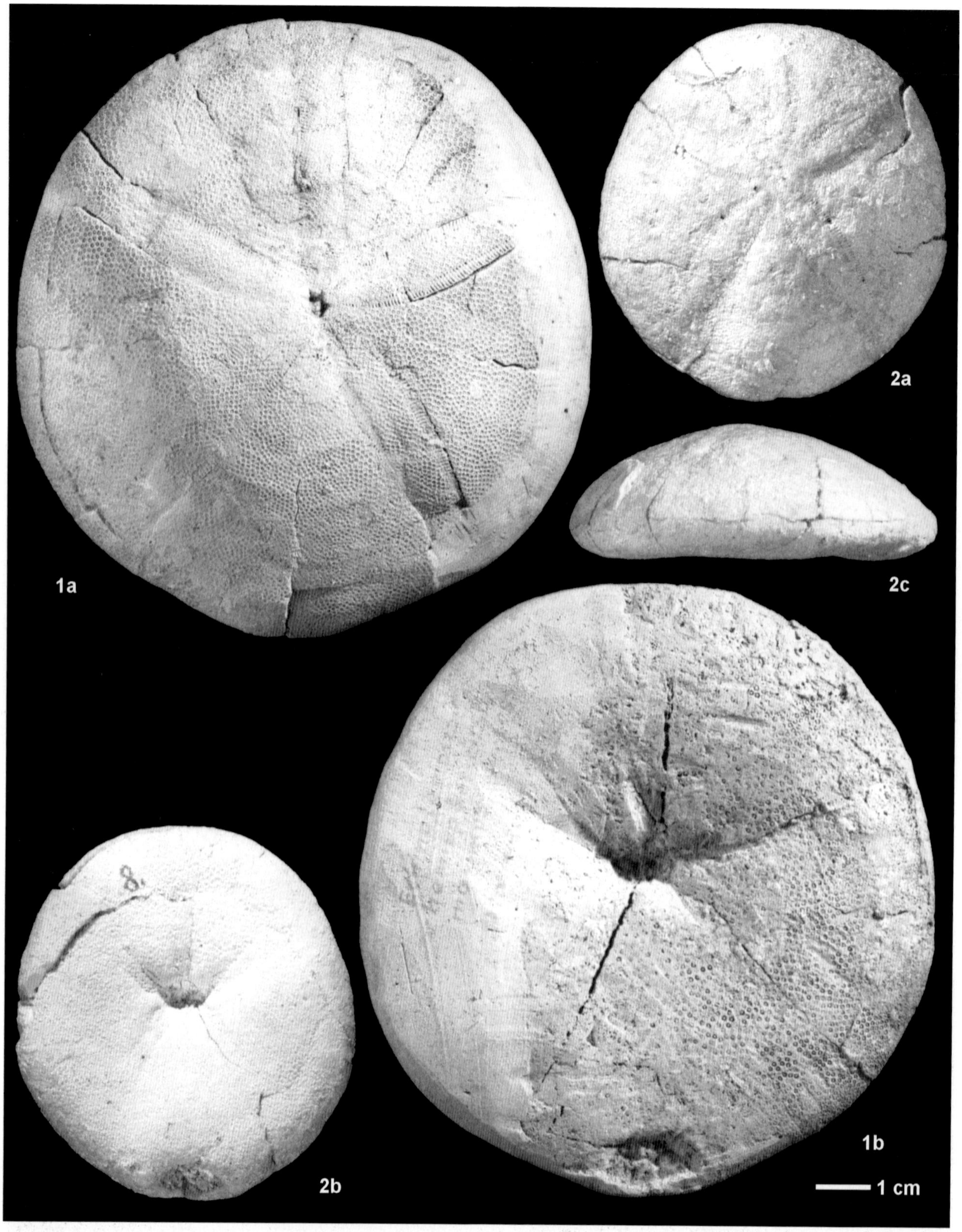

Echinolampas barcinensis **Lambert, 1906:** 1a-b: Mannersdorf, NÖ (NHMW 2002z0181/0007; left margin reconstructed); 2a-c: Müllendorf, Bgld (NHMW 1997z0178/1743)

Figures 1a and 1b coated with ammonium chloride; a: aboral, b: oral, c: left lateral view.

Echinolampas barcinensis **Lambert, 1906:** 1a-c: Mannersdorf, NÖ (NHMW 1978/2020/21; a: aboral, b: oral, c: left lateral view) ➪

1a
1b
1c
1 cm

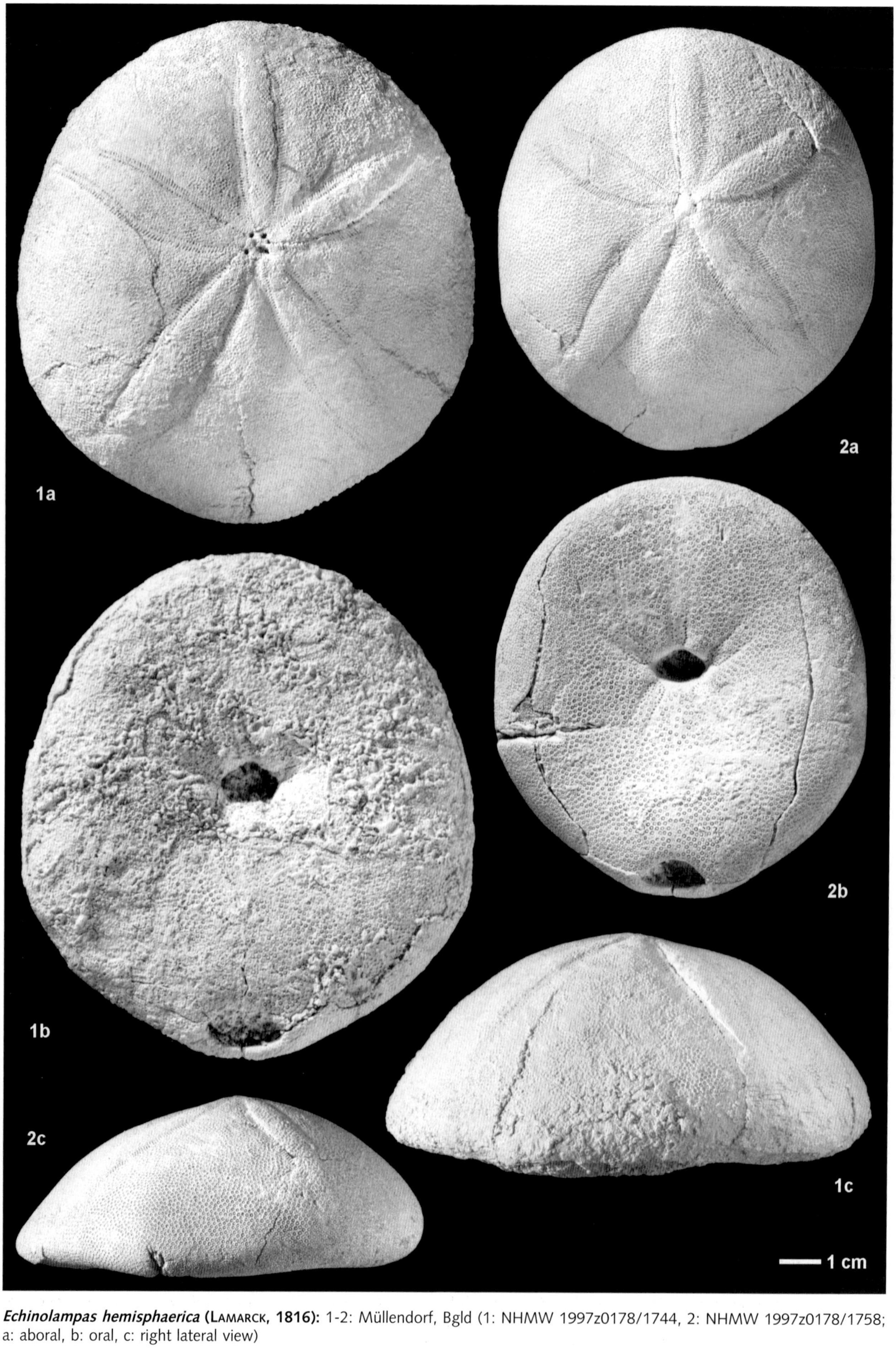

Echinolampas hemisphaerica **(Lamarck, 1816):** 1-2: Müllendorf, Bgld (1: NHMW 1997z0178/1744, 2: NHMW 1997z0178/1758; a: aboral, b: oral, c: right lateral view)

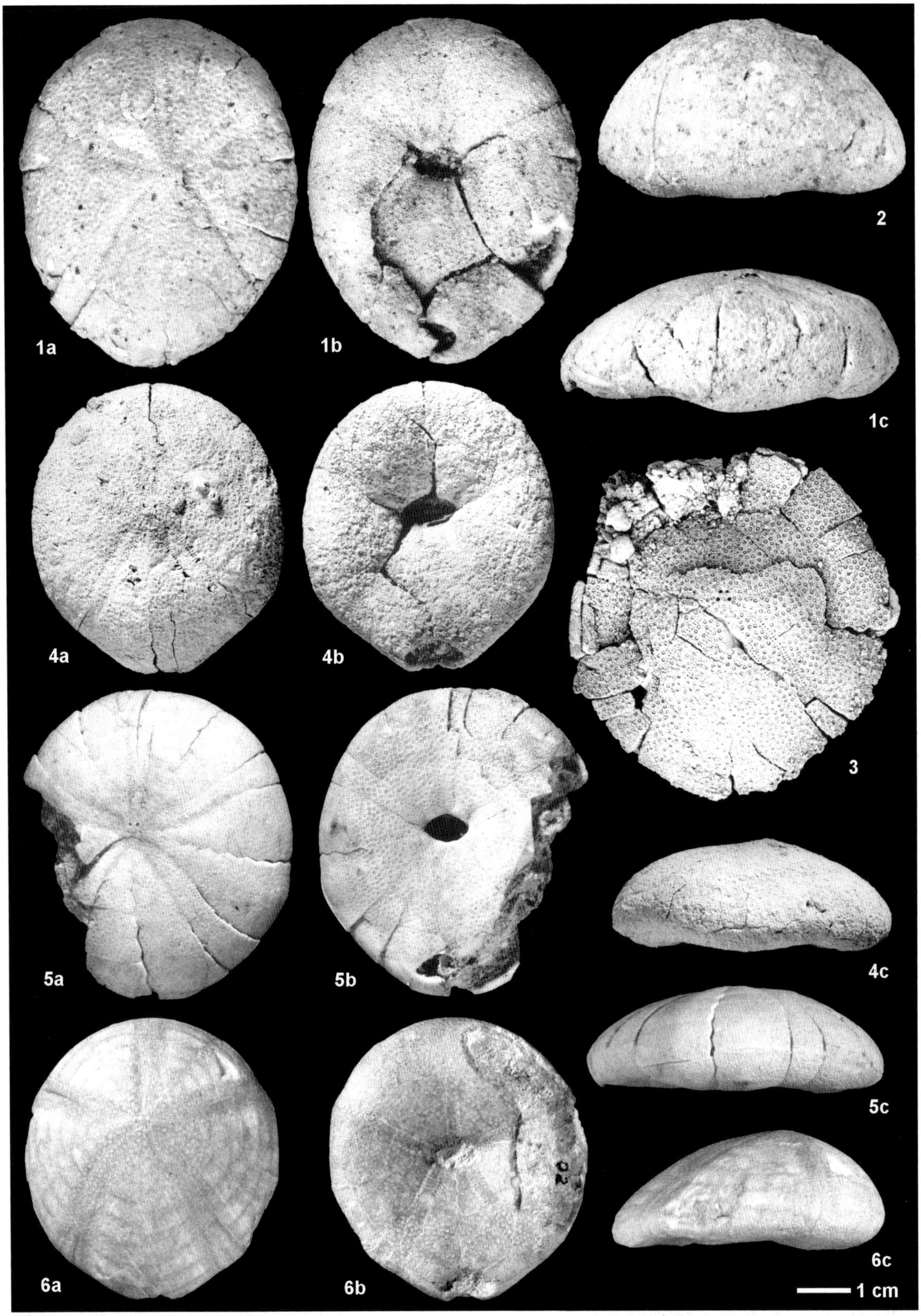

Echinolampas manzonii **Pomel, 1883:** 1-3: Hartl Fm., Hartl hill, Eisenstadt, Bgld [1: Wanzenböck coll. W4, 2: W5 (mirrored for easier comparison), 3: W6]; ***Echinolampas*** **aff.** ***manzonii*** **Pomel, 1883:** 4a-c: Hartl Fm., Hartl hill, Eisenstadt, Bgld (NHMW 1997z0178/1717B); 5-6: Stotzing, Bgld (5: Wanzenböck coll. W17, 6: W7)

Figures 3 and 4a-c coated with ammonium chloride; a: aboral, b: oral, c: right lateral view

Echinolampas sayni **LAMBERT, 1913:** 1-6: Grübern, near Maissau, NÖ (1: NHMW 1914.VII.20/4, 2: NHMW 1914.VII.20/2, 3: NHMW 2003z0035/0001, 4: NHMW 1914.VII.20/5, 5-6: NHMW 2003z0035/0002)

Figures 3a and 3b coated with ammonium chloride; a: aboral, b: oral, c: right lateral view

Echinolampas sayni **LAMBERT, 1913:** 1a-c: Gauderndorf, NÖ (NHMW 1850.IX.83, probable syntype of *E. angustistellata* LAUBE, 1869)

Echinolampas schultzi **KROH, nov. nom.:** 2-3: Gauderndorf, NÖ (2: NHMW 1997z0178/511, 3: NHMW 2003z0036/0001)

a: aboral, b: oral, c: lateral view (1c and 3c: left lateral; 2c: right lateral)

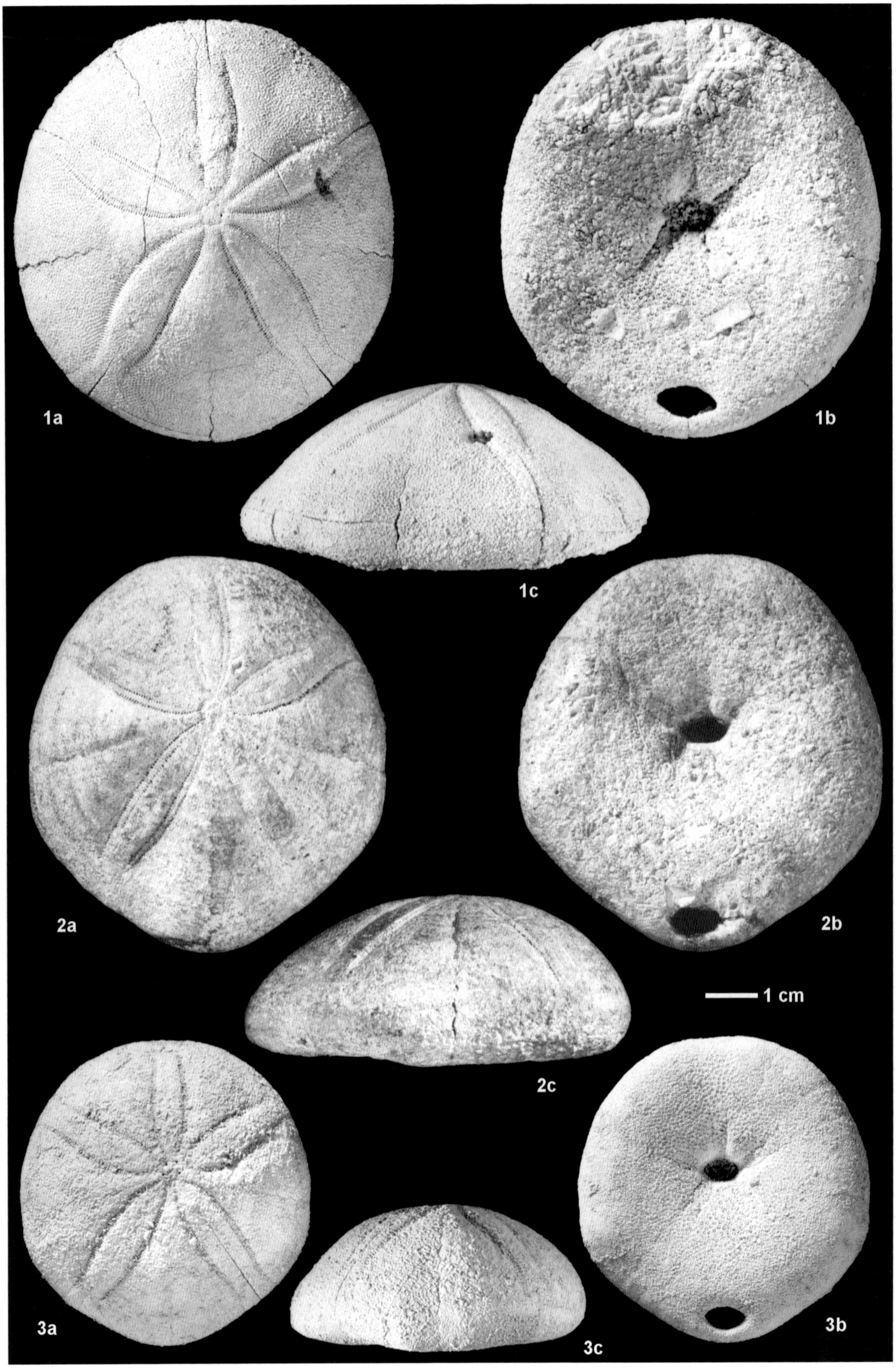
1a
1b
1c
2a
2b
2c
1 cm
3a
3b
3c

Echinolampas schultzi **Kroh, nov. nom.:** 1a-b: pathologic specimen (? disease), Gauderndorf, NÖ (NHMW 2003z0036/0007); 2a-c: juvenile specimen, Gauderndorf, NÖ (NHMW 1914.VII.20/1); ***Echinolampas*** **sp. 1:** 3a-c: Baden, NÖ (NHMW 2003z0034/0001)

Figures 2a-c coated with ammonium chloride; a: aboral, b: oral, c: right lateral view

Echinolampas schultzi **Kroh, nov. nom.:** 1: between Gauderndorf and Kleinmeiseldorf, NÖ (NHMW 2003z0036/0003); 2: Eggenburg (train station), NÖ (neotype, NHMW 2003z0040/0001); 3: Eggenburg, NÖ (NHMW 1972/1551/122a)

Figures 1a-c and 3a-c coated with ammonium chloride; a: aboral, b: oral, c: right lateral view

Conolampas elegans **(Airaghi, 1900):** 1-2: sandpit Mayer, Stotzing, Bgld (both Wanzenböck coll., 1: oral view of W13, 2: dorsal view of W14)

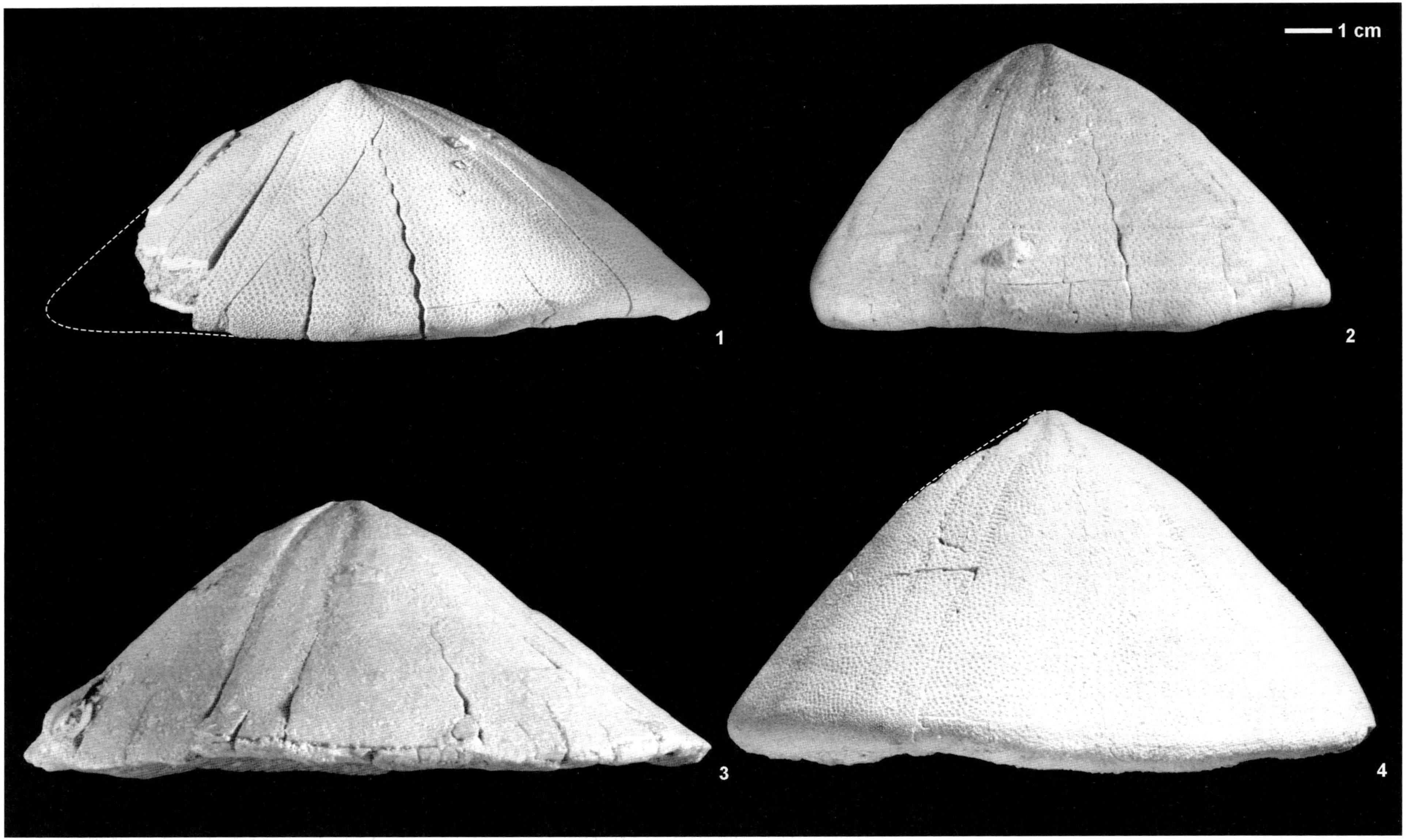

Conolampas subpentagonalis **(GREGORY, 1891):** 1, 3: sandpit Mayer, Stotzing, Bgld (1: NHMW 2003z0008/0001, 3: WANZENBÖCK coll. W9)
Conolampas elegans **(AIRAGHI, 1900):** 2, 4: sandpit Mayer, Stotzing, Bgld (2: NHMW 2003z0008/0003, 4: WANZENBÖCK coll. W14)

1 cm
1a
1b

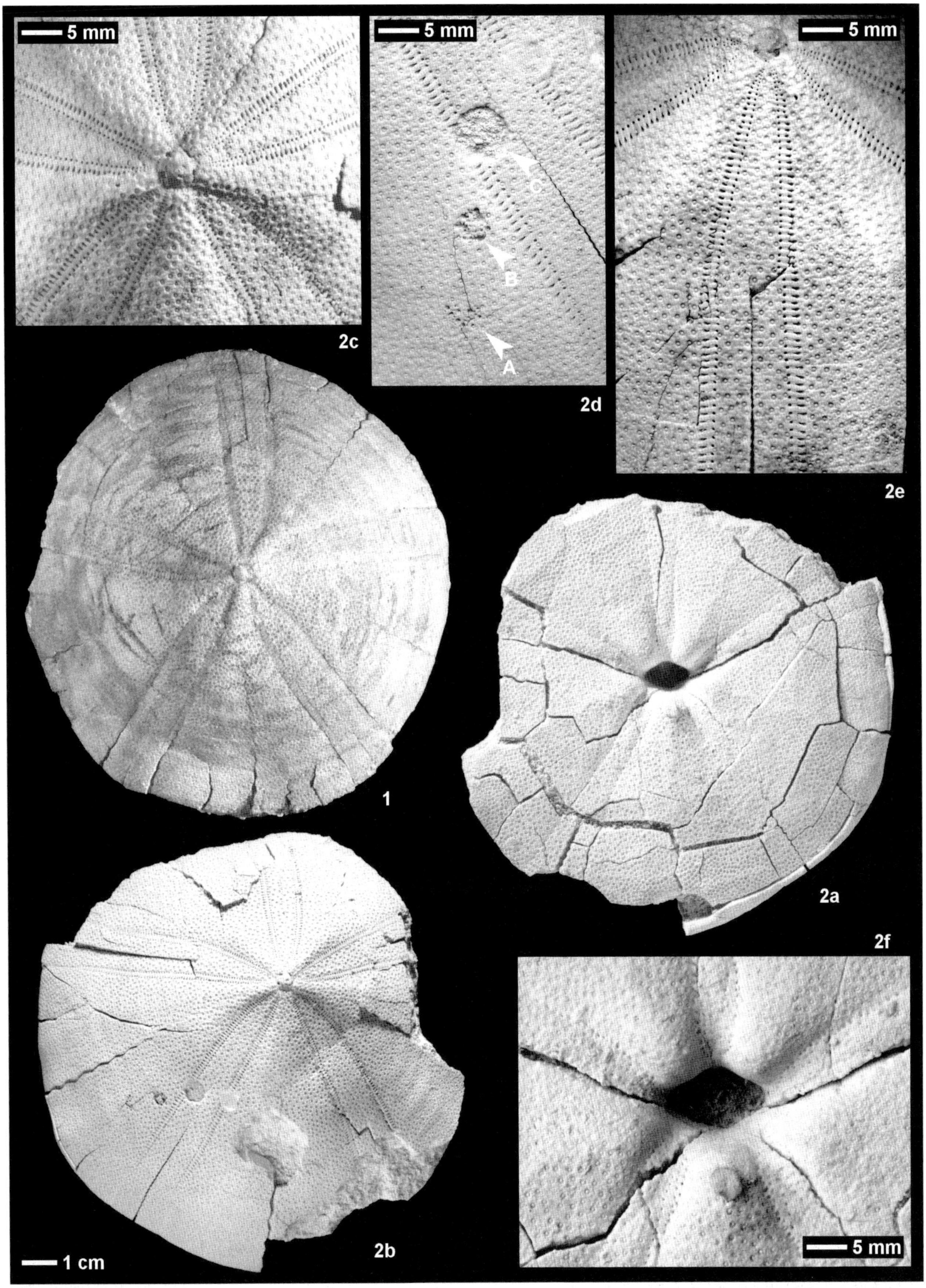

Conolampas subpentagonalis **(Gregory, 1891):** 1: sandpit Mayer, Stotzing, Bgld (Wanzenböck coll. W10); 2: sandpit Mayer, Stotzing, Bgld [NHMW 2003z0008/0001, 2a: aboral view, 2b: oral view, 2c: apical disc, 2d: predatory borings (presumably by cassid or ficid gastropods) in aboral interambulacrum 4 b and poriferous zone Va (A: slight etching traces; B: stronger traces, but test not penetrated; C: penetration), 2e: adapical petal IV, 2f: close up of the floscelle]

Conolampas subpentagonalis **(Gregory, 1891):** 1a-b: sandpit Mayer, Stotzing, Bgld (Wanzenböck coll. W9)

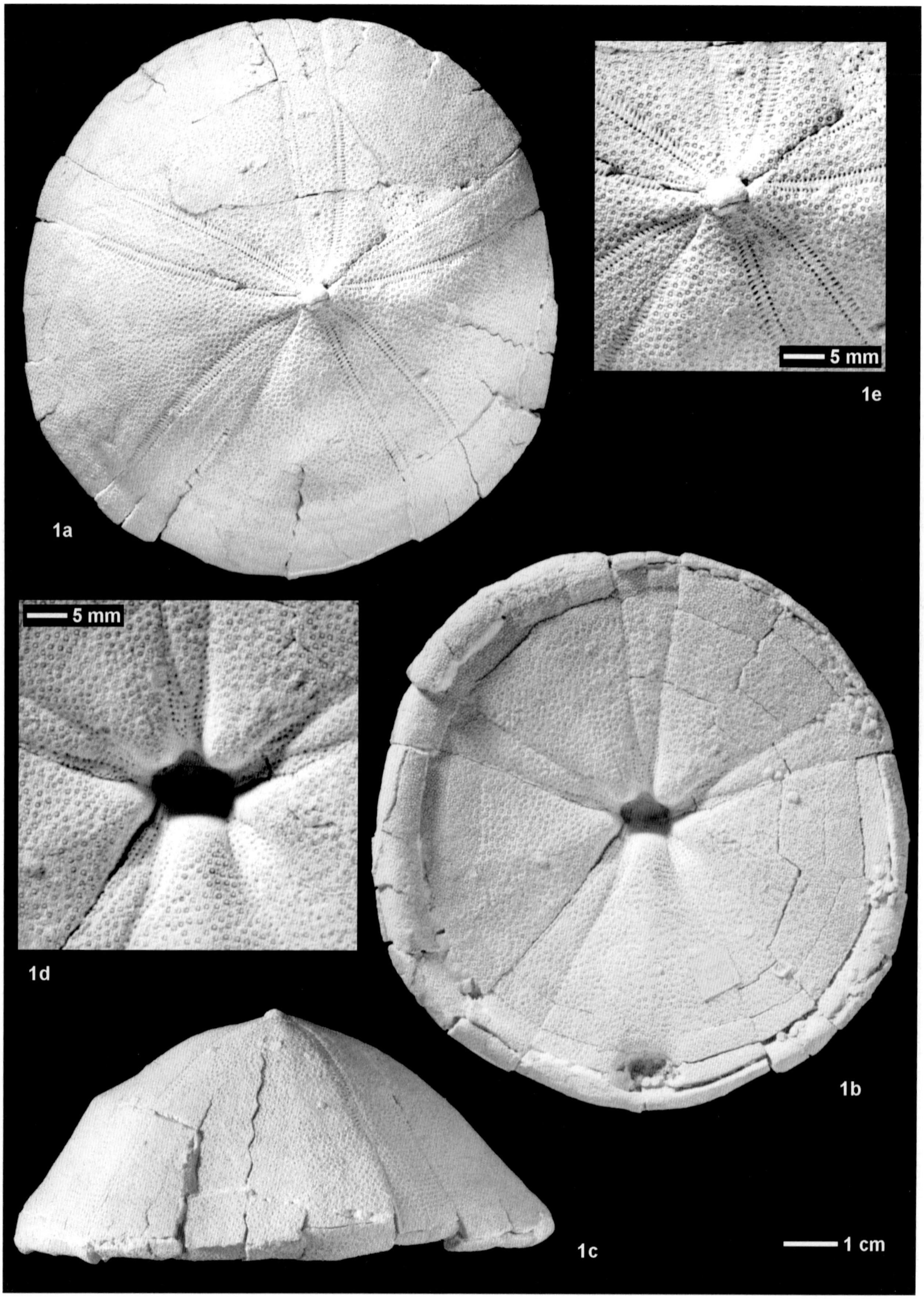

***Conolampas* cf. *subpentagonalis* (Gregory, 1891):** 1a-d: Hornstein, Bgld (NHMW 1970/1396/898)

Figures 1a-d coated with ammonium chloride; a: aboral, b: oral, c: right lateral view, d: close up of the floscelle, e: apical disc

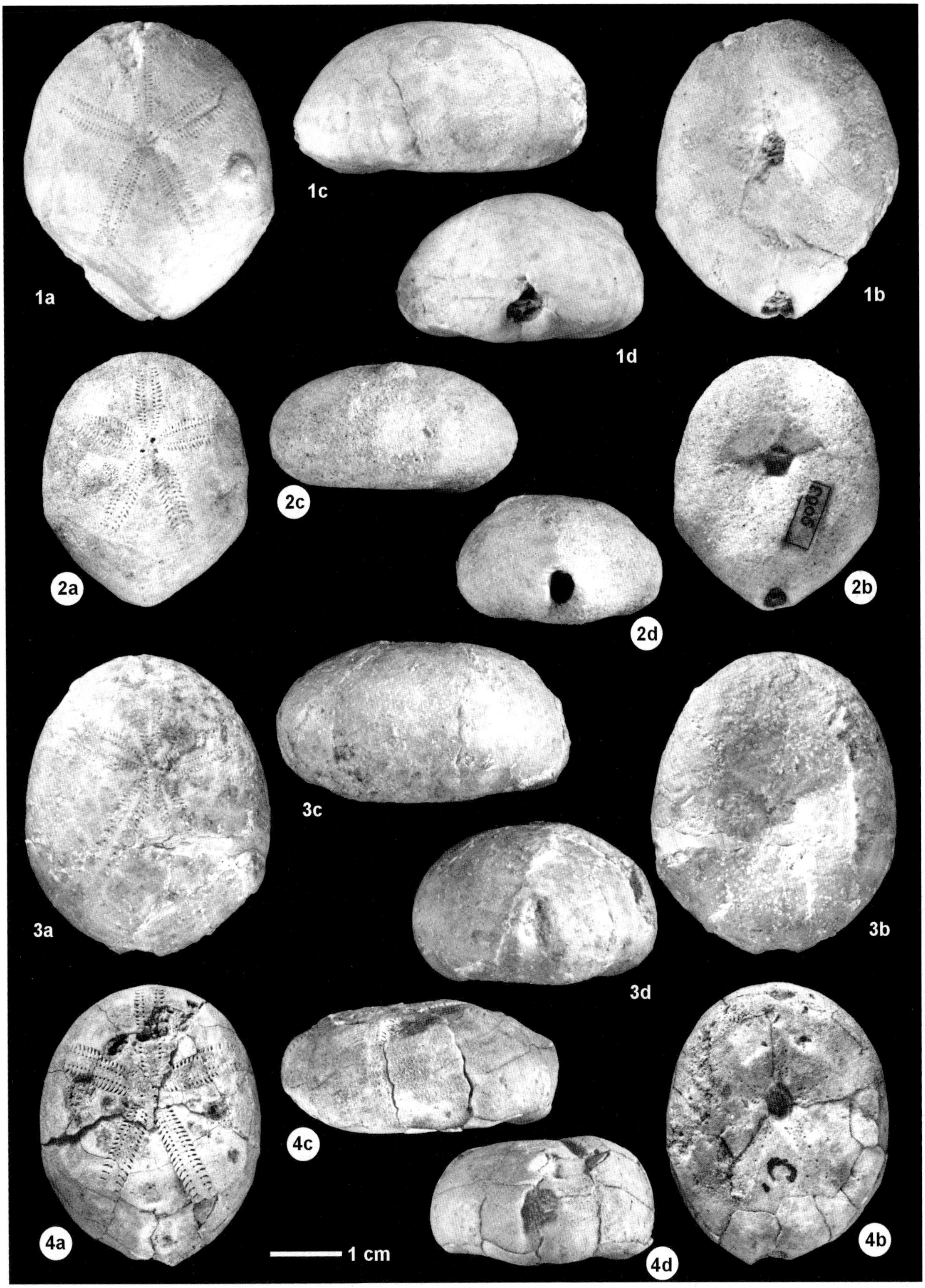

Pliolampas vassalli **(Wright, 1855):** 1a-d: sandpit Mayer, Stotzing, Bgld (Wanzenböck coll. W25); 2a-d: Podjarków, near Lwów, western Ukraine (NHMW 1859.XLV.557)
Studeria corsica **? (Cotteau, 1877):** 3a-d: Lafarge quarry, Retznei, Styria (NHMW 2002z0173/0001); 4a-d: Monte Balistro, Corsica (NHMW 1857.XV.124)

a: aboral, b: oral, c: left lateral view (except 1c: right lateral view), d: posterior view

Ditremaster scillae **(Wright, 1855):** 1a-c: Sievering, Vienna (holotype of *Hemiaster rotundus* Laube, 1869; GBA 1871/3/1)
Pericosmus latus **(Desor in Agassiz & Desor, 1847b):** 2a-b, 3: sandpit Mayer, Stotzing, Bgld (2: NHMW 2004z0093/0001, 3: NHMW 2004z0093/0002)
Schizaster (S.) eurynotus **Sismonda, 1841:** 4: sandpit Mayer, Stotzing, Bgld (NHMW 2004z0093/0009); 5: Müllendorf, Bgld (NHMW 2004z0094/0001)

a: aboral, b: right lateral, c: oral view; scale bar at top margin valid for figs. 1a-c, scale bar at right hand margin valid for figs. 2-5

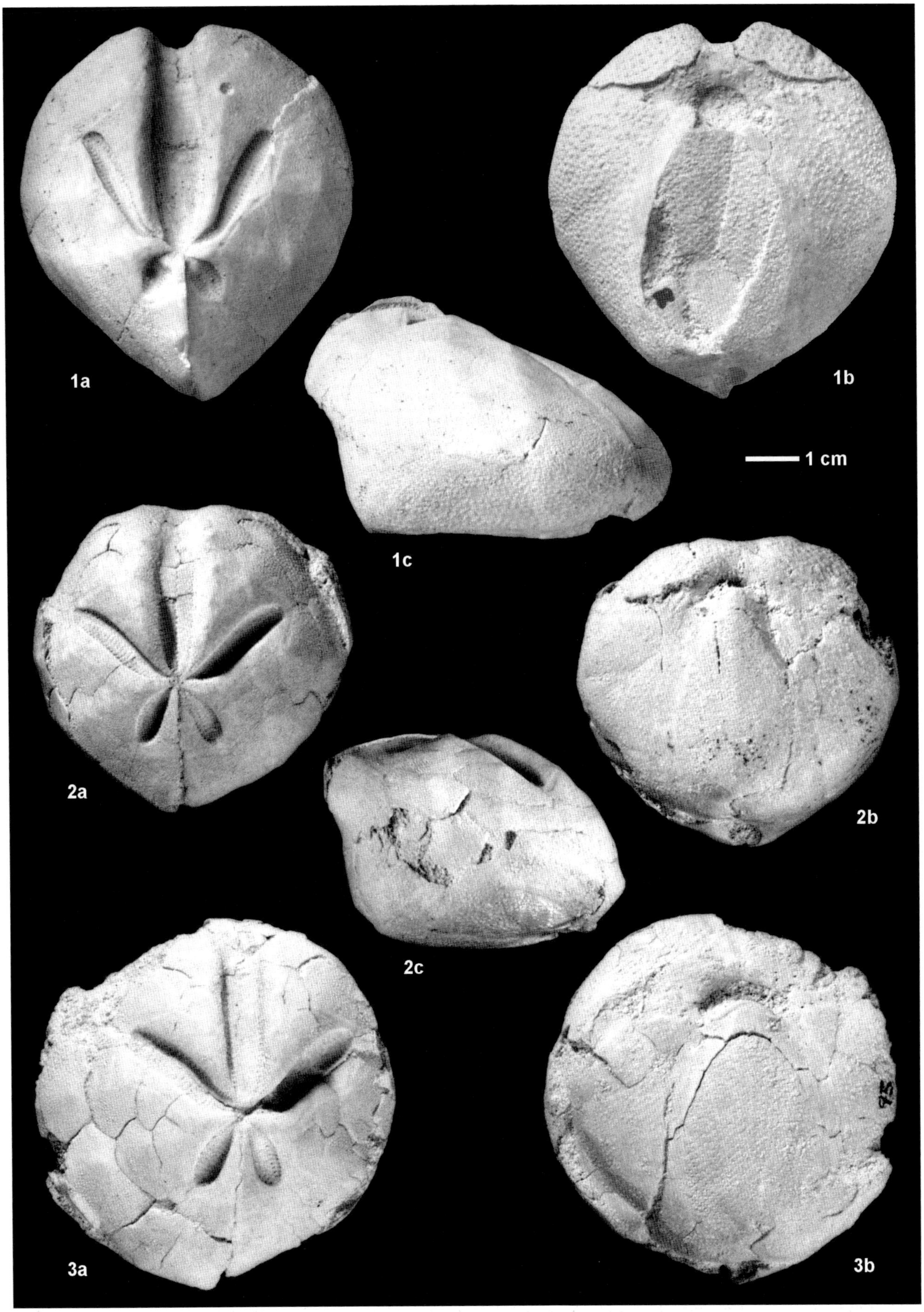

Schizaster* (*S.*) *eurynotus **SISMONDA, 1841:** 1a-c: sandpit Mayer, Stotzing, Bgld (WANZENBÖCK coll. W31)
Aliaster cotteauii **(WRIGHT, 1855):** 2a-c, 3a-b: sandpit Mayer, Stotzing, Bgld [WANZENBÖCK coll. W20 (figs. 2a-c), W21 (figs. 3a-b)]

a: aboral, b: oral, c: right lateral view

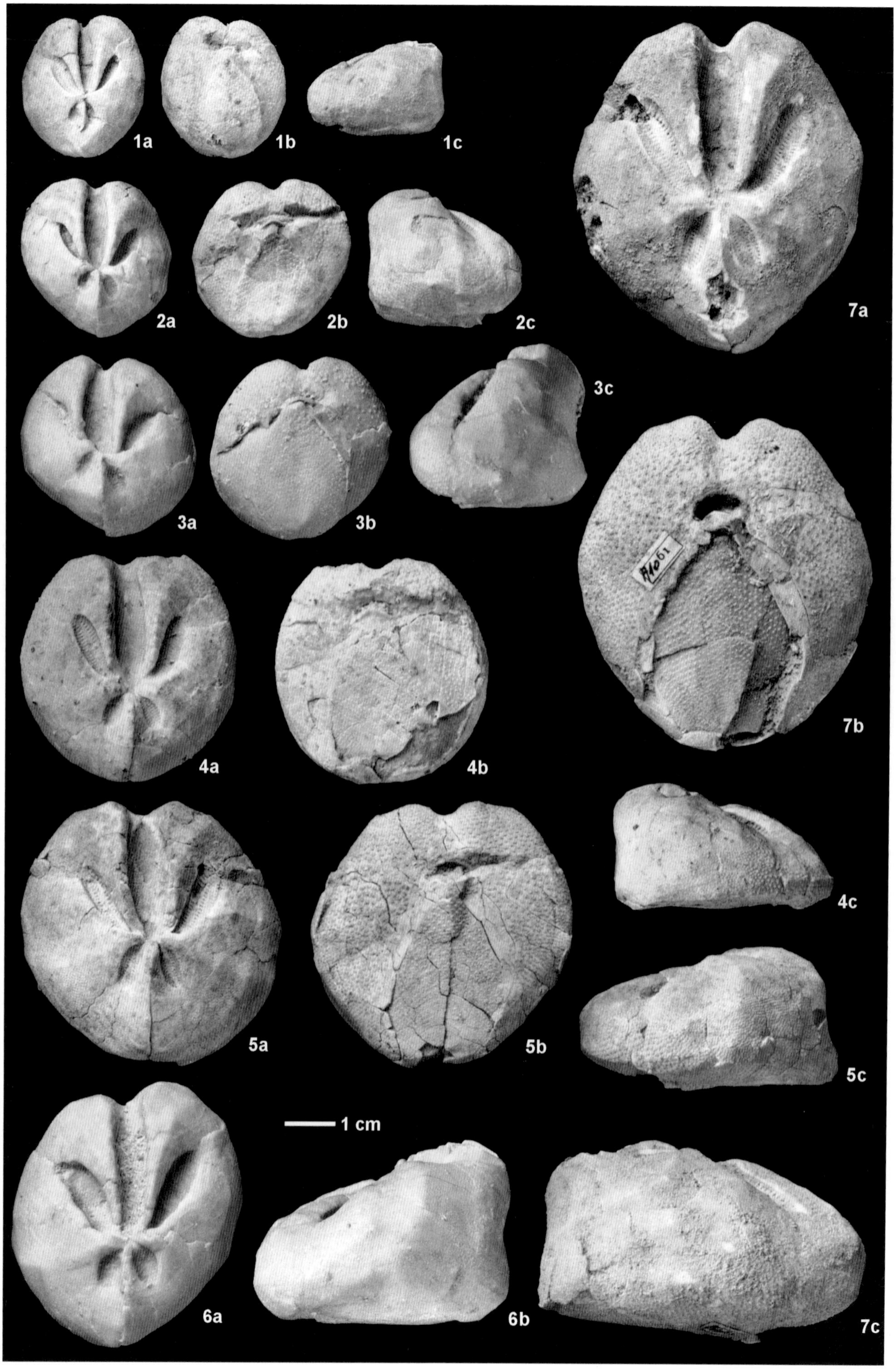
1a
1b
1c
2a
2b
2c
3a
3b
3c
4a
4b
4c
5a
5b
5c
6a
6b
7a
7b
7c
1 cm

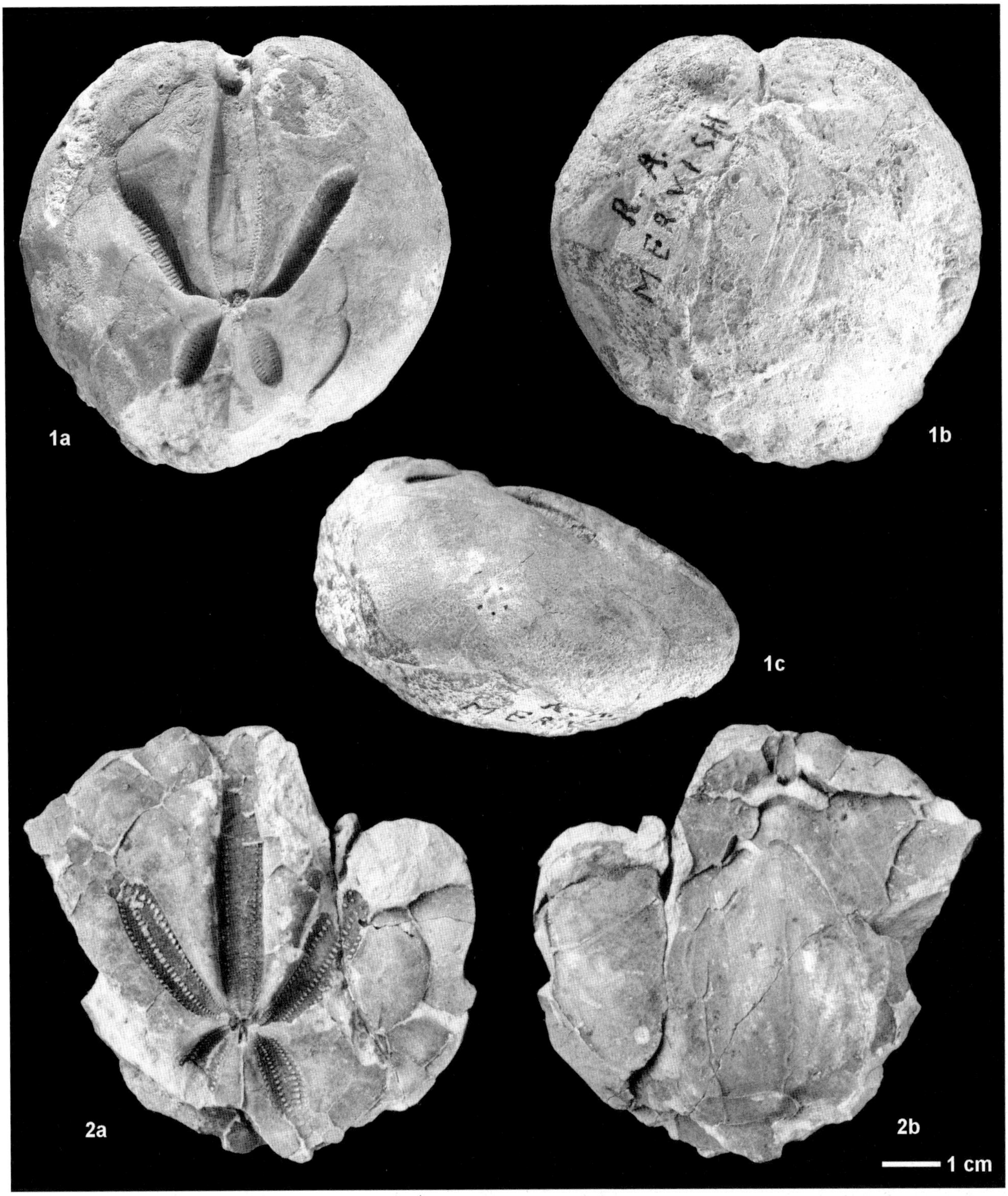

***Schizaster* (*S.*) *eurynotus* Sismonda, 1841:** 1a-c: Mörbisch, Bgld (holotype of *Schizaster leithanus* Laube, 1869, GBA 1871/3/2)
***Schizaster* (*S.*) *laubei* Hoernes, 1875a:** 2a-b: Wetzelsdorf in der Weststeiermark, Stmk (NHMW 2004z0096/0002)

a: aboral, b: oral, c: right lateral view

***Schizaster* (*S.*) *karreri* Laube, 1869:** 1a-c, 2a-c, 4a-c: Wagna, Stmk (1: NHMW 2004z0001/0036, 2: NHMW 2004z0001/0037, 4: NHMW 2004z0001/0035); 3a-c, 6a-b: Lafarge quarry Rosenberg, Retznei, Stmk (3: NHMW 2004z0001/0034, 6: NHMW 2004z0001/0033); 5a-c, 7a-c: Kalksburg, Vienna (5: NHMW 2004z0001/0032a; 7: NHMW 1866.XLVII.37).

a: aboral, b: oral, c: lateral view (except fig. 6b)

***Schizaster* (*S.*) *laubei* Hoernes, 1875a:** 1a-c: Ottnang, OÖ (lectotype; NHMW 1854.II.149b; a: aboral, b: oral, c: right lateral view); 2a-b: natural cast of the aboral surface of the same specimen photographed with two different angles of lighting to reveal different structures (NHMW 1854.II.149c)

***Linthia* ? *hlinnensis* (SENEŠ, 1955):** 1a-b: Ottnang, OÖ (GBA 1875/01/77)]
***Linthia* ? *summesbergeri* nov. sp.:** 2-7: Ebelsberg Fm., Weikerlsee, Linz, OÖ [2: holotype, NHMW 2003z0026/1187a; 3: natural internal cast (specimen lost, corresponding cast preserved under NHMW 2003z0026/1196); 4: paratype, NHMW 2003z0026/1190; 5: paratype, NHMW 2003z0026/1178; 6-7: paratype, natural internal mould (2003z0026/1213) and corresponding cast (NHMW 2003z0026/1212)]

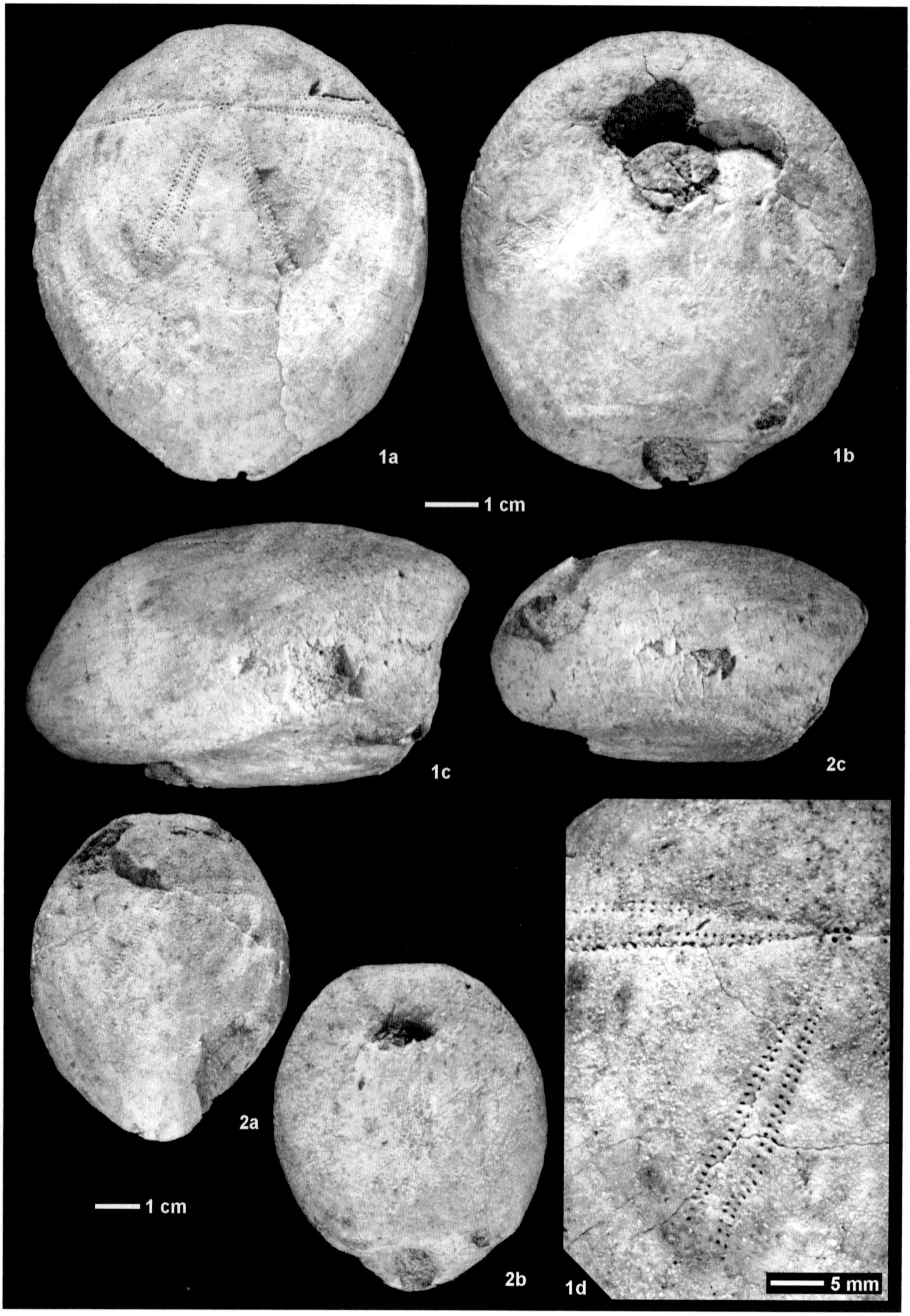

Prenaster fuchsi **(LAUBE, 1871):** 1a-d, 2a-c: Gauderndorf, NÖ (1: lectotype, NHMW 1866.XX.18; 2: paralectotype, NHMW 1866.XX.19)

a: aboral, b: oral, c: left lateral view, d: detail of the petals; central scale bar valid for figs. 1a-c, 2c, lower left scale bar valid for figs. 2a-b

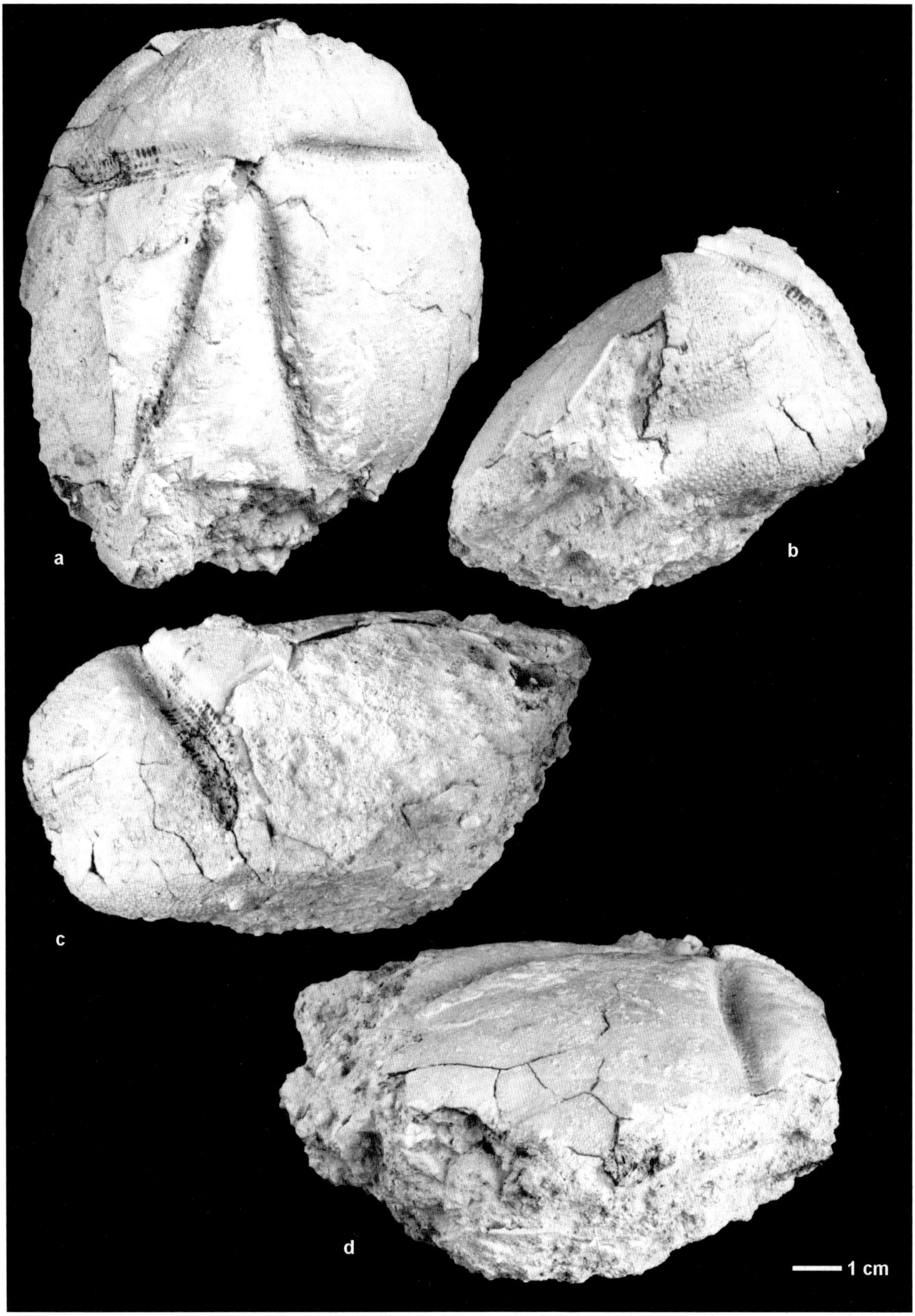

Pseudobrissus corsicus **(COTTEAU in LOCARD, 1877):** a-d: Wöllersdorf, NÖ (KNÖPPEL coll.)

a: aboral, b: anterior, c: left lateral, d: right lateral view

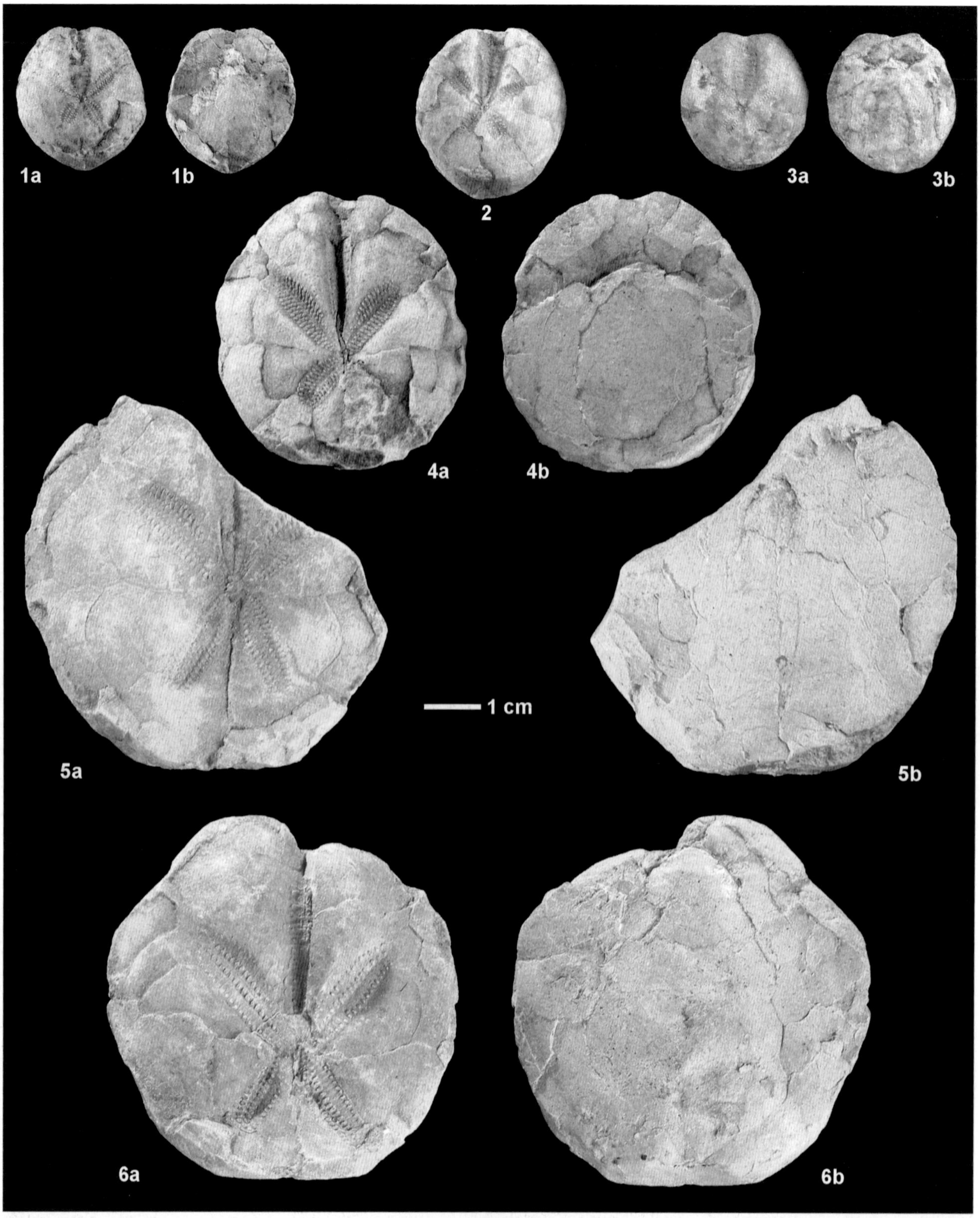

Brissopsis ottnangensis **Hoernes, 1875:** 1-6: lectotype (4a-b) and paralectotypes (1-3, 5-6), Ottnang, OÖ (1: GBA 1875/1/78j; 2: GBA 1875/1/78k; 3: GBA 1875/1/78c; 4: GBA 1875/1/78a; 5: GBA 1875/1/78h; 6: GBA 1875/1/78g)

a: aboral, b: oral view

Brissopsis ottnangensis **Hoernes, 1875:** 1a-c: Retznei, Styria, (NHMW 2004z0098/0002; a: dorsal, b: anterior, c: left lateral view); 2: Devínska Nová Ves, Slovak Republic (SNM z24177); 3-4: Devínska Nová Ves, Slovak Republic, (3: SNM z24181; 4: SNM z24174; in both specimens the inner surface of the test was fixed with glue to enable removing the sediment covered external surface); 5: Kalksburg, Vienna (NHMW 1981/55/12); 6: Haller Schlier, Pfarrkirchen, OÖ (NHMW 1977/1891a); 7: Devínska Nová Ves, Slovak Republic (NHMW 2004z0095/0008; inside of test infiltrated with resin to allow exposing the oral surface retaining the spination)

central "5 mm" scale bar valid for all figs. except fig. 6

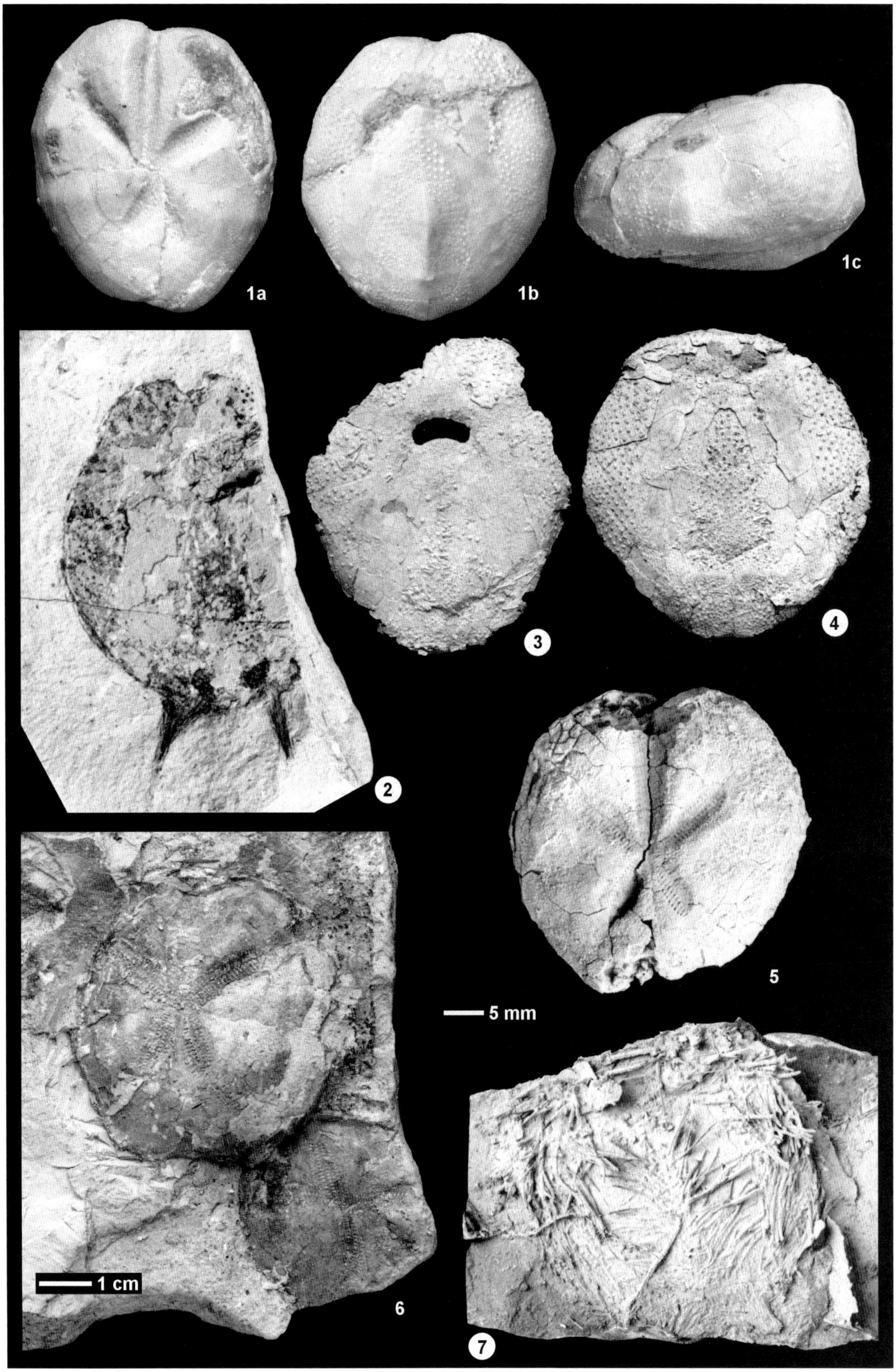
1a
1b
1c
2
3
4
5
5 mm
1 cm
6
7

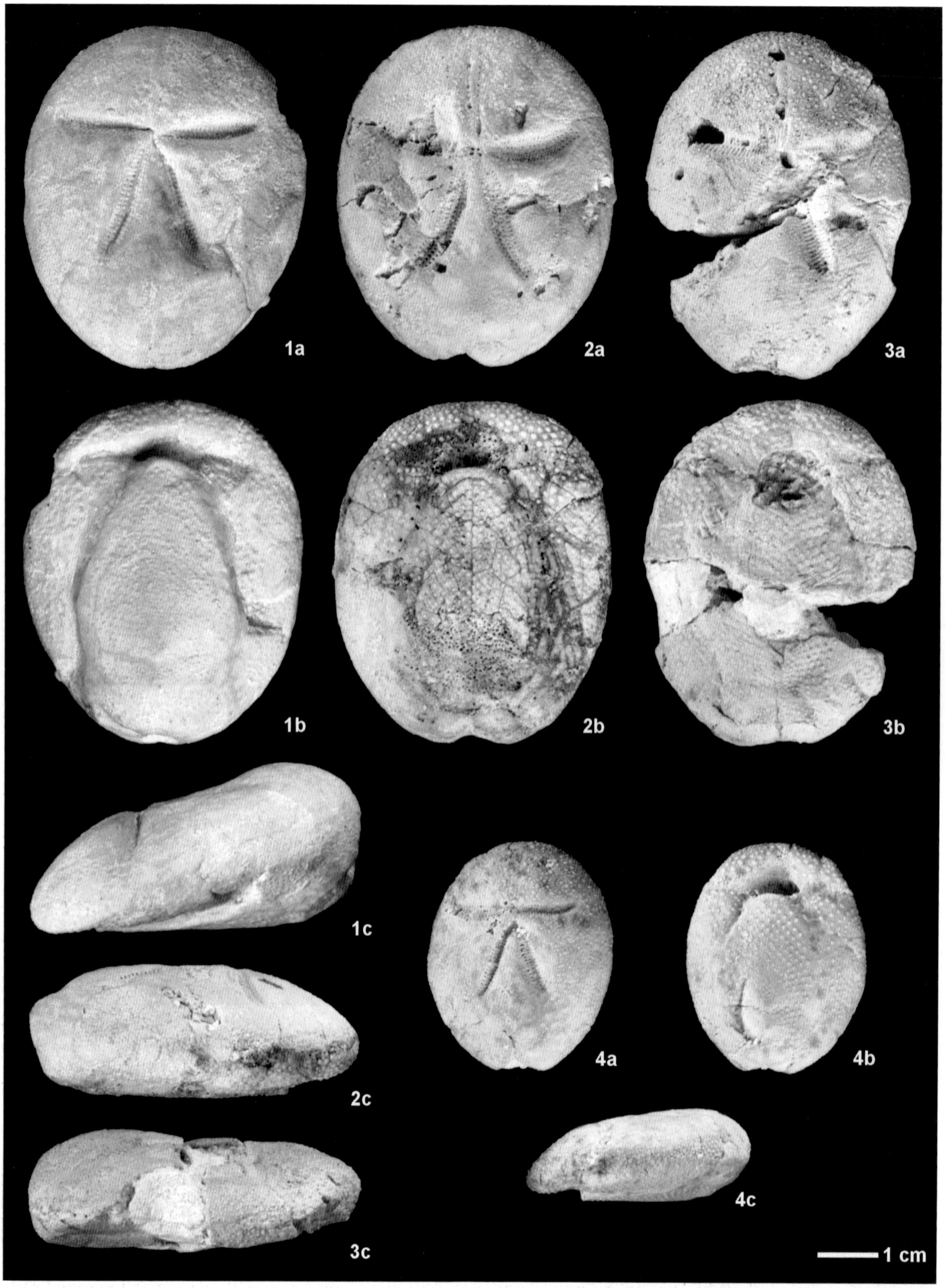

Brissus abeli **(REIDL, 1941):** 1-4: Müllendorf, Bgld (1: NHMW 1997z0178/1742a; 2: 1997z0178/1742b; 3: cast of paratype, NHMW 2002z0180/0006; 4: 1997z0178/1742d)

a: aboral, b: oral, c: lateral view

Meoma locardi **(Cotteau in Locard, 1877):** 1a-b: "Gallenegg", near Zagorje ob Savi (formerly Sagor, Krain), Slovenia (NHMW C.1967, a: aboral, b: oral view); 2: Trbovlje (formerly Trifail), Slovenia (GBA, no inv. no.)

1a
1b
1 cm
1c
3
2c
2a
2b
2d
5 mm

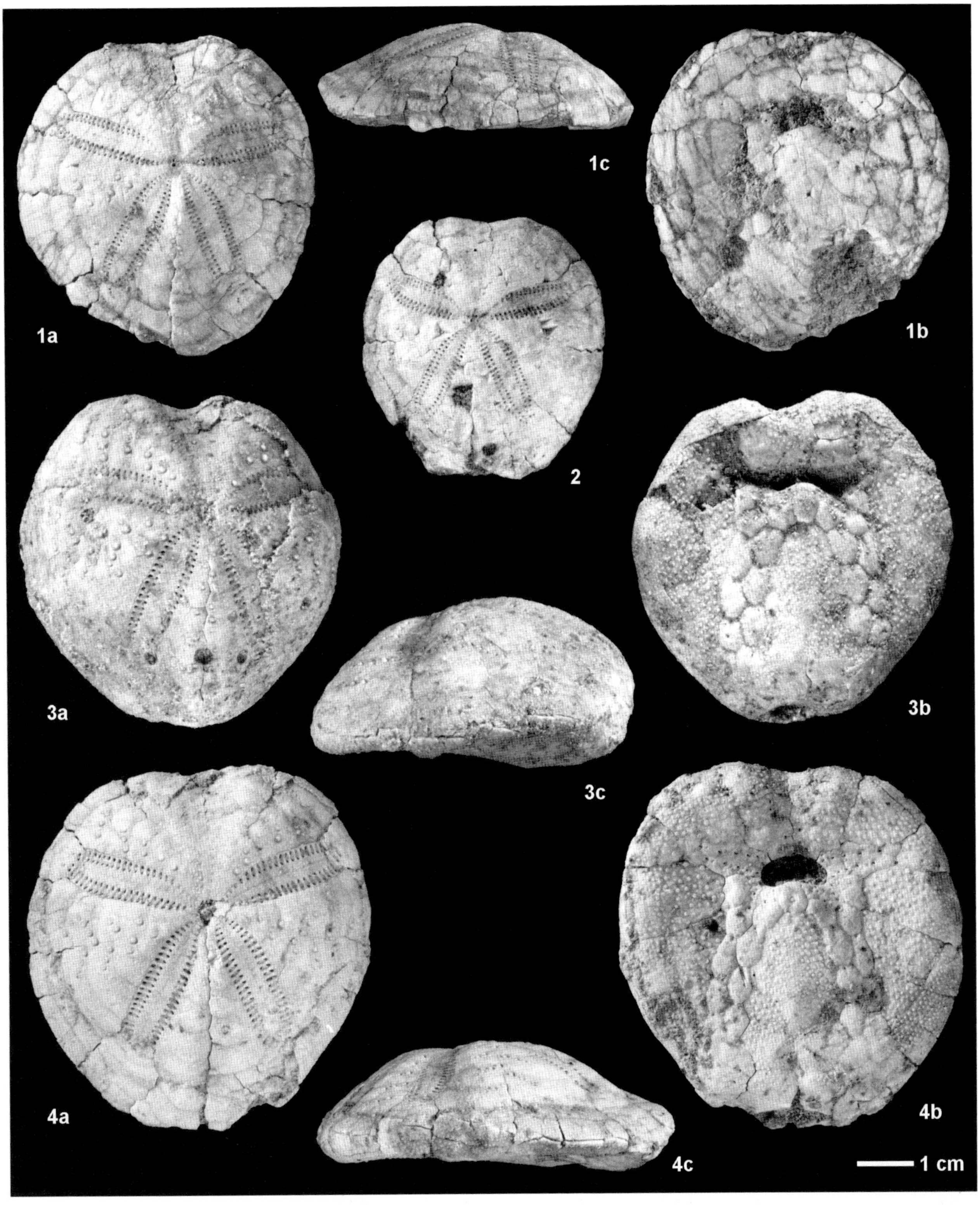

Spatangus austriacus **LAUBE, 1869:** 1-4: Grübern, near Maissau, NÖ (1a-c: NHMW 1914.VII.19e; 2: NHMW 1914.VII.19j; 3a-c: NHMW 2002z0159/0005; 4a-c: NHMW 1914.VII.19d)

a: aboral, b: oral, c: lateral view

⇦ ***Spatangus austriacus*** **LAUBE, 1869:** 1a-c, 2a-d: Bayersdorf, near Maissau (1: lectotype, NHMW 1866.I.1286; 2: paralectotype, NHMW 26.IV.726); 3: Grübern, near Maissau, NÖ (AO of SCHULTZ, 1998, NHMW 1914.VII.19c)

a: aboral, b: oral, c: lateral view; "1 cm" scale bar valid for all figs. except fig. 2d (detail of petals)

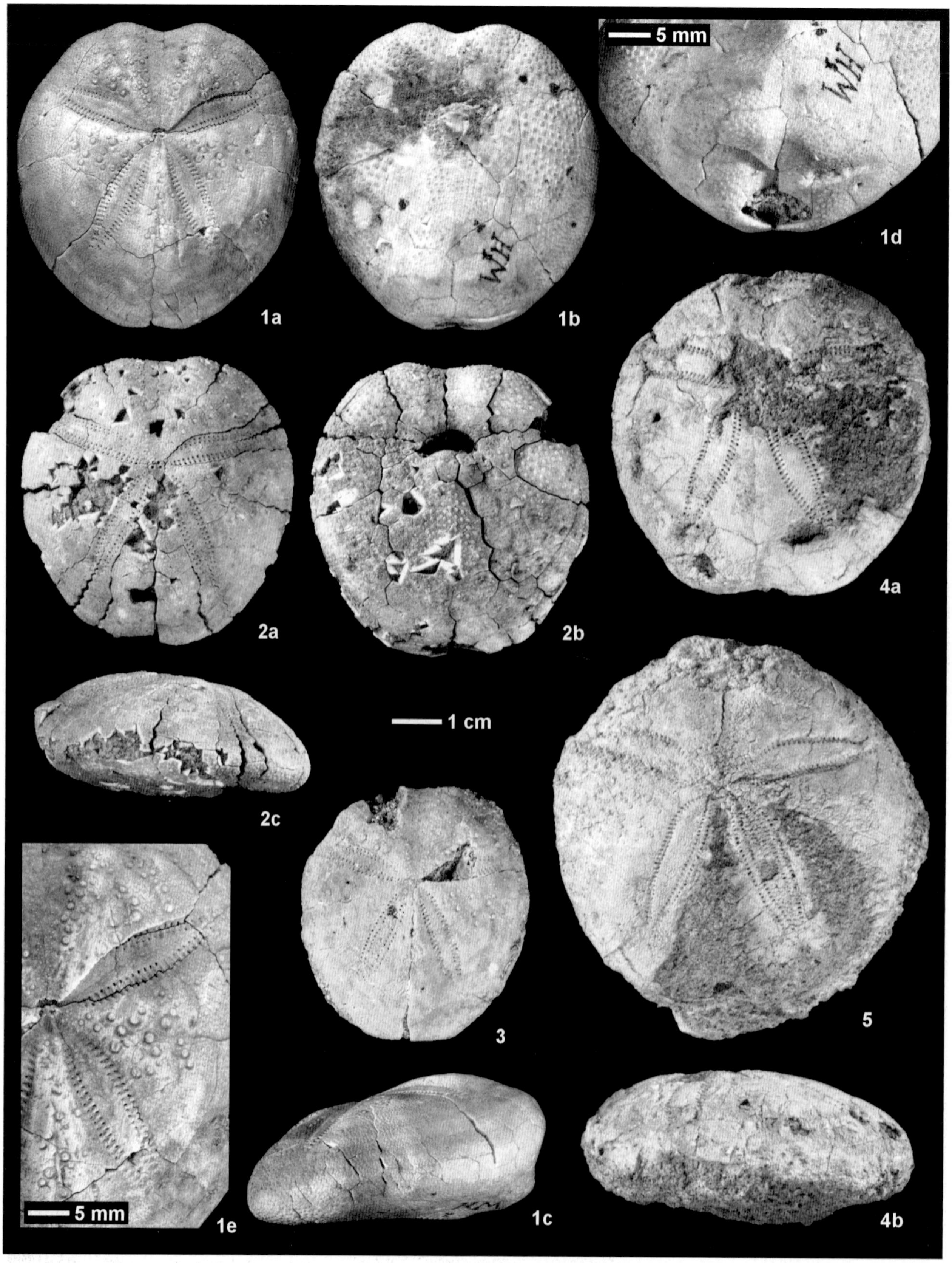

***Spatangus* cf. *austriacus* Laube, 1869:** 1a-e: Maria Dreieichen, NÖ (NHMW 1860.L.205); 2a-c: Grübern, near Maissau, NÖ (NHMW 1914.VII.19k); 3: Bayersdorf, near Maissau, NÖ (NHMW 1934.I.72)
***Spatangus* sp. 1:** 4a-b: Eggenburg-Schindergraben, NÖ (NHMW 592/1964); 5: Eggenburg, NÖ (NHMW 2002z0168/0001)

a: aboral, b: oral, c: lateral view, d: subanal fasciole, e: detail of petals; "1 cm" scale bar valid for all figs. except fig. 1d-e

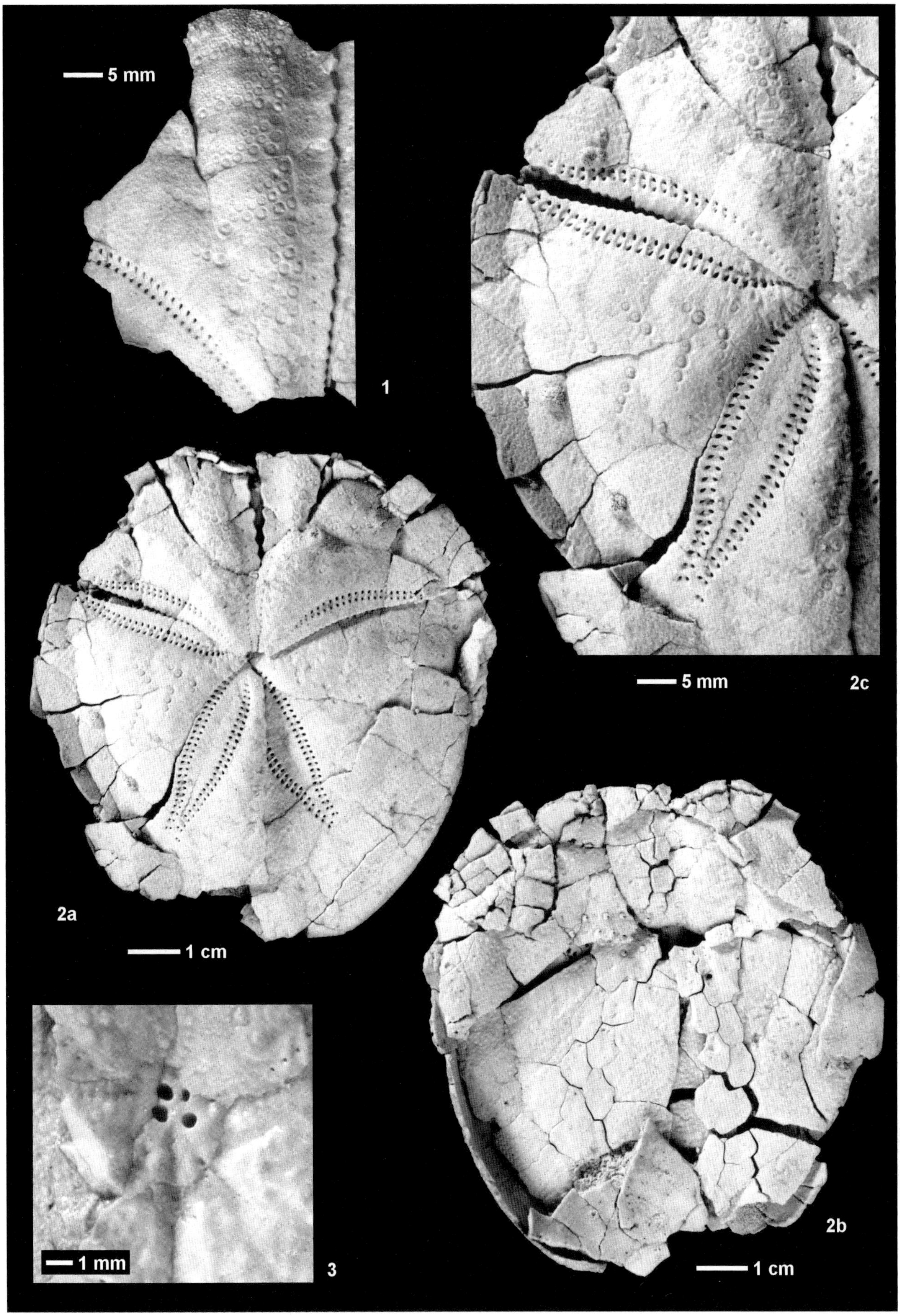

***Spatangus* sp. 2:** 1-3: Kummer quarry, St. Margarethen, Bgld (1: NHMW 1976/1844/45; 2a-c: NHMW 1986/138/130; 3: apical disc of NHMW 1976/1844/54)

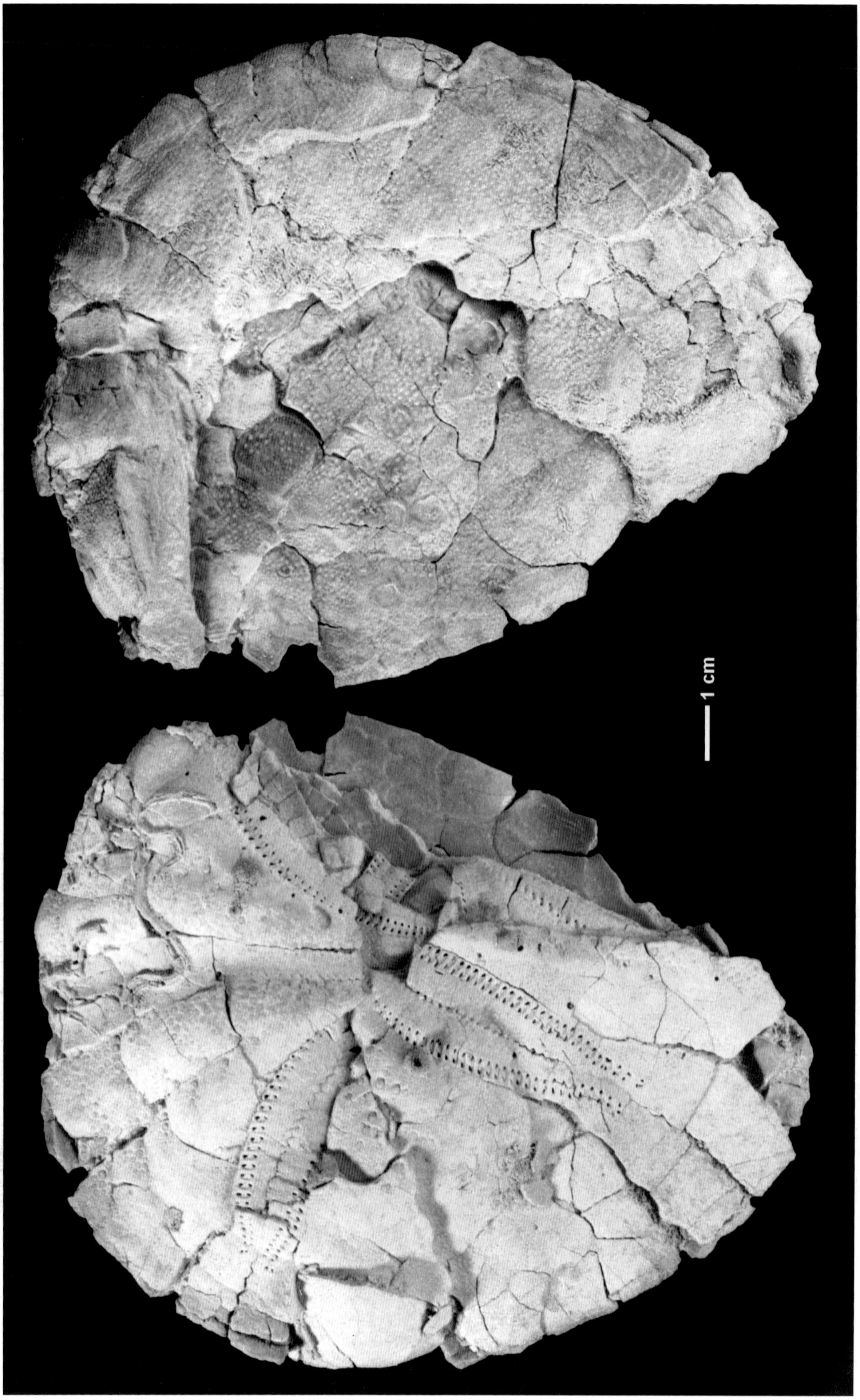

***Spatangus* sp. 2:** a-b: Kummer quarry, St. Margarethen, Bgld (NHMW 1992/151/3; a: aboral, and b: oral view)

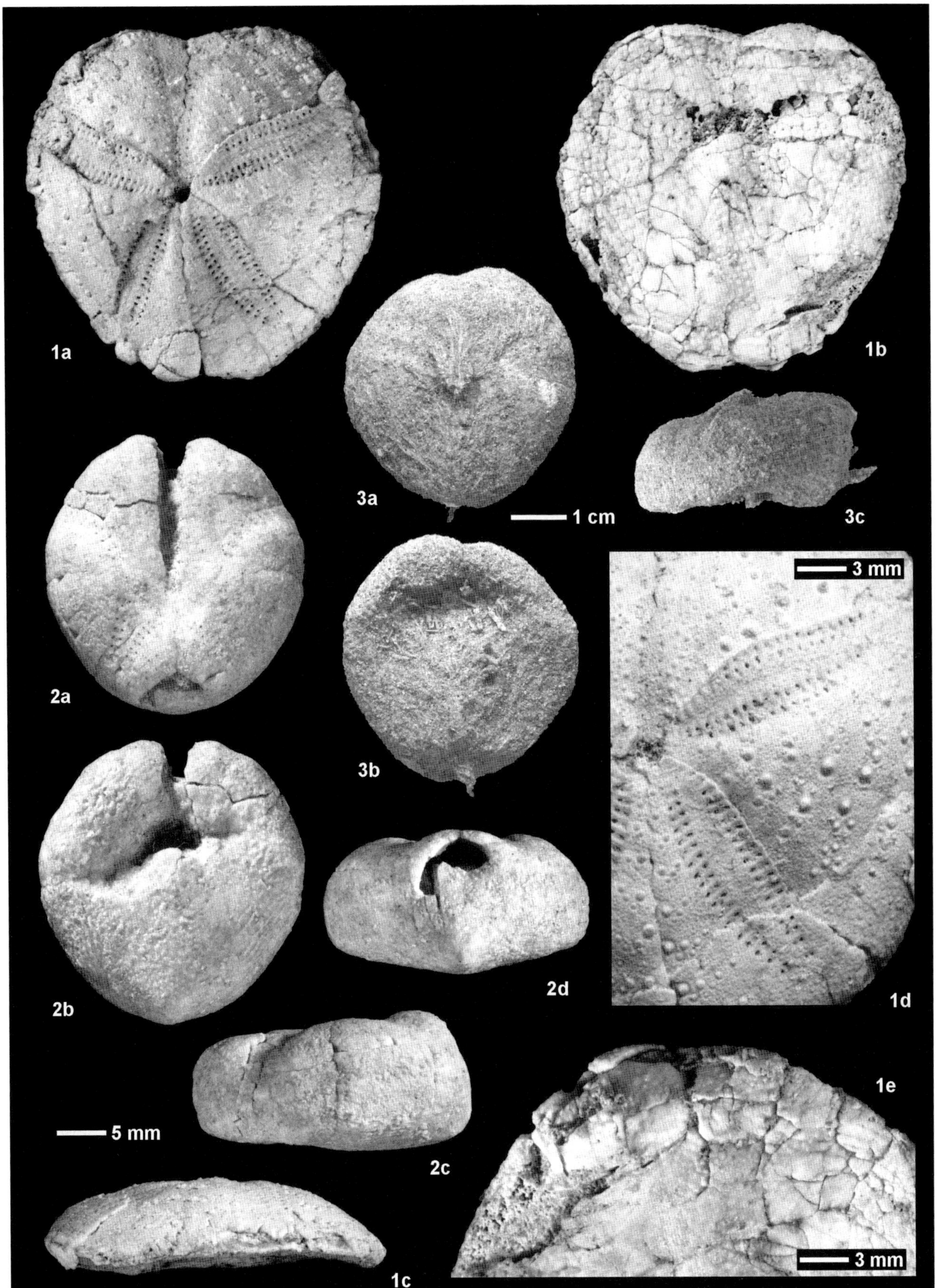

Mariania deydieri **(COTTEAU, 1897):** 1a-e: Grübern, near Maissau, NÖ (NHMW 1998z0048/0102; d: detail of petals, e: subanal area)
Echinocardium depressum **(AGASSIZ in AGASSIZ & DESOR, 1847b):** 2a-d: Florianibründl, near Eggenburg, NÖ (NHMW 2002z0171/0001; d: posterior view)
Echinocardium **sp.:** 3a-c: Gösing, NÖ (KM no inv. no.)

a: aboral, b: oral, c: lateral view; "5 mm" scale bar valid for figs. 1a-c and 2a-e, "1 cm" scale bar valid for figs. 3a-c

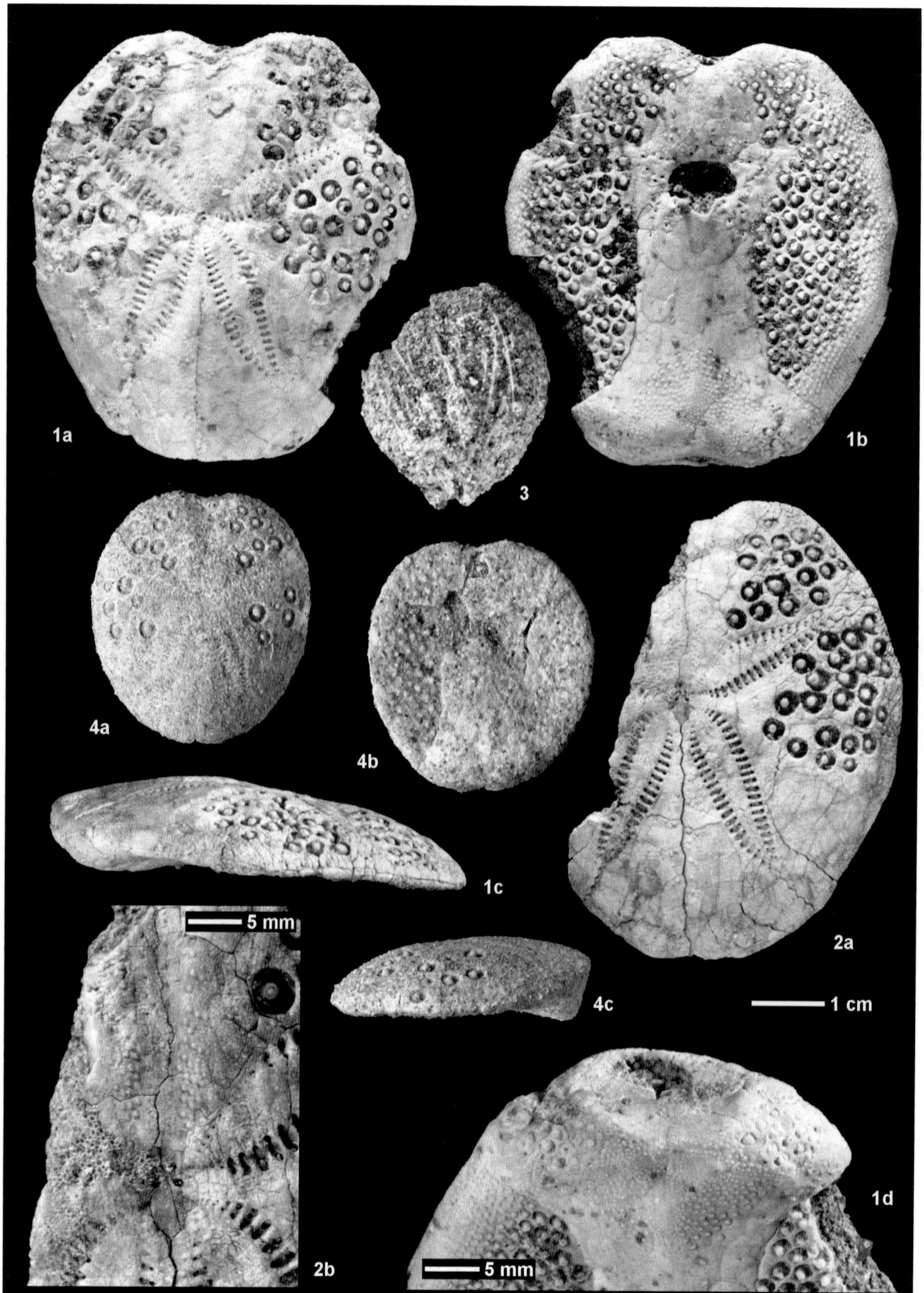

Hemipatagus ocellatus **(Defrance, 1827):** 1a-d, 2a-b: Grübern, NÖ (1: NHMW 1914.VII.22b, 2: NHMW 1914.VII.22a); 3, 4a-c: Gösing, NÖ (3: NHMW 1998z0048/0063; 4: KM without inv. no.)

a: aboral, b: oral, c: lateral view, d: oblique view of the subanal fasciole; "1 cm" scale bar valid for all figs. except figs. 1d and 2b

Hemipatagus ocellatus **(DEFRANCE, 1827):** 1a-c, 2a-c: Kremserberg, near Eggenburg, NÖ [syntypes of *Spatangus* (*Maretia*) *pernata* SCHAFFER, 1912; 1: KM F/0127; 2: KM F/0128]
Spatangoida indet.: Ebelsberg Fm., vicinity of Linz, OÖ [NHMW 2003z0026/1380; accumulations of spatangoid spine fragments, fish remains, mollusc debris and larger foraminifera, interpreted as feeding remains (? regurgitated material)]

"1 cm" scale bar valid for all figs. except figs. 2c and 3